NANOTECHNOLOGY IN PLANT PATHOLOGY

a division of
NIPA GENX ELECTRONIC RESOURCES & SOLUTIONS P. LTD.
New Delhi-110 034

About the Author

Dr. N.G. Ravichandra, Professor of Plant Pathology and Scheme Head, All India Co-ordinated Research Project (Nematodes), Department of Plant Pathology University of Agricultural Sciences, GKVK, Bengaluru, Karnataka, India, is involved in teaching, research and extension activities. He has been a major advisor to the Post-Graduate students majoring in Plant Pathology for M.Sc., (Agri.) and Ph.D. Degree Programmes.

Dr. N.G. Ravichandra has authored nine text books in Plant Pathology & Nematology, has 260 scientific research papers published in National and International journals, contributed chapters to twelve text books, edited five books and also authored ten books in Kannada language.

Dr. N.G. Ravichandra has identified and reported a new species of root-knot nematode (*Meloidogyne enterolobii*) on Guava and a new host, Lac plant (*Flemingia semialalta* L.) for root-knot nematode (*Meloidogyne incognita*) from Karnataka. These are the new and first reports from India.

Dr. N.G. Ravichandra has been bestowed with three National awards for his research accomplishments and contributions. He has been recognized as the International Editor and Reviewer for three prestigious International Journals and National Editor for the eminent 'Indian Phytopathology' Journal. He has developed eight technologies for the management of root-knot nematodes infecting tomato, brinjal and rice, which have been included in the Package of Practices of University of Agricultural Sciences, Bengaluru, Karnataka, India.

Dr. N.G. Ravichandra has operated research projects funded by the Department of Bio-Technology & Indian Council of Agricultural Research, New Delhi, India as Principal Investigator and Co-Principal Investigator; participated and presented original research papers on various aspects of plant diseases in National and International conferences; undergone advanced training programs sponsored by the Department of Biotechnology & Indian Council of Agricultural Research, New Delhi, India; an active life member of professional societies including Nematological Society of India, Indian Phytopathological Society, Society of Mycology & Plant Pathology, National Environmental Science Academy, Institute of Agricultural Technology, Glacier Journal Research Foundation and Institute of Scholars

NANOTECHNOLOGY IN PLANT PATHOLOGY

N.G. Ravichandra

Professor of Plant Pathology & Scheme Head, AICRP (Nematodes)
Department of Plant Pathology, College of Agriculture
University of Agricultural Sciences, GKVK
Bangaluru-560 065, Karnataka, India

a division of

NIPA GENX ELECTRONIC RESOURCES & SOLUTIONS P. LTD.

New Delhi-110 034

a division of

NIPA GENX ELECTRONIC RESOURCES & SOLUTIONS P. LTD.

101,103, Vikas Surya Plaza, CU Block
L.S.C.Market, Pitam Pura, New Delhi-110 034
Ph : +91 11 27341616, 27341717, 27341718
E-mail:newindiapublishingagency@gmail.com
www: www.nipabooks.com

For customer assistance, please contact
Phone: + 91-11-27 34 17 17 Fax: + 91-11- 27 34 16 16
E-Mail: feedbacks@nipabooks.com

ISBN: 978-93-90591-87-9

Composed and Designed by NIPA.

To

The University of Agricultural Sciences, Bengaluru

Acknowledgement

My respectful thanks to,

Dr. S. Rajendra Prasad, Hon'ble Vice Chancellor, University of Agricultural Sciences, GKVK, Bengaluru, Karnataka, India

Dr. Y.G. Shadakshari, Director of Research, University of Agricultural Sciences, GKVK, Bengaluru

Dr. Umarao, Project Coordinator, AICRP (Nematodes), Indian Agricultural Research Institute, New Delhi

My thanks to Dr. T. Narendrappa, Professor & University Head, Dr. Y.M. Somashekhara, Professor & Head, Dr. A. Nagaraja, Dr. N. Nagaraju, Dr. H.K. Ramappa, Dr. H.A. Prameela, and Mr. B.R. Nagendra, Department of Plant Pathology, UAS, GKVK, Bengaluru

I would like to express my deep sense of gratitude to my beloved parents Sri. N. Gurushankar and Smt. Parvathamma. I am grateful to my in-laws for their affection and encouragement.

I wish to record my appreciation and affection to my wife, Ms. Deepthi and daughter, Ms. Anusha for their care, patience, keen interest and affection throughout the preparation of the manuscript.

My heartfelt thanks to my dear brother-in-law Mr. B. Sudesh Kumar and my beloved sister Ms. N.G. Tara, for their sustained moral support, affection and special care. I fondly acknowledge their keen interest, continued encouragement and blessings being showered on me.

My special thanks to NIPA, New Delhi, India, who deserves commendation for their professional contribution.

Preface

Nanotechnology is one of the most fascinating and rapidly advancing sciences that has grown exponentially in the past decade. The plant disease diagnosis pathogen detection and residual analysis have become much more precise and quick with the use of nanomaterials and nanosensors. Such an accurate technology helps to design a proper strategy that may modify crop environment to adversely affect crop pathogens, leading to high-quality monitoring and management of plant diseases. As global demand for food production escalates against a changing climate, nanotechnology could sustainably mitigate many challenges in disease management.

Under various degree programmes at Agricultural and Horticultural Universities, nanotechnology is prescribed either as an exclusive subject with specific syllabus or is being discussed in relation to plant diseases. This book covers the syllabus prescribed by the Indian Council of Agricultural and Research, New Delhi, India for nanotechnology aspects of Plant Pathology. There is a need for a comprehensive book on the significance of nanotechnology in plant pathology. This book has been prepared to furnish the latest information on this fast developing topic.

The book "Nanotechnology in Plant Pathology" explicates essential aspects and is conveniently divided into nine unique chapters, covering the latest information and developments in nanotechnology with respect to plant diseases. The chapters include Introduction to nanotechnology, Categories and characterization of nanoparticles, Nanotechnology in plant disease diagnosis, Nanoparticles in plant disease management, Nanotechnology and pesticides, Nanotechnology in pesticides delivery, Regulations and risks in nanotechnology, Nanotoxicity and Nanosafety. Quality diagrams at appropriate places, convincing tables and suitable graphs / illustrations have been furnished for an impressive reading. Under each chapter, a separate bibliography providing the list of references cited has also been included for additional reading.

The contents of this book reflecting an extensive literature search will be useful, particularly for teachers, researchers and students in Plant Pathology the State Departments of Horticulture & Agriculture, Commercial protection centers, Administrators, Policy planners and all those who are interested and concerned with the use and application of nanoparticles in plant pathology.

The author appreciates receiving suggestions and constructive criticism that would improve the quality of the book.

Bengaluru

2022 **N.G. Ravichandra**

Contents

1

Introduction

The term nanotechnology, buzzword of present day science owes its origin from the Greek word 'nano' literally meaning dwarf. When it is expressed in terms of dimension one nanometer equals to one billionth of a meter (1nm = 10^{-9} m). A nanometer is a thousandth of a thousandth of a thousandth of a meter, about 60,000 times smaller than a human hair in diameter or the size of a virus, a typical sheet of paper is about 100,000 nm thick, a red blood cell is about 2,000 to 5,000 nm in size, and the diameter of DNA is in the range of 2.5 nm. Therefore, nanotechnology deals with matter that ranges from one-half the diameter of DNA up to 1/20 the size of a red blood cell.

"Imagine you take a large boulder and determine its mass and surface area to volume ratio. Now you take a hammer and you smash the rock. The accumulated mass of the resulting smaller pieces remains the same, but the surface [area to volume [ratio] goes way up. Now take another hammer, and smash all of those small pieces into tiny pieces. They still have the same [combined] mass, but the surface area goes through the roof. That's what nanoparticles are" (Ryan Otter, *Biologist,* the Middle Tennessee State University).

The subject nanotechnology deals with manufacturing, study and manipulation of matter at nano-scale (or atomic scale) in the size range of 1-100 nm which may be called as 'nanoparticles' (Rajan, 2004). Any material when attenuated at nanometer scale (less than 100 nm) exhibits new properties that are entirely different from its bulk counterpart due to small size and high surface to volume ratio. It is shown in terms of higher plasticity at high temperature, higher hardness, breaking strength and toughness at low temperature, higher chemical reactivity and surface energy and high mobility in the body of an organism including cellular entry.

Nanotechnology refers to controlling, building and restructuring materials and devices on the scale of atoms and molecules. A nanometer (nm) is one- billionth of a meter. To get the sense of the nano scales, the width of the human hair is 80,000 nm and the smallest things visible with the naked human eye are 10,000 nm across. At nano scales, the basic rules of chemistry and physics are not applicable. One example of this technology is the carbon nanotube discovered in 1991, which is only a few nanometers in diameter but can conduct electricity better than copper, 100 times stronger than steel but only one sixth of its weight. Novel agricultural systems can make excellent use of nanotech-enabled "smart" devices that can perform a dual role of being a preventive and early warning system. These devices can identify plant related health issues even before they become visible to the farmers and simultaneously provide remedial measures. These nanotech systems can also be used to monitor the delivery of pesticides. User-friendly and eco-friendly nano delivery systems for pesticides have started to find their place in the market. These can allow the use of pesticides with the absolute minimum risk of environmental damage. Companies have implemented nanoemulsions in commercial pesticide products.

Nanotechnology is the manipulation or self-assembly of individual atoms, molecules or molecular clusters into structures to create materials and devices with new or vastly different properties (Sunkara V Manorama, 2004). Nanotechnology has the potential to revolutionize the agricultural and food industry with new tools for the molecular treatment of diseases, rapid disease detection, enhancing the ability of plants to absorb nutrients etc. Smart sensors and smart delivery systems will help the agricultural industry combat plant pathogens. In the near future, nanostructured catalysts will be available which will increase the efficiency of pesticides and herbicides, allowing lower doses to be used. Nanotechnology will also protect the environment indirectly through the use of alternative (renewable) energy supplies, and filters or catalysts to reduce pollution and clean-up existing pollutants. Nanotechnology is the understanding and control of matter at dimensions between approximately 1 and 100 nanometers, or nanoscale. Unusual physical, chemical, and biological properties can emerge in materials at the nanoscale. These properties may differ in important ways from the properties of bulk materials and single atoms or molecules. Encompassing nanoscale science, engineering, and technology, nanotechnology involves imaging, measuring, modeling, and manipulating matter at this length scale. The sizes of organisms and different biosubjects are given in table 1.

Table 1: Size of different organisms and biomolecules on the micro- and nanometric scale

S.No.	Nature of organisms & different biosubjects	Size range (μm)	Size on nanometric scale (nm)
1	Plant, animal cell	10–100	10 000–10 0000
2	Bacteria	≤1–10	≤1000–100
3	Virus	0.03–0.1	30–100
4	Simple molecules (proteins, DNA turns)	0.001–0.01	1–10
5	Atoms (DNA 'base')	0.0001–0.001	0.1–1

For comparison, the wavelength of visible light is between 400 and 700 nm. A living cell has dimensions of microns (thousands of nanometers) (Table 2).

Table 2: Comparisons of scale from macro to molecular

Size (μm)	Examples	Terminology
0.1 – 0.5	Individual chemical bonds	Molecular/atomic
0.5 – 1.0	Small molecules, pores in zeolites	Molecular
1 - 1000	Proteins, DNA, mesopores,inorganic nanoparticles	Nano
103 - 104	Microfluidic channels, MEMS, Devices on a silicon chip, living cells	Micro
> 104	Normal bulk matter	Macro

Nanotechnology-based products and its applications in agriculture include nano-fertilizers, nano-herbicides, nano-pesticides, recalcitrant contaminants from water, nano-scale carriers, nanosensors, veterinary care, fisheries and aquaculture, detection of nutrient deficiencies, preservation, photocatalysis, nanobarcode, quantum dots etc. This fast growing technology is already having a significant commercial impact, which will certainly increase in the future. While nanotechnologies offer many opportunities for innovation, the use of nanomaterials in food and agriculture has also raised a number of safety, environmental, ethical, policy and regulatory issues. Nanotechnology is emerging as the sixth revolutionary technology in the current era after the Industrial Revolution of Mid 1700s, Nuclear Energy Revolution of the 1940s, The Green Revolution of 1960s, Information Technology Revolution of 1980s and Biotechnology Revolution of the 1990s.

A survey by Salamanca Buentella et al. (2005) predicted several nanotechnology applications for agricultural production for developing countries within next 10 years. These included nanoforms zeolites for slow release and efficient dosage of water and fertilizers for plants; drugs for livestock, nanocapsules and herbicide delivery; nanosensors for soil quality and for plant health monitoring, nanosensors for pests detection; nanomagnets for removal of soil contaminants and nanoparticles for new pesticides, insecticides, and insect repellents.

Nanotechnology is the design, fabrication and utilization of materials, devices and systems through control of matter on the nanometer length scale and exploitation of novel phenomena and properties (physical, chemical, biological) at that length scale. It is now more properly labeled as "molecular nanotechnology" (MNT) or "nano-scaleengineering". By taking advantage of quantum-level properties, nanotechnology allows for unprecedented control of the material world, at the nanoscale, providing the means by which systems and materials can be built with exact specifications and characteristics, allowing materials to be lighter, stronger, smarter, cheaper, cleaner and more precise. Nanotechnology has the potential to advance agricultural productivity through genetic improvement of plants, delivery of genes and drug molecules to specific sites at cellular levels, and nano-array based gene- technologies for gene expressions in plants and animals under stress conditions. The potential is increasing with suitable techniques and sensors being identified for precision agriculture, natural resource management, early detection of pathogens and contaminants in food products, smart delivery systems for agrochemicals like fertilizers and pesticides, smart systems integration for food processing, packaging and other areas like monitoring agricultural and food system security (Abhay Kumar, 2003).

Further developments in nanotechnology in this sector can be expected to become the main economic driving forces in the long run and benefit consumers, producers, farmers, ecosystems, and the general society at large. Several metallic nanoparticles such as silver and copper have the property of antimicrobial activity. Polymer-based copper nano-compounds have been investigated with the antifungal activity against plant infecting fungi. Efficacy of silica-silver nanoparticles was studied against *Rhizoctonia solani, Pythium ultimum, Botrytis cinerea, Magnaporthe grisea* and *Colletotrichum gloeosporioides* for their control. Similarly, nano-based products were used for the control of pumpkin disease, the powdery mildew and it was demonstrated that the infecting pathogens on the leaves causing the disease were vanished within three daysafter spraying the product (Shafqat Saeed et al., 2014).

'Nanotechnology', is a brain child of late Richard Feynman, Nobel laureate in Physics in the year 1965. Nanotechnology, the frontier science of materials having unique properties than their macroscopic or bulk counter parts, has wider applications in material sciences and biology as well (Amarendra N. Misra et al., 2013). The ability of nano-materials to work at the molecular level, atom by atom, to create large structures with fundamentally new molecular organization is the key to its applications. Nanotechnology will leave no field untouched by its ground breaking scientific innovations. The agriculture and food industry is no exception. Nanotechnology has provided new solutions to problems in plants, plant products and post-harvest technology in enhancing the quality of plant products.

Nanoparticles show properties sharp contrast to their bulk in many respects which is utilised for its use in 'nanotechnology' viz., small size (1-100nm), large surface to volume ratio, chemically alterable physical properties, change in the chemical and physical properties with respect to size and shape, structural sturdiness in spite of atomic granularity, enhanced or delayed particles aggregation depending on the type of the surface modification, enhanced photoemission, high electrical and heat conductivity and improved surface catalytic activity (Amarendra N. Misra et al., 2013).

Nanoparticle: A nanoparticle (or nanopowder or nanocluster or nanocrystal) is a microscopic particle with at least one dimension less than 100 nm.

1.1 Major characteristics of nanoparticles

- Nanoparticle research is currently an area of intense scientific research, due to a wide variety of potential applications in biomedical, optical, and electronic fields.
- Nanoparticles are of great scientific interest as they are effectively abridge between bulk materials and atomic or molecular structures.
- A bulk material should have constant physical properties regardless of its size, but at the nano-scale this is often not the case.
- Size-dependent properties are observed such as quantum confinement in semiconductor particles, surface plasmon resonance in some metal particles and superparamagnetism in magnetic materials.
- The properties of materials change as their size approaches the nanoscale and as the percentage of atoms at the surface of a material becomes significant.
- For bulk materials larger than one micrometre the percentage of atoms at the surface is minuscule relative to the total number of atoms of the material.
- The interesting and sometimes unexpected properties of nanoparticles are not partly due to the aspects of the surface of the material dominating the properties in lieu of the bulk properties.
- Nanoparticles exhibit a number of special properties relative to bulk material.
- For example, the bending of bulk copper (wire, ribbon, etc.) occurs with movement of copper atoms/clusters at about the 50 nm scale.

- Copper nanoparticles smaller than 50 nm are considered super hard materials that do not exhibit the same malleability and ductility as bulk copper.
- The change in properties is not always desirable.
- Ferroelectric materials smaller than 10 nm can switch their magnetisation direction using room temperature thermal energy, thus making them useless for memory storage.
- Suspensions of nanoparticles are possible because the interaction of the particle surface with the solvent is strong enough to overcome differences in density, which usually result in a material either sinking or floating in a liquid.
- Nanoparticles often have unexpected visible properties because they are small enough to confine their electrons and produce quantum effects.
- For example gold nanoparticles appear deep red to black in solution.
- Nanoparticles have a very high surface area to volume ratio.
- This provides a tremendous driving force for diffusion, especially at elevated temperatures.
- Sintering can take place at lower temperatures, over shorter time scales than for larger particles.
- This theoretically does not affect the density of the final product, though flow difficulties and the tendency of nanoparticles to agglomerate complicates matters.
- The large surface area to volume ratio also reduces the incipient melting temperature of nanoparticles.

1.2 Nanotechnology Research, Development and Innovation in India

Nanotechnology has been heralded as a revolutionary technology by many scholars worldwide. Being an enabling technology, it has the potential to open up new vistas in the field of R&D in various multiple disciplines and have wide domain of sectoral applications, ranging from healthcare/medicines, electronics, textiles, agriculture, construction, water treatment, and food processing to cosmetics. Much of these applications are very much pertinent for a developing country like India. In this context, the government has been playing a pioneering role in fostering and promoting nanotechnology R&D in India since early 2000s.

The government has played a pioneering role in promoting nanotechnology Research and Development in India. It has taken many initiatives to foster and promote R&D in India through several of its departments. Many schemes/ programmes have been launched for infrastructure and human resource development (Amit Kumar, 2014). The government has also undertaken many bilateral/multilateral/regional initiatives with many countries to promote nanotechnology research, development and innovation (RDI). It has also encouraged PPP model to encourage nano-based product development.

The 9th Five-Year Plan (1998-2002) had mentioned for the first time that national facilities and core groups were set up to promote research in frontier areas of S&T which included superconductivity, robotics, neurosciences and carbon and nano materials. In 2001-2002, the DST set up an Expert Group on "Nanomaterials: Science and Devices". The Government identified the need to initiate a Nanomaterials Science and Technology Mission (NSTM) in the 10th Five Year Plan (2002- 07) after taking into consideration the developments in nanotechnology.

Subsequently, the National Nanoscience and Nanotechnology Initiative (NSTI) was launched in October, 2001 under the aegis of the Department of Science and Technology of the Ministry of Science. The motive of launching NSTI in 2001 was to create research infrastructure and promote basic research in nanoscience and nanotechnology. It focused on various issues relating to infrastructure development, basic research and application oriented programmes in nanomaterial including drugs/drug delivery/gene targeting and DNA chips. Nanotechnology was heralded as a revolutionary technology with applications in almost every aspect of life. Overwhelmed by the promising prospects of nanotechnology applications and in order to further enhance the visibility of India in nano science and technology, a Nano Science and Technology Mission (NSTM) was envisioned to give desired thrust to research and technology development in this area. The Eleventh Five-Year Plan (2007- 2012) categorically mentioned projects to create high value and large impact on socio-economic delivery involving nano material and nano devices in health and disease.

Accordingly, on 3 May 2007, a Mission on Nano Science and Technology (Nano Mission) was launched by the DST to foster, promote and develop all aspects of nanoscience and nanotechnology which have the potential to benefit the country. The Mission is steered by a Nano Mission Council (NMC) under the Chairmanship of Prof. CNR Rao. The technical programmes of the Nano Mission are also being guided by two advisory groups, viz. the Nano Science Advisory Group (NSAG) and the Nano Applications and Technology

Advisory Group (NATAG). The primary objectives of the Nano-Mission include Infrastructure Development for Nano Science and Technology Research; Public Private Partnerships and Nano Applications and Technology Development Centres; Human Resource Development; International Collaborations and Academia-Industry partnerships to be nurtured under these programmes (DST 2008).

In the Twelfth Five Year Plan (2012-2017) too, the government gave its approval for continuation of the Mission on Nano Science and Technology (Nano Mission) in its Phase-II at a total cost of Rs. 650 crore. The Nano Mission, in this new phase, would make greater effort to promote application- oriented R&D so that some useful products, processes and technologies also emerge. It will continue to be anchored in the Department of Science and Technology and steered by a Nano Mission Council chaired by an eminent scientist (PIB 2014).

The main departments which have been involved in nanotechnology RDI in India are discussed below.

i. Department of Science and Technology (DST): DST is the nodal agency in the Indian nanotechnology innovation system. It has since 1980s launched may programmes/schemes to foster R&D on miniature-scale and on nano-scale. Some such major programmes/ schemes are as follows (Amit Kumar, 2014).

- Intensification of Research in High Priority Areas (IRHPAS): A programme launched by the DST during the sixth Five Year Plan (1980-1985).
- The committee on Emerging Technologies was set up in 1997 to fund research for three years. Science and Engineering Research Council (SERC) also initiated a programme on nanocrystalline material.
- The National Programme on Smart Materials (NPSM), which is a five year programme funded for Rs. 90 crore , was launched jointly by the five government departments, viz. Defense Research and Development Organisation (DRDO), Council of Scientific and Industrial Research (CSIR), Department of Space DOS, DST and Ministry of Information Technology (MIT) in the year 2000.
- Nano Science and Technology Initiative (NSTI), launched by the DST, in 2001, focused on various issues relating to infrastructure development, basic research and application oriented programmes in nanomaterial including drugs/drug delivery/gene targeting and DNA chips. It gave way to Nano Mission in 2007.

- The Nano Science and Technology Mission (NSTM) – an umbrella programme – was launched in the year 2007 to promote R&D in this emerging area of research in a comprehensive fashion.

For developing physical infrastructure, the DST has established Units on Nano Science, Centres for Nanotechnology, Centre of Computational Materials Science and Thematic Units of Excellence on Computational Material Science in various universities and government research centres/laboratories across the country. For nurturing the Public-Private-Partnership (PPP) activities, the DST had sanctioned a few nanotechnology projects (DST 2008). The notable among them are as follows.

Nano Functional Materials Technology Centre (NFMTC) at the Institute of Technology Madras, Chennai: The Centre is working on cost effective method for the production of oxide ceramic powders of nano size; consolidation and sintering of nanocrystalline oxide powders for the production of bulk ceramics; nanocrystalline diamond (NCD) films/coatings on die-inserts and plugs to increase wear-resistance and durability; cost effective production of large scale and highly pure random and aligned carbon nanotubes (CNT); and nanostructured multidrug-delivery system for hard tissue applications and CNTs for laser based treatment of cancer by photodynamic therapy.

- Development of high performance rubber nanocomposites for tyre engineering: This project being implemented at Mahatma Gandhi University, Kottayam envisages the development of novel technologies in tyre engineering band on nanosize fillers in collaboration with Apollo Tyres.
- Research programme on Smart and Innovative Textiles (SMITA) at the Indian Institute of Technology, Delhi.
- Support to Nanotechnology Business Incubator (NBI) at NCL: The NBI has nurtured activities by seven start-up companies on items like– computational modelling of flow and chemical processes, therapeutic potential of biotechnologically engineered antibodies, ocular and maxillofacial implants, and 12 start-up companies are under incubation presently as Resident Incubates on items like maxillo-facial surgery, organic chemical synthesis, etc., (DST 2013).

As far as the international collaborations in the field on nanoscience and nanotechnology research, development and innovation are concerned, the DST has launched bilateral joint research projects with more than 25 countries and multilateral research projects with regional/multilateral bodies such as EU, BRICS and IBSA. The DST has been actively promoting international collaborations through bilateral joint research-projects over a period of time. The investment made by DST has remarkably increased almost 10 times in the period 2006-2012. These international collaborative projects cover a range of nanoscience and nanotechnology basic and applied research in areas having diverse applications such as Healthcare/medicine/drug delivery; Sensors; Energy storage in batteries; Solar energy; Fuel cells; Other areas of use of nano-materials such as thin films, carbon nanotubes, nano composites, etc.

ii. Department of Biotechnology (DBT): The DBT is basically active in the area of nano biotechnology R&D. It has been engaged in promoting interdisciplinary research, fostering innovations and promoting development of translational research in various areas of nano biotechnologies including developing new therapeutics, diagnostics for early disease detection and imaging, design and development of tissue specific drug delivery, medical devices, and fabricating sensors for detection of chemicals and pathogens in food and crop.

In the biotechnology sector, the government, while realising the importance of nanotechnology applications in biotechnology, sees nano-bio industries as a new-generation biotech industries (GoI 2012). In the period around 2007, the Department of Biotechnology (DBT) initiated a programme on nanobiotechnology. Since then, this Programme has promoted basic R&D in the following areas (DBT 2008).

- Nanotechnology for food/agriculture: weed utility, nanosensor for crop protection, pesticide delivery vehicles, nanocides, smart packaging, sensors for detecting pathogens and chemicals in food and crop, etc.
- Nanotechnology for animal husbandry: biodegradable nanoparticles for drug delivery, etc.
- Nanotechnology for environment management: biosysnthesis of nanoparticles, treatment of industrial effluent, waste management, etc.
- Nanotechnology for healthcare/medicine/drug delivery: Drug delivery system, disease diagnosis, cancer and TB therapy, scaffolds, medical devices, implants and imaging, etc.
- Nanotechnology in other allied areas such as bioengineering, water filtration, etc.

Department of Electronics and Information Technology (DeITy): The DeITy has established Centre for Materials for Electronics Technology (C-MET) at Pune, Hyderabad and Trissur. These centres are involved in nanotechnology R&D activities, particularly in nanomaterials. They provide matrials, fabricating facilities, prototyping facilities, technology transfer and consultancy services for the interested entities including industries.

iii. Some of the key schemes/policies that have been launched in this direction are as follows.

- DeITy started the Nanotechnology Development Programme (NDP) in 2004 to create infrastructure and promote basic research in the areas of nanoelectronics and nanometrology in India. The mission was to develop knowledge base in the electronics materials and their processing technology for the Indian industry (DeITy 2006).
- A special package was initiated to fuel further growth of semiconductor manufacturing. This Special Incentive Package Scheme (SIPS) for Semiconductor Fabrication and Micro and Nanotechnology Manufacture Industry was launched by the DeITy in 2007 to attract investments for setting up semiconductor fabrication and other micro and nanotechnology manufacturing industries in India. It has received very positive response from prospective investors (DeITy 2011).
- In the whole scheme of creating globally competitive Electronics System Design and Manufacturing (ESDM) industry in India, the nano-electronics has been identified as a key element to meet the country's needs and serve the international market (DeITy 2013). The National Policy on Electronics (NPE) launched in 2012, talks about the need for a separate sector specific strategy.
- A Nanoelectronics Innovation Council has also been set up recently by DeITy in 2012. The scope of the Council includes the entire electronics systems hardware infrastructure enabled by nanoelectronics technology.
- Electronics Development Fund, involving US$ 2 billion, has been proposed by DeITy to promote innovation, product development, commercialisation of products in electronic system design and manufacturing (ESDM), nanoelectronics and IT sectors.

iv. Department of Industrial Policy and Promotion (DIPP): DIPP established a Nano-Manufacturing Technology Centre (NMTC) and Academy of Excellence for Advanced Manufacturing Technology (AEAMT) at the Central Manufacturing Technology Institute (CMTI) in Bangalore, Karnataka.

v. Department of Industrial and Scientific Research: The DISR through its autonomous body, the CSIR provides for projects funding as well as various grants and fellowships and help in S&T and human resource development in the basic and applied sciences including that in nanoscience and nanotechnology.

The CSIR has 39 laboratories established all across the country. They are engaged in R&D in biological, chemical, physical and engineering fields. Nano-related R&D is also being carried out in some of these labs.

To sum up, the development of nanotechnology in India has been mainly conceived and continued on the premise that this new and emerging technology has huge potential to help the country address societal challenges such as provision of drinking water, healthcare, etc., and simultaneously achieve economic gains through growth in the nanotech-based industrial sector. There have been various concerted efforts since early 2000s by the government to foster and promote nanotechnology in India. The Plan documents and various initiatives taken by various departments/ministries have placed much emphasis on this technology (Amit Kumar, 2014). In addition to this, there have been efforts to establish a regulatory framework at national level for addressing the risk and safety aspects of nanotechnology. The world over, there have been serious efforts taking place in this regard and this is the high time that India too come up with a framework to ensure safety of humans and environment and minimise any unintended consequences out of the use of nanotechnology. The multilateral/bilateral cooperation play a major role in promoting cutting-edge basic R&D by providing Indian scientists access to sophisticated equipment/facilities in advanced countries. India's early involvement with various international/ inter-governmental organizations, such as International Standards Organization (ISO), Organization for Economic Cooperation and Development (OECD), and IRGC, for the development of standards, safe lab practices and risk governance is quite significant.

1.3 Nanotechnology in Agriculture

Agricultural nanotechnology has emerged in the late 1990s and is developed and applied all over the world. However, this technology has not developed so fast in different sectors of agriculture. This has not even found its market to the expected scale but has the potential to improve agricultural production. This agro nanotechnology can be utilized for developing healthy seeds that can improve plant germination, growth, yield, and quality. This technology has the potential to increase the storage period for vegetables and fruits. Organic pesticides and fertilizers can be developed by the proper use of agro nano technology. It has found its use in genetic engineering and plant breeding. However, the application of nanotechnology provides numerous advantages related to food safety and quality, at the same time it may present a potential risk not only to human health, but can affect animals and the environment as well as nanoparticles may have toxicological effects on biological systems (Wani Khursheed Ahmad and Kothari Richa, 2018).

In the field of agriculture, nanotechnology has been used to heighten the crop production with quality enrichment by improving farming systems as schematically shown in fig.1 (Yifen Shang et al., 2019). In order to achieve food security, advanced nano-engineering is a handy tool for boosting crop production and assuring sustainability. Nanotechnology helps to improve agricultural production by increasing the efficiency of inputs and minimizing relevant losses. Nanomaterials offer a wider specific surface area to fertilizers and pesticides. In addition, nanomaterials as unique carriers of agrochemicals facilitate the site-targeted controlled delivery of nutrients with increased crop protection. Due to their direct and intended applications in the precise management and control of inputs (fertilizers, pesticides, herbicides), nanotools, such as nanobiosensors, support the development of high-tech agricultural farms. The integration of biology and nanotechnology into nonosensors has greatly increased their potential to sense and identify the environmental conditions or impairments.

The emergence of engineered nanomaterials and their actions within the frame of sustainable agriculture have revolutionized world agriculture canvass dramatically by novelty, fast growth and enormity to meet the projection of global food demand. In sustainable agriculture, the protection of the environment from pollution is the crucial target for trade, and nanomaterials provide an assurance of better management and conservation of inputs to plant production. The potential of nanomaterials encourages a new green revolution with reduced farming risks. However, there are still huge gaps in our knowledge of the uptake capacity, permissible limit and the ecotoxicity of different nanomaterials. Therefore, further research is urgently needed to unravel the behavior and fate of altered agriculture inputs and their interaction with biomacromolecules present in living systems and environments.

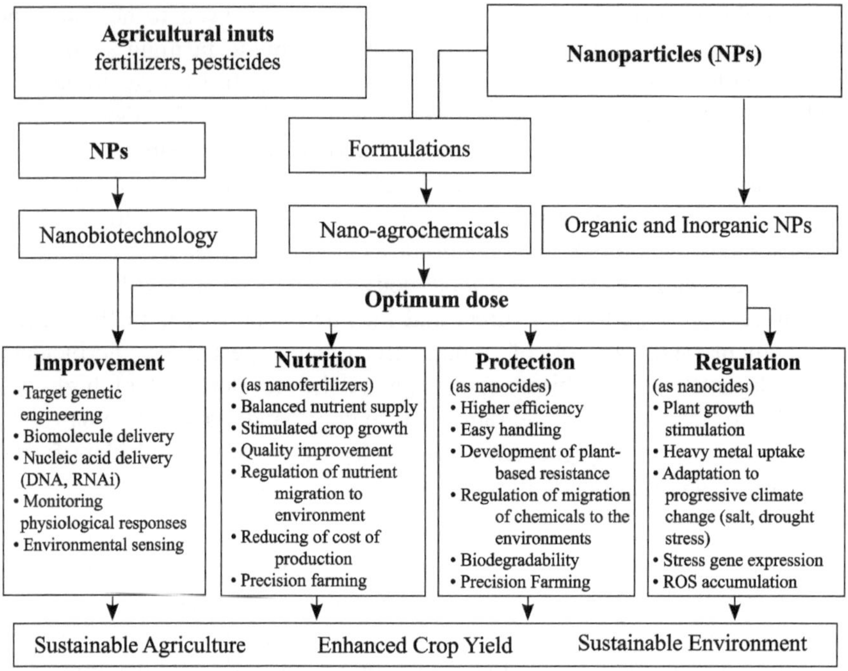

Fig. 1: Overview of potential applications of nanomaterials in sustainable agriculture production

In agriculture, some of the world's largest makers of pesticides, fertilizers, and other farm inputs and technologies are betting on nanotechnology to bring unprecedented precision to crop and livestock production (Touseef Hussain, 2017). Potential applications of nanotechnology in crop protection include controlled release of encapsulated pesticide, fertilizer and other agrochemicals in protection against pests and pathogens, early detection of plant disease and pollutants including pesticide residues by using nanosensors. The potential applications of nanomaterials in crop protection, helps in the development of efficient and potential approaches for the management of plant pathogens. Nanotechnology has emerged as one of the most innovative scientific field in agriculture.

It provides opportunity to develop improved systems for monitoring environmental conditions and delivering nutrients or pesticides as appropriate, improve our understanding of the biology of different crops and thus potentially enhance yields or nutritional values. Precision farming-Nanotechnology application here makes farming more targeted and scientific. Precision farming makes use of computers, global satellite positioning systems, and remote sensing devices to measure various parameters. Accurate information through applications of Nanotechnology for real time monitoring of soil conditions, environmental changes and diseases and plant health issues. Nanoscale devices with novel properties could be used to make agricultural systems "smart".

Nanotechnology has the potential to revolutionize the agricultural and food industry with new tools for the molecular treatment of diseases, rapid disease detection, enhancing the ability of plants to absorb nutrients etc. Smart sensors and smart delivery systems will help the agricultural industry combat viruses and other crop pathogens. In the near future nanostructured catalysts will be available which will increase the efficiency of pesticides and herbicides, allowing lower doses to be used. However, currently there is a limited understanding of human health and safety risks associated with this technology (Touseef Hussain, 2017). The potential application of nanocides in different agricultural applications needs further research investigation with respect to synthesis, toxicology and its effective application at field level. Nanotechnology has got applicability in various fields, but research and development is still at bench-top scale. Great efforts are required in commercialization of nanocides for agricultural applications, which requires proper protection needs, testing priorities, risk assessment and regulatory guidance at global level. Controlled use of the technology will open opportunities for developing new materials and methods that will enhance our ability to develop faster, more reliable and more sensitive analytical systems. Overall the scenario presents us with the view that nanotechnology is here to stay.

Nanotechnology, the process to generate, manipulate, and deploy nanomaterials, represents an area holding significant promise for the agricultural scenario (Vandana Ghormade et al., 2011). A wide variety of materials are used to make NPs, such as metal oxides, ceramics, silicates, magnetic materials, semiconductor quantum dots (QDs), lipids, polymers, dendrimers and emulsions. Polymers display controlled release of ingredients, a character useful for developing polymeric NPs as agrochemical carriers. Metal nanoparticles display size dependent properties such as magnetism (magnetic NPs), fluorescence (QDs) or photocatalytic degradation (metal oxide NPs) that have biotechnological applications in sensor development, agrochemical degradation and soil remediation (Table 3).

Table 3: Applications of nanotechnology in agriculture

Application	Nanoparticles
Pesticide delivery	
a. Chemical	
Avermectin	Porous hollow silica (15 nm)
Ethiprole or phenylpyrazole	Poly-caprolactone (135 nm)
Gamma cyhalothrin	Solid lipid (300 nm)
Tebucanazole/chlorothalonil	Polyvinylpyridine and polyvinylpyridine-co- styrene (100 nm)
b. Biopesticides	
Plant origin: nanosilica	Nanosilica (3–5 nm)
essential oil encapsulation	Solid lipid (200–294 nm)
Microorganisms: Lagenidium giganteum cells in emulsion	Silica (7–14 nm)
Microbial product: absorption of Myrothrecium verrucaria enzyme complex	Chitosan/kaolin (250–350 nm)
c. Controlled delivery	
NPK controlled delivery	Nano-coating of sulfur (100 nm layer) Chitosan (78 nm)
Genetic material delivery DNA	Gold (10–15 nm), Gold (5–25 nm), Starch (50–100 nm)
Double stranded RNA	Chitosan (100–200 nm)
d. Pesticide sensor	
Carbofuran/triazophos	Gold (40 nm)
DDT	Gold (30 nm)
Dimethoate	Iron oxide (30 nm), zirconium oxide (31.5 nm)
Organophosphate	Zirconium oxide (50 nm)
Paraoxon	Silica (100–500 nm), Carbon nanotubes
Pyrethroid	Iron oxide (22 nm)
e. Pesticide degradation	
Lindane	Iron sulfide (200 nm)
Imidacloprid	Titanium oxide (30 nm)

Potential applications of nanotechnology in agriculture include delivery of nanocides–pesticides encapsulated in nanomaterials for controlled release; stabilization of biopesticides with nanomaterials; slow release of nanomaterial assisted fertilizers, biofertilizers and micronutrients for efficient use; and field applications of agrochemicals, nanomaterials assisted delivery of genetic material for crop improvement. Nanosensors for plant pathogen and pesticide detection, and NPs for soil conservation or remediation are other areas in agriculture that can benefit from nanotechnology. Enzyme immobilization for nanobiosensor using nanomaterials involves the high value low volume application of enzymes (Kim et al., 2006).

While in recent years the enormous progress of nanotechnology in biomedical sciences has revolutionized therapeutic and diagnostic approaches, the comprehension of nanoparticle-plant interactions, including uptake, mobilization and accumulation, is still in its infancy. Deeper studies are needed to establish the impact of nanomaterials (NMs) on plant growth and agro-ecosystems and to develop smart nanotechnology applications in crop improvement. Sanzari Ilaria et al., (2019) provided a short overview of NMs employed in plant science and concisely describe key NM-plant interactions in terms of uptake, mobilization mechanisms, and biological effects.

Metallic nanoparticles, such as gold (Au), and silver (Ag) NPs, have been widely introduced in plant science for different applications. Their chemical synthesis is quite costly and requires the use of hazardous chemicals. However, greener approaches based on the use of plant extract as well as ionizing radiation chemistry in aqueous solutions have been developed. Also oxidized NMs, such as MgO, CaO, ZnO, and TiO2 NMs, have been widely proposed, thanks to their superior electrical, catalytic and light absorption properties. Over the recent years, the interest in polymeric nanomaterials is predominantly increasing due to their biocompatibility, low-cost synthesis and capability to response to external stimuli. Core/Shell NPs are also available and can be manufactured with a variety of combination of materials such as inorganic/inorganic, inorganic/organic, organic/inorganic, and organic/organic materials.

Nanogels (NGs) are a new category of NM with a growing interest in the nanotechnology community. They have excellent physicochemical properties, colloidal stability, high encapsulation capacity of biomolecules (bioconjugation), and stimuli-responsiveness (pH, temperature, etc.) (Sanzari Ilaria et al., 2019). NGs are defined as nano-sized ionic and non-ionic hydrogels made of synthetic or natural polymeric chains, chemically or physically cross-linked. NGs possess a high water content (70–90% of the entire structure), a high degree of porosity and high load capacity. The most common NGs are chitosan, alginate, poly (vinyl alcohol), poly (ethylene oxide), poly (ethyleneimine), poly (vinylpyrrolidone), poly(N-isopropylacrylamide). NGs with hybrid structures, made of polymeric or non-polymeric materials can be obtained. Hybrid NGs have been classified in nanomaterial– nanogel, which are synthesized by incorporation of nanosized materials such as magnetic or carbonaceous nanoparticles, and polymer–nanogel composites, which include interpenetrated networks (IPNs), copolymer, and core-shell particles. The main advantage of IPNs and copolymer NGs relies on their stimuli-responsiveness, whereas core-shell NGs are more promising for encapsulating biomolecules and drug delivery.

On the base of consolidated evidence reported in cell and animal models, soft materials, like nanogels, and polymeric nanostructures should be further exploited as favorable candidates to develop new strategies for controlled release of biomolecules and plant genome editing. Owing to their safe profile, high loading capacity and excellent cargo protection from degradation polymeric and hydrogel-based NPs have shown undeniable advantages in drug delivery. Moreover, this kind of NMs has been elegantly employed to achieve a controlled (spatial and temporal) release of cargos triggered by external stimuli (e.g., UV, NIR, acoustic waves etc.) in cell and animal models. These outstanding results suggest that the huge potential of soft nanomaterials remains almost unexplored in plants. Besides a few successful attempts for agrochemicals delivery, more efforts are needed to design strategies and smart tools based on polymeric or hybrid materials for applications in plants. Of course, a careful analysis of manufacturing scalability and cost-effectiveness needs to be undertaken before the extensive use of polymeric nanomaterials in agriculture.

The delay in plant nanotechnology might be overcome by encouraging the activation of multidisciplinary approaches for the design and the synthesis of smart nanomaterials. To this aim, joint collaborative initiatives, merging complementary professional competencies such those of plant biologists, geneticists, chemists, biochemists, and engineers, may disclose new horizons in phytonanotechnology.

Nanomaterials and Agriculture: Nanomaterials often have chemical, physical, or biological properties that are different from those of their larger counterparts and due to their different properties, nanomaterial may pose different safety issues than their larger counterparts. There has been main attention in using nanotechnology in agriculture and the food system due to great potential as it can improve the quality of different products, also, with the rapid advancement of nanotechnology since the last decade of last century, controlled preparation of Nanomaterials with desired morphology and size, and newly established concepts and methodology have underpinned the solid bases to solve the unsolved questions in nutrient uptake. Direct applications of nanotechnology in agricultural production and products generally embrace delivery of various agrochemicals, the study of plant disease mechanisms and genomes improvement (Waleed Fouad Abobatta, 2018).

Nanomaterials have been recently reported to have a potential role in plant protection by: i) increasing the dispersion and wettability, and the affinity to pesticide target (Stefania Mura et al, 2013); ii) providing better penetration of herbicides through cuticles and tissues of weeds; iii) altering viral capsids to achieve different configurations and deliver specific nucleic acids, enzymes or antimicrobial peptides acting against the parasites. The use of nanostructured catalysts could improve the efficiency of pesticides and herbicides, allowing lower doses to be used. A potential development of new nanomaterials is towards the building of agrochemicals containing different nano-substances for different functions (plant protection, fertilization, hormones, etc.) and encapsulated separately to avoid interactions between them and degradation. In this way, several substances could be applied with the same treatment to the crop, and the nanocapsules could be regulated to release their load according to the characteristic of every loaded substance. Some studies showed the potential use of nanotechnologies to detect microbes, contaminants, toxic pollutants and for monitoring soil conditions, crop growth, plant microenvironment and enzyme/substrate interactions. Nanotechnology can also be applied for remediation purposes through the construction of nano-photocatalysts able to degrade organic pesticides and industrial pollutants in harmless and useful products. Filters or catalysts could also be created to reduce pollution or clean-up existing pollutants.

There are various positive effects of nanomaterials in Agriculture. In nano-world, materials less than 100-nanometer size behave completely different, the rules that manage the behavior of the elements of our known world start to give way to the rules of quantum mechanics, and everything changes. There are various advantages that nanotechnologies offer due to the unique functional properties of nanoparticles and materials including the higher solubility of nanoparticles in suspension; the higher surface area and particle size of the nanoparticles, which facilitates penetration of seed coats and subsequently emerging roots; better bioavailability of molecules to the seed radicals; the properties of nanomaterial for agricultural applications. The model Nanomaterials for agricultural applications are supposed to have the following properties; providing actual concentration and controlled release of fertilizers or pesticides in response to certain conditions (TiO_2 Nano particles used as plant fertilizer for Mung bean to enhance crop production) ; improved targeted activity; lower eco harmful with safe and relaxed transport.

Nanomaterials usage in agriculture: Nanomaterials have many usages in all stages of agricultural production, in different forms and various procedures such as:nano-fertilizer for balance crop nutrition; crop improvement (Zinc Nano fertilizer used to enhance crop production of *Pennisetum americanum*); plant protection ingredients (pesticides, fungicides, weedicides); weed management; nano pesticides; nano sensors; post-harvest technology; bioprocessing (bio synthesized) nanoparticles for agricultural use; bio sensors for Aqua culture; nano biotechnology (Analysis of gene expression and regulation); monitoring the identity and quality of agricultural produce; precision agriculture:

Precision agricultural techniques might be used to promote increase crop yields but not damage soil and water, decrease nutrients loss due to leaching and emissions, in addition to enhance nutrients long-term incorporation by soil microorganisms; Seed technology; Water management; Plant growth regulators; Soil management; Agricultural engineering aspects; Food technology. Some applications of nanotechnology in agriculture include Crop improvement; Increase efficient fertilizers and pesticides; Soil management; Plant disease detection; Analysis of gene expression and Regulation and Post-Harvest Technology.

Nanotechnology offers a new scientific approach to break this yield barrier and may improve our understanding of the biology of different crops. This has the potential to enhance yields and nutritional values, and improve systems for monitoring environmental conditions and delivering nutrients or pesticides as appropriate. In addition, it can offer solutions to meet the challenges of food security and of environmental remediation. The use of nanotechnology in agriculture shows promise. However, research into most applications is still ongoing, such as in the field of crop breeding and production of more selective, effective and easier-to-dose plant protection products (Stefania Mura et al., 2013).

Some studies demonstrated the role of nanomaterials in enhancing seed germination rates in tomato, spinach and rice and consequently the crop dry matter yield, at least in the early stages of growth. However, several concerns remain on the possible phytotoxicity of some types of nanomaterials (nano-sized Zn, nano-Al, La_2O_3, Gd_2O_3, nano-Ag), depending on their composition, concentration, size and physical chemical properties and on the susceptibility of the plant species. Plant production could also be increased through the development of nanomaterials that can be used as a coating layer to enable the slow release of traditional fertilizers.

Another field of application of nanotechnology is particle farming. This yields NPs for industrial use by growing plants in defined soils. For example, research has shown that alfalfa plants grown in gold rich soil absorb gold NPs through their roots and accumulates them in their tissues. The gold NPs can be mechanically separated from the plant tissue following harvest. The formation of NPs of an alloy of gold-silver-copper using plants was also reported. Industry has shown a great deal of interest in this area (Stefania Mura et al., 2013).

Nanosilver: A growing use of nanosilver is being found in a number of consumer products, including food and health food, water, and food contact surfaces and packaging materials. Indeed, the use of nanosilver as an antimicrobial, antiodourant, and a (proclaimed) health supplement has already surpassed all other nanomaterials currently in use in different sectors. Most current uses of nanosilver relate to health food and packaging applications, but its use as an additive to prepare antibacterial wheat flour was the subject of a rather recent patent application.

Nanosilica: Amorphous nanosilica is known to be used in food contact surfaces and food packaging applications. Amorphous silica has been used for many years in food applications, such as in clearing processes of beers and wines, and as a free-flowing agent in powdered soups. The conventional bulk form of silica is a permitted food additive (SiO_2 INS 551), but the material may not have been tested with a focus on nanosilica. Porous silica is used in nanofiltration to remove undesired components in food and beverages such as the bitter taste in some plant extracts.

Nanotitanium dioxide: The conventional bulk form of titanium dioxide has already been approved as an additive for food use [titanium dioxide (TiO_2) INS 171], but the conventional form may also contain a nanosized fraction. Nanotitanium dioxide is used in a number of consumer products (e.g. paints, coatings) and its use may extend to foodstuffs. For example, a patent (US Patent US5741505) describes the potential application of nanoscale inorganic coatings directly on food surfaces to provide a barrier to moisture and oxygen, and thus improve shelf life and/or the flavour impact of foods. The materials used for the nanocoatings, intended to be applied in a continuous process as a thin amorphous film of 50 nm or less, include titanium dioxide (along with silicon dioxide and magnesium oxide).

Nanoselenium: Nanoselenium is being marketed as an additive to a green tea product, with a number of (proclaimed) health benefits resulting from enhanced uptake of selenium.

Nanocalcium: Nanocalcium salts are the subject of patent applications (Sustech GMBH & Co, Düsseldorf, Germany; http://www.nanomat.de) for intended use in chewing gums. Nanocalcium and nanomagnesium salts are also available as health supplements.

Nano-iron: Nano-iron is available as a health supplement. Nano-iron is also used in the treatment of contaminated water, where it is claimed to decontaminate water by breaking down organic pollutants and killing microbial pathogens. Iron nanoparticle technology has received considerable attention for its potential application in groundwater treatment and site remediation.

Agribusiness sector: Thinking strategically will become increasingly important for successful agribusinesses of the future with respect to nanotechnology application (Vincent Sabourin and Alpha Ayande, 2017).

Nanotechnology could be used to enhance the possibilities of developing conventional and stranded agribusiness resources. Nanotechnology can make the industry considerably greener and competitive, with its current growth rate of 25% (US$ 1.08billion) annually. The opportunity for application of nanotechnology in agriculture is prodigious. Nanotechnology, focusing on special properties of materials emerging from nanometric size has the potential to revolutionize the agricultural and food sectors, biomedicine, environmental engineering, safety and security, water resources, energy conversion, and numerous other areas.

Metal based nanoparticles in agriculture: With the fast-developing nanotechnology, metal based nanoparticles (NPs) production and application are increased significantly. These metal based NPs can enter agricultural land through both direct and indirect pathways (Hao Chen, 2018). The physical chemical properties of both metal based NPs (e.g. size, surface charge, surface coating) and soil matrix (e.g. pH, ionic strength, mineral composition, dissolved organic matter) all play important roles in determining the mobility, transformation and potential risks of metal based NPs in plant and soil system. NPs can be accumulated to plant roots and translocated to other parts of the plants. The properties of both plant and metal based NPs are playing critical roles to this process. Systematic research of metal based NPs in environmentally relevant concentrations and conditions is needed for the future study. Despite increasing concern about NPs in the environment and subsequent ecosystem effects, little information available about the quantity of NPs entered agriculture soil system thorough direct or indirect input. Research of the fate, transport, bioavailability and potential environmental risks of metal based NPs are still in early stage.

Metal based NPs may have both positive and negative effect on plant. As toxicity occurs at all physiological to biochemical and molecular levels, beneficial effect also displays in accelerating growth, photosynthetic yield, fruit yields and pest control. However, the mechanism for NPs phytotoxicity and the NPs translocation pathways inside plants are still largely unknown. The interaction of metal based NPs with soil, plant root and soil microbial in root rhizosphere should also be emphasized in future studies to truly reflective the effect of metal based NPs in natural environmental conditions.

Nanodiffusion in the environment: Nanoparticles are not a human invention and have always existed naturally. For example, NPs have been found in 10,000-year old glacial ice cores, and there is evidence of natural NP formation in sediments at the Cretaceous-Tertiary boundary. If we consider atmospheric dust alone, estimates indicate approximately one billion metric tons per year produced globally and even with a fraction of this as ultrafine particles, this would be millions of tons of natural NPs. If we compare this against production estimates for manufactured nanomaterials of the order of a few thousand tons per year for each major type of material, it is clear that exposure to natural NPs vastly outweighs any anthropogenic production. There are several mechanisms that create NPs in the environment and these can be either geological or biological. Geological mechanisms include physicochemical weathering, authogenesis/neoformation (e.g., in soils), and volcanic activity. These geological processes typically produce inorganic NPs. Biological mechanisms typically produce organic nano-molecules, although some organisms can produce mineral granules in cells.

Nanomaterials in soil: Nanoscience is of crucial importance to the soil sciences because many natural compounds of the soils are nanoparticulate or have nanoscale features. At the nanoscale, particle-particle interactions are either dominated by weak Van der Waals forces, stronger polar and electrostatic interactions, or covalent interactions. Depending on the viscosity and polarisability of the fluid, particle aggregation is determined by the interparticle interaction. Modifying the surface layer can enhance or hinder the tendency of a colloid to coagulate. In liquids, particle charge can be stabilised by electrochemical processes at surfaces. The details of nanoparticle–nanoparticle interaction forces and nanoparticle–fluid interactions are of key importance to describe physical and chemical processes, and the temporal evolution of free NPs.

Different nanomaterials can be found in soil. Nanomaterials that only exist in the nanoparticle range and mineral NPs that are minerals in nanosize range but also exist at larger sizes (Stefania Mura et al., 2013). Different nanomaterials are present in soils as nanoscale aggregates of natural organic matter, bacterial appendages, organic and inorganic particles as zeolites, humic and fulvic acids, allophane and imogolite, hydroxides and aluminosilicate, clay minerals, amorfous substance and other nanominerals as ferrihydrite, Fe (hydr)oxides (hematite, goethite, maghemite, lepidocrocite, magnetite), Mn (hydr)oxides, uraninite (UO_2) and swelling clays (montmorillonite). In particular, different classes of inorganic materials such as silicates (clay and mica), oxides/hydroxides (MnO), carbonates ($CaCO_3$), phosphates, metal sulphides (ZnS) or organic materials as cell fractions, macromolecules or biocolloids (bacteria) can be found in soils. Manufactured or engineered NPs can also be found in soils. These can be released in the environment either unintentionally (corrosion or combustion of products) or intentionally in different forms: metal oxides [TiO2, zinc oxide (ZnO)], semiconductor materials (quantum dots), carbon derived nanomaterials (carbon nanotubes), zero valent metals (zero valent iron), nanopolymers. These classes of compounds come from the different industries (e.g. fullerene from electronic and optic, CeO_2 from automotive, AgNPs from medical devices, ZnO from solar cells, TiO_2 from cosmetics and painting), and significant amounts of these nanomaterials are deposited in soil.

Nanotechnology could have several applications in soil science. As discussed above, the use of nanofertilisers can facilitate nutrient transport to the rhizosphere when needed, and in more suitable amounts and composition, thereby improving efficiency of use. NPs locked onto the roots can also enhance elemental uptake. Nanomaterials, such as hydrogels and zeolites, were reported to be useful to improve water-holding capacity of the soil and to absorb environmental contaminants.

Development of plant protection products: Sustainable agriculture calls for minimal use of agrochemicals in order to protect the environment (Yurii A Krutyakov et al., 2016). It has caused an increase in the rate of nanoparticles use, in particular silver nanoparticles (AgNPs) due to their safety for mammals, unique biological activity and a broad spectrum of action against fungal and bacterial pathogens. Until now the use of AgNPs dispersions in the agricultural sector has been essentially limited due to many factors decreased their stability (mixing with other pesticides, presence of electrolytes). A versatile synthesis of polyampholyte surfactant (tallow amphopolycarboxyglycinate) stabilized AgNPs was presented. Unique aggregation behavior (via dynamic light scattering and UV–vis spectroscopy) and biocidal activity of obtained silver colloids were observed. AgNPs were characterized by exclusively high aggregative stability in the presence of coagulating agents $NaNO_3$ and $NaSO_4$ (up to 1 M), during drying/redispergation, and frost/defrost cycles. The dispersion of AgNPs showed high biocidal activity (EC_{50} is ten times lower than commercial species ones) with respect to *Phytophthora infestans* and phytopathogenic fungi. This points to the possibility of successful application of silver preparations within agriculture with the goal of partial reduction of the use of toxic and expensive synthetic antibiotics and pesticides.

Insect pest management: The potential uses and benefits of nanotechnology are enormous. These include management of insect pests through the formulations of nanomaterials-based insecticides. Traditional strategies like integrated pest management used in agriculture are insufficient, and application of chemical pesticides have adverse effects on animals and human beings apart from the decline in soil fertility. Therefore, nanotechnology would provide green and efficient alternatives for the management of insect pests in agriculture without harming the nature (Mohamed Ragaei and Al-kazafy Hassan Sabry, 2014). This art is focused on traditional strategies used for the management of insect pests and potential of nanomaterials in insect pest control as modern approaches of nanotechnology.

Nanopesticide formulations increase the solubility of poorly soluble active ingredient and helps in releasing the active ingredient slowly (Ashish Khandelwal and Ritika Joshi, 2018). The bioavailability of poorly water-soluble agrochemicals can be increased through the use of additives or by nanoparticulate formation of agrochemicals.

1.4 Potential risks of Nanotechnology

There are some negative effects of nanomaterials on biological systems and the environment caused by nanoparticles, like chemical hazards on edible plants after treatment with high concentration of Nano Silver, also, in some cases, nanomaterial generated free radicals in living tissue leading in DNA damage, therefore nanotechnology should be carefully evaluated before increasing the use of the nano agromaterials (Waleed Fouad Abobatta, 2018).

Nanotechnology considers a novel key to growing agricultural production through implementing nutrient efficiency, improve plant protection practices, also, nanotechnology may have real solutions for various agriculture problems like improved crop varieties, plant protection, detect diseases and monitor plant growth. Nanotechnology offers generous visions for the development agricultural sector through advanced applications and the probability of products and increases global crops production volume to feed the world population in next decades. Promising results and applications are already being developed in the areas of nano nutrients, implement crop productivity, protect plants (herbicides and pesticide), nano-packing and Nano sensors.

While nanopesticides are an attractive technological advancement, many of the approaches and assumptions used to assess the risks of conventional chemicals may not be appropriate for engineered nanoparticles such as some used in nanopesticides (Virginia Tressider, 2015).

1.5 Nanotechnology in Plant Pathology

Plant diseases, as already known, are the major factors that limit crop productivity. The first step towards managing a plant disease under greenhouse and field conditions is to correctly identify the pathogen. Nano-diagnostic kit equipment can easily and quickly detect potential serious plant pathogens, allowing plant pathologist to help farmers in preventing epidemic diseases (Kamel Ahmed Abd-Elsalam, 2015). The application of nano-technology for quicker, cheaper, and more precise diagnostic procedures of plant diseases was deal along with information on nano-technology that can offer green and eco-friendly alternatives for plant disease management. Apart from being ecofriendly, fungi are used as bio-manufacturing units, which will provide an added benefit in being easy to use, as compared to other microbes. Nanotechnological application in plant pathology is still in the early stages. However, nano-fungicides, nano-pesticides and nano-herbicides are being used extensively in agriculture practices. Remote activation and monitoring of intelligent nanodelivery systems can assist agricultural growers in the future to minimize fungicides and pesticides use. Nanoparticle-mediated gene transfer would be useful for improvement of crops resistant to pathogens and pest. Such an accurate technology may also help to design a proper integrated disease management system which may modify crop environments to adversely affect plant pathogens.

Engineered nanoparticles are materials between 1 and 100 nm and exist as metalloids, metallic oxides, nonmetals, and carbon nanomaterials and as functionalized dendrimers, liposomes, and quantum dots. Their small size, large surface area, and high reactivity have enabled their use as bactericides/fungicides and nanofertilizers (Wade Elmer and Jason C. White, 2018). Nanoparticles can be designed as biosensors for plant disease diagnostics and as delivery vehicles for genetic material, probes, and agrichemicals. In the past decade, reports of nanotechnology in phytopathology have grown exponentially. Nanomaterials have been integrated into disease management strategies and diagnostics and as molecular tools. Most reports summarized herein are directed toward pathogen inhibition using metalloid/metallic oxide nanoparticles as bactericides/fungicides and as nanofertilizers to enhance health. The use of nanoparticles as biosensors in plant disease diagnostics was also reviewed. As global demand for food production escalates against a changing climate, nanotechnology could sustainably mitigate many challenges in disease management by reducing chemical inputs and promoting rapid detection of pathogens.

An advanced approach adopted to protect crops and maintain long-term sustainability is nanotechnology, which acts on the concept of "minimal usage with maximum effect." Nanoformulations of pesticides, nanosensors, and the generation of nanoparticle-mediated resistant crop varieties are a part of the spectrum of agricultural nanoapplications. Agrochemicals for plant protection that are encapsulated with polymer-based nanomaterials (nanocapsules, nanospheres, micelles, and nanogels), lipid-based nanomaterials (liposomes, solid lipid nanoparticles), porous inorganic nanomaterials (mesoporous silica nanoparticles, porous hollow nanoparticles), and clay-based nanomaterials (clay, layered double hydroxides) for controlled release ensure better solubility, stability, and long-term effect of the agrochemicals (Sunita Ojha et al., 2018).

Silver, zinc oxide, copper oxide and magnesium oxide nanoparticles have shown promising antimicrobial activity against phytopathogens. Usability of nanobased materials also lies in gene delivery, infection detection, pesticide/ nutrient level measurement, etc., which will be helpful in effective farming and crop protection.

Potential applications of nanotechnology in crop protection include controlled release of encapsulated pesticide, fertilizer and other agrochemicals in protection against pests and pathogens, early detection of plant disease and pollutants including pesticide residues by using nanosensors (Ashish Khandelwal and Ritika Joshi, 2018). The potential applications of nanomaterials in crop protection, helps in the development of efficient and potential approaches for the management of plant pathogens.

Chitosan: In search of natural antimicrobials to avoid harmful synthetic chemicals, chitosan and chitosan nanoparticles are found to be more effective against plant pathogens like *Fusarium solani.* Inhibitory effect was also influenced by particle size and zeta potential of chitosan nanoparticles. The chitosan therefore could be formulated and applied as a natural antifungal agent in nanoparticles form to enhance its antifungal activity.

Silver nanoparticles: The application of silver nanoparticles as antimicrobial agents is because of its economical production and multiple modes of inhibitory action to microorganisms. Silver nanoparticles are the most studied and utilized nanoparticles in bio-system because of its strong inhibitory and antimicrobial activities . Nano silver colloid is more adhesive on bacterial and fungal cell surface; hence act as better fungicide because of its well dispersed and stabilized silver nanoparticles solution.

Copper nanoparticle: The growth of phytopathogenic fungus Fusarium graminearum inhibited due to strong synergistic effect between chitosan and copper which results due to complexation of copper with chitosan nanogels. These nanohydrogels due to its biodegradable natures are included as a new generation of copper-based bio-pesticides and it could also be developed into an efficient input based delivery system particularly for copper based fungicides as plant protectant.

Zinc nanoparticle: The nano-ZnO derived from zinc nitrate showed fungal cell wall deformity due to synthesis of hydroxyl and superoxide radicals and death due to high energy transfer in fungal pathogen *Aspergillus fumigatus.* ZnO nanoparticles prevent the conidiophores and conidial development and cause deformation of fungal hyphae which ultimately leads to the death of fungal hyphae.

Nanotechnology is a promising field of interdisciplinary research and its practical applications into agriculture industry is receiving attention nowadays due to the potential benefits that nanomaterials (NMs) can guarantee in several respects such as pests management. Plant diseases caused by bacteria, fungi, insects, nematodes, phytoplasmas, and viruses cause economic losses by reducing attainable yields, product quality, and/or shelf life (Angelo Mazzaglia et al., 2017).

Existing plant disease management relies predominantly on toxic pesticides that are potentially harmful to humans and the environment. Nanotechnology can offer advantages to pesticides, like reducing toxicity, improving the shelf-life, and increasing the solubility of poorly water-soluble pesticides, all of which could have positive environmental impacts. A review by Elizabeth A. Worrall et al., (2018) explores the two directions in which nanoparticles can be utilized for plant disease management: either as nanoparticles alone, acting as protectants; or as nanocarriers for insecticides, fungicides, herbicides, and RNA-interference molecules. Despite the several potential advantages associated with the use of nanoparticles, not many nanoparticle-based products have been commercialized for agricultural application. The scarcity of commercial applications could be explained by several factors, such as an insufficient number of field trials and underutilization of pest–crop host systems.

Traditional plant protection strategies often prove insufficient and application of chemicalbased pesticides has negative effects on animals and human beings apart from causing decline in soil fertility. Nanotechnology would provide green and efficient alternatives for the management of plant diseases without harming the Nature, while the most promising strategies, in this scenario, are the use of micro and nanotechnology to promote a more efficient assembly and then release of specific and environmental sustainable active principles. The broad range of nanotechnology applications in agriculture also includes nanopesticides (NPs) for the control of crop pests and diseases.

The use of nanoparticles to protect plants can occur via two different mechanisms:, nanoparticles themselves providing crop protection, or nanoparticles as carriers for existing pesticides or other actives, such as double-stranded RNA (dsRNA), and can be applied by spray application or drenching/soaking onto seeds, foliar tissue, or roots. Nanoparticles, as carriers, can provide several benefits, like enhanced shelf-life, improved solubility of poorly water-soluble pesticides, reduced toxicity, and boosting site-specific uptake into the target pest. Another possible nanocarrier benefit includes an increase in the efficacy of the activity and stability of the nanopesticides under environmental pressures (UV and rain), significantly reducing the number of applications, thereby decreasing toxicity and reducing their costs.

Nanotechnology in the agricultural sector is not fully reaching the market. Most of the developed nanoparticle-based pesticides are at the very early stage of development; hence, the efficacy and toxicity of the nanopesticides on soil and environment needs be studied in more detail. Of the many studies of pesticides loaded onto nanoparticles, only two of them have conducted field trials; *chlorfenapyr* loaded onto silica nanoparticles with additional emulsions was applied to *Brassica chinese*, and imidacloprid loaded onto sodium alginate nanoparticles. Another issue with the early stages of research on nanoparticles in plant protection is the current lack of long-term trials. Only a few studies have included extended time trials in their studies; for example, RNAi/ nanoparticle topical delivery platform called BioClay to protect plants against viruses 20 days post-spray application. Azadirachtin in either zinc oxide or chitosan nanoparticles in a jar of ground nuts infested with the groundnut bruchid insect over 180 days was studied (Elizabeth A. Worrall et al., 2018).

Nanopesticides are an emerging technological advancement where, in relation to pesticide use, there is a lack of a clear definition of what is, and what is not a nanopesticide, by regulatory bodies. Unlike conventional pesticides, the effects of nanopesticides may be dependent on the uptake, bioavailability, concentration, and toxicity of the nanoparticles, as well as the ratio of the active bound to them. Also, there is limited data concerning the issue of pesticide resistance and how the addition of nanoparticles could conceivably reduce its incidence.

Climate change will negatively affect crop production by exacerbating the incidence of disease and decreasing the efficacy of conventional approaches to disease control. Nanotechnology is a promising new strategy for plant disease management that has many advantages over conventional products and approaches, such as better efficacy, reduced input requirements, and lower eco-toxicity. Studies on crop plants using various nanomaterials (NMs) as protective agents have produced promising results. A review by Lin Fu et al., (2020) focused on the use of NMs in disease management through three different mechanisms: as antimicrobial agents; as biostimulants that induce plant innate immunity; and as carriers for active ingredients such as pesticides, micronutrients, and elicitors. The potential benefits of nanotechnology are considered, together with the role that NMs might play in future disease management and crop adaptation measures.

Nanotechnology for the control of plant diseases is a promising technique in plant pathology either by providing controlled delivery of functional molecules or as diagnostic tool for disease detection, an important step in plant disease treatment (Sharon et al 2010). Several studies were conducted using Nanosized particles to control fungal pathogens including such as *Pythium ultimum, Magnaporthe grisea, Colletotrichum gloeosporioides, Botrytis cinerea* and, *Rhizoctonia solani*, as well as bacteria including *Bacillus subtilis, Azotobacter chrococuum, Rhizobium tropici, Pseudomonas syringae* and *Xanthomonas compestris* pv. *vesicatoria* (Park et al., 2006).

Nanoparticles of defined concentrations could be successfully used for the control of various plant diseases caused by several phytopathogens like silver, alumina-silicate, TiO_2, carbon nanotubes etc. Pesticides are used on large scale and most of the pesticides are resistant to biodegradation and are found to be carcinogenic in nature even at trace levels. Conventional methods of pesticide removal are disadvantageous due to their inherent time consumption or expensiveness. Nanoparticles alleviate both of these drawbacks and hence, they can be effectively utilized for the mineralization of pesticides (Anita Tanwar and Sushil, 2019).

Private companies are investigating whether intentionally manufactured nano-size active ingredients can give increased efficacy or greater penetration of useful components in plants. Indeed, decreasing the particle size and consequently increasing the surface area can be beneficial for parameters like the rate of solubility, the coverage of a leave surface etc. In particular, the use of nano materials as carriers and delivery systems for active ingredients is being researched (in particular solid-liquid formulations) (Claudia Parisi et al., 2014).

There are over 3,000 patents worldwide for potential agrochemical usage of nanotechnology but they are most likely patents with broad claims, filed with the scope of guarantee freedom to operate in the field in case of future commercial developments. In reality, today very few, if any, intentionally manufactured nano-sized formulations exist on the market. Agrochemical large companies are constantly exploring the possibilities offered by nanotechnology, among other innovative technologies. Additionally, nanoparticles usually present higher costs compared to particles of bigger sizes (Claudia Parisi et al., 2014). These costs are normally compensated by higher returns in the medical and pharmaceutical sectors, but so far the agricultural applications require too high investments. Research will anyway continue in the agro-chemical sector to evaluate potential future advantages.

Nanomaterials are carriers of agrochemicals. They facilitate site targeted delivery of various nutrients needed for better growth and high productivity of plants. Another application of NPs in agriculture includes application of nanobiosensors in the crop protection and nanodevices for genetic manipulation of plant. However, some reports reflect the negative impact of NPs on the environment. The researchers are working on NMs to minimize the negative impacts on the environment (Ajey Singha et al., 2015).

Several studies have reported that nanoparticles (NPs) can be used to suppress pathogens which increased crop growth. Chemically synthesized Cu NPs demonstrated higher pathogenic fungal inhibition in comparison to the fungicide bavistin (Ajey Singha et al., 2015). Effective concentrations were reported for inhibition of colony formation by silver compounds on *Bipolaris sorokiniana* was greater than that on *Magnaporthe grisea*. Silver compounds inhibited colony formation of *B sorokiniana* by 50% at optimum concentration. Significant reduction in mycelial growth was observed from spores incubated with silver NPs. Silver NPs greatly reduced the number of germinating fragments relative to the control at 24 hour incubation of spores with a 2.5 ppm solution of NPs.

ZnO NP was found to be a good antibacterial agent. ZnO NP reduced the cell number of *Salmonella typhimurium* and *Staphylococcus aureus* to zero within 8 and 4 hour. This antibacterial activity of ZnO NP can be used in food system as preservative agent. The nanosilver and nano TiO_2 with a good potential may be used for removing of the bacterial contaminants in the tobacco plant. Ag NP exposure causes toxicity to bacteria. Ag NP treatment can prevent replication and protein synthesis.

Nanobiotechnology has the potential to revolutionize diverse sectors including medicine, agriculture, food, textile and pharmaceuticals. Disease diagnostics, therapeutics and crop protection strategies are fast emerging using nanomaterials preferably nanobiomaterials. It has potential for development of novel nanobiomolecules which offer several advantages over conventional treatment methods. RNA nanoparticles with many unique features are promising candidates in disease treatment. The miRNAs are involved in many biochemical and developmental pathways and their regulation in plants and animals (Chaudhary et al., 2019). These appear to be a powerful tool for controlling various pathological diseases in human, plants and animals, however there are challenges associated with miRNA based nanotechnology. Several advancements made in the field of miRNA therapeutics make it an attractive approach, but a lot more has to be explored in nanotechnology assisted miRNA therapy. The miRNA based technologies can be employed for detection and combating crop diseases as well. Despite these potential advantages, nanobiotechnology applications in the agricultural sector are still in its infancy and have not yet made its mark in comparison with healthcare sector.

Nanotechnology may be developed and deployed for real time monitoring of a wide variety of fertilizers, herbicide, pesticide, insecticide, heavy metals, organic pollutants and pathogens (Ivo Iavicoli et al., 2017). The increased application of pesticides in various agricultural activities has made it necessary to explore the unique chemical and physical properties of nanomaterials to develop innovative sensors for pesticide residue detection. These "nanosensors," intended as analytical devices having at least one sensing dimension no > 100 nm, fabricated for monitoring physico-chemical properties in places otherwise difficult to reach, may offer greater sensitivity, low detectionlimits, selectivity, fast detection rates, and more portability than conventional detection techniques. Nanomaterials may also offer improved detection limits for bacterial, viral and fungal pathogen determination in plants.

1.6 Disease Diagnosis

Traditional molecular diagnostic methods are widely used in laboratories all over the world to identify plant pathogenic organisms with high degree of sensitivity and specificity. The common disadvantage of the traditional diagnostic methods is that they are time consuming and lack high sensitivity. Consequently, developing low-cost methods which improve the accuracy and rapidity of plant pathogens diagnosis is needed (Veeranna et al., 2017).

Current technologies require a relatively large amount of target tissue and rely on multiple assays to accurately identify distinct plant. Still, most of these procedures cannot be applied in the field (on-site detection) or in developing countries where incomes are small. Furthermore, the high price and short shelf halflife of some molecular biology reagents, such as enzymes and primers, limit the application of traditional molecular methods in developing countries. Nanotechnology may have actual solutions against many agriculture problems like plant disease control. Nano-based materials will be presented which will increase the efficacy of fungicides and pesticides, allowing only minor doses to be used. Moreover, nanodiagnostic and microfluidics offer novel tools to improve the sample preparation step that remains difficult to integrate in a miniaturized platform. The signal amplification approaches could challenge those of target amplification. Quick on-site detection of plant pathogens using, nano based kits, nanosensor, nanobiosensors, nanobarcodes and other portable diagnostic systems will also help the agricultural and food industry to manage different plant diseases.

Modern nanomolecular techniques are used for monitoring or understanding of pathogen population genetics, plant microbe interactions and gene transfer between pathogens and even the host. Furthermore, nanoparticles such as nano-sized silica-silver have recently been applied as antimicrobial and antifungal agents. Additionally, nanomaterials can be used for mycotoxin detection and detoxication, increasing plant resistance, plant disease forecasting and nano-molecular diagnostics of plant pathogens. Nano particles and quantum dots (QDs) have emerged as essential tools for fast detection of a particular biological marker with extreme accuracy. Nano diagnostic kit like multiplex dipstick immunoassay which can easily and quickly detect potential serious plant pathogens viz., Fusarium toxins, zearalenone, T-2 and HT-2 toxins, deoxynivalenol and fumonisins in wheat, oats and maize allowing experts to help farmers in the prevention of epidemics. Furthermore, nanopore DNA sequencing tools, QDs, nanostructured platforms for the detection of aflatoxin-B1 by means of functionalization of cMWCNTs/ITO electrode surface with monoclonal aflatoxin-B1 antibodies (anti-AFB1) using electrochemical technique (Veeranna et al., 2017). Nanostructural layer biosensors are fabricated for the detection of Aspergillus niger fungi through changes observed in the electrical resistance of samples. Nanofabricated imaging and LAMP have the potential to raise sensitivity, specificity and speed of the pathogen detection, and can be used for high-quality monitoring and crop protection.

Nano-phytopathology is a cutting-edge science which uses nanotechnology for detecting, diagnosing and controlling plant disease and their pathogens at an early stage, owing to crop protection from epidemic diseases (Touseef Hussain, 2017). The modern plant pathologist strives to apply his knowledge in nano phytopathology to enable understanding of controlling factors of plant diseases and to develop and/or evaluate eco-friendly diagnostic measurements. Modern nano molecular techniques are used for monitoring or understanding of pathogen population genetics, plant microbe interactions and gene transfer between pathogens and even the host. Furthermore, nanoparticles such as nano sized silicasilver have recently been applied as antimicrobial and antifungal agents. Additionally, nanomaterials scan be used for mycotoxin detection and detoxication, increasing plant resistance, plant disease forecasting and nano-molecular diagnostics of plant pathogens.

Nanoparticles such as gold nanoparticles, magnetic nanoparticles and quantum dots, are most widely used for molecular detection purpose. Gold nanoparticles are widely used in rapid immune diagnosis, due to their unique physiochemical properties, including small-sized effect, surface effect and catalytic effect, and can also react with nucleic acid through an Au-S covalent bond and show high application value in DNA identification and detection. The bio-barcode assay is an ultrasensitive method of amplification and detection of proteins or nucleic acids. DNA biobar coded tests employ oligo nucleotide-modified magnetic gold nanoparticles (AuMNPs) for signal amplification and for simple separation of a target protein from the sample. The large b-DNAto-recognition agent ratio affords a means of substantial signal amplification. It is also promising by allowing the quick detection of numerous protein targets at low-attomolar concentrations.

Delivery of pesticides or medicines is either provided as "preventative" treatment, or is provided once the disease causing organism has multiplied and symptoms are evident in the plant (Srilatha, 2011). In this context, nanotechnologies offer a great opportunity to develop new products against pests. Nanotechnology improves their performance and acceptability by increasing effectiveness, safety, patient adherence, as well as ultimately reducing health care costs. Nanoscale devices are envisioned that would have the capability to detect and treat an infection, nutrient deficiency, or other health problem, long before symptoms were evident at the macro-scale. This type of treatment could be targeted to the area affected with a greater awareness of the hazards associated with the use of synthetic organic insecticides, there has been an urgent need to explore suitable alternative products for pest control .The broad application of Molecular Biology revolutionized the field of diagnostics. Today, Nanomaterials have been designed for a variety of biomedical and biotechnological applications, including biosensors, enzyme encapsulation.

Nanoparticles technology has emerged as a strategy to tackle developing new materials and selecting appropriate materials for each specific treatment, other factors need to be optimally selected in order to design better targeted Nano particles. These factors include the particles size, shape, sedimentation, drug encapsulation efficacy, desired drug release profiles, distribution in the body, circulation and cost. Specifically, application of Nanoparticles technology in plant pathology targets specific agricultural problems in plant–pathogen interactions and provide new ways for crop protection.

Bibliography

Ajey Singha , N. B. Singha,, Imtiyaz Hussaina , Himani Singha , and S.C. Singh Plant-nanoparticle interaction: An approach to improve agricultural practices and plant productivity , 2015. Intl. J. Pharmaceu. Sci. Invention, 4: 25-40

Amit Kumar, 2014. Nanotechnology Development in India- An Overview. Discussion Papers, RIS (Research and Information System for Developing Countries), 40 pp.

Angelo Mazzaglia, Elena Fortunati, Josè Maria Kenny, Luigi Torre and Giorgio Mariano Balestra, 2017. Nanomaterials in Plant Protection. In : Prof. Monique A. Axelos Prof. Marcel H. Van de Voorde Dr. Dr. h.c. (Edrs.). Wiley Online Library.

Anita Tanwar and Sushil, 2019. Role and effects of nanotechnology used in pesticides and agriculture field. Proc., AIP Conference , 2142pp

Ashish Khandelwal and Ritika Joshi, 2018. "Synthesis of Nanoparticles and their Application in Agriculture". Acta Sci. Agri., 2. : 10-13.

Chaudhary, V., Jangra, S. & Yadav, N.R., 2018. Nanotechnology based approaches for detection and delivery of microRNA in healthcare and crop protection. J . Nanobiotechnol., 16: 40

Claudia Parisi, Mauro Vigani and Emilio Rodríguez-Cerezo, 2014. Proceedings of a workshop on "Nanotechnology for the agricultural sector: from research to the field". JRC Scientific and Policy Reports, Institute for Prospective Technological Studies Contact information, Edificio Expo. c/o Inca Garcilaso, Spain.

DBT. 2008. Annual Report 2007-08. Department of Biotechnology, New Delhi.

DeITy. 2006. Annual Report 2005-06. Department of Electronics and Information Technology, New Delhi

DeITy. 2011. Annual Report 2010-11. Department of Electronics and Information Technology, New Delhi

DeITy. 2013. Annual Report 2012-13. Department of Electronics and Information Technology, New Delhi.

DST. 2001. Annual Report 2000-01. Department of Science and Technology (DST), Government of India, New Delhi.

DST. 2008. Annual Report 2007-08. Department of Science and Technology, Government of India. New Delhi.

DST. 2013. Annual Report 2012-13. Department of Science and Technology. Government of India, New Delhi.

Elizabeth A. Worrall, Aflaq Hamid , Karishma T. Mody , Neena Mitter and Hanu R. Pappu, 2018. Nanotechnology for Plant Disease Management. Agronomy, 8 ; 2-24

GOI. 2002. 10th Five-Year Plan 2002-2007. Planning Commission. Government of India. New Delhi: SAGE Publications India Pvt Ltd.

GOI. 2007. 11th Five-Year Plan 2007-2012. Planning Commission, Government of India. New Delhi: SAGE Publications India Pvt Ltd.

Hao Chen, 2018. Metal based nanoparticles in agricultural system: behavior, transport, and interaction with plants. Chem. Speciation & Bioavailability, 30:1, 123-134,

Ivo Iavicoli, Veruscka Leso, Donald H. Beezhold and Anna A. Shvedova., 2017. Nanotechnology in agriculture: Opportunities, toxicological implications, and occupational risks. Toxicol. Appl .Pharmacol., 329: 96–111.

Kamel Ahmed Abd-Elsalam, 2015. Applications of nano-biotechnological strategies in phytopathology. Journal of Biotechnology & Biomaterials, World Bio Summit & Expo November 02-04, 2015 Dubai, UAE. DOI: 10.4172/2155-952X.C1.046

Lin Fu, Zhenyu Wang, Om Parkash Dhankher and Baoshan Xing, 2020.Nanotechnology as a new sustainable approach for controlling crop diseases and increasing agricultural production, Journal of Experimental Botany, 71: 507–519

Mohamed Ragaei and Al-kazafy Hassan Sabry, 2014. Nanotechnology for Insect Pest Control. Intl. J. Sci. Env. & Tech., 3: 528 – 545

PIB. 2014. "Continuation of the Mission on Nano Science and Technology in the 12th Plan Period". Press Information Bureau (PIB), 20 February.

Sanzari Ilaria, Leone Antonietta, Ambrosone Alfredo, 2019. Nanotechnology in Plant Science. To Make a Long Story Short . Fronti. Bioengg.& Biotech., 7: 120.

Srilatha B., 2011. Nanotechnology in Agriculture. J. Nanomedic Nanotechnol., 2:123. doi:10.4172/2157-7439.1000123

Stefania Mura, Giovanna Seddaiu, Fabio Bacchini, Pier Paolo Roggero and Gian Franco Greppi, 2013. Advances of Nanotechnology in Agro-Environmental Studies. Italian J. Agron., DOI: 10.4081/ija.2013.e18

Sunita Ojha, Deepika Singh, Arghya Sett, Hasnahana Chetia, Debajyoti Kabiraj and Utpal Bora, 2018. Nanomaterials in Plants, Algae, and Microorganisms. Nanotechnol. in Crop Protn., 1: 345-391

Touseef Hussain, 2017 . Nanotechnology: Diagnosis of Plant Diseases. Agri Res & Tech: Open Access J 10(1): ARTOAJ.MS.ID.555777

Touseef Hussain, 2017. Nanocides: smart delivery system in agriculture and horticultural crops. Adv Plants Agric Res., 6:175

Veeranna, D., Singh, S., Kumar, R. and Jeevan, B., 2017. Nanotechnology as diagnostic tool for plant disease detection. BioQuest, Vol. 1, No. 1

Vincent Sabourin and Alpha Ayande, 2017. Commercial Opportunities and Market Demand for Nanotechnologies in Agribusiness Sector. J. Technol. Manag. Innov. 10 :40-52

Virginia Tressider, 2015. Nanopesticides: a promising new pesticide solution? ECOS Bull., Issue 14.

Wade Elmer and Jason C. White, 2018. The Future of Nanotechnology in Plant Pathology. Ann. Rev.Phytopath., 56: 111-136

Waleed Fouad Abobatta, 2018. Nanotechnology Application in Agriculture. Acta Scientific Agriculture, 2 : 99-102.

Wani Khursheed Ahmad and Kothari Richa, 2018.. Agricultural Nanotechnology: Applications and Challenges. Annals Pl. Sci., 7: 2146-2148,

Yifen Shang, Md. Kamrul Hasan, Golam Jalal Ahammed, Mengqi Li, Hanqin Yin, Jie Zhou, 2019. Applications of Nanotechnology in Plant Growth and Crop Protection: A Review. Molecules, 24: 2558

Yifen Shang, Md. Kamrul Hasan, Golam Jalal Ahammed, Mengqi Li, Hanqin Yin, Jie Zhou, 2019. Applications of Nanotechnology in Plant Growth and Crop Protection: A Review. Molecules, 24: 2558

Yurii A Krutyakov, Alexey A Kudrinskiy, Pavel M Zherebin, Alexey D Yapryntsev, Marina A Pobedinskaya, Sergey N Elansky, Albert N Denisov, Dmitry M Mikhaylov and Georgii V Lisichkin, 2016. Tallow amphopolycarboxyglycinate-stabilized silver nanoparticles: new frontiers in development of plant protection products with a broad spectrum of action against phytopathogens. Mater. Res. Express, 3: 3075403.

2

Categories and Characterization of Nanoparticles

It all started after the famous lecture, "There is plenty of room at the bottom" given by Richard Feynman on December 29, 1959. Nanomaterials are intermediate between macroscopic solid and of atomic and molecular systems. Nanomaterials have certain properties which make them different from that of the bulk materials, including large fraction of surface atoms, high surface energy, spatial confinement, and reduced imperfections. Nanotechnology deals with various structures of matter having dimensions of the order of a billionth of a meter (Ray Unmesha. 2018). From the advent of nanotechnology, people realized that certain materials can exhibit different properties based on its size and shape.

2.1 Nanoscience v/s Nanotechnology

Nanoscience and nanotechnology are the study and application of extremely small things and can be used across all the other science fields, such as chemistry, biology, physics, materials science, and engineering. The terms nanoscience and nanotechnology are often used interchangeably, but they are in fact two very different things. It's hard to imagine just how small nanotechnology is. One nanometer is a billionth of a meter, or 10^{-9} of a meter.

Here are a few examples: There are 25,400,000 nanometers in an inch; A sheet of newspaper is about 100,000 nanometers thick; On a comparative scale, if a marble were a nanometer, then one meter would be the size of the Earth.

Nanoscience and nanotechnology involve the ability to see and to control individual atoms and molecules. Everything on earth is made up of atoms, the food, the clothes, the buildings and houses and our own bodies. Although modern nanoscience and nanotechnology are quite new, nanoscale materials were used for centuries. Alternate-sized gold and silver particles created colors in the stained glass windows of medieval churches hundreds of years ago. The artists back then just didn't know that the process they used to create these beautiful works of art actually led to changes in the composition of the materials they were working with. Today's scientists and engineers are finding a wide variety of ways to deliberately make materials at the nanoscale to take advantage of their enhanced properties such as higher strength, lighter weight, increased control of light spectrum, and greater chemical reactivity than their larger-scale counterparts.

Nanoscience: Nanoscience is the study of structures and materials on an ultra-small scale and the unique and interesting properties these materials demonstrate. Nanoscience is cross disciplinary, meaning scientists from a range of fields including chemistry, physics, biology, medicine, computing, materials science and engineering are studying it and using it to better understand our world.

Nanotechnology: Nanotechnology (also sometimes called molecular manufacturing), on the other hand, is the design, production and application of structures, devices and systems at the nanoscale. So essentially one is studying nanomaterials and their properties and the other is using those materials and properties to create something new or different.

2.2 Characterization of nanoparticles

Nanoparticle characterization is a broad and complex discipline. However, measurement and standardization often lag behind the rapid development of new nanomaterials and their applications. Furthermore, the demands on reproducibility and quality control increase steadily as new materials leave the discovery stage and move into commercial applications. Nanoparticles exist in various chemical compositions ranging from micelles to metal (oxide)s, from synthetic polymers to large biomolecules. Each of these materials features a completely different chemistry, which can be analyzed by a variety of methods including optical spectroscopy, X-ray fluorescence and absorbance, Raman spectroscopy, and solid-state NMR. However, often the behavior of nanoparticles is largely governed by their nanometer dimensions. As such, throughout nanoparticle characterization, the investigation of size, shape, surface charge and porosity is a fundamental step for fully understanding and predicting their behavior.

Nanoparticles are of widespread use due to an emerging and prevailing understanding of their potential impact on particulate additives, human health and environmental sustainability, and due to the increasing use and production of man-made nanoparticles in the environment. These are used and created by many different processes in many different applications. Measuring and characterizing its properties poses an interesting analytical challenge to its kind of process and development. It is challenging to assess morphological characteristics and physiochemical properties of nanoparticles. It is not only necessary to properly characterize the nanoparticles, but also to evaluate their biological effect (Afolalu, et al., 2019). To do this, it is necessary to explore and evaluate all relevant exposure routes. In order to cope and compensate with the modification of its properties, it seems reasonable to characterize nanoparticles materials by its surface charge, functional groups and catalytic activity. Different classification methods find application in the analysis of the Nanoparticles' various chemical and physical properties.

2.2.1 Morphological characterizations

The morphological characterizations of NPs usually garner prodigious attention because of the influence morphology has over most of the Nanoparticles' properties. Different methods exist to characterize NPs for studying their morphology, but techniques that entail the use of microscopy such as SEM, TEM and Polarized Optical Microscopy (POM) constitute the most substantial methods. Thc Scanning Electron Microscope method depends on the electron examining rule, and it gives at nanoscale level all accessible data about the NPs. Writing is widely accessible, where individuals utilized this method to discover the characteristics of their nanomaterials, as well as the scattering of NPs in the mass or lattice (Afolalu, et al., 2019). The SWNTs which are scattered in the polymer lattice made up of nylon-6 and poly (butylene) terephthalate (PBT) was discovered using this method. A similar group additionally gave investigation of their materials using the POM, and it was found that the materials had spherulities which were star like shaped, whose size reduced with the increase in the amount of SWNTs present in the lattice. TEM micrographs have been produced showing different characteristics of gold NPs, arranged using different techniques. TEM also provides basic data on materials with at least two layers, such as in the Co3O4 NPs where the empty quadrupolar shell structure is made visible by TEM. In Li-particle batteries, it was found that these NPs can be used efficiently as anode. Permeable multishell structure prompts shorter Li+ spreading distance with satisfactory space abrogated to cradle the volume extension, high cycling performance, more noteworthy rate limit and explicit limit.

2.2.2 Particle size and surface area characterization

The size of NPs can be estimated using a variety of techniques, which include TEM, AFM, XRD, SEM, and dynamic light dispersing (DLS). While the first 4 gives a better estimate of size than the DLS, the NPs size can be estimated at incredibly low dimensions only with the zeta potential size analyzer/DLS. In one study, the DLS method was used to study silica NPs size variation with serum protein absorption rate. The results showed that the size of the NPs expanded with protein layer. But in the event of agglomeration and hydrophilicity, DLS may demonstrate inability to accurately measure during which the Differential Centrifugal Sedimentation (DCS) high-resolution technique is considered (Afolalu et al., 2019). In addition to DCS, a comparatively fresh and unique technique for tracking nanoparticles (NTA) is available for the analysis of proteins, DNA and other organic systems. NPs in liquid media are analyzed and visualized in NTA strategy by comparing the rate of Brownian motion to the size of the NP thereby allowing the size distribution profile of the nanoparticle to be located in a fluid medium with diameters ranging from 10 to 1000 nm. The NTA technique was found to produce very good results when contrasted with DLS and was also found to be very accurate with far better peak resolution when used in sizing monodisperse and polydisperse samples.

The polymer and protein sample concentration and particle size suspended differentially sized NPs were performed and an overview was provided for further study on the effect of data evaluation and experimental parameters. Large areas of nanomaterials offer exceptional opportunity for several uses and BET is the top way to decode NP surfaces. This operation is based upon the Brunauer–Emmett–Teller (BET) hypothesis and the adsorption and disorption rules. This is why nitrogen gas is used regularly.

2.2.3 Optical characterizations

The optic property is a major feature governing photo-catalytic application of NPs, which means that photochemists must understand the techniques employed for optical characterization which find their root in the law of beer-lambert and the principles of light to reveal the mechanism behind photochemical proceedings (Afolalu et al., 2019). The techniques of optical characterization are used to gather information on NPs reflectivity, absorption, luminescence and phosphorescence properties. NPs with various colours, particularly semiconductor and metallic NPs, are commonly known and are therefore best used in photographic applications. Knowledge of the absorption and reflection values of these documents comes from an understanding of each application's fundamental mechanism.

The well-known optical instruments include the UV, photoluminescence and none ellipsometers and are relevant for the identification of optical properties of NPs. The optical instruments are also known as UV-visibles. The UV / Diffuse Reflectance Spectrometer (DRS) is majorly arranged contraction capable of evaluating optical transmission, intake and reflection. The past two are useful to each other, while the last referred (DRS) is an exceptional use of strategy for precedents usually sold. The procedure for validating np bandgaps and distinctive nanomaterials is particularly commendable. The photoactivity and conductance of the materials are essential for packaging the bandgap. The carbon nanodot-carbon nitride (C3N4) was regarded for the metal-free water part as a photocatalyst. The photographic boundaries of this material are explicitly associated with the estimate of the UV–VIS Spectroscopy (2.74–2.77 eV) bandgap.

Furthermore, this framework also provides for the movement of absorption in the event of a doping event, a composite action or a heterostructure of NP materials. The association of nanocomposites $LaFeO_3$, MMT and $LaFeO_3$/MMT analyzed their range of electromagnet UV absorption in order to investigate their optical properties. their assortment. Strong red movement checked whether a nanocomposite occurred when the MMT and $LaFeO_3$ NPs were immaculate. $LaFeO_3$/MMT and $LaFeO_3$ showed a quite wide osmosis range of between 400 and 620 nm in their band gap. Photocatalysis of these driving forces for sunlight is made wide in this possession.

2.3 Physicochemical properties

The optical and electronic characteristics of NPs are becoming more and more essential between wards. For example, good metal NPs have subordinate visual characteristics and show a robust band clear of UV's, a band that is missing in the mass metal scope. The outcome of this excitation band is the Limited Surface Plasma Resonance (LSPR) when the repeat photon event is compatible with the complete conductive electron's excitation (Afolalu et al., 2019). The LSPR zenite spectrum is defined to be slim depending on the dimensions, shape and interparticle which differentiates the NPs equally from one type of dielectric and adjacent features, including substrates, solvents, and adsorbents. The mean free route for Ag and silver is approximately 50 nm, higher than the NP density of these products. As such, the LSPR in these NP's locations standing resonance state, instead of dispersing light participation, is typical of the mass.

Specific properties (Mechanical) of NPs allow researchers to search for new method and applications in multiple key fields, including tribology, surface design, nano-manufacturing and nanomaterials. The mechanical watchfulness of the NPs can be determined by examining certain mechanical parameters, such as flexible parts, strength, strain, connection and tension. In addition to these parameters, coagulation and petroleum additionally contribute to the mechanical features of NPs. When they stand out from microparticles and their mass fabrics, NPs demonstrate distinct mechanical characteristics. Furthermore, the complexity of the force between NPs and the outer layer in a lobed up or lobed touch checks if the NPs indented on the surface of the match scheme or bent when the touch pressure is usually important. This fundamental data may show how the contact procedure is performed by the NPs. Normal inspections of the mechanical characteristics and its involvement with any form of surface are essential in enlightening surface efficiency and boosting the material removal.

Nanofluids are generated by distributing strong particles of a nanometric scale into liquids like water, ethylene glycol or oils. Nanofluids are dependent on displaying preferential characteristics compared to those of usual warmth trading liquids and liquids that are assessed to contain minor particles. Since the light trade happens outside the cells, using the particles with important complete surface region is appealing. A safety suspension is also produced by the huge difficult and quick surface area. From early on, it has been shown that nanofluids in water or ethylene, including CuO or Al_2O_3 NPs, demonstrate hot conductivity.

Nanoscale materials often present properties different from their bulk counterparts, as their high surface-to-volume ratio results in an exponential increase of the reactivity at the molecular level. Such properties include electronic, optical and chemical properties, while the mechanical characteristics of the nanoparticles (NPs) may also differ extensively (Stefanos Mourdikoudis et al., 2018). This enables them to be an object of intensive studies due to their academic interest and the prospective technological applications in various fields. Nanoscience and nanotechnology are still undergoing constant growth, and the scientific community is rather aware that there may be certain differences between the way analytical characterization methods operate for nanomaterials, in comparison with their more 'traditional' modes of use for more 'conventional' (macroscopic) materials.

Fundamental properties of nanoparticles are strong function of their size and shapes. Surfactants are also equally important, since not only it stabilizes the nanoparticles but also makes nanoparticles compatible for various biomedical applications. The different size and shapes of the nanoparticles certainly have an influence and important place in clinical applications.

Nanoparticles consist of three layers: the surface layer, the shell layer, and the core. The surface layer usually consists of a variety of molecules such as metal ion, surfactants, and polymers. Nanoparticles may contain a single material or maybe consist of a combination of several materials. Nanoparticles can exist as suspensions, colloids, or dispersed aerosols depending on their chemical and electromagnetic properties (Ananya Madal, 2018). The properties of nanoparticles are dependent their size. For instance, copper nanoparticles than are smaller than 50 nm are super hard materials and do not exhibit the properties of malleability or ductility of bulk copper.

Other changes that are dependent on the size of nanoparticles are superparamagnetism exhibited by magnetic materials, quantum confinement by semiconductor Q-particles, and surface plasmon resonance in some metal particles. Research has also demonstrated that absorption of solar radiation in photovoltaic cells is much higher in nanoparticles than it is in thin films of continuous sheets of bulk material. This is because nanoparticles are smaller and can absorb greater amount of solar radiation. Nanoparticles exhibit enhanced diffusion at elevated temperatures due to their high surface area to volume ratio. This property of nanoparticles allows sintering to take place at lower temperatures than in the case of larger particles. While this diffusion property exhibited by nanoparticles may not affect the density of the product, it can lead to agglomeration.

The principal parameters of nanoparticles are their shape (including aspect ratios where appropriate), size, and the morphological sub-structure of the substance. Nanoparticles are presented as an aerosol (mostly solid or liquid phase in air), a suspension (mostly solid in liquids) or an emulsion (two liquid phases). In the presence of chemical agents (surfactants), the surface and interfacial properties may be modified. Indirectly such agents can stabilise against coagulation or aggregation by conserving particle charge and by modifying the outmost layer of the particle. Depending on the growth history and the lifetime of a nanoparticle, very complex compositions, possibly with complex mixtures of adsorbates, have to be expected.

In the typical history of a combustion nanoparticle, for example, many different agents are prone to condensation on the particle while it cools down and is exposed to different ambient atmospheres. Complex surface chemical processes are to be expected and have been identified only for a small number of particulate model systems. At the nanoparticle - liquid interface, polyelectrolytes have been utilised to modify surface properties and the interactions between particles and their environment. They have been used in a wide range of technologies, including adhesion, lubrication, stabilization, and controlled flocculation of colloidal dispersions

At some point between the Angstrom level and the micrometre scale, the simple picture of a nanoparticle as a ball or droplet changes. Both physical and chemical properties are derived from atomic and molecular origin in a complex way.

2.4 Nanoparticle - nanoparticle interaction

At the nanoscale, particle-particle interactions are either dominated by weak Van der Waals forces, stronger polar and electrostatic interactions or covalent interactions. Depending on the viscosity and polarisability of the fluid, particle aggregation is determined by the interparticle interaction. By the modification of the surface layer, the tendency of a colloid to coagulate can be enhanced or hindered. For nanoparticles suspended in air, charges can be accumulated by physical processes such as glow discharge or photoemission. In liquids, particle charge can be stabilised by electrochemical processes at surfaces. The details of nanoparticle - nanoparticle interaction forces and nanoparticle, fluid interactions are of key importance to describe physical and chemical processes, and the temporal evolution of free nanoparticles. They remain difficult to characterize due to the small amount of molecules involved in the surface active layer. Both surface energy, charge and solvation are relevant parameters to be considered.

Due to the crucial role of the nanoparticle, nanoparticle interaction and the nanoparticle – fluid interaction, the term free nanoparticle can be easily misunderstood. The interaction forces, either attractive or repulsive, crucially determine the fate of individual and collective nanoparticles. This interaction between nanoparticles resulting in aggregates and/or agglomerates may influence on their behaviour. In gas suspensions, aggregation is crucially determined by the size and diffusion, and coagulation typically occurs faster than in the liquid phase as the sticking coefficient is closer to unity than in liquids.

Nanoparticles have different analytical requirements than conventional chemicals, for which chemical composition and concentration are sufficient metrics. Nanoparticles have other physical properties that must be measured for a complete description, such as size, shape, surface properties, crystallinity, and dispersion state. The bulk properties of nanoparticles are sensitive to small variations in these properties, which has implications for process control in their industrial use. These properties also influence the health effects of exposure to nanoparticles of a given composition. An additional challenge is that sampling and laboratory procedures can perturb the nanoparticles› dispersion state, or bias the distribution of their other properties. In environmental contexts, many methods cannot detect low concentrations of nanoparticles that may still have an adverse effect. A high background of natural and incidental nanoparticles may interfere with detection of the target engineered nanoparticle, as it is difficult to distinguish the two. Nanoparticles may also be mixed with larger particles. For some applications, nanoparticles may be characterized in complex matrices such as water, soil, food, polymers, inks, complex mixtures of organic liquids such as in cosmetics, or blood.

There are several overall categories of methods used to characterize nanoparticles. Microscopy methods generate images of individual nanoparticles to characterize their shape, size, and location. Electron microscopy and scanning probe microscopy are the dominant methods. Because nanoparticles have a size below the diffraction limit of visible light, conventional optical microscopy is not useful. Electron microscopes can be coupled to spectroscopic methods that can perform elemental analysis. Microscopy methods are destructive and can be prone to undesirable artifacts from sample preparation, or from probe tip geometry in the case of scanning probe microscopy. Additionally, microscopy is based on single-particle measurements, meaning that large numbers of individual particles must be characterized to estimate their bulk properties (Hassellöv Martin, ct al., 2008).

Spectroscopy, which measures the particles' interaction with electromagnetic radiation as a function of wavelength, is useful for some classes of nanoparticles to characterize concentration, size, and shape. X-ray, ultraviolet–visible, infrared, and nuclear magnetic resonance spectroscopy can be used with nanoparticles. Light scattering methods using laser light, X-rays, or neutron scattering are used to determine particle size, with each method suitable for different size ranges and particle compositions.

2.4.1 Size and dispersion

Dispersion is the degree to which particles clump together into weakly bound agglomerates (pictured) or strongly bound aggregates. Particle size is the external dimensions of a particle, and dispersity is a measure of the range of particle sizes in a sample. If the particle is elongated or irregularly shaped, the size will differ between dimensions, although many measurement techniques yield an equivalent spherical diameter based on the surrogate property being measured. Size can be calculated from physical properties such as settling velocity, diffusion rate or coefficient, and electrical mobility. Size can also be calculated from microscope images using measured parameters such as Feret diameter, Martin diameter and projected area diameters; electron microscopy is often used for this purpose for nanoparticles. Size measurements may differ between methods because they measure different aspects of particle dimensions, they average distributions over an ensemble differently, or the preparation for or operation of the method may change the effective particle size.

The nanoscale is the dimensional range of approximately 1 to 100 nanometres. Cells and bacteria are measured in micrometres, and electronic components on a silicon chip are usually around 1 micrometre in size (Chennupati Jagadish and Amanda Barnard, 2020).

To reach the nanoworld one has to go smaller again. A nanometre (nm) is 10^{-9}, which is one-thousandth of a micrometre, or one-billionth of a metre. This is the scale at which we measure atoms and the molecules they make. To have some idea of the nanoscale, 10 hydrogen atoms laid side by side measure a nanometre across, a strand of DNA is 2.5 nm in diameter, while a red blood cell is about 7,000 nm wide. Need a few more examples? A human hair is between 50,000 and 100,000 nanometres thick, a single sheet of paper is around 75,000 nanometres thick, while a pinhead is around a million nanometres wide (Fig. 2). If each person on Earth was the size of a nanometre, everyone on the planet would fit into one Hot Wheels matchbox car. So nano is super, super tiny.

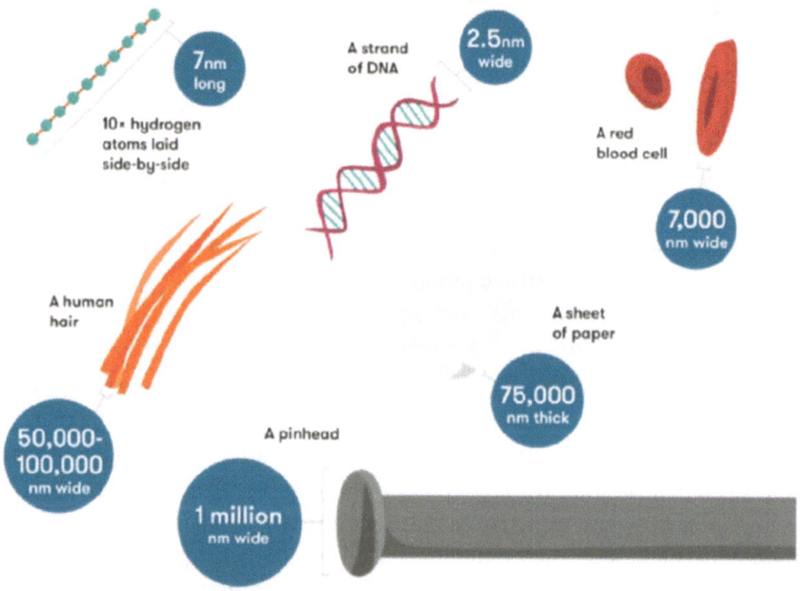

Fig. 2: Nanoworld

2.4.2 Shape

Morphology refers to the physical shape of a particle, as well as its surface topography, for example, the presence of cracks, ridges, or pores. Morphology influences dispersion, functionality and toxicity and has similar considerations as size measurements (Figs.3-5). Evaluation of morphology requires direct visualization of the particles through techniques like scanning electron microscopy, transmission electron microscopy, and atomic force microscopy. Several metrics can be used, such as sphericity or circularity, aspect ratio, elongation, convexity, and fractal dimension. Because microscopy involves measurements of single particles, a large sample size is necessary to ensure a representative sample, and orientation and sample preparation effects must be accounted for.

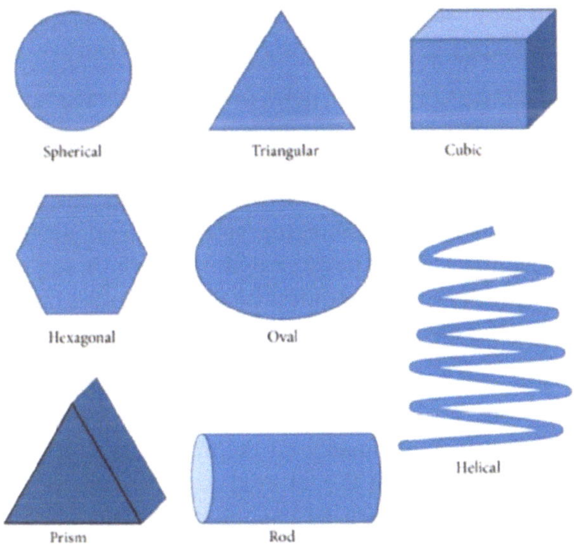

Fig. 3: Shapes of nanoparticles

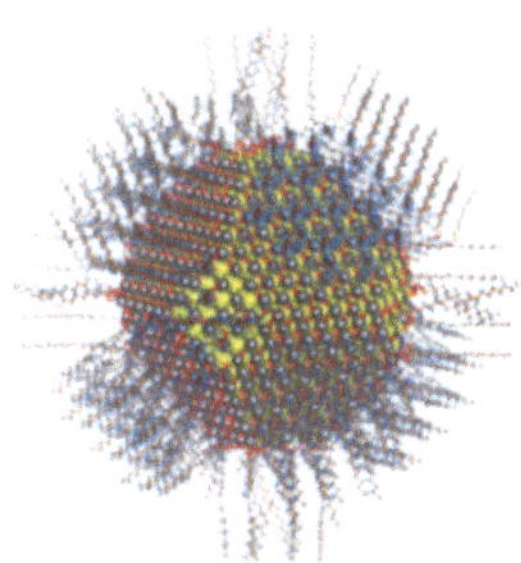

Fig. 4: The surface of a nanoparticle

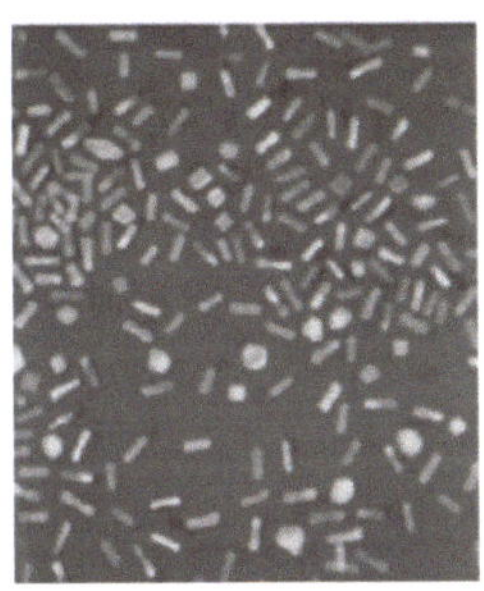

Fig. 5: Size and shapes of nanoparticles

2.4.3 Chemical composition and crystal structure

Bulk chemical composition refers to the atomic elements of which a nanoparticle is composed, and can be measured in ensemble or single-particle elemental analysis methods. Ensemble techniques include atomic absorption spectroscopy, inductively coupled plasma optical emission spectroscopy or inductively coupled plasma mass spectrometry, nuclear magnetic resonance spectroscopy, neutron activation analysis, X-ray diffraction, X-ray absorption spectroscopy, X-ray fluorescence, and thermogravimetric analysis. Single-particle techniques include time-of-flight mass spectrometry, as well as utilizing elemental detectors such as energy-dispersive X-ray analysis or electron energy loss spectroscopy while using scanning electron microscopy or transmission electron microscopy (Stefaniak Aleksandr, 2017).

The arrangement of elemental atoms in nanoparticles may be organized into a crystal structure or may be amorphous. Crystallinity is the ratio of crystalline to amorphous structure. Crystallite size, the size of the crystal unit cell, can be calculated through the Scherrer equation. Generally, crystal structure is determined using powder X-ray diffraction, or selected area electron diffraction using a transmission electron microscope, though others such as Raman spectroscopy exist. X-ray diffraction requires on the order of a gram of material, whereas electron diffraction can be done on single particles.

Surface area: Although surfaces or, more precisely, the surface atomic and electronic structure, determine the way materials interact with their environment, the influence of surface chemistry on the bulk of the material is generally considered to be small (Jürgen Biener et al., 2009). However, in the case of high surface area materials such as nanoporous solids, surface properties can start to dominate the overall material behavior. This allows one to create new materials with physical and chemical properties that are no longer determined by the bulk material, but by their nanoscale architectures. Several examples ranging from nanoporous gold to surface engineered carbon aerogels that demonstrate the tuneability of nanoporous solids for sustainable energy applications were discussed.

Surface area is an important metric for engineered nanoparticles because it influences reactivity and surface interactions with ligands. The surface area to volume ratio for a material or substance made of nanoparticles has a significant effect on the properties of the material. Firstly, materials made up of nanoparticles have a relative larger surface area when compared to the same volume of material made up of bigger particles. Specific surface area refers to the surface area of a powder normalized to mass or volume. Different methods measure different aspects of surface area. Direct measurement of nanoparticle surface area utilizes adsorption of an inert gas such as nitrogen or krypton under varying conditions of pressure to form a monolayer of gas coverage. The number of gas molecules needed to form a monolayer and the cross-sectional area of the adsorbate gas molecule are related to the «total surface area» of the particle, including internal pores and crevices, using the Brunauer–Emmett– Teller equation. Organic molecules can be used in place of gasses, such as ethylene glycol monoethyl ether.

It means that the surface area to volume ratio increases as the radius of the sphere decreases and vice versa. It also means that when a given volume of material is made up of smaller particles, the surface area of the material increases. Therefore, as particle size decreases, a greater proportion of the particles are found at the surface of the material. For example, a particle of size 3 nm has 50% of its particles on the surface; at 10 nm, 20% of its particles are on the surface; and at 30 nm, 5% of its particles are on the surface. Therefore, materials made of nanoparticles have a much greater surface area per unit volume ratio compared with the materials made up of bigger particles. This leads to nanoparticles being more chemically reactive. As chemical reactions occur between particles that are on the surface, a given mass of nanomaterial will be much more reactive than the same mass of material made up of large particles. This means that materials that are inert in their bulk form are reactive when produced in their nanoparticle form.

Measuring the surface area of nanoparticles is a key to understanding nanoparticle properties, behavior, applications, and hazards. There are several indirect measurement techniques for airborne nanoparticles, which do not account for porosity and other surface irregularities and therefore may be inaccurate. Real-time diffusion chargers measure the "active surface area", the area of the particle that interacts with the surrounding gas or ions and is accessible only from the outside. Electrical mobility analyzers calculate the spherical equivalent diameter, which can be converted using geometric relationships. These methods cannot discriminate a nanoparticle of interest from incidental nanoparticles that may occur in complex environments such as workplace atmospheres. Nanoparticles can be collected onto a substrate and their external dimensions can be measured using electron microscopy, then converted to surface area using geometric relations.

Surface chemistry and charge : One can exploit this complexity to design tunable interface controlled materials with high surface area such as nanoparticles, nanowires and low-density nanoporous materials (Jürgen Biener et al., 2009). Their surface-to-volume ratio increases with deceasing feature size and for small enough feature sizes, their properties are no longer dominated by the bulk of the material but by surface atoms. Based on coordination, one can differentiate between three types of surface atoms which, in the order of decreasing coordination, are terrace atoms, step edge atoms, and kink sites. Simple geometrical considerations reveal that the curved surfaces, which are typical for most high surface area materials, are dominated by the more under coordinated step edge and kink site surface atoms. It is indeed the under coordinated which gives rise to new properties. For example, surface stress is the consequence of an electronic relaxation by transferring electronic charge into in-plane bonds. This surface stress induces a pressure in the bulk that then further affects the chemical, physical and mechanical properties of the material. For example, theoretical studies have shown that tensile strain can make gold less noble by increasing the ability of gold to bond simple adsorbates more strongly. Thus high surface area materials open a new dimension in material design for a multitude of technological important areas, including energy storage and conversion, sensing and catalysis. Beyond morphology, the properties of high surface area materials can be further tuned by surface engineering.

Surface chemistry refers to the elemental or molecular chemistry of particle surfaces. No formal definition exists for what constitutes a surface layer, which is usually defined by the measurement technique employed. For nanoparticles a higher proportion of atoms are on their surfaces relative to micron-scale particles, and surface atoms are in direct contact with solvents and influence their interactions with other molecules. Some nanoparticles such as quantum dots may have a core–shell structure where the outer surface atoms are different than those of the interior core.

The surface charge of nanoparticles is suggested as one of the key parameters which decide their biological impact. Surface charge generally refers to the charge from adsorption or desorption of protons on hydroxylated sites on a nanoparticle surface. Surface charge is difficult to directly measure, so the related zeta potential is often measured instead, which is the potential at the double layer's slipping plane, which separates mobile solvent molecules from those that remain attached to the surface. Zeta potential is a calculated rather than measured property, and is a function of both the nanoparticle of interest and its surrounding medium, requiring a description of the measurement temperature; the composition, pH, viscosity, and dielectric constant of the medium; and value used for the Henry function to be meaningful. Zeta potential is used as an indicator of colloidal stability, and has been shown to be predictive of nanoparticle uptake by cells (Stefaniak Aleksandr, 2017). Zeta potential can be measured by titration to find the isoelectric point or through electrophoresis including laser Doppler electrophoresis.

One aspect of NP structure that may influence toxicity is the identity and charge of ligand molecules used to functionalize the NP surface. These surface chemistries have the potential to increase or decrease negative biological impacts, yet their impacts are poorly understood (Bozich et al., 2014). Surface charge is the major parameter determining cellular uptake efficiency, although other factors such as aggregation/agglomeration, protein corona formation, and compositional elements can also influence the cellular uptake partly or indirectly (Soyeon Jeon et al., 2018).

Although positive charge appears to improve the efficacy of imaging, gene transfer, and drug delivery, a higher cytotoxicity of such constructs has been reported. A review by Eleonore Fröhlich (2012) summarized findings on the role of surface charge on cytotoxicity in general, action on specific cellular targets, modes of toxic action, cellular uptake, and intracellular localization of NPs. Effects of serum and intercell type differences are addressed. Cationic NPs cause more pronounced disruption of plasma-membrane integrity, stronger mitochondrial and lysosomal damage, and a higher number of autophagosomes than anionic NPs. In general, nonphagocytic cells ingest cationic NPs to a higher extent, but charge density and hydrophobicity are equally important; phagocytic cells preferentially take up anionic NPs. Cells do not use different uptake routes for cationic and anionic NPs, but high uptake rates are usually linked to greater biological effects. The different uptake preferences of phagocytic and nonphagocytic cells for cationic and anionic NPs may influence the efficacy and selectivity of NPs for drug delivery and imaging.

2.4.4 Solubility

Solubility is a measurement of the degree to which material dissolves from a nanoparticle to enter solution. Material dissolved as part of a solubiity test can be quantified using atomic absorption spectroscopy, inductively coupled plasma optical emission spectroscopy, and inductively coupled plasma mass spectroscopy, with the last being generally the most sensitive. Two related concepts are biodurability, the rate of dissolution in a biological fluid or surrogate, and biopersistence, the rate at which a material is cleared from an organ such as the lung by physical and chemical dissolution processes. Analytical techniques for solubility quantitatively measure total elemental concentration in a sample, and do not discriminate between dissolved or solid forms. Therefore, a separation process must be used to remove the remaining particles. Physical separation techniques include size exclusion chromatography, hydrodynamic chromatography and field flow fractionation. Mechanical separation techniques utilize membranes and/or centrifugation. Chemical separation techniques are liquid–liquid extraction, solid–liquid extraction, cloud point extraction, and the use of magnetic nanoparticles.

The solubility of polymer nanoparticles is a complex phenomenon dependent on solvent–solute and solute–solute interactions. Contrary to phase separation in standard polymerization reactions, which is a well established research area, the relationship between the solubility of polymer nanoparticles and the resulting diameter of the nanoparticles has been largely overlooked (Todd Cowen et al., 2018). Importance of separating spinodal-character phase separation from binodal-character phase separation in polymer nanoparticle synthesis was also asserted. In order to study the dependence of water solubility and hydration behavior of nanoparticles on their surface polarity, Zhu et al, (2016) designed polar nanoparticles with varying surface polarity by assigning atomic partial charge to the surface of C60. The water solubility of the nanoparticle is enhanced by several orders of magnitude after the introduction of surface polarity.

The use of different methods for the characterization of NPs was described (Stefanos Mourdikoudis et al., 2018). These techniques are sometimes exclusive for the study of a particular property, while in other cases they are combined discuss all these techniques in a comparative way, considering factors such as their availability, cost, selectivity, precision, non-destructive nature, simplicity and affinity to certain compositions or materials. The techniques are analysed in depth, despite their big number presented herein. There are microscopy-based techniques, which provide information on the size, morphology and crystal structure of the nanomaterials. Other techniques are specialized for certain groups of materials, such as the magnetic techniques. Examples of these techniques are SQUID, VSM, FMR, and XMCD. Many other techniques provide further information on the structure, elemental composition, optical properties and other common and more specific physical properties of the nanoparticle samples (Table 4). Examples of these techniques include X-ray, spectroscopy and scattering techniques.

Table 4: Characterization of nanoparticles using techniques

Technique	Main information derived
X-Ray Deffraction	Crystal structure, composition, crystalline grain size
X-Ray Absorption Spectroscopy (X-Ray Absorption Fine Structure, X-Ray Absorption near Edge Structure)	X-ray absorption coefficient (element-specific) – chemical state of species, interatomic distances, Debye– Waller factors, also for non-crystalline NPs
Small-Angle X-Ray Scattering	Particle size, size distribution, growth kinetics
X-Ray Photoelectron	Electronic structure, elemental composition, oxidation states,
Spectroscopy	ligand binding (surface-sensitive)
Fourier Transform Infrared Spectroscopy	Surface composition, ligand binding
Nuclear Magnetic Resonance	Ligand density and arrangement, electronic core structure, atomic composition, influence of ligands on NP shape, NP size
Brunauer–Emmett–Teller	Surface area
Thermal Gravimetric Analysis	Mass and composition of stabilizers
Low-Energy Ion Scattering	Thickness and chemical composition of self-assembled monolayers of NPs

UV-Vis Spectroscopy	Optical properties, size, concentration, agglomeration state, hints on NP shape
Photoluminescence spectroscopy	Optical properties – relation to structure features such as defects, size, composition
Dynamic Light Scattering	Hydrodynamic size, detection of agglomerates
Nanoparticle Tracking Analysis	NP size and size distribution
Differential Centrifugal Sedimentation	NP size and size distribution
Inductively Coupled Plasma Mass Spectrometry	Elemental composition, size, size distribution, NP concentration
Secondary Ion Mass Spectrometry, Time of Light -Spectrometry	Chemical information (surface-sensitive) on functional group, molecular orientation and conformation, surface topography, MALDI for NP size
Resonant Mass Measurement	Please check the relevant parts of the manuscript
Microelectro-Mechanical System	Magnetization saturation, magnetization remanence, blocking temperature
Superconducting Quantum Interference Device Magnetometry	
Vibrating Sample Magnetometry	Similar to SQUID through M–H plots and ZFC-FC curves
Mössbauer Spectroscopy	Oxidation state, symmetry, surface spins, magnetic ordering of Fe atoms, magnetic anisotropy energy, thermal unblocking, distinguish between iron oxides
Ferromagnetic Resonance	NP size, size distribution, shape, crystallographic imperfection, surface composition, M values, magnetic anisotropic constant, demagnetization field
Technique	Main information derived
X-ray Magnetic Circular Dichroism	Site symmetry and magnetic moments of transition metal ions in ferro- and ferri-magnetic materials, element specific
Magnetic Susceptibility, Magnetophoretic Mobility	Please check the relevant parts of the manuscript
Super Paramagnetic Relaxometry	Core properties, hydrodynamic size distribution, detect and localize superparamagnetic NPs
Transmission Electron Microscopy	NP size, size monodispersity, shape, aggregation state, detect and localize/quantify NPs in matrices, study growth kinetics
High-Resolution TEM	All information by conventional TEM but also on the crystal structure of single particles. Distinguish monocrystalline, polycrystalline and amorphous NPs. Study defects
Liquid TEM	Depict NP growth in real time, study growth mechanism, single particle motion, superlattice formation
Cryo-TEM	Study complex growth mechanisms, aggregation pathways, good for molecular biology and

	colloid chemistry to avoid the presence of artefacts or destroyed samples
Electron Diffraction	Crystal structure, lattice parameters, study order–disorder transformation, long-range order parameters
Scanning Transmission Electron Microscopy	Combined with HAADF, EDX for morphology study, crystal structure, elemental composition. Study the atomic structure of hetero-interfaces
Aberration-corrected (STEM, TEM)	Atomic structure of NP clusters, especially bimetallic ones, as a function of composition, alloy homogeneity, phase segregation
Electron Energy Loss Spectroscopy	Type and quantity of atoms present, chemical state of atoms, collective interactions of atoms with neighbors, bulk plasmon resonance
Electron tomography	Realistic 3D particle visualization, snapshots, video, quantitative information down to the atomic scale
SEM-HRSEM, T-SEM-EDX	Morphology, dispersion of NPs in cells and other matrices/supports, precision in lateral dimensions of NPs, quick examination– elemental composition
Electron Backscatter Diffraction	Structure, crystal orientation and phase of materials in SEM. Examine microstructures, reveal texture, defects, grain morphology, deformation
Atomic Force Microscopy	NP size and shape in 3D mode, evaluate degree of covering of a surface with NP morphology, dispersion of NPs in cells and other matrices/ supports, precision in lateral dimensions of NPs, quick examination– elemental composition
Technique	Main information derived
Magnetic Force Microscopy	Standard AFM imaging together with the information of magnetic moments of single NPs. Study magnetic NPs in the interior of cells. Discriminate from non-magnetic NPs

The acquisition of a full picture of the variety of features that are associated with a nanomaterial requires typically the use of numerous techniques, often needing to use more than one of them for evaluating well and completely even a single property. Of course, there are challenges in the scientific community for the further improvement of the accuracy and resolution of many techniques.

Nanoparticles can be characterized in two ways. The first method of which are methods aiming to determine the physical properties of the nanoparticle itself, such as size, shape, monodispersity, or crystal structure. Other characterization methods aim to determine chemical characteristics of the particle, including the presence and method of bonding of ligands or other conjugated molecules, and the zeta potential.

2.5 Physical characterization of nanoparticles

The physical characterizations of the prepared metallic and bimetallic Pt-alloy nanoparticles were studied by various analyses (Prashant et al., 2018). The size and size distribution were studied by using the dynamic light scattering principle using the particle size distribution analyzer (Malvern Zeta Sizer Nano Series, Nano-S90, Malvern Instruments, UK).

A list of techniques commonly employed to characterize physical aspects of nanoparticles include UV-Visible Spectroscopy, Transmission Electron Microscopy (TEM), Dynamic Light Scattering (DLS), Differential Centrifugal Sedimentation (DCS), Nanoparticle Tracking Analysis (NTA), Scanning Electron Microscopy (SEM) and X-Ray Diffraction (XRD).

As the field of colloidal science continues to expand, tools for rapid and accurate physiochemical characterization of colloidal particles will become increasingly important. Particle Scattering Diffusometry (PSD), a method that utilizes dark field microscopy and the principles of particle image velocimetry to measure the diffusivity of particles undergoing Brownian motion was studied (Katherine N. Clayton et al., 2016). PSD measures the diffusion coefficient of particles as small as 30 nm in diameter and is used to characterize changes in particle size and distribution as a function of small, label-free, surface modifications of particles. The rapid sizing of particles using three orders-of- magnitude less sample volume than current standard techniques was determined and use PSD to quantify particle uniformity. Furthermore, PSD is sensitive enough to detect biomolecular surface modifications of nanometer thickness.

Nanoparticles exhibit improved physical, chemical and biological properties based on specific characteristics such as size distribution and morphology. Due to their size nanoparticles have higher surface to volume ratio having distinct properties as compared to the bulk counterpart offering many new developments in biosensors, nano fertilizers, nano pesticides and many more (Ishrat Fatima et al., 2019). Metal oxide nanoparticles are of great interest due to their safe, cheap, environment friendly, easy synthesis procedure and technological applications in physics as well as medical and biology. To synthesize nanoparticles of various shapes, size and composition depending on specific requirements, numerous methodologies has been developed such as precursor from liquid, solid or gas phase.

Sol-gel method allows synthesizing nano particles of high purity and better homogeneity compared to the traditional process of oxide fusion. Gas phase synthesis of nano particles is based on nucleation and growth in an inert gas environment and the processing is mainly by laser ablation, electro spray, plasma spray and spray pyrolysis. Laser ablation is an attractive method due to its ability to offer narrow size distribution and a low level of impurities. Synthesis parameters such as laser wave length, energy, pulse width and ablation time can affect the characteristics of nano particles.

2.6 Chemical characterization of nanoparticles

Nanomaterials have a large surface area to volume ratio. The size and structure of nanomaterials largely depend on the salt and surfactant additives, reactant concentrations, reaction temperatures, and solvent conditions used during their synthesis. In engineered nanomaterials, size is a critical factor that controls the circulation and navigation of nanomaterials in the bloodstream, penetration across the physiological drug barriers, site- and cell-specific localization and even induction of cellular responses (AZoNano, 2020). The importance of the size/shape of nanoparticles must be assessed based on individual cases, due to the wide differences in the behavior of different nanomaterials. The shape of nanomaterials affects cellular uptake, biocompatibility, and retention in tissues and organs. Further, the disposition and translocation of nanomaterials in the organism may be dependent on their shape, size, and state of agglomeration.

Understanding the physicochemical properties of nanomaterials is essential in evaluating toxicological and ecological hazards. Chemical properties that are important in characterizing materials include composition, structure, molecular weight, boiling and melting points, vapor pressure, octanol-water partition coefficient, water-solubility, reactivity, and stability. In addition, information on the formulation and manufacture of materials is important in understanding its product variability, purity, performance, and use. It is important to chemically identify and characterize nanomaterials.

Nanomaterials commonly consist of at least two of the following units: metallic, semiconducting and organic particles or molecules (Ping-Chang Lin et al., 2014). Additionally, nanomaterials are generally coated with polymers or biorecognition molecules to improve biocompatibility and selective targeting of biologic molecules. A common feature of all nanomaterials is their large ratio of surface area to volume, which may be orders of magnitude greater than that of macroscopic materials. Still, the final size and structure of nanomaterials depend on the salt and surfactant additives, reactant concentrations, reaction temperatures, and solvent conditions used during their synthesis.

Some nanomaterials that are nanomedicines or considered to be potential nanomedicines are generally split into several categories based on the types of nanomaterials or the application areas, such as drug delivery, drugs and therapies, in vivo imaging, in vitro diagnostics, biomaterials and implants.

Advances in nanotechnology have opened up a new era of diagnosis, prevention and treatment of diseases and traumatic injuries. Nanomaterials, including those with potential for clinical applications, possess novel physicochemical properties that have an impact on their physiological interactions, from the molecular level to the systemic level. There is a lack of standardized methodologies or regulatory protocols for detection or characterization of nanomaterials.

To study the internal chemical structural details, chemical characterization is done.

i. Optical Spectroscopy: Optical Absorption Spectroscopy (OAS), Photoluminescence (PL), Fourier Transform Infrared Spectroscopy (FTIR) and Raman Spectroscopy.

ii. Electron Spectroscopy: Energy Dispersive X-ray Spectroscopy (EDS), X-ray Photoelectron Spectroscopy (XPS), Auger Electron Spectroscopy (AES) and Ultraviolet photoelectron spectroscopy (UPS).

iii. Ionic Spectrometry

iv. Rutherford Backscattering Spectrometry (RBS)

v. Secondary Ion Mass Spectrometry (SIMS).

Since nanoparticles can be made from a wide variety of materials, chemical characterization methods can vary significantly. Some most common techniques are Gel electrophoresis, $^{1}H/^{13}C$ nuclear magnetic resonance spectroscopy (NMR), Fourier-transform infrared spectroscopy (FTIR), Surface enhanced Raman spectroscopy (SERS), Biofunctionality Testing (e.g. immunoblotting), X-ray photoelectron spectroscopy and Inductively couple plasma mass spectrometry (ICP-MS).

The chemical composition of nanoparticles range from a single raw material to multiple ones, assembled into multilayered and composite objects.

Structural characterization techniques: Variation in the size of nanoparticle determines the optical properties of the material. Similarly, for different applications of the nanomaterial, the information about the shape, lattice constants and crystallinity are important. To know the size, shape, lattice constants and crystallinity of the material, structural characterization is done. The techniques include X-ray diffraction techniques- Electron microscopy: Scanning Electron Microscopy (SEM), Transmission Electron Microscopy (TEM)/High Resolution (HR)TEM with Selected Area Electron Diffraction(SAED), Small Angle X-ray Scattering (SAXS), Environmental Transmission Electron Microscopy (ETEM) and Scanning Probe Microscopy

(SPM), Atomic Force Microscopy (AFM), Scanning Tunneling Microscopy (STM), Dynamic Light Scattering.

2.7 Surface characterization of nanomaterials and nanoparticles

The challenges inherently associated with understanding nanomaterials and the roles surface and interface characterization methods can play in meeting some of the challenges were analyzed (Donald R. Baer et al., 2013).

It is equally valuable for researchers to understand how characterization approaches (surface and otherwise) can help to minimize synthesis surprises and to determine how (and how quickly) materials and properties change in different environments. Appropriate application of traditional surface sensitive analysis methods (including x-ray photoelectron and Auger electron spectroscopies, scanning probe microscopy, and secondary ion mass spectroscopy) can provide information that helps address several of the analysis needs. In many circumstances, extensions of traditional data analysis can provide considerably more information than normally obtained from the data collected. Less common or evolving methods with surface selectivity (e.g., some variations of nuclear magnetic resonance, sum frequency generation and low and medium energy ion scattering) can provide information about surfaces or interfaces in working environments (operando or *in situ*) or information not provided by more traditional methods.

NPs are not simple molecules itself and therefore composed of three layers i.e., The surface layer, which may be functionalized with a variety of small molecules, metal ions. surfactants and polymers; The shell layer, which is chemically different material from the core in all aspects and The core, which is essentially the central portion of the NP and usually refers the NP itself. Owing to such exceptional characteristics, these materials got immense interest of researchers in multidisciplinary fields. Mesoporousity imparts additional characteristics in NPs. The NPs can be employed for drug delivery, chemical and biological sensing, gas sensing, CO_2 capturing and other related applications (Ibrahim Khan et al., 2019).

Nanoparticles are unlike conventional chemicals in that their chemical composition and concentration are not sufficient metrics for a complete description, because they vary in other physical properties such as size, shape, surface properties, crystallinity, and dispersion state. Nanoparticles are characterized for various purposes, including nanotoxicology studies and exposure assessment in workplaces to assess their health and safety hazards, as well as manufacturing process control.

Nanoparticles measure less than 100 nanometers in at least one of their external dimensions, and often have properties different from the bulk versions of their component materials, which make them technologically useful (Anonymous, 2013).

2.8 General applications of nanoparticles

Nanoparticles are widely used in bioimaging applications due to their ability to produce varying intensity of colors in solutions by changing the thickness of the nanoshell, the aspect ratio, and the percentage of gold. For instance, a 20-nm gold produces a wine red color solution while a 20-nm platinum produces a yellowish-gray solution. Some nanoparticles such as gold nanoparticles also have the advantage of melting at much lower temperatures (~300 °C for 2.5 nm size) than their bulkier counterparts (gold slabs melt at ~ 1064 °C). Nanoparticles have also been used in cosmetic technology. For instance, zinc oxide particles have been found to have superior UV blocking properties compared to its bulk substitute. Therefore, it is used in the preparation of sunscreen lotions (Ananya Madal, 2018).

As the most prevalent morphology of nanomaterials used in consumer products, nanoparticles have an enormous range of potential and actual applications. Table 5 summarizes the most common nanoparticles used in various product types available on the global markets.

Table 5: General applications of nanoparticles

No.	Industrial sectors	Nanoparticles
1.	Agriculture	Silver, silicon dioxide, potassium, calcium, iron, zinc, phosphorus, boron, zinc oxide and molybdenum
2.	Automotive	Tungsten, disulfidesilicon dioxide, clay, titanium dioxide, diamond, copper, cobalt oxide, zinc oxide, boron nitride, zirconium dioxide, tungsten, γ-aluminium oxide, boron, palladium, platinum, cerium (IV) oxide, carnauba, aluminium oxide, silver, calcium carbonate and calcium sulfonate
3.	Construction	Titanium, dioxidesilicon dioxide, silver, clay, aluminium oxide, calcium carbonate calcium silicate hydrate, carbon, aluminium phosphate cerium (IV) oxide and calcium hydroxide
4.	Cosmetics	Silver, titanium dioxide, gold, carbon, zinc oxide, silicon dioxide, clay, sodium silicate, kojic acid and hydroxy acid
5.	Electronics	Silver, aluminum, silicon dioxide and palladium
6.	Environment	Silver, titanium dioxide, carbonmanganese oxide, clay, gold and selenium
7.	Food	Silver, clay, titanium dioxide, gold, zinc oxide, silicon dioxide, calcium, copper, zinc, platinum, manganese, palladium and carbon
8.	Home appliance	Silver, zinc oxide, silicon dioxide, diamond and titanium dioxide

9.	Medicine[	Silver, gold, hydroxyapatite, clay, titanium dioxide, silicon dioxide, zirconium dioxide, carbon, diamond, aluminium oxide and ytterbium trifluoride
10.	Petroleum	Tungsten, disulfidezinc oxide, silicon dioxide, diamond, clay, boron, boron nitride, silver, titanium dioxide, tungsten, γ-aluminium oxide, carbon, molybdenum disulfide and γ-aluminium oxide
11.	Printing	Toner, deposited by a printer onto paper or other substrate
12.	Renewable energies	Titanium, palladium, tungsten disulfide, silicon dioxide, clay, graphite, zirconium (IV) oxide-yttria stabilized, carbon, gd-doped-cerium (IV) oxide, nickel cobalt oxide, nickel(II) oxide, rhodium, sm-doped-cerium (IV) oxide, barium strontium titanate and silver
13.	Sports and fitness	Silver, titanium dioxide, gold, clay and carbon
14.	Textile	Silver, carbon, titanium dioxide, copper sulfide, clay, gold, polyethylene terephthalate and silicon dioxide

2.9 Applications of nanotechnology in plant protection

In the era of climate change, global agricultural systems are facing numerous, unprecedented challenges. In order to achieve food security, advanced nano-engineering is a handy tool for boosting crop production and assuring sustainability. Nanotechnology helps to improve agricultural production by increasing the efficiency of inputs and minimizing relevant losses. Nanomaterials offer a wider specific surface area to fertilizers and pesticides. In addition, nanomaterials as unique carriers of agrochemicals facilitate the site-targeted controlled delivery of nutrients with increased crop protection. Due to their direct and intended applications in the precise management and control of inputs (fertilizers, pesticides, herbicides), nanotools, such as nanobiosensors, support the development of high-tech agricultural farms. The integration of biology and nanotechnology into nonosensors has greatly increased their potential to sense and identify the environmental conditions or impairments (Fig. 6) (Yifen Shang et al., 2019).

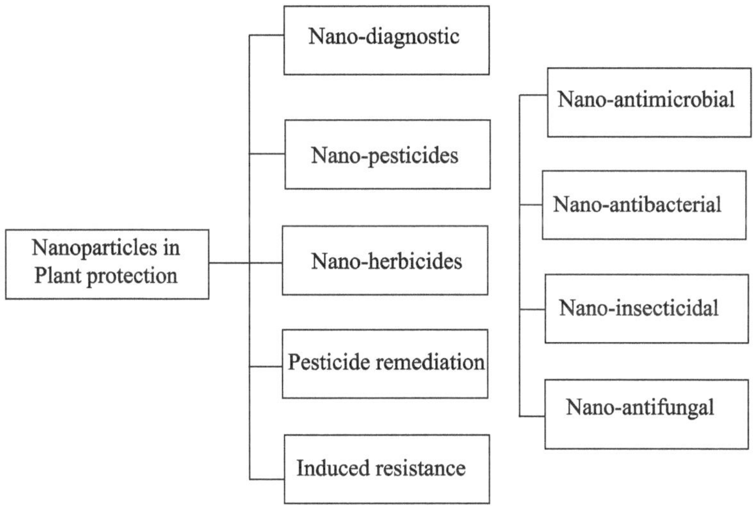

Fig. 6: Uses of nanoparticles in plant protection

Nanoparticles can be used for multiple plant protection purposes,such as pathogen detection (nanodiagnostics), pest control (against microbial pathogens, fungi, bacteria and insects), weed control, pesticide remediation, induced resistance and so on. The assistance of nanotechnology in plant protection products has exponentially increased to achieve higher crop production. In general, conventional crop protection practices involve applications of large-scale and over-dose of fungicides, herbicides and insecticides. Among the applied pesticides, more than 90% are either lost in the environment or unable to reach the target sites essential for effective pest control. This not only increases the expenses of crop production, but also causes the depletion of environmental systems.

It is to be noted that the presence of active ingredients in minimum effective concentration of a formulation at the target sites is essential for assurance of better protection of plants from pest invasion and subsequent crop loss. In this connection, the development of new plant protection formulation has long been a very cognitive field of agricultural research. One such technology is nanoformulation or encapsulation of pesticides which has revolutionized the plant protection sector. Nanoformulation of pesticides contains a very tiny number of particles that act as active ingredients of pesticides, whereas other engineered nano-structures also have useful pesticidal properties (Yifen Shang et al., 2019). The nanoencapsulation of pesticides is the coating of active ingredients of pesticides with another material of various sizes at nano- range where encapsulated materials are referred to as internal phase of the core material (pesticides) and capsulation materials are known as external phase, i.e., the coating nanomaterials.

Nanoformulations or encapsulations of pesticides facilitate the persistence or controlled release of active ingredients in root zones or inside plants without compromising effectiveness. On the other hand, conventional formulations of pesticides or herbicides not only limit water solubility of pesticides, but also injure other organisms, leading to increased resistance to target organisms. In contrast, nanoformulations assist to overcome the above-mentioned limitations. Nanoformulations of pesticides boost crop yields by increasing pesticide efficacy through regulating transport potential of pesticide. Moreover, nanomaterials in pesticide formulation show some useful properties like increased stiffness, permeability, thermal stability, solubility, crystallinity and also biodegradability essential for sustainable agro-environmental system. More importantly, the timely and controlled release of active ingredients reduces the total amount of pesticides required for pest and disease control, an important feature of integrated pest management (IPM). Furthermore, sustainable agriculture demands minimum use of agrochemicals, so as to protect environmental depletion and others non-target species. Moreover, the minimum use of pesticides reduces the cost involved in crop production.

It has been estimated that on a global scale annual agricultural crop losses account for 2000 billion US dollars due to plant diseases, insect pests and weeds, and only in United States for pathogen control efforts simply through fungicide applications cost exceeds $600 million (Yifen Shang et al., 2019). Fascinatingly, in such circumstances, the use of NPs is being reported as efficient alternative to directly suppress pathogen infection and activity, leading to the increased crop growth and yield. For example, halloysites a types of clay nano tubes used as cost-effective carriers of pesticides in agriculture. These nano tubes not only show extended release period of active ingredients (AI), but also provide assurance of better contact with minimum environmental effect.

In addition, many investigations have provided evidence that nanoformulations of pesticides facilitate the widening of plant-based systemic acquired resistance (SAR) against pests. For examples, silica nanosphere formulations can increase ability of pesticides to penetrate through the plant and reach the cell sap, thereby exerting systemic effect to control chewing or sucking type insects like aphid. Latex fabricated bioactive AuNPs reduced the catalytic potential of trypsin, a vital insect protease and thereby bio-controlling devastating insects. This catalytic inhibition might be due to the interaction of metallic NPs with proteins via covalent interactions, electrostatic interactions or binding to -SH group of amino acid (Yifen Shang et al., 2019).

Inorganic NPs, such as ZnO, Cu, SiO_2, TiO_2, CaO, MgO, MnO and AgNPs play important roles in various arena of plant protection including microbial activity and bacterial diseases. For example, ZnO nanoparticles have recently been shown to provide effective growth control of *Fusarium graminearum, Penicillium expansum, Alternaria alternate, F. oxysporum, Rhizopus stolonifer, Mucor plumbeus* and *A. flavus* as well as pathogenic bacteria *Pseudomonas aeruginosa* Nano-Cu application was found to be more effective against *Phytophthora infestans* compared to currently available non-nano Cu formulations in tomato. Besides, Si, and TiO_2 have been found promising to suppress crop diseases directly, through antimicrobial activity. MONPs inhibit the development of fungal conidia and conidiophores which cause ultimate death of fungal hyphae. Likewise, weeds are also considered as serious threat to global agricultural production as they compete with crops for their nutrient, water and light. However, the application engineered nanomaterials containing herbicides provide the solution in eco-friendly manner. Germination, root and shoot lengths, fresh and dry weights, and photosynthetic pigments with total protein significantly decreased in weeds exposed to SiO_2 nanoparticles.

Thus, the nanomaterials in pesticides, fungicides and herbicides have a tremendous scope in sustainable agricultural development.

2.10 Morphology and broad categories of nanoparticles

The morphology of nanoparticles helps serve their various purposes such as long carbon nanotubes being used to bridge an electrical junction. Amorphous particles usually adopt a spherical shape or nanospheres and anisotropic microcrystlaline whiskers correspond to their particular crystal shape. Small nanoparticles usually form clusters. These may be of various shapes like rods, fibers, and cups etc. The study of fine particles is called micromeritics. Controlling the morphology of nanoparticles is of key importance for exploiting their properties for their use in several emerging technologies. Optical filters and bio-sensors are among the many applications that use optical properties of gold nanoparticles and it requires anisotropy of the particle shape as larger shapes produce greater plasmon losses.

Nanoparticles can be classified into different types according to the size, morphology, physical and chemical properties. Some of them are carbon-based nanoparticles, ceramic nanoparticles, metal nanoparticles, semiconductor nanoparticles, polymeric nanoparticles and lipid-based nanoparticles.

Nanomaterials can be nanoscale in one dimension (eg. surface films), two dimensions (eg. strands or fibres), or three dimensions (eg. particles). They can exist in single, fused, aggregated or agglomerated forms with spherical, tubular, and irregular shapes. Common types of nanomaterials include nanotubes, dendrimers, quantum dots and fullerenes.

Nanostructured materials are classified as Zero dimensional, one dimensional, two dimensional, three dimensional nanostructures. Nanomaterials are materials which are characterized by an ultra fine grain size (< 50 nm) or by a dimensionality limited to 50 nm. Nanomaterials can be created with various modulation dimensionalities, viz., zero (atomic clusters, filaments and cluster assemblies), one (multilayers), two (ultrafine-grained overlayers or buried layers) and three (nanophase materials consisting of equiaxed nanometer sized grains) as shown in the figures.

i. **Zero dimensional:** Materials wherein all the dimensions are measured within the nanoscale (no dimensions, or 0D, are larger than 100 nm). The most common representation of zero-dimensional nanomaterials are nanoparticles (Fig. 7). Nanoparticles can be amorphous or crystalline; be single crystalline or polycrystalline; be composed of single or multi-chemical elements; exhibit various shapes and forms ; exist individually or incorporated in a matrix.

ii. **One-dimensional nanomaterials:** One dimension that is outside the nanoscale. This leads to needle like-shaped nanomaterials. 1-D materials include nanotubes, nanorods, and nanowires. 1-D nanomaterials can be amorphous or crystalline; single crystalline or polycrystalline ; chemically pure or impure; standalone materials or embedded in within another medium; metallic, ceramic, or polymeric

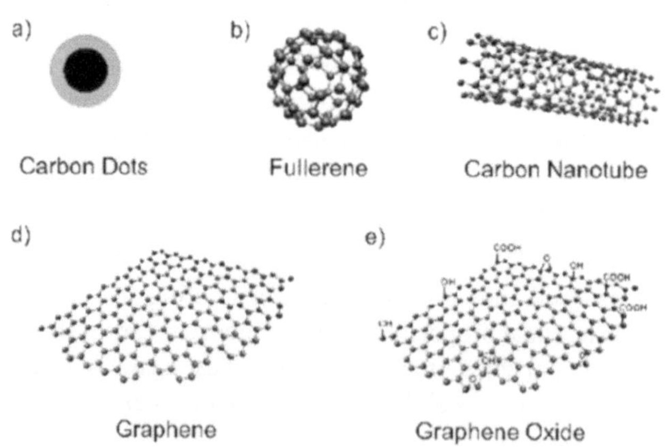

Fig. 7: Zero dimensional carbon nanomaterials, a) carbon dots, and b) fullerene. One dimensional carbon nanomaterial, c) carbon nanotube. Two dimensional carbon nanomaterial, d) graphene, and e) graphene oxide

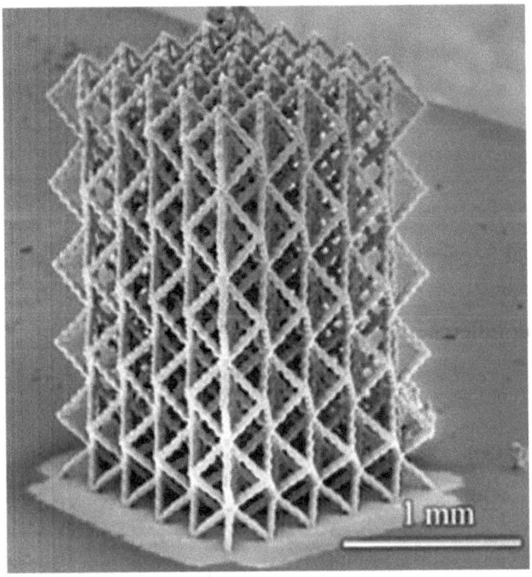

Fig. 8: 3D Microstructure fabricated from nanoparticles

iii. **Two-dimensional nanomaterials:** Two of the dimensions are not confined to the nanoscale. 2-D nanomaterials exhibit plate-like shapes (Fig. 7). Two-dimensional nanomaterials include nanofilms, nanolayers, and nanocoatings. 2-D nanomaterials nanomaterials can be: amorphous or crystalline; made up of various chemical compositions; used as a single layer or as multilayer structures ; deposited on a substrate; integrated in a surrounding matrix material ; metallic, ceramic, or polymeric

iv. **Three-dimensional nanomaterials:** Bulk nanomaterials are materials that are not confined to the nanoscale in any dimension. These materials are thus characterized by having three arbitrarily dimensions above 100 nm (Fig. 8). Materials possess a nanocrystalline structure or involve the presence of features at the nanoscale. In terms of nanocrystalline structure, bulk nanomaterials can be composed of a multiple arrangement of nanosize crystals, most typically in different orientations. With respect to the presence of features at the nanoscale, 3-D nanomaterials can contain dispersions of nanoparticles, bundles of nanowires, and nanotubes as well as multinanolayers.

Types and examples of Nanomaterials: Nanomaterials (gold, carbon, metals, meta oxides and alloys) with variety of morphologies (shapes) are depicted in Fig. 9.

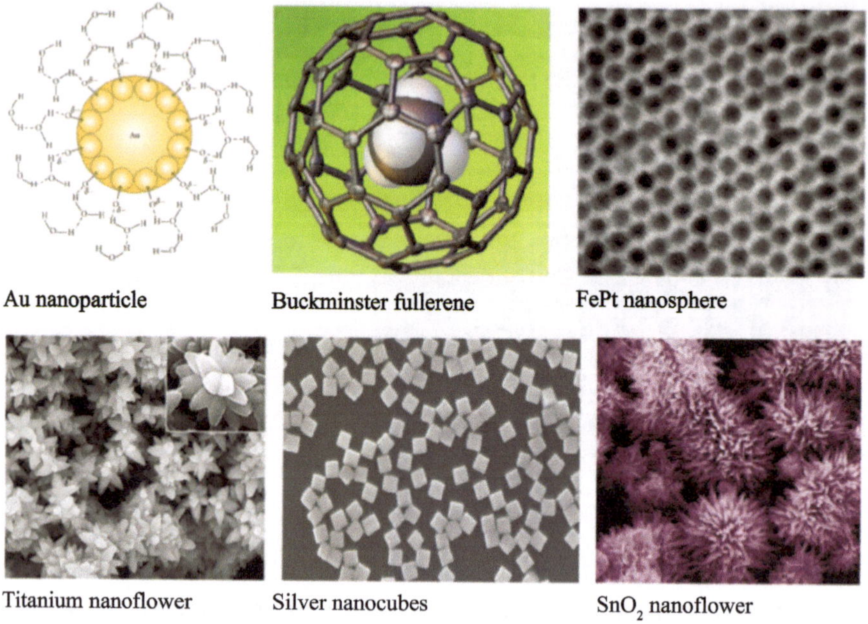

Fig. 9: Shapes of nanomaterials

Organic and inorganic nanoparticles: Organic and inorganic nanoparticles are materials of two or more dimensions, with a size in the range of 1–100 nm. Nanoparticles are prepared with organic polymers (organic nanoparticles) and/or inorganic elements (inorganic nanoparticles).

Organic nanoparticles are materials of two or more dimensions, with a size in the range of 1–100 nm. The chemical composition and the shape of a nanoparticle also influence its specific properties. The main groups of organic Nano carriers are liposomes, micelles, protein/peptide based and dendrimers (Fig. 10). Organic nanoparticles, nanocrystals and nanobeads are of major interest in material and life sciences. Biopolymer nanoparticles are offering numerous advantages that embrace the simplicity of their preparation from well-understood biodegradable, biocompatible polymers and their high stability in biological fluids during storage. Several types of polymers have been tested as potential drug delivery systems; including nanoparticles, dendrimers, capsosomes and micelles.

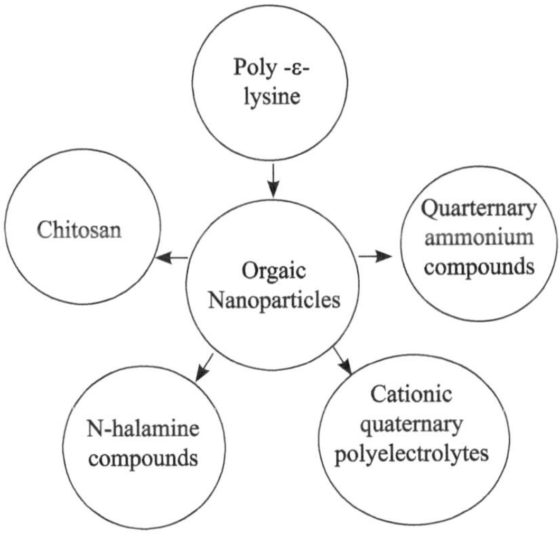

Fig. 10: Types of organic nanoparticles

Biopolymer nanoparticles are offering numerous advantages which embrace the simplicity of their preparation from well-understood biodegradable, biocompatible polymers and their high stability in biological fluids during storage. Since the emergence of Nanotechnology in the past decades, the development and design of organic and bioorganic nanomaterials has become an important field of research. And several types of polymers have been tested and are used in drug delivery systems; including nanoparticles, dendrimers, capsosomes and micelles. Researchers have found, the synthesized polymers even serves as a good carrier and plays a vital role in carrying a drug. And in other hand they are used in food industries too for food package purposes.

There are thousands of organic chemicals are in present in various pharmaceutical to consumer product and are being used in dyes, flavoring agents. It can be explained in organic compounds ranging in diameter from 10 to 1μm. Ultrafine particles are the same as nanoparticles and between 1 and 100 nanometers in size, fine particles are sized between 100 and 2,500 nanometers, and coarse particles cover a range between 2,500 and 10,000 nanometers. Physical and chemical methods for the synthesis of nanoparticles of complex organic compounds (mechanical grinding, laser ablation, solvent replacement, reduction in solution, ion association, etc.) were analyzed (A Yu Utekhina and Gleb B Sergeev, 2011) Emphasis was placed on a method that employs supercritical fluids and on cryochemical synthesis. Physicochemical properties of nanoparticles of organic compounds and the prospects of their use were considered.

Inorganic nanoparticles cover a broad range of substances including elemental metals, metal oxides, and metal salts. Silver nanoparticles (AgNPs) are used in many products as a bactericide, whereas gold nanoparticles (AuNPs) are explored for many possible applications because of their catalytic activity. Both silver and gold nanoparticles possess the so-called surface plasmon as a result of the collective oscillation of the electrons, which gives them excellent optical properties and a large enhancement of the electric field on their surface. Quantum dots (QDs) are semiconductor nanocrystals with all three dimensions falling in the 1–10 nm size range. Semiconductor quantum dots are commercially available and offer a viable alternative to fluorescently labeled particles, whilst iron oxide nanoparticles have been approved for human use in magnetic resonance imaging (MRI) applications as contrast enhancers (Fig. 11) (Juan J.Giner-Casares et al., 2016).

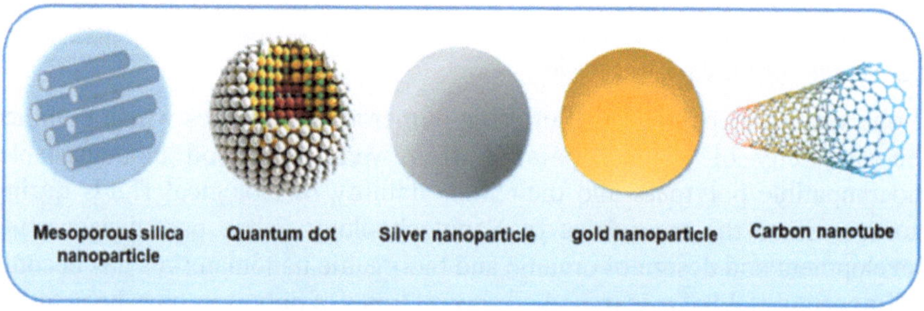

Fig. 11: Types of inorganic nanoparticles

In the field of modern material science, inorganic nanoparticle has been developed based upon their unique physical properties, particularly in biotechnology. Based upon these factors of inorganic nanoparticles they have certain physical properties that mainly include size- dependent optical, magnetic, electronic and catalytic properties. Bio related application are involved for the preparation of these interesting nanoparticles like iron oxides, gold, silver, silica, quantum dots etc (Heera and Shanmugam, 2015). Novel physical properties mainly related because of their size approaches nanometer scale dimension.

Magnetic nanoparticles: Magnetic nanoparticles belong to the group of nanotechnology-based materials with an impact in fields of analytical chemistry, biosensing, and nanomedicine. Magnetic effects are caused by movements of particles that have both mass and electric charges. These particles are electrons, holes, protons, and positive and negative ions (Fig.12) (Wilczewska et al., 2012). A spinning electric-charged particle creates a magnetic dipole, so-called magneton. In ferromagnetic materials, magnetons are associated in groups. A magnetic domain (also called a Weiss domain) refers to a volume of ferromagnetic material in which all magnetons are aligned in the same direction by the exchange forces. This concept of domains distinguishes ferromagnetism from paramagnetism. The domain structure of a ferromagnetic material determines the size dependence of its magnetic behavior. When the size of a ferromagnetic material is reduced below a critical value, it becomes a single domain. Fine particle magnetism comes from size effects, which are based on the magnetic domain structure of ferromagnetic materials.

Magnetic nanoparticles show remarkable new phenomena such as high field irreversibility, high saturation field, superparamagnetism, extra anisotropy contributions, or shifted loops after field cooling. These phenomena arise from narrow and finite-size effects and surface effects that dominate the magnetic behavior of individual nanoparticle. The magnetization behavior of these particles above a certain temperature, i.e., the blocking temperature, is identical to that of atomic paramagnets (superparamagnetism) except that large susceptibilities and, thus, an extremely large moment are involved (Abolfazl Akbarzadeh et al., 2012).

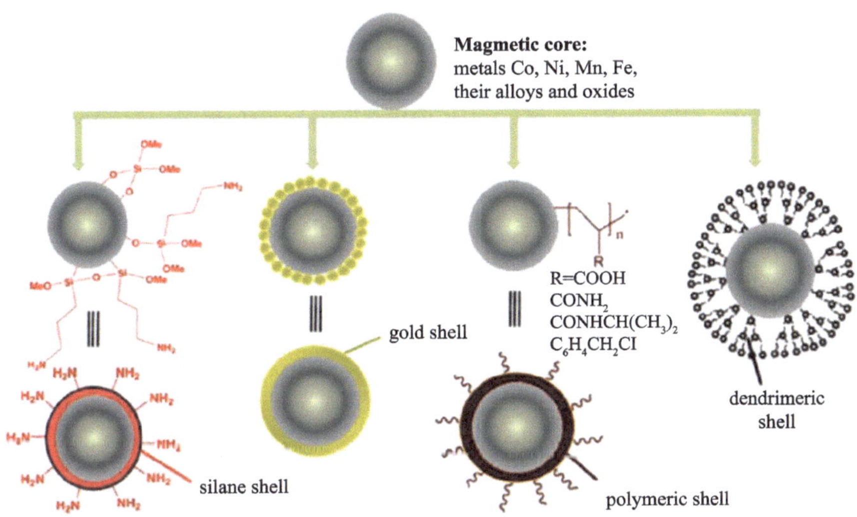

Fig. 12: Magnetic nanoparticles with various shells

The four applications they have described for magnetic nanoparticles and microparticles aided in diagnosis and treatment of diseases in the following years include magnetic separation of biological entities contributed to the development of diagnostics, magnetic nanocarriers contributed to drug delivery, while, radio frequency-controlled magnetic nanoparticles provided a new approach for cancer treatment, in addition to magnetic resonance imaging (MRI) applications (Jiri Kudr et al., 2017). Nanomedicine introduced numerous novel materials for drug delivery and applications in molecular imaging where accurate detection is dependent on molecular signatures. Magnetic particles are also applicable in the microscopic manipulation of nano- and micro-sized objects.

The stability of ferrite nanoparticles in a solution can be increased by modifying their surface using surfactants, or derivatives of phosphoric acid or silicon. However, the magnetic property of nanoparticles can also be of disadvantage in certain situations. For example, ferroelectric materials smaller than 10 nm can switch their magnetization direction using room temperature thermal energy, thus making them unsuitable for memory storage (Ananya Madal, 2018).

Carbon-based nanoparticles : Carbon-based nanoparticles include two main materials, viz., carbon nanotubes (CNTs) and fullerenes. CNTs are nothing but graphene sheets rolled into a tube. These materials are mainly used for the structural reinforcement as they are 100 times stronger than steel. CNTs can be classified into single-walled carbon nanotubes (SWCNTs) and multi-walled carbon nanotubes (MWCNTs). CNTs are unique in a way as they are thermally conductive along the length and non-conductive across the tube. Fullerenes are the allotropes of carbon having a structure of hollow cage of sixty or more carbon atoms. The structure of C-60 is called Buckminsterfullerene, and looks like a hollow football. The carbon units in these structures have a pentagonal and hexagonal arrangement. These have commercial applications due to their electrical conductivity, structure, high strength, and electron affinity.

Carbon nanoparticles can be used for the delivery of drugs at a cellular level. One research team used carbon nanoparticles created through basic means to deliver infusions of drugs to a pig. The conclusion from that experiment is that carbon nanoparticles could be used to deliver a variety of drugs to treat cancer. Well-known classic examples of carbon allotropes are diamond, amorphous carbon and graphite (Fig.13). More recently discovered allotropes are buckminster fullerenes or buckyballs, carbon nanotubes and of course graphene. Nanoparticles used in drug delivery system include Chitosan, Alginate, Xanthan gum, Cellulose, Liposomes, Polymeric micelles, Dendrimers and Inorganic nanoparticles.

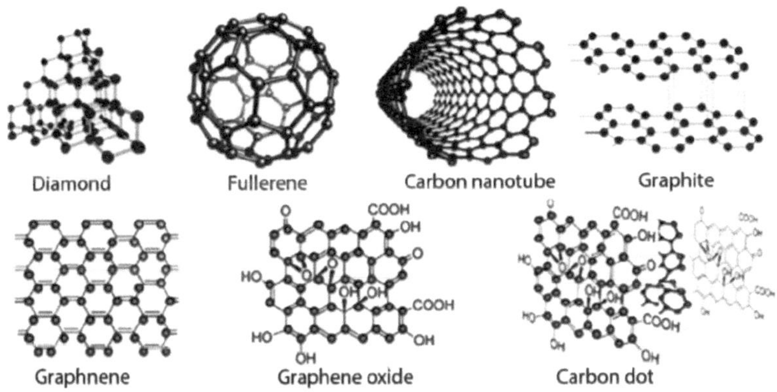

Fig. 13: Carbon nanoparticles

Ceramic nanoparticles: Ceramic nanoparticle is a type of nanoparticle that is composed of ceramics, which are generally classified as inorganic, heat-resistant, nonmetallic solids that can be made of both metallic and nonmetallic compounds. The material offers unique properties. Ceramic nanoparticles are inorganic solids made up of oxides, carbides, carbonates and phosphates. These nanoparticles have high heat resistance and chemical inertness (Fig.14). They have applications in photocatalysis, photodegradation of dyes, drug delivery, and imaging. By controlling some of the characteristics of ceramic nanoparticles like size, surface area, porosity, surface to volume ratio, etc, they perform as a good drug delivery agent. These nanoparticles have been used effectively as a drug delivery system for a number of diseases like bacterial infections, glaucoma, cancer, etc.

Metal nanoparticles: Metal nanoparticles are prepared from metal precursors. These nanoparticles can be synthesized by chemical, electrochemical, or photochemical methods. In chemical methods, the metal nanoparticles are obtained by reducing the metal-ion precursors in solution by chemical reducing agents. These have the ability to adsorb small molecules and have high surface energy. These nanoparticles have applications in research areas, detection and imaging of biomolecules and in environmental and bioanalytical applications. For example gold nanoparticles are used to coat the sample before analyzing in SEM. This is usually done to enhance the electronic stream, which helps us to get high quality SEM images (Fig. 15).

Fig. 14: Ceramic nanoparticles

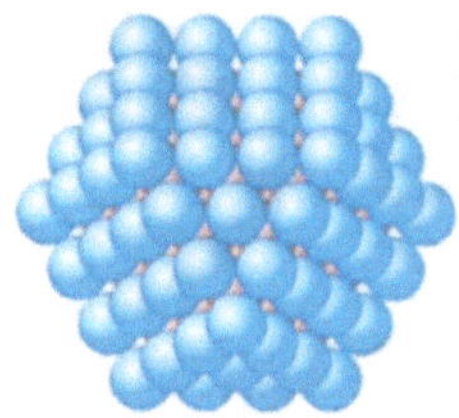

Fig. 15: Metal nanoparticle

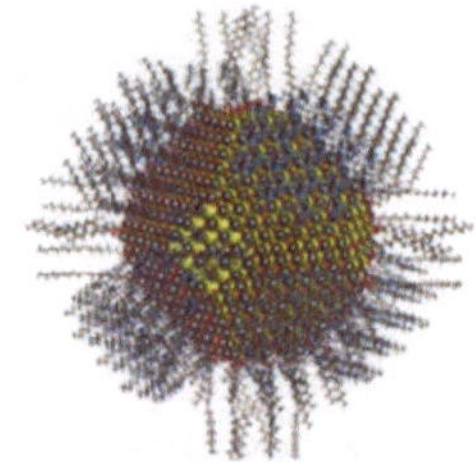

Fig. 16: Semiconductor nanoparticle

Semiconductor nanoparticles : Semiconductor nanoparticles have properties like those of metals and non-metals. They are found in the periodic table in groups II-VI, III-V or IV-VI. These particles have wide band gaps, which on tuning shows different properties (Fig 16). They are used in photocatalysis, electronics devices, photo-optics and water splitting applications. Some examples of semiconductor nanoparticles are GaN, GaP, InP, InAs from group III-V, ZnO, ZnS, CdS, CdSe, CdTe are II-VI semiconductors and silicon and germanium are from group IV.

Lipid-based nanoparticles: Lipid nanoparticles are generally spherical in shape with a diameter ranging from 10 to 100nm. It consists of a solid core made of lipid and a matrix containing soluble lipophilic molecules. The external core of these nanoparticles is stabilized by surfactants and emulsifiers (Fig.17). These nanoparticles have application in the biomedical field as a drug carrier and delivery and RNA release in cancer therapy. Thus, the field of nanotechnology is far from being saturated and it is, as the statistic says, sitting on the staircase of an exponential growth pattern. It is basically at the same stage as the information technology was in the 1960s and biotechnology in the year of 1980s. Thus it can easily be predicted that this field would witness a same exponential growth as the other two technological field witnessed earlier.

Dendrimers: These nanomaterials are nanosized polymers built from branched units. The surface of a dendrimer has numerous chain ends, which can be tailored to perform specific chemical functions. This property could also be useful for catalysis. Also, because three-dimensional dendrimers contain interior cavities into which other molecules could be placed, they may be useful for drug delivery. Dendrimers are repetitively branched molecules (Fig. 18) (Astruc et al., 2010).

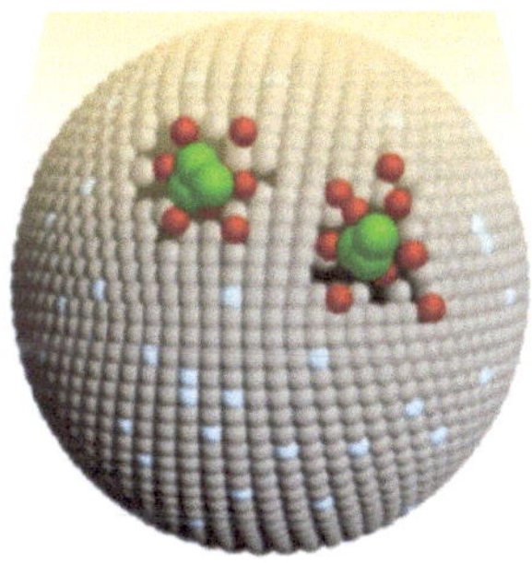

Fig. 17: Lipid-based nanoparticle

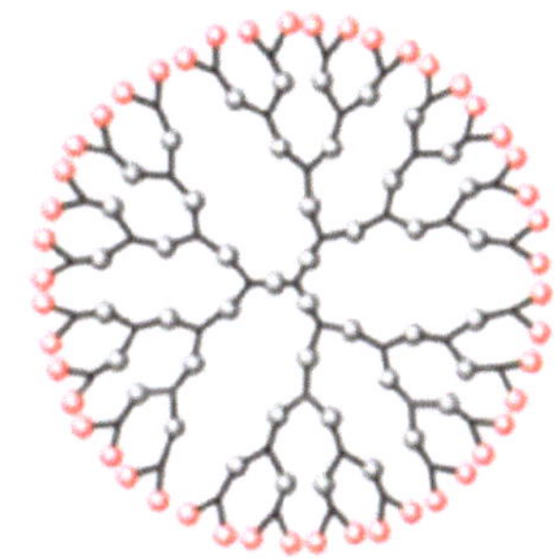

Fig. 18: Dendrimer

The name comes from the Greek word δένδρον (dendron) which translates to "tree". Synonymous terms for dendrimer include arborols and cascade molecules. However, dendrimer is currently the internationally accepted term. A dendrimer is typically symmetric around the core, and often adopts a spherical three-dimensional morphology. The word dendron is also encountered frequently. A dendron usually contains a single chemically addressable group called the focal point or core. The difference between dendrons and dendrimers is illustrated in the top figure, but the terms are typically encountered interchangeable.

Composites: Composites combine nanoparticles with other nanoparticles or with larger, bulk-type materials. Nanoparticles, such as nanosized clays, are already being added to products ranging from auto parts to packaging materials, to enhance mechanical, thermal, barrier, and flame-retardant properties (Fig. 19). Nano-composite materials have been extensively used in many engineering applications such as aerospace engineering, mechanical engineering, electrical engineering, and bio- and chemical engineering owing to the fact that their mechanical properties are considerably higher than that of the original homogeneous material (without inclusions). Understanding and being able to predict the effective mechanical properties of nano-composites hold the key for designing and developing such novel materials. There have been several micromechanical models proposed to predict the effective material properties of nano-composites (Alexander Vakhrushev, 2020).

Fullerenes: The fullerenes are a class of allotropes of carbon which conceptually are graphene sheets rolled into tubes or spheres. These include the carbon nanotubes (or silicon nanotubes) which are of interest both because of their mechanical strength and also because of their electrical properties. The first fullerene molecule to be discovered, and the family's namesake, buckminsterfullerene (C_{60}), was prepared in 1985 by Richard Smalley, Robert

Curl, James Heath, Sean O'Brien, and Harold Kroto at Rice University. The name was a homage to Buckminster Fuller, whose geodesic domes it resembles.

Fullerenes have since been found to occur in nature. More recently, fullerenes have been detected in outer space (Cami et al., 2010). The family is named after buckminsterfullerene (C_{60}), the most famous member, which in turn is named after Buckminster Fuller. A buckyball, more correctly, a spherical fullerene, is a molecule of carbon atoms forming a hollow spherical "shell." Many sizes and configurations are possible (and stable!), but due to Carbon's chemical properties they always have a few common characteristics: the faces are rings of 5 or 6 carbon atoms, and each carbon atom bonds to 3 others. The most commonly occurring one, buckminsterfullerene, is made with 60 carbon atoms in the shape of a soccer ball (Fig. 19).

Fig. 19: Buckminster fullerene

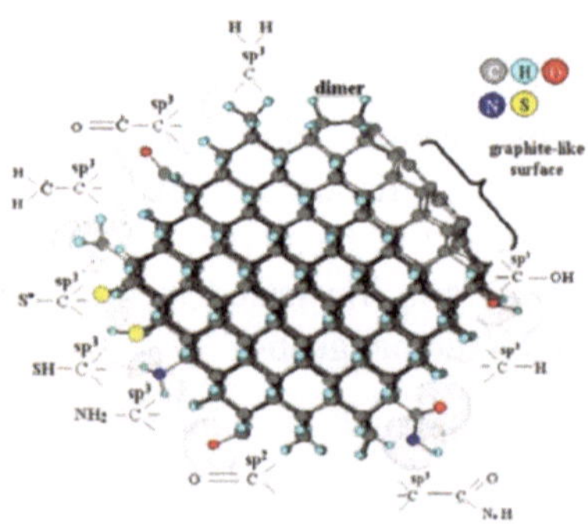

Fig. 20: Nanodiamond

Fig. 21: Onion-like carbons

For the past decade, the chemical and physical properties of fullerenes have been a hot topic in the field of research and development, and are likely to continue to be for a long time.

Nanodiamond: Nanodiamonds are carbon nanoparticles with truncated octahedral architecture that are about 2–8 nm in diameter and can deliver a wide range of therapeutics, including small molecules, proteins, and nucleic acids. Because of their inexpensive, large-scale synthesis, potential for surface functionalization, and high biocompatibility, nanodiamonds are widely investigated as a potential material in biological and electronic applications and quantum engineering (Mochalin et al., 2011).

Diamond, spherical and truncated octahedron diamond with predominant 3-bonded carbon (Fig.20) is one of the hardest materials known to date and regarded as the king of all gemstones because of its scientific qualities in hardness, chemical corrosiveness, thermal expansion and conductivity, electrical insulation, and biocompatibility. Nanodiamond (ND) is a cubic structural diamond, with structure and properties as that of a diamond and is 5nm in diameter on average. Nanodiamond contains a variety of diamondbased materials at nanoscale (the length scale of approximately 1–100 nm) including purephase diamond films, diamond particles, and their structural assemblies. Nanodiamonds have a wide range of applications in electrochemical coatings, polymer compositions, antifriction coatings, polishing, lubricants, biosensors, imaging probes, implant coatings and drug carriers (Ratna Kumari and Sobha, 2013).

Onion-like carbons (OLCs): The onion-like carbons are polyhedral nanoparticles with well aligned concentric rings and show a high degree of symmetry. Carbon onions are quasi- spherical nanoparticles, possess a hollow core encased with three to eight closed graphitic shell structures (Fig.21). The outer diameters are in the range of 20–100 nm. They can serve as nanocapsules for drug delivery systems. The external graphite acts as a template for the attachment of desirable functional groups and protects the substances present inside. Carbon onions exhibit unique zero-dimensional spherical or concentric shell structures with small (<10nm) diameters. They often go by other names such as carbon nano-onions or onion-like carbon (OLC). We synthesize carbon onions by annealing (usually in vacuum) nanodiamond particles. Because of their structure, high conductivities, and dispersion properties, they are often used as conductive additives in electrochemical energy storage devices, such as battery and supercapacitor electrodes, where an increase in the rate performance (power) of the device is necessary (McDonough and Gogotsi, 2013).

However, carbon onions can be more than just an additive. When investigated as the main active material in microsupercapacitors, devices containing carbon onion electrodes showed stack capacitances 4 orders of magnitude higher than that of electrolytic capacitors, an impressive performance in the field of ultracapacitors. Additionally, carbon onions have also been explored with ionic liquid electrolytes, widening the operating voltage range and enabling use in low-temperature regimes. Carbon onions have also been coated with redox active molecules or functionalized with electrochemically active functional groups, improving the overall electrochemical performance. Because of each of these qualities and performance ranges, carbon onions could be expanded into many other applications, such as textile supercapacitors, antennas, biomedical devices, sensors, and much more.

Nanofibers: Nanofibers are fibers with diameters in the nanometer range. Nanofibers can be generated from different polymers and hence have different physical properties and application potentials. Examples of natural polymers include collagen, cellulose, silk fibroin, keratin, gelatin and polysaccharides such as chitosan and alginate (Khajavi et al., 2016). Examples of synthetic polymers include poly(lactic acid) (PLA), polycaprolactone (PCL), polyurethane (PU), poly(lactic-co-glycolic acid) (PLGA), poly(3-hydroxybutyrate-co-3-hydroxyvalerate) (PHBV), and poly(ethylene-co-vinylacetate) (PEVA). Polymer chains are connected via covalent bonds. The diameters of nanofibers depend on the type of polymer used and the method of production. All polymer nanofibers are unique for their large surface area-to-volume ratio, high porosity, appreciable mechanical strength, and flexibility in functionalization compared to their microfiber counterparts. Nanofibers have many possible technological and commercial applications. They are used in tissue engineering, drug delivery, cancer diagnosis, lithium-air battery, optical sensors and air filtration.

They may be cylindrical or conical, filamentous carbon nanostructures from a quasi-onedimensional (1D) filament (Fig.22). Their diameter ranges from a few to hundred nanometers, where as lengths range from less than a micrometer to millimeters. The morphological structure is often divided into plate CNFs, ribbon-like CNFs, herringbone CNFs (Rassaei et al., 2007). These are used for electrode materials, as anodes for rechargeable lithium-ion batteries.

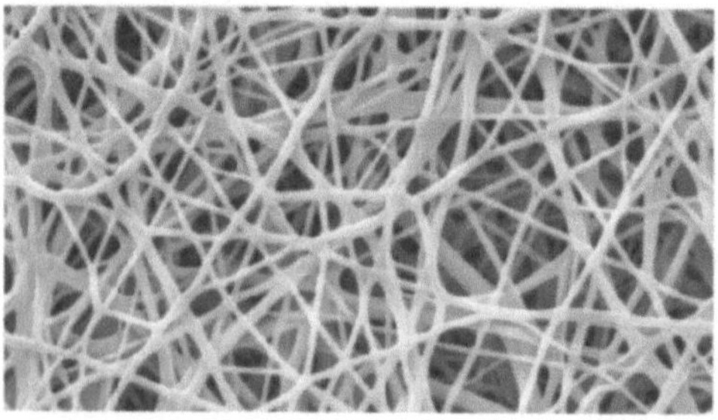

Fig. 22: Nanofiber

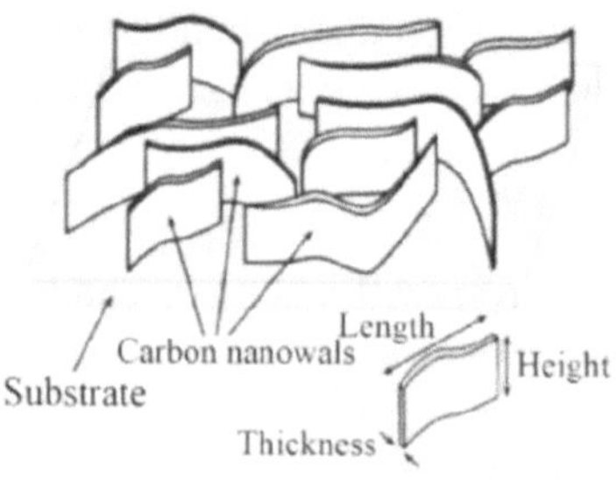

Fig. 23: Carbon nanowalls

Nanowalls: These are also referred to as " nanoflakes". Carbon nanowalls (CNWs) consist of vertically aligned graphene sheets standing on the substrates, form two dimensional wall structure with large surface areas and sharp edges (Fig.23). The thickness of CNWs ranges from a few nm to a few tens nm. CNWs can also be used as a template for loading other nanomaterials and the resulting hybrid nanostructures are used as photocatalysts. The optical properties of carbon nanowall ilms in the visible range have been studied and reported for the first time. Depending on the film structure, ultra-low total reflectance up to 0.13% can be reached, which makes the CNW films a promising candidate for the black body-like coating and thus for a wide range of applications as a light absorber.

Yolk/Shell nanoparticles: Core/shell nanoparticles were first reported in the early 1990s with a simple spherical core and shell structure, but the area is gradually diversifying in multiple directions such as different shapes, multishells, yolk/shell etc., because of the development of different new properties of the materials, which are useful for several advanced applications (Rahul Purbia and Santanu Paria, 2015). Among different sub-areas of core/ shell nanoparticles, yolk/shell nanoparticles (YS NPs) have drawn significant attention in recent years because of their unique properties such as low density, large surface area, ease of interior core functionalization, a good molecular loading capacity in the void space, tunable interstitial void space, and a hollow outer shell. The YS NPs have better properties over simple core/shell or hollow NPs in various fields including biomedical, catalysis, sensors, lithium batteries, adsorbents, DSSCs, microwave absorbers etc., mainly because of the presence of free void space, porous hollow shell, and free core surface.

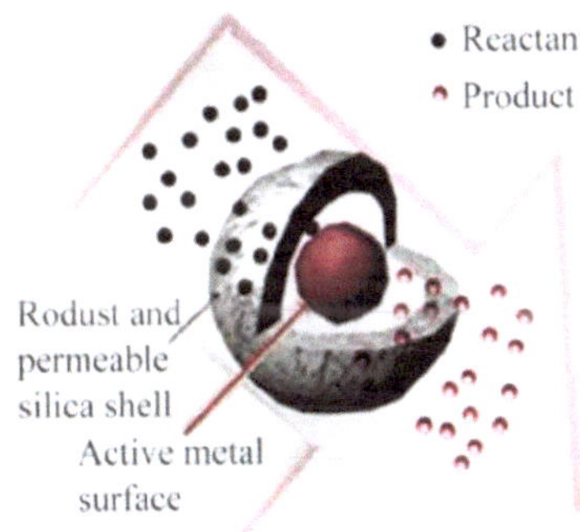

Fig. 25: Yolk / shell nanoparticles

Yolk/shell nanoparticles (YSNs) consist of a nanoparticle core surrounded by a hollow shell and are named after an egg with a shell, interstitial space and the core - the yolk (Fig. 25). Yolk-shell structures have been prepared with various types of movable cores, such as gold, SiO_2 and magnetic Fe_3O_4. Yolk/ shell nanoparticles are one of the important nanomaterials for a wide variety of applications such as catalysis, delivery, lithium-ion batteries and biosensors due to their tailorability and functionality in both the cores and hollow shells (Jian Liu, et al., 2011). Due to its tiny size they provide a large surface area. These nanostructures reduce drug risks; by breathable pores the drug enters into the cell and do not flow.

Nanosized silver: Nanosilver particles (NSPs), are among the most attractive nanomaterials, and have been widely used in a range of biomedical applications, including diagnosis, treatment, drug delivery, medical device coating, and for personal health care. Silver nanoparticles are nanoparticles of silver of between 1 nm and 100 nm in size (Graf Christina et al., 2003). While frequently described as being 'silver', some are composed of a large percentage of silver oxide due to their large ratio of surface to bulk silver atoms. Numerous shapes of nanoparticles can be constructed depending on the application at hand. Commonly used silver nanoparticles are spherical, but diamond, octagonal, and thin sheets are also common. Their extremely large surface area permits the coordination of a vast number of ligands. The properties of silver nanoparticles applicable to human treatments are under investigation in laboratory and animal studies, assessing potential efficacy, toxicity and costs.

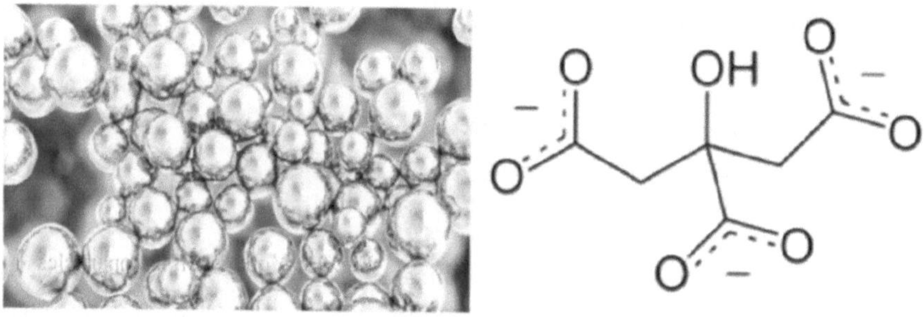

Fig. 26: Nanosilver particles Structure

Silver (Ag) is known to have antimicrobial activity both in ionic or nanoparticle forms. The powerful antimicrobial effect of silver especially in unicellular microorganisms is believed to be brought about by enzyme inactivation (Kim et al., 1998) (Fig.26). Nano silver whose antimicrobial effect has been tested against many disease causing pathogens of animals and plants is the most studied and utilized nanoparticle. Silver is also an excellent plant growth stimulator. Antifungal effect of nano silver colloids (average diameter of 1.5 nm) was studied against the powdery mildew pathogen of rose caused by *Sphaerotheca pannosa* var. *rosae*. Silver is now an accepted agrochemical replacement' and maximum no. of patents are filed for 'nano silver' for preservation and treatment of diseases in agriculture field (Sharon et al., 2010). Application of silver in management of plant diseases has been tested with reference to two fungal pathogens of cereals and rice blast.

In vitro assays indicated that silver both in ionic and nanoparticle forms inhibited colony growth of both the pathogens but *M. grisea* was comparatively more sensitive to silver application. When tested in vivo with perennial ryegrass (*Lolium perenne*) silver ions and nanoparticles brought significant reduction in disease severity when applied 3 hours prior to pathogen inoculation. In another study, silver nano particles synthesized extracellularly by *Alternaria alternata* were found to cause significant enhancement in the antifungal action of the triazole fungicide fluconazole against *Candida albicans*, *Phoma glomerata* and *Trichoderma* sp.(Gajbhiye et al. , 2009). However, no significant enhancement was observed with respect to the fungi *Phoma herbarum* and *Fusarium semitectum*.

Nano silica: Silicon dioxide nanoparticles, also known as silica nanoparticles or nano-silica, are the basis for a great deal of biomedical research due to their stability, low toxicity and ability to be functionalized with a range of molecules and polymers. Nano-SiO_2 is a white fluffy powder composed of high purity amorphous silica powder. Because of its small particle size, nano-SiO_2 had the advantages of large specific surface area, strong surface adsorption, large surface energy, high chemical purity and good dispersion. The properties of silica include both chemical and physical properties such as hardness, color, melting and boiling point, and reactivity. Silica under normal conditions of temperature and pressure is a solid, crystallized mineral.

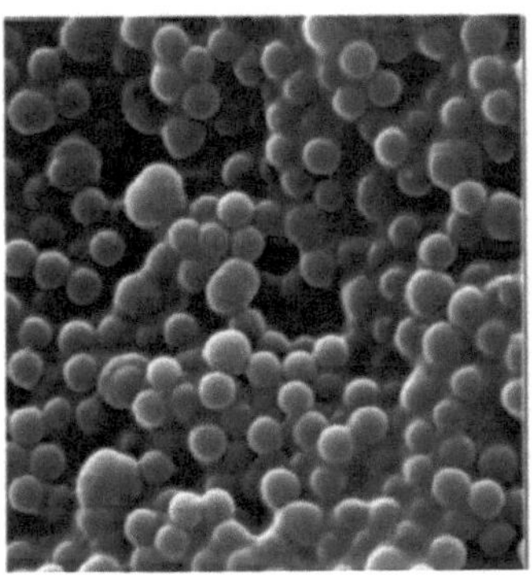

Fig. 27: SEM images of silica nanoparticles

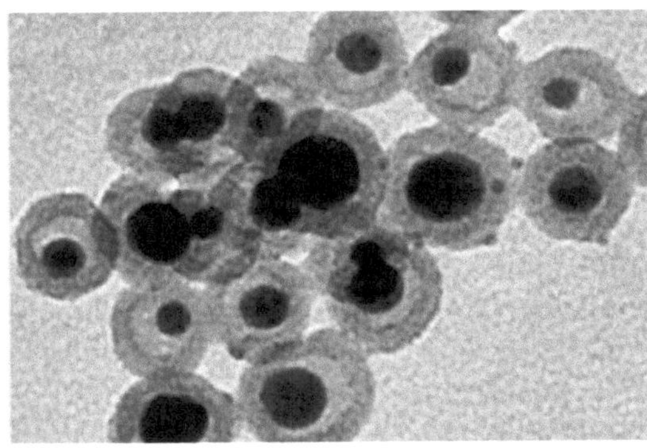

Fig. 28: Nano silica silver

Silicon increases disease resistance and stress resistance in plants. It also stimulates the physiological activity and growth of plants (Chowdappa and Shivakumar Gowda, 2013) (Fig.27). Honeycomb mesoporous silica nanoparticle (MSN) system with 3nm pores have been used to deliver DNA and chemic also into plant cells and intact leaves. They loaded the gene and its chemical inducer into the MSN systemand capped the ends with gold nanoparticles and studied the release pattern of chemicals and induction of gene expression in the plants under controlled-release conditions. Their study showed an application of silica nanoparticles in target-specific delivery of proteins, nucleotides and chemicals in plant biotechnology.

Nano silica silver: Silicon (Si) is known to be absorbed into plants to increase disease resistance and stress resistance. Silica itself has no direct effect on pathogenic microorganisms and has no effect on diseases. So a new composition of nano-sized Silica- Silver for control of various plant diseases has been developed. Alumino-Silicate nanotubes sprayed on plant surfaces are also developed and are easily picked up in insect hairs. Insects actively groom and consume pesticide-filled nanotubes. These are biologically more active and relatively more environmentally-safe pesticides. Mesoporous Silica nano particles can deliver DNA and chemicals into Plants thus, creating a powerful new tool for targeted delivery into plant cells (Amarendra N. Misra et al., 2013). It has been successfully used to introduce DNA and chemicals into Arabidopsis, tobacco and corn plants. The other advantage is that with the mesoporous nanoparticles, one can deliver two biogenic species at the same time.

Silver is a great antimicrobial agent, the metal's ions permeabilize bacterial membranes and interrupt bacterial metabolism, but are relatively (although debately) safe to human cells. Nano silica silver is an uniform mix of silver nanoparticles (4-10 nm in size) encased in hollow porous silica shells that were 13-20 nm in diameter. Under an electron microscope, the hollow void between the silver and the silica made the particles resemble little rattles,nanorattles (Fig. 28) (Anonymous, 2014). A nano-silica–AgNPs composite material is proposed as a novel antifouling adsorbent for cost-effective and ecofriendly water purification (Sujoy K. Das et al., 2013). Fabrication of well-dispersed AgNPs on the nano-silica surface, designated as NSAgNP, has been achieved through protein mediated reduction of silver ions at ambient temperature for development of sustainable nanotechnology. The coated proteins on AgNPs led to the formation of stable NSAgNP and protected the AgNPs from oxidation and other ions commonly present in water. The NSAgNP exhibited excellent dye adsorption capacity both in single and multicomponent systems, and demonstrated satisfactory tolerance against variations in pH and dye concentration.

The adsorption mainly occurred through electrostatic interaction, though π–π interaction and pore diffusion also contributed to the process. Moreover, the NSAgNP showed long-term antibacterial activity against both planktonic cells and biofilms of Gram-negative *Escherichia coli* and *Pseudomonas aeruginosa.* The antibacterial activity of AgNPs retarded the initial attachment of bacteria on NSAgNP and thus significantly improved the antifouling properties of the nanomaterial, which further inhibited biofilm formation. Scanning electron and fluorescence microscopic studies revealed that cell death occurred due to irreversible damage of the cell membrane upon electrostatic interaction of positively charged NSAgNP with the negatively charged bacterial cell membrane. The high adsorption capacity, reusability, good tolerance, removal of multicomponent dyes and *E. coli* from the simulated contaminated water and antifouling properties of NSAgNP will provide new opportunities to develop cost-effective and ecofriendly water purification processes.

Copper nanoparticle: A copper nanoparticle is a copper based particle 1 to 100 nm in size. Like many other forms of nanoparticles, a copper nanoparticle can be formed by natural processes or through chemical synthesis (Heiligtag, et al., 2013). These nanoparticles are of particular interest due to their historical application as coloring agents and their modern-day biomedical ones. Copper nanoparticles display unique characteristics including catalytic and antifungal/antibacterial activities that are not observed in commercial copper. First of all, copper nanoparticles demonstrate a very strong catalytic activity, a property that can be attributed to their large catalytic surface area. With the small size and great poro sity, the nanoparticles are able to achieve a higher reaction yield and a shorter reaction time when utilized as reagents in organic and organometallic synthesis.In fact, copper nanoparticles that are used in a condensation reaction of iodobenzene attained about 88% conversion to biphenyl, while the commercial copper exhibited only a conversion of 43%.

Copper-based fungicides produce highly reactive hydroxyl radicals which can damage lipids, proteins, DNA, and other biomolecules. It plays an important role in disease prevention and treatment of large variety of plants (Fig.29). Complexation of copper with chitosan nanogels was shown to have strong synergistic effect between c hitosan and copper in inhibiting the growth of *Fusarium graminearum*. Because of its bio-compatibility, these nanohydrogels are included as a new generation of copper-based bio-pesticides and it could also be developed into an efficient delivery system for copper based fungicides for plant protection (Chowdappa and Shivakumar Gowda, 2013). Low melting point soda-lime glass powder containing copper nanopartic les showed efficient antimicrobial activity against gram-positive, gram-negative bacteria, yeast and fungi, key reason for the increased antimicrobial activity is because of inhibitory synergistic effect of the Ca_{2+} lixiviated from the glass.

The figure shows a transmission electron microscopy image of aged copper nanoparticles. These nanoparticles are oxidized giving a core shell morphology with a metallic copper (Cu) core and a fully oxidized copper oxide (CuO) shell (see cartoon representation). Additionally, there appear to be larger particles in the image from particle coalescence. These aged nanoparticles show different chemical behavior compared to new Cu nanoparticles and fully oxidized CuO nanoparticles.

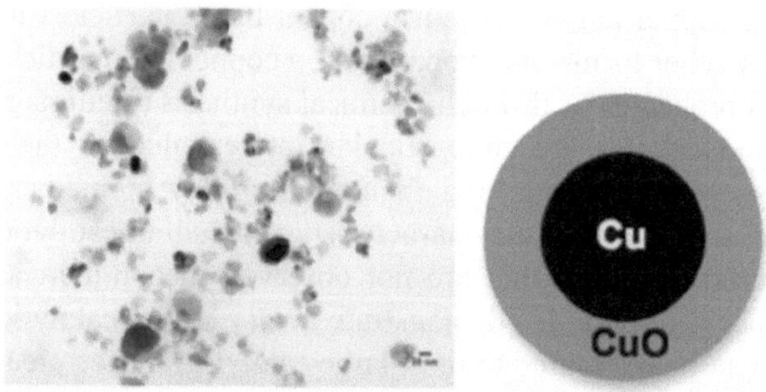

Fig. 29: Copper nano particles

(Image courtesy: Imali A. Mudunkotuwa, Grassian research group, University of Iowa)

Nanocomposites: Nanocomposites are materials that incorporate nanosized particles into a matrix of standard material. The result of the addition of nanoparticles is a drastic improvement in properties that can include mechanical strength, toughness and electrical or thermal conductivity. Nanocomposite is a multiphase solid material where one of the phases has one, two or three dimensions of less than 100 nanometers (nm) or structures having nano- scale repeat distances between the different phases that make up the material. The idea behind Nanocomposite is to use building blocks with dimensions in nanometre range to design and create new materials with unprecedented flexibility and improvement in their physical properties (Tian Zhiting et al., 2013). Nanocomposites are found in nature, for example in the structure of the abalone shell and bone. The use of nanoparticle-rich materials long predates the understanding of the physical and chemical nature of these materials.

Their structure is usually more complicated than that of microcomposites, and that is especially valid for polymer/layered silicate nanocomposites. Besides the usually assumed individual silicate platelets and tactoids, layered silicate nanocomposites may contain also large particles and a silicate network can also form in them at large extent of exfoliation. Polymer nanocomposite materials, consisting of a polymer matrix embedded with nanoscale fillers or additives that reinforce the inherent properties of the matrix polymer, play a key role in many industrial applications. Understanding of the relation between thermodynamic interactions and macroscopic morphologies of the composites allow for the optimization of design and mechanical processing (Fig.30) (Dai et al., 2019).

Chowdappa and Shivakumar Gowda (2013) used chitosan silver nanopartic le (chitosan -Ag Np) composite (size distribution from 10 nm to 15 nm) for inhibition of conidial germination in *Colletotrichum gloeosporioides*. They observed complete inhibition of spore at 100µg/ml concentration of the composite, but chitosan alone did not show significant inhibition at the same concentration. They observed the shrinking of spores and other morphological changes in the spores treated with CS-Ag nanocomposite.

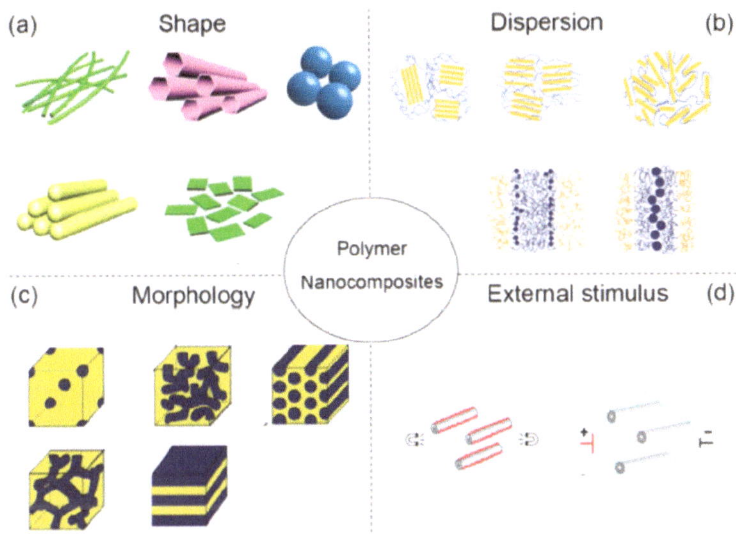

Fig. 30: Nanocomposites

Preparation and antifungal activity of composite films of pullulan and Ag nanoparticles (NP) against Aspergillus niger as a model system has been reported. These composite films showed strong inhibitory action on fungal sporulation, which was confirmed by disruption of the spores cells when observed under SEM. Silver in an ionic state exhibits high antimicrobial activity. A new nano-sized Silica-Silver composite has been developed for control of various plant diseases (Chowdappa and Shivakumar Gowda, 2013). Composite showed good antifungal activity where pathogens disappeared from the infected leaves within three days of spraying and the plants remained healthy thereafter. An attempt was done to determine the effective concentration of composites and also used it effectively for suppression of growth of many pathogens. Nano composites showed 100% growth inhibition of *Pythium ultimum, Magnaporthe grisea, Colletotrichum gloeosporioides, Botrytis cinerea* and *Rhyzoctonia solani* at 10 ppm concentration. Whereas, *Bacillus subtilis, Azotobacter chrococuum, Rhizobium tropici, Pseudomonas syringae* and *Xanthomonas compestris* pv. *vesicatoria* showed 100% growth inhibition at 100 ppm concentration.

Polymeric nanoparticles: They are sub-micron (1 to 1000nm) colloidal particles comprising active pharmaceutical ingredients encapsulated within or adsorbed to macromolecular substances, polymer. (Fig. 31) (Carina I.C.Crucho and Maria Teresa Barros, 2017). Polymers are the most common materials for constructing nanoparticle-based drug carriers Intense interest lies in the potential of polymeric NP s to revolutionize modern medicine. To determine the ideal nanosystem for more effective and distinctly targeted delivery of therapeutic applications, particle size, morphology, material choice, and processing techniques are all research areas of interest. Utilizations of polymeric NP s include drug delivery techniques such as conjugation and entrapment of drugs, prodrugs, stimuli-responsive systems, imaging modalities, and theranostics. Cancer, neurodegenerative disorders, and cardiovascular diseases are fields impacted by NP technologies that push scientific boundaries to the leading edge of transformative advances for nanomedicine (Brittany L. Banik et al., 2016). Polymeric nanoparticle is a type of nanoparticle. The dispersion of preformed polymers and the polymerization of monomers are two strong strategies mainly involved for preparation (Prasad Rao and Kurt Geckeler, 2011); 10 1000nm it is the range of size involved with solid particles. Polymer-based nanoparticles can improve the efficacy, solubility, toxicity, bioavailability and pharmacokinetic profile of a drug molecule.

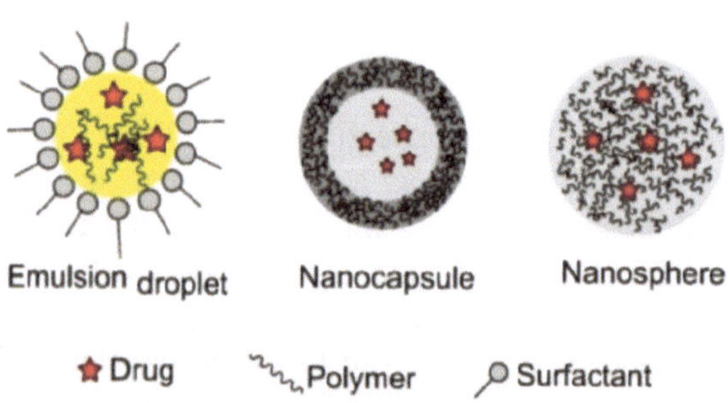

Fig. 31: Polymeric nanoparticles

Polymeric nanoparticles are organic based nanoparticles. Depending upon the method of preparation, these have structures shaped like nanocapsular or nanospheres. A nanosphere particle has a matrix-like structure whereas the nanocapsular particle has core-shell morphology. In the former, the active compounds and the polymer are uniformly dispersed whereas in the latter the active compounds are confined and surrounded by a polymer shell. Some of the merits of polymeric nanoparticles are controlled release, protection of drug molecules, ability to combine therapy and imaging, specific targeting and many more. They have applications in drug delivery and diagnostics. The drug deliveries with polymeric nanoparticles are highly biodegradable and biocompatible.

Liposome and nanoparticle: A liposome is a tiny bubble (vesicle), made out of the same material as a cell membrane. Liposomes can be filled with drugs, and used to deliver drugs for cancer and other diseases. Membranes are usually made of phospholipids, which are molecules that have a head group and a tail group. Nanoscale drug delivery systems using liposomes and nanoparticles are emerging technologies for the rational delivery of chemotherapeutic drugs in the treatment of cancer. Their use offers improved pharmacokinetic properties, controlled and sustained release of drugs and, more importantly, lower systemic toxicity (Fig. 32) (Yogeshkumar Malam and Alexander M Seifalian, 2009). A liposome has an aqueous solution core surrounded by a hydrophobic membrane, in the form of a lipid bilayer; hydrophilic solutes dissolved in the core cannot readily pass through the bilayer. These liposomes work to deliver drug by diffusion rather than by direct cell fusion. Liposomes are one of the methods based upon the different types of nanoparticles. Structure of liposomes consists of one or more phospholipid bilayers and they are sphere-shaped vesicles to carry compound of interest. Today liposomes have been useful in the field of reagent and tool in various scientific disciplines. Since many features involved in liposome they made their own way in the market (Heera and Shanmugam, 2015). Cosmetic and pharmaceutical industries numerous molecules act as a carrier, and in the field of Food and farming industries liposomes involved in encapsulation to grow delivery system that can entrap unstable compounds.

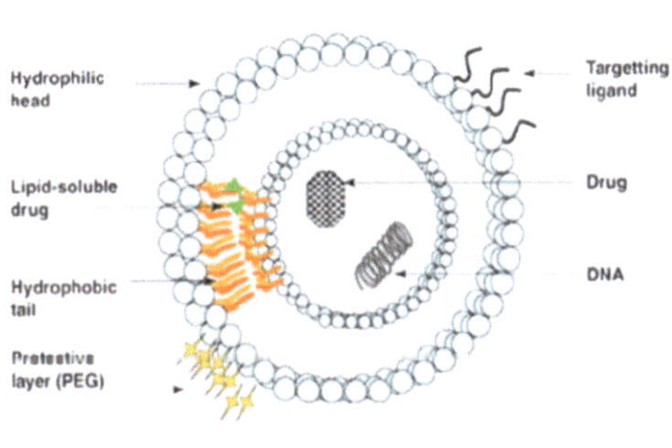

Fig. 32: Liposome and nanoparticle

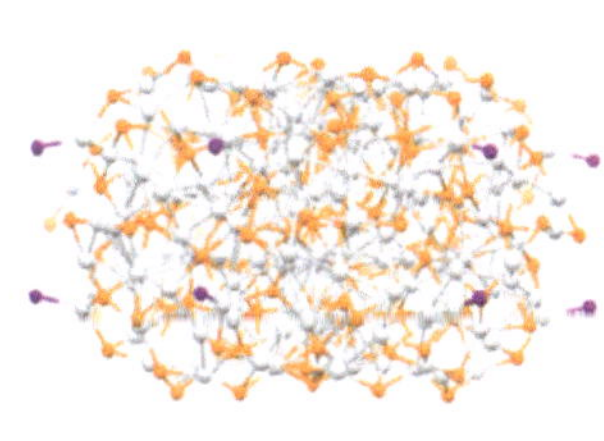

Fig. 33: Nanocrystal

Nanocrystal: A nanocrystal is a type based upon material particle having at least one dimension smaller than 100 nanometres and mainly composed of atoms in either a single or poly-crystalline arrangement (Jens-Uwe et al., 2008). Nanocrystals are aggregates of around hundreds or thousands of molecules that combine in a crystalline form, composed of pure drug with only a thin coating comprised of surfactant or combination of surfactants. When embedded in solids, nanocrystals may exhibit much more complex melting behaviour than conventional solids and may form the basis of a special class of solids. They can behave as single-domain systems (a volume within the system having the same atomic or molecular arrangement throughout) that can help explain the behaviour of macroscopic samples of a similar material without the complicating presence of grain boundaries and other defects (Fig. 33) (Talapin, 2012). Semiconductor nanocrystals having dimensions smaller than 10 nm are also described as quantum dots.

Nanotube: A nanotube is a nanometer scale tube like structure. Nanotubes are members of the fullerene structural family. Their name is derived from their long, hollow structure with the walls formed by one-atom-thick sheets of carbon called graphene. These sheets are rolled at specific and discrete ("chiral") angles and the combination of the rolling angle and radius decides the nanotube properties; for example, whether the individual nanotube shell is a metal or semiconductor. Nanotubes are categorized as single-walled nanotubes (SWNTs) and multi-walled nanotubes. Carbon nanotubes are large molecules of pure carbon that are long and thin and shaped like tubes, about 1-3 nanometers (1 nm = 1 billionth of a meter) in diameter, and hundreds to thousands of nanometers long. As individual molecules, nanotubes are 100 times stronger-than-steel and one-sixth its weight. Carbon nanotubes are composed of carbon atoms linked in hexagonal shapes, with each carbon atom covalently bonded to three other carbon atoms. Carbon nanotubes have diameters as small as 1 nm and lengths up to several centimeters. Although, like buckyballs, carbon nanotubes are strong, they are not brittle. The nanotube is empty prior to filling, and the drugs enter from one end only. The other end of the nanotube is capped and far enough away from encapsulation that it has no effect on encapsulation (Fig.34) (Duangkamon Baowan et al, 2017).

Nano gold: Nano gold is tiny particles of gold that are so small they're measured in nanometers. A nanometer is a billionth of a meter, smaller than the wavelength of light. When gold is nano-sized, it has some surprising properties. For example, nano gold can look red, orange, or even blue. Nanotechnology takes advantage of special properties at the nanoscale to create new materials and devices. Gold nanoparticles can be used as markers to indicate the presence of specific strands of DNA.

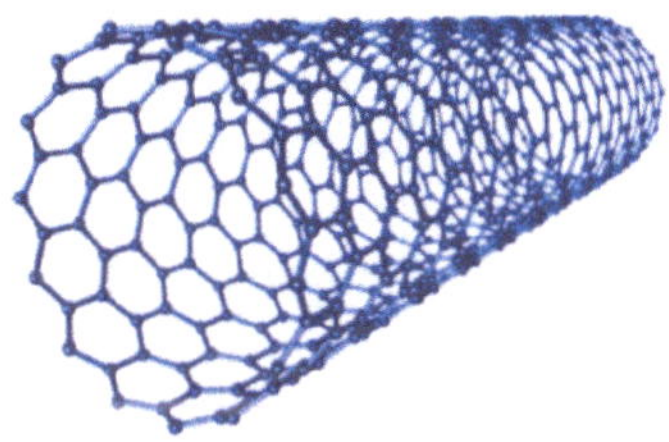

Fig. 34: Carbon nanotube

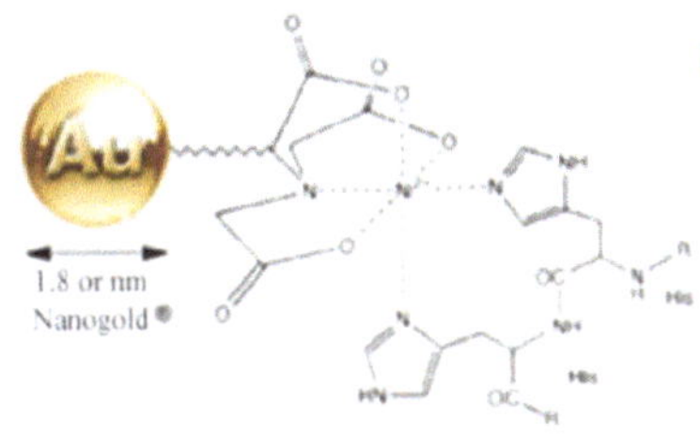

Fig. 35: Nano gold structure

Gold nanoshells, tiny spheres of glass covered with a thin layer of gold, are being tested as a part of a new cancer therapy. Gold nanoparticles are available in sizes ranging from 5 nm to 400 nm in diameter with numerous surface functionalities in a variety of solvent compositions.

A strong absorption band in the visible region is shown with gold nanoparticles when the frequency of the electromagnetic field is resonant with the coherent electron motion, which is called surface plasmon resonance absorption (Fig.35) (Sara Biglari et al., 2014). Interaction with the electric field results in a polarization of the electrons with respect to and relative to the ionic core of a nanoparticle. The dipole oscillations of the free electrons with respect to the ionic core of a spherical nanoparticle result in the so-called surface plasmon absorption. When the frequency of the electromagnetic field becomes resonant with the coherent electron motion, a strong absorption band around 520 nm in the spectrum is observed, which is the origin of the observed brilliant color of the nanoparticles in solution. The extracellular synthesis of nano gold particles (NGP) by *Fusarium oxysporum* and actinomycete *Thermomonospora* sp., respectively has been reported. The same group has reported the intracellular synthesis of gold nanoparticles by fungus *Verticillium* sp. as well. The reaction of $AuCl^{4-}$ ions with the extract of geranium leaves and an endophytic fungus, *Colletotrichum* sp., present in the leaves, leads to the formation of AuNPs. Gold nanoparticles-modified electrodes are used in the assembly of electrochemical DNA biosensors. They constitute useful analytical tools for sequence-specific DNA diagnosis and detection due to their inherent advantages of low cost, sensitivity and rapidity of response. Pathogens, bacteria and viruses can be detected via their unique, respective nucleic acid sequences.

Karnal bunt (KB) of wheat incited by *Tilletia indica* is an economically important quarantined fungal pathogen. Similarity in teliospore configuration makes it difficult to differentiate KB teliospores from the teliospores of other bunt fungi using conventional approaches. In order to determine the correct identity of KB teliospore, an infectious entity, it is essential to develop specific diagnostic probes and high quality of immunological reagents against infectious entities. The immuno-dipstick assay was developed by Sadhna Singh et al., (2010), which was quite sensitive to detect the antigens of even five teliospores. For on-site testing of KB, a nano-gold based lateral flow immuno-dipstick assay (LFID) was also developed using anti-teliospore antibodies. For development of LFID, colloidal gold was prepared and conjugated with anti-teliospore antibodies which were further characterized.

Nanocapsule: A 'Nanocapsule' is a nanoparticle that consists of a shell and a 'space', in which desired substances may be placed is a nanoscale shell made from a nontoxic polymer (Fig.36). They are vesicular systems made of a polymeric membrane which encapsulates an inner liquid core at the nanoscale. For example, polyelectrolyte (PE) capsules prepared by means of the layer-by-layer technique with tailored properties, fulfill the requirements for nano-organized systems in a satisfactory manner. With these shells, material can be delivered to predefined organs. Nanocapsules are prepared using molecules called phospholipids, which are hydrophobic (water-repellant) on one end and hydrophilic (water-loving) on the other. When such molecules are placed in an aqueous environment, they can spontaneously form capsules in which the hydrophobic portions are inside, protecting them from contact with water. In particular these nanocapsules are used for the delivery of fragile biological substances.

Nanocapsules have many uses, including promising medical applications for drug delivery, food enhancement, nutraceuticals, and for self-healing materials. The benefits of encapsulation methods are for protection of these substances to protect in the adverse environment, for controlled release, and for precision targeting (Ezhilarasi et al., 2012). Nanocapsules can potentially be used as MRI-guided nanorobots or nanobots, although challenges remain. The typical size of the nanocapsule used for various applications ranges from 10-1000 nm. However, depending on the preparation and use of the nanocapsule, the size will be more specific. Table 6 displays how nanocapsules exhibit different traits based on the method by which they were prepared. Nanocapsule types vary by size, drug concentration and active substance release time (Mora-Huertas et al., 2010).

Table 6: Different traits of Nanocapsules

	Mean size (nm)	Drug concentration in diluted dispersion (mg/ml)	Drug concentration in concentrated dispersion (mg/ml)	Active substance release time (90%) (min)
Nanoprecipitation	250	0.002–0.09	0.15–6.5	750
Emulsion-diffusion	425	~0.2	50	60
Double emulsification	400	2–5	20–50	45
Emulsification	300	~0.24	12	>2000
Coacervation				

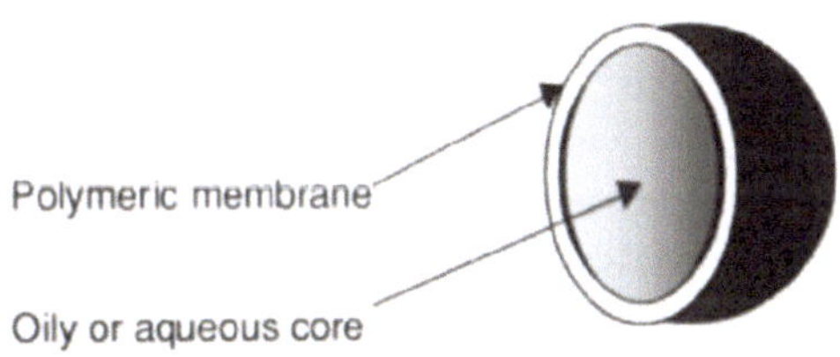

Fig. 36: Nanocapsule

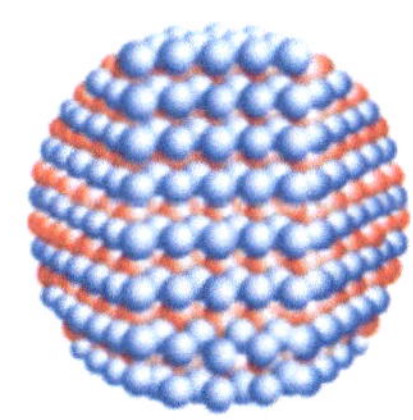

Fig. 37: Quantum dots

Nanocapsule structure consists of nanovesicular system that is formed in a core-shell arrangement. The shell of a typical nanocapsule is made of a polymeric membrane or coating. The type of polymers used is of biodegradable polyester, as nanocapsules are often used in biological systems. The core of a nanocapsule is composed of an oil surfactant that is specifically selected to coordinate with the selected drug within the polymeric membrane. The specific oil used must be highly soluble with the drug, and non-toxic when used in a biological environment. The oil-drug emulsion must have low solubility with the polymer membrane to ensure that the drug will be carried throughout the system properly and be released at the proper time and location. When the proper emulsion is obtained, the drug should be uniformly dispersed throughout the entire internal cavity of the polymeric membrane.

Quantum dots (QDs) : These are man-made nanoscale crystals that can transport electrons. When UV light hits these semiconducting nanoparticles, they can emit light of various colors. These artificial semiconductor nanoparticles have found applications in composites, solar cells and fluorescent biological labels. Nanoparticles of semiconductors, quantum dots, were theorized in the 1970s and initially created in the early 1980s. If semiconductor particles are made small enough, quantum effects come into play, which limit the energies at which electrons and holes (the absence of an electron) can exist in the particles. As energy is related to wavelength (or color), this means that the optical properties of the particle can be finely tuned depending on its size. Thus, particles can be made to emit or absorb specific wavelengths (colors) of light, merely by controlling their size. Quantum dots are artificial nanostructures that can possess many varied properties, depending on their material and shape. For instance, due to their particular electronic properties they can be used as active materials in single-electron transistors.

These are minute semiconductor structures on which the next generation of lasers are based upon. Quantum dots are typically only 10 to 50 nm across and contain just a few hundred or thousand atoms. The properties of a quantum dot are not only determined by its size but also by its shape, composition, and structure, for instance if it's solid or hollow (Fig. 37). A reliable manufacturing technology that makes use of quantum dots' properties, for a wide-ranging number of applications in such areas as catalysis, electronics, photonics, information storage, imaging, medicine, or sensing ,needs to be capable of churning out large quantities of nanocrystals where each batch is produced according to the exactly same parameters. Because certain biological molecules are capable of molecular recognition and self-assembly, nanocrystals could also become an important building block for self-assembled functional nanodevices.

Quantum dots often described as 'artificial atoms,' exhibit discrete energy levels, and their bandgap can be precisely modulated by varying the size. QDs are inorganic semiconductor nanocrystals, typically composed of a cadmium selenide (CdSe) core and a zinc sulphide (ZnS) shell and whose excitons (excited electron-holepairs) are confined in all three dimensions, giving rise to characteristic fluorescent properties. QDs are extremely photostable, bright and are characterized by broad absorption profiles, high extinction coefficients and narrow and spectrally tunable emission profiles (Ana R. Santos et al., 2010).

Quantum dots have found applications in composites, solar cells (Grätzel cells) and fluorescent biological labels (for example to trace a biological molecule) which use both the small particle size and tuneable energy levels. Advances in chemistry have resulted in the preparation of monolayer-protected, high-quality, monodispersed, crystalline quantum dots as small as 2 nm in diameter, which can be conveniently treated and processed as a typical chemical reagent. In the language of materials science, nanoscale semiconductor materials tightly confine either electrons or electron holes. Quantum dots are sometimes referred to as artificial atoms, emphasizing their singularity, having bound, discrete electronic states, like naturally occurring atoms or molecules. It was shown that the electronic wave functions in quantum dots resembles the ones in real atoms. By coupling two or more such quantum dots an artificial molecule can be made, exhibiting hybridization even at room temperature (Cui Jiabin et al., 2019).

2.11 Bionanoparticles

Bionanoparticles possess unique multifunctional and self-assembling properties, particularly, protein cages like plant viruses or ferritin but also other well-defined self-assembling protein structural motifs are valuable building blocks with great potential in (bio-) nanotechnology. A steeply increasing number of research works present promising results and applications in biomedicine, diagnostics and analytics as well as nanoelectronics. However, the use of bionanoparticles for hybrid and soft protein–polymer composite materials has not received high attention yet (Fig. 37a) (Günther Jutz and Alexander Böker, 2011). The use of biomolecules is considered as an obvious step in the synthesis and construction of next-generation nanomaterials and devices. A whole new branch termed bionanotechnology seeks for scientific as well as economic breakthroughs in the development of bio-inorganic nanomaterials with novel properties for computation and nanotechnology, new methods in diagnosis and analytics or new drugs and drug delivery systems. There are several reducing agent are used for the synthesis of nanoparticles, which impart harmful effect on the environment and humans. These chemicals cause several hazardous waste in the environment. These are the reasons for that the chemical and physical method of nanoparticle synthesis is less common in implementation (Kumar et al., 2015).

The biogenic method of nanoparticles production is simple, eco-friendly and is of great interest due to possible application in catalysis, medicine and nano- optoelectronics. Biologically fabricated nanostructures offer substantially different properties like good adhesion, tribologically good properties (like lubrication - due to this it is able to interact with surfaces), optical and electrical properties. It is possible to get the nanoparticles from simple microbes and bacteria up to more evolved plants. For the synthesis of metal and metal sulphide nanoparticles, fungi are the most efficient biological agents. *Verticillium* sp. and *Fusarium oxysporum* are used for the synthesis of metal nanoparticles, as they can reduce the metal ions extracellularly.

Fig. 37a: Bionanoparticle

An extracellular enzyme, hydrogenase in the *F. oxysporum* shows excellent redox properties and it can act as an electron shuttle in metal reduction. The electron shuttle or other reducing agents (e.g. hydrochinones) released by microorganisms are capable of reducing ions to nanoparticles. When challenged with silver nitrate in aqueous medium the fungal members *Aspergillus flavus, Aspergillus fumigatus* and *Phanerochaete chrysosporium* produce stable silver nanoparticles (Ratna Kumari and Sobha, 2013).

Biosynthesis of metal NPs : Nanotechnology is a revolutionary field just at its onset, the trend in the next decades being its integration with the green chemistry approach. Several strategies involving exhaustive strain selection, cultivation modes, recombinant gene expression, metabolic engineering, protein re-design and re-engineering, and predictive modeling will allow to create nanobioreactors, a new nanobiotechnology arena with a high potential impact in many fields. In vitro biosynthesis of Ag-NPs was studied in several microorganisms, showing different NP profiles and properties (Ana Cauerhff and Guillermo R. Castro, 2013). These authors reported the synthesis of Ag-NPs using 24 hrs to 3 days old microbial supernatants of *Bacillus cereus, Bacillus subtilis, Escherichia coli, Enterobacter cloacae, Klebsiella pneumoniae, Lactobacillus acidophilus, Staphylococcus aereus Pseudomonas aeruginosa, Candida albicans* and *Aspergillus niger*. The Ag-NPs ranged from 28.2 to 122 nm in diameter with an average of 52.5 nm. The general mechanism was described as non-specific and based on the presence of hydroquinones, molecules with high redox properties, which could act as electron shuttles in metal reductions. The largest plasmon resonance peak was observed in the supernatant of *K. pneumoniae* after 5 min of incubation in the presence of $AgNO_3$. Besides, energy dispersive spectroscopy (EDS) and transmission electron microscopy (TEM) analysis demonstrated the presence of Ag-NPs but also nanocrystals, which can be attributed to the insoluble salt formation in the presence of several anions in the medium (e.g. Cl-1, SO4 -2 and MoO4 -2). In addition, EDS spectra showed the presence of sodium, calcium, iron, and chloride in the Ag-NPs (Shahverdi et al. 2007). Additionally, the NP size dispersion increases by the presence of impurities in the solution. Both, contamination of NPs by other elements and changes in the NP size are serious limitations for specific therapeutic applications.

In the case of metal NPs, a selected wild-type *Bacillus licheniformis* Ag-resistant was grown in the presence of $AgNO_3$ in the medium. Silver NPs were detected intracellularly and the proposed mechanism involved the cytoplasmatic NADH-dependant nitrate reductase and Ag^o formation followed by the nucleation into Ag-NPs of 50 nm average size. In a further work, the presence of NADH- dependant nitrate reductase was confirmed and optimized for the production of AgNPs. Also, intracellular biosynthesis of manganese oxide NPs was described in batch culture of *Bacillus* sp.. The maximum oxidation rate of Mn+2 was reached at stationary growth phase, and the NP was characterized as MnO_2. The manganese transport into the bacterial cells is still unknown, but the authors speculated with the induction of cellular mechanism during the exponential growth phase to reduce the toxicity of manganese forming less toxic molecules intracellularly.

2.12 Categories of nanoparticles for plant disease /pest management:

2.12.1 Nanoparticles as protectants

Nanoparticles are materials that range between 10 to 100 nanometers (nm), and can be designed with unique chemical, physical, and biological properties, to distinctively differ from those of their molecular and bulk counterparts. Nanoparticles alone have the potential to be directly applied to plant seeds, foliage, or roots for protection against pest and pathogens, such as insects, bacteria, fungi, and viruses (Elizabeth A.Worrall et al., 2018). Metal nanoparticles such as silver, copper, zinc oxide, and titanium dioxide have been intensively researched for their antibacterial and antifungal properties, and are known for their antiviral properties. This section gives a brief overview and an update on current literature reviews pertaining to the individual nanoparticles that already exist. Recently, silver nanoparticles have increased in popularity, due to "green synthesis" production in plants, bacteria, fungi, or yeast. Silver nanoparticles have shown antifungal inhibition of *Alternaria alternata, Sclerotinia sclerotiorum, Macrophomina phaseolina, Rhizoctonia solani, Botrytis cinerea*, and *Curvularia lunata* by well diffusion assay. When silver nanoparticles were sprayed onto bean leaves, complete suppression of sun-hemp rosette virus was observed. Faba bean plants challenged with bean yellow mosaic virus and sprayed with silver nanoparticles, produced remarkably better results when the nanoparticles were applied 24 h post-infection, compared to spray application before infection, or simultaneously at the time of inoculation. Silver nanoparticles have shown immense potential for plant disease management against fungal and bacterial pathogens, but there are significant hurdles associated to them, such as production, toxicity, and soil interaction.

Other commonly used metal nanoparticles include copper, titanium dioxide, and gold. Nanoparticles, such as copper and titanium dioxide, are more commonly being utilized for fertilizer with little research into plant disease management. The research covering titanium dioxide, silver, and copper nanoparticles for antimicrobial properties, and aluminum nanoparticles as an insecticidal dust, has been highlighted. Furthermore, titanium dioxide nanoparticles in fertilizers have produced protection from bacteria and inactivation of viruses. Introducing poly-dispersed gold nanoparticles via a mechanical abrasive was seen to melt and dissolve the Barley yellow mosaic virus particles conferring resistance to the plant.

Chitosan is another popular nanoparticle that has favorable biological properties, such as biodegradability, biocompatibility, non-allergenicity, and antimicrobial activity, with low toxicity to animals and humans. Chitosan nanoparticles induce viral resistance in plant tissues by protecting them against infections caused by the mosaic virus of alfalfa, snuff, peanut, potato, and cucumber. Chitosan nanoparticles have shown antimicrobial properties, such as controlling *Fusarium crown*, root rot in tomato, Botrytis bunch rot in grapes, and *Phyricularia grisea* in rice, but are less effective against bacteria (Elizabeth A.Worrall et al., 2018). Potential mechanisms were summarized that lead to the antimicrobial effects of chitosan, such as agglutination, disruption of the cell membrane, inhibition of H^+-ATPase activity, inhibition of toxin production and microbial growth, inhibition of the synthesis of messenger RNA and proteins, and blockage of nutrient flow. Antiviral effects have been observed in beans against Bean mild mosaic virus and, in tobacco, against Tobacco mosaic virus and Tobacco necrosis virus. Chitosan has also shown effectiveness against oleander aphid (*Aphis nerii*), cotton leafworm (*Spodoptera littoralis*), root-knot nematode (*Meloidogyne javanica*), and nymphs of the spear psylla (*Cacopsylla pyricola*). Chitosan holds promise as both nanoparticles alone and has shown to have an enormous potential as a nanocarrier.

2.12.2 Nanoparticles that act as carriers

Nanoparticles are also commonly used as carriers to entrap, encapsulate, absorb, or attach active molecules to develop effective agricultural formulations. The common nanoparticles which have been used as carriers for insecticides, fungicides, herbicides, and RNAi-inducing molecules are summarized below (Elizabeth A.Worrall et al., 2018).

i. Silica nanoparticles: Silica nanoparticles are the mostly used nanomaterials to build pH nanosensor. Silica nanoparticles are optically transparent, of low toxicity, inert to pH, and (some are) degradable. More attractively, their surface can be easily functionalized via the well-established silane technology. Colloidal silica (silicon dioxide) nanoparticles are amorphous materials and are generally spherical in shape. They can be made to have a broad range of sizes and their surface chemistry easily modified to target a variety of applications. Regardless of size, dried silica nanoparticles are a white powder.

In their nonporous form, silica nanoparticles are known for being absorbent and abrasive, whereas mesoporous silica nanoparticles have important applications in drug delivery and nanomedicine (Fig.38) (Zhigang Xu, et al., 2017).

They can be easily synthesized with a controlled size, shape, and structure, making them highly advantageous delivery vehicles. They are commonly produced in a spherical shape with pore-like holes; for example, porous hollow silica nanoparticles (PHSNs) or mesoporous silica nanoparticles (MSNs). PHSN and MSN commonly load the pesticide into the inner core to protect the active molecules and, therefore, provide a sustained release. The shell structure of PHSNs protects the active molecules inside the nanoparticles against degradation by UV light. The literature suggests that silicon has already been used to enhance plant tolerance against various abiotic and biotic stresses and, therefore, silica nanoparticles seem to be the natural choice for the development of agri-products for pests.

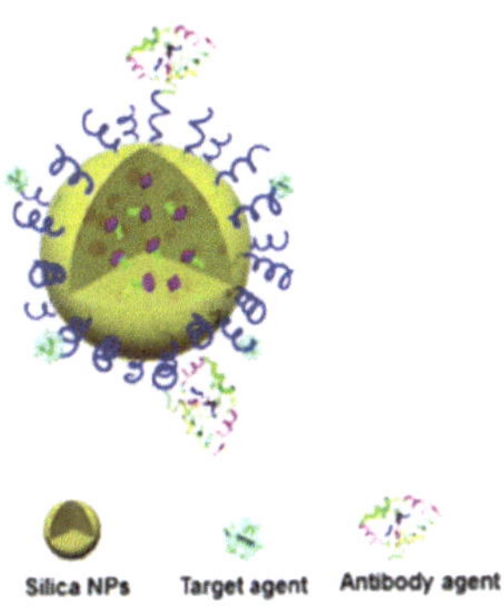

Fig. 38: Silica nanoparticle

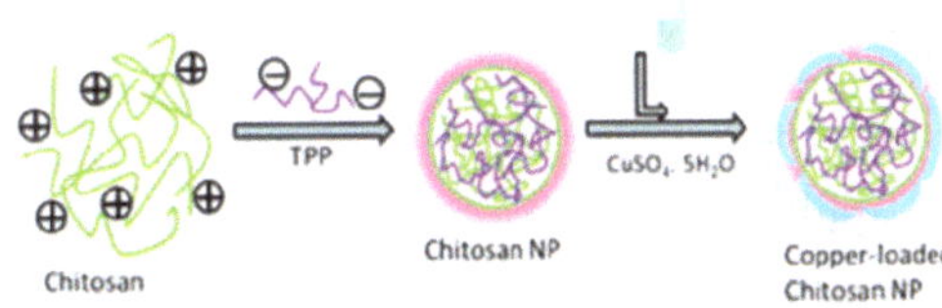

Fig. 39: Copper-loaded chitosan nanoparticle

ii. **Chitosan nanoparticles:** Chitosan is a biodegradable, biocompatible polymer regarded as safe for human dietary use and approved for wound dressing applications . Chitosan has been used as a carrier in polymeric nanoparticles for drug delivery through various routes of administration. Chitosan has chemical functional groups that can be modified to achieve specific goals, making it a polymer with a tremendous range of potential applications. These have low solubility in aqueous media, due to their hydrophobic properties. As a result, chitosan is commonly mixed with a copolymer, organic and inorganic, to improve its solubility. The particle size of the copper-loaded chitosan nanoparticles (CuCNPs) is in the range required for drug delivery applications and cancer targeting (<200 nm).. the superior anticancer effect of CuCNPs is attributed to the generation of a higher mitochondrial ROS level compared to that of the control (Fig. 39) (Jin-wei Ai et al., 2017).

Chitosan has reactive amine and hydroxyl groups, allowing modification, graft reactions, and ionic interactions, enabling improvement of chitosan properties. Chitosan adheres well to the epidermis of leaves and stems, prolonging the contact time and facilitating the uptake of the bioactive molecules. Chitosan has been used as a carrier in polymeric nanoparticles for drug delivery through various routes of administration. Chitosan has chemical functional groups that can be modified to achieve specific goals, making it a polymer with a tremendous range of potential applications. Nanoparticles (NP) prepared with chitosan and chitosan derivatives typically possess a positive surface charge and mucoadhesive properties such that can adhere to mucus membranes and release the drug payload in a sustained release manner. Chitosan-based NP have various applications in non-parenteral drug delivery for the treatment of cancer, gastrointestinal diseases, pulmonary diseases, drug delivery to the brain and ocular infections which will be exemplified in this review. Chitosan shows low toxicity both *in vitro* and some *in vivo* models (Munawar, A. Mohammed et al., 2017).

iii. Solid lipid nanoparticles (SLNs): They are nano-sized colloidal carriers composed of lipid, surfactants, and drugs in appropriate ratios. SLNs are another antimicrobial drug delivery platform in which the small size and lipid core provide some unbeatable properties that make them superior over other existing drug carriers. They are similar to emulsions, and are composed of lipids that are solid at room temperature. An advantage of SLNs is that they provide a matrix to entrap lipophilic active molecules without the use of organic solvents. Furthermore, SLNs can also provide controlled release of various lipophilic components, due to decreased mobility of the active in the solid matrix. Surfactants are used to stabilize the SLN when dispersed into water. Their main drawbacks are that they have a low loading efficiency, and the active may leak out of the structure during storage.

SLNs are typically particulate systems with mean diameters ranging from 50 nm up to 1000 nm and used for various drug delivery applications. The small size of SLNs offers several advantages like larger surface, high drug loading capacity, interaction with a target site up to molecular level, and enhanced bioavailability of the drug. The lipid core facilitates the incorporation of a wide variety of drugs that get dissolved, dispersed, or entrapped in SLNs. In addition, the solid lipid core in SLNs reduces the "drug mobility" and hence reduces the chances of its degradation. Development of solid lipid nanoparticles is one of the emerging fields of lipid nanotechnology (Mashaghi et al., 2013) with several potential applications in drug delivery, clinical medicine and research, as well as in other discipline. Due to their unique size-dependent properties, lipid nanoparticles offer the possibility to develop new therapeutics.

The ability to incorporate drugs into nanocarriers offers a new prototype in drug delivery that could hold great promise for attaining the bioavailability enhancement along with controlled and site specific drug delivery.

Solid lipid nanoparticles are at the forefront of the rapidly developing field of nanotechnology with several potential applications in drug delivery, clinical medicine and research, as well as in other varied sciences. They are nano-sized colloidal carriers composed of lipid, surfactants, and drugs in appropriate ratios (Fig.40) (Vijay Mishra et al., 2018). SLNs are another antimicrobial drug delivery platform in which the small size and lipid core provide some unbeatable properties that make them superior over other existing drug carriers. Due to their unique size-dependent properties, lipid nanoparticles offer the possibility to develop new therapeutics. The ability to incorporate drugs into nanocarriers offers a new prototype in drug delivery that could be used for secondary and tertiary levels of drug targeting. Hence, solid lipid nanoparticles hold great promise for reaching the goal of controlled and site specific drug delivery and hence have attracted wide attention of researchers (Mukherjee et al., 2009). For controlling the drug delivery in 1990's solid lipid nanoparticles played a dominant role. There are certain alternate carrier systems to emulsions, liposomes and polymeric nanoparticles as a colloidal Carrier system.

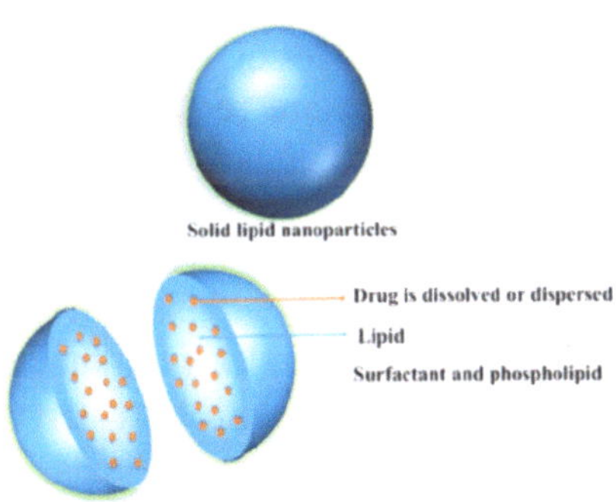

Fig. 40: Solid lipid nanoparticle

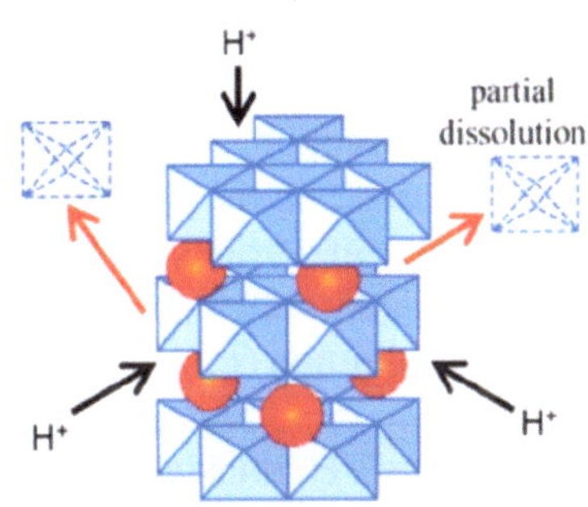

Fig. 41: Layered double hydroxide

iv. Layered Double Hydroxides (LDHs): These are a class of ionic solids characterized by a layered structure with the generic layer sequence [AcB Z AcB]n, where c represents layers of metal cations, A and B are layers of hydroxide (HO^-) anions, and Z are layers of other anions and neutral molecules (such as water) (Fig.41) (Toshiyuki Hibino, 2015). Lateral offsets between the layers may result in longer repeating periods. The intercalated anions (Z) are weakly bound, often exchangeable; their intercalation properties have scientific and commercial interest (Khan Aamir I. and O'Hare Dermot, 2002).

LDHs occur in nature as minerals, as byproducts of metabolism of certain bacteria, and also unintentionally in man-made contexts, such as the products of corrosion of metal objects.

LDHs are clays that form into hexagonal sheets with layers of the active molecules trapped in the interlayer space. LDH nanoparticles breakdown under acidic conditions such as the addition of moisture and carbon dioxide from the atmosphere. Positively charged delaminated LDH lactate nanoparticles have been shown to facilitate the transport of biologically active materials across the plant cell wall barrier. Silica, chitosan, SLN, and LDH nanoparticles have been utilized as common carriers for plant disease management (Elizabeth A.Worrall et al., 2018). The individual studies using these nanoparticles, and less frequently used nanoparticle types, are discussed in-depth in the sections categorized by insecticide, fungicide, herbicide, or RNAi.

2.12.3 Insecticides, fungicides and herbicides

This section gives a brief outline on the types of pesticides and their classification system, designed to aid in their correct usage, and the prevention of pesticide resistance. Insecticides can be classified into systemic (absorbed by the plant) or contact insecticides (direct contact required), and can be classified into at least 55 different chemical classes, grouped according to physiological functions affected: nerve/muscle, growth, respiration, midgut, or unknown/non- specific]. Importantly, there are 30 modes of action (MoA) (including multiple sites and unknown groups) which are required to be applied, in rotation, to prevent insecticide resistance. Fungicides work by contact (barrier on the plant that is not absorbed) or by penetration (absorbed by the plant at various levels of systemic movement). Spray coverage of fungicides is extremely important, since systemic movement within the whole plant is uncommon (Elizabeth A.Worrall et al., 2018). Fungicides are split into 49 Fungicide Resistance Action Committee (FRAC) codes with 14 different MoA groups]. Herbicides can be selective by killing only target weeds, or non-selective by killing all vegetation. Spray application of herbicides can be applied at pre-plant (non- selective applied to the soil before planting), pre-emergence (before weed seedlings emerge through the soil) or post-emergence (weed seedlings already emerged from the soil). There are 23 Herbicide Resistant Action Committee (HRAC) groups of herbicides. Rotation within these groups is critical to prevent pesticide resistance.

Nanoparticles as carriers for insecticides: Loading insecticides into nanoparticles first started in the early 2000s. Since then, conventional insecticides (27 studies) and bioactive compounds with insecticidal properties (13 studies) have been conducted with a range of nanoparticles. These studies have explored eight different MoA and a range of essential oils (not included in Insecticide Resistance Action Committee (IRAC) classification).

Slow release of active molecules could also potentially decrease the toxicity of the insecticides. To date, four studies have tested if sustained release of insecticides from nanoparticles could decrease toxicity. For example, when testing for cytotoxicity of imidacloprid loaded onto sodium alginate nanoparticles, it was claimed "that these experiments clearly reveal that at this concentration loaded-pesticide nanoformulation is very less toxic than original pesticide". However, only a small difference was observed, and no statistical analysis was conducted.

Nanoparticles as carriers for fungicides : Starting in 1997, initial studies on nanofungicides were conducted on incorporating fungicides into solid wood (Elizabeth A.Worrall et al., 2018). Since then, conventional fungicides (20 studies) and biocides with antifungal properties (6 studies) have been conducted with a wide range of nanoparticles. Nine FRAC groups, and a range of essential oils not included in the fungicide groups, were studied. The most commonly investigated nanoparticle carriers were polymer mixes, silica, and chitosan. A wide range of fungi were used to check the efficiency of the nanofungicide. However, few plants were tested, nor were there many toxicity studies. Similar to insecticides, nanoparticles were exploited to improve low-water-solubility issues, decrease volatilization, and improve stability while providing a slow sustained release.

Nanosized bacterial ghosts, non-denatured empty cell envelopes of Gram-negative bacteria were used to increase adherence to the leaf surface and improve the low water-solubility of tebuconazole. *Pectobacterium cypripedii* was picked as the nanosized bacterial ghost, due to its ability to adhere to plants. Of the six plants tested (rice, soya, cabbage, cotton, barley, and corn), fluorescently labelled ghosts without a loaded fungicide adhered best to rice leaves (55% remaining), and soya leaves the lowest (10% remaining), when treated with heavy simulated rain in glasshouse conditions. All six plants were compared with ghost-loaded tebuconazole or with two tebuconazole commercial treatments (WP 25 and EW 250) against a wide range of fungi. Plants that were not subjected to rainfall showed similar or higher protection to the commercial products with one exception. When the plants were subjected to heavy rain 1 h post-treatment, and challenged with fungus, the majority of groups did not perform as well as the commercial products with one exception. However, when washed 24 h post-treatment, ghost-loaded tebuconazole was equal to, or greater than, WP 25 treatments, but EW 250-treated controls were generally more efficient.

Pyraclostrobin, another low-water-soluble fungicide, was loaded onto chitosan–lactide copolymer nanoparticles at different concentrations. Three and five days post-application, they found that the nanofungicide was either similar to, or less efficient at preventing inhibition of *C gossypii*, when compared to commercial pyraclostrobin. However, an increase in inhibition was observed at 7 days post-treatment, compared to active alone]. In another trial, kaempferol (another low-soluble fungicide) loaded onto lecithin/chitosan showed 67% inhibition efficacy after 60 days storage on a *Fusarium oxysporum*-infected Petri dish.

In a series of studies, tebuconazole and chlorothalonil were loaded into a wide range of nanoparticles to improve the low solubility of the fungicides for deployment into solid woods (Elizabeth A.Worrall et al., 2018). All four studies tested the mass lost due to decay from *Gloeophyllum trabeum* in southern yellow pine wood over 50–55 days (later trials also included birch, Trametes versicolor, KATHON 930, and the insecticide chlorpyrifos). Hydrophobic chlorothalonil encapsulated into nanoparticles were not as efficient, due to the size and stability of the formulation, but with an increase in concentration, minimal decay of the wood due to fungal infection was achieved. Nanoparticle-loaded tebuconazole was able to drastically decrease the amount of fungicide delivered (from ca. 2 kg active/m^3 to 0.2 kg/m^3) with negligible change in the amount of decay. Birch wood was harder to treat due to its smaller pit pores. Chlorothalonil and tebuconazole, encapsulated with a surfactant-free method, created smaller median particle diameters and a more stable aqueous suspension, ultimately increasing uptake into the wood. Negligible decay (5% or less mass loss) was achieved with drastically reduced concentrations compared to industry standards for both fungicides in southern pine exposed to *G. trabeum* and birch wood exposed to *T. versicolor*, clearly showing that the smaller, more stable surfactant-free nanoparticles afforded more resistance to fungal decay.

Essential oils with fungicidal properties evaporate too quickly for large scale commercial use. Five individual essential oil components were encapsulated into MSN to show higher antifungal activity 14 days post-Aspergillus niger-infection, when compared to the bulk essential oil components. Similarly, SLNs have also been used to stabilize *Zataria multiflora* essential oil, providing protection against six fungi.

Leaching, the movement of water and chemicals through soil, is a major pesticide issue, but not many studies have been conducted with soil. The fungicide metalaxyl was loaded onto MSNs, and observed leaching in soil between free metalaxyl (76% release) and encapsulated metalaxyl (11.5%) within a period of 30 days. When tested in water, the encapsulated metalaxyl had an increased release rate of 47% compared to the 11.5% seen in the soil, highlighting the importance of testing within the farming environment. Two different types of nanoparticles, solid lipid or polymeric were tested the cytotoxicity of carbendazim and/or tebuconazole loaded onto the nanoparticles. A decrease in toxicity with the nanoparticle-loaded pesticides was observed in preosteoblast and fibroblast mouse cell lines. In the soil leaching experiments, the addition of nanoparticles decreased the release rate in soil layer release experiments when compared to the commercial formulation. Ebrafish was treated for 96 h with azoxystrobin and difenoconazole loaded into poly(butylene succinate) and poly(lactic acid) shells, and observed reduced toxicity compared to other formulations.

Slow release of the active molecules was achieved with nanosized calcium carbonate carrying validamycin. Validamycin encapsulated in nanoparticles showed lower efficiency than validamycin alone, over a period of 1–7 days, in an *in vitro* assay against *Rhizoctonia solani*. However, two weeks later, the nanoparticle formulation showed slightly better results than the active alone, highlighting the efficacy of the nanoformulation over longer periods of time. An increase in the fungal inhibition rate was noted when carbendazim-loaded polymeric nanoparticles were tested against *Fusarium oxysporum* and *Aspergillus parasiticus*, compared to carbendazim alone. Phytotoxicity studies confirmed that the nanoformulated carbendazim was safer for germination and root growth of *Cucumis sativa, Zea mays* and *Lycopersicum esculentum* seeds.

MSNs were used to load pyrimethanil, and studied the uptake in cucumber plants over 48 days using high performance liquid chromatography tandem mass spectrometry. The uptake and the dosage of pesticide-loaded MSN in leaves had almost no effect on the distribution and dissipation rate in cucumber plants. Their study has shown that pyrimethanil-loaded MSN has a minimal risk of accumulating in the edible part of the plant.

Nanoparticles as carriers for herbicides : Most of the herbicide nanocarrier studies primarily focused on reducing the environmental impact caused by herbicides (Elizabeth A.Worrall et al., 2018). When compared with insecticide and fungicide nanocarrier studies, herbicide research clearly prioritizes reducing the non-target toxicity of the herbicides. Development of nanoparticle-based herbicides has also included a wider variety of nanoparticles. For example, montmorillonite clay layers coated with a pH dependent polymer, core–hollow shell manganese carbonate, nanosized tubular halloysite and platy kaolinite, amino-activated iron (II, III) oxide magnetic nanoparticles and nanosized rice husks, have been used to date.

It was also found that the leaching effects in soil were reduced when the herbicide was bound to the nanosized rice husks. Polymeric nanoparticles containing ametryn, atrazine, or simazine were slightly less toxic than herbicide alone, when subjected to genotoxicity studies using human lymphocyte and *Allium cepa* cell cultures.

Bibliography

A Yu Utekhina and Gleb B Sergeev, 2011. Organic nanoparticles. Russian Chem. Rev., 80 : 219 Abolfazl Akbarzadeh, Mohamad Samiei and Soodabeh Davaran, 2012. Magnetic nanoparticles: preparation, physical properties, and applications in biomedicine. Nanoscale Res Lett., 7:144.

Afolalu, S. A., Soetan, S. B., Ongbali, S. O., Abioye, A. A. and Oni, A. S., 2019. Morphological characterization and physio-chemical properties of nanoparticle – review. 1st International Conference on Sustainable Infrastructural Development. IOP Conf. Series: Materials Science and Engineering 640, 012065

Alexander Vakhrushev , 2020. Head of Department, Russian Academy of Sciences, Institute of Mechanics, Ural Branch, Moscow, Russia

Ana Cauerhff and Guillermo R. Castro, 2013 . Bionanoparticles, a green nanochemistry approach. Electronic J. Biotechnol., ISSN: 0717-3458

Ananya Madal, 2018. Properties of Nanoparticles. Bull. News-Med. Life science

Anonymous, 2013. Current Strategies for Engineering Controls in Nanomaterial Production and Downstream Handling Processes". U.S. National Institute for Occupational Safety and Health: 1–3, 47–49, 57–58. November 2013.

Anonymous, 2014. Nanoparticles of silica-encased silver nanoparticles may improve biomedical implants The American Ceramic Society, Westerville

Astruc D, Boisselier E, Ornelas C (April 2010). "Dendrimers designed for functions: from physical, photophysical, and supramolecular properties to applications in sensing, catalysis, molecular electronics, photonics, and nanomedicine". Chemical Reviews, 110:1857–959. AZoNano, 2020. Nanomaterials - Chemical Indentification and Characterization of Nanomaterials, EPA Nanotechnology White Paper EPA 100/B-07/001

Bozich Jared, Lohse Samuel Torelli, Marco Murphy, Catherine Hamers, Robert Klaper and Rebecca. 2014. Surface chemistry, charge and ligand type impact the toxicity of gold nanoparticles to Daphnia magna. Environ. Sci. Nano., 1: 260-270

Brittany L. Banik, Pouria Fattahi and Justin L. Brown, 2016. Polymeric nanoparticles: the future of nanomedicine. WIREs Nanomed. Nanobiotechnol, 8:271–299.

Cami, J.,Bernard-Salas, J., Peeters , E. and Malek, S. E., 2010. "Detection of C60 and C70 in a Young Planetary Nebula". Science. 329: 1180–1182

Carina I.C.Crucho and Maria TeresaBarros, 2017. Polymeric nanoparticles: A study on the preparation variables and characterization methods. Mat. Sci. & Engg., C, 80: 771-784

Chennupati Jagadish and Amanda Barnard, 2020. Nanoscience: thinking big, working small. Bull. Australian Acad. Sci.,

Cui Jiabin, Panfil Yossef, E., Koley Somnath, Shamalia Doaa, Waiskopf Nir, Remennik Sergei, Popov Inna, Oded Meirav and Banin Uri, 2019. "Colloidal quantum dot molecules manifesting quantum coupling at room temperature". Nature Communications. 10 : 5401.

Dai, X., Hou, C., Xu, Z., Yang, Y., Zhu, G., Chen, P., Huang, Z.and Yan, L.T., 2019. Entropic Effects in Polymer Nanocomposites. Entropy, 21: 186.

Donald R. Baer, Mark H. Engelhard, Grant E. Johnson, Julia Laskin, Jinfeng Lai, Karl Mueller, Prabhakaran Munusamy, Suntharampillai Thevuthasan, Hongfei Wang, Nancy Washton,

Alison Elder, Brittany L. Baisch, Ajay Karakoti, Satyanarayana V. N. T. Kuchibhatla and DaeWon Moon, 2013. Surface characterization of nanomaterials and nanoparticles: Important needs and challenging opportunities. J. Vac. Sci. Technol. A., 31 (5): 050820.

Duangkamon Baowan, Barry J.Cox, Tamsyn A.Hilder, James M.Hill and NgamtaThamwattana, 2017. Chapter 8 - Nanotubes as Drug Delivery Vehicles. P. 211-241. In : Modelling and Mechanics of Carbon-Based Nanostructured Materials. Micro and Nano Technologies. William Andrew Publ.

Eleonore Fröhlich, 2012. The role of surface charge in cellular uptake and cytotoxicity of medical nanoparticles. Int. J. Nanomedicine., 7: 5577–5591

Elizabeth A. Worrall , Aflaq Hamid , Karishma T. Mody, Neena Mitter and Hanu R. Pappu, 2018. Nanotechnology for Plant Disease Management. Agronomy , 8 : 2-24

Ezhilarasi, P. N.; Karthik, P.; Chhanwal, N.; Anandharamakrishnan, C. (2012). "Nanoencapsulation Techniques for Food Bioactive Components: A Review". Food and Bioprocess Techn., 6: 628–47.

Graf Christina, Vossen, Dirk L.J., Imhof Arnout, van Blaaderen and Alfons, 2003. A General Method To Coat Colloidal Particles with Silica. Langmuir, 19 : 6693–6700.

Günther Jutz and Alexander Böker, 2011. Bionanoparticles as functional macromolecular building blocks – A new class of nanomaterials. Polymer, 52: 211-232

Hassellöv Martin, Readman, James W., Ranville, James F. and Tiede, Karen, 2008. "Nanoparticle analysis and characterization methodologies in environmental risk assessment of engineered nanoparticles". Ecotoxicology, 17: 344–61

Heera, P. and Shanmugam, S., 2015. Nanoparticle Characterization and Application: An Overview. Int.J.Curr.Microbiol.App.Sci., 4: 379-386

Heiligtag Florian J. and Niederberger Markus, 2013. The fascinating world of nanoparticle research. Materials Today, 16: 262–271

Ibrahim Khan, Khalid Saeed and Idrees Khan, 2019. Nanoparticles: Properties, applications and toxicities. Arabian J. Chem., 12:908-931

Ishrat Fatima, Muhammad Aslam, Muhammad Waqas Yonas, Muhammad Haroon, Muhammad Saqib Mushtaq, Faheem Abbas and Muhammad Usman Tahir, 2019. Synthesis and characterization of nanoparticles for agricultural applications. The Int. J. Global. Sci., 2:, 32-37

Jin-wei Ai, Wen Liao and Zhi-Long Ren, 2017. Enhanced anticancer effect of copper-loaded chitosan nanoparticles against osteosarcoma. RSC Adv., 7 : 15971-15977

Jiri Kudr, Yazan Haddad, Lukas Richtera, Zbynek Heger, Mirko Cernak, Vojtech Adam and Ondrej Zitka, 2017. Magnetic Nanoparticles: From Design and Synthesis to Real World Applications. Nanomaterials (Basel), 7: 243

Juan J.Giner-Casares, MalouHenriksen-Lacey and MarcCoronado-Puchau, 2016. Inorganic nanoparticles for biomedicine: where materials scientists meet medical research. Materialstoday, 19: 19-28

Jürgen Biener , Arne Wittstock , Theodore F. Baumann , Jörg Weissmüller , Marcus Bäumer and Alex V. Hamza, 2009. Surface Chemistry in Nanoscale Materials. Materials 2 : 2404-2428 Katherine N. Clayton, Janelle W. Salameh, Steven T. Wereley, and Tamara L. Kinzer-Ursem, 2016. Physical characterization of nanoparticle size and surface modification using particle scattering diffusometry. Biomicrofluidics , 10: 054107

Khajavi, R., Abbasipour, M. and Bahador, A., 2016. "Electrospun biodegradable nanofibers scaffolds for bone tissue engineering". J Appl Polym Sci., 133 : 1-19

Khan Aamir I. and O'Hare Dermot , 2002. Intercalation chemistry of layered double hydroxides: recent developments and applications. J. Mat. Chem., 12 : 3191-3198

Mashaghi, S., Jadidi, T., Koenderink, G. and Mashaghi, A., 2013. Lipid Nanotechnology. Int. J. Mol. Sci. 14: 4242-4282

McDonough, K. and Gogotsi, Y., 2013. Carbon Onions: Synthesis and Electrochemical Applications, Interface, Fall 2013, 61-66

Mochalin, V. N, ; Shenderova, O. Ho, D. and Gogotsi, Y., 2011. "The properties and applications of nanodiamonds". Nature Nanotechnol., 7: 11–23

Mora-Huertas, C.E., Fessi, H., and Elaissari, A., 2010. "Polymer-based nanocapsules for drug delivery". Intl. J. Pharmaceu., 385 : 113–42.

Mukherjee, S., Ray, S. and Thakur, R.S., 2009. Solid lipid nanoparticles: a modern formulation approach in drug delivery system. Indian J. Pharm. Sci., 71:349-358.

Munawar, A. Mohammed, Jaweria T. M. Syeda, Kishor M. Wasan and Ellen K. Wasan, 2017. Pharmaceutics, 9: 53

Ping-Chang Lin, Stephen Lin, Paul C. Wang and Rajagopalan Sridhar, 2014. Techniques for physicochemical characterization of nanomaterials. Biotechnol., Adv., 32: 711–726.

Prashant L. Suryawanshi, Rajesh Kumar Polagani, Shirish H. Sonawane, 2018. Synthesis of Nanomaterials for Fuel Cell Applications, p.205-226. In: Nanomaterials for Green Energy-Micro and Nano Technologies. Bharat A. Bhanvase, Vijay B. Pawade, Sanjay J. Dhoble, Shirish H. Sonawane, Muthupandian Ashokkumar (Edrs.), Elsevier.

Rahul Purbia and Santanu Paria, 2015. Yolk/shell nanoparticles: classifications, synthesis, properties, and applications. Nanoscale, 7: 19789-19873

Ray Unmesha. 2018. What are the Different Types of Nanoparticles?. AZoNano, , https://www.azonano.com/article.aspx?ArticleID=4938.

Shahverdi, A.R., Minaeian, S., Shahverdi, H.R., Jamalifar, H. and Nohi, A.A.,2007. Rapid synthesis of silver nanoparticles using culture supernatants of Enterobacteria: A novel biological approach. Process Biochem., 42 : 919-923

Soyeon Jeon , Jessica Clavadetscher , Dong-Keun Lee , Sunay V. Chankeshwara , Mark Bradley and Wan-Seob Cho, 2018. Surface Charge-Dependent Cellular Uptake of Polystyrene Nanoparticles. Nanomaterials, 8: 1028-1039

Stefaniak Aleksandr B., 2017. Principal Metrics and Instrumentation for Characterization of Engineered Nanomaterials. In : Metrology and Standardization of Nanotechnology. Mansfield, Elisabeth; Kaiser, Debra L.; Fujita, Daisuke; Van de Voorde, Marcel (Eds.). Wiley-VCH Verlag. pp. 151–174

Stefanos Mourdikoudis,, Roger M. Pallares and Nguyen T. K. Thanh ,2018. Characterization techniques for nanoparticles: comparison and complementarity upon studying nanoparticle properties. Nanoscale, 10 : 1-64

Sujoy K. Das, Md. Motiar R. Khan T. Parandhaman, Fathima Laffir, Arun K. Guha, G. Sekaran and Asit Baran Mandal, 2013. Nano-silica fabricated with silver nanoparticles: antifouling adsorbent for efficient dye removal, effective water disinfection and biofouling control. Nanoscale, 5 : 5549-5560

Talapin, D. V., 2012. "Nanocrystal solids: A modular approach to materials design". MRS Bull., 37: 63–71.

Tian Zhiting, Hu Han and Sun Ying,2013. "A molecular dynamics study of effective thermal conductivity in nanocomposites". Int. J. Heat Mass Transfer., 61: 577–582.

Todd Cowen, Kal Karim and Sergey A. Piletsky, 2018. Solubility and size of polymer nanoparticles. Polymer Chem., 36 : 12-28

Toshiyuki Hibino, 2015. Layered double hydroxide–agarose composites for water treatment: carbonate contamination during the drying process. Appl. Clay Sci., 10 : 116-117.

Vijay Mishra , Kuldeep K. Bansal ,Asit Verma ,Nishika Yadav ,Sourav Thakur ,Kalvatala Sudhakar and Jessica M. Rosenholm · 2018. Solid Lipid Nanoparticles: Emerging Colloidal Nano Drug Delivery Systems. Pharmaceutics, 10 : 191

Wilczewska, A. Z., Niemirowicz, K., Markiewicz, K. H. and Car, H., 2012. Nanoparticles as drug delivery systems. Pharmacol. Rep., 64 : 1020-1037}.

Yifen Shang, Md. Kamrul Hasan, Golam Jalal Ahammed, Mengqi Li, Hanqin Yin, Jie Zhou, 2019. Applications of Nanotechnology in Plant Growth and Crop Protection: A Review. Molecules, 24: 2558

Yogeshkumar Malam and Alexander M.Seifalian, 2009. Liposomes and nanoparticles: nanosized vehicles for drug delivery in cancer. Tre.Pharmacol.Sci., 20: 592-599

Zhigang Xu,, Xiaoqian Ma,, Yong-E. Gao,, Meili Hou,, Peng Xue,, Chang Ming Li, and Yuejun Kang, 2017. Multifunctional silica nanoparticles as a promising theranostic platform for biomedical applications. Mat. Chem. Front., 5: 6847

Zhu, J., Ou, X., Su, J. and Li, J., 2016. The impacts of surface polarity on the solubility of nanoparticle. J Chem Phys., 145 (4):044504.

Shahverdi, A.R., Minaeian, S., Shahverdi, H.R., Jamalifar, H. and Nohi, A.A.,2007. Rapid synthesis of silver nanoparticles using culture supernatants of Enterobacteria: A novel biological approach. Process Biochem., 42 : 919-923.

3

Nanotechnology in Plant Disease Diagnosis

Biological diagnostic methods rely on the transmission and amplification of the pathogen in indicator hosts that develop symptoms. These methods test the biological properties of pathogens and early successes in the control of intracellular pathogens were largely due to the diagnosis of the causal agents by biological tests. The sensitivity of biological/serological tests is usually determined by the minimum number of pathogen propagules required to initiate infection. Inoculation with fewer particles could lead to false negative results. Biological techniques for disease diagnosis and pathogen detection are usually highly accurate but too slow and not amenable to large-scale application. Non-biological diagnosis or component analysis depends on the recognition of components specific to the pathogen (Choudhary Prassan et al., 2018). Electron microscopy, nucleic acid hybridization and polymerase chain reaction (PCR) amplification are examples of component analysis. Amidst the serological and molecular detection methods of disease diagnostics, Nanotechnology (coined by Taniguchi in 1974) is one of the most fascinating and rapidly advancing sciences and possesses potential to revolutionize many disciplines of science, technology, medicine and agriculture.

3.1 Nanodetection of plant pathogens

Nanosensors can be potentially used in the crop protection, disease diagnosis, pathogen detection and residual analysis with more precision. Significance of nanoparticles is being recognized worldwide and several countries have set up committees to monitor nanotechnological advancements and to harness its benefit for mankind. The Indian Government has also set up a Nano research project during the eleventh five year plan with multiple varsities and institutes taking up the technology. Many of these techniques were first developed in the medical sciences before being applied to the plant sciences and, in the present situation, to plant pathology. The disease diagnosis, pathogen detection and residual analysis may become much more quick and precise with the use of Nanosensors. These aspects can be attributed to the ultra small size of the particle and there high degree of specificity/reactivity.

Nanoparticles can be used as biomarkers or as a rapid diagnostic tool for detection of phytobacteria, viruses and fungi. Nanoparticles based sensors might offer improved detection limits in detecting viral pathogens in plants (Choudhary Prassan et al., 2018). Multiplexed detection of pathogen DNA (dendrimer-like DNA-based) with fluorescence-intensity-coded nanobarcodes, which contain a built- in code and a probe for molecular recognition have been reported. Nanoparticles can either be directly modified for use in pathogen detection or for a diagnostic tool to detect compounds indicative to a diseased condition. A variety of biosensors have been developed and described in the recent literature for mycotoxin analysis some of which have the potential of multi-array mycotoxins identification. Recent breakthroughs in nanotechnology enable the preparation of various nanoparticles and nanostructures with few technical hurdles. Nanoparticles display fascinating electronic and optical properties and can be synthesized using different types of materials for electronics and sensing applications.

Nano-chips are types of microarrays that contain fluorescent oligo capture probes through which the hybridization can be detected. The nanochips are known for their sensitivity and specificity in detecting single nucleotide change in bacteria and viruses A fluorescence silica nanoparticle was utilized in combination with antibodies to detect *Xanthomonas axonopodis* pv. *vesicatoria* that causes bacterial spot disease in plants. nanogold based immunosensors was used that could detect *Tillitia indica* using Surface Plasmon Resonance (SPR). Application of SPR sensor in the seed certification and plant quarantines may prove highly effective and accurate in the detection of infection in plants.

A mouse monoclonal antibody and a SPR sensor was used for label free detection of urediniospores from *Puccinia striiformis* f. sp. *tritici* (Pst).

Completion of the resultant assay took approximately 45 minutes with a detection level of 3.1×10^5 urediniospores/ml. Skottrop et al., (2007) also pre-incubated a murine monoclonal antibody with sporangia of *P infestans.* The separation of bound and unbound antibody was permitted by a 5 minute at 37°C centrifugation (1500g). Free antibody was then passed over a surface containing bound goatntimouse polyclonal antibody, allowing quantization of free antibody. The limit of detection was 2.2×10^6 sporangia/mL, and no cross-reactivity was seen when other fungal strains, such as *Melampsora euphorbia* and *Botrytis cinerea* that were assayed in parallel. Dunne et al., 2007 also developed a SPR-based inhibition assay that incorporated monomeric and dimeric scFv antibody fragments for permitting the detection of between 390 and 12,000 ppb and 190 and 24,000 ppb of aflatoxins (naturally occurring mycotoxins produced by several strains of *Aspergillus* spp.), AFB1 (aflatoxin B1) with monomeric and dimeric scFvs respectively.

Gold nanoparticles are excellent markers to be used in biosensors as several optical or electrochemical procedures can be modified to identify pathogens. A number of nanoparticle-based experiments have been performed to develop biomolecular detection with DNA or protein functionalized gold nanoparticles, which are used as the target-specific probes (Choudhary Prassan et al., 2018). The ability of gold nanoparticles was highlighted to act as fluorescence quenchers and, therefore, it could be used to solve major drawbacks in molecular biology experiments. For example, a DNA oligonucleotide could be synthesized, fluorescently labeled at its 5' end and conjugated at the 3' end with gold nanoparticles. These oligonucleotides can be used in diagnostic procedures, particularly in cases where DNA analysis cannot be circumvented, and in the diagnosis of the phytoplasma associated with the flavescence doree (FD) of grapevine. The gold nanoparticles efficiently quench the fluorescence of light harvester polymers, such as polyfluorene, and will open new perspectives in the development of the optical performances of nanobiotransducers for diagnostic purposes. Furthermore, a diagnostic probe made of a specific oligonucleotide bearing a fluorescein at its 5' end and 2-nm gold particles at its 3' end act as a Nanobiotransducer in DNA hybridization. It produces a stronger fluorescence signal when hybridized to target DNA.

Nanowire based approaches have been used for direct, label free detection and extreme sensitivity and good selectivity which could provide a road ahead for integrated, high throughput, multiplexed DNA detection for genetic screening and biothreat detection. Two-terminal silicon nanowires was used in which the surfaces of the silicon nanowire devices were modified with peptide nucleic acid receptors designed to recognize wild type versus the mutation site in the cystic fibrosis transmembrane receptor gene.

Nano-phytopathology is a cutting-edge science which uses nanotechnology for detecting, diagnosing and controlling plant disease and their pathogens at an early stage, owing to crop protection from epidemic diseases (Touseef Hussain, 2017) The modern plant pathologist strives to apply his knowledge in nano phytopathology to enable understanding of controlling factors of plant diseases and to develop and/or evaluate eco-friendly diagnostic measurements. Modern nano molecular techniques are used for monitoring or understanding of pathogen population genetics, plant microbe interactions and gene transfer between pathogens and even the host. Furthermore, nanoparticles such as nano sized silicasilver have recently been applied as antimicrobial and antifungal agents. Additionally, nanomaterials scan be used for mycotoxin detection and detoxication, increasing plant resistance, plant disease forecasting and nano-molecular diagnostics of plant pathogens. The bio-barcode assay is an ultrasensitive method of amplification and detection of proteins or nucleic acids. DNA biobar coded tests employ oligo nucleotide-modified magnetic gold nanoparticles (AuMNPs) for signal amplification and for simple separation of a target protein from the sample. The large b-DNAto-recognition agent ratio affords a means of substantial signal amplification.

It is also promising by allowing the quick detection of numerous protein targets at low-attomolar concentrations and nucleic acids at high-zeptomolar levels under optimized conditions. The concept of the bio-barcode assay is unique and represents a potential alternative to the PCR technique.

3.2 Nanosensors for plant disease forecasting

This is a management method used to forecast the chances or severity of plant diseases and to help farmers make cost-effective decisions for controlling diseases (Bogue, 2008). Presently research is being carried out by using nanosensors to improve pathogen detection methods in crop systems. Many electronic companies have been examining the use of electrical conducting polymers such as polyaniline, polythiophene and polypyrolle. These polymers can also be used to fabricate sensors that can detect molecular signals with very low intensity of spoilage and food-borne pathogens within minutes. The bio-nanosensors have the potential of increased sensitivity and therefore a significantly reduced response-time to discover potential disease problems. Such bio-analytical nanosensors were utilized to detect and quantify minute amounts of contaminants such as viruses, bacteria, fungi, toxins and other bio-hazardous substances in the agriculture and food systems (Choudhary Prassan et al., 2018). Therefore, these biosensors may have a huge impact on the precision farming methods. Nano-sensors can be linked to a GPS for real-time monitoring of disease and distributed throughout the field to monitor soil conditions and crop health.

The union of biotechnological and nanotechnological approaches in bio-sensors will be used to construct equipment with increased sensitivity, allowing an earlier response to ecological changes and disease prevalence. Nanosensors will allow us to identify plant diseases before visible symptoms appear and thus facilitate their control. Precision farming will allow improved agriculture production by providing precise data, helping growers to make better decisions. Nanofabricating images help plant pathologists detect early signs of disease and the rapidly expanding use of diagnostic imaging technologies continues to push the boundaries of plant pathologist diagnostic capabilities. Nanotechnology offers us the opportunity to precisely tune and control the chemical and physical properties of contrast materials in order to overcome concerns with toxicity, useful imaging time, tissue specificity and signal strength (Choudhary Prassan et al., 2018).

3.3 Nanodiagnosis and Nano-Phytopathology

The extension/integration of molecular diagnostics on a nanoscale is a promising technology for identifying pathogens. Nanomolecular diagnostic is the use of nanobiotechnology to diagnose plant diseases and this can be termed as nanodiagnostics.In particular, several nanodevices and nanosystems are used for sequencing single molecules of DNA (Mohammad A. Khiyami et al., 2014). Assays with the use of nano-size devices to investigate DNA sequences and diagnose disease are becoming faster, more flexible and more sensitive. It is noteworthy that new detection techniques involving nano-biosensors for pathogen identification will likely be a cornerstone to this trend. During the 1980s and 1990s, phytopathologists relied on visual assessment to identify plant diseases. Identifying plant pathogens via conventional techniques may take several days and therefore researchers need rapid detection tools that can provide results within a few hours. To develop such detection tools, phytopathologists are constantly working with nanotechnologists. Nano and/ or phytopathology researchers are attempting to develop an easy assay that is moveable and accurate and does not need any difficult method for procedure, so that farmers can use the mobile laboratory themselves to detect specific diseases.

Newly developed nanomaterials with special nanoscale characteristics could present a tremendous breakthrough in pathogen and contaminant detection. Nanotechnology is also driving the development of lab-on-a-chip systems for detecting toxicity in waters, observing nutrients in irrigation water and controlling the quality in food production. Nano-phytopathology is a cutting- edge science which uses nanotechnology for detecting, diagnosing and controlling plant disease and their pathogens at an early stage, owing to crop protection from epidemic diseases. The modern plant pathologist strives to apply his knowledge in nanophytopathology to enable understanding of controlling factors of plant diseases and to develop and/or evaluate eco-friendly diagnostic measurements.

Modern nanomolecular techniques are used for monitoring or understanding of pathogen population genetics, plant- microbe interactions and gene transfer between pathogens and even the host. Furthermore, nanoparticles such as nanosized silica-silver have recently been applied as antimicrobial and antifungal agents. Additionally, nanomaterials can be used for mycotoxin detection and detoxication, increasing plant resistance, plant disease forecasting and nano-molecular diagnostics of plant pathogens. Potential applications of nanophytopathology are shown in Fig.42 (Mohammad A. Khiyami et al., 2014).

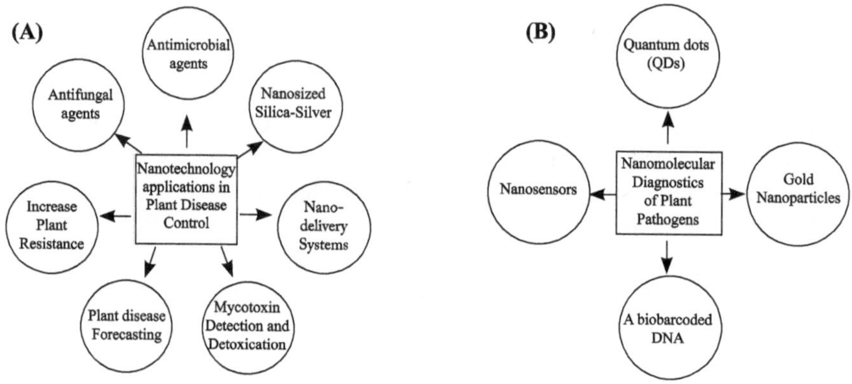

Fig. 42: Potential applications of nanotechnology in plant pathology: (a) plant disease control and (b) detection of plant pathogens

3.4 Nanodiagnosis of plant diseases

Nanodiagnostic using nanodevices and nanosystems are faster, sensitive, and flexible for the detection/identification of crops diseases compared with traditional diagnostic methods; for these reasons the current chapter will give information about nanotechnology applied to diagnosis of diseases caused by fungi, bacteria, nematodes and viruses in some important economic crops (Álvarez et al., 2018).

Early detection of plant diseases has prompted Nanotechnologists to look for a nano solution for protecting the food and agriculture from bacteria, fungi and viral agents through autonomous nanosensors linked into a GPS system for real-time monitoring throughout the field to monitor soil conditions and crop (Amarendra N. Misra et al., 2013) The union of biotechnology and nanotechnology in sensors will create equipment of increased sensitivity, allowing an earlier response to environmental changes and diseases. There is an urgent need of ultrasensitive diagnostic tool which can detect the molecular defects, be it at genomic or biochemical level, rapidly. Bio-systems are endowed with functional nanometric devices such as enzymes, proteins, and Nucleic acids, which detectable vital processes in plants. Disease diagnosis is difficult mainly because of the extremely low concentrations of biochemicals and also due to the presence of very low amount of detectable virus and many fungal or bacterial infections.

Traditional methods of identification based on visual symptoms and culturing in laboratory are time-consuming, labour-intensive, costly and have very low sensitivity and accuracy (Kashyap et al., 2017). The immunological and nucleic acid-based detection techniques enhanced the sensitivity, accuracy and speed but require highly skilled manpower, sophisticated and heavy instrumentation facility. Recent advances in development of functional nanoparticles (electronic, optical, magnetic, or structural) that can be covalently linked to biological molecules such as peptides, proteins, and nucleic acids opened way to develop low cost, accurate, rapid and on-site detection system by integration of molecular diagnostics and nanotechnology.

Following list provides nanotechnology-based systems for detection and diagnosis of plant pathogens.

Nanotechnology system	Application	References
Fluorescent silica nanoparticles (FSNP) combined with antibody molecules	Fluorescence silica nanoprobe as a biomarker for rapid detection of *Xanthomonas axonopodis* pv. *vesicatoria,* responsible for bacterial spot disease in tomatoes and peppers	Yao et al. (2009)
Copper nanoparticles-modified gold electrode	Electrocatalytic oxidation of phytohormone salicylic acid at copper nanoparticles-modified gold electrode and its detection in oilseed rape infected with fungal pathogen *Sclerotinia sclerotiorum*	Wang et al. (2010)
Nanogold particles	Surface plasmon resonance based immunosensor for Karnal bunt (*Tilletia indica*) diagnosis based on the experience of nanogold based lateral flow immune-dipstick test	Singh et al. (2010)
Thioglicolic acid modified cadmium-telluride QD	Development of a quantum dots FRET-based biosensor for efficient detection of *Polymyxa betae*, a vector of beet necrotic yellow vein virus (BNYVV) responsible for *Rhizomania* disease in sugar beet	Safarpour et al. (2012)
Thioglicolic acid-modified cadmium-telluride quantum dots (CdTe-QD)	Detection of *Candidatus Phytoplasma aurantifolia* with a quantum dots FRET-based biosensor	Rad et al. (2012)
CuO nanoparticles and nanostructural layer biosensors	Synthesis of CuO nanoparticles and fabrication of nanostructural layer biosensors for detecting *Aspergillus niger* fungi	Etefagh et al. (2013)
Modified QD that contained carboxylic groups	Development of a fluorescence resonance energy transfer (FRET)-based DNA biosensor for detection of synthetic oligonucleotide of *Ganoderma boninense*,	Bakhori et al. (2013)

	an oil palm pathogen	
Nanoribbon	Polypyrrole nanoribbon-based chemiresistive immunosensors for viral plant pathogen detection	James (2013)
Nanorod	Direct detection of orchid viruses using nanorod- based fibre optic particle plasmon resonance immunosensor	Lin et al. (2014)
TiO_2 or SnO_2 nanoparticles on-screen-printed carbon electrodes	Electrochemical detection of p-ethylguaiacol, a fungi infected fruit volatile using metal oxide nanoparticles	Fang et al. (2014)
Nanowire	Plant diseases detection using nanowire as biosensor transducer	Ariffin et al. (2014)
Silver nanoparticle	Development of a helicase-dependent isothermal amplification (HDA) in combination with on-chip hybridization for the detection of selected Phytophthora species	

Biosensor, quantum dots (QD), nanostructured platforms, nanoimaging and nanopore DNA sequencing tools have the potential to raise sensitivity, specificity and speed of detection, facilitate high-throughput analysis, and can be used for large-scale monitoring and on-site detection of plant pathogens.

Nanotechnology, nano particles and quantum dots (QDs) have emerged as essential tools for fast detection of a particular biological marker with extreme accuracy (Mohammad A. Khiyami et al., 2014). Biosensor, QDs, nanostructured platforms, nanoimaging and nanopore DNA sequencing tools have the potential to raise sensitivity, specificity and speed of the pathogen detection, facilitate high- throughput analysis, and to be used for high-quality monitoring and crop protection. Furthermore, nanodiagnostic kit equipment can easily and quickly detect potential serious plant pathogens, allowing experts to help farmers in the prevention of epidemic diseases. Such an accurate technology may help to design a proper integrated disease management system which may modify crop environments to adversely affect crop pathogens.

Detection and identification of plant diseases could be achieved by both direct and indirect methods. Direct approaches typically involve analysis of plant pathogens (e.g., bacteria, oomycetes, fungi, and viruses) or biomolecular markers (e.g., nucleic acids, proteins, and carbohydrates) isolated from infected plant tissues (Zheng Li et al., 2020). Indirect diagnosis recognizes plant diseases through changes in physiological or histological indices such as leaf surface temperature or humidity, spectroscopic features of plant tissues, morphology, growth rate, and emissions of volatile organic compounds (VOCs). A wide variety of spectroscopic, electrochemical or molecular technologies could serve as direct or indirect detection methods.

Several mainstream categories of nanoscale techniques for agricultural diagnosis were summarized, in both direct or indirect fashions, including microneedle patches, nanopore sequencing platform, plant wearables, and nanoparticle or array-based sensors Zheng Li et al., (2020). The development of nanodiagnostic tools for AgBio science is in its infancy.

3.4.1 Nanodiagnostic kits

Plant pathologists plan to study this 'lab in a box' very soon. This refers to packing sophisticated measuring devices, reagents, power supply and other features that now take up laboratory space into a parcel no larger or heavier than a briefcase (Goluch et al., 2006). A briefcase-sized kit is transferred to a field where crops are growing to search for pathogens that could infect and reduce the yield. This is a quick and precise procedure. Nanodiagnostic kit equipment can easily and quickly detect potential serious plant pathogens, allowing experts to help farmers in prevention of disease epidemic from breaking out. For example, 4mycosensor is a tetraplex competitive antibody-based assay in a dipstick format for the real-time detection of zearalenone ZEA, T-2/HT-2, DON and FB1/FB2 mycotoxins on the same single strip in corn, wheat, oat and barley samples at or below their respective European maximum residue limits (MRLs). The proposed immunoassay protocol is fast, cheap, easy-to-use and suitable for the purpose of quick screening of mycotoxins in cereals.

Plant pathologists plan to study this 'lab in a box' very soon. This refers to packing sophisticated measuring devices, reagents, power supply and other features that now take up laboratory space into a parcel no larger or heavier than a briefcase. A briefcase-sized kit is transferred to a field where crops are growing to search for pathogens that could infect and reduce the yield. This is a quick and precise procedure. Nanodiagnostic kit equipment can easily and quickly detect potential serious plant pathogens, allowing experts to help farmers in prevention of disease epidemics from breaking out. For example, 4mycosensor is a tetraplex competitive antibody-based assay in a dipstick format for the real-time detection of ZEA, T-2/HT-2, DON and FB1/FB2 mycotoxins on the same single strip in corn, wheat, oat and barley samples at or below their respective European maximum residue limits (MRLs). The proposed immunoassay protocol is fast, cheap, easy-to-use and suitable for the purpose of quick screening of mycotoxins in cereals.

Nucleic acid- and antibody-based molecular assays are gold-standard methodologies for the diagnosis of plant diseases, but the analyzing procedures are complex and laborious. The prominent physical or chemical properties of nanomaterials have enabled their use as innovative and high-performance diagnostic tools for numerous plant pathogens and other important disease biomarkers. Engineered nanomaterials have been incorporated into traditional laboratory molecular assays or sequencing technologies that offer notable enhancement in sensitivity and selectivity. Meanwhile, nanostructure-supported noninvasive detection tools combined with portable imaging devices (e.g., smartphones) have paved the way for fast and on-site diagnosis of plant diseases and long-term monitoring of plant health conditions, especially in resource-poor settings.

3.4.2 Bio-barcoded DNA (b-DNA)

With the rapid development of nanotechnology, the bio-barcode assay (BCA), as a new diagnostic tool, has been gradually applied to the detection of protein and nucleic acid targets and small-molecule compounds. BCA has the advantages of high sensitivity, short detection time, simple operation, low cost, good repeatability and good linear relationship between detection results. However, bio-barcode technology is not yet fully formed as a complete detection system, and the detection process in all aspects and stages is unstable. Therefore, studying the optimal reaction conditions, optimizing the experimental steps, exploring the multi-residue detection of small- molecule substances, and preparing immuno-bio-barcode kits are important research directions for the standardization and commercialization of BCA. The main theme of this review was to describe the principle of BCA, provide a comparison of its application, and introduce the single-residue and multi- residue detection of macromolecules and single-residue detection of small molecules. We also compared it with other detection methods, summarized its feasibility and limitations, expecting that with further improvement and development, the technique can be more widely used in the field of stable small-molecule and multi-residue detection (Yuanshang Wang et al., 2019).

The bio-barcode assay is an ultrasensitive method of amplification and detection of proteins or nucleic acids. DNA bio-barcoded tests employ oligonucleotide-modified magnetic gold nanoparticles (AuMNPs) for signal amplification and for simple separation of a target protein from the sample. The large b-DNA-to-recognition agent ratio affords a means of substantial signal amplification. It is also promising by allowing the quick detection of numerous protein targets at low-attomolar concentrations (Goluch et al., 2006) and nucleic acids at high-zeptomolar levels under optimized conditions .The concept of the bio-barcode assay is unique and represents a potential alternative to the PCR technique.

DNA barcoding has proven to be a versatile tool for plant disease diagnostics in the genomics era. As the mass parallel and next generation sequencing techniques gained importance, the role of specific barcodes came under immense scrutiny. Identification and accurate classification of phytopathogens need a universal approach which has been the main application area of the concept of barcode. The present review entails a detailed description of the present status of barcode application in plant disease diagnostics. A case study on the application of Internal Transcribed Spacer (ITS) as barcode for *Aspergillus* and *Fusarium* spp. sheds light on the requirement of other potential candidates as barcodes for accurate identification. The challenges faced while barcoding novel pathogens have also been discussed with a comprehensive outline of integrating more recent technologies like meta-barcoding and genome skimming for detecting plant pathogens (Choudhary et al., 2021).

3.4.3 Quantum dots (QDs)

QDs are semiconductor nanoparticles that fluoresce when stimulated by anexcitation light source. Furthermore, QDs are inorganic fluorophores presenting major advantages over traditional organic fluorophores used as markers on nucleic acids or proteins for visual detection (Choudhary Prassan et al., 2018). The mycosynthesis of semiconductor nanomaterials was first reported in unicellular yeast, which was shown to be capable of producing cadmium sulphide (CdS) crystallites in response to a cadmium salt stress. Different microbes have also been used for the biosynthesis of CdS; however, few studies have focused on its luminescent properties. Very luminescent CdSe QDs were created by the fungus, *F oxysporum* when incubated with a mixture of CdCl2 and SeCl4 at room temperature. A proficient myco-mediated synthesis of highly fluorescent CdTe QDs was accomplished by the *F. oxysporum* isolates when reacted with a mixture of CdCl2 and TeCl2 at ambient conditions. Description of these biosynthesized CdTe nanoparticles was performed by transmission electron microscopy (TEM) and selected-area electron diffraction (SAED) An effective biosynthesis method to prepare easily harvested biocompatible cadmium telluride (CdTe) QDs with tunable fluorescence emission using yeast cells was developed.The confocal images of the yeast cells show obvious green emission from the synthesized CdTe QDs. The CdTe QD units were spread only in the cytoplasm and nucleus of yeast cells, while none were found in the cell membrane.

Biocompatible semiconductor crystals are composed of a nucleus and a shell allowing the binding of ligands and thus the attachment of this fluorescent marker to the target pathogen. A hybridization-based biosensor known as the quartz-crystal microbalance biosensor is a good example. The quartz- crystal microbalance vibrates under electric stimulus, and a change in mass due to the attachment of any compound to its surface can be detected through a reduction in the frequency of its vibration. When a nucleic acid probe is attached to the surface of a quartz-crystal microbalance biosensor and then exposed to complementary PCR product, a hybrid forms and causes the resonance frequency of the quartz-crystal microbalance biosensor to decrease dramatically. This system can be combined with fast PCR protocols to decrease the time for specific environmental detection of plant pathogens (Sharon et al., 2010).

Recent developments in molecular detection technology, nanosensors, and a combination of both resulting in more convenient, effective and specific assays have opened the door to greater use of these tests for detecting plant pathogens. These assays will help to free growers, crop consultants, and plant-health-care professionals from having to rely excessively on symptomatology and/or time-consuming diagnostic procedures, and permit early detection of pathogens. However, they should be viewed as management tools, to be used in conjunction with other diagnostic procedures, knowledge of the crop, and understanding of the biology of the pathogen and the ecology of the disease. The portable diagnostic equipment, nanoparticles-based, bio-barcoded DNA sensor, and the QD have potential applications in the multiplex detection of plant pathogens and toxigenic fungi. To date, mobile diagnostic assays have been developed to rapidly detect plant disease and may be used to prevent epidemics.

These nano-based diagnostic kits not only increase the speed of pathogen detection but also increase the accuracy of the diagnosis.

During the next decade, nanodevices, which can make thousands of measurements speedily and very cheaply, will become available. Future prospects in plant disease diagnostic will continue in miniaturization of biochip technology to the nanoscale range. Nanophytopathology can be applied as a tool to understand plant pathogen interactions, providing new methods for crop protection. Specific nanodevices and DNA nanodevices could enable accurate tracking, detection and diagnosis of plant pathogens in the early stages of plant disease.

Engineered nanoparticles exist as metalloids, metallic oxides, nonmetals and carbon nanomaterials and as functionalized dendrimers, liposomes, and quantum dots. Nanomaterials have been integrated into disease management strategies and diagnostics and as molecular tools. Most reports summarized herein are directed toward pathogen inhibition using metalloid/metallic oxide nanoparticles as bactericides/fungicides and as nanofertilizers to enhance health. The use of nanoparticles as biosensors in plant disease diagnostics was also reviewed (Elmer Wade and White Jason, 2018). As global demand for food production escalates against a changing climate, nanotechnology could sustainably mitigate many challenges in disease management by reducing chemical inputs and promoting rapid detection of pathogens.

Nanodiagnostic and microfluidics offer novel tools to improve the sample preparation step that remains difficult to integrate in a miniaturized platform. The signal amplification approaches could challenge those of target amplification. Quick on-site detection of plant pathogens using, nanobased kits, nanosensor, nanobiosensors, nanobarcodes and other portable diagnostic systems will also help the agricultural and food industry to manage different plant diseases. Nanoparticles such as nano-sized silica-silver have recently been applied as antimicrobial and antifungal agents (Veeranna et al., 2017). Additionally, nanomaterials can be used for mycotoxin detection and detoxication, increasing plant resistance, plant disease forecasting and nano-molecular diagnostics of plant pathogens.

Nano particles and quantum dots (QDs) have emerged as essential tools for fast detection of a particular biological marker with extreme accuracy. Nano diagnostic kit like multiplex dipstick immunoassay which can easily and quickly detect potential serious plant pathogens viz., Fusarium toxins, zearalenone, T-2 and HT-2 toxins, deoxynivalenol and fumonisins in wheat, oats and maize allowing experts to help farmers in the prevention of epidemics. Furthermore, nanopore DNA sequencing tools, QDs, nanostructured platforms for the detection of aflatoxin-B1 by means of functionalization of cMWCNTs/ITO electrode surface with monoclonal aflatoxin-B1 antibodies (anti-AFB1) using electrochemical technique. Nanostructural layer biosensors are fabricated for the detection of Aspergillus niger fungi through changes observed in the electrical resistance of samples. Nanofabricated imaging and LAMP have the potential to raise sensitivity, specificity and speed of the pathogen detection, and can be used for high-quality monitoring and crop protection.

Nanodiagnostic kit equipment can easily and quickly detect potential serious plant pathogens, allowing experts to help farmers in the prevention of epidemic diseases. A review by Mohammad A. Khiyami et al., (2014) dealt with the application of nanotechnology for quicker, more cost-effective and precise diagnostic procedures of plant diseases. Such an accurate technology may help to design a proper integrated disease management system which may modify crop environments to adversely affect crop pathogens.

3.4.4 Portable Genome Sequencer (Nanopore sequencing system)

Many companies are exploring the idea of using nanopore technology. There are two main challenges that must be addressed in any nanopore sequencing: how to distinguish the nucleotides as the strand passes through the nanopore and how to control the speed of the DNA strand as it passes through the nanopore (Mohammad A. Khiyami et al., 2014). Even with these unique challenges the nanopore sequencing platform is thought to be simple and straight forward because theoretically very long reads can be generated from a low quantity of nucleic acid. The protein nanopore and enzyme were designed to control a single strand of DNA, and as the DNA goes through the nanopore a direct electronic analysis is conducted. The protein nanopore is inserted in a polymer bilayer membrane across the top of a microwell. Each microwell has a sensor chip that measures the ionic current as the single molecule passes through the nanopore (Fig.43).

However, the speed at which the DNA strand travels through the nanopore is too fast for accurate identification. IBM and Roche together are developing a new sequencing technology described as 'DNA transistor' which could potentially record the nucleotide sequence as the template is pulled through the nanopore sensor. Portable genome sequencer (MinION) was able to sequence 10 kb of a single sense and anti-sense DNA strand and will make next-generation sequence (NGS) within the reach of many groups and research environments.

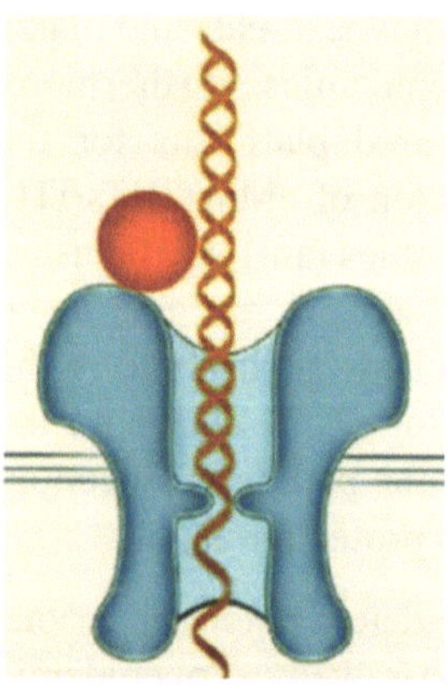

Fig. 43: Diagram of a DNA molecule travelling through a protein nanopore

Nanopore platform implemented within current diagnostic equipment has the potential of analysing the entire genome in minutes instead of hours. Nanotechnology can be applied to plant science research in order to analyse plant genomics, gene function and pathogens detection as well as improvement of crop species.

3.4.5 Loop-Mediated Isothermal Amplification (LAMP-PCR)

Loop-mediated isothermal amplification (LAMP) of DNA is a simple, cost effective, and rapid method for the specific on-site detection method of genomic DNA using a set of six oligonucleotide primers with eight binding sites hybridizing specifically to different regions of a target gene, and a thermophilic DNA polymerase from *Geobacillus stearothermophilus* for DNA amplification.

There are several detection devices varying from gold nanoparticles tagged with short fragments of DNA to multicolour optical coding for biological tests that have been achieved by embedding different-sized QDs into polymeric microbeads. The thiol-modified oligonucleotides (Au-nanoprobes) technique discards the use of electrophoretic analysis of LAMP-PCR products, thus increasing speed, specificity of results and decreasing costs (Kaewphinit et al. 2013). LAMP has the potential to implement early detection of plant pathogens at a local level (on a farm) instead of in a laboratory or less well-resourced settings by molecular methods for monitoring inoculum levels in the air and is currently developing models to predict when and where a pathogen will first occur in the agricultural fields.

For instance, LAMP-PCR of DNA is an easy, commercial and quick method for the specific detection of genomic DNA using a set of six oligo primers with eight binding sites hybridizing, specifically to various regions of a target gene and a thermophilic DNA polymerase from G. stearothermophilus for DNA amplification of *Fusarium graminearum*. Therefore, in the LAMP technique, diagnostics can be performed without examination of the amplification product during or after the amplification. Predominantly, positive reactions will be expected to show bright green fluorescence, while a negative reaction would remain light orange. This would be the expected result when SYBR Green I dye is added, meaning that the LAMP amplicons will be visualized directly and seen with the naked eye or under ultraviolet trans-illumination.

Positive results were obtained with *F. graminearum* isolates, while all reactions with other Fusarium and fungal species included for specificity remained negative, with a response identical to the negative control (Mohammad A. Khiyami et al., 2014). This methodology was used to detect plant pathogenic fungi to rhizospheric microorganisms such as *Phytophthora sojae, P. ramorum* and *P. kernoviae, Aspergillus flavus* and *A. parasiticus,* and toxigenic Fusarium.

3.4.6 Nanoparticle-based sensors

Imaging techniques, such as thermographic and hyperspectral imaging, have been used in the field for indirect detection of plant disease. However, many of them have notable limitations, such as susceptibility to parameter changes of the environment and lack of specificity for disease subtypes or strains (Zheng Li et al., 2020). Rapid progresses in chemo- or biosensing technology have led to a broad range of successful applications, such as quality assessment of medicinal and industrial products. Recently, several nanoparticle-based chemo- or biosensors have been proposed and commercialized for agricultural diagnosis. Depending on transduction mechanisms of the designed sensory interactions, the analytes could be recognized by characteristic optical or electrical outputs of the nanosensors. The sensor's detection specificity could be enhanced by selective chemical interactions or by the use of biospecific recognition elements such as DNA oligos, antibodies, aptamers, and enzymes. The detection sensitivity could be improved by the use of surface-enhanced optical properties (e.g., surface plasmon resonance (SPR)) or electron-conductive nanoscale substrates such as carbon-based nanomaterials (e.g., graphene or carbon nanotube) as transducers. In this section, we will highlight three main classes of nanosensors or nanobiosensors, including metal or metalloid nanoparticles, quantum dots, and array-based nanosensors.

A list of representative nanosensors or biosensors used for plant disease diagnosis, their detection mechanisms, and associated performance is summarized below.

Target	Sensor component	Sensor Fabrication	Detection mechanism	Sensitivity	Specificity	
P. syringae	Au NP[a]-ss DNA	Nanoparticle functionalization and DNA hybridization	Electrochemistry	214 pM	Unresponsive to *B.cinerea and F. oxysporum*	Low
P. ramorum and *P. lateralis*	Ag NP-ss DNA	Nanoparticle functionalization and DNA hybridization	SERS[b]	N/A	Unresponsive to *P. lateralis*	Low
Aflatoxins	Ag NR[c]	Surface functionalization	SERS	5×10^{-5} M for Aflatoxin B1	Discriminable among Aflatoxin B1, B1, G1 and G2	Low
B. thuringiensis and *B. subtilis*	Pt NP-IgG[d]	Surface immunological functionalization	MALDI-TOF MS[e]	~10^2 CFU	Discriminable between two bacteria	Low
X. campestris	Si NP-Rubpy-IgG	Surface immunological functionalization	Fluorescence quenching	3×10^2 CFU	N/A	Moderate
R. solanacearum	Au NP-ss DNA	Surface functionalization	Colorimetry	7.5 ng	No specific band revealed by PCR	Low
F. oxysporum	CdSe/ZnS-MPA[f] QD[g]	Surface functionalization	Fluorescence	25 μg mL^{-1}	N/A	High
Arabidopsis thaliana	CdSe-PEI[h] QD	Surface functionalization	Fluorescence	10 μg mL^{-1}	N/A	High

V. mali	CD[i]	Surface functionalization	Fluorescence	pH range of 2.13–9.34	No interference from ions	Low
F. avenaceum	CD	Surface functionalization	Fluorescence	N/A	Discriminable between *P. aeruginosa* and *F. avenaceum*	Low
3′-Diphosphate-5′-diphosphate	Tb(III)-CD	Metal ligation	Fluorescent ratiometry	50 nM	Unresponsive to analogues including ATP, CTP, and UTP	Moderate
Citrus tristeza	CdTe QD-CD	Surface immunological functionalization	FRET[j]	520 ng mL^{-1}	N/A	High
Citrus tristeza	CdTe QD-Rd[k]	Surface immunological functionalization	FRET[j]	520 ng mL^{-1}	N/A	High
p-Ethylguaiacol	TiO_2 and SnO_2 NP	Surface functionalization	Electrochemistry	35–62 nM	Slightly responsive to p-ethylphenol, unresponsive to 3-octanone and 1-octen-3-ol	Low

Ethylene	Cu(I) tris (pyrazolyl) borate-SWCNT[l]	Ball-milling	Chemiresistor	0.5 ppm	Unresponsive to common solvents except acetonitrile, THF, and acetaldehyde	Low
Ethylene	Pd(II)-pH dye- SiO2 microspheres	Ultrasonic spray	Colorimetry	0.17 ppm	Unresponsive to alcohols or	Moderate
Terpene vapor	Au NP@ sol–gel	One-pot hydrolysis	LSPR	N/A	Discriminable among cis-jasmone, α-pinene, limonene, and γ-terpiene	Low
Various VOCs[m]	Au NR-cysteine and nanoporous dyes	Surface functionalization	LSPR	0.4–1.7 ppm for hexenal; 1.8–5.2 ppm for phenol	Discriminable among over ten VOCs	Low

a Nanoparticle. b Surface-enhanced Raman spectroscopy. c Tris(bipyridine)ruthenium(II) chloride. d Immunoglobulin G. e Matrix-assisted laser desorption/ionization-time of flight mass spectroscopy. f 3-Mercaptopropionic acid. g Quantum dot. h Polyethylenimine. i Carbon dot. j Förster resonance energy

Gold nanoparticles as biosensor: Utilization of nanomaterials or nanoparticles in biosensors allows the development and use of some novel signal detection procedures and equipment. Different strategies such as antibody–antigen, adhesion–receptor, antibiotic and complementary DNA sequence recognitions have been discovered for a specific detection between target phytopathogenic cells and bio-functionalized nanomaterials. Gold nanoparticles are excellent markers to be used in biosensors as several optical or electrochemical procedures can be modified to identify pathogens. A number of nanoparticle-based experiments have been performed to develop biomolecular detection with DNA- or protein-functionalized gold nanoparticles, which are used as the target-specific probes. Several nanobiosensors for the molecular diagnosis of food-borne pathogens and agro-terrorism agents can be found in recent studies. These detection methods include conductive polymer nanowires, carbon nanotubes, nanoporous silicon and gold nanoparticles.

The ability of gold nanoparticles to act as fluorescence quenchers was highlighted and could be used to solve major drawbacks in molecular biology experiments (Dubertret et al. 2001). For example, a DNA oligonucleotide could be synthesized, fluorescently labeled at its 5′ end and conjugated at the 3′ end with gold nanoparticles. These oligonucleotides can be used in diagnostic procedures, particularly in cases where DNA analysis cannot be circumvented, and in the diagnosis of the phytoplasma associated with the flavescence dorée (FD) of grapevine. The gold nanoparticles efficiently quench the fluorescence of light harvester polymers, such as polyfluorene, and will open new perspectives in the development of the optical performances of nanobiotransducers for diagnostic purposes. Furthermore, a diagnostic probe made of a specific oligonucleotide bearing a fluorescein at its 5′ end and 2-nm gold particles at its 3′ end acts as a nanobiotransducer in DNA hybridization. It produces a stronger fluorescence signal when hybridized to target DNA.

3.4.7 Nanostructured platform for mycotoxin detection

Nanostructured platform for mycotoxin detection and detoxification mycotoxins are secondary metabolites produced by pathogenic filamentous fungi causing disease and death in humans and other animals. Mycotoxicoses are diseases caused by the ingestion of mycotoxins by humans and animals, mostly through the consumption of polluted food. They can induce various different biological effects, i.e. carcinogenic, mutagenic, teratogenic, estrogenic, immunotoxic, nephratoxic and neurotoxic effects. Avariety of biosensors have been developed and described in the recent literature for mycotoxin analysis some of which have the potential of multi-array mycotoxins identification (Mohammad A. Khiyami et al., 2014). The main role of nanosensors is to decrease the time for fungal pathogen detection. Food spoilage can be detected by nanosensors such as an array of thousands of nanoparticles designed to be visualized in different colours in contact with food pathogens.

An immunosensor technique coupled with a flow injection system can be used for the quick, sensitive and discriminatory quantification of zearalenol in corn silage samples. (Panini et al., 2010). Horseradish peroxidase (HRP) biosensors have been applied without any pre-treatment to determine Ochratoxin A OTA in spiked beer samples, and OTA isolated from roasted coffee. Additionally, the performance of the HRP biosensors has been revealed yielding average revivals of 103% and 99%, respectively in both cases of OTA as mentioned earlier. The ozonization and adsorption efficiency of modified nanodiamonds to decrease the content of aflatoxin-B1 was examined. Silica and clays are most efficient in combination with smaller sized water molecules and smaller mycotoxins such as aflatoxins and ochratoxins. However, clays are less efficient in binding the larger mycotoxins such as fumonisin and deoxynivalenol (vomitoxin) because the distance among clay layers is not sufficient to accommodate the larger molecules. By using nano-sized clay, the space between the layers of clay has been prolonged 10 times. As a result, the nanoclay can bind the whole family of mycotoxins. A rapid enzyme-linked immunosorbent assay (ELISA) method was investigated by reducing the coating, blocking and competition time required in usual ELISA method using superparamagnetic nanoparticles. This method was effective for detecting aflatoxin M1 (AFM1) in milk in a linear working range of 4–250 ng/L.

A nanostructured cerium oxide film-based immunosensor was also developed for the detection of food-borne mycotoxins. A moveable machine has recently been developed that can concurrently identify various bacterial, fungal toxins and pathogens in stored food. Obviously biosensors can be an exciting alternative to the conventional techniques for the detection of mycotoxins and pathogens in food (Biswal et al., 2012).

3.4.8 Nanobiosensors for plant pathogen detection

Biosensor could be defined as a biological sensor i.e. a device which combines a biological component with a physicochemical detector component for the detection of an analyte. Recent advancement in the field of nanotechnology facilitates the material science-based approach to nanoscale synthesis of nanomaterials. According to the behaviour and role of biosensors, there are different types such as photometric biosensor, biological biosensors, electrochemical biosensors etc (Srivastava Priyanka et al., 2013).

Nanosensors are emerging as promising tools for the applications in the agriculture and food production. They offer significant improvements in selectivity, speed and sensitivity compared to traditional chemical and biological methods (Enisa Omanović-Mikličanina et al., 2016). Nanosensors can be used for determination of microbes, contaminants, pollutants and food freshness (Joyner and Kumar 2015). The nanosensors used in food analyses combine knowledge of biology, chemistry and nanotechnology and may also be called nanobiosensors.

Characteristics for an ideal nanobiosensor : An ideal nanobiosensor should be, highly specific for the purpose of the analyses i.e. a sensor must be able to distinguish between analyte and any "other" material ; stable under normal storage conditions; specific interaction between analytes should be independent of any physical parameters such as stirring, pH and temperature ; reaction time should be minimal; the responses obtained should be accurate, precise, re-producible and linear over the useful analytical range and also be free from electrical noise ; the nanobiosensor must be tiny, biocompatible, non-toxic and non-antigenic ;should be cheap, portable and capable of being used by semi-skilled operators (Manju Kumari Choudhary et al., 2015).

There is strong interest for developing new biosensing systems for early detection of plant diseases with high sensitivity and specificity at the point-of-care. Mohga Khater et al., (2017) revised here the recent advancement in the development of advantageous biosensing systems for plant pathogen detection based on both antibody and DNA receptors. The use of different nanomaterials such as nanochannels and metallic nanoparticles for the development of innovative and sensitive biosensing systems for the detection of pathogens at the point-of-care is also shown. Plastic and paper-based platforms have been used for this purpose, offering cheap and easy-to-use really integrated sensing systems for rapid on-site detection.

Biosensor platforms based on nanomaterials

Recent breakthroughs in nanotechnology enable the preparation of various nanoparticles and nanostructures with few technical hurdles. Nanoparticles display fascinating electronic and optical properties and can be synthesized using different types of materials for electronics and sensing applications (Yi Fang and Ramaraja P. Ramasamy, 2015). For biosensing application, the limit of detection and the overall performance of a biosensor can be greatly improved by using nanomaterials for their construction. The popularity of nanomaterials for sensor development could be attributed to the friendly platform it provides for the assembly of bio-recognition element, the high surface area, high electronic conductivity and plasmonic properties of nanomaterials that enhance the limit of detection. Various types of nanostructures have been evaluated as platforms for the immobilization of a biorecognition element to construct a biosensor. The immobilization of the biorecognition element, such as DNA, antibody and enzyme, can be achieved using various approaches including biomolecule adsorption, covalent attachment, encapsulation or a sophisticated combination of these methods. The nanomaterials used for biosensor construction include metal and metal oxide nanoparticles, quantum dots, carbon nanomaterials such as carbon nanotubes and graphene as well as polymeric nanomaterials.

Nanoparticles have been utilized with other biological materials such as antibody for detecting *Xanthomonas axonopodis* that causes bacterial spot disease. Gold nanoparticle-based optical immunosensors have been developed for detection of karnal bunt disease in wheat using surface plasmon resonance (SPR). In addition to single probe sensors, nano-chips made of microarrays which contain fluorescent oligo probes were also reported for detecting single nucleotide change in the bacteria and viruses with high sensitivity and specificity based on DNA hybridization (López et al., 2009). Fluorescent silica nanoparticles (FSNPs) combined with antibody as a biomarker have been studied as the probe, which successfully detected plant pathogens such as *Xanthomonas axonopodis* pv. *vesicatoria* that causes bacterial spot disease in Solanaceae plants.

Quantum dots (QD) have also been used for biosensor construction for disease detection. Due to their unique and advantageous optical properties, they have been used for disease detection using fluorescence resonance energy transfer (FRET) mechanism, which describes energy transfer between two light-reactive molecules. Many QD-FRET-based sensors have been developed for phytodisease detection such as the witches' broom disease of lime (WBDL) caused by *Candidatus Phytoplasma aurantifolia* (*Ca P aurantifolia*). The immunosensor constructed showed a high sensitivity and specificity of 100%, and a detection limit of 5 *ca P aurantifolia*/µL. Additionally, QD-FRET was also reported to detect the disease vectors (Yi Fang and Ramaraja, P.Ramasamy, 2015.). For example, Rhizomania, which is the most destructive disease in sugar beet, is caused by beet necrotic yellow vein virus (BNYVV). *Polymyxa betae* (Keskin), the only known vector of BNYVV, for transmission of the virus to the plants was successfully reported to be detected by QD-FRET- based sensor.

In addition to QD-based sensors, the use of other novel materials for sensor fabrication have been explored in order to attain high sensitivity and low limits of detection. Gold nanoparticles are widely used nanomaterials due to their high electroactivity and electronic conductivity for electron transfer. Recently, nanomaterial-based electrochemical sensors have been reported for plant disease detection by our group. The application of gold nanoparticle (AuNP) modified electrode has been reported by our group for the electrochemical detection of methyl salicylate, a key plant volatile organic compound released by plants during infections. Moreover, in addition to gold nanoparticles, semiconductive metal oxide nanoparticles have also been reported for VOC detection due to its advantages such as low cost, suitability for electron conduction for amperometric signal and the ease at which to obtain a desired size and shape. In addition to the detection of plant VOC which is indicative of a particular disease, nanoparticles can also be used for detection of compounds released by the pathogen itself. Different types of phytopathogens, phytobacteria, viruses and fungi have been detected through nanoparticle based amperometric biosensors.

3.4.9 Nanowire as Biosensor Transducer

Nanowire biosensors are a class of sensors or measuring devices, of which the major sensing components are made of wires or fibers in nanometer scale formed by biological molecules, such as DNA molecules, polypeptides, fibrin proteins, and filamentous bacteriophages, etc. (Zhang et al., 2013) A bionanowire is a one-dimensional fibril-like nanostructure, with the diameter constrained to tens of nanometers or less and unconstrained length.

Nanowires are considered as key nanomaterials because of their electrical controllability for accurate measurement, and chemical-friendly surface for various sensing applications (Zongjie Wang, et al., 2016). Advances in the performance and functionality of nanowire sensors are also surveyed to highlight recent progress in this area. These advances include the improvements in reusability, sensitivity in high ionic strength solvent, long-term stability, and self-powering. Overall, with the advantages of ultra-sensitivity and the ease of fabrication, it is expected that nanowires will contribute significantly to the development of biological and medical sensors in the immediate future.

Among recent research advances exploiting new nanomaterials for biomolecule analysis, silicon nanowires (SiNWs), which are configured as field-effect transistors (FETs), have emerged as one of the most promising and powerful platforms for label-free, real-time, and highly-sensitive electrical detection of proteins, as well as, many other biological species (Gengfeng Zheng and Charles M. Lieber, 2013). A detailed protocol was described for realizing SiNW biosensors for protein detection that includes SiNW synthesis, FET device fabrication, surface receptor functionalization, and electrical sensing measurements. Moreover, incorporating both p-type and n-type SiNWs in the same sensor array provides a unique means of internal control for sensing signal verification. The nanowire based biosensor application is a most reliable sensor nowadays because of advantages towards detecting biological molecule especially plant diseases (Shahrul et al., 2013).

Biosensors could be a thrilling alternative to the conventional systems for the detection of toxins and pathogens. Biosensors are reliable which allow the specific detection of target analyte at minimum cost. Recent advancement in the field of nanotechnology has been used to develop highly sensitive bio-sensor's chips by using multidisciplinary approach (Srivastava Priyanka, et al., 2013).

The QD-based Fluorescence Resonance Energy Transfer (FRET) genosensor, most widely used in the biomolecule detection fields, and QD-based nanosensor for Rev-RRE interaction assay were presented as examples by Sungyeap Hong and Cheolho Lee (2018). A QCM-based genosensor was reported for the detection of *Cymbidium Mosaic Potexvirus* (CymMV) and *Odontoglossum Ringspot Tobamovirus* (ORSV), a number of biosensors for the detection of plant viruses combining various bioconjugates with different interface elements have been developed.

The SPR biosensors have been extensively studied as biosensors for the detection of plant viruses; the potato virus Y (PVY), apple stem pitting virus (ASPV), maize chlorotic mottle virus (MCMV), barley stripe mosaic virus (BSMV) and cowpea mosaic virus (CPMV). Citrus Tristeza Virus (CTV) was the most extensively researched. The sensor, labeled with a CTV coat protein-derived antibody on the CdTe quantum dot, is a representative example (Sungyeap Hong and Cheolho Lee, 2018). In fact, using antibody derived from CTV coat protein, alternative types of biosensors were reported to enhance the specificity and sensitivity of fluorescence resonance energy transfer-based biosensor.

There is a growing urgency for highly sensitive and selective detection techniques, usable for a wide number of applications, including disease diagnosis as well as preventive measures. Nanobiosensors represent a new promising tool for virus detection (Krejcova et al., 2015). This review gives a brief survey of the issue of viral detection, comprising diagnostics of target structure of viruses such as nucleic acids or proteins. This review covers different detection principles, methods of fabrication, and applications of virus biosensors.

To detect pathogens and toxins in food, fibre-optic biosensor has commonly used tapered fibre-optic probes coated with antibodies. An approach has been developed for rapid determination of *Escherichia coli* using a flow-injection system. Electrochemical measurement of K3Fe (CN)6, reduced by microbial metabolism, allowed the quantitative determination of bacteria and fungi in 20 minutes. A new biosensor system for the rapid diagnosis of soil-borne diseases, consisting of two biosensors was developed (Hashimoto et al., 2008). The system was constructed using equal quantities of two different microbes, each individually immobilized on an electrode.

Fluorescent silica nanoparticles (FSNP) combined with antibody molecules successfully detected plant pathogens such as *Xanthomonas axonopodis* pv. *vesicatoria* which causes bacterial spot disease in tomatoes and peppers. Copper oxide (CuO) nanoparticles and nanolayers were synthesized by sol–gel and spray pyrolysis methods, respectively. Both CuO nanoparticles and nanostructural layer biosensors were used for detecting the *A. niger* fungi.

Various nano-inspired biosensors have been reported that range from detection of plant infections (fungal, viral and bacterial), abiotic stress, metabolic content, phytohormones, mi RNAs, genetically modified (GM) plants to transcriptional and genetically encoded biosensors in very short span of time (Vinay Kumar and Kavita Arora, 2020). Considering nano-inspired-biosensors based applications, these are expected to play a major role in combating growing challenges in the form of plant disease detection, management and genetic transformation for plant biotechnology, agriculture and food production enhancement. A novel microbial detection technology based on NPs is being developed. Silica-based NPs (60 nm) were filled with a fluorescent dye and conjugated to an antibody specific to a surface antigen of the microbe of interest. Detection of a single bacterial cell was possible using this technique (Zhao et al., 2004). This method has potential for sensitive detection of plant pathogens.

Nanoscale "Smart" biosensors can provide security in food manufacturing, processing, and shipping of products through the detection of pathogens and contaminants. These sensory systems detect biological molecules such as sugars or proteins in foods to detect pathogens and other contaminants (Amarendra N. Misra et al., 2013). Nanosensors are small, portable, sensitive with real-time monitoring, specific, quantitative, reliable, accurate, reproducible, robust and stable which can overcome the deficits of present sensors.

Imaging techniques, such as thermographic and hyperspectral imaging, have been used in the field for indirect detection of plant disease (Zheng Li et al., 2020). However, many of them have notable limitations, such as susceptibility to parameter changes of the environment and lack of specificity for disease subtypes or strains. Rapid progresses in chemo- or biosensing technology have led to a broad range of successful applications, such as quality assessment of medicinal and industrial products. Recently, several nanoparticle-based chemo- or biosensors have been proposed and commercialized for agricultural diagnosis. Depending on transduction mechanisms of the designed sensory interactions, the analytes could be recognized by characteristic optical or electrical outputs of the nanosensors. The sensor's detection specificity could be enhanced by selective chemical interactions or by the use of biospecific recognition elements such as DNA oligos, antibodies, aptamers, and enzymes. The detection sensitivity could be improved by the use of surface-enhanced optical properties (e.g., surface plasmon resonance (SPR)) or electron-conductive nanoscale substrates such as carbon-based nanomaterials (e.g., graphene or carbon nanotube) as transducers. In this section, we will highlight three main classes of nanosensors or nanobiosensors, including metal or metalloid nanoparticles, quantum dots, and array-based nanosensors. A list of representative nanosensors or biosensors used for plant disease diagnosis, their detection mechanisms, and associated performance is summarized in table 7.

Table 7: Summary of nanoparticle-based sensors for plant pathogen or disease marker detection

Target	Sensor component	Sensor Fabrication
Pseudomonas syringae	Au NP-ssDNA	Nanoparticle functionalization and DNA hybridization
P. ramorum and *P. lateralis*	Ag NP-ssDNA	Nanoparticle functionalization and DNA hybridization
Aflatoxins	Ag NR	Surface functionalization
Bacillus thuringiensis and *B. subtilis*	Pt NP-IgG	Surface immunological functionalization
Xanthomonas campestris	Si NP-Rubpy-IgG	Surface immunological functionalization
Ralstonia solanacearum	Au NP-ssDNA	Surface functionalization
Fusarium oxysporum	CdSe/ZnS-MPA[f]	Surface functionalization
Verticillium mail	CD	Surface functionalization
F.avenaceum	CD	Surface functionalization
Citurs citriza	CdTe QD-CD	Surface immunological functionalization

Nanosensors with immobilized bioreceptor probes selective for target analyte molecules are termed as nanobiosensors. These are very sensitive, precise, quantitative, accurate, reproducible and robust, consequently can be employed to detect and quantify minute amounts of contaminants such as viruses, bacteria, fungi, toxins and other biohazardous substances in the agriculture and food systems (Kashyap et al., 2017). These sensors can be linked to GPS and distributed throughout the field for real-time monitoring of disease, soil conditions and crop health in precision farming. Nanosensors will allow us to identify plant diseases before visible symptoms appear and thus facilitate better disease management and enhance agriculture production and productivity.

The nanomaterials transduce the chemical binding event on their surface lead to change in conductance of the nanowire in an extremely sensitive, real-time and quantitative fashion. By employing this concept, nanowire was used as biosensor transducer for the detection of Cucumber Mosaic Virus (CMV) and Papaya Ring Spot Virus (PRSV). These results indicated that the nanowires are a good candidate material for fabricating nanoscale biosensors for making remote-controlled nanbiosensors for future applications in crop health care testing, disease diagnostics and environmental monitoring.

Advantages of nanobiosensors : Measurement of more variables , Greater sensitivity, Simple to use and less sample material required, Faster detection rates, Portable, Multi-analyte and high throughput, Works at atomic scale with highest efficiency and increased surface to volume ratio (Manju Kumari Choudhary et al., 2015).

Disadvantages of nanobiosensors : Nano-scale quantification is still lacking and nanobiosensors are still under infancy stage.

3.4.10 Array-based nanosensors

Among the state-of-the-art nanosensors, one format capable of multiplexing and discrimination of different analytes is the array-based sensor assemblies composed of several chromophores or synthetic nanomaterials as an analog to the mammalian olfactory system. A common form of the array-based sensor is also known as the "electronic nose" (e-nose), which incorporates electronic transducers instead of chemical sensors in the array (Zheng Li et al., 2020). The array-based approach is particularly beneficial in distinguishing highly similar analyte mixtures due to its cross-reactivity and ability to measure molecular fingerprints. The multidimensional output data of sensor arrays can be analyzed by supervised or unsupervised chemometrics, which provides highly precise and predictive results.Alternatively, array-based chemical sensors can provide optical readout (e.g., colorimetric or fluorescent), which is easier to analyze and interpret compared with e-nose signals. In addition, chemical sensor arrays are easy to fabricate, based on specific chemical interaction, which brings in much better chemical specificity, extremely cost-effective, and more resistant to environmental interference such as humidity and temperature than most e-nose devices. Chemical sensor arrays can be easily prepared by embedding various analyte-responsive dyes such as Lewis acid-base colorants, Brønsted acidic or basic colorants, solvatochromic or vapochromic dyes, and redox indicator colorants in hydrophobic nanoporous substrates such as modified silica sol–gels.

E-noses made of metal oxides or conductive polymer coatings have drawn considerable attention as multiplexed gas sensors for tracing VOC biomarkers in plant stress events ranging from pathogen infection, pest invasion, to physical wounding. Particular emphasis is placed on the detection of ethylene, one of the most important phytohormones, using chemiresistive or colorimetric method. Of special interest, is the design of a nanoplasmonic sensor array comprising AuNPs and a molecularly imprinted sol–gel (AuNPs@ MISG) for selective detection of terpenes.

Recently, a smartphone-integrated VOC sensing platform was developed that utilized plasmonic nanoparticles for early detection and differentiation of tomato late blight (within 2 days of infection) from the other fungal pathogens that lead to similar symptoms of tomato foliage. The gold nanoparticles were functionalized with chemical ligands that were reactive to specific leafy VOCs, and printed on paper strips to generate unique patterns of color changes for each VOC. The results were then scanned by a smartphone microscope for digital pattern recognition and differentiation. The described sensor platform will be not only useful for fingerprinting plant pathogens of interest, but also well-suited for distinguishing various abiotic stresses, such as mechanical damages, drought, and nutrition deficiency. Future research may be focused on improving sampling protocols that do not require pre-collection and concentration of VOCs prior to data analysis.

3.4.11 Nanostructured platforms-based diagnostics

The development of single-molecule sequencing technology has brought about impressive advancements in the detection accuracy and sensitivity of probing DNA structure variations for both clinical and agricultural studies (Zheng Li et al., 2020). As a third-generation sequencing (TGS) technique, the nanopore sequencing platform uses a motor protein to transport a single- stranded DNA or RNA through either a protein or a solid-state nanopore to generate nucleotide-dependent electronic current signals which enables facile analysis of pathogenic polynucleotides. Compared to earlier generations of sequencing techniques, nanopore sequencing platform takes advantages of real-time and long reads of sequence data, fast running times, scalable genome mapping, and small sample loadings.

A large number of latest studies using nanopore sequencing platforms have involved in the plant pathogen diagnosis. For example, Chalupowicz et al. (2019) have successfully set up a standard protocol using a handheld sequencing system developed by Oxford Nanopore Technologies (named "MinION") for diagnosis of different plant bacteria, viruses, fungi, and phytoplasma, including *P. digitatum* in lemon and *S. lycopersicum* in tomato. The total assay time is within 2 h, and the results are comparable to conventional diagnostic procedures (e.g., PCR and ELISA). Despite some challenges remaining in the present equipment such as high per-read error rate and poor discriminatory ability among similar sequences, continuous breakthroughs in nanopore technology will lead to the creation of more powerful sequencing platforms.

Advancementofnanotechnologyandbiotechnologyhasexpandedtheutilization of nanostructure as a novel sensing platforms, owing to their ultrahigh surface area-to-volume ratio, size dependent electrical properties and possibility of device miniaturization (Kashyap et al., 2017). Nanomaterials, such as carbon nanomaterials (carbon nanotubes and graphene), nanowires, nanocomposites, and nanostructured metal oxide nanoparticles are being used in the designing of nanosensing platforms for pathogen and mycotoxin determination. Other types of nanostructure platforms are based on microfluidics systems, which can also be used to detect pathogens efficiently in real time and with high sensitivity. A major advantage of such systems is their miniature format and their potential to detect compounds of interest in minute sample volumes with rapidity.

An array of thousands of nanoparticles to be visualized in different colours in contact with food pathogens was designed. NanoSiO_2 and chitosan-based nanobiocomposite film was fabricated on an indium tin oxide (ITO) substrate to co-immobilize r-IgGs and BSA for ochratoxin-A (OTA) detection. Horseradish peroxidase (HRP) biosensors have also been applied without any pre-treatment to determine OTA (*A. ochraceus* and *Penicillium viricatum*) in spiked beer samples and roasted coffee. Magnetic nanoparticleswere used to improvise electrochemical immunosensor for the detection of ultratrace quantities of aflatoxin-M1 (AFM1 (up to 0.01 ppb) produced by *Aspergillus flavus* in foodstuffs. A 'lab-on-chip' strategy integrating an electrokinetic magnetic bead-based electrochemical immunoassay on a microfluidic chip was developed for the quick, sensitive and discriminatory quantification of zearalenone produced by *Fusarium* sp.

Silica and clays are most efficient in combination with smaller-sized water molecules and smaller mycotoxins such as aflatoxins and ochratoxins. However, clays are less efficient in binding the larger mycotoxins such as fumonisin and deoxynivalenol (vomitoxin) because the distance among clay layers is not sufficient to accommodate the larger molecules. By using nanosized clay, the gap between the layers of clay has been prolonged ten times. As a result, the nanoclay can bind the whole family of mycotoxins. A rapid enzyme-linked immunosorbent assay (ELISA) was improvised by using superparamagnetic nanoparticles. The assay was effective for detecting aflatoxin-M1 (AFM1) with a detection limit of 4-250 ng l^{-1}. Cysteamine-functionalized gold nanoparticles (C-AuNP) along with aflatoxin-B1 antibodies (aAFB1) were immobilized on a 4-mercaptobenzoic acid-based self-collected monolayer on a gold electrode (MBA/Au) to fabricate a BSA/aAFB1-CAuNP/MBA/Au immunoelectrode. These electrodes were used to detect AFB1 in the range of 10–100 ng l^{-1} . Recently, a moveable machine has been developed that can concurrently identify various bacterial, fungal toxins and pathogens in stored food. From all these reports, it seems that nanostructured platforms can be an exciting alternative to the conventional techniques for the detection of mycotoxins and pathogens spoiling food and agricultural crops.

With recent advances in the technology of micro-electro-mechanical systems (MEMS), more and more implantable and wearable electronics using conductive nanomaterials have emerged as sensor components for long-term and on-demand monitoring of plant VOCs or other biomarkers. Plant wearable becomes a new frontier of crop diagnostics, which relies on ultrathin and ultra-lightweight nanosensor design to attach flexible sensor devices directly on plant tissues such as leaves for continuous monitoring. A roll-to-roll fabrication method of a graphene-based wearable sensor was developed that can monitor water evaporation from plant leaves (Zheng Li et al., 2020).

3.4.12 Nano-chips

Nano-chips were identified as a type of microarrays containing oligo capture probes capable of detecting the hybridization. Nanochips can detect single nucleotide changes of bacteria and viruses with high sensitivity and specificity. A successful attempt was made to detect *Xanthomonas axonopodis* pv. *vesicatoria*, which causes bacterial spot disease in Solanaceae plants, by using fluorescence silica nanoparticles in conjunction with antibodies (Ammar Ahmad, 2018). Moreover, Karnal bunt disease detection in wheat was accomplished by using nano-gold based immunesensors and Surface Plasmon Resonance (SPR).

Using physiologically active, complex, polyfunctional, multicomponent nano systems, nano chips in the pre-sowing rice treatment nanotechnology allows precise seeding; enhancing field seed germination; reducing plant morbidity rate considerably or even deleting disease incidents; increasing plant adaptability to unfavorable environmental conditions; expanding crop yield; improving products quality by not using toxic chemical means of plant protection; intensifying competitive ability of agricultural products as well as assuring environmental safety in crop cultivation regions (Voropaeva et al., 2014). Nutrition micro- and mesoelements, as also intense activity plant growth regulators and other physiologically active substances, depending on the crop variety, cultivation conditions and possible risks, can be added to such pre-sowing treatment (nano) technology.

3.4.13 Nanofabrication techniques

Nanofabrication techniques have been used in creating artificial plant parts such as stomata and xylem vessel which are then used to study the infection process and behavior of pathogens inside host plant for example, *Uromyces appendiculatus* (fungus causing rust disease of bean), *Colletotrichum graminicola* (fungus causing anthracnose in corn) and *Xylella fastidiosa* (xylem limited bacterium causing Pierce's disease of grapevine) (Meng et al., 2005). Such study would help breeder to look for specific stomatal characters to prevent entry inside host through stomata or leaf surface characters to prevent appressorium formation before penetration of fungi or vascular characters to obstruct the movement of vascular pathogens like bacteria, fungi etc. In other words, it would help formation of proper breeding strategy to screen for or to develop disease resistant crop plants.

The advancement in nanofabrication and characterization tools have enabled studies of physical, chemical and biological interactions between plant cell organelles and various disease causing pathogens. A better understanding of plant pathogenic mechanisms such as flagella motility and biofilm formation will lead to improved treatment strategies to control the diseases and protect production (Cursino et al., 2009). For example, spatial and temporal studies of plant pathogenic xylem inhabiting bacteria have traditionally been conducted by monitoring changes in bacterial populations through destructive sampling techniques of tissues at various distances from inoculation sites. This approach seriously limits the information that can be obtained regarding colonization, biofilm development, and subsequent movement and re-colonization at new areas, primarily because the same region or sample site cannot be followed temporally. Micro-fabricated xylem vessels with nano-size features have been shown very useful in gaining an appreciation of the mechanisms and kinetics of bacterial colonization in xylem vessels such that novel disease control strategies may be developed (Hongda Chen and Rickey Yada, 2011).

Surface engineering techniques for cellular micropatterning are emerging as important tools to clarify the effects of the microenvironment on cellular behavior, as cells usually integrate and respond the microscale environment, such as chemical and mechanical properties of the surrounding fluid and extracellular matrix, soluble protein factors, small signal molecules, and contacts with neighboring cells. Furthermore, recent progress in cellular micropatterning has contributed to the development of cell-based biosensors for the functional characterization and detection of drugs, pathogens, toxicants, and odorants. In this regard, the ability to control shape and spreading of attached cells and cell-cell contacts through the form and dimension of the cell-adhesive patches with high precision is important. Commitment of stem cells to different specific lineages depends strongly on cell shape, implying that controlled microenvironments through engineered surfaces may not only be a valuable approach towards fundamental cell-biological studies, but also of great importance for the design of cell culture substrates for tissue engineering (Otsuka, 2010).

The rapidly expanding use of diagnostic imaging technologies continues to push the boundaries of plant pathologist diagnostic capabilities. Using these techniques, plant doctors are able to diagnose crop diseases earlier and more precisely. Nanotechnology offers us the opportunity to precisely tune and control the chemical and physical properties of contrast materials in order to overcome concerns with toxicity, useful imaging time, tissue specificity and signal strength. In the 'mesoscopic' size range of 5–100 nm diameter, nanoparticles also have large surface areas and functional groups for conjugating to multiple diagnosis tools (Mohammad A. Khiyami et al., 2014). Thus, progress in the field of nano-scale contrast agents will play a key role in the continued enhancement of our diagnostic imaging capabilities in the coming years.

For instance, electron beam and photolithography techniques were used to fabricate topographies that mimic leaf surface features as well as the internal plumbing of plants, and then nano-imaging technologies were used to study how bacteria and fungi invade and colonize the leaf. Lithography was used to nanofabricate a pillared surface on silicon wafers. This lawn of miniature pillars was between 1.4 and 20 μm wide and spaced various distances apart. It was used to examine the movement across the surface by the fungus that mimicked some of the characteristics of the host plant. Images of the *Colletotrichum graminicola* crawling across the nanofabricated surface assisted the researchers to determine that the fungus needs to make a minimum contact of at least 4.5 μm before it starts to develop appressoria.

Nanophytopathology, hence, can be applied as a tool to understand plant–pathogen interactions, providing new methods for crop protection. Specific nanodevices and DNA nanodevices could enable accurate tracking, detection and diagnosis of plant pathogens in the early stages of plant disease. The portable diagnostic equipment, nanoparticle-based, bio-barcoded DNA sensor and the QD have potential applications in the multiple detection of plant pathogens and toxigenic fungi. To date, mobile diagnostic assays have been developed to rapidly detect plant disease and may be used to prevent epidemics. These nano-based diagnostic kits not only increase the speed of pathogen detection but also increase the accuracy of the diagnosis. Additionally, the combination of nanotechnology with microfluidic systems has been effectively applied in molecular plant pathology and can be adapted to detect specific pathogens and toxins.

The advent of new innovative nanotechnologies in fields of plant pathology, nano-fabrication, and characterization of tools have allowed researchers to easily understand physical, chemical, and biological interactions between plant cells and pathogens (Ammar Ahmad, 2018).

Spatial and temporal studies of plant pathogenic xylem inhabiting bacteria have been conducted by monitoring changes in bacterial populations through destructive sampling techniques of tissues at various distances from inoculation sites (Hoch, 2007). The development of several microfabricated fluidic devices was completed in which various xylem-inhabiting plant pathogenic bacterial species were studied. The devices were constructed from molded polymerized polydimethylsiloxane (PDMS) sandwiched between glass microscope slides and coverglasses fitted with fluidic access ports.

3.4.14 Smart Nano-dust

Smart dust' technology is an idea that thousands of tiny sensors could be scattered like invisible eyes, ears and noses across the farm fields and battle fields. Autonomous nanosensors were developed that are in the size of a match-stick head (Vedpriya Arya, 2019). Using silicon-etching technology these 'motes' or (smart dust sensors) would feature onboard power supply computational abilities and the ability to detect and then communicate with the other motes in the vicinity. In this way, the individual motes would self organize into computer networks and capable of relaying data using wireless technology. The new radio telescope of the LOFAR (Low Frequency Array) project is based on tens of thousands of antennas that are connected to each other with a large ICT infrastructure.

Uniform scattering and distribution of innumerable number of nano-sensors over farmlands like dust will act in the future as smart eyes, ears, and noses of farming industrial complex. Through utilization of smart dust, many parameters such as temperature, humidity, nutrient deficiencies, insect infestations, and disease infestations can be sensed, detected, and monitored. Smart nano-dust particles or nano-sensors, which are connected to global positioning system (GPS), would be able to operate autonomously, communicate wirelessly information they collect, and sense in real time as well as respond rapidly to any variations in aforementioned parameters (Ammar Ahmad, 2018). This complex intelligence network of smart particles or sensors can warn farmers in advance to devise ways to deal with any aberrations in environment. Additionally, employment of smart dust systems can help farmers evaluate the amount of pollutants in environment.

3.4.15 Phage Biosensors

Biosensors that use phages or bacteriophages as biorecogntion elements are called phage biosensors, and are environmentally more robust than immunosensors, low in cost, easy to produce, and widely used to detect pathogens. They contain viruses that infect bacteria and display peptides or proteins on their surface, that is, bacterial host strains, through the specific receptor molecules present on the surface of the bacterial cell. Phage biosensors can be stored for longer times with minimal loss of binding affinity (Rajesh Kumar Saini et al., 2017).

Application of nanomaterials adds further advantages to such systems, such as, further miniaturization, measurement of more variables, greater sensitivity, less sample material required, faster detection rates read- outs in real time and application of novel detection methodologies (e.g. electronic, colorimetric, fluorometric, and mass changes) (Vandana Ghormade et al., 2011).

Currently uni-molecular and array type of nanomaterial based biosensors are being developed for detection of pesticides.

A zirconium oxide NPs (~50 nm) based immunoassay was developed for sensitive detection (0.02 nM) of organophosphate pesticides using phosphorylated enzyme acetylcholinesterase as a potential biomarker (Wang et al. 2009). Guo et al. (2009) developed a gold NPs (40 nm) based portable one-step strip assay allowing simultaneous screening for two pesticides carbofuran and triazophos within a short time (8–10 min) without any equipment.

3.4.16 Aptamer-based Biosensors

Aptamers exhibit many advantages as recognition elements in biosensing when compared to traditional antibodies. They are small in size, chemically stable and cost effective (Shiping Song et al., 2008). More importantly, aptamers offer remarkable flexibility and convenience in the design of their structures, which has led to novel biosensors that have exhibited high sensitivity and selectivity. Recently, the combination of aptamers with novel nanomaterials has significantly improved the performance of aptamer-based sensors. In view of the unprecedented advantages brought by aptamers, it is expected aptamer-based biosensors to find broad applications in biomedical diagnostics, environmental monitoring and homeland security.

Selection technique may discover binding between specific nucleic acid sequences and nonspecific nucleic acid targets with great specificity and affinity. The newly developed technique is called SELEX, while the resulting DNA or RNA oligonucleotides are termed aptamers. Aptamers can be considered as nucleic acid analogues of Abs; they can bind with high affinity and specificity to a broad range of targets, such as small molecules, proteins viruses, or cells and they exhibit a great affinity for a great range of target analytes, which include metal ions, pathogenic microorganisms, and proteins. Another important property of aptamers is that they have numerous benefits over Abs, such as their reproducibility and accurate chemical production greater stability than Abs, no requirement of immunization of animal hosts ability to undergo reversible denaturation, easy modification with new functional groups without any effect on their activity for their immobilization by incorporating reporter molecules (e.g., fluorophores or enzymes) and functional groups, ability to bind to selected nucleic acids, and ability to work in extreme conditions, whereas Abs are only stable under physiological conditions and also have a higher- surface density of receptors (Rajesh Kumar Saini et al., 2017). Due to their many advantages, numerous aptamer-based biosensors have been developed for the detection of a wide range of targets. Currently, nucleic acid aptamers have been selected for more than 150 targets including small molecules, such as cocaine, aspartame, growth factors peptides, toxins, viral proteins, cells and bacteria.

The development of aptasensors for mycotoxins determination in food and agricultural products over the last 11 years was analyzed (Guo Xiaodong et al. 2020). In addition, challenges and future trends for the selection of aptamers toward various mycotoxins and aptasensors for multi-mycotoxins analyses were summarized.

A fast, sensitive and reliable electrochemical sensing platform for the detection of airborne soybean rust spores was presented (Vadim Krivitsky et al, 2021).

3.5 Nanodetection of Plant viruses

Recent advances in nanotechnology have led to a new class of biological detection method that combines magnetic enrichment and optical detection. Although the method has been investigated by some researchers, up to now there is little report on using UCNPs to detect the target. It is predicted that techniques involving novel nanoparticle labels may become one of the effective approaches for the screening of targets in the near future (Mingzhe Zhang et al., 2013). A simple and convenient method has been put forward to simultaneously detect multiple viruses in plant samples based on antibody-labeled UCNPs coupled with immunomagnetic separation.

A sensitive, specific and rapid method for the detection of three different kinds of plant viruses including tomato ringspot virus (ToRSV), bean pod mottle virus (BPMV) and arabis mosaic virus (ArMV) was demonstrated using novel upconversion nanoparticles (UCNPs) as a fluorescence marker coupled with immunomagnetic separation Magnetic nanoparticles (MNPs, ~100 nm) were coated with different antibodies were employed to capture and enrich the target viruses. Then antibody-conjugated UCNPs as signal probes were added to form sandwich complexes. This was followed by a fluorescence measurement using a 980 nm laser. This method not only avoids the difficulty in simultaneous detection of multiple independent organic fluorphores that require distinct excitation wavelengths, and auto-fluorescence of biological samples due to the higher-energy excitation of QDs but also amplifies detection signal by UCNPs-tags together with easy separation of samples by magnetic forces, demonstrating the potential to be used for detecting virus in the field of environmental safety and other fields.

Several nanomaterials such as quantum dots, carbon nanotubes, graphene oxide, silica, and metal nanoparticles are used as biosensors for the detection of the pathogenic virus (Priyom Bose, 2020). Several nanoparticles can be used in virus detection. The use of nanoparticles as a marker for the rapid and accurate detection of infectious agents in small sample volumes helps in early detection at an affordable cost.

Plant viruses such as Cucumber mosaic cirus, *Pantoea stewartii*, Plum pox vius, Prunus necrotic ring sopt virus, Citrus tristeza virus, Potato virus x are reported to be detected using immuno assays (Vinay Kumar and Kavita Arora, 2020). Optical DNA hybridization biosensors have also been reported that provided PCR independent visual nucleic acid detection offering simplicity, high sensitivity and quick results . Recently, a nano inspired biosensor, in the series of DNA based diagnostics, the localized surface plasmon resonance (LSPR) of gold nanoparticles (AuNPs) was reported to be used to develop a colorimetric nano-biosensing system to detect the unamplified Tomato yellow leaf curl virus (TYLCV) genome in infected plants.

A review enlists thirteen biosensors (Table 8) that were developed for plant viruses and brings in most of the information in one platform about use of quantum dots and magnetic nanoparticles through immunomagnetic separation and FRET (fluorescence resonance energy transfer) phenomina through surface plasmon resonance and optical measurements (Hong and Lee, 2018). This is interesting to note that most of the biosensors reported are immuno based that make use of optical modes of detection rather than electrochemical detection having detection limit reaching as low as to ng mL^{-1}.

These biosensors warrant implications of nano materials and their composites for improvisation in the performance of biosensors for plant.

Table 8: Surface plasmon resonance (SPR) and quantum dot (QD) based plant virus biosensors

Virus	Nano and biomolecular Conjugate	Limit of Detection	Mode of detection/ nanomaterial
PVY	Monoclonar antibody	0.31 mg/mL	SPR
MCMV	Anti-MCMV antibody	1 ng/mL	SPR
BSMV	Specific oligonucleotide from RNA	14.7 ng/mL	SPR
ASPB	DNA aptmer from coat protein	500 mg/mL	SPR
CPMV	Monoclonal antibody	16 mg/mL	SPR
CPMV	Surface immobilized CPMV CdSc-ZnS core	–	FRET
CaMV	23 mer derived from CaMV 35S PbS nanoparticle	4.38×10^{-12} mol/L	FRET
CTV	Antibody to CTV coat protein-InP	2 nM for antibody	Electrochemcial
CTV	CTV-CP antibody-CdTe	220 ng/mL	electrochemical
CTV	AuNPs-CTV-CP/Ads-CTV-CP antibody, AuNP/QD	130 ng/mL	FRET
CTV	CTV-CP antibody-CdTe	198 ng/mL	FRET
GVA	Grapevine virus A type proteins ZnO films	1 pg/mL – 10 ng/ mL	FRET
ToRSV, BPMV & ArMV	Antibody, Fe_2O_4/SiO_2 MNPs and SiO_2/UCNPs (upconversion nanoparticles)	100 ng/mL	FRET

Plant virus-based systems, in particular, are among the most advanced and exploited for their potential use as bioinspired structured nanomaterials and nano-vectors. Plant viruses have a size particularly suitable for nanoscale applications and can offer several advantages. In fact, they are structurally uniform, robust, biodegradable and easy to produce. Moreover, many are the examples regarding functionalization of plant virus-based nanoparticles by means of modification of their external surface and by loading cargo molecules into their internal cavity. This plasticity in terms of nanoparticles engineering is the ground on which multivalency, payload containment and targeted delivery can be fully exploited. (Lico, et al., 2013).

Cowpea mosaic virus (CPMV) is a plant virus that has been developed for multiple biomedical and nanotechnology applications, including immunotherapy (Chao Wang et al., 2019). Two key platforms are available: virus nanoparticles (VNPs) based on the complete CMPV virion, including the genomic RNA, and virus-like nanoparticles (VLPs) based on the empty CPMV (eCPMV) virion. CPMV particles were uniquely able to boost populations of potent antigen- presenting cells, such as tumor-infiltrating neutrophils and activated dendritic cells. These results will facilitate the development of CPMV and eCPMV as immunotherapeutic vaccine platforms with tailored.

Rapid detection of pathogens is crucial to minimize adverse health impacts of nosocomial, foodborne, and waterborne diseases. Gold nanoparticles are extremely successful at detecting pathogens due to their ability to provide a simple and rapid color change when their environment is altered. Mohit, S. Verma et al., (2015) reviewed general strategies of implementing gold nanoparticles in colorimetric biosensors, highlighting how gold nanoparticles have improved conventional genomic analysis methods by lowering detection limits while reducing assay times, later focused on emerging point-of-care technologies that aim at pathogen detection using simpler assays. These advances will facilitate the implementation of gold nanoparticle-based biosensors in diverse environments throughout the world and help prevent the spread of infectious diseases.

Vasily G. Panferov, et al., (2018) developed a highly-sensitive lateral flow immunoassay (LFIA) for rapid detection of potato leafroll virus (PLRV), an infectious agent of one of the most widespread potato diseases. The proposed LFIA combines the formation of sandwich immune complexes with gold nanoparticles (GNP) as labels and silver enhancement. The enhancement stage was realized using mixture of silver lactate and hydroquinone and subsequent addition of chloride- containing buffer to stop silver reduction. LFIA with silver enhancement was 15 times more sensitive (detection limit 0.2 ng/mL; 15 min) compared with conventional LFIA (detection limit 3 ng/mL; 10 min). The enhanced LFIA detected PLRV in leaves' extracts of infected potato in dilutions higher than enzyme-linked immunosorbent assay.

3.6 Nanodetection of fungal infections

Fungal infections in plants amount to 80% of infections cuased in plants. Fungi are group of versatile eukaroytic carbon-heterotropic organisms that can maintain symbiotic/saparophytic/pathogenic relation with plants (Vinay Kumar and Kavita Arora, 2020). Fungi act as decomposers and recyclers of organic materials while each plant is attacked by fungal pathogens, however, only few lead to infections. Damage caused by fungal pathogens to plants is tremendous therefore, timely detection of fungal infections is extremely important. Potato blight, downy mildew of grapes and bushels of wheat are some most common examples of fungal infections in crops.

Pseudocerocospora fijiensis is a hemibiotrophic fungus which is known to infect banana plants and leads to Black Sigatoka (leave streak disease) which leads to decrease in productivity and quality of banana. A surface plasmon resonance based immunosensor was reported to detect the desired fungus in real leaf extract samples using polyclonal antibody against cell wall protein HF1 of *P. fijiensis* immobilized onto gold chip in lateral flow assay giving linear response for HF1 in 39.1 to 122 μg per mL with detection limit 11.7 μg per mL, sensitivity 0.0021 units reflectance per ng per mL.

Yellow rust disease in wheat is caused by fungus *Puccinia striiformis* which may impact the quality and production by almost 100%. For timely detection of infection at initial stage of crop development is extremely important from agriculture point of view (Mahlein et al., 2013).

Most important potato infection causing fungi include *Rhizocotonia solani, Spongospora subterranea, Synchytrium endobioticum, Alternaria solani, Collectorichum coccodes, Fusarium* spp etc. Disease diagnosis of six fungal pathogens were reported in easy and convenient qPCR based microarray in open microreactors (6 × 8) that offered multiplex field diagnostics of six fungal pathogens of potato crops viz *A. solani, A. alternata, R. solani, C. coccodes, S. subterranea, Fusarium* (Luna-Moreno, et al., 2019). This method offered detectionl limit that varied between 0.6 and 43.5 pg of DNA of target species with standard deviation of 0.3–0.7%.

Late blight in potatoes and tomatoes caused by *Phytophthora infestans* was visually detected using a novel method that used integration of universal primer mediated asymmetric PCR with gold nanoparticle (AuNP)-based lateral flow biosensor upto 0.1 pg per μL genomic DNA (Zhan et al., 2018).

Fungal infections are very severe and can contaminate seed, soil, weeds while spreading through wind and water, therefore, availability of sensitive method is extremely important. Most of the reports for fungal detection mainly make use of gold nanoparticle and the detection limit is limited to sub pg mL^{-1} range. This is very less compared to existing biosensors for various applications. Mendes et al., (2009) developed a biosensor that can detect *Phakopsora pachyrhizi*, the causal agent of rust in soybean.

Micro needle patches: A new technique that uses microneedle patches to collect DNA from plant tissues in one minute, rather than the hours needed for conventional techniques was developed (Anonymous, 2019). DNA extraction is the first step in identifying plant diseases, and the new method holds promise for the development of on-site plant disease detection tools.

It is exciting to see the new application of microneedle patch technology in agriculture and plant science. The microneedle technique successfully extracted pathogen DNA from all field-collected infected tomato leaves in a recent blind test. DNA extraction has been a significant hurdle to the development of on- site testing tools.

3.7 Nanodetection of bacterial infections

Most of the bacteria are generally known to be associated with plants are beneficial or saprotropic i.e., do not harm the cells, however, there are approximately 100 known bacterial species that are known to cause disease in plants belonging to tropical or subtropical regions (Vinay Kumar and Kavita Arora, 2020). Some common examples include *Agrobacterium, Erwinia, Proteobacteria, Pseudomonas, Xanthomonas* etc whose infection leads to loss of production and may result in symptoms such as galls and overgrowths, wilts, leaf spots, specks and blights, soft rots, as well as scabs and cankers (Lertanantawong et al., 2018) Among the visible symptom based plant disease diagnosis, a lateral flow immunoassay (LFIA) based test strips were reported to detect *Ralstonia solanacearum* which is a quarantine organism that causes potato brown rot using enlarging gold nanoparticles (GNPs), antibodies against the pathogen and tetrachnoroauric (III) anion reduction (amplification solution). This method reported detection limit upto 3×10^4 cells per mL in potato tuber extracts in 3 min for GNP size enhancement and 15 min including sample preparation time.

An electrochemical method has been described for detection of endophytic bacteria infected (through roots) Agave tequilana plants using electrochemical detection of H2O2 produced by leaves of host plant tissues to destroy the bacterial infection (Lima et al. 2018).

An acoustic based biosensor (the quartz crystal microbalance) that could sense three out of the most reported plant pathogenic bacteria, *Ralstonia solanaceraum, Pseudomonas syringae* pv. *tomato* and *Xanthomonas campestris* pv. *vesicatoria* (Papadakis et al., 2015).

Environmental monitoring and medical diagnosis recently improved by the development of novel sensors. The analytes may be detected using sensors exploiting electrical, chemical, electrochemical, optical, magnetic or vibrational signals (Valeria Scala et al., 2018). The sensitivity and specificity could be improved respectively by the use of nanomaterial matrices as transducers (such as metal and metal oxide nanoparticles, quantum dots, carbon nanomaterials) and by the use of bio-recognition elements (such as DNA, antibody, enzymes etc.). Various types of nanostructures have been evaluated as platforms for the immobilization of a bio-recognition element to construct a biosensor.

Developing quick and sensitive molecular diagnostics for plant pathogen detection is challenging. Herein, a nanoparticle based electrochemical biosensor was developed for rapid and sensitive detection of plant pathogen DNA on disposable screen-printed carbon electrodes (Lau et al., 2017). This 60 min assay relied on the rapid isothermal amplification of target pathogen DNA sequences by recombinase polymerase amplification (RPA) followed by gold nanoparticle-based electrochemical assessment with differential pulse voltammetry (DPV). This method was 10,000 times more sensitive than conventional polymerase chain reaction (PCR)/gel electrophoresis and could readily identify *P syringae* infected plant samples even before the disease symptoms were visible. On the basis of the speed, sensitivity, simplicity and portability of the approach, it is believed that the method has potential as a rapid disease management solution for applications in agriculture diagnostics.

Fluorescent silica nanoprobe as a biomarker for detection has attracted much attention in the field of nano-biotechnology recently but no further research applications using fluorescent silica nanoparticles (FSNP) combined with antibody molecules reported to detect pathogen detection (Yao et al., 2009). The fluorescence silica nanoprobe biomarker will have potential for rapid diagnosis applications on plant diseases.

3.8 Nanodetection of nematode infections

Abamectin (Abm) is a biopesticide with a strong activity against a wide variety of phytonematodes. However, Abm's poor mobility in the soil compromises its nematicide performance because of the limited zone of protection surrounding the growing root system of the plant. In a study, Abm's soil physical chemistry was manipulated by encapsulating Abm within the Red clover necrotic mosaic virus (RCNMV) to produce a plant virus nanoparticle (PVN) delivery system for Abm (Jing Cao et al., 2015). The transmission electron microscopic and dynamic light scattering characterization of Abm-loaded PVN (PVN^{Abm}) indicated the resultant viral capsid integrity and morphology comparable to native RCNMV. In addition, the PVN^{Abm} significantly increased Abm's soil mobility while enabling a controlled release strategy for Abm's bioavailability to nematodes. As a result, PVN^{Abm} enlarged the zone of protection from *Meloidogyne hapla* -root knot nematodes in the soil as compared to treating with free Abm molecules. Tomato seedlings treated with PVN^{Abm} had healthier root growth and a reduction in root galling demonstrating the success of this delivery system for the increased efficacy of Abm to control nematode damage in crops.

Measuring the temperature of objects at a nanometer-scale has been a long challenge, especially in living biological samples, because of the lack of precise and reliable nanothermometers. An international team of researchers has realized a quantum technology to probe temperature on a nanometer-scale, and have observed a 'fever' in tiny nematode worm (*Caenorhabiditis elegans*) under pharmacological treatment (Anonymous, 2020).

Quantum sensing is a technology that exploits the ultimate sensitivity of fragile quantum systems to the surrounding environment. High-contrast MRIs are examples of quantum spins in fluorescent diamonds and are some of the most advanced quantum systems working at the forefront of real-world applications. Applications of this technique to thermal biology were introduced seven years ago to quantify temperatures inside cultured cells. However, they had yet to be applied to dynamic biological systems where heat and temperature are more actively involved in biological processes. The research team decorated the surface of the nanodiamonds with polymer structures and injected them to *C elegans* worms, one of the most popular model animals in biology. They needed to know the base "healthy" temperature of the worms. Once inside, the nanodiamonds moved quickly but the team's novel quantum thermometry algorithm successfully tracked them and steadily measured the temperature. A fever was induced within the worms by stimulating their mitochondria with a pharmacological treatment. The team's quantum thermometer successfully observed a temperature increase in the worms.

3.9 miRNA based nanotechnology

RNA nanoparticles with many unique features are promising candidates in disease detection and treatment. The miRNAs are involved in many biochemical and developmental pathways and their regulation in plants and animals. These appear to be a powerful tool for controlling various pathological diseases in human, plants and animals, however there are challenges associated with miRNA based nanotechnology. Several advancements made in the field of miRNA therapeutics make it an attractive approach, but a lot more has to be explored in nanotechnology assisted miRNA therapy. The miRNA based technologies can be employed for detection and combating crop diseases as well. A review by Chaudhary et al., (2018) focused on current knowledge on miRNA based detection and treatment of various disease using RNA nanotechnology.

MiRNAs are key governors of altered gene expression under many pathological conditions in plants, animals and humans, and are characterized by altered miRNA levels. Therefore, precise and sensitive assessment of miRNA can lead to early and correct disease diagnosis and make these small noncoding RNA transcripts a valuable biomarker.

In the absence of target miRNA, the split DNAzymes form an inactive DNAzyme motif with their substrate through partial pairing at the end of each strand and the fluorescence is quenched. Inside the cells, the target miRNA binds with both of the two halves of split DNAzymes forming the active secondary structure in the catalytic cores which can cleave the substrates resulting in the rupture of the substrate and recovery of fluorescence. A dual-mode electronic biosensor was developed for miRNA-21 detection based on gold nano-particle-decorated MoS2 nano-sheet which may prove to be potential candidate in biosensing field from nucleic acid to protein detection. A sensor for the detection of specific miRNA sequences was developed based on graphene quantum dots (GQDs) and ssDNA-UCNP@SiO_2. Carbonaceous nanomaterials such as nanosized graphene oxide and carbon nanoparticles have also been used in miRNA detection which is based on their fluorescence quenching properties.

3.10 Nanobiosensors for pesticide detection

Due to constant application, pesticides enter the food chain through various ways, such as water, air, and soil, and severely affect the environment, as well as human health because they are carcinogenic and cytotoxic in nature. Over the past decade, considerable attention has been given to the development of rapid, sensitive, specific, and economically viable methods for pesticide detection to ensure food safety and security (Rajesh Kumar Saini, et al., 2017). Nanotechnology is a fast-growing field that enables the development of materials with brand new properties that can be used as inexpensive, portable nanosensors (by combining a biological recognition mechanism with a physical transduction technique) for rapid and real-time detection of pesticides at decentralized locations. This review provided a critical evaluation of recent advances in this area and newer trends in biosensors, such as tailor-designed biomolecules that include aptamers and molecularly imprinted polymers, optical, electrochemical, and piezoelectric materials, and so forth. A conclusive remark will also highlight the existing challenges and future prospects in this new area of agricultural nanotechnology.

They are highly sensitive chemical or physical sensors that can detect a single virus, as well as a low concentration of substance that could be potentially harmful. A nanosensor consists of a biosensitive layer (biological recognition element, i.e., antibody (Ab), enzyme, protein, DNA, etc., covalently attached to the transducer, at least one of their sensing dimensions being not greater than 100 nm) that produces a physiochemical change when the target analyst interacts with the bioreceptor and converts this change into an electrical signal.

Nanosensors can be classified on the basis of receptors molecules, structure, and applications into specificity for the target species, sensitivity to changes in target species concentrations, fast response time, extended lifetime of at least several months and small size (miniaturization) with the possibility of low-cost (Rajesh Kumar Saini et al., 2017). For pesticides detection, numerous immunosensors have also been designed that utilize transducers that work on optical, electrochemical, piezoelectric, and mechanical methods. Abs combined with nanomaterials are also used in biosensors. An amperometric immunosensor was designed for the detection of the pesticide atrazine, which contained carbon paste solid polymer electrolyte comprised of polyaniline (PANI), a conducting polymer/poly(vinyl sulfonic acid) (PVS). Designing of several conductometric immunosensors has also been attempted for environmental analysis.

Bibliography

Álvarez S.P., Tapia M.A.M., Medina J.A.C., Ardisana E.F.H. and Vega M.E.G.,2018. Nanodiagnostics Tools for Microbial Pathogenic Detection in Crop Plants. In: Exploring the Realms of Nature for Nanosynthesis. Prasad R., Jha A., Prasad K. (Edrs.) Nanotechnology in the Life Sciences. Springer, Cham.

Amarendra N. Misra , Meena Misra and Ranjeet Singh, 2013. Nanotechnology in Agriculture and Food Industry- Review Paper. Int. J. Pure Appl. Sci. Technol., 16 : 1-9

Ammar Ahmad S., 2018. Nanotehcnologies associated to floral resources in agri-food sector. Acta Agronomica, 67: 146-159

Anonymous, 2019. New microneedle technique speeds plant disease detection.. Science Daily. North Carolina State University.

Anonymous, 2020. "Real-time nanodiamond thermometry probing in-vivo thermogenic responses. Sci. Adv. Nanowerk News , Osaka City University.

Avinash Marwal and Gaur, R.K., 2019. Nanophytovirology: An emerging field for disease management. Plant Diseases-Current Threats and management Trends. Sniezana Topolovee-Pintaric (Edrs). DOI: 10.5772/intechopen.86653

Biswal, S.K., Nayak, A.K., Parida, U.K. and Nayak, P.L., 2012. Applications of nanotechnology in agriculture and food sciences. Int. J. Sci. Innovations Discov., 2:21–36.

Bogue, B., 2008. Nanosensors 2008 - A review of recent progress. Emerald Group Publishing Limited, UK.

Chalupowicz, L., Dombrovsky, A., Gaba, V., Luria, N., Reuven, M., Beerman, A., Lachman, O., Dror, O., Nissan, G. and Manulis-Sasson, S., 2019. Diagnosis of plant diseases using the Nanopore sequencing platform. Pl.Path., 68: 229-238

Chao Wang, Veronique Beiss, Nicole F. Steinmetz, 2019. Cowpea Mosaic Virus Nanoparticles and Empty Virus-Like Particles Show Distinct but Overlapping Immunostimulatory Properties. J. Virol., 93 (21) e00129-19

Chaudhary, V., Jangra, S. and Yadav, N.R., 2018. Nanotechnology based approaches for detection and delivery of microRNA in healthcare and crop protection. J Nanobiotechnol., 16, 40

Choudhary Prassan, Singh Vineeta, Goswami Sanjay and Chakdar Hillol., 2018. Plant disease diagnostics at nanoscale..P 366-372. In. Agricultural, Allied Sciences & Biotech. for Sustainability of Agriculture, Nutrition & Food Security National Bureau of Agriculturally Important Microorganisms, Kaithal, India

Choudhary Prassan, Singh Bansh, Chakdar Hillol and Saxena Anil, 202). DNA barcoding of phytopathogens for disease diagnostics and bio-surveillance. World J. Microbiol.& Biotechnol.,37. 10.1007/s11274-021-03019-0.

Dubertret B, Calame M, Libchaber AJ , 2001. Single-mismatch detection using gold-quenched fluorescent oligonucleotides. Nat Biotechnol. 19: 365-70.

Elmer Wade and White Jason,2018. The Future of Nanotechnology in Plant Pathology. Ann. Rev. Phytopathol., 56. 111-133.

Enisa Omanović-Mikličanina , Mirjana Maksimović, 2016. Nanosensors applications in agriculture and food industry. Bull. Chemists and Technologists of Bosnia and Herzegovina, 47 : 59-70

Fang, Yi and Ramaraja, P. Ramasamy, 2015. Current and Prospective Methods for Plant Disease Detection. Biosensors, 5 : 537-561

Gengfeng Zheng and Charles M. Lieber, 2013. Nanowire Biosensors for Label-Free, Real-Time, Ultrasensitive Protein Detection. Methods Mol. Biol., 790: 223–237.

Goluch, E.D., Nam, J.M., Georganopoulou, D.G., Chiesl, T.N., Shaikh, K.A., Ryu, K.S., Barron, A.E., Mirkin, C.A. and Liu, C.,2006. A bio-barcode assay for on chip attomolar-sensitivity protein detection. Lab Chip 6: 1293-1299.

Guo Xiaodong, Wen Fang, Zheng Nan, Saive Matthew, Fauconnier Marie-Laure and Wang Jiaqi, 2020. Aptamer-Based Biosensor for Detection of Mycotoxins. Fron.in Chem.,8 : 195

Hashimoto, Y., Nakamura, H., Asaga, K. and Karube, I., 2008. A new diagnostic method for soil-borne disease using a microbial biosensor. Microbes Environ., 23:35-9.

Hoch, H.C., 2007. Use of nano-fabricated surfaces for studying colonization & dispersal of bacteria in water conducting plant vessels. USDA, N Y Agricultural Expt Station (N/A), Geneva,NY 14456

Hong, S. and Lee, C., 2018. The current status and future outlook of quantum dot-based biosensors for plant virus detection. Plant Pathol. J., 34 : 85-92

Hong, S. and Lee, C., 2018. The current status and future outlook of quantum dot-based biosensors for plant virus detection. Plant Pathol. J., 34 : 85-92

Julie A. Willoughby, 2015. Development of Abamectin Loaded Plant Virus Nanoparticles for Efficacious Plant Parasitic Nematode Control . ACS Appl. Mater. Interfaces , 7: 9546–9553

Kaewphinit, T., Santiwatanakul, S. and Chansiri, K., 2013. Colorimetric DNA based biosensor combined with loop-mediated isothermal amplification for detection of mycobacterium tuberculosis by using gold nanoprobe aggregation. Sens Transducers, 142:123–128.

Kashyap, P.L., Kumar, S. and Srivastava, A.K., 2017. Nanodiagnostics for plant pathogens. Environ.Chem. Lett., 15: 7–13

Krejcova, L., Michalek, P., Merlos Rodrigo, M., Heger, Z., Krizkova, S., Vaculovicova, M., Hynek, D., Adam, and Kizek, R.,2015. Nanoscale virus biosensors: state of the art. Nanobiosensors in Disease Diagnosis. 4:47-66

Lau, H., Wu, H. and Wee, E., 2017. Specific and Sensitive Isothermal Electrochemical Biosensor for Plant Pathogen DNA Detection with Colloidal Gold Nanoparticles as Probes. Sci Rep., 7: 388-396 .

Lertanantawong, B., Krissanaprasit, A., Chaibun, T., Gothelf, K.V. and Surareungchai, W., 2019. Multiplexed DNA detection with DNA tweezers in a one-pot reaction. Mater. Sci. Energy Technol., 2 : 503-508

Lico, C., Schoubben, A., Baschieri, S., Blasi, P. and Santi, L., 2013. Nanoparticles in Biomedicine: New Insights from Plant Viruses. Curr. Med. Chem., 20 : 3471–87

Lima, A.S., Prieto, K.R., Santos, C.S., Paula Valerio, H.,. Garcia-Ochoa, E.Y., Huerta- Robles, A.., Beltran-Garcia, M.J., Di Mascio, P. and Bertotti, M., 2018. In- vivo electrochemical monitoring of H2O2 production induced by root-inoculated endophytic bacteria in Agave tequilana leaves. Biosens. Bioelectron., 99: 108-114

López, M.M., Llop, P., Olmos, A., Marco-Noales, E., Cambra, M. and Bertolini, E., 2019. Are molecular tools solving the challenges posed by detection of plant pathogenic bacteria and viruses? Curr. Issues Mol. Biol. , 11:13–46

Luna-Moreno, D. , Sánchez-Álvarez, A. , Islas-Flores, I. , Canto-Canche, B., Carrillo-Pech, M., Villarreal-Chiu, J.F. and Rodríguez-Delgado , M., 2019. Early detection of the fungal banana black sigatoka pathogen Pseudocercospora fijiensis by an SPR immunosensor method. Sensors (Basel), 23 : 465

Mahlein, T. Rumpf, P. Welke, H.W. Dehne, L. Plumer, U. Steiner, E.C. Oerke, 2013. Development of spectral indices for detecting and identifying plant diseases. Remote Sens. Environ., 128 :. 21-30

Manju Kumari Choudhary, Manvendra Singh and Vinod Saharan,2015. Application of Nanobiosensors in Agriculture. Popular Kheti, 3: 130-135.

Mendes,R.K., Carvalhal, R.F., Machado, S.D.R., Kubota, L.T., . Surface Plasmon resonance immunosensor for early diagnosis of Asian rust on soybean leaves. Biosensors and Bioelectronics, 24: 2483-2487

Mingzhe Zhang, Wujian Chen, Xi Chen, Yongjiang Zhang, Xiaojia Lin, Zhiyi Wu and Mingfu Li, 2013. Multiplex Immunoassays of Plant Viruses Based on Functionalized Upconversion Nanoparticles Coupled with Immunomagnetic Separation. J.Nanomat., vol. 2013, Article ID 317437, 8 pages

Mohammad A. Khiyami, Hassan Almoammar, Yasser M. Awad, Mousa A. Alghuthaymi, Kamel A. Abd-Elsalam, 2014. Plant pathogen nanodiagnostic techniques: forthcoming changes? Biotechnol. Biotechnollgical Equip., 28 : 775–785

Mohammad Hasan Dad Ansari, Santosh Lavhale, Raviraj M. Kalunke, Prabhakar L. Srivastava, Vaibhav Pandit, Subodh Gade, Sanjay Yadav, Peter Laux, Andreas Luch, Donato Gemmati, Paolo Zamboni, and Ajay Vikram Singh, 2020. Recent Advances in Plant Nanobionics and Nanobiosensors for Toxicology Applications. Curr. Nanosci.,, 16: 27-41 Mohga Khater, Alfredo de la Escosura-Muñiz and Arben Merkoçi, 2017. Biosensors for plant pathogen detection. Biosensors and Bioelectronics, 93: 72-86

Mohit, S., Verma, Jacob L. Rogowski , Lyndon Jones and Frank X. Gu , 2015 . Colorimetric biosensing of pathogens using gold nanoparticles. Biotechnol. Adv., http:// dx.doi.org/10.1016/j.biotechadv.2015.03.003.

Panini, N.V., Bertolino, F.A., Salinas, E., Messina, G.A., and Raba, J., 2010. Zearalenone determination in corn silage samples using an immunosensor in a continuous-flow/ stopped- flow systems. Biochem. Engg. J., 51:7–13

Papadakis, G., Skandalis, N., Dinopoulou, A., Glynos, P. and Gizel, E., 2015. Bacteria murmur: Application of an acoustic biosensor for plant pathogen detection. PLos One, 10(7):e0132773

Priyom Bose, 2020. The Role of Nanotechnology in Virus Detection. Editorial Feature, AZO NANO, Apr.28,2020.

Rajesh Kumar Saini, Laxmi P.Bagri and Anil Kumar Bajpai, 2017. Smart nanosensors for pesticide detection. New Pesticides and Soil Sensors, Academic Press, 519-559

Seon-Yeong Kwak, Min Hao Wong, Tedrick Thomas Salim Lew, Gili Bisker, Michael A. Lee, Amir Kaplan, Juyao Dong, Albert Tianxiang Liu, Volodymyr B. Koman, Rosalie Sinclair, Catherine Hamann, and Michael S. Strano, 2017.. Nanosensor Technology Applied to Living Plant Systems. Ann. Rev.Anal.Chem., 10: 113-140

Shahrul, A.B , Tijjani, A., U. Hashim and Faridah Salam, 2013. Plant Diseases Detection Using Nanowire as Biosensor Transducer. Adv. Ma. Res., 832:113-117

Sharon, M., Choudhary, A.K., Kumar, R.J.,2010. Nanotechnology in agricultural diseases and food safety. J. Phytol., 2:83-92

Shiping Song, Lihua Wang, Jiang Li, Chunhai Fan, Jianlong Zhao, 2008. Aptamer-based biosensors, TrAC Tre. Anal. Chem.,, 27: 108-117

Spicer, C.D., Jumeaux,C., Gupta,B., Stevens,M.M., 2018. Peptide and proteinnanoparticle conjugates versatile platforms for biomedical applications. Chem. Soc. Rev., 47: 3574,DOI:10.1039/C7C500877E

Srivastava Priyanka, Pandey Shashank, M. K.. Muhammad Aslam , Singh Prashant and Singh Krishan Pal, 2013. Nanobiosensors: Diagnostic Tool for Pathogen Detection, Int. Res. J. Biological Sci., 2 : 76-84

Sungyeap Hong and Cheolho Lee, , 2018. The Current Status and Future Outlook of Quantum Dot-Based Biosensors for Plant Virus Detection. The Pl. Pathol. J., 34:85-92

Touseef Hussain, 2017. Nanotechnology: Diagnosis of Plant Diseases. Agri Res & Tech: Open Access J 10 (1): ARTOAJ.MS.ID.555777

Vadim Krivitsky, Eran Granot, Yoav Avidor, Ella Borberg, Ralf T. Voegele and Fernando Patolsk, 2021. Rapid Collection and Aptamer-Based Sensitive Electrochemical Detection of Soybean Rust Fungi Airborne Urediniospores. ACS Sens., 6 : 1187–1198

Valeria Scala, Nicoletta Pucci and Stefania Loreti, 2018. The diagnosis of plant pathogenic bacteria: a state of art. Frontiers In Bioscience, Elite, 10 : 449-460,

Vasily G. Panferov, Irina V. Safenkova, Nadezhda A. Byzova, Yuri A. Varitsev, Anatoly V. Zherdev and Boris B. Dzantiev, 2018. Silver-enhanced lateral flow immunoassay for highly-sensitive detection of potato leafroll virus. Food & Agril. Immunol., 29: 445-457

Vedpriya Arya, 2019. Nano-approach towards Sustainable Agriculture and Precision Farming. Intl. J. Adv. Engg. Manage. & Sci. (IJAEMS), 5 :639-647

Veeranna D, Singh S, Kumar R, Jeevan B., 2017. Nanotechnology as diagnostic tool for plant disease detection. BioQuest 1 : 17-18

Vinay Kumar and Kavita Arora, 2020. Trends in nano-inspired biosensors for plants. Mat.Sci for Energy Tech., 3: 255-273

Voropaeva , N., Karpachev , V., Varlamov , V.and Figovsky, O., 2014. Influence of efficient, multicomponent, polyfunctional, physiologically active (nano) chips with herbicide activity on rice crop growth, development, yield and on weed growth inhibition. Intl. Lett. Chem. Phys. & Astron., 7 : 62-68

Yao, K. S., Li, S. J., Tzeng, K. C., Cheng, T. C., Chang, C. Y., Chiu, C. Y., Liao, C. Y., Hsu, J. J., & Lin, Z. P.,2009. Fluorescence Silica Nanoprobe as a Biomarker for Rapid Detection of Plant Pathogens. Adv. Mat. Res., 79–82 : 513–516

Yuanshang Wang, Maojun Jin, Ge Chen, Xueyan Cui, Yudan Zhang, Mingjie Li, Yun Liao, Xiuyuan Zhang, Guoxin Qin, Feiyan Yan, A.M. Abd El-Aty, Jing Wang,2019. Bio-barcode detection technology and its research applications: A review. J.Adv. Res.,20 : 23-32,

Zhan, F., Wang, T., Iradukunda, L. and Zhan , J., 2018. A gold nanoparticle-based lateral flow biosensor for sensitive visual detection of the potato late blight pathogen, Phytophthora infestans. Anal Chim. Acta., 1036: 153-161

Zhang, X.E., Men, D. and Wei, H., 2013. Nanowire Biosensors. In: Encyclopedia of Biophysics. Roberts G.C.K. (Ed.) Springer, Berlin, Heidelberg. https://doi.org/10.1007/978-3-642-16712-6_718

Zheng Li, Tao Yu, Rajesh Paul, Jingyuan Fan, Yuming Yang and Qingshan We, 2020. . Agricultural nanodiagnostics for plant diseases: recent advances and challenges. Nanoscale Adv., 2 : 3083-3094

Zongjie Wang, Suwon Lee, Kyo-in Koo and Keekyoung Kim, 2016. Nanowire-Based Sensors for Biological and Medical Applications. IEEE Transactions on Nanobioscience, 15 :186-199.

4

Nanoparticles in Plant Disease Management

Agri-nanotechnology is a new branch of biology that has originated due to the compatibility of nanosized inorganic and organic particles with biological functions. Based on enhanced effectiveness, the new age drugs are nanoparticles of polymers, metals, or ceramics, which can facilitate several biological applications. Plant diseases are caused by pathogens including fungi, bacteria, mycoplasma, viruses and viroids, nematodes, parasitic plants, and protozoa. The variation in life cycle of these disease-causing organisms produces a variety of responses to micro- and macroclimate, and variation in temperature can lead to differential symptom expression (Sharma Rishu et al., 2016). In the recent times, the increased incidence and severity of several plant diseases is the resultant of wide-scale application of intensive farming and extensive use of high-yielding varieties. The conventional plant disease control measures add toxic chemicals to the environment in addition to the increased cost of food production. And, with the changing environmental conditions, it has been predicted that a large number of new causal organisms of many dreadful diseases will emerge fast due to climate change. Moreover, a number of new physiological races and isolates of the existing pathogens will make them more virulent and destructive. In the changing scenario of climate change, an enormous change in host–pathogen interaction is predicted. Genomics, molecular breeding, and nanotechnology are highly promising emerging techniques to develop disease-resistant crop varieties. Climate change is affecting agriculture due to global warming with an average temperature increase of 0.74 °C since past 100 years, and atmospheric CO_2 concentration has increased from 280 ppm in 1750 to 400 ppm in 2013. These sudden changes will have deep adverse effect on various crops and microorganisms on the earth.

Nanoparticles have wide potential in plant disease diagnosis and as an ecofriendly mode of disease management. Nanosensor, mini-detection instruments could play a vital role in pathogen detection and management of various plant diseases (Rishu Sharma et al., 2016). Study identifies existing efficacy of anti-microbial activities of silver, siliconsilver, Zinc- iron-magnesium oxides, copper, validamycin, chitosan and iron nano- particles. Seven different types of gaps were identified for use of these particles in pathogen managements and these were pertains to pathogen types, hypothetical mode of actions, application of these particles at phyllosphere and rhizosphere, environmental stability and economics of nano-particles (Manish Mathur et al., 2018). The potential application of nanomaterials in crop protection helps in the development of efficient and potential approaches for the management of plant pathogens. Several studies were conducted using nano-sized particles to control fungal pathogens such as *Pythium ultimum, Magnaporthe grisea, Colletotrichum gloeosporioides, Botrytis cinerae* and *Rhizoctonia solani*, as well as pathogenic bacteria including *Bacillus subtilis, Azotobacter chrococuum, Rhizobium tropici*, *Pseudomonas syringae* and *Xanthomonas compestris* pv. *vesicatoria.*

Engineered nanoparticles (NPs) (1–100 nm) that have demonstrated activity in suppressing plant diseases are metalloids, metallic oxides, nonmetals, and carbon nanoparticles (Wade Elme et al. 2018). NPs have been integrated into disease management strategies as bactericides/fungicides and as nanofertilizers to enhance plant health. Although there are reports of over 18 different NPs of single element and carbon nanomaterials affecting disease and/or plant pathogens, only Ag, Cu, and Zn have received much attention thus far. Some NPs act directly as antimicrobial agents while others function more in altering the nutritional status of the host and thus activate defense mechanisms. For example, NPs of Ag and Cu can be directly toxic to microorganisms. Other NPs of B, Cu, Mn, Si, and Zn appear to function in host defense as fertilizers. As demand for food production increases against a warming climate, nanoparticles will play a role in mitigating the new challenges in disease management resulting in a reduction in active metals and other chemical inputs.

4.1 Disease occurrence and prior warning

Nanotechnology provides efficient tools for early detection of plant diseases by diagnostic tools in managing insects and pathogens by targetted controlled delivery of agri-based chemicals (Sharma Rishu et al., 2016). Detecting and diagnosing a plant disease at an early stage is vital and has thus tempted scientists to look for a 'nanosolution' for protecting food and agriculture from bacteria, fungus, and viral agents so that tons of food can be protected from possible outbreak. Diagnosis involves pathogen detection and control. An accurate and before-hand detection of pathogens would affect positively in protecting crops from diseases by timely application of pesticides in the field.

4.2 Application of nanoparticles in management of plant diseases

The nanoparticle are used to protect plants in two different mechanisms, viz. nanoparticle act as carrier for pesticides can be applied by spray application and nanoparticle themselves providing protection to crop (Pulugu Sai et al. 2020). Nanoparticle act as carrier can provide several benefits like increases shelf life of pesticides, it increases the solubility of poorly water soluble pesticides, increases the site specific uptake into target pest and reduces toxicity. Nanoparticles can also be directly applied to plant seed, foliage roots for protection against pest and pathogen. Metal nanoparticles such as silver copper, zinc oxide and titanium oxide have been intensively researched for their antibacterial and antifungal properties.Silver nanoparticles have shown antifungal inhibition of *Alternaria alternata, Sclerotinia sclerotiorum Rhizoctina solanai* (Krishnaraj et al., 2012). When poly-dispersed gold nanoparticles introduced into plant through mechanical abrasive was seen to melt and dissolve the Barley yellow mosaic virus particles conferring resistance to plant.

Plant diseases can be managed using nanotechnology by controlled release of encapsulated pesticide, and other agrochemicals in protection against pests and pathogens. The potential application of nanomaterials in crop protection helps in the development of efficient and potential approaches for the management of plant pathogens (Gopal et al. 2011). Nanoparticles remain bound to the cell wall of pathogens and cause deformity due to high-energy transfer leading to its death. Nanotechnological application in plant pathology targets specific agricultural problems in host–pathogen interactions and could provide new avenues for crop protection.

Nanosized particles have been studied to control various fungal pathogens such as *Pythium ultimum, Magnaporthe grisea, Colletotrichum gloeosporioides Botrytis cinerae*, and *Rhizoctonia solani*, as well as pathogenic bacteria including *Azotobacter chrococuum, Bacillus subtilis, Pseudomonas syringae Rhizobium tropici*, and *Xanthomonas compestris* pv. *vesicatoria* (Sharma Rishu et al., 2016). The two most important criteria in disease management are met by nanocomposites, i.e., maximum efficacy with minimal ecological impact and less toxicity on humans. Some of the NPs that have entered into the arena of controlling plant diseases are nanoforms of carbon, silver, silica and alumino-silicates. Scientists are concentrating on carbon nanotubes (CNTs). Also, the use of nanomaterials including copper, zinc, titanium magnesium, gold, alginate, and silver has been developed, but silver NPs (Nano-Ag) have been proved to be the extraordinarily effective exhibiting antimicrobial efficacy against pathogens, viz., eukaryotic microorganisms bacteria, and viruses . An ecofriendly fungicide is underdevelopment that uses nanomaterials to liberate its pathogen-killing properties, only when it is inside the targeted pathogen. Various nanoparticles are used for controlling plant diseases as have been enlisted in Table 9.

Table 9: Use of nanoparticles for managing plant pathogens

Nanomaterial	Application	Functions
Thiamine dilauryl sulfate (TDS) nanoparticles	Antifungal activity of TDS NPs (258.6 nm) against *Colletotrichum gloesporoides* associated with pepper anthracnose	NPs at 100 ppm conc. Showed 80 % growth inhibition of *Colletotrichum gloesporoides* compared to control. TDS NPs penetrated inside the hyphal cell membrane and destructed the cells
Validamycin loaded with nanosized calcium carbonate	Controlled release of validamycin-loaded nanosized calcium carbonate (50–200 nm) against Rhizoctonia solani	The formulation showed better germicidal efficacy against *Rhizoctonia solani* compared to conventional technical validamycin when studied after 7 days. The nanoformulation showed the release of validamycin up to two weeks
Chitosan NPs (CS NPs)	Efficacy of CS NPs on fungal growth and chili seed quality	CS NPs at a conc. Of 0.6 % (w/v) significantly delayed mycelia growth of *Rhizopus* spp., *Colletotrichum* spp. (*C. capsici, C. gloesporoides*) and *Aspergillus niger* when compared to contro
Nanocopper	Antibacterial activity against *Xanthomonas axanopodis* pv. *punicae*, causing bacterial blight of pomegranate	Nanocopper inhibited the growth of X. axonopodis at 0.2 ppm, i.e., >10,000 times lower than that usually recommended for copper oxychloride
Light-activated nanoscale formulations of TiO_2	Nanoscale formulations of TiO_2 with Ag and Zn on *Xanthomonas perforans* toward the control of bacterial spot of tomato	Compared to control, TiO_2/Ag and TiO_2/Zn had high photocatalytic activity against *X. perforans* and the combination significantly reduced bacterial spot severity without causing any adverse effects on tomato yield
Copper nanoparticles	Cu-based NPs (11– 55 nm) were tested on tomato against *Phytophthora infestans* for their antifungal activity	The synthesized Cu-based NPs were more effective than the commercial agrochemicals at lower concentrations and drastically reduced the active ingredient rate. The particles were not showed phytotoxicity on treated plants

DNA-directed silver nanoparticles on graphene oxide	Antibacterial activity of nanoparticles against *Xanthomonas perforans*	Nanoparticles at 100 ppm concentration severely reduced the bacterial spot disease compared to untreated tomato transplants in greenhouse condition without causing phytotoxicity
Light-activated nanoparticle formulation of titanium dioxide with Zinc	Nanocomposite for management of bacterial leaf spot on *Rosa noare*	Field applications of TiO2/Zn nanoformulation at H 500–800 ppm significantly reduced bacterial spot severity compared with the untreated control and other commercial bactericides
Chitosan-based nanoparticles (chitosan, chitosan- saponin, and Cu–chitosan nanoparticles)	In vitro evaluation of chitosan-based nanoparticles against *Alternaria alternata, Macrophomina phaseolina* and *Rhizoctonia solani*	Cu–chitosan nanoparticles were most effective at 0.1 % concentration and showed 89.5, 63.0 and 60.1 % growth inhibition of *A. alternata*, *M.phaseolina* and *R. solani*, respectively, and were effective in controlling spore germination

Nano-scale platforms, biological sensors, miniature detection devices and nano-sensors could play a significant role in pathogen detection and disease management in the future. various nano-particles utilized by Manish Mathur et al., (2018) in disease management and the possibility of large scale adaptability of nano-particles by integrating into present practices and avoiding crop losses reviewed. Study identifies existing efficacy of anti-microbial activities of silver, siliconsilver, Zinc-iron-magnesium oxides, copper, validamycin, chitosan and iron nano-particles. Seven different types of gaps were identified for use of these particles in pathogen managements and these were pertains to pathogen types, hypothetical mode of actions, application of these particles at phyllosphere and rhizosphere, environmental stability and economics of nano-particles.

4.2.1 Carbon Nanotubes

CNTs arc nanostructures, which are allotropes of carbon with extraordinary mechanical, electrical, thermal, optical, and chemical properties. A nanotube can be visualized as a hexagonal network of hexagonal carbon atoms which upon rolling give a cylindrical structure (Sharma Rishu et al., 2016). CNTs have many applications, especially in the fields of nanotechnology, electronics, and architecture. CNTs can be used during seed germination to deliver desired molecules to provide them protection against diseases. Among the carbon-based (CB) NPs, the impact of fullerene C70 and fullerol in agricultural sciences has been extensively studied and it has been found that these two types of CB NPs get readily accumulated in plants.

Encapsulation of pesticides on CNT-g-PCA hybrid material was confirmed via TEM analysis. Experiments indicated that new the CNT-g-PCA-EP hybrid material in comparison with bulk pesticide had a superior toxic influence on *Alternaria alternata.*

The addition of surface functional groups to multi-walled carbon nanotubes (MWCNTs) expands their application in engineering, materials, and life science (Xiuping Wang et al., 2017). The antifungal activities of MWCNTs was explored with different surface groups against *Fusarium graminearum.*

4.2.2 Nanosilver

Silver can affect various microorganisms and their biological processes and also inhibits protein expression. The use of nanosized silver particles as antimicrobial agents has become more common as technological advances make their production more economical. Silver has potential applications in management of plant diseases. Since silver exhibits multiple modes of inhibitory action to microorganisms, it may be used for controlling various plant pathogens in a relatively safer way as compared to synthetic fungicides. Coreduction method using thermal treatment was used for controlled synthesis of Ag NPs. The initial reduction was performed using sodium borohydride (NaBH4) at 60 °C (stage I), which induced generation of a large number of Ag NPs. The new silver nuclei are formed by the reduction of silver cations (Sharma Rishu et al., 2016).

Nano silver is the most studied and utilized nano particle for Bio-system. It has long been known to have strong inhibitory and bactericidal effects as well as a broad spectrum of antimicrobial activities (Madhuri Sharon et al,., 2010). Silver nanoparticles, which have high surface area and high fraction of surface atoms, have high antimicrobial effect as compared to the bulk silver. The antifungal effectiveness of colloidal nano silver (1.5 nm average diameter) solution was studied against rose powdery mildew caused by *Sphaerotheca pannosa* Var *rosae.* The nano silver colloidal solution of concentration of 5000 ppm was diluted in 10 ppm of 500 kg and sprayed at large area of 3306 m2 polluted by rose powdery mildew. Two days after the spray more than 95% of rose powdery mildew faded out and did not recur for a week. Nano silver colloid is a well dispersed and stabilized silver nano particle solution and is more adhesive on bacteria and fungus, hence are better fungicide. No wonder that maximum patents are filed for 'Nano silver for preservation and treatment of diseases in agriculture field.' This popularity of nano silver has caused concern about regulating and classifying the nano silver as pesticide (Anderson 2009). In May 2008, the International Center for Technology

Assessment (ICTA) submitted a petition to EPA requesting that it regulate nano-silver used in products as a pesticide under the Federal Insecticide, Fungicide and Rodenticide Act (FIFRA).

Silver is now an accepted agrochemical replacement. It eliminates unwanted microorganisms in soil and hydroponics systems. It is being used as foliar spray to stop fungi, moulds, rot and several other plant diseases. Moreover, silver is an excellent plant-growth stimulator. There are literally thousands of other essential uses for this odorless, nearly tasteless and colorless, totally benign, powerful, non-toxic disinfectant and healing agent.

NPs formed at the initial stage subsequently participated in the growth process, where trisodium citrate (TSC)-mediated reduction of unspent Ag ions was favored at the higher temperature, i.e., 90 °C, prevailing in stage II. It inhibited colony formation of both *Bipolaris sorokiniana* and *Magnaporthe grisea* (Sharma Rishu et al., 2016). Silver ions and NPs effectively reduced leaf spot and gray leaf spot on perennial ryegrass (*Lolium perenne*) without noticeable phytotoxicity. The inhibition efficiency of fungus colony exists as long as silver ions do not get neutralized with chloride ions. Silver has strong bactericidal and inhibitory effects. Silver NPs, which have high surface area and high fraction of surface atoms, have high antimicrobial effect as compared to the bulk silver.

The antifungal activity of silver nanoparticles was evaluated against sclerotium-forming phytopathogens, *Rhizoctonia solani, Sclerotia sclerotiorum*, and *S. minor* during a study conducted and the data demonstrated that the nanoparticles strongly inhibited the fungal growth and sclerotial germination growth (Nikunj Patel et al., 2014). It is also suggested that nanometer-sized silvers possess different properties, which might come from morphological, structural and physiological changes. It was also shown that the nano-particles efficiently penetrate into microbial cells, which implies that lower concentrations of nano-sized silvers would be sufficient for microbial control. This would be efficient, especially for some organisms that are less sensitive to antibiotics due to the poor penetration of some antibiotics into cells.

A study observed that silver nanoparticles disrupt transport systems, including ion efflux. The dysfunction of ion efflux can cause rapid accumulation of silver ions, interrupting cellular processes at their lower concentrations such as metabolism and respiration by reacting with molecules. Also, silver ions are known to produce reactive oxygen species (ROS) via their reaction with oxygen, which are detrimental to cells, causing damage to proteins, lipids, and nucleic acids. Most fungi have shown a high inhibition effect at a 100 ppm concentration of silver nanoparticles. In most cases, inhibition increases as the concentration of AgNPs is increased. It happens due to the high density at which the solution was able to saturate and cohere to fungal hyphe and to deactivate plant pathogenic fungi. upon treatment with Ag, DNA loses its ability to replicate, resulting in inactivated expression of ribosomal subunit proteins, as well as certain other cellular proteins and enzymes essential to ATP production (Nikunj Patel et al., 2014). It has also been hypothesized that Ag+ primarily affects the function of membrane-bound enzymes, such as those in the respiratory chain.

The microscopic data revealed that silver nanoparticle- treated hyphae were seriously damaged on hyphal walls, resulting in the plasmolys is of hyphae (Min et al., 2009).

The biocidal efficacy of ZnO and $ZnTiO_3$ nanopowders was determined against *Aspergillus niger.* DNA-directed silver NPs grown on graphene oxide (GO) composites effectively decreased cell viability in culture and on plants of *Xanthomonas perforans* causing bacterial spot of tomato in Florida, while the pathogen has developed resistance to Cu fungicides. These compounds (Ag@dsDNA@GO) show excellent antibacterial activity in culture at a very low concentration of 16 ppm with higher adsorption rate. Severity of tomato bacterial spot is significantly reduced by application of Ag@dsDNA@GO at 100 ppm in greenhouse when compared to untreated and showed no phytotoxicity.

Several lines of evidence support the enhanced efficiency of silver nano-particles on antimicrobial activity. Silver nano-particles are highly reactive as they generate Ag+ions while metallic silver is relatively un-reactive. It was also shown that the nano-particles efficiently penetrate into microbial cells, which implies lower concentrations of nano-sized silver would be sufficient for microbial control. Study suggested that silver nanoparticles can significantly delay mycelial growth of *C. gloeosporioides* in a dose dependent manner in vitro (Manish Mathur et al., 2018). Antifungal efficiency of silver nano-particles was observed at 24 h after inoculation, suggesting that direct contact of silver with spores or germ tubes is critical in inhibiting disease development.

Antifungal properties of silver nano-particles, silver ions, acrylate paint and cotton fabric impregnated with silver nano-particles were assessed against *Aspergillus niger, Aureobasidium pullulans* and *Penicillium phoeniceum* (Manish Mathur et al., 2018). Bioassay of elemental and nano-sulphur against *A niger* showed that nanosulfur was more efficient than its elemental structure. Three types of silver nanoparticles (AgNPs) were applied at concentrations of 10, 25, 50, and 100 ppm concentration for controlling *Alternaria laternata,*

A. brassicicola, A. solani, Botrytis cinerea, Cladosporium cucumerinum, Corynespora cassiicola, Cylindrocarpon destructans, Didymella bryoniae, Fusarium spp., *Glomerella cingulata, Monosporascus cannonballus, Pythium aphanidermatum, P spinosum* and *Stemphylium lycopersici* presented on various hosts. They reported that WA-CV-WB13R types of AgNPs showed maximum inhibition of most fungi on PDA with 100 ppm of AgNPs.

Cavity slide experiment was conducted to quantify antagonistic potential of AgNPs against *Bipolaris sorokiniana* causing spot blotch disease in wheat with their conidial concentration of 2,4 and 10 µg/ml (Mishra et al.,2014) and observed that 2, 4 and 10 µg/ml concentrations of bsAgNPs accounted for complete inhibition of conidial germination, whereas in the absence of bsAgNPs, conidial germination was 100%. Dose dependent study of nano-silver on *Xanthomonas campestris* pv. *campestris* (causing cabbage black rot) showed significant reduction of cabbage black rot in the pot experiment (Manish Mathur et al., 2018).

Sunhemp rosette virus treated with solutions of the silver nano-particles at a concentration of 50 mg/L. After 30 sec, a 10 µL droplet was deposit on a carbon coated nickel TEM grid and exposed to a 2.5% solution of PBS/ glutaraldehyde vapours for 30 min. Leaves inoculated with (SHRV) showed complete suppression of the disease (Jain and Kothari , 2014).

Silver nanoparticles may be synthesid using plant extracts or microbes viz., fungi, bacteria, yeast and actinomycetes (Table 10) (Sahayaraj and Rajesh, 2011).

Table 10: Antimicrobial activities of bionanoparticles synthesized using silver with botanicals and microbes

Biological entity	Test microorganisms
a. Using silver with botanicals	
Argimone mexicana	*Escherichia coli* and *Pseudomonas aeruginosa, Aspergillus flavus*
Aloe Vera	*E. coli*
Caulerpa scalpelliformis	*Xanthomonas campestris* pv. *malvacearum*
Caulerpa veravalensis	*X. campestris* pv.*malvacearum*
Desmodium triflorom	*Staphylococcus* sp., *E. coli, Bacillus subtilis*
Cinnamon zeylanicum	*E. coli* BL-21
Gliricidia sepium	*S. aureus, E. coli, P. aeruginosa, Klebsiella pneumoniae*
Ocimum sanctum & *Vitex negundo*	*S. aureus, Vibrio cholerae,Proteus vulgaris, P. aeruginosa.*
Padina pavonica	*X. campestris* pv.*malvacearum*
Sagassum wightii	*S. aureus, B. rhizoids, E. coli, P. aeruginosa*
Vernonia cinerea	*X.campestris* pv.*malvacearum*

Trianthema decandra	*E. coli* , *P. aeruginosa*
Solanus torvum	*P. aeruginosa, S. aureus, Aspergillus flavus* , *A. niger*
Svensonia hyderabadensis	*A.niger, Fusarium oxysporum, Curvularia lunata, Rhizopus arrhizus*
b. Silver with various microbes	
Amylomyces rouxii KSU-09	*Shigella dysenteriae typeI,Staphylococcus aureus, Citrobacter* sp., *Escheeichia coli, Pseudomonas aeruginosa, Bacillus subtilis, Candida albicans* and *Fusarium oxysporum*
Alternaria alternata	*Phoma glomerata, P.herbarum, F. semitectum, Trichoderma* sp. and *C.albicans*
Aspergillus sp.	*Staphylococcus* sp.
Euphorbia hirta	*S. aureus, E. coli, K.pneumoniae, B. cereus* and *P. aeruginosa*
Klebsiella pneumoniae	*S. aureus, E. coli*
Fusarium acuminatum	*S. aureus, Salmonella typhi,Staphylococcus epidermidis, E. coli*
Phytophthora infestans	*E. coli, S. typhi, K. pneumoniae,P. vulgaris,B. subtilis, S. aureus*
Trichoderma viride	*S. typhi, E. coli, S. aureus, Micrococcus luteus*

Mechanism of silver nanoparticles on bacteria: The mechanisms behind the activity of nano silver on bacteria are not yet fully elucidated, the three most common mechanisms of toxicity proposed up to now include uptake of free silver ions followed by disruption of ATP production and DNA replication; formation of Reactive Oxygen Species (ROS) and direct damage to cell membranes (Sahayaraj and Rajesh, 2011).

Uptake of free silver ions: Antimicrobial properties of ionic silver are well known; thus, it is likely that eluted ions from silver nanoparticles are accountable for at least a part of their antibacterial properties. At sub-micromolar concentrations, $Ag+$ interacts with enzymes of the respiratory chain such as NADH dehydrogenase ensuing in the uncoupling of respiration from ATP synthesis. Ionic silver also binds with transport protein leading to proton leakage, inducing collapse of the proton motive force. Ionic silver hampers the uptake of phosphate and thus causes the efflux of intracellular phosphate.

4.2.3 Silicon nano-particles

Silicon (Si) is known to be absorbed into plants and to increase disease resistance and stress resistance (Manish Mathur et al., 2018). Aqueous silicate solution, used to treat plants, is reported to exhibit excellent preventive effects on pathogenic microorganisms causing powdery mildew or downy mildew in plants thus used for treating diseased plants. It promotes the physiological activity and growth of plants and induces disease and stress resistance in plants. But, since silica has no direct disinfection effects on pathogenic microorganisms in plants, it does not exhibit any effect on established diseases. Further, the effects of silica significantly vary with the physiological environment and thus, it is not registered as an agricultural chemical. As mentioned above, silver is known as a powerful disinfecting agent. It kills unicellular microorganisms by inactivating enzymes having metabolic functions in the microorganisms by oligodynamic action.

A new composition of nano-sized silica-silver for control of various plant diseases has been developed by Park et al., (2006), which consists of nano-silver combined with silica molecules and water soluble polymer, prepared by exposing a solution. DNAdirected silver nano-particles grown on graphene oxide (GO) composites effectively decreased cell viability in culture and on plants of *Xanthomonas perforans* causing bacterial spot of tomatoes.

4.2.4 Nanosilica–Silver composite

The role silicon (Si) plays in increasing the disease and stress resistance has been recorded (Sharma Rishu et al., 2016). Aqueous silicate solution has been reported to treat diseased plants to prevent them from pathogens causing powdery and downy mildew (Brecht et al. 2004). Moreover, it promotes the physiological activity and growth of plants and induces disease and stress resistance in plants. But, since silica has no direct disinfection effects on pathogenic microorganisms in plants, it does not exhibit any effect on established diseases. Further, the effects of silica significantly vary with the physiological environment, and thus, it is not registered as an agricultural chemical.

The recent addition to this kitty is nanosized silica–silver which is capable of controlling many plant diseases have been developed, which consist of nanosilver combined with silica molecules and water soluble polymer, prepared by exposing a solution. The 'effective concentration' of nanosized silica–silver in suppressing the fungal growth of many pathogens causing plant diseases was studied and reported that few pathogens showed cent percent (100%) growth suppression at 10 ppm. Chowdappa and Gowda (2013) observed an inhibition of conidial germination in *Colletotrichum gloeosporioides* with chitosan–silver NP (chitosan–Ag Np) composite (size distribution from 10 to 15 nm).

Nano Silica-silver composite is known to be absorbed into plants to increase disease resistance and stress resistance (Madhuri Sharon et al., 2010). Aqueous silicate solution, used to treat plants, is reported to exhibit excellent preventive effects on pathogenic microorganisms causing powdery mildew or downy mildew in plants, Moreover, it promotes the physiological activity and growth of plants and induces disease and stress resistance in plants. But, since silica has no direct disinfection effects on pathogenic microorganisms in plants, it does not exhibit any effect on established diseases. Further, the effects of silica significantly vary with the physiological environment and thus, they are not registered as an agricultural chemical.

Silver in the form of a metal or oxide, is stable in the environment, but because of its low antimicrobial activity it is used in relatively increased amount, which is not very desirable.

Application of Ag@dsDNA@GO at 100 ppm on tomato transplants in a greenhouse experiment significantly reduced the severity of bacterial spot disease compared to untreated plants, giving results similar to those of the current grower standard treatment, with no phytotoxicity.

Nanoalumino-Silicate : Many of the big pesticide industries are now formulating pesticides which are extremely efficient, biologically active, and environmentally safe at nanoscale. The use of alumino-silicate nanotubes with active ingredients is one such type, which has proved advantageous when sprayed on plant surfaces because of its easy pick by insect hairs and lastly these nanotubes get consumed by them leading to their death. Leading chemical companies are now formulating efficient pesticides at nano scale. One of such effort is use of Alumino-Silicate nanotubes with active ingredients. The advantage is that Alumino-Silicate nanotubes sprayed on plant surfaces are easily picked up in insect hairs. Insects actively groom and consume pesticide-filled nanotubes. They are biologically more active and relatively more environmentally-safe pesticides (Madhuri Sharon et al., 2010).

Leading chemical companies are now formulating efficient pesticides at nano scale. One of such effort is the use of aluminum-Silicate nanotubes with active ingredients (Nikumj Patel et al., 2014). The advantage is that Alumino- Silicate nanotubes sprayed on plant surfaces are easily picked up in insect hairs. Insects actively groom and consume pesticide-filled nanotubes. They are biologically more active and relatively more environmentally- safe pesticides.

Mesoporous silica nanoparticles: Mesoporous silica nanoparticles (MSNs) are thermally stable nanoparticles with tractable porosity, having the tendency to deliver DNA and chemicals into plants by targetted release (Sharma Rishu et al., 2016). The spherical porous silica NP systems with particles having arrays of independent porous channels were developed. A honeycomb-like structure is formed by channels which can be used to fill chemicals. These NPs have a unique 'capping' strategy that seals the chemical inside. They have also demonstrated that the caps can be chemically activated to pop open and release the cargo inside the cells where it is delivered. This potential of NPs for controlled delivery efficiency can be even more increased by controlling time. Plant cells have rigid cell walls and need to be modified with chemical coatings. This technique has been successfully used in Arabidopsis, maize, and tobacco plants to inject DNA and chemicals for desired results.

Mesoporous silica nano particles can deliver DNA and chemicals into Plants thus, creating a powerful new tool for targeted delivery into plant cells. Lin's research group has developed porous, silica nanoparticles systems that are spherical in shape and the particles have arrays of independent porous channels. The channels form a honeycomb-like structure that can be filled with chemicals or molecules (Madhuri Sharon et al., 2010). These nanoparticles have a unique "capping" strategy that seals the chemical inside. They have also demonstrated that the caps can be chemically activated to pop open and release the cargo inside the cells where it is delivered. This unique feature provides total control for timing the delivery. Plant cells have rigid cell wall. Hence to penetrate it they had to modify the surface of the particle with a chemical coating. It has been successfully used to introduce DNA and chemicals in to Arabidopsis, tobacco and corn plants. The other advantage is that with the mesoporous nanoparticles, one can deliver two biogenic species at the same time.

4.2.5 Chitosan nanoparticles

Chitosan nanoparticles have hydrophobic properties due to this property it is low soluble in aqueous solution. As a result, chitosan is commonly mixed with organic and inorganics to improve solubility (Pulugu Sai et al., 2020). The chitosan inhibition mode against fungi is defined by these three mechanism , viz., it is proposed that chitosan can penetrate the cell wall of fungus and bind to its DNA and hinder mRNA synthesis and in turn, impact the production of vital enzymes and protin, chitosan chelates with metal ions, which are suggested as a possible antimicrobial action mode and the positively charged chitosan interact with negative charge of phospholipid components of fungi membrane, which alter cell permeability of plasma membrane and lead to leakage of cellular components, which subsequently leads to death ofcell. Chitosan nanoparticle exhibited antifungal activity in vitro and could protect the finger millet plants from blast disease caused by *Pyricularia grisea.* Chitosan nanoparticles were reported to be effective against plant pathogenic fungi and bacteria effecting tomatos. The growth inhibitory effects was maximum in *F. oxysporum* fallowed by *P. capsici*, *Xanthomonas compestris* pv. *vesicatoriaas* well as *Erwnia carotovora* was inhibited.

Chitosan polymer and chitosan nano-particles at a concentration of 0.6% w/v significantly delayed mycelial growth of *Rhizopus* sp. *Colletotrichum capsici, C. gloeosporioides*, and *Aspergillus niger* when compared with 0.15% w/v captan, 0.2% w/v chitosan nanoparticles, and the control (PDA). The chitosan NP (0.5, 1.0, 2.0, and 4.0 mg /ml-1) was evaluated on *Fusarium equiseti* and *Curvularia lunata* present on Latropha curcas, and their study revealed that Chitosan completely inhibited sporulation of *C. lunata* and spore germination of *F. equiseti.* Inoculation with *F. equiseti* and *C. lunata* reduced seed germination of *J. curcas* by 20 and 26.6 %, respectively. However, application of chitosan before inoculation inhibited pathogenic activity. Therefore, chitosan did not affect seed germination and caused inhibitory effects on *F. equiseti* and *C. lunata.* Chowdappa et al., (2013) observed an inhibition of conidial germination in *Colletotrichum gloeosporioides* with chitosan-silver nanoparticle (chitosan- Ag Np) composite (size distribution from 10 nm to 15 nm).

4.2.6 Iron nano-particles

"Smart Delivery Systems" for agriculture can possess timely controlled, spatially targeted, self-regulated, remotely regulated, preprogrammed, or multi-functional characteristics to avoid biological barriers to successful targeting (Manish Mathur et al., 2018). In order to develop smart treatment-delivery system in plant, Cucurbita pepo plants were treated with carboncoated Fe nano-particles *in vitro*. The magnetic core consisting of Fe nano-particles allow themselves to be guided to a place of interest in the body (affected part) of an organism using small magnets that create a magnetic field. The carbon coating provides biocompatibility and acts as a surface for adsorption where various types of molecules of interest (drug/DNA/chemical/enzyme) can be adsorbed.

The antifungal activity of Magnesium oxide (MgO) and iron oxide (FeO) nanoparticles prepared by bio safe method was evaluated for *Penicillium expansum, Aspergillus niger, Alternaria alternata, Mucor plumbeus, Penicillium chrysogenum, Trichothecium roseum* and *Rhizoctonia solani.* It was observed from the study that all the concentrations of nanoparticles brought about significant inhibition in the spore germination and mycelial growth of all the rot causing fungi (Koka et al., 2019). However, the highest inhibition in the germination of all the test fungi was observed at higher concentrations followed by lower concentrations of nanoparticles. It was observed from the present study that MgO and FeO nanoparticles showed significant antimycotic activity against all the tested fungal pathogens. However, highest reduction in spore germination was observed against *Mucor plumbeus* whereas least reduction of spore germination was observed against Aspergillus niger at different concentration of nanoparticles of MgO respectively. Likewise, the maximum inhibition in the fungal growth was observed against *Alternaria alternata* and least inhibition in zone of fungal growth due to MgO nanoparticle was found against *Mucor* plumbeus respectively. Similarly, highest inhibition in spore germination was found against *Penicillium expansum* and least inhibition in spore germination was found against *Aspergillus niger* at different concentrations of nano FeO. Likewise, the maximum inhibition in the fungal growth was found against *Rhizoctonia solani* and *Trichothecium roseum* and least inhibition in zone of fungal growth was found against *Penicillium expansum* and *Mucor plumbeus* at 0.1mg/ml, 0.25mg/ml and 0.5 mg/ml concentrations of FeO nanoparticles respectively. Activity index was recorded highest against *P. chrysogenum* (0.97) in case of MgO and against T. roseum (0.97) in case of FeO nanoparticles respectively.

4.2.7 Gold nanoparticles

DNA-gold nanoparticle probes hold promise as a new generation of biosensors in the detection of pathogenic microorganisms. In this technique, the gold nanoparticle oligonucleotideprobes are hybridized with the complementary DNA, which stabilizes gold nanoparticles against aggregation (retaining thenative pink color of the colloidal gold). In the absence of complementary DNA, the solution turns purple because theaggregation of gold nanoparticles leads to the shift in absor-bance peak toward a longer wavelength (Pulugu Sai et al., 2020). Gold nanoparticles feature agglomeration related to color production that was used in the detection of pesticide. Developing a color signal facilitated easy visual detection when gold nanoparticles marked antibodies bound to the pesticide residues. The gold nanoparticle-based dipstick technique suited the detection of numerous toxins in environmental and food samples and can be used for rapid examination of pesticides on the site .Gold nanoparticle-based optical immunosensors have been developed for detection of Karnal bunt disease in wheat.

A nanoparticle based electrochemical biosensor was developed for rapid and sensitive detection of plant pathogen DNA on disposable screen-printed carbon electrodes. This 60 min assay relied on the rapid isothermal amplification of target pathogen DNA sequences by recombinase polymerase amplification (RPA) followed by gold nanoparticle-based electrochemical assessment with differential pulse voltammetry (DPV) (Lau et al., 2017). This method was 10,000 times more sensitive than conventional polymerase chain reaction (PCR)/gel electrophoresis and could readily identify *P. syringae* infected plant samples even before the disease symptoms were visible. On the basis of the speed, sensitivity, simplicity and portability of the approach, it is believed that the method has potential as a rapid disease management solution for applications in agriculture diagnostics.

Antibacterial activity of gold nanoparticles was evaluated against Gram negative test organism *Pseudomonas* sp. In sterile condition, prior sterilized and dried cotton cloth (2 × 2 cm) was immersed in flask containing FocJT1and FocJT1 augmented with HAuCl4 for one hour at room temperature (Janki et al., 2013). The cloth was collected and dried. From overnight grown culture of *Pseudomonas* sp., 0.1 mL was spread on PDA plates and allowed to dry for 10 min. Au nanoparticle impregnated cloth was kept in the centre of inoculated plate and incubated at 37°C for 24 hours. Plates were observed for zone of inhibition.

4.2.8 Zinc nanoparticles

In agriculture ZnO is primarily used as a micronutrient fertilizer however, its antimicrobial properties are well known (Pulugu Sai et al., 2020). Mechanism of action of nano-ZnO derived from zinc nitrate on important pathogen *Aspergillus fumigatus* demonstrated that it made cell wall deformity by hydroxyl and superoxide radicals mediated in fungal and finally led to death due to high energy transfer. ZnO nanoparticles (ZnO NPs) were also reported to be effective against two postharvest pathogenic fungi (*Botrytis cinerea* and *Penicillium expansum*), thus contributing in agriculture and food safety application. ZNo nanoparticles prevented the development of conidiophores and conidia of *P expansum*, which eventually lead to death of fungal hyphae.

Zinc oxide (ZnO), Magnesium oxide (MgO) and Titanium oxide (TiO_2): The antifungal activity of zinc oxide and magnesium oxide nanoparticles prepared by bio-safe method was evaluated for *Alternaria alternate, Fusarium oxysporum, Rhizopus stolonifer* and *Mucor plumbeus* (Wani and Shah, 2012). It was observed from the study that all the nanoparticles at different concentrations brought about significant inhibition in the germination of spores of *Alternaria alternata, Fusarium oxysporum, Rhizopus stolonifer* and *Mucor plumbeus*. However, the highest inhibition in the germination of all the test fungi was observed at higher concentrations followed by lower concentrations of nanoparticles. The nanoparticles of MgO at highest concentration were found most effective in reducing the spore germination followed by nanoparticles of ZnO at the same concentration.

The potential biocidal efficacy of ZnO and $ZnTiO_3$ nano-powders against the fungus *Aspergillus niger* was assessed (Manish Mathur et al., 2018).

$ZnTiO_3$ nano-powder showed higher growth inhibition efficiency than ZnO. Nanosuspension (0.0 to 0.5 ml) of Zinc and Magnesium oxide was used against *Alternaria alternata, Fusarium oxysporum, Rhizopus stolonifer* and *Mucor plumbeus.* They concluded that Nano-Mgo at highest concentration was most effective in reducing the spore germination followed by nano-Zno. Also, 30, 15, 7.5, 3.75, 1.87 and 0.938 mg/ml concentration of zinc nano rods were used towards controlling bacterial strains of *Pectobacterium carotovorum* subsp. *Wasabiae, P. atrosepticum, Dickeya chrysanthemi, D. solani* and *D. dianthicola.* They found that 30 mg/ml ZnO suspension concentration showed largest inhibition zone. Light Activated Nanoscale Formulations of TiO_2 significantly reduced disease incidence of Bacterial Spot of Tomato (*Xanthomonas perforans*).

4.2.9 Copper nano-particles

Antibacterial properties of nano-copper (20, 16, 8, 4, 2, 0.5, 0.4, and 0.2 ppm concentrations of nano-copers) against pomegranate bacterial blight (*Xanthomonas axonopodis* pv. *punicae*) was studied and reported that nano-copper suppressed xap growth at 0.2 ppm, i.e., >10,000 times lower than that usually recommended for Cu-oxychloride (Manish Mathur et al., 2018). Green synthesized copper nanoparticles were evaluated against *Colletotrichum capsici* where Carbendazim 50 WP @ 500 ppm and copper oxychloride 50 WP @ 2500 ppm served as standard checks. The mycelia inhibition of *Colletotrichum capsici* caused by copper nanoparticles was studied on PDA medium. CuNP-M @ 1000 ppm showed highest mycelial inhibition of 99.78% followed by 93.75% at 500 ppm and CuNP-E @ 1000 ppm compared to standard fungicides, carbendazim 50 WP @ 500 ppm (72.82%) and copper oxychloride 50 WP @ 2500 ppm (85.85%) (Krishnan and Shivanand Iliger et al., 2021). The CuNP-M @ 500 ppm were significantly superior to carbendazim 50 WP @ 500 ppm and copper oxychloride 50 WP @ 2500 ppm, but was statistically at par with CuNP-E @ 1000 ppm. This shows effectiveness of much lower concentration of copper nanoparticles compared to conventional fungicides. In detached fruit method, nanoparticles applied before inoculation of pathogen showed better results with regard to incubation period, lesion number and lesion size than after inoculation of pathogen. The study revealed a simple, convenient, non-toxic and cost-efficient technique for the synthesis of nanoparticles and their effectiveness against Colletotrichum capsici. CuNP-M first time synthesized and evaluated against *Colletotrichum capsici* performed better than CuNP-E.

Copper nanoparticles varying in sizes from 5 to 295 nm have been synthesized using leaf extracts and latex from plants, and using bacteria and fungi. Biosynthesized copper nanoparticles show good antimicrobial activity inhibiting the growth of pathogenic bacteria and pathogenic fungi. Copper nanoparticles enhance the germination and growth of some plants at lower concentrations, whereas high concentrations result in retarded growth. Studies carried on in vitro antifungal activity of copper nanoparticles reported the maximum antifungal activity against *Curvularia lunata* MTCC 2030 followed by *Alternaria alternata* MTCC6572, *Fusarium oxysporum* MTCC1755 and *Phoma destructive* DBT66 (Kasana et al., 2017). Copper nanoparticles synthesized using Citrus medica also demonstrated inhibitory activity against plant pathogenic fungi, *Fusarium culmorum, F. oxysporum* and *F. graminearum*. In another study, copper nanoparticles showed maximum antifungal activity against *Fusarium equiseti* with 25 mm zone of inhibition followed by *F. oxysporum* and *F. culmorum.* Copper nanoparticles of 20–50 nm at a concentration of 450 ppm could inhibit 93.98% growth of the Fusarium sp after 9 days of incubation Copper-based fungicide has been used in disease prevention and treatment in many plant species.

4.2.10 Validamycin nano-particles

Controlled release of validamycin loaded nano-sized calcium carbonate (50 to 200 nm) against *Rhizoctonia solani* was tested (Manish Mathur et al., 2018). Nano-encapsulation of thiamine dilauryl sulfate (TDS), a vitamin B1 derivative, (at 50 and 100 ppm concentrations for 0 to 24 hours) for inhibition of the spore germination and mycelia growth of *Fusarium oxysporum* f. sp. *raphani.* Study showed highest mycelia growth inhibition activity with 100 ppm TDS. Entrapping TDS solution in the nano size of 150 nm or smaller increases the surface area acting on the spores of *F. oxysporum*, as well as the spore penetration capability of TDS, thus generating a control effect greater than that of general pesticides acting on mycelia as reported.

Nano-sized calcium carbonate (nano-CC) was studied in terms of acting as a carrier for a pesticide. Nano-CC was prepared by reaction of calcium chloride and sodium carbonate by the reversed-phase microemulsion method and then loaded with the pesticide validamycin (Qian et al., 2011). The resulting material was characterized by X-ray diffraction analysis and scanning electron microscopy. The loading efficiency, sustained-release performance, germicidal efficacy, and stability also were investigated. The size of the loaded nano-CC can be adjusted to between 50 to 200 nm by varying the water/surfactant molar ratio from 30/1 to 10/1, and the loading efficiency can be increased to about 20% by increasing the size of the nano-CC. The material displayed better germicidal efficacy against *Rhizoctonia solani* compared to conventional technical validamycin after about 7 days, and the time of the release of validamycin was extended to 2 weeks. Given the loading efficiency, stability, sustained-release performance and good environmental compatibility of the material, the method for its preparation may be extended to other hydrophilic pesticide.

4.3 Successful applications of nanoparticles in plant disease management

Plant pathologists across the globe are working toward developing an effective solution for food and agricultural products protection from various pathogens. Thus, to integrate nanomethods with existing ones is utmost important to protect crops. Another upcoming field is myconanotechnology, which promotes the synthesis of NPs from fungi with contractible shape and size. Myconanotechnology with potential applications provides exciting waves of transformation in agriculture and captivates microbiologists to develop techniques for assurance in food security using green chemistry.

Nanotechnological approaches have potential to curb crop losses in a specific frame of time (Sharma Rishu et al., 2016). Nanotechnology applications present significant opportunities to more proficiently and unhazardous treat fungicides, herbicides, and fertilizers, by controlling their release. Now, efforts are being made to develop disease management strategies which are relatively less hazardous to humans and animals with minimum use of synthetic fungicides. The antifungal effect of silver NPs has received only minor attention. Silver NPs have strong inhibitory potential against fungi Botrytis cinerea causing gray mold. Kirby– Bauer disk diffusion method was used to evaluate the effect of fluconazole and silver NPs for their antifungal activity against three fungal pathogens including *Phoma glomerata, Phoma herbarium*, and *Fusarium semitectum.* Ag_2S nanocrystals on amorphous silica particles showed antifungal activity against *Aspergilus niger.* Silver ions and NPs were evaluated to determine the antifungal action on *Bipolaris sorokiniana* and *Magnaporthe grisea.*

The laboratory and field evaluations of NPs and silver ions evidenced for decreased disease development of phytopathogenic fungi. The antifungal effects of silver NPs were evalauted especially on sclerotia-forming phytopathogenic fungi. The antifungal activity of silver NPs was assessed against filamentous ambrosia fungi in South Korea. The effect of silver NPs on plant pathogenic spores of *Fusarium culmorum* was studied. The silver NPs were also found to exhibit antifungal activity against *Fusarium oxysporum.* Silver NPs strikingly decreased the number of germinating fragments and sprout length, relative to the control. *Botrytis cinerea* growth was restricted using zinc oxide NPs causing deformed mycelial pattern and similar inhibition pattern was observed in *Penicillium expansum* leading to death of fungal mats

(Sharma Rishu et al., 2016) The mycosynthesized silver NPs may be nontoxic to human and animals than synthetic fungicides. Moreover, in addition to the toxicity that NPs may cause on algae, plants, and fungi, they may also have some positive effects.

The antifungal activity of the silver NPs was evaluated on *Colletotrichum gloeosporioides*, which is responsible for causing fruit anthracnose. Silver NPs significantly reduced the mycelia growth of *C. gloeosporioides* in a dose-dependent manner. Antifungal properties of silver NPs, silver ions, acrylate paint, and cotton fabric impregnated with silver N were assessed against *Aspergillus niger, Aureobasidium pullulans,* and *Penicillium phoeniceum* (Khaydarov et al. (2011). Bioassay of elemental and nanosulfur against *Aspergillus niger* showed that nanosulfur was more efficient than its elemental structure. 'Smart Delivery Systems' for agriculture can possess timely controlled, spatially targeted, self-regulated, remotely regulated, preprogramed, or multifunctional characteristics to avoid biological barriers to successful targeting. In order to develop smart treatment–delivery system in plant, Zucchini (*Cucurbita pepo*) plants were treated with carbon-coated Fe NPs in vitro (Sharma Rishu et al., 2016). The magnetic cores consisting of Fe NPs allow themselves to be guided to a place of interest in the body (affected part) of an organism using small magnets that create a magnetic field. The carbon coating provides biocompatibility and acts as a surface for adsorption where various types of molecules of interest (drug/DNA/chemical/enzyme) can be adsorbed. González-Melendi et al. (2008) were the first to report the penetration and transport of NPs inside whole plant.

Plants are responsive to various stress conditions and they combat it through physiological changes, viz., induction of systemic defense. A sensitive electrochemical sensor was developed by harnessing this indirect stimulus, and the use of gold electrode with copper NPs has been demonstrated to check the levels of salicylic acid and pathogen detection. An extensive research on similar sensors and sensing techniques, however, needs to be expanded for detecting pathogens, their by-products, or monitoring physiological changes in plants. Nanodispersed formulations can be prepared in a simple cost-effective manner and are suited for developing new forms of fungicidal materials. Among the most widely used nanoformulation for the treatment of diseases in turf and ornamental plants, soil, and seeds is Syngenta's Banner MAXX™. It is commercialized as a microemulsion concentrate formulation, providing excellent tank mix compatibility and stability (Sharma Rishu et al., 2016). Another nanoformulation 'Nano-5' is a marketed product and acts as a natural mucilage organic solution to control many plant pathogens (Table 11).

Table 11: Prevention of plant diseases with nanoformulation Nano-5

Plant disease	Mode of application	Killing time
Gray mold, blast, Fusarium wilt, early blight	Spray Nano-5 on leaf surface every 3 days	1–2 h
Late blight, Phytophthora diseases, southern blight, white root rot, blister blight of tea, rust	Apply to roots twice	1–2 h
Sclerotinia rot, ergot, powdery mildew, Fusarium root rot, downy mildew	Spray on leaf surfaces once in 5–7 days	Stops infection within 1–2 h
Bakanae disease, white rust, leaf blight, soft rot	Apply to roots twice h	Stops infection within 1–2
Bacterial wilt, leaf spot, rot, brown leaf spot, black rot	Apply to roots twice	Stops infection within 1–2 h
Mosaic, ring spot, transitory yellowing, Tristeza virus, Exocortis viroid	Spray on leaf surfaces and to roots once in 3 days	Complete control in 7 days

Resistance in plants would help in management of the above-mentioned agents to overcome the problem of economic loss. Nanoparticle-mediated plant transformation has the potential for genetic modification of plants for further improvement. In particular, employment of nanoparticles in agriculture could target specific problems in plant protection, pathogen detection, and deciphering plant–pathogen interactions and offers new methods for plant disease management (Mahendra et al., 2012). For example, introduction of resistance genes in plant cells using nanotechnological approaches for developing resistant varieties will curtail the expenses incurred on agrochemicals. Nanophytopathology can be applied as a tool to understand plant–pathogen interactions, which will provide new methods for crop protection.

Various research groups focused their interest on the usage of eco-friendly pesticides. Similar to chemical pesticides, nanoparticlebased pesticides and herbicides are being explored for the application of the antimicrobial agents to protect crops from various diseases. Extensive studies on nanoparticle-based systems may eliminate the intensive use of pesticides in the agricultural sector. The antifungal properties of nanoparticles can help to formulate nanoparticle- based pesticides. Among the different inorganic nanoparticle-based antimicrobial agents, silver has been extensively studied by many researchers because of its several advantages over other nanoparticles such as copper, zinc, gold, ZnO, Al_2O_3 and TiO_2. The use of nanosilver has been studied recently against *Colletotrichum gloeosporioides.* Other nanoparticles (Fe, Cu, Si, Al, Zn, ZnO, TiO_2, CeO_2, Al_2O_3 and carbon nanotubes) have been reported to have some adverse effects on plant growth apart from the antimicrobial properties (Alaa Y. Ghidan and Tawfiq M. Al Antary, 2019). Sometimes, nanoparticles also have an effect on the growth of useful soil bacteria, such as *Pseudomonas putida* KT2440.

Both the concentrations of nickel nanoparticles (50 and 100 ppm) inhibited the fungal mycelial growth on solid media, and the inhibitions were significant ($p \leq 0.05$) over control. Nickel nanoparticles at 100 ppm concentration inhibited the mycelial growth of *F oxysporum* f. sp. *lactucae* and *F oxysporum* f. sp. *lycopersici* by 60.23 and 59.77%, respectively, over control. The inhibitory effects of nickel nanoparticles were also assessed in liquid medium, and the results were similar with solid media. In the liquid media, the fresh mycelial weight of the tested fungal pathogens decreased significantly and more than 50% reduction was recorded with the use of nickel nanoparticles at the concentration of 100 ppm. The results revealed that mycelial growth of tested pathogens was inhibited in a concentration dependent manner. These results suggest that using Ni nanoparticle solution can significantly increase the surface areas acting on the mycelia of *Fusarium* and mycelial growth.

Nickel nanoparticles at the concentration of 100 ppm decreased the number of spore development by 81.40 and 74.60% in *F. oxysporum* f. sp. *lactucae* and *F. oxysporum* f. sp. *lycopersici,* respectively. The conidial germination was negatively affected by nickel nanoparticles. The inhibitory effect of Ni nanoparticles on spore germination could be due to their fungicidal effect. These results agreed with the results obtained from other workers on antifungal effects of different metal nanoparticles against some pathogenic fungi such as silver nanoparticles and zinc nanoparticles against copper nanoparticles. Inhibitory effect of Ni nanoparticles could be due to producing of extracellular enzymes from fungi as survival agents caused by stress of toxic materials or could be due to large surface areas and small sizes to penetrate into the cell membrane of pathogen and work in the cytosols.

The antifungal activity of zinc oxide (ZnO) and magnesium oxide (MgO) nanoparticles prepared by bio-safe method was evaluated for *Alternaria alternata, Fusarium oxysporum, Rhizopus stolonifer* and *Mucor plumbeus* (Wani and Shah, 2012). It was observed from the study that all the nanoparticles at different concentrations brought about significant inhibition in the germination of spores of *Alternaria alternata, Fusarium oxysporum, Rhizopus stolonifer* and *Mucor plumbeus.* However, the highest inhibition in the germination of all the test fungi was observed at higher concentrations followed by lower concentrations of nanoparticles. The nanoparticles of MgO at highest concentration were found most effective in reducing the spore germination followed by nanoparticles of ZnO at the same concentration. Antifungal effect of silver nanoparticles on some pathogenic fungi has also been reported by some workers. It is observed a significant reduction in mycelial growth and spore germination incubated with silver nanoparticles. Inhibitory effect was found to be dependant over incubation and the concentration of silver nanoparticles.

Nanofungicides based on silver nanoemulsion tackle fungal pathogen problems in agriculture and hold promise in the future. World over scientists have been propagating this branch of science through their research, but very little is done to bring this technology at farmers level (Nikunj Patel et al., 2014). They addressed various problems associated with the existing methods of fungal phytopathogen control and also the emphasis has been laid upon the use of AgNPs as Ag nanoemulsion to control phytopathogens in the field and application of AgNPs along with the biofertilizers to supplement their effect and promote the plant growth in a better way.

The recent advances in plant disease management using nanoparticles themselves as protectants, as well as nanoparticles as carriers for insecticides, fungicides, herbicides and dsRNA for RNA-interference (RNAi)-mediated protection were discussed (Elizabeth A. Worral et al., 2018). For years, nanotechnology has shown the potential to create a strong footprint for developing effective formulations. However, the authors believe that agricultural nanotechnology research is not reaching its potential. due to the scarcity of commercial applications.

4.4 Types of nanoparticles for plant disease management

4.4.1 Nanoparticles as Protectants

Nanoparticles are materials that range between 10 to 100 nanometers (nm), and can be designed with unique chemical, physical, and biological properties, to distinctively differ from those of their molecular and bulk counterparts. Nanoparticles alone have the potential to be directly applied to plant seeds, foliage, or roots for protection against pest and pathogens, such as insects, bacteria, fungi, and viruses. Metal nanoparticles such as silver, copper, zinc oxide, and titanium dioxide have been intensively researched for their antibacterial and antifungal properties, and are known for their antiviral properties. This section gives a brief overview and an update on current literature reviews pertaining to the individual nanoparticles that already exist. Silver nanoparticles have shown antifungal inhibition of *Alternaria alternata, Sclerotinia sclerotiorum, Macrophomina phaseolina, Rhizoctonia solani, Botrytis cinerea*, and *Curvularia lunata* by well diffusion assay. When silver nanoparticles were sprayed onto bean leaves, complete suppression of sun-hemp rosette virus was observed. Faba bean plants challenged with bean yellow mosaic virus, and sprayed with silver nanoparticles, produced remarkably better results when the nanoparticles were applied 24 h post-infection, compared to spray application before infection, or simultaneously at the time of inoculation. Silver nanoparticles have shown immense potential for plant disease management against fungal and bacterial pathogens, but there are significant hurdles associated to them, such as production, toxicity, and soil interaction (Elizabeth A. Worral et al., 2018).

Other commonly used metal nanoparticles include copper, titanium dioxide, and gold. Nanoparticles, such as copper and titanium dioxide, are more commonly being utilized for fertilizer with little research into plant disease management. The research covering titanium dioxide, silver, and copper nanoparticles for antimicrobial properties, and aluminum nanoparticles as an insecticidal dust, has been highlighted. Furthermore, titanium dioxide nanoparticles in fertilizers have produced protection from bacteria and inactivation of viruses. Introducing poly-dispersed gold nanoparticles via a mechanical abrasive was seen to melt and dissolve the Barley yellow mosaic virus particles conferring resistance to the plant.

Chitosan is another popular nanoparticle that has favorable biological properties, such as biodegradability, biocompatibility, non-allergenicity, and antimicrobial activity, with low toxicity to animals and humans. Chitosan nanoparticles induce viral resistance in plant tissues by protecting them against infections caused by the mosaic virus of alfalfa, snuff, peanut, potato, and cucumber. Chitosan nanoparticles have shown antimicrobial properties, such as controlling Fusarium crown, root rot in tomato, Botrytis bunch rot in grapes, and *Phyricularia grisea* in rice, but are less effective against bacteria. Potential mechanisms were summarized that lead to the antimicrobial effects of chitosan, such as agglutination, disruption of the cell membrane, inhibition of H+ -ATPase activity, inhibition of toxin production and microbial growth, inhibition of the synthesis of messenger RNA and proteins, and blockage of nutrient flow (Elizabeth A. Worral et al., 2018). Antiviral effects have been observed in beans against Bean mild mosaic virus and, in tobacco, against Tobacco mosaic virus and Tobacco necrosis virus.

Nanomaterials as unique carriers of agrochemicals facilitate the site-targeted controlled delivery of nutrients with increased crop protection. Due to their direct and intended applications in the precise management and control of inputs (fertilizers, pesticides, herbicides), nanotools, such as nanobiosensors, support the development of high-tech agricultural farms. The integration of biology and nanotechnology into nonosensors has greatly increased their potential to sense and identify the environmental conditions or impairments (Yifen Shang et al., 2019). The remarkable antimicrobial potential of various metal nanoparticles may be applied for the development of novel nanoantimicrobials (antiviral, antibacterial, fungicidal, and pesticidal agents, etc.), which can be used as an effective tool for the management of plant diseases.

4.4.2 Nanoparticles act as Carriers

Smart nano-based pesticides are designed to efficiently delivery sufficient amounts of active ingredients in response to biotic and/or abiotic stressors that act as triggers, employing targeted and controlled release mechanisms (Camara, et al. 2019). The advantages of controlled release for the delivery of agrochemicals include reductions of phytotoxicity, leaching losses, volatilization, drift, and soil degradation, as well as increased safety during application. However, the release of active agents from loaded particles mainly occurs by means of passive diffusion, capsule erosion, or osmotic pressure, resulting in poor control of pesticide release. Hence, stimuli-responsive systems are promising candidates for improving controlled release properties, promoting site-specific and smart release of pesticides in response to biotic or abiotic stimuli.

The advantages of controlled release for the delivery of agrochemicals include reductions of phytotoxicity, leaching losses, volatilization, drift, and soil degradation, as well as increased safety during application. However, the release of active agents from loaded particles mainly occurs by means of passive diffusion, capsule erosion, or osmotic pressure, resulting in poor control of pesticide release. Hence, stimuli-responsive systems are promising candidates for improving controlled release properties, promoting site-specific and smart release of pesticides in response to biotic or abiotic stimuli. Prospecting new responsive carriers that respond to environmental stimuli including diseases could be an important strategy for improving the resistance of crops to the stresses caused by these environmental factors.

The design of formulations using stimuli-sensitive carriers is a strategy to improve the controlled release system, where the macromolecules present an active response to small signals or modifications in the surrounding environment, resulting in changes in their physicochemical properties favoring the release of loaded substances (Milani et al. 2017). When used in agrochemical formulations, these sensitive carriers can allow the release of active compounds in response to abiotic or biotic stimuli such as light, pH, temperature, magnetic field, enzymes, among others. Stimuli-controlled delivery systems can be prepared using natural polymers (such as alginate, cyclodextrin, starch, chitosan, carboxymethylcellulose, and ethylcellulose) and silicas due to their wide availability and biodegradability. Recently, carriers based on pillararenes and curcubit[n]urils were studied in pesticide and herbicide delivery due to their responsiveness to environmental stimuli. The use of stimuli-sensitive carriers in nanoformulations can enable the release of different compounds in an effective manner and at defined times, hence increasing bioefficacy, while reducing side effects, the dosage required, and the number of applications. The use of stimuli-responsive particles is a promising technology that is still in an early stage of development and clearly needs further exploitation to extend its use in agroindustry. Therefore, the aim of this review is to discuss the current state of development of stimuli-responsive carrier systems, where the mechanisms of release of active ingredients are activated by different triggers, in response to biotic and abiotic factors, as well as to highlight the perspectives and commercial challenges for the use of smart delivery systems in the agricultural sector.

The development of nanomaterials and related technologies has provided new ways of creating intelligent nano-pesticides. Nanomaterials have important applications in areas such as pharmacology and biology because their physical and chemical properties are different from those of macroscopic materials. Nanomaterials feature unique surface properties, a small size, and quantum size effects. Nano-microcapsules or nanospheres have been prepared using light-sensitive, thermo-sensitive, humidity-sensitive, and enzyme- and soil pH-sensitive high polymer materials to deliver pesticides. These nanostructured materials are prepared via processes such as adsorption, coupling, encapsulation, and embedding (Camara et al., 2019). Such formulations can protect pesticides' active ingredients and enhance stability, control the release of core materials, reduce or obscure odors, decrease volatility, and isolate components. In recent years, the application of nanomaterials and technologies in the field of pesticides has made considerable strides. In particular, targeted delivery and controlled release of nanomaterials can improve pesticide utilization and reduce residue and pollution. Pesticide nanocapsule formulations have slow release and protection performance and, due to their small size, improvable pesticide droplet ductility, wettability, and target adsorption when spraying fields; these methods provide efficient and environmentally friendly advantages. While progress has been made in research on pesticide nanocapsule formulations, preparation technology warrants further study. Popular pesticide microcapsule preparation methods are shown in Table 12.

Table 12: Main pesticide microcapsule preparation methods

Method	Preparation Process	Example
Interfacial polymerization	Two reactive monomers are dissolved in two different solvents. When one solvent is dispersed in another solvent, the two monomers undergo a polycondensation reaction at the phase interface to form microcapsules	Natural pyrethrin nanocapsules
In situ polymerization	Two or more water-soluble monomers are polymerized to form a water-insoluble dipropylthiocarbamate, polymer and are deposited on the surface of the core material for coating	S-ethyl acetochlor, atrazine, methotrexate
Emulsion polymerization	A solvent-insoluble monomer is dispersed in a solvent to form a uniform emulsion via mechanical agitation, high-speed shearing, and vigorous shaking with a surfactant (an emulsifier). Then, the polymerization reaction is initiated to form the polymer to achieve core material encapsulation	Abamectin nanocapsule suspension], natural pyrethrin nanocapsules
Membrane emulsification	The dispersion phase enters the continuous phase through a shirasu porous glass membrane under inert gas pressure, and the continuous phase breaks into droplets on the membrane surface by the shear forces of the SPG membrane and droplets.	Chlorantraniliprole nanocapsules, avermectin nanocapsules
Solvent evaporation	The wall material and core material are dispersed in the organic phase, added to the solution immiscible with the wall material, and the wall material is precipitated to form the microcapsule by heating and evaporating the solvent	Spinosad nanocapsules
Nano- precipitation	The interfacial interaction between solvent and non-solvent disperses the polymer and drug from the oil phase into the aqueous phase. This material can quickly wrap the drug and obtain nanocapsules through precipitation	Pyrazole azoxystrobin nanocapsules, azoxystrobin microcapsules
Double coacervation	Two oppositely charged water-soluble polymers form a wall around the water-insoluble pesticide active ingredient, which is a spontaneous liquid- to-liquid separation caused by electrostatic interactions	Azoxystrobin microcapsules

Nanostructured materials can also respond to external stimuli such as temperature, light, humidity, and pH to achieve an environmental response and controlled release of pesticides. Therefore, intelligent microcapsules that can respond to the external environment have attracted considerable attention.

Although environmentally-responsive nano-drugs have been rapidly developed and applied in the medical field, the application of these materials to pesticides has only recently begun.

In pesticides field, although simple slow release has been achieved, materials showing responsive release based on environmental changes have yet to be developed. The pesticide field also requires continued systematic research, development, and improvement in environmentally responsive, targeted, controlled-release pesticides. In particular, effective systems that are responsive to internal bio-stimulation present a great challenge in this field. Pesticide-responsive controlled-release agents effectively improve pesticide utilization and substantially reduce pesticides' negative effects on agricultural products and the environment. Agents should have the ability to respond intelligently to the external environment and should have a trigger release of active ingredients to effectively control harmful organisms. Microcapsule pesticide formulations are an important application of this technology in pesticide formulation preparation. Compared with conventional pesticide formulations, these systems have advantages such as long lifetimes, reduced environmental effects, and lower toxicity. These features may also improve rain resistance. In recent years, development prospects in areas related to pesticide- controlled release have drawn much attention (Zhao et al., 2017).

If a pesticide microcapsule can be properly applied, the pesticide will be released only at the target tissue. This process allows the pesticide to be used more effectively while ensuring that affected plant tissues receive the optimum therapeutic effect, thus reducing environmental pollution. Therefore, the application of microcapsule materials in pesticides and the development of new environmentally intelligent and responsive pesticide formulations represent new directions in pesticide formulation.

Nanoparticles are also commonly used as carriers to entrap, encapsulate, absorb, or attach active molecules to develop effective agricultural formulations.

The common nanoparticles which have been used as carriers for insecticides, fungicides, herbicides and RNAi-inducing molecules, are summarized below.

a. Silica nanoparticles: They can be easily synthesized with a controlled size, shape and structure, making them highly advantageous delivery vehicles. They are commonly produced in a spherical shape with pore-like holes; for example, porous hollow silica nanoparticles (PHSNs) or mesoporous silica nanoparticles (MSNs). PHSN and MSN commonly load the pesticide into the inner core to protect the active molecules and, therefore, provide a sustained release. The shell structure of PHSNs protects the active molecules inside the nanoparticles against degradation by UV light. The literature suggests that silicon has already been used to enhance plant tolerance against various abiotic and biotic stresses and, therefore, silica nanoparticles seem to be the natural choice for the development of agri-products for pests. A novel hydrophilic delivery system through loading Abamectin with porous silica nanoparticles (Abam-PSNs) was developed in order to improve the chemical stability, dispersity, and the controlled release of Abamectin (Wang et al., 2014). These results suggest that Abam-PSNs can significantly improve the performance of controllable release, photostability, and water solubility of Abamectin by changing the porous structure of silica nanoparticles, which is favorable to improve the bioavailability and reduce the residues of pesticides.

A novel delivery system was developed for post-loading a biopesticide to mesoporous silica nanoparticles (MeSiNPs) functionalised with N-rich derived polymer (Olivija Plohl et al., 2021). The system showed a fast initial release behaviour in the first 300 min, which was continued by slow release behaviour and started to reach equilibrium after 10 days. It has been shown that the MeSiNPs@bPEI showed an excellent potential for use in agrochemical applications, not only as a nanocarrier for controlled and slow pesticide release, but may also act as a soil improver by providing Si and N as key nutrient sources.

Mesoporous silica nanoparticles (MSNs) with particle diameters of ~110 nm and pore sizes of ~3.7 nm were synthesized via a liquid crystal templating mechanism. A water-soluble chitosan (CS) derivative (N-(2-hydroxyl) propyl-3-trimethyl ammonium CS chloride, HTCC) was successfully capped on the surface of pyraclostrobin-loaded MSNs (Lidong Cao et al., 2016). HTCC-capped MSNs released faster than bare MSNs in the initial stage. Pyraclostrobin-loaded HTCC-capped MSNs with half doses of pyraclostrobin technical demonstrated almost the same fungicidal activity against *Phomopsis asparagi* which obviously reduced the applied pesticide and enhanced the utilization efficiency. Therefore, HTCC-decorated MSNs demonstrated great potential as nanocarriers in agrochemical applications.

Hollow porous silica nanospheres (HSNs) have unveiled as a potential carrier for pesticide delivery. However, HSNs mostly suffer from postsynthesis loading process of pesticides because of the tiny mesopores onto shell (Md Nuruzzaman et al., 2020). To eradicate this disadvantage, it was hypothesized that developing a large through hole or pore opening on shells (>10 nm) could be effective for the postsynthesis loading of active molecules onto HSNs using a simple immersion method.

Mesoporous silica nanoparticles (MSNs) have become popular nanomaterials in smart delivery systems. Although research progress in the application of MSNs as pesticide carriers has been achieved, multifunctional MSNs endowed with bright luminescent centers facilitating the tracking of MSNs in biological systems and versatile structural properties possessing a high drug loading capacity and regulable release are still highly desirable. Cao et al., (2018) reported a fluorophore-free method to endow MSNs with stable fluorescence and a double-shelled hollow structure; they were prepared by a selective-etching strategy and subsequent annealing treatment. This research seeks to develop a novel nanocarrier platform for the potential application of pesticides in sustainable plant protection.

The synthesis and evaluation of the direct antifungal activity of mesoporous silica nanoparticles (MSN) was described compared to metalaxyl (recommended fungicide) against *Alternaria solani* under laboratory and greenhouse conditions (Aly Derbalah, et al., 2018). The structural features of MSN such as high porosity, small particle size and suitable shape contributed to its high antifungal efficacy against *Alternaria solani*. Laboratory synthesized MSN showed marked increase in tomato growth parameters compared to untreated control. This study presented promising results of the use of MSN as an effective and safe alternative of fungicides for managing tomato early blight.

The surface characteristics of MSN support this point of view since the positively charged MSN interacts with the protein thiol groups (-SH) on the fungal cell surface to cause cell lysis. Finally, MSN may function against *A. salani* by increasing tomato plant resistance, physiological characteristics and its self-defense against the disease by absorbing silicon (Si) into plants.

b. Chitosan nanoparticles : They have low solubility in aqueous media, due to their hydrophobic properties. As a result, chitosan is commonly mixed with a copolymer, organic and inorganic, to improve its solubility Chitosan has reactive amine and hydroxyl groups, allowing modification, graft reactions, and ionic interactions, enabling improvement of chitosan properties. Chitosan adheres well to the epidermis of leaves and stems, prolonging the contact time and facilitating the uptake of the bioactive molecules (Elizabeth A. Worral et al., 2018).

Chitosan-based agronanochemicals exhibit highly controlled release behavior that subsequently increases its bioavailability with high circulation and retention time in the plant tissue (higher half-lives, t1/2) (Farhatun Najat Maluin and Mohd Zobir Hussein, 2020). Thus, the controlled release of active ingredients in agrochemicals aims to address the problems associated with the excessive usage of agrochemicals by reducing the quantiies and frequency application in the field. The agrochemical release from the chitosan matrix can be triggered by two types of stimuli, viz., biotic stress, such as the presence of plant pathogens (fungi and bacteria), nematodes, insects, pest and weeds; and abiotic stress factors, such as pH, temperature, salinity, flooding, drought and other environmental factors.

The release mechanism upon stimulus-response is through pore diffusion, surface desorption, capsule swelling and degradation. The diffusion-controlled release of methomyl at pH 6.0 with t1/2 of 36–70 h has been obtained by its encapsulation into carboxymethyl chitosan and azidobenzaldehyde (Farhatun Najat Maluin and Mohd Zobir Hussein, 2020). The release of the insecticide acetamiprid from nanocapsules of chitosan-alginate was reported to be pH-dependent, in which half of the acetamiprid was released after 36 h at pH 7.0 and 4.0 compared to only 24 h needed to release the same amount at pH 10.

The mechanism of actions of chitosan against pathogens : The antimicrobial action of chitosan on pathogens (bacteria, fungi and virus) relies on several mechanisms, viz., the positive charge of the protonated chitosan enables electrostatic interactions with the negative charge of the pathogen surface; the cell damage and leakage of the pathogen, hence increases its membrane permeability and subsequently results in cell death ; chitosan then chelates the essential elements (including metal ions, minerals and nutrients) for the growth of pathogens, thus, preventing the normal growth of pathogens; DNA/RNA interaction of pathogens with the penetrated chitosan leads to the inhibition of the mRNA syncretization and pathogen reproduction; and the deposition of chitosan on the microbial surface of pathogens forms a barrier to extracellular transport of the essential nutrients and metabolites from entering the cell, hence, inhibits the normal growth of pathogens (Farhatun Najat Maluin and Mohd Zobir Hussein, 2020).

The efficiency of the mechanism in action can be enhanced by the small size of the chitosan nanoformulations due to their high surface area that comes in contact with the pathogens. The reduction in gray mold incidence in pre- and post- harvest with a 2-fold increase of PAL upon treatment with 1% chitosan on grapes has been reported. Similarly, rice and wheat treated with chitosan induced PAL production and reduced disease incidence. The enzyme polyphenol oxidase helps to catalyze the phenolic substances to lignin biosynthesis. Hence, the increase of these enzymes can be interpreted as an increase in lignin formation that contributes to the building up of cell wall structure, hence, forming a barrier for the penetration of pathogens (Li and Zhu., 2013).

Previous studies have also reported on the accumulation of defense-related secondary metabolites, including phytoalexins, phenolic compounds, lignin and callose in plants treated with chitosan. Phytoalexins are toxins that have antimicrobial and antioxidant properties produced upon infection by pathogens, the results of metabolites' reactions against the disease. Interestingly, plants incorporated with chitosan were also found to be able to elicit phytoalexins. Moreover, upon chitosan treatment, plants exhibit pathogenesis-related proteins, including chitinase and β-1,3-glucanase (Xing et al., 2015). In response to pathogen attack, plants induce proteins and peptides with antimicrobial properties to protect themselves. Interestingly, chitosan also can act as an elicitor in inducing the pathogenesis-related proteins and improve the plant resistance against pathogens. Chitinase and β-1,3-glucanase can act as a catalytic converter in the hydrolysis of chitin and β-D-glucans that can be found in the fungal cell wall and insect exoskeletons. Hence consequently, it degrades the fungal cell wall and stops the fungal growth on the host plant.

Biodegradable chitosan-lactide copolymer (CPLA) was prepared and characterized as hydrophobic pesticide-pyraclostrobin carrier (Lei Xu et al., 2014). Nano-precipitation method was used to fabricate pyraclostrobin-loaded nanoparticles. Loading content (LC), encapsulation efficiency (EE), size, and size distribution of the resulting nanoparticles were investigated.

Although fungicides could be the best solution in combating fungal infections in crops, however, the phytotoxic level of fungicides to the crops should be tested first to ensure that it is safe for the crops. Moreover, nanocarrier systems of fungicides could play a significant role in the advancement of crop protection. For this reason, chitosan was chosen in the present study as a nanocarrier for fungicides of hexaconazole and/or dazomet in the development of a new generation of agronanofungicides with a high antifungal potent agent and no phytotoxic effect. Hence, the encapsulation of fungicides into the non-toxic biopolymer, chitosan was aims to reduce the phytotoxic level of fungicides (Farhatun Najat Maluin et al., 2020).

Polymer nanocapsules carrying a fungicide, hexaconazole were developed through ionotropic gelation method utilizing chitosan and tripolyphosphate (TPP). The nanocapsules were characterized by photon correlation spectroscope (PCS), transmission electron microscope (TEM), and Fourier transform infra- red (FTIR) spectroscope (Chauhan et al., 2017).

Solid lipid nanoparticles (SLNs): These are similar to emulsions, and are composed of lipids that are solid at room temperature. An advantage of SLNs is that they provide a matrix to entrap lipophilic active molecules without the use of organic solvents. Furthermore, SLNs can also provide controlled release of various lipophilic components, due to decreased mobility of the active in the solid matrix. Surfactants are used to stabilize the SLN when dispersed into water . Their main drawbacks are that they have a low loading efficiency, and the active may leak out of the structure during storage.

The introduction of Solid Lipid Nanoparticles (SLN) was in late 1991 to be a substitute transporter system to what known old colloidal carriers, like emulsions, liposomes, and polymeric micro- and nanoparticles (Ahmed A.A Alsaad et al., 2020). SLN has merits and potentialities of classical structures except avoiding some of their common and known disadvantages.

Solid lipid nanoparticles (SLNs) represent a transport system alternative to conventional colloidal carriers. SLNs combine the advantages of traditional and modern systems if some of their major difficult proceedings are eliminated. SLN production techniques include drug incorporation, drug loading and release capacity, with a special focus on drug release methods.

With the aim to establish a novel nanocarrier system with higher payload and higher photo-protection for deltamethrin (active ingredient) compared to solid lipid nanoparticles (SLN), nanostructured lipid carriers (NLC) were prepared by combining hot homogenization and sonication with different ratios of corn oil (liquid lipid) and beeswax (solid lipid) (Nguyen et al., 2012).

Carbendazim (MBC) (methyl-2-benzimidazole carbamate) and tebuconazole (TBZ) ((RS)-1-(4-chlorophenyl)-4, 4-dimethyl-3- (1H-1,2,4-triazol-1-ylmethyl) pentan-3-ol) are widely used in agriculture for the prevention and control of fungal diseases. Solid lipid nanoparticles and polymeric nanocapsules are carrier systems that offer advantages including changes in the release profiles of bioactive compounds and their transfer to the site of action, reduced losses due to leaching or degradation and decreased toxicity in the environment and humans (Campos et al., 2015). Two types of nanoparticle were prepared as carrier systems for a combination of TBZ and MBC, then investigated the release profiles of the fungicides as well as the stabilities and cytotoxicities of the formulations. Both nanoparticle systems presented high association efficiency (>99%), indicating good interaction between the fungicides and the nanoparticles. The release profiles of MBC and TBZ were modified when the compounds were loaded in the nanoparticles and cytotoxicity assays showed that encapsulation of the fungicides decreased their toxicity. These fungicide systems offer new options for the treatment and prevention of fungal diseases in plants.

c. Layered double hydroxides (LDHs): These are clays that form into hexagonal sheets with layers of the active molecules trapped in the interlayer space. LDH nanoparticles breakdown under acidic conditions like the addition of moisture and carbon dioxide from the atmosphere. Positively charged delaminated LDH lactate nanoparticles have been shown to facilitate the transport of biologically active materials across the plant cell wall barrier. Silica, chitosan, SLN, and LDH nanoparticles have been utilized as common carriers for plant disease management. The individual studies using these nanoparticles, and less frequently used nanoparticle types, are discussed in-depth in the sections categorized by insecticide, fungicide, herbicide, or RNAi. The use of host inorganic sources for the encapsulation of agrochemicals in silicate clays and zeolites, bring LDH to be subject of several studies (Luíz Paulo Figueredo Benício et al., 2015). Synergistic effect benefits of the interaction between inorganic sources and agrochemicals may be used with environmental hazard reduction.

Layered double hydroxides (LDHs) have been known for many decades as catalyst and ceramic precursors, traps for anionic pollutants, catalysts and additives for polymers, but their successful synthesis on the nanometer scale a few years ago opened up a whole new field for their application in nanomedicine (Katharina Ladewig et al., 2009).

Soil-compatible layered double hydroxides (LDHs) were hybriduzed with natural antibiotic substances (Park et al, 2010). The cinnamate-LDH hybrid was synthesized by a typical coprecipitation method. Cinnamic acid was analyzed by high performance liquid chromatography which was operated at 280 nm with C18 column. Its controlled release property was evaluated in a cultivated soil as well as a simulated soil solution. Its antifungal activity was examined against the growth of *Phytophyhora capsici* in a potato dextrose agar medium and a red pepper seedling, respectively.

Its antifungal activity against the growth of *P. capsici* in a potato dextrose agar medium and a red pepper seedling definitely shows that the hybrid is very effective in controlling the root rot of red pepper. This study demonstrated that the hybridization of natural antibiotic substances with layered double hydroxides could be a fascinating alternative for green formulation of pesticides.

4.4.3 Nanoparticles and RNAi in plant protection

The discovery of the RNAi pathway has heralded new and innovating approaches for the management of pests and pathogens. RNAi, a conserved eukaryotic mechanism, is involved in the growth, development, and host defense against viruses and transposons, that can also be hijacked to target insects, fungi, viruses, and weeds (Elizabeth A. Worral et al., 2018). In plants, RNAi is triggered by dsRNA, which is processed into small-interfering RNA (siRNA) by Dicer-like (DCL) enzymes. These siRNAs are incorporated into a RNA-induced silencing complex (RISC); siRNAs then direct the RISCs through base pairing to degrade the pathogen RNA, by preventing it from being used as a translation template. Since its discovery, the RNAi pathway has emerged as a powerful tool to combat plant pests and pathogens by genetic modification. However, genetically modified organisms are controversial and highly regulated in most countries; thus, research into new dsRNA delivery methods is being undertaken.

Despite the numerous potential advantages of nanotechnology in plant protection, it has not yet made its way commercially into our diseased fields. Many countries across the globe are pursuing Research & Devclopment for nanotechnological application in agriculture to nullify the toxic effects of chemicals used in field. In the future, nanoscale devices with novel properties could be used to make agricultural systems 'smart.' A recent study by toxicologists reported that nanomaterials may be toxic but still therc are no regulations to their exposure till date (Sharma Rishu et al., 2016).

Topical application of pathogen-specific double-stranded RNA (dsRNA) for virus resistance in plants represents an attractive alternative to transgenic RNA interference (RNAi). However, the instability of naked dsRNA sprayed on plants has been a major challenge towards its practical application (Mitter et al., 2017). dsRNA that can be loaded on designer, non-toxic, degradable, layered double hydroxide (LDH) clay nanosheets was demonstrated. Once loaded on LDH, the dsRNA does not wash off, shows sustained release and can be detected on sprayed leaves even 30 days after application.

RNA nanoparticles with many unique features are promising candidates in disease treatment. The miRNAs are involved in many biochemical and developmental pathways and their regulation in plants and animals. These appear to be a powerful tool for controlling various pathological diseases in human, plants and animals, however there are challenges associated with miRNA based nanotechnology (Chaudhary et al., 2018). Several advancements made in the field of miRNA therapeutics make it an attractive approach, but a lot more has to be explored in nanotechnology assisted miRNA therapy. The miRNA based technologies can be employed for detection and combating crop diseases as well. Despite these potential advantages, nanobiotechnology applications in the agricultural sector are still in its infancy and have not yet made its mark in comparison with healthcare sector. This review provided a platform to discuss nature, role and use of miRNAs in nanobiotechnology applications.

Magnesium oxide nanoparticles (nMgO): Engineered nanoparticles have provided a basis for innovative agricultural applications, specifically in plant disease management. In a interdisciplinary study, by conducting comparison studies using macroscale magnesium oxide (mMgO), (Chen Juanni et al., 2020) evaluated the fungicidal activity of MgO nanoparticles (nMgO) against soilborne *Phytophthora nicotianae* and *Thielaviopsis basicola* for the first time under laboratory and greenhouse conditions. *In vitro* studies revealed that nMgO could inhibit fungal growth and spore germination and impede sporangium development more efficiently than could macroscale equivalents. Indispensably, direct contact interactions between nanoparticles and fungal cells or nanoparticle adsorption thereof were found, subsequently provoking cell morphological changes by scanning electron microscopy/energy-dispersive spectrometry (SEM/EDS) and transmission electron microscopy (TEM). In addition, the disturbance of the zeta potential and accumulation of various modes of oxidative stress in nMgO-exposed fungal cells accounted for the underlying antifungal mechanism. In the greenhouse, approximately 36.58 and 42.35% decreases in tobacco black shank and black root rot disease, respectively, could testify to the efficiency by which 500 μg/ml of nMgO suppressed fungal invasion through root irrigation (the final control efficiency reached 50.20 and 62.10%, respectively) when compared with that of untreated controls or mMgO. This study will extend our understanding of nanoparticles potentially being adopted as an effective strategy for preventing diversified fungal infections in agricultural fields.

Deeper insight into the antifungal mechanisms of nMgO against phytopathogenic fungi was investigated in comparison with those of macroscale MgO (mMgO) (Chen Juanni et al., 2020). The direct-contact damage process was evaluated as a possible antifungal mechanism. Magnesium oxide with the advantages of non- toxicity, environmental friendliness, ease of availability and biocompatibility with human cells, are recognized as safe disinfection agents by the U.S. Food and Drug Administration without any harmful byproducts; thus, they hold great promise in both medical therapeutics and water decontamination (Chen Juanni et al., 2020). *In vitro* studies have demonstrated that nMgO could be applied as a microbicide against gram-positive (*Bacillus subtilis* and *Staphylococcus aureus*) and Gram-negative (*Escherichia coli*) bacteria and fungal pathogens. Antibacterial activity has been shown to be affected by the size, pH, concentration, and form of the nMgO. Unlike for agrichemicals, these toxicity mechanisms are most likely due to direct physical and chemical inactivation during the interaction, avoiding the dissolution of vegetative fungal spores through the secretion of amino acids and malic acid.

Several studies have suggested that the formation of reactive oxygen species (ROS) and its accumulation in cells seem to be the underlying mechanism of metal nanoparticles against bacterial pathogens, particularly because the generated ROS directly destroy the cell multiplication capacity (Chen, et al., 2014). It is generally believed that microorganism disinfection relies on the direct interaction between nanoparticles and biological cells. nMgO have enormous potential as antibacterial agents, effectively suppressing agricultural bacterial and fungal diseases caused by *Ralstonia solanacearum*, a medical and foodborne pathogen. nMgO could act as an underlying alternative to fungicides for tobacco black shank and root black rot management, even possibly controlling other plant pathogens. Nevertheless, future investigations are needed to evaluate how the nanoparticles affect the toxic activities of pathogens from the view of the whole genome and clarify the control efficacy under field conditions, as well as the related microbial response mechanism in rhizospheric soils.

The antibacterial mechanisms of magnesium oxide nanoparticles (MgONPs) were tested against the *Ralstonia solanacearum in vitro* and *in vivo* for the first time (Lin Cai et al., 2018). *R. solanacearum* has contributed to catastrophic bacterial wilt, which has resulted in the world-wide reduction of tobacco production. The results demonstrated that MgONPs possessed statistically significant concentration-dependent antibacterial activity, and the minimum inhibitory concentration (MIC) and minimum bactericidal concentration (MBC) were measured as 200 and 250 μg/mL, respectively. Additional studies, aimed at understanding the toxicity mechanism of MgONPs, indicated that physical injury occurred to the cell membranes, along with decreased motility and biofilm formation ability of *R. solanacearum*, due to the direct attachment of MgONPs to the surfaces of the bacterial cells, which was observed by scanning electron microscopy (SEM) and transmission electron microscopy (TEM). Reactive oxygen species (ROS) accumulation could also be an important reason for the antibacterial action, inducing DNA damage. The toxicity assessment assay under greenhouse conditions demonstrated that the MgONPs had exerted a large effect on tobacco bacterial wilt, reducing the bacterial wilt index.

By conducting comparison studies using macroscale magnesium oxide (mMgO), the fungicidal activity of MgO nanoparticles (nMgO) was evaluated against soilborne *Phytophthora nicotianae* and *Thielaviopsis basicola* for the first time under laboratory and greenhouse conditions (Juanni Chen et al, 2020). *In vitro* studies revealed that nMgO could inhibit fungal growth and spore germination and impede sporangium development more efficiently than could macroscale equivalents. This study extends the understanding of nanoparticles potentially being adopted as an effective strategy for preventing diversified fungal infections in agricultural fields. The effects of magnesium oxide nanoparticles (MgO NP) were studied on disease resistance in tomato plants against *Ralstonia solanacearum*, as well as its antibacterial activity (Imada et al., 2016). The treated plants showed very little inhibition of bacterial wilt. In contrast, when roots were drenched with a MgO NP suspension prior to inoculation with the pathogen, the incidence of disease was significantly reduced.These results indicated that MgO NP induces systemic resistance in tomato plants against *R. solanacearum.*

Copper (Cu) bactericides are often ineffective due to the presence of Cu-tolerant strains. Magnesium oxide (MgO) is an effective alternative to Cu bactericides against *Xanthomonas* spp. However, the effects of particle size on bactericidal activity and fruit elemental levels are unknown. In this study, nano (20 nm) and micron (0.3 and 0.6 μm) size MgO particles were compared for efficacy (Liao, et al., 2019). Nano MgO had significantly greater in vitro bactericidal activity against Cu-tolerant *X. perforans* than micron MgO at 25–50 μg/ml. In field experiments nano and micron MgO applied at 200 and 1,000 μg/ml were evaluated for disease control. Nano MgO at 200 μg/ml was the only treatment that consistently reduced disease severity compared to the untreated control. Inductively Coupled Plasma Optical Emission Spectroscopy revealed that nano MgO applications did not significantly alter Mg, Cu, Ca, K, Mn, P and S accumulation compared to fruits from the untreated plots. It was demonstrated that although both nano MgO and micron MgO had bactericidal activity against Cu-tolerant strains *in vitro*, only nano MgO was effective in bacterial spot disease management under field conditions.

4.5 Modes of actions of nano-particles

Nano-particles can be applied directly into the soil or these particles can be utilized as carriers of some chemicals like pheromones, SAR inducing chemical, polyamine synthesis inhibitors or even (Manish Mathur et al., 2018).

Nanometer-sized silvers possess different properties, which might come from morphological, structural, and physiological changes (Nel et al., 2006). Indeed, several lines of evidence support the enhanced efficiency of silver nano-particles on antimicrobial activity. Silver nano-particles are highly reactive as they generate Ag+ ions while metallic silver is relatively un-reactive It was also shown that the nano-particles efficiently penetrate into microbial cells, which implies lower concentrations of nano-sized silver would be sufficient for microbial control. This would be efficient, especially for some organisms that are less sensitive to antibiotics due to the poor penetration of some antibiotics into cells (Manish Mathur et al., 2018). Silver has strong bactericidal and inhibitory effects. Silver nano-particles, which have high surface area and high fraction of surface atoms, have high antimicrobial effect as compared to the bulk silver. A previous study observed that silver nano-particles disrupt transport systems including ion efflux. The dysfunction of ion efflux can cause rapid accumulation of silver ions, interrupting cellular processes at their lower concentrations such as metabolism and respiration by reacting with molecules. Also, silver ions are known to produce reactive oxygen species via their reaction with oxygen, which is detrimental to cells, causing damage to proteins, lipids, and nucleic acids.

Silver nanoparticles can significantly delay mycelial growth of *Colletotricchum gloeosporioides* in a dose dependent manner in vitro. Antifungal efficiency of silver nano-particles was observed at 24 h after inoculation, suggesting that direct contact of silver with spores or germ tubes is critical in inhibiting disease development. Moreover, antifungal efficiency of silver was also observed at 5 days after inoculation, suggesting that silver nanoparticles could have penetrated the plant cell wall and inhibited the disease development. AgNP severely damaged *Magnaporthe grisea* hyphal walls, resulting in the plasmolysis of hyphae. Considering many cellular effects of silver ions, silver nanoparticle-mediated collapse in pathogen hyphae is probably not only by damaging hyphal walls, but also other cellular effects. The preventative and post-inoculation application of the silver nano-particles effectively reduced disease severity on plants at all concentrations. A mechanism of this antifungal activity is suggested by the direct effect on germination and infection process in the fungi. *M. grisea* can cause foliar disease and reproduce as asexual conidia.

Inhibition efficiency of AgNP positively associates with concentrations of AgNPs. They hypothesized that this could be due to the high density at which the solution was able to saturate and cohere to fungal hyphe and to deactivate plant pathogenic fungi. Ag NPs synthesized thorough Aspergillus terreus (KC462061) inhibited the growth of *A. flavus* by disturbing cellular functions which caused deformation in fungal hyphae. AgNPs cause decrease in spores number, abnormality and hypertrophy, these special effects lead to destroyed and damaged of spores (Othman et al., 2014). Dose dependent impact of nano-silver on *Xanthomonas campestris* pv. *campestris* could destroy the cell membrane and increase the cell conductivity of the tested bacteria. AgNPs was synthesized by using *Serratia* sp. and evaluated these AgNPs against spot blotch disease in wheat caused by *Bipolaris sorokiniana*. They hypothesized that the treatment of bsAgNPs enhanced lignifications which could have worked as a hindrance against pathogen attack in plants treated with pathogen and bsAgNPs in combination (B4), while least lignin deposition in pathogen challenged plants (BC) favoured pathogen attack. Leaf-spot disease caused by *Xanthomonas perforans* (Cu resistant) can be inhibited by DNA-directed silver (Ag) nanoparticles (NPs).

Numerous studies on the mechanism of action of TiO_2 identified three possible modes of action, including direct oxidation of coenzyme A, which inhibits cell respiration leading to cell death; change in cell permeability; and cell wall damage followed by cytoplasmic membrane damage. Plant cells might also be expected to be susceptible to this chemical effect. The physical size of plant cells and, in particular, the thickness of their cell walls may exceed the lethal radius of diffusion of the free radicals generated by TiO_2, affording selectivity compared with their pathogens. *X perforans* has a cell wall thickness of $\approx$10 nm and overall cellular dimensions of roughly 500 by 1,500 nm. Titanium NPs was used to understand the interaction between *Bacillus amyloliquefaciens*, a plant growth promoting bacterium and the host plant *Brassica napus* for providing protection against *Alternaria brassicae.*

The antifungal activity of nano-copper was studied against plant pathogenic fungi Cioffi et al., (2004). The antifungal activity of three nano-composites performed on *Saccharomyces cerevisiae* yeast was screened using a twostep protocol. Antifungal activity correlates with the electrothermal atomic absorption spectroscopy (ETAAS) analysis of the copper released by the nano-composites in a yeastfree culture broth. Result demonstrated that the releasing properties of such nanocomposites can also be controlled by a proper modulation of the CuNPs loading.

Increased activity of defence enzymes in leaves of chitosan treated turmeric plants may play a role in restricting the development of disease symptoms. The eliciting properties of chitosan make chitosan a potential antifungal agent for the control of rhizome rot disease of turmeric. Increase in chitinase and chitosanase activity may play a role in enhanced resistance in turmeric plants against *P. aphanidermatum* infection (Anusuya and Sathiyabama, 2014). They applied chitosannano-particles to induce antifungal hydrolyses in turmeric plant found correlation between reduction on incidence of rhizome-rot and enhanced activity of defense enzymes such as peroxidases, polyphenol oxidases, protease inhibitors and ß-1, 3- glucanases.

4.6 Limitations and Gaps

Despite the numerous potential advantages of nanotechnology in plant protection it has not yet made its way commercially in our diseased fields. Firstly, Nanotech products require high initial investments and secondly large scale field use is a pre requisite for its application. And there are numerous reports of nano-materials biosynthesis from plant pathogens. Nanotechnology can play as a catalyst for enhancing agricultural growth rate. Many countries across the globe are pursuing Research and Development for nanotechnological application in agriculture to nullify the toxic effects of chemicals used in field. The broad implication of nanotechnology for society can be grouped into two categories, namely environmental, health and safety implication and societal dimensions.

In the future, nano-scale devices with novel properties could be used to make agricultural systems "smart". A rapidly growing body of toxicological studies suggest that a number of nano-materials may be toxic (Kahru and Dubourguier, 2010), yet there are currently no regulations that limit nano-material exposures. It is widely recognized that the environmental impacts of nanomaterials need to be understood, although a number of laboratory trials have measured acute toxicity and sub lethal effects of engineered nano-particles on organisms. Some of these nanotechnologies may help reverse environmental degradation or may replace the more toxic technologies currently in use. Others may end up substituting a new problem for an old one. It is important to develop the scientific tests and models that can distinguish the alternatives. Though, exact a priori predictions of all environmental impacts from controlled laboratory conditions cannot be determined. The toxicity of nanomaterials has to be clearly understood before its commercialization and field application. The potential application of nano-materials in different agricultural applications needs further research with respect to synthesis, toxicology and its effective application at field level. In the field of agriculture, there are still many possibilities to explore with new nano-products and techniques. Barring the miniscule limitations, nano-materials have a tremendous potential in making crop protection methodologies cost-effective and environmental friendly.

Majority of the studies conducted for pathogens belonging to phylum Ascomycota (includes *Aspergillus niger, Aureobasidium pullulans, Botrytis cinerea, Cladosporium cucumerimum, Colletotrichum gloeosporioides, Corynespora cassiicola, Curvularia lunata, Cylindrocarpon destructans, Didymella bryoniae, Fusarium* Spp, *Magnaporthe grisea, Monosporascus cannonballus, Penicillium phoeniceum, Saccharomyces cerevisiae, Sclerotinia sclerotiorum* and *Sphaerotheca pannosa* var. *rosae*. Few studies on genera belonging to Proteobacteria like *Dickeya chrysanthemi, Pectobacterium carotovorum, Xanthomonas campestris* pv. *campestris* and *Xanthomonas perforansi* and Zygomycota genera like *Mucor plumbeus* and *Rhizopus stolonifer* were conducted, while *Pythium aphanidermatum* (Oomycetes) and *Rhyzoctonia solani* (Basidiomycota) were studied in one representative studies each (Manish Mathur et al., 2018).

These trends suggested that majority of the research works focused on fungi that producing specialized external asexual conidia and thus, forcing to conclude that nano-particles might choosing the easiest site in their target pathogen. However, such generalized statement need to explore. Gaps were also pertains to implication of this novel technology for management of obligate pathogens likes rust fungi and *Phytophthora*, facultative saprophytes like *Ustilago* spp., bacteria like *Pseudomonas* and ds DNA and RNA viruses, phytoplasma etc. Most of the suggested modes of actions of nano-particles are hypothetical therefore; efforts should be directed toward exploration of their suppression behaviour at physiological, biochemical and cellular levels. Further role of these highly reactive particles as elicitors for inducing host resistance also need exploration for different host-pathogen complex.

At field level, applications of nano-particles or nano coated materials at phyllosphere or at rhizosphere and their subsequent behaviour on these applied sites should be figure out. As far as present status is concern, in most research these particles were applied as curative one, however, research required for production of some preventive nano-coated material so that epidemic avoidance can be attempted. The use of Carbon Nano Tubes have enhanced the plant growth in tomato, while another study using carbon nano-tubes depicted that it had inhibitory effect on root elongation in tomato whereas in onion and cucumber it showed enhancement in root elongation (Canas et al., 2008). Thus, large-scale application of CNTs still needs review and further experimentation. Some other studies have also depicted the toxic effect of multi-walled carbon nano-tubes (MWCNT's) in plant cells and application of MWCNTs was found to be responsible for accumulation of reactive oxygen species (ROS) and subsequently decreased cell proliferation and cell death. Based on the positive as well as negative effects of CB NPs it can be stated that the response of plants or plant cells to NPs varies with the plant species, stages of growth and the nature of the NPs. Further research on nano-sciences is needed to revealthe most efficient and useful NPs combinations for betterment of agriculture.

4.7 Green synthesis of nanoparticles

The conventional methods of developing nanoparticles NPs are costly and produce very toxic product, so the need of hour is to reduce the risk of toxicity in the environment from the different chemicals used in the physical and chemical methods. The alternate approaches found to develop NPs is "green synthesis". Presently, green synthesis is the watch word for the combination of NPs by plants or their metabolites. This innovation is particularly compensating as far as decreasing the poisonous quality caused by the conventionally integrated NPs (Aman Gour and Narendra Kumar Jain, 2019), who discussed the perspectives by which metal particles can be integrated from green methods in the perspective of green methods utilized in the NPs combination. In the green strategies, plant metabolites and natural substances are utilized to orchestrate the NPs for the pharmaceutical and other applications. Some characterization methods are also reviewed along with applications of NPs. The surging popularity of green methods has triggered synthesis of NPs using different sources, like bacteria, fungi, algae, and plants, resulting in large-scale production with less contamination. Green synthesis is an ecofriendly and biocompatible process, generally accomplished by using a capping agent/stabilizer (to control size and prevent agglomeration), plant extracts, yeast, or bacteria (Ahmad et al., 2019).

Nanoparticles are expected to play a vital role in the management of future plant diseases, and they are expected to provide an environmentally friendly alternative to traditional synthetic fungicides. In a study, silver nanoparticles (AgNPs) were green synthesized through the mediation by using the endophytic bacterium *Pseudomonas poae* strain CO, which was isolated from garlic plants (*Allium sativum*) (Ezzeldin Ibrahim et al., 2020). Following a confirmation analysis that used UV–Vis, the in vitro antifungal activity of the biosynthesized AgNPs was examined with the size of 19.8–44.9 nm, which showed strong inhibition in the mycelium growth, spore germination, the length of the germ tubes and the mycotoxin production of the wheat Fusarium head blight pathogen, *Fusarium graminearum*. Furthermore, the microscopic examination showed that the morphological of mycelia had deformities and collapsed when treated with AgNPs, causing DNA and proteins to leak outside cells. The biosynthesized AgNPs with strong antifungal activity were further characterized based on analyses of X-ray diffraction, transmission electron microscopy, scanning electron microscopy, EDS profiles, and Fourier transform infrared spectroscopy. The results indicated that the biosynthesized AgNPs may have a great potential in protecting wheat from fungal infection.

Biosynthesized metallic NPs such as silver, copper, gold, and zinc, have been reported to act against both Gram-positive and Gram-negative bacteria such as *Bacillus subtilis*, *E coli* and *Staphylococcus aureus*. These NPs showed inhibitory actions against some pathogenic fungi including *Aspergillus niger*, *Fusarium oxysporum*, *A fumigatus*, and inhibitory effects against other pathogenic microbes (Md. Arshad Ali et al., 2020). The silver NPs have received more attention due to "green synthesis" from plants, bacteria, fungi or yeasts. The MgO and MnO_2 NPs synthesized using chamomile flower extract showed activity against *Acidovorax oryzae* strain RS-2, a bacterium causing bacterial brown stripe disease of rice. Nanotechnology has potential prospects in plant pathogenic management in various ways. The most commonly used method is the direct application of NPs to seeds or foliar spray against the plant pathogens, which are more greatly suppressed compared to chemical pesticides. In another method, nanomaterials are used as carrier materials of chemical active ingredients for controlled release of pesticides. The green synthesized ZnONPs and TiO_2NPs with lemon fruit extract at room temperature showed their antibacterial activities against *Dickeya dadantii*, a causal bacterium of sweet potato stem and root rot disease.

Metal nanoparticles such as silver, gold and copper have been widely synthesized by employing physical, chemical and biological methods. The bio-based protocols are now turning into viable alternative to conventional physical and chemical methods, as they appear to be simple, low cost and ecofriendly (Abdulrhman A. Almadiy and Gomah E. Nenaah, 2018). To avoid the use of risky and toxic solvents, many researchers affirmed the use of biological reducing agents such as bacteria, fungi and plants, that are natural, low-cost, and ecofriendly for producing silver nanoparticles for various nanotechnological applications. The upsurge in green synthesis routes of nanomaterials permit better shape and size control, crystal growth and stabilization of the produced nanomaterials. AgNPs synthesized through green procedures using plant extracts showed antimicrobial activity against pathogenic microorganisms, but so far, there has been limited research on the use of AgNPs against plant pathogenic fungi.

A study was aimed to determine the antifungal activity of α-chaconine, and αsolanine, two major steroidal alkaloids of potato, when tested individually or as mixtures against *A. alternata, R. solani, B. cinerea,* and *F. oxysporum* f. sp. *lycopersici*, four of the important pathogenic fungi infecting plants. Large-scale production of nanoparticles from diverse fungal strains has great potential since they can be grown even in vitro (Siddiqi et al, 2016).

Physical synthesis of nanoparticle gives low yield, while as chemical methods are toxic to the environment. However, biological synthesis of nanoparticles is advantageous than physical and chemical methods because of rapid synthesis, better control over size and shape characteristics, less toxicity, cost-effectiveness, and eco-friendly approach (Mehrajud Din Talie et al., 2020). Various plants, bacteria, fungi, algae, and viruses have been used for biological synthesis of nanoparticles. However, the most widely accepted approach is using bacteria and fungi because they are easy to manipulate and handle.

Several methods have been applied for the management of fungal and bacterial pathogens but these have some limitations. However, synthesized nanoparticles have great potential to be used as antifungal and antibacterial agents because they are considered as alternate, cost effective, and eco-friendly management strategy for the control of pathogenic microbes. Various mushrooms, namely, *Volvariella volvacea, Pleurotus sajorcaju, Pleurotus florida, Ganoderma lucidum, Ganoderma applanatum, Agaricus bisporus, Fomes fomentarius, Helvella* sp., and *Microporus xanthopus* have been assessed for the synthesis of silver and other nanomaterials. Silver nanoparticles (AgNPs) have become increasingly popular as antifungal, antibacterial, antioxidant, and antiinflammatory agents.Considering the present environmental scenario and other safety measures, there is widespread interest in developing cost effective, benign, and eco-friendly approach for the synthesis of nanoparticles. Therefore, the present study entitled "Green synthesis of AgNPs using *Helvella leucopus* Pers. and their antimycotic activity against fungi causing fungal rot of Apple" was carried out 1st time from Kashmir, India. These synthesized AgNPs were characterized and evaluated for their antifungal activity against fungal rot pathogens such as *Penicillium chrysogenum, Aspergillus niger*, and *Alternaria alternata* with respect to spore germination and reduction in zone of inhibition.

The pathogenic fungi can develop different defenses to fungicides which mean they can adapt to fungicides by mutations, leading to the loss of fungicide efficacy. For many decades, copper ions have been used as pesticides, fungicides, and fertilizers. For example, some copper salts have been used for reducing mildew in grapes. However, the use of copper ions presents some disadvantages, such as instability in field conditions, high solubility in water, and formulations with high concentrations of copper salts. Therefore, it is necessary to develop long-term efficient, eco-friendly, and cost-effective fungicides. The green synthesis of fungicides, which consist of copper nanoparticles (Cu-NPs) prepared in aqueous media was proposed (Nicolaza Pariona et al., 2019).. Through in vitro experiments, the antifungal efficacy against *Fusarium solani, Neofusicoccum* sp., and *Fusarium oxysporum* was investigated. Although the antifungal activity differs for each fungal species, it was found that the Cu-NPs induce strong morphological changes in the mycelium. Additionally, the damage of the cell membranes of the pathogens was revealed by microscopic observations. For the three evaluated fungi, fluorescence microscopy demonstrated the intracellular generation of reactive oxygen species in the mycelium. This work proved that the green-synthesized Cu-NPs are potential fungicides against *F. solani, Neofusicoccum* sp., and *F. oxysporum*.

Seed-borne fungus *Penicillium duclauxii* was examined in this study to investigate its capability of synthesizing silver nanoparticles (Ag-NPs) (Khalid S.Almaary et al., 2020). *In vitro* experiments were conducted using corn-grain contaminating fungal isolate. Ag-NPs detection and characterization were assayed by the aid of spectroscopic techniques. Spectroscopy (energy dispersive), X-ray diffraction, transmission electron-microscope and optical absorption dimensions were employed. Ag-NPs with biosynthesized were used to test invitro against *Bipolaris sorghicola*; the cause of target leaf spot disease on sorghum plants. The myco-synthesis of Ag NPs using *Penicillium duclauxii* was proved in this study. Moreover, *Bipolaris sorghicola* was successfully inhibited by such Ag NPs *in vitro.*

In order to reduce the excessive reliance on the toxic chemical fungicides, a study was aimed to isolate the total potato glycoalkaloids (TPAs), and the two steroidal alkaloids α-chaconine and α-solanine from potato (Abdulrhman A. Almadiy and Gomah E. Nenaah, 2018). Their structures were characterized using physical and spectroscopic methods including (UV, IR, 1H, 13C-- NMR, 2D 1H1H COSY, HMBC and NOESY).

An investigation by Ashraf et al., (2020) invovled the green synthesis of silver nanoparticles (AgNPs) from the aqueous leaf extract of *Melia azedarach* by following a microwave-assisted method to control *Fusarium oxysporum*, the causal agent of tomato wilt. Efficient inhibition of *F. oxysporum*, the causal agent of tomato wilt, was achieved after exposure to MLE-AgNPs both in vivo and in vitro. *In vitro* studies exhibited repressed fungal mycelial growth with 79–98% inhibition as compared to the control. Significant increases in growth parameters of tomato seedlings were observed after treatment with biosynthesized nanoparticles as compared to *F. oxysporum*-infected plants grown without them under greenhouse conditions.

Synergistic biological activities: Nanoparticles are attributed to green nanotechnology methods to know the synergistic biological activities. To accomplish this phenomenon, a study was aimed to synthesize copper oxide nanoparticles (CuO NPs) by using *Syzygium alternifolium* stem bark, characterized those NPs using expository tools and to elucidate high prioritized antimicrobial and anticancer activities (Yugandhar et al., 2017). Synthesized particles exhibited a color change pattern upon synthesis and affirmed its respective broad peak at 285 nm which was analyzed through UV–vis spectroscopy. Antimicrobial studies of particles showed highest inhibitory activity against *E coli* and *T harzianum* among bacterial and fungal strains, respectively.

Biogenic synthesis of silver nanoparticles employing fungi offers advantages, including the formation of a capping from fungal biomolecules, which provides stability and can contribute to biological activity. In this work, silver nanoparticles were synthesized using *Trichoderma harzianum* cultivated with

(AgNP-TS) and without enzymatic stimulation (AgNP-T) by the cell wall of *Sclerotinia sclerotiorum* (Guilger-Casagrande et al., 2019). The nanoparticles were evaluated for the control of *S. sclerotiorum.* Low levels of cytotoxicity and genotoxicity were observed. This study opened perspectives for further exploration of fungal biogenic nanoparticles, indicating their use for the control of *S. sclerotiorum* and other agricultural pests.

Iron oxide nanoparticles were synthesized through the green chemistry method. It involved the use of plant extract for synthesis of metal oxide nanomaterials (Momanyi Kerubo Rachael and Rajiv, 2019). The highest zone of inhibition was obtained in *S. rolfsii* (16.5 ± 0.5)at 50 μg/ml of concentration of TFeONPs. Biosynthesized TFeONPs were capable of inhibiting the growth of fungal pathogens such as *S rolfsii* and *F oxysporum* and they might be employed as antifungal agents in agriculture to control fungal infections.

Green synthesis of silver nanoparticles was carried out using aqueous leaf extract of *Acalypha indica.* The formation of nanosilver was confirmed by UV-Vis Spectroscopy (Rajyasri Ghosh et al., 2018). The antifungal property of silver nanoparticles was evaluated against pathogenic fungus *Colletotrichum capsici* both *in vitro* and in field inoculation experiment. Application of silver nanoparticles produced significant inhibition of growth of fungal mycelia as well as conidial germination *in vitro.* Results of the present study highlighted the eco-friendly approach of plant mediated synthesis of nanosilver and its potential application in the field as an alternative to chemical fungicides for disease management.

A novel and facile approach to effectively synthesize and stabilize copper nanoparticles (CuNPs) was presented using the sugars dextrose, dextrin and β-cyclodextrin (Debajyoti Ray *et al,* 2015). This work provided a biocompatible method to arrest the growth of a potent rice pathogen. This has widespread implications in areas where rice is the main food crop.

A study was focused on the extracellular synthesis of silver nanoparticles (AgNPs) using culture supernatant of an agriculturally important bacterium, *Serratia* sp. BHU-S4 and demonstrates its effective application for the management of spot blotch disease in wheat (Mishra et al., 2014). The bsAgNPs exhibited strong antifungal activity against *Bipolaris sorokiniana,* the spot blotch pathogen of wheat.

Mechanisms of action of green nanoparticles against phytopathogens: Although green synthesized metallic NPs have been studied for their potentials against phytopathogens, the exact mode of actions of NPs is not completely understood (Md. Arshad Ali et al., 2020). To date, some mechanisms have been reported, such as protein dysfunction (e.g., oxidation of cysteine in Fe-binding site, destruction of Fe–S cluster, exchange of catalytic metal, and exchange of structural metal), production of reactive oxygen species (ROS) and antioxidant depletion, impaired membrane function (e.g., membrane damage, loss of membrane potential), interference with nutrient uptake (e.g., inhibition of Fe (III) transporter gene expression) and genotoxicity (e.g., double-strand breaks). These mechanisms may not operate individually, but function in combination against various phytopathogens.

The adhesion of NPs with microbial cell membrane occurs due to the electrostatic attraction between the negatively charged cell membrane of microbes and NPs with positive or low negative charges. The morphological structures of the membrane are disturbed by the NPs and the membrane depolarization causes the disruption of membrane permeability and respiratory actions, and ultimately damages the cell structures, leading to cell death. This disruption of cell structure leads to the leakage of internal cell content including proteins, enzymes, DNA, and metabolites. In addition, the NPs may cause irregular pits on the microbial cell wall that facilitate NPs' entry into periplasmic space and inside the cells. NP actions on membrane damage and pit formation on the cell surface can be observed by using TEM and SEM. The green synthesized AgNPs affects *Acidovorax oryzae* strain RS-2 using TEM and showed highly ruptured cell walls, leakage of cytoplasmic and nucleic contents, swollen structure leading to bacterial death. Biosynthesized AgNPs were applied to *Fusarium graminearum* strain PH-1 and antifungal activities of distortion of hyphae and damaging cell walls were observed by both SEM and TEM. Similar effects of AgNPs on several other phytopathogenic fungi such as *Alternaria alternata, Botrytis cinera* and *Trichosporon asahii* were also observed (Xia, et al., 2016).

The toxicity of NPs may be caused by the formation of ROS. Free radicals can damage the cell wall and various biomolecules such as proteins, lipids and DNA. DNA damages, such as mutations, deletions, single-strand breaks, double-strand breaks, and cross-linking with proteins, may occur.

In order to achieve the green control of plant diseases with the green-synthesized NPs, the major aspects to be considered, viz., AgNPs is the major metallic NPs tested on plant pathogens and shows toxicities to plants and plant-associated microbes. Less phytotoxic metals, such as Zn, Fe, Mg, and Mn, should be tested ; Minimum inhibitory concentration of metallic NPs on pathogens should be determined in vitro and in vivo ; Metallic NP effects on plant growth and development at the working concentration should be determined ; Metallic NPs at the working concentration in vivo should be less toxic to other plant-associated microbes than the target pathogens; Metallic NP effects on plant microbiota should be determined and ; The dynamic of the metallic NPs and their effects on soil microbiota should be determined. Few studies on plant disease control in fields by metallic NPs have been done. Future studies should determine the efficiency of metallic NPs in fields along with the assessment of their potential risks to environments and human health (Md. Arshad Ali et al., 2020).

4.7.1 Green and chemically synthesised nanoparticles

The nanoparticles are classified into different classes such as inorganic nanoparticles, organic nanoparticles, ceramic nanoparticles and carbon base nanoparticles. The inorganic nanoparticles are further classified into metal nanoparticles and metal oxide nanoparticles (Irfan Ijaz et al., 2020). Similarly carbon base nanoparticles are classified into Fullerene, Carbon nanotubes, Graphene, Carbon nanofiber and carbon black. Nanoparticles are also classified on the basis of dimension such as one dimension nanoparticles, two-dimension nanoparticles and three-dimension nanoparticles. The nanoparticles are synthesized by using two approaches like top-down approach and bottom-up approach. In this review chemical, physical and green synthesis of nanoparticles was reported. The synthesized nanoparticles are synthesized using different qualitative and quantitative techniques. The Qualitative techniques include Fourier Transform Infrared Spectroscopy (FT-IR), UV-Vis spectrophotometry, Scanning electron microscope (SEM), X.ray diffraction (XRD) and Atomic Force Microscopy (AFM). The Quantitative techniques include Transmission Electron Microscopy (TEM), High Angle Annular Dark-Field Imaging (HAADF) and Intracranial pressure (ICP).

Synthesis and characterisation of silver nanoparticles and their effect on antifungal efficacy of common agricultural fungicides was reported (Pragati Jamdagni et al., 2018). Silver nanoparticles were synthesised using biological and chemical reduction methods employing *Elettaria cardamomum* leaf extract and sodium citrate, respectively. Nanoparticles were then characterised using UV-Visible spectroscopy, X-ray diffraction (XRD), transmission electron microscopy, and dynamic light scattering (DLS). While XRD assigned particles size of 31.86 nm for green and 41.91 nm for chemical silver nanoparticles with the help of the Debye-Scherrer formula, DLS specified monodisperse nature of both suspensions. Nanoparticles were tested individually and in combination with fungicides (carbendazim, mancozeb, and thiram) against fungal phytopathogens.

A study on combined effect of zinc oxide nanoparticles (ZnO) with common agricultural fungicides was conducted. Nanoparticles were synthesized using chemical reduction process and characterized through UV-Visible spectroscopy, X-ray diffraction (XRD), Dynamic Light Scattering (DLS) and Transmission Electron Microscopy (TEM). Synthesized nanoparticles were observed to be in size range of 12-63 nm as confirmed by TEM micrograph. DLS established polydisperse nature of nanoparticles and provided effective hydrodynamic diameter of 76.15 nm, confirming the hypothesis of particles being in nano range (Pragati Jamdagni et al., 2018).. Nanoparticles were tested against fungal phytopathogens, namely *A. alternata, A. niger, B. cinerea, F. oxysporum* and *P. expansum*. Nanoparticles used in the study exhibited good antifungal activity. In summary, the study reports synthesis of ZnO nanoparticles with good antimicrobial potential and amelioration of activity of fungicides was observed when used in combination with nanoparticles.

Since green mode of nanoparticles (NPs) synthesis is simple, advantageous and environment friendly relative to chemical and physical procedures, various plant species have been used to fabricate copper and copper oxide nanoparticles (Cu/CuO-NPs) owing to the presence of phytochemicals which often act as capping as well as stabilizing agent (Siddiqi et al., 2020). A comparison of the traditional/conventional method of fabrication of NPs with those of green protocols has also been made. Some misconception of copper chemistry has also been critically discussed in terms of oxidation and reduction reactions.

In materials science, "green" synthesis has gained extensive attention as a reliable, sustainable, and eco-friendly protocol for synthesizing a wide range of materials/nanomaterials including metal/metal oxides nanomaterials, hybrid materials, and bioinspired materials (Singh et al., 2018). As such, green synthesis is regarded as an important tool to reduce the destructive effects associated with the traditional methods of synthesis for nanoparticles commonly utilized in laboratory and industry. The fundamental processes and mechanisms of "green" synthesis approaches were summarized, especially for metal and metal oxide [e.g., gold (Au), silver (Ag), copper oxide (CuO), and zinc oxide (ZnO)] nanoparticles using natural extracts. Importantly, we explored the role of biological components, essential phytochemicals (e.g., flavonoids, alkaloids, terpenoids, amides, and aldehydes) as reducing agents and solvent systems. The stability/toxicity of nanoparticles and the associated surface engineering techniques for achieving biocompatibility are also discussed. Applications of such synthesized products to environmental remediation in terms of antimicrobial activity, catalytic activity, removal of pollutants dyes, and heavy metal ion sensing were coverd.

Antifungal activity: The antifungal activity of the synthesized Cu-NPs was evaluated by measuring the mycelial radial growth for all the treatments for an incubation period of 6 days. It can be observed that the antifungal activity of the Cu-NPs depends on the pathogenic fungi. Similar results have been reported for the antifungal activity of copper nanoparticles against different fungi. Higher concentrations of Cu-NPs in the growth media promote greater IRG (smaller colony areas), Similarly, the Cu-NPs change the morphology, color, form, texture, and density of the fungal mycelium (Nicolaza Pariona et al., 2019). For example, at 0.1 mg mL^{-1}, the IRG was 55%, while the other two fungi species were not affected at this concentration. The sensitivities of the fungal species to Cu-NPs were different, this also has been reported for Cu-NPs of 8 nm in size against *F. solani* and other three fungal species.

Neofusicoccum sp. exhibited high tolerance to Cu-NPs at low concentrations. In contrast, concentrations as high as 0.5 mg mL^{-1} inhibited the growth of the fungus with IRG greater than 95%. High antifungal activity against to *F. oxysporum* has also been reported for Cu-NPs of 3–10 nm in size with a cetyl trimethylammonium bromide coating. In contrast, it has been reported that microparticles of copper-containing metal–organic frameworks at concentrations as high as 0.5 mg mL^{-1} caused 50% of inhibition of *F. oxysporum*. The great antifungal activity of the Cu-NPs could be attributed to both adequate particle size and the coating used in the synthesis process.

Nanotechnology has enabled the fungicide active agents (hexaconazole) to be encapsulated into chitosan nanoparticles with the aim of developing a fungicide nanodelivery system that can transport them more effectively to the target cells (*Ganoderma*) (Maluin et al., 2019). A pathogenic fungus, *Ganoderma boninense* (*G boninense*), is destructive to oil palm whereby it can cause significant loss to oil palm plantations located in the Southeast Asian countries, especially Malaysia and Indonesia. In regard to this matter, a series of chitosan nanoparticles loaded with the fungicide, hexaconazole, was prepared using various concentrations of crosslinking agent sodium tripolyphosphate (TPP). The resulting particle size revealed that the increase of the TPP concentration produced smaller particles. In addition, the *in vitro* fungicide released at pH 5.5 demonstrated that the fungicide from the nanoparticles was released in a sustainable manner with a prolonged release time up to 86 h. On another note, the *in vitro* antifungal studies established that smaller particle size leads to lower half maximum effective concentration (EC50) value, which indicates higher antifungal activity against *G. boninense*.

Three different fungi were isolated from samples of citrus. *Alternaria alternata* was the most abundant fungal pathogen isolated from citrus leaf and fruit spots followed by *Alternaria citri* and *Penicillium digitatum*. Silver nanoparticles were tested as a potent fungicide that may replace the dangerous cytotoxic fungicides (Abdelmalek and Salaheldin, 2016). Results revealed that Silver nanoparticle, 150 ppm, showed potent antifungal activity against the isolated fungi that will open the gate of new generation of less harmful fungicides replacing the cytotoxic dangerous currently used fungicides.

The antifungal effect of nanoparticles of cobalt and nickel ferrite against phytopathogenic fungi was reported (Parul Sharma et al., 2017). As a proof of concept, it was also shown how such nanoparticles can be used as fungicides in plants. The developed cobalt and nickel ferrite nanoparticles ($CoFe_2O_4$ and $NiFe_2O_4$) were successfully tested for antimycotic activity against three plant-pathogenic fungi: *Fusarium oxysporum, Colletotrichum gloeosporioides*, and *Dematophora necatrix*. In addition, it was also observed that these ferrite nanoparticles reducedd the incidence of Fusarium wilt in capsicum. The study suggested that nanoparticles of $CoFe_2O_4$ and $NiFe_2O_4$ could be used as an effective fungicide in plant disease management.

The antifungal activity of metal oxide nanomaterials (zinc oxide (ZnO), magnesium oxide (MgO), and ZnO:MgO and $ZnO:Mg(OH)_2$ composites) prepared under different conditions of synthesis was evaluated against strains of *Colletotrichum gloeosporioides* obtained from papaya and avocado (Susana C. De la Rosa-García et al., 2018). All nanoparticles (NPs) at the tested concentrations significantly inhibited the germination of conidia and caused structural damage to the fungal cells. According to the radial growth test, the fungal strain obtained from avocado was more susceptible to the NPs than the strain obtained from papaya. The effect of the tested NPs on the fungal strains confirmed that these NPs could be used as strong antifungal agents against *C. gloeosporioides* to control anthracnose in tropical fruits.

4.8 Nanobiofungicides: The next-generation of fungicides?

Application of pesticides, which encompasses insecticides, fungicides, herbicides, nematicides etc., being used for plant defence mechanisms, embarrasses pest organisms. Although, pesticides eliminate the problem of pests, the use of synthetic pesticides has resulted in unsustainable management of our soil resources. This can be explained as due to the development of resistance by the pest organisms on continuous exposure to the pesticides, thus posing a challenge which leads to development of new classes of pesticides (Bhattacharyya Atanu et al., 2016). These pesticides, apart from targeting the pest organisms, causes undesirable effects to all matrices of the environment- viz, soil, water, air, biota etc., Hence, the need for the development of ecofriendly pesticides becomes immediate inevitability. However, there is no one single method for efficient command on insect pests. Among the various classes of pesticides, fungicides form a major group of domineering plant diseases of fungal origin, either by inhibiting the growth of the fungi or by complete biocidal activity. The significance of fungicides is due to the fact that fungal diseases stand first in crop thrashing world wide. Currently, apart from the existing synthetic fungicides in the market, biofungicides occupies an unique position in controlling target diseases of fungal origin. An advanced technology in the field of biopesticides is the employment of engineered nanomaterials. These nanomaterials are more reactive can conjugate with biofungicides forming covalent bonds. These unique properties of nanomaterials are exploited to manage the plant-pest chain.

Nano-based antimicrobials are a recent addition to the fight against fungal pathogens, replacing the poisonous elements like heavy metals and toxins with a viable alternative. The expression nanofungicides is used to describe any fungicide formulation that intentionally includes entities in the nanometer size range (here we include entities up to 100 nm), is designated with a "nano"prefix (e.g., nanohybrid , nanocomposite), and/or is claimed to have novel properties associated with the small size (Abd-Elsalam and Alghuthaymi, 2015). On this basis, nanofungicides include a wide variety of products which are described in detail consequently. Several formulation types of nanopesticides including organic ingredients (e.g., active ingredient(s), polymer-based inorganic silica-based nanoparticles and titanium dioxide, nanoemulsions and nanoclays in various forms (e.g., particles, micelles). Nevertheless, before a broad-spectrum use of nanotechnology-based products in plant protection, non-targeted effects and public concern should be reduced or counterd. Remote activation and monitoring of intelligent nano-delivery systems can assist agricultural growers of the future to minimize fungicides and pesticides use. Intelligent use of chemicals on the nano scale can be a suitable solution for this problem. These materials are used into the part of plant that was attacked by disease or pest. Also these carriers in nano scale have self-regulation, this means that the medication on the required amount only be delivered into plant tissue.

Nanotechnological application in plant pathology is still in the early stages. More nanophytopathological studies on physiology of host and pathogen, interaction, infection process and disease diagnosis will help in developing new disease management strategy to develop nanopesticides that are less harmful to the environment than conventional formulations (Kah et al., 2013).

The bonding between safe antimicrobial agents such as biocontrol agents, GRAS subestances, essential oil, biopolymer and curpic, and sulfur NPs could result in synergistic effect. The effect of combination between different active ingredients may increase antimicrobial activity, reduce pesticide usage and delay the development of fungal resistance. The antimicrobial activities of the inorganic nanoparticles such as S, Ag, CuO, MgO and ZnO were investigated separately or combined with biopolymer in previous studies (Rai and Ingle, 2012).

Nanobiocide a product prepared by mixing several bio-based chemicals was reported to eliminate fungus *Magnaporthe grisea*, the causal agent of rice blast disease.

Few researchers have studied the combined antimicrobial effect of inorganic NPs with bioorganic pesticides in the field of plant protection, let alone demonstrated synergistic effect (Kah et al., 2013).

Nanoformulations are viewed to be safer and environment friendly option for plant disease management, but high toxicity of nanoparticles inadvertently released in the environment may pose greater threat to man and other organisms. The ecotoxicological effects of nanomaterials on plant, and soil microorganisms have been widely explored.

There is a need for phytoxicity study on seed systems exposed to different concentration of nanoparticles dispersion to check phytotoxicity end point such as root length, percentage of germination, adsorption and aggregation of nanoparticles (Kumari et al., 2012). Nanofungicides can be prepared in a simple cost-effective manner are suitable for formulating new types of biohybrids nanocide materials would be use as a new environment friendly antimicrobial against different fungal pathogenic organisms in plant.

The synthesis of silver nanoparticles by a biological method using *Bacillus amyloliquefaciens* MH046937 as a new source was attaempted (Abd El Aty et al., 2020). For attaining maximum biosynthesis of nano-silver an orthogonal array of five-four level factors was applied to determine optimum conditions for maximum bacterial biomass production. The biosynthesized silver nanoparticles are characterized using UV-vis spectrophotometer, Transmission electron microscope and Dynamic Light Scattering (DLS). AgNPs were tested for its antibacterial and antifungal potential against Gram- positive, Gram-negative bacteria strains and fungi strains. In-vitro studies revealed good antifungal effects of AgNPs against phytopathogenic *Fusarium oxysporum* (33 mm IZD) and *F. solani* (26 mm IZD). In the greenhouse experiment, AgNPs caused a remarkable enhancement in the *Vicia faba* growth as; the fresh dry weight of root and shoot, shoot length and stem diameter in soil invaded with *F. oxysporum*, *F. solani* and *Rhizoctonia solani* compared with the control.

The local nano-MgO synthesizer strain was isolated from *Ocimum sanctum* plant and deposited in GenBank as endophytic *Streptomyces coelicolor* strain E72 (Shahira H. EL-Moslamy, 2018). Its intracellular metabolic fraction that contains 7.2μg/μl of carbohydrate, 6.3g/l of protein and 5.2 nmol/hr/ml of nitrate reductase used to produce multi-surface shaped nano-MgO with diameter ~25nm. These results are expected to gather sufcient knowledge to discover and develop a new cheap and eco-friendly nano-MgO as an extremely strong antimicrobial agent used in biomedical applications.

Eco-friendly alternative management strategies were developed by Kamel A. Abd-Elsalam and Alexei R. Khokhlov (2017) including the use of heavy nanometal-tolerant *Trichoderma* strains and the main effective material in conventional fungicides (curpic, sulfur, phosphorus and zinc) for controlling plant diseases. Studies on bioefficacy and the mechanism of the nanocides against the most important plant diseases in Egypt were evaluated. Nanofungicides represent an emerging technological development that could offer a range of benefits including increased efficacy, durability, and a reduction in the amounts of active ingredients that need to be used.

Essential oil components (EOCs) are known for their antifungal properties; however, their high volatility limits their application as antimicrobial agents. Strategies used for controlling the volatility of EOCs include encapsulation or loading into porous materials (Mousa Alghuthaymi et al., 2020). This study evaluated the in vitro antifungal activity of selected EOCs (carvacrol, cinnamaldehyde, eugenol and thymol) against the fungus *Aspergillus niger* when loaded into MCM 41 and β cyclodextrin (β CD). Carvacrol and thymol in Mobil Composition of Matter No. 41 (MCM 41) displayed remarkable enhanced antifungal properties in comparison to the pure or βCD encapsulated EOCs. In fact, carvacrol and thymol were able to maintain antifungal activity and inhibit fungal growth for 30 days, suggesting better applicability of these EOCs as natural preservatives.

The efficiency of four different nanoformulations (F1, F2, F3, and F4) and a former formulation (F0) from the fungus *Talaromyces flavus* were investigated in reducing the incidence of cotton Verticillium wilt in greenhouse conditions (Laleh Naraghi and Maryam Negahban, 2019). Results showed decreased percentages of disease severity in treatments of nanoformulations (85-97%) and the former formulation (95%) in comparison to the infected control.

4.9 Nanophytovirology

Plant virus particularly spherical virus is considered to be the naturally occurring nanomaterials (Alaa Y. Ghidan and Tawfiq M. Al Antary, 2019). It is essential to develop therapy practices to strengthen the management of viral diseases in economically important crops. Interactions among nanoparticles, plant phytopathogens, and plant responses have attracted increasing attention for antiviral activity; moreover, the influence of nanoparticles on plants supports the idea that nanoparticles can be employed as a new antiviral strategy in agriculture (Lin Cai et al., 2019).

Zinc oxide nanoparticles (ZnONPs) and silica nanoparticles (SiO_2NPs) were fabricated to control *Tobacco mosaic virus* (TMV) infection *in vitro* and *in vivo.* TMV particles exhibited substantial aggregation and breakage *in vitro* after pretreatment for 2 h with the NPs (ZnONPs or SiO_2NPs). Interestingly, the NPs also increased the dry weight and fresh weight of tobacco plants compared with the control. It was found that NPs can also be absorbed by the leaves and transported throughout the plant, implying a possible synergistic nutritional provision and means of immune regulation. This strategy of developing NPs as new antiviral materials in agriculture broadened the avenues for applying and researching viral materials.

The potential antiviral property of MeNPs makes them a powerful option for controlling these histological agents. It is crucial to determine the dosage of NPs, the application intervals, their effect as a biostimulant, and the clarification of the mechanisms of action, which are not fully understood. A review by Marcela Vargas-Hernandez et al., (2020) was focused on discussing the ability of metal nanoparticles and metal oxides to control viruses that affect agriculture through an exhaustive analysis of the characteristics of the particles and their interaction processes for a possibly beneficial effect on plants.

Because the synthesis methods largely determine the geometry of the nanoparticles, these affect the physicochemical properties of NPs, such as the size, dispersity, type of extrinsic and/or intrinsic defects, morphology, and crystal structure. The size of the NPs influences their entry into cells and interactions with biomolecules. On the other hand, the cellular uptake efficiency of nanostructures is dependent on their size, surface charge, and shape. The diversity of the characteristics of the metal nanoparticles relates to their toxicity (Marcela Vargas-Hernandez et al., 2020). As such, MeNPs are proposed as a promising antiviral agent.

Antiviral activity of metallic nanoparticles: The virus multiplication cycle involves several events, viz., the entry into the cells, the synthesis of the genome and viral proteins, the assembly of viral components and the viral output of infected cells. Consequently, if a compound can inhibit one or more stages of the virus cycle, it can be considered an antiviral agent. Due to the ability of MeNPs to intervene in some steps of virus replication, they have been studied for antiviral therapy in recent decades. Studies have been developed along two lines, the antiviral activity of functionalized MeNPs or pure or non-functionalized nanoparticles. Regarding the latter, the main compounds evaluated are gold and silver, as well as some metal oxide nanoparticles, such as iron oxide, zinc oxide, and titanium dioxide. Notably, several studies have demonstrated the antiviral activity of metal nanoparticles against human viruses.

The mechanisms of action used by silver nanoparticles against viruses are the interaction with gp120, competition for the binding of the virus to the cell, inactivation of the virus particles before entry, interaction with double-stranded DNA, and binding with viral particles. Gold/copper sulfide core-shell nanoparticles (Au/CuSNPs) induce viral capsid protein degradation and capsid damage. Furthermore, the antiviral effect of metallic nanoparticles is due to a synergistic effect with other chemical compounds. Selenium nanoparticles (SeNPs) enhance the anti-EV71 activity of oseltamivir in the human astrocytoma cell model.

Antiviral activity of MeNPs in plants : Studies have focused on the use of nanoparticles for the management of plant diseases; however, research aimed at combatting viral diseases is still in its beginning stages (Marcela Vargas-Hernandez et al., 2020). The antiviral mechanisms of MeNPs are not well understood but current available studies could provide evidence of the mechanisms involved. The antiviral activity of MeNPs has been demonstrated both *in vitro* and *in vivo* with different plants and is effective against positive- and negative-sense single-stranded RNA viruses (Table 13). Many authors describe the interaction of MeNPs with the surface of the virus as one step of the antiviral mechanism. The antiviral activity of ZnONPs and SiO_2NPs against tobacco mosaic virus (TMV) *in vitro* was repoted.

Table 13: Antiviral activity of metal and metal and metal oxide nanoparticles.

Nanoparticle	Size	Plant	Pathogen	Effect
Silver nanoparticles (AgNPs)	10–20 nm	*Cymopsis tetragonaloba* (SHRV)	Sunhemp rosette virus	Complete suppression of the disease
AgNPs	77 nm	*Vicia faba*	Bean yellow mosaic virus (BYMV)	Decrease in virus concentration, percentage of infection, and disease severity Reduction in lesions on infected leaves
Schiff base nanosilver NPs	18–20 nm.	*Nicotiana tabacum*	Tobacco mosaic virus (TMV)	Reduction of the harm of TMV to tobacco
AgNPs	12 nm	*Solanum tuberosum*	Potato virus Y (PVY)	Resistance to virus infection
Graphene oxide-silver NPs (GO-AgNPs)	3050 nm	*Lactuca sativa*	Tomato bushy stunt virus (TBSV)	Decrease in virus concentration, infection percentage, and disease severity
AgNPs	12.6 ± 5 nm	*Solanum tuberosum* L. cv. *Spunta*	Tomato spotted wilt virus (TSWV)	Decrease in TSWV infectivity and produces an inhibitory effect in local lesions
		Solanum lycopersicum	Tomato mosaic virus (ToMV)	Reduction in disease severity and virus infection
		Solanum lycopersicum	Potato virus Y (PVY)	Reduction in disease severity and virus infection
Gold NPs (AuNPs)	30 to 50 nm	*Hordeum vulgare*	Barley yellow mosaic virus (BaYMV)	Dissociation of virus particle in vitro
AuNPs	3.151 to 31.67 nm	*Hordeum vulgare*	Barley yellow dwarf virus-PAV (BYDV-PAV)	Destroys gold barley yellow dwarf virus and eliminates virus infectivity hazards
Titanium dioxide NPs (TiO2NPs)	~3–5 µm	*Vicia faba* L. Fabaceae	Broad bean stain virus (BBSV)	Reduction in disease severity
TiO2NPs	20 nm	*Nicotiana* benthamiana	Turnip mosaic virus (TuMV)	Effectively limits viral infection and replication
TiO2NPs	20 nm	*N. benthamiana*	Turnip mosaic virus (TuMV)	Effectively limits viral infection and replication
Zinc oxide NPs (ZnONPs)	18 nm	N. *benthamiana*	Tobacco mosaic virus (TMV)	Suppression of the speed of TMV invasion

In vivo studies of the effectiveness of MeNPs against viral infections in the Solanaceae, Asteraceae, Poaceae, Fabaceae, and Cucurbitaceae families have also been conducted. The application of MeNPs in plants interferes with the reproduction of viruses through different mechanisms. Furthermore, nanoparticles affect the diseases caused by viruses. CeO2NPs inhibits the reproduction of viruses by penetrating the TMV-infected leaf segment of *Datura stramonium* and *Nicotiana tabacum* through the vascular system and inhibits reproduction. Graphene-based silver nanocomposites exhibit potent activity against tomato bushy stunt virus (TBSV) in lettuce, decreasing the virus concentration and disease severity. The antiviral activity of the NPs was observed in both the foliar application and the soil supplementation. Cucumis sativus infected with papaya ringspot virus (PRSV) and treated with SiO2NPs via soil supplementation produced a significant reduction in disease symptoms (Elsharkawy and Mousa, 2015).

Application of TiO2NPs using sprays and soil drenches after 24 h of broad bean strain virus (BBSV) inoculation in faba bean plants causes a dramatic decrease in BBSV. The antiviral effect and the responses of the plants depend on the treatment time, that is, whether they are applied before, along with, or after the viral infection. The effect of AgNPs synthesized by *Bacillus pumilus*, *Bacillus persicus*, and *Bacillus licheniformis* had on yellow bean mosaic virus was studied. A post-infection treatment 24 h after virus exposition prevented all negative symptoms caused by the virus. However, plants treated with the AgNPs and infected simultaneously showed weak BYMV symptoms, and a pre-infection treatment 72 h before inoculation showed no effect on the virus concentration and disease severity.

Chenopodium amaranticolor plants showed the most significant antiviral effect against tomato spot wilt virus (TSWV) when the AgNPs were sprayed 24 h after inoculation. A decrease in virus concentration and disease percentage for potato plants sprayed with AgNPs 24 h after inoculation with tomato bushy stunt virus (TBSV) was reported (Marcela Vargas-Hernandez et al., 2020). On the other hand, *Cymopsis tetragonaloba* leaves sprayed with AgNPs and sunhemp rosette virus (SHRV) showed complete suppression of the disease by inactivating virus replication. A mixture of AgNPs and salicylic acid (SA) applied before 3 and 7 days of virus infection shows a synergistic antiviral effect against TBSV compared to individual AgNPs and SA treatments. The above shows the ability of NPs mixed with other compounds to present a synergistic effect. Some authors suggest that MeNPs activate the immune system of plants (Derbalah and Elsharkawy, 2019). The use of MeNPs as antivirals in plants is due not only through the inactivation of viruses and inhibition of multiplication but also the activation of plant defense mechanisms that leads to plant immunity and growth response, in which the antioxidant system, resistance genes, and plant hormones participate.

Effect of MeNPs as biostimulants in virus-infected plants: Plant biostimulants are substances and materials, other than nutrients and pesticides, that when applied to a plant, seeds, or growing substrates in specific formulations, can modify the physiological processes of plants in a way that provides potential benefits to growth, development, and defense responses. Depending on the nanoparticle and the state of the plant, NPs provide a positive (eustress) or negative (distress) impact. Studies on healthy tobacco plants treated with SiO_2, ZnO, and Fe_2O_3 NPs increased tobacco growth (Marcela Vargas-Hernandez et al., 2020). Therefore, this biostimulant effect can cause eustress in healthy and also in stressed plants, for example plants infected with a virus, since it counteracts the destructive effects they cause.

The use of nanoparticles has many advantages due to its low cost, high stability, incorporation of both hydrophilic and hydrophobic substances, secure handling, and they are renewable; the physicochemical properties can also be controlled. The current rate of development of antiviral agents is promising due to the constant appearance of emerging or re-emerging viruses. Due to the above, the use of metal nanoparticles is an exciting opportunity for novel antiviral therapy uses in crops; however, it is essential to take into account its toxicity and manage the application doses. Many of the MeNPs and MoNPs do not present toxicity to humans at the concentrations used as antimicrobials, and many of these concentrations are also those used to activate the antiviral activity, which is an advantage for their large-scale use. Furthermore, the concentrations at which they show antiviral activity do not have negative impacts on plants and sometimes even show a positive or eustress effect.

Despite the advantages that potentially harmful aspects of the MeNPs present must be considered. Fate should be monitored as well as their potential toxicity and levels of overdosing. Plants can absorb and translocate MeNPs to different parts depending on their bio-ability, concentration, solubility, and exposure time (Siddiqi and Husen, 2017). It is essential to take the doses and the time of application into account and monitor the phenological status of the plant to determine whether they are displaying. any negative effects. It is also necessary to establish the process of accumulation of the nanoparticles, especially in the edible parts of the plant, and even the effect of MeNPs at the nutraceutical level, such that they are not toxic and may even have a beneficial impact for consumers. Although there are no reports related to the time in which MeNPs remain in the plants, nanoparticles can accumulate in fruits and seeds. Studies have reported the transgenerational effects of CeO_2NPs in second-generation seedlings. Those were grown with seeds from parental plants treated with 10 mg/L CeO_2NPs showing more cerium compared to non-treated plants. MeNPs are taken up and accumulate in plants that are food for insects, animals, or humans entering the food chain, and some of them can damage some biological entities of the food chain. Studies have reported that crickets fed NP CeO_2- exposed zucchini leaves accumulate more Ce than bulk exposed insects. It was indicated that the accumulation of CeO_2NPs in zucchini results in trophic transfer and can contaminate the food chain.

MeNPs such as ZnO, TiO_2, CeO_2, and Fe_3O_4 NPs affect soil enzyme activities and change soil bacterial communities (Marcela Vargas-Hernandez et al., 2020). Besides, CeO2NPs have been found to compromise soil fertility. Because of this, it is a critical point for FDA approval of the detailed knowledge of the biodegradation rate. Some polymer encapsulated metal oxide nanoparticles have a degradation time in different media variables from a few days to months. Knowing the biodegradation time of nanoparticles and the damage that may be caused to the potentially susceptible biota, the generation of the application management program to reduce the potential toxicity in the biological communities involved should be an important commitment.

It has been reported that nanoparticle formulations are 10–100 times more effective than traditional formulations to contend with similar problems. Some of these formulations can be used at the rate of 50 mL of nanoemulsions or 8 g per hectare (Keswani, 2020). Studies have reported antiviral activity of MeNPs at a maximum concentration of 200 μg/mL; according to these data, it was established that the amount that could be used for virus control in agriculture is approx. 20–40 g/ha for each application. The large-scale application of nanoparticles in agriculture requires mass production. The low-cost mass production of nanoparticles has gained much attention in industrial and academic circles. Physical methods for the synthesis of nanoparticles are expensive, require strong material, toughness, and strength, and are difficult to facilitate on a large scale, while chemical methods are used in mass production; they could have a negative impact on the environment. An alternative to the use of nanoparticles that harm the environment is an environmentally friendly green approach for their synthesis. The production of MeNPs through biogenesis has advanced in aspects of profitability, and eco-friendly since bacteria, fungi, and plants are involved. However, there are other aspects to improve, green synthesis using plant extracts can be comparable to conventional methods, but mass production requires highly reactive substances or energy-consuming (Peralta-videa et al., 2016).

The ability to act by interfering with any of the stages of the infection to prevent the induction of virus symptoms, to be selective to avoid damaging the host, to have a broad spectrum, and have no harmful effects on the environment would be the requirements to finally achieve a curative antiviral chemical. One of the most promising strategies for the effective use of antivirals is to interfere with virus recognition and entry into the host (Galdiero et al., 2011). The use of nanoparticles as compounds to control the virus can be an alternative to antiviral compounds. On the other hand, studies have reported that the antiviral activity of MeNPs is higher when applied in cultures before infection. Therefore, the use of MeNPs is proposed as a preventive control that can be applied to early-stage plants, either to seeds or seedlings. These types of particles present a wide range of infectious diseases. Preventive control will not only avoid infection by viral pathogens but also by fungi or bacteria (Sánchez-López et al., 2020).

The use of NPs has excellent potential to deal with the problem of virus-caused diseases in agriculture. It is necessary to establish precise complementary methodologies such that the technology is ready to be used without risks to the environment or consumers. Among this complementary knowledge, it is required to prescribe doses, the stage of phenological development of the plant to make the applications, and the precise types of NPs that produce the greatest advantages.

4.10 Nanophytomycology

A series of inorganic and organic nanomaterials have been developed and proven to exhibit prominent antifungal properties on phytopathogenic fungi in vitro, and some of them still exerted their toxicity effects under greenhouse and field conditions. Several nanoparticles including TiO_2, CuO, Zn, ZnO, carbon nanomaterials, Al, and Si nanoparticles have been reported to display toxicity toward phytopathogenic bacteria and fungi, decreasing the disease incidence (Chen et al., 2020).

The antifungal activity of nMgO against *Phytophthora nicotianae* and *Thielaviopsis basicola, in vitro* and in a greenhouse compared with that of mMgO revealed that in comparison with mMgO, nMgO had a significantly higher inhibitory effect on spore germination, sporangium formation, and hyphal development. All the toxicity behaviors may be mostly attributed to nanoparticle–cell direct contact, classically observed using SEM/EDS and TEM technologies, and the subsequent oxidative stress. Additionally, investigations in greenhouse experiments that irrigated the application of nanomaterials found that nMgO reduced fungal disease occurrence when compared with untreated controls. These results proposed a new view that nMgO could act as an underlying alternative to fungicides for tobacco black shank and root black rot management, even possibly controlling other plant pathogens. .

Metal nanoparticles can act as potential fungicides when compared to the conventional chemical fungicides. A study was based on utilization of the myco-nanoparticles synthesized from endophytic fungi isolated from *Solanum nigrum* (Akther and Hemalatha, 2019). The mycosilver nanoparticles are characterized by SEM, TEM and EDAX analysis. The results of the study exhibited the silver nanoparticles were of different morphological shapes including spherical, cylindrical and loosely agglomerated with an average size of 2–50 nm. Further, the plant pathogenic fungi including *Fusarium graminearum, Fusarium udum, Rhizoctonia solani* and *Aspergillus niger* were treated with mycosilver nanoparticles to test the efficacy to control plant pathogens and showed broad spectrum antifungal activity against the phytopathogens by inhibiting the radial growth.

For the control of some plant fungal diseases, biogenic nanoparticles have proved to be effective nanomaterial-based fungicides. Biosynthesis of silver nanoparticles was performed by using phytopathogenic fungi (*Alternaria* sp.) isolated from banana cultivated soil. *Alternaria* sp. can grow very fast and produce high enough bioactive compounds (Theint Theint Win et al., 2020). Synthesized AgNPs have demonstrated remarkable potential for the use of antifungal compounds to combat plant diseases.

The fungicidal properties of nano-size silver colloidal solution were used as an agent for antifungal treatment of various plant pathogens. WA-CV-WA13B, WA-AT-WB13R, and WA-PR-WB13R silver nanoparticles (AgNPs) were used at concentrations of 10, 25, 50, and 100 ppm (Kim et al., 2012). Eighteen different plant pathogenic fungi were treated with these. The results indicated that AgNPs possess antifungal properties against these plant pathogens at various levels. Treatment with WA-CV-WB13R AgNPs resulted in maximum inhibition of most fungi. Results also showed that the most significant inhibition of plant pathogenic fungi was observed on PDA and 100 ppm of AgNPs.

The green synthesis of fungicides, which consist of copper nanoparticles (Cu-NPs) was prepared in aqueous media. Through *in vitro* experiments, the antifungal efficacy against *Fusarium solani, Neofusicoccum* sp., and *Fusarium oxysporum* was investigated (Nicolaza Pariona et al., 2019). Although the antifungal activity differs for each fungal species, it was found that the Cu-NPs induce strong morphological changes in the mycelium.

As an alternative to synthetic fungicides, the use of silver nanoparticles (Ag NPs) as antifungal agents has become more widespread, as scientific progress make their production more cost-effective. One of the applications that Ag NPs can be used in is plant diseases management. Silver nanoparticles have a high antimicrobial activity and are broadly utilized for several disinfection purposes including water and materials' sanitization for medical purposes. There have been comparatively few studies on using silver against plant pathogenic fungi. In this study, silver nanoparticles (Ag NPs) were used at concentrations of 0.0, 0.0002, 0.0005, 0.0007, 0.0009, 0.0014 and 0.0019 mol/L (Abdallah Mohamed Elgorban et al., 2016). Various concentrations of Ag NPs have antifungal properties to control *R. solani* AGs. The obtained results also revealed that strong inhibition of *R. solani* AGs was noticed on CDA at all concentrations. Silver is well known for its antimicrobial potential. However, in the form of nanoparticles, the antimicrobial activity of silver can be much superior. The antimicrobial potential of AgNPs was evaluated at different concentrations (25-100 ppm), against three pathogenic plant fungus plant (*Alternaria solani, Corynespora cassiicola* and *Fusarium* spp.), in two different fungal media in term of inhibition of radial growth (Pankaj K. Tyagi et al., 2020). Up to 100% inhibition for *A. solani* and *Fusarium* spp. and 85% inhibition for *Corynespora cassiicola* was observed at 100 ppm AgNPs concentration on potato dextrose agar (PDA). Further, exposure of AgNPs on *Drosophila melanogaster* confirmed that Bio-AgNPs are nontoxic as compared to CH-AgNPS. It was concluded that Bio-AgNPs are safe to use due to their nontoxic nature.

The inhibitory affects of different concentrations of high and low molecular weight of chitosan (CH-HMw and CH-LMw, respectively) and chitosan nanoparticles (CH-HMw-NPs and CH-LMw-NPs, respectively) against linear growth and spores/sclerotia germination of the 18 phytopathogenic fungi of tomato, potato and green bean were evaluated *in vitro* (Mahmoud Abdel-Aziz, 2019). The studies showed that CH-HMw-NPs and CH-LMw-NPs were highly effective against all tested pathogenic fungi compared with CH-HMw and CH- LMw. Furthermore, CH-HMw-NPs and CH-LMw-NPs at concentrations 0.1% and 0.05% showed completely inhibited (100%) the mycelial growth of all tested pathogens. Meanwhile, Chitosan HMw and LMw at 1.0% caused 100% reduction of all tested pathogenic fungi. *Rhizoctonia solani* and *Fusarium solani* followed by *F oxysporum* and *Macrophomina phaseolina* were highly sensitive to all concentrations of chitosan than *Alternaria solani, Sclerotium rolfsii* and *Phytophthora infestans*. It was concluded that chitosan NPs could be a new development for the generation of chitosan based bio-nanopesticides against fungal diseases exploited for delivery of agrochemicals.

4.11 Nanophytobacteriology

In recent years, nanotechnology has been increasingly applied to the development of novel antimicrobials for the management of bacterial plant diseases. In particular, significant development in nanomaterials synthesis, such as polymeric, carbon-based and metallic, has attracted researchers' attention towards applications in managing plant diseases caused by bacteria. Toxicity considerations, including negative environmental effects, have also led to the redesign of nanomaterials by tuning the size and shape and by surface modification, leading to increased antimicrobial activity and decreased ecological toxicit (Ocsoy et al., 2013).

Many studies have shown that nanoparticles that can be used to play an important role in the management of plant disease and promotion of plant growth (Kasprowicz et al., 2015). For example, metallic nanoparticles such as zinc oxide nanoparticles and MgO nano-flowers showed high antibacterial activity against bacterial leaf blight disease. Furthermore, silver nanoparticles (AgNPs) showed strong inhibition on Gram-positive and Gram-negative bacteria. Particularly, biological methods have been more and more applied in synthesize of the nanoparticles, which are more safe and eco-friendly compared to the traditional physical and chemical methods.

Bacterial spot caused by *Xanthomonas perforans* is a major disease of tomatoes, leading to reduction in production by 10–50%. While copper (Cu)- based bactericides have been used for disease management, most of the *X. perforans* strains isolated from tomatoes in Florida and other locations worldwide are Cu-resistant. DNA-directed silver (Ag) nanoparticles (NPs) grown on graphene oxide (GO) was developed. These Ag@dsDNA@GO composites effectively decreased *X. perforans* cell viability in culture and on plants. At the very low concentration of 16 ppm of Ag@dsDNA@GO, composites showed excellent antibacterial capability in culture with significant advantages in improved stability, enhanced antibacterial activity and stronger adsorption properties (Ocsoy et al., 2013). Application of Ag@dsDNA@GO at 100 ppm on tomato transplants in a greenhouse experiment significantly reduced the severity of bacterial spot disease compared to untreated plants, giving results similar to those of the current grower standard treatment, with no phytotoxicity.

The biosynthesis of silver nanoparticles (AgNPs) through the use of endophytic bacteria is a safe replacement for the chemical method. A study by Ezzeldin Ibrahim et al., (2019) was aimed to synthesize AgNPs using endophytic bacterium *Bacillus siamensis* strain C1, which was isolated from the medicinal plant *Coriandrum sativum*. The synthesized AgNPs at 20 μg mL^{-1} showed a strong antibacterial effect against the pathogen of rice bacterial leaf blight and bacterial brown stripe, while an inhibition zone of 17.3 and 16.0 mm was observed for *Xanthomonas oryzae* pv. *oryzae* (Xoo) strain LND0005 and *Acidovorax oryzae* (Ao) strain RS-1, respectively.

Furthermore, the synthesized AgNPs significantly inhibited bacterial growth, biofilm formation and swimming motility of Xoo strain LND0005 and Ao strain RS-1. In addition, the synthesized AgNPs significantly increased root length, shoot length, fresh weight and dry weight of rice seedlings compared to the control. Overall, this study suggests that AgNPs have the potential to protect rice plants from bacterial infection and plant growth promotion.

The development of multidrug resistant bacteria is becoming a severe problem, causing adverse effects on plants, animals, and public health (Behiry et al., 2018) and resulting in millions of infections and thousands of deaths on a global scale. Although a number of new antibiotics have been produced in the last couple of decades, the growth in antimicrobial resistance is faster than the current rate of developments in effective antibiotics production. A significant decrease of over 50% in approved antibiotics in the last two decades has made it more alarming and compels researchers to search for novel, appropriate, and efficient antimicrobial agents that are cost effective.

Bacterial canker of tomato caused by the bacterial pathogen *Clavibacter michiganensis* subsp. *michiganensis* (Cmm) is a major limiting factor for tomato production worldwide. Currently there exists no resistant variety of tomato to bacterial canker; only cultural and chemical controls are available (Asma Noshad et al., 2020). This study synthesized AgNPs (silver nanoparticles) via a green chemistry route and investigated their bactericidal potential against bacterial canker of tomato (BCT). AgNPs were prepared using mycellial aqueous extract of agriculturally beneficial fungi *Pythium oligandrum*. AgNPs treated seeds showed a normal germination rate with normal seedling growth. An in-vitro study found that the prepared AgNPs caused the maximum inhibition of the bacterial pathogen. In the greenhouse the introduction of AgNPs significantly prevents and inhibits the bacterial pathogen Cmm on tomato plants. These results suggested that this process is a strong candidate for industrial scale production of AgNPs. These particles act as an inhibitor and broad spectrum antibacterial agent against cmm, and hence offer a new and eco-friendly alternative in BCT control.

This bactericidal potential of AgNPs makes it a potential candidate to eliminate rather than merely inhibit the spread of infection, and many research studies support it (Kumari et al., 2009). But since the target bacteria was Cmm it was hypothesized that it could be an option to inhibit bacterial canker. Hence, based on the results obtained in the in-vitro trial, a greenhouse experiment was designed to observe the effect of AgNPs not only on seed germination but also on the mature tomato plant. Seedlings exposed to both concentrations of AgNPs showed full germination after 12 days. Some studies in the past considered a threshold value of AgNPs toxicity of 5 mg/L–1000 mg/L, whereas here 0.088 mg/L, 0.176 mg/L, and 0.44 mg/L were used which as the lowest possible concentration in this study (Asma Noshad et al., 2020).

Antibacterial properties of zinc nanoparticles against *Pseudomonas aeruginosa* was reported as the novel material (Khan and Rizvi, 2014). Copper nanoparticles were found effective against *Bacillus subtilis* and *P. aeruginosa.* The effect of the nanomaterial TiO_2 on the phyllosphere microbial community was investigated. Analysis results by PCR denaturing gradient gel electrophoresis revealed a TiO_2-induced change in the community structure of microorganisms (Shuang Wang, et al., 2009). An epiphytic bacterium, *Bacillus cereus*, was chosen to study the role of the superoxide dismutase-encoding genes, sodA-1 and sodA-2, and its survival from TiO_2 photocatalysis. These results showed that the expression of sodA-1 and sodA-2 was induced by photocatalytic oxidation, with a higher induction observed in sod A-2. In addition, compared with wild-type *B. cereus* 905, a reduced bacterial population was observed in a sodA-1 and sodA-2 double deletion mutant strain KOS on a cucumber leaf surface sprayed with TiO_2. Because the phyllosphere is considered as one of the major habitats for microorganisms, and substantial areas of the earth are covered with leaves, the results of this work provides information of the potential impact of photocatalytic nanomaterial in the environment.

Effect of TiO_2 concentrations on bacterial populations : The 2-week-old cucumber plants were grown in a greenhouse covered with a PEP-UV2702AD plastic film (Shanghai HiTeC Plastics Co., China). Solar UV-A and UV-B radiation to the leaves was monitored by a UV sensor (YK34UV, Taiwan Taina Co., China). The plot in the greenhouse was 60 m long, 8 m wide, and was divided into fifteen separated blocks. The experimental design was a randomized, complete block with three replications for each of the five TiO_2 treatments tested. The treatments were devised according to the dosage of TiO_2 (0, 0.002, 0.02, 0.2, and 2 mg mL.

TiO_2 in water was sonicated for 20 min before the experiment so that the particles could be uniformly dispersed. To simulate a durative deposit with TiO_2 on the leaf surface, newly expanded leaves of each plant were sprayed with various concentrations of the TiO_2 suspensions once a week from the early growth stage. Eighty plants were treated with a 100-mL suspension per block, each time. The experiment lasted 30 days and sampling was performed a day after the last TiO_2 treatment. The trials were conducted in the greenhouse located at Shangzhuang Agricultural Experimental Station, China Agricultural University, Beijing, China. To collect leaf samples, fully expanded young leaves (30 g per replication) were chosen randomly from the plant canopy on July 30, 2007, using sterile scissors to nip individual leaves into sterile plastic bags containing washing buffer. These bags were transported to the laboratory on ice and sonicated for 10 min in an ultrasonic cleaning bath. The resulting bacterial suspension was centrifuged for DNA extraction or plated for viable count. Viable counts of the total population size of leaf-associated bacteria were determined by serial dilution of cultures and plating on nutrient agar (1% w/v). Results are representative of at least three independent experiments

Effect of TiO_2 on the foliar bacterial community : DGGE profiles and cluster analysis showed that similar community structures were observed in the control (0 mg mL. TiO_2) and low-level TiO_2 treatment (0.002 mg mL . TiO_2) (Shuang Wang, et al., 2009). However, an obvious TiO_2-induced change in the community structure was observed at a TiO_2 concentration of 0.02 mg m. The total culturable leaf surface population sizes in the phyllosphere community were significantly reduced with an increase in the TiO_2 concentration to 0.02 mg mL. The TiO_2 affected the foliar bacterial community. To further understand the differences between various treatments, dominant 16S rRNA DGGE bands obtained from microorganisms on cucumber leaves were isolated, cloned, and sequenced.

4.12 Nanophytonematology

A plethora of chemical, physical and biological techniques continues to evolve leading to the production of noble metal nanoparticles. Alongside, biological organisms including plant, fungi and bacteria are an ideal source for green synthesis of nanoparticles with desired shape and size. Some of these nanoparticles also have nematicidal properties, which apply to numerous genera of plant parasitic nematodes and also to plant pathogenic fungi and bacteria (Bhau et al., 2016). Plant parasitic nematodes are major agricultural pests causing crop losses worth hundreds of billions dollars annually worldwide. Traditional control measures depend upon highly toxic nematicides. In theory, seed treatment for nematode control is optimal, but is largely ineffectual due to poor rhizosphere delivery. Active ingredients of various nanoparticles have also shown evidence of being potentially effective nematicides, which makes these nanoparticles a suitable noble source to control nematode infection in plants. Although very limited reports are available on the use of nanoparticles to control plant nematodes, very encouraging reports are there and research in this area is getting lot of attention.

New approaches by green silver nanoparticles (GSN) were used to control plant-parasitic nematode to avoid chemical nematicide hazards (Kamal

Fouad Abdellatif et al., 2016). Analyses of the incorporation of different concentrations of nanoparticles on two different algae (*Ulva lactuca* and *Turbinaria turbinata*) were carried out. Furethermore, the effect of GSN on the eggplant DNA profile was studied using RAPD and EST molecular markers. GSN (17 $mg.mL^{-1}$) obtained from *U. lactuca* was more effective in reducing second- stage juveniles (J2s) of *M. javanica* (69.44%) population in soil. All treatments improved eggplants growth parameters. *Meloidogyne javanica* reproduction and growth were affected. All treatments were comparable (K) to chemical control (Vydate®24%L); resulted in significant reduction in number of galls, number of female nematodes, and number of egg-masses per root system. The same finding was noticed on the number of nematode juveniles tow (J2s) in soil compared with those of (M) control (Table 14). Change in DNA profile using of both RAPD and EST markers was noted. GSN did not cause any phototoxicity in eggplants under treatment.

Table 14: Efficacy of GSN on reproduction parameters of *M. javanica* in eggplant

Treatments	($mg.100\ mL^{-1}$)	Galls/ root	Females/ root	Egg masses/ root	Juveniles (J2s)/ 250 soil
Turbinaria turbinata	A (17)	53 ed	46.25b	58.25b	1116c
	B (12.75)	99cd	55b	74.5b	1140c
	C (8.5)	95.75bcd	67.25b	72b	1308c
	D (4.25)	96.25bcd	106.25b	130.25b	1894b
Ulva lactuca	E (17)	51e	54b	58.5b	1023c
	F (12.75)	62.75edc	65.25b	72.5b	1127.5c
	G (8.5)	86.5bcde	57.25b	91b	1219c
	H (4.25)	114b	73.5b	111b	1916b
Chemical control	K(2 ml Vydate®)	45.75e	47.5b	57.5b	948c
Control (Nematode)	L (Infected)	438 a	298.5a	278.5 a	3348a
Control	M (uninfected)	-	-	-	-

Means connected with the same letter(s) within a column are not significantly different ($P \leq 0.05$) according to Duncan's Multiple Range tests.

Growth parameters of eggplants, weight and length of shoots and roots and number of leaves, are listed in Table 15. All treatments showed improvement in plant growth. However, no remarkable increase in root weight parameters was noticed.

Table 15: Growth reactions of eggplant treated with application of GSN synthesized by marine algae, and chemical nematocides in the presence and absence of root-knot nematode

Treatments	(mg.100 mL^{-1})	Shoot		Root		Leave
		length (cm)	weight (g)	length (cm)	Weight (g)	Number
Turbinaria	A (17)	26.5abc	10.79bcd	10.75a	3.16ab	8ab
turbinata	B (12.75)	28 bcd	11.73cd	11ab	5.17ab	7.75ab
	C (8.5)	30.5 cd	14.77e	10a	4.19ab	9.75bc
	D (4.25)	26 ab	10.24bc	9.75a	4.08ab	7.75ab
	E (17)	25.5ab	11.17cd	11.75 a	4.05ab	25.5ab
Ulva lactuca	F (12.75)	28bcd	9.92bc	9.5a	2.87ab	28ab
	G (8.5)	28.25abc	9.91bc	11ab	3.7ab	28.25ab
	H (4.25)	25.5ab	8.16b	9a	2.93ab	25.5a
Chemical control	K(2 mL)	31.5 d	15.66 e	11ab	3.81ab	11.5c
Control	Vydate®	20.5 a	5.72 a	10.5a	6.08a	8ab
(Nematode)	L (Infected)	27.75 abcd	13.26 de	13.5 b	1.58b	6 a
Control	M (uninfected)					

Means connected with the same letter(s) within a column are not significantly different ($P \leq 0.05$) according to Duncan's Multiple Range tests.

In order to have effective management, nanoparticles offer a lot of potentials to effectively control nematodes population through its targeted approach, nanoparticles mediates higher penetration of chemicals pesticides with a minimum concentration of chemical used, as there penetrates easily through the cell wall and produces maximum effect. Gold nanoparticles kill nematodes itself, but when used along with chemicals nematicides their efficiency increased significantly (Thakur and Shirkot, 2017). A direct application of nanoparticles in soil and exposed parts of plants may also affect the on target organisms such as significant symbionts, solubilizers and nanoparticles may act as carriers of unwanted chemicals, concentrated intermediates of various pesticides/insecticides etc. Thus to reckon the scope and applications of nanoparticles in plant disease management the effect can be divided into two perspectives direct effect of nanoparticles on nematodes and use nanoparticles in formulating nematicides. Due to their ultra-sub microscopic size nanoparticles gain the high degree of reactivity and sensitivity and thus have potential to prove very useful in control the PPNs, in addition, nematicidal residue analysis. As chemical and physical properties of nanoparticles vary greatly as compared to larger form, thus it has become imperative to evaluate the effect of nanoparticles on nematodes to harness the beneficial effects of this technology in the plant protection, especially against phytonematodes.

Gold, Silver NP possess nematicidal activity that may provide an alternative to high-risk synthetic nematicides or inconsistent biological control agents. High frequency (biweekly) and high application doses (\$ 90.4 mg/m^2) of Gold, Silver NP may be required to achieve effective field efficacy for root-knot nematode. Combining AgNP with an irrigation system such as fertigation or tank-mixture with compatible chemicals that supplement the AgNP nematicidal effect may increase applicability of AgNP. Further understanding of the mechanism in the nematicidal action of Silver NP also warrants improvement of Gold, Silver NP efficacy.

Cyanobacterium-based silver nanoparticles are considered not only as an efficient nano-nematicide but also as a bio-stimulant material for plant growth. They could be employed as a part of an integrated program for controlling some plant diseases. Silver nanoparticles (Ag-NPs) were biosynthesized from aqueous extract of the cyanobacterium, Nostoc sp. PCC7524. Full characterization of the biosynthesized Ag-NPs was monitored by UV-vis spectroscopy, transmission electron microscopy, X-ray diffraction pattern, Zeta sizer and Fourier transform infrared spectroscopy (Hamed et al., 2019). *In vitro* assay against *Meloidogyne javanica* showed that Ag-NPs significantly decreased egg hatching of *M. javanica* at different applied concentrations (3, 6, 12, 25, and 50%, v/v). Fifty percent of Ag- NPs induced the highest reduction percent (94.66%). Moreover, Ag-NPs and $AgNO_3$ significantly increased the percentages of larval mortality of the second-stage juveniles (J2) with concentration and time-dependent responses. Ag-NPs or AgNO3 at 2.4 ml/l, 24 h, completely inhibited the growth of J2 compared to 23% inhibition using aqueous cyanobacterial extract (ACE). *In vivo* effect of Ag-NPs on faba bean-infected plant under greenhouse conditions was achieved by treating soil with three different concentrations of 1, 2, and 3 ml/kg soil over two consecutive seasons. Ag-NPs significantly reduced root galling from 39.6 to 78.7% and J2 population in the soils from 32.2% to 86.7% in the 2018 season and from 21.9 to 78.1% and 40.0 to 81.0% in the 2019 season, respectively. Moreover, 3 ml/kg soil of Ag-NP treatment showed statistically comparable effects to that of vydate nematicide but with remarkable enhancement of faba bean growth parameters as compared to those of vydate or $AgNO_3$ treatments in the two seasons. The considerable *in vitro* and *in vivo* nematicidal potential of the cyanobacterium-based Ag-NPs, besides their bio-stimulant effect on plant growth, makes them feasible for the biological control of *M. javanica.*

The role of copper/iron based nanoparticles was studied as bioactive agents against the root knot nematode, *Meloidogyne* spp; along with the plant enhancement properties on the nematode infested host (Gkanatsiou Ch et al., 2019). Same sized PEGylated Cu, CuFe and $CuFeO_2$ NPs, were solvothermally synthesized characterized and evaluated against *M. incognita* and *M. javanica* nematodes, in terms of paralysis and biological cycle arrest in tomato plants. Comparisons were made against the commercial nematicide fosthiazate. All the tested compositions revealed nematicidal activity, but CuFe NPs were found best effective exhibiting the lowest EC50 value calculated at 0.03 μg ai g/soil. Additionally, a fresh shoot and root weight increase was evident in tomato plants treated with CuFe NPs.

The nematicidal activity of the NPs was determined in the descending order of CuFe (CuO) > Cu (CuO) = $CuFeO_2$ (Cul+), with the last two compositions revealing similar efficacy. Copper salts have already been reported of nematicidal activity when used in combination with maleimide derivatives, while metal complexes have been found effective both against the root-knot nematodes and *Fusarium oxysporum* . In specific copper sulphate, principally used as a fungicide but also for treating copper-deficient soils, has been reported to exhibit an EC50 value of 280 μg/mL against *M. javanica* while an EC50 value of 48.6 μg/mL against *M. incognita* was reportrd. Application of a mixture of maleic acid and copper sulfate reduced RKN disease on tomato by 51.72% in a pot experiment, suppressed gall formation on melon and decreased nematode population density in the soil. Regarding copper based NPs, in previous studies, against plant pathogenic bacteria, among NPs of the same sizes with varyin compositions such as Cu (CuO), Cu_2O (Cul+) and CuO (Cu_2+) NPs, metallic copper has been found the most potent (Rahul et al., 2014).

Abamectin (Abm) is a biological and high-efficiency pesticide against *Meloidogyne incognita.* In this study, a powerful method, flash nanoprecipitation (FNP), was adopted to successfully produce Abm-loaded nanoparticle suspensions with high drug loading capacity (>40%) and encapsulation efficiency (>95%), where amphiphilic block copolymers (BCPs) poly(lactic- co-glycolic acid)-b-poly(ethylene glycol) (PLGA-b-PEG), poly (D, L-lactide)- b-poly(ethylene glycol) (PLA-b-PEG), or poly(caprolactone)-b-poly(ethylene glycol) (PCL-b-PEG) were used as the stabilizer to prevent the nanoparticles from aggregation (Zhinan Fu et al., 2019). The effect of the drug-to-stabilizer feed ratio on the particle stability was investigated. Moreover, the effect of the BCP composition on the morphology of Abm-loaded nanoparticles for controlling *M. incognita* was discussed. Notably, spindle-like nanoparticles were obtained with PCL-b-PEG as the stabilizer and found significantly more efficient (98.4% mortality at 1 ppm particle concentration) than spherical nanoparticles using PLGA-b-PEG or PLA-b-PEG as the stabilizer.

Overall, the lethal concentration assays indicated that spindle-like Abm-loaded nanoparticles prepared using PCL-b-PEG as the stabilizer by FNP have higher nematicidal efficiency than spherical Abm-loaded nanoparticles using PLGA-b-PEG or PLA-b-PEG as the stabilizer. In addition, the Abm- loaded nanoparticles caused irreversible paralysis in *M. incognita*, and the nematode mortality increased with the increasing concentration of Abm-loaded nanoparticles. Thus, the nematode control effectiveness can be improved by using increased concentration of Abm-loaded nanoparticles.

The eco-nematicide activity of freshwater green macroalga, *Cladophora glomerata* aqueous extract and their synthesized silver nanoparticles (Ag-NPs) against root-knot nematode, *Meloidogyne javanica* was investigated on tomato plants. The formation and chemical structure of Ag-NPs was examined (Ghareeb et al., 2020). The results indicated that GSNPs from *C. glomerata* exhibited the highest eco-nematicide activity in the laboratory bioassay on egg hatchability and juveniles (J2S) mortality of *M. javanica* compared with the chemical commercial nematicide Rugby 60% (Table 16). Also, results showed a significant reduction in galls number, egg masses, females per root system/plant, and mortality of juveniles. The results of PAL and PPO enzyme expression for the control plants remained relatively stable, while the plant inoculated with nematode *M. javanica* as well as the activity of genes in scope was increased from 14 to 28 Days after Nematode Inoculation (DANI). These activities were improved in inoculated plants and treated with *C. glomerata* extract and their green syntheses of Ag-NPs and the other plants treated with Rugby 60% (4 mL/L). The greatest activities of the three enzymes were evident after 14 days after the nematode inoculation. It can be concluded that the green synthesized nanoparticles using *C. glomerata* could be used as potent nematicides against *M. javanica* which induces the immune system to defend against nematode infection.

The toxicity of three nanoparticles, silver (AgNP, 20 nm), silicon oxide (SiO_2NP, 11-14 nm) and titanium oxide (TiO_2NP, 20 nm), to *Meloidogyne incognita*, was studied inlaboratoryand pot experiments (Abbas Salahi Ardakani, 2013). Results showed 100% J2 immobility and mortality in treatments of 800, 400 and 200 mg ml^{-1} of AgNP; the LC50 for AgNP was 100 mg ml^{-1}. J2 mortality was 48, 16 and 7.3% in 100, 50 and 25 mg ml^{-1} of AgNP, respectively.

Table 16: The effect of *C. glomerata* extract and their silver nanoparticles on the numbers of nematode galls, egg masses, females/root system, and the number of second-stage juveniles (J_2S)/250g soil tomato plant infected with *M. javanica* (Mj) after 60 days.

Treatment	Con	Nematode parameters and the decrease in galls over control (D%)ta Extract							
		Galls	D %	Egg Masses	D %	Females/Root	D %	J_2S/250 gm Soil	D %
M. javanica Alone	–	$178^a \pm 5.8$	–	$154.2^a \pm 2.4$	–	$165.5^a \pm 5.2$		$185^a \pm 6.4$	–
M. javanica+Rugby 60%	–	$11.8^c \pm 0.6$	93.4	$8.1^c \pm 0.6$	95	$21.24^c \pm 07$	87	$10.0^b \pm 0.6$	94.6
M. javanica + *C. glomera-*	–	$29.4^b \pm 0.7$	83.0	$10.2^b \pm 0.7$	93.5	$33^b \pm 0.6$	80	$13.0^c \pm 0.6$	93.7
M. javanica+GSNPs	S	$12.2^c \pm 0.6$	93.1	$6.2^d \pm 0.6$	96	$18.3^d \pm 0.7$	89	$5.0^d \pm 0.5$	97.3

Data are means of ten replicates; means with the same letter(s) in each column are not significantly different at ($P \leq 0.05$).

J2 mortality was 4.3 and 2% in 800 and 400 mg ml^{-1} of TiO NP, respectively. No J2 mortality was seen in all other treatments after 24 h exposure. In a pot experiment, all treatments of AgNP and 0.02% TiO_2NP completely controlled *M. incognita*. Treatments of 0.02, 0.01 and 0.005% of AgNP as well as 0.02% of TiO_2NP were toxic to tomato plants and significantly reduced tomato root and stem length and fresh weights in comparison to control.

A greenhouse experiment was conducted to evaluate the effect of silver nanoparticles (AgNPs) at concentrations of 0.25, 0.5 and 1 mM as a potential nematicide on *M. incognita* infecting tomato (Ahmed Hammad Nour El-Deen and Bahig Ahmed El-Deeb, 2018). Results indicated that application of AgNPs improved plant growth and reduced nematode infection in comparison to silver nitrate and control treatments. The highest increment of fresh weight as well as the lowest numbers of galls and egg-masses was obtained when tomato plants was treated with AgNP produced by ginger extract at 1 mM.

Silver nanoparticles (AgNP) were evaluated as a nematicidal substance in laboratory and screenhouse experiments. Second infective juveniles (IJ2) of *Meloidogyne incognita* were exposed to AgNP in water at with the consternations of 20, 40, 200, 500, and 1500 ppm/ml. the concentration of 200 ppm caused 52% mortality at the third day, while 500 ppm caused 51% mortality after one day and 64%, 82% after the second and the final day, respectively (Entsar H. Taha, 2016). The most effective concentration was 1500 ppm which caused 89%, 93, and 96.5 %, respectively. In the screen house experiment, all the concentrations of AgNP inhibited the nematode growth (gall and egg formation and final population) and eggs hatchability. However the high concentrations of 200 ppm, 500 ppm, and 1500 ppm were more significant in their effect. The application of AgNP didn't show toxic effect on the plant growth or the free living nematodes. It was concluded that, the laboratory assays to the nematicidal effect of AgNP, and the screenhouse evaluation demonstrated that, its nano silver may be a more safety alternative method to control root knot nematodes.

In vitro evaluation for the nematicidal efficiency of copper nanoparticles (CuNPs) was conducted against, *Meloidogyne incognita*. CuNPs were prepared according to the chemical reduction method; physicochemical characterization of CuNPs was done using UV-Vis spectroscopy, Dynamic Light Scattering and Transmission Electron Microscopy (Elwy A. Mohamed et al., 2019). When second stage juveniles (J2) of *M. incognita* were incubated in soil saturated with CuNPs (spherical shape; 100 nm diameter) for 3 days, it was found that J2 mortality was directly proportional to the concentration of CuNPs and 0.2 g/L was sufficient to cause 100% mortality. Statistical analysis showed that all mortalities caused by treatment with CuNPs at different concentrations were statistically significant compared with non-treated control. Conclusively, this paper may provide a potential alternative nematicide against *M. incognita.*

The effectiveness of leaf extracts of *Conyza dioscoridis* and *Urtica urens* that were prepared as silver nanoparticles (Ag-NP) as nematicidal alternatives was investigated (Atef Mohamed Khedr Nassar et al., 2018). Results revealed that silver nanoparticles of *C. dioscoridis* and *Urtica urens* extracts had great nematicidal activity against the 2nd stage juvenile and eggs of *Meloidogyne incognita.* Also, the Ag-NP showed similar nematicidal effect to the reference nematicides. The GC-MS analysis revealed the increase of certain metabolites due to the formulation of the Ag-NPs. Aromadendrene, 1-hydroxy-1,7-dimethyl-4-isopropyl-2, 7-cyclodecadiene, 6-epi-shyobunol, 4-hexy-lacetophenone, β-isocomene, caryophyllene, β- and α-selinene, α-cadinol, berkheyaradulen (It's a sesquiterpene hydrocarbon that is detected in many plant species; Chrysanthemum boreale, chamomile oil, and many others,) and bis-(2-ethylhexyl)phthalate were increased more than 2.5-folds in the Ag-NP compared to the extract. Therefore, the green synthesis of metal nanoparticles might be safe, effective and affordable nematicide alternatives.

The physicochemical properties of SiC nanoparticles were characterized and investigated their impact on the multicellular animal model, *Caenorhabditis elegans* and the plant parasitic nematode *Meloidogyne incognita* (Al Banna et al., 2018). The X-ray diffractogram patterns and chemical analyses confirmed that the content is SiC. The bioassays indicated that SiC nanoparticles did not affect hatching of larvae of both nematodes, whereas they affected dramatically the survival of the first stage larvae (L1) of C. elegans, but not the second stage juveniles J2s of *M. incognita.* There was no effect of SiC on length and width of the dead L1; however, internal organs, particularly the intestine, exhibited black color indicating uptake of the SiC nanoparticles.

Sulfur nanoparticles (S-NPs) were prepared using rosemary (*Rosmarinus officinalis*) leaves aqueous extract and sodium thiosulfate pentahydrate ($Na_2S_2O_3.5H_2O$) at room temperature. The S-NPs were characterized by UV–visible spectrophotometer, Fourier transform infrared (FT-IR), Scanning electron microscopy (SEM), Transmission electron microscopy (TEM), and X-ray diffraction (XRD) techniques (Luma S. Al Banna et al., 2020). The sulfur nanoparticles are crystalline in nature with average size of 40 nm. The morphology of S-NPs could be controlled by tuning the amount of rosemary leaves aqueous extract and sulfur ions. The cytotoxic effects of the S-NPs on hatching of second stage root-knot and mortality of second stage of root-knot nematode (*Meloidogyne javanica*) were tested. These results confirmed that the S-NPs synthesized using rosemary leaves aqueous extract and could possibility be used as nematicidal to manage *M. javanica* infestation.

A bioinspired nanoparticle technology was developed using the coat protein (CP) assemblies of Tobacco mild green mosaic tobamovirus (TMGMV) to deliver nematicides to roots reduce dosage and improve soil mobility efficacy and safety (Raghavan Charudattan, 2019). The CP's zwitterionic nature allows the loading and release of nematicides. The TMGMV-delivered nematicides (TDN) are efficacious against *Caenorhabditis elegans* in liquid cultures and the TMGMV-assisted delivery enhances the nematicide's soil mobility compared to the free nematicide and contemporary synthetic nanoparticles.

4.13 Can nanoparticles replace chemicals?

Nanotechnology has entered into the world of pesticides and is capable of pest control by overcoming the use of pesticides such as insecticides, fungicides, and herbicides as in the present times. With new nanoscale techniques of mixing and harnessing genes, genetically modified plants would be replaced by 'atomically modified' plants (Rishu Sharma et al., 2016). Pesticides can be more precisely packaged to knock out unwanted pests. The crux of nanotechnology remains in compressing the materials to nanometer range by keeping the size small. The application of silicon is integrated into field management of pests and disease due to its non-residue property. A reduction in pest infestation and increased resistance has been observed in wheat cultivars after silicon application. Application of nanosilica to the tomato plants may minimize the problems caused by Spodoptera littoralis leading to development of average resistance and in turn affecting the feeding preference of this pest. Lastly, it deeply impacts the longevity and fecundity of the insect, thus decreasing the insect population density and ultimately decreasing the yield losses. Though the use of NPs can control the pest populations, further study needs to validate their viability and environmental impacts in comparison with the chemical pesticides. Potential adaptations for dealing with impacts of climate change on plant diseases are achieved by applying genetic changes, laws, and economic changes and mitigation practices. The amount of GHG can be decreased by adopting new cropping systems and crops that reduce the GHG production by emitting less nitrous oxide.

Nanoparticles could be used for site-targeted delivery of nanoparticles to avoid damage to plant parts that might happen by various conventional agrochemical application methods and also to support environmental protection by avoiding soil and water pollution (Jha et al. 2009). Nanoparticle-mediated plant genetic engineering and use of nanosensors in agriculture are frontier areas to increase crop production by avoiding pest and disease incidences, which will be a boon to farmers. Some of the NPs that have entered into the arena of controlling plant diseases are nanoforms of carbon, silver, silica, and alumino-silicates. Nanoparticles with similar mode of action to chemical pesticides can be used as carrier of active ingredients of pesticides, host defense-inducing chemicals to the target pathogens. Nanoparticles may hit/target virus particles due to their ultra-small size and may open a new field of virus control in plants. Plant ion concentrations are affected by environmental conditions. Plants absorb essential and nonessential elements under specific growing environments. An uptake of these ions above certain concentrations can cause toxicity. Plants produce natural nanomaterials under certain necessary for their growth (Rishu Sharma et al., 2016).

Understanding the genetic constraints for pathogen and plant adaptation will allow better mechanistic models by studying ecology and genetic interaction in order to mitigate impacts of climate change. The possible changes in precipitation, temperature, concentration of CO_2, CH_4, nitrous oxide, and O_3 are expected to have significant effect on crops and pathogen development, survival, and spread. The changes in climate have affected various pathogens differently; for example, with the increased CO_2, the fecundity of Colletotrichum gloesporoides increases. Increased CO_2 has led to increased pathogen load in prairie. Similarly, under drought conditions observed plant structural changes, leaf inhibition and stomatal closure. With the changed temperature and precipitation pattern, there is an expansion of Phythophthora cinnamoni in Europe (Bergot et al. 2004).

To sum up, the occurrence and development of new pathogenic races is a continuing problem, and the use of chemicals to control pests is expensive and not always effective. In recent years, the use of nanomaterials has been considered as an alternative solution to control plant pathogens. Agricultural practices usually include the systematic application of a wide array of active compounds at variable dosages and frequencies, which represent a wide range of selective regimes. Metal oxide nanoparticles have controlled the green peach aphid. Magnesium hydroxide, bionanoparticles synthesized were the best control to *Myzus persicae* compared to other synthetic nanoparticles. Zinc oxide nanoparticles synthesized using aqueous Punica granatum peel extract was tested for its potential antimicrobial activity against some selected microbes. Also this research work determined the effect of synthesized ZnONPs on green peach aphid and antibacterial efficacy against standard strains of Gram-positive *Staphylococcus aureus* and Gram-negative *Escherichia coli.* Other nanoparticles (Fe, Cu, Si, Al, Zn, ZnO, TiO_2, CeO_2, Al_2O_3 and carbon nanotubes) have also been reported to have some adverse effects on plant growth. Sometimes nanoparticles also have an adverse effect on the growth of useful soil bacteria, such as *Pseudomonas putida* KT2440. Both the concentrations of nickel nanoparticles (50 and 100 ppm) inhibited the fungal mycelial growth on solid media, and the inhibitions were significant over control. Nickel nanoparticles at 100 ppm concentration inhibited the mycelial growth of *F. oxysporum* f. sp. *lactucae* and *F. oxysporum* f. sp. *lycopersici*. Green methods for synthesizing nanoparticles with plant extracts are advantageous as it is simple, convenient, environment friendly and require less reaction time.

Nanomaterials prepared by eco-friendly and green methods may increase agriculture potential for improving the fertilization process, plant growth regulators, pesticides delivery of active component to the desired target sites, treatment of wastewater and also enhancing the absorption of nutrients in plant. In addition, they minimize the amount of harmful chemicals that pollutes the environment. Hence, this technology helps in reducing the environmental pollutants. Nanotechnology has recently gained attention due to wide applications in different fields such as in agriculture medicine and environment. The large surface area offered by the tiny nanoparticles, which have high surface area, makes them attractive to address challenges not met by different control methods. Nanotechnology applications are currently being researched, tested and in some cases already applied across the entire spectrum of food technology, from agriculture to food processing, packaging and food supplements. They are with unique chemical, physical, and mechanical properties. In recent years, agricultural waste products have attracted attention as source of renewable raw materials. Insecticide resistance is one of the best examples of evolution occurring on an ecological time scale. The study of insecticide resistance is important, because it leads to a better understanding of evolutionary mechanisms operating in real time. The development of insecticide resistance in pest insects has been an increasing problem for agriculture and public health. Agricultural practices usually include the systematic application of a wide array of active compounds at variable dosages and frequencies.

4.14 Nanoparticles in the management of post harvest plant diseases

Nanomaterial treatments have demonstrated promising results and they are being investigated to reduce the utilization of synthetic fungicides to manage post-harvest rot in fruits and vegetables. Sergio Ruffo Roberto et al., (2019) reviewed a wide range of nanomaterials used in the management of post-harvest decays related to each selected fruit crop including grape, citrus, banana, apple, mango, peach, and nectarine. Other examples also used are apricot, guava, avocado, papaya, dragon, pear, longan, loquat, jujubes, and pomegranate fruits. In the field of post-harvest technology of fruits and vegetables, the nanotechnology approach is helpful for: controlling post-harvest diseases; introducing a new invention for packaging films; preventing impact of gases and unsafe rays; improving packaging appearance; and helping for labeling fresh products using the multiple chips (nanobiosensors). Nanotechnology was used to produce nanomaterials for use as antifungal agents in several commodities, including fruits and vegetables. Several nanomaterials have shown potential in post-harvest management and have been developed to control diseases in citrus, grapes banana, apple, mango, peach, and nectarine.

Nanoparticles having particular characteristics are useful in postharvest technologies. It can improve the post-harvest shelf life by controlling the growth, development and spread of disease causing microorganisms. With the development of new generation of packaging coverage's (Films), increasing strength, quality and packaging beauty and using the nanobiosensors for labelling the food products has proved beneficial for controlling postharvest losses (Singh, 2020).

i. Citrus: Clay-chitosan nanocomposite (CCNC) was prepared at different ratios of clay/chitosan and was tested against *Penicillium digitatum* on 'Valencia Late' sweet orange (*Citrus sinensis* L. Osb.). In in vitro tests, a complete inhibition of the pathogen was achieved at 20 $\mu g{\cdot}mL^{-1}$ for clay/ chitosan (1:0.5), clay/chitosan (1:1), and clay/chitosan (1:2). In in vivo trials, a complete inhibition of the disease was obtained, and a high reduction (around 70%) of lesion development was reported for clay/chitosan at 1:2 ratio at 20 $\mu g{\cdot}mL^{-1}$. Scanning electron microscopy showed that treatment with this nanocomposite caused severe collapse and irregular branching of hyphae in the apical part. In addition, CCNC caused a degradation of DNA of the tested pathogen at 20 $\mu g{\cdot}mL^{-1}$ (Sergio Ruffo Roberto et al., 2019). Silver nanoparticles, with an average diameter of 10 ± 5 nm were tested in vitro against *Alternaria alternata, P. digitatum* and *Alternaria citri* isolated from citrus fruits. This nanoparticle at 150 $mg{\cdot}L^{-1}$ showed a good antifungal efficacy against the three tested pathogenic fungi. The results of this study confirmed silver nanoparticles as promising nanomaterials, even if compared to iprodione or difenoconazole fungicides at the same concentration. To understand the mechanism in which silver nanoparticles works, it has been mentioned that it may cooperate with phosphorus and sulphur and their interaction may damage the DNA and proteins, and be consequential in cell death.

In 'Valencia Late' sweet orange (*C. sinensis* L. Osb.), copper nanoparticles (48 nm) were tested against *P digitatum* and *Fusarium solani.* The *in vitro* tests showed that a complete inhibition of *P digitatum* was recorded at 20 μg·mL^{-1} for copper nanoparticles. The percentage of reduction in colony diameter of *P. digitatum* was 43%, 60%, and 100% for copper nanoparticles, respectively. A complete inhibition of *F. solani* was obtained at 60 μg·mL^{-1} copper nanoparticles. The in vivo experiments demonstrated that copper nanoparticles completely inhibited green mold, in a direct action. In addition, when copper nanoparticles and *P. digitatum* were added into separate wounds, the reduction was 88.5% for copper nanoparticles. For fusarium rot, in direct action, copper nanoparticles completely inhibited the rot development. The mechanism by which copper nanoparticles acted against the pathogen was explained by degradation of fungal DNA. In fact, earlier investigations related to the mechanism of copper nanoparticles on plant pathogenic fungi have not been enough to explain the precise action. Indeed, it gives the impression to suppose that the small-size range of copper nanoparticles may possibly add to its antifungal effect, given that it can simply penetrate pathogen cell membrane and lead to cell death. The ZnO nanoparticles (<50 nm) at concentrations from 0 mM to 15 mM were tested against a number of fungal contaminants, including *A. alternata* (Sardella et al., 2017). The pathogen was inhibited by ZnO nanoparticles at concentrations higher than 6 mM. In the presence of ZnO nanoparticles, clear morphological aberrations in the fungal structures were detected.

On lemon fruits, TiO_2 nanoparticles (7 nm) applied on polypropylene film were prepared to control *Penicillium expansum*. The fruits were enveloped in the prepared film and treated with irradiation by UVA during two-weeks storage period at 25 °C. The authors confirmed a noteworthy decrease of decay in lemons wrapped in prepared coated film as compared to the control. They also referred this efficacy to the accumulation of ROS from the UVA-irradiated TiO_2 surfaces that suppress fungus development. A summary of recent application of different nanomaterials against post-harvest disease of citrus is mentioned in Table 17 (Sergio Ruffo Roberto et al., 2019).

Table 17: Recent application of different nanomaterials used as alternative control means to manage post-harvest disease of citrus.

Fruit/Cultivar	Scientific Name	Nanomaterial	Objective	Size (nm)
Valencia Late orange	*Citrus sinensis* L. Osb.	Clay-chitosan nanocomposite	Antifungal	NM *
Lemon	*Citrus limon* L. Osb.	Chitosan-clay nanocomposite	Coating	NM *
Thomson navel orange	*C. sinensis* L. Osb.	Chitosan-clay nanocomposite	Edible coating	NM *
Tangerine	*Citrus tangerine* Hort. ex Tanaka	Chitosan/ montmorillonite	Coating/ antifungal	NM *
Washington navel orange	*C. sinensis* L. Osb.	Silver nanoparticles	Antifungal	10 ± 5
Valencia Late	*C. sinensis* L. Osb.	Copper nanoparticles	Antifungal	48
-	*C. sinensis* L. Osb.	ZnO nanoparticles	Antifungal	<50
Lemon	*C. limon* L. Osb.	TiO_2 nanoparticles	Coating/ antifungal	7

NM * = size of the nanomaterials was not mentioned.

ii. Grapes: For 'Flame Seedless' table grape (*Vitis vinifera* L.), five safe nanomaterials were prepared and characterized, including silica, chitosan, and copper nanoparticles (NPs) and their combination to be used against gray mold (Sergio Ruffo Roberto et al., 2019). The authors summarized that nanomaterials, i.e. silica and chitosan, are promising compounds for management of gray mold of table grape cv. Flame Seedless. The mechanism of chitosan and silica NPs was explained by asymmetrical branching of the pathogen hyphae in the apical section and the failure of linearity and irregular mycelia growth detected by scanning electron microscopy (SEM). Also, the two nanomaterials exhibited a genotoxicity supported by degradation of pathogen DNA.

Combined applications of chitosan nanoparticles/silica nanoparticles showed a synergistic effect against gray mold. Under artificial conditions, chitosan/ silica nanocomposite at 1% reduced the incidence of decay by 59% as compared to control. Moreover, when the formulated nanocomposite was applied in the field, as a preharvest application, it reduced Botrytis mold development by 59% and 83%, for 'Italia' and 'Benitaka' table grapes, respectively, giving results even better than the individual chitosan or silica nanoparticles alone. The authors concluded that chitosan nanoparticles and chitosan/silica nanocomposite were able to preserve 'Italia' table grape bunches from mass loss, while no negative effect of those nanomaterials was observed in terms of fruit quality. To explain the mechanism of chitosan/silica nanocomposite several approaches were studied, including ROS, ATP content, and MMP of the pathogen spores. In particular, higher increase in fluorescent spores in presence of this formulated nanocomposite at 0.6% was detected as compared to the control. Also, it reduced the ATP content after 5, 15, and 30 min of incubation by 34%, 38%, and 35%, respectively as compared to control. In addition, MMP decreased, resulting in a loss of mitochondria function (Youssef et al., 2019). In addition, silica-silver nanosized (1–5 nm) was examined at 0.3–100 μg·mL^{-1} concentration against ten plant pathogenic fungi and bacteria, including *B. cinerea*, showing a complete inhibitory effect at 10 μg·mL^{-1}.

A coating based on pullulan and polymeric nanocapsules (153.9 nm) containing thyme (*Thymus vulgaris* L.) was prepared and applied to increase the shelf life of table grapes. Grapes coated with the formulated nanocapsules maintained their color, firmness, titratable acidity, and soluble solids content for longer as compared to untreated grapes. In grapes stored for 6 months, gray mold incidence was lower in coated fruits as compared to untreated ones. This formulated coating avoided the quick evaporation of volatile compounds and increased their persistence on the grape surfaces, thus delaying grape ripening and extending their shelf life (Sergio Ruffo Roberto et al., 2019). The antifungal efficacy in SiO_2/Ag_2S nanocomposites (300 nm) was reported for the first time against *Aspergillus niger.* The results showed that SiO_2/Ag_2S nanocomposites suppress the fungal development and sporulation. The authors refereed this effect to the presence of Ag_2S nanophases, and they observed a synergistic effect due to Ag_2S antifungal centres and the SiO_2 surfaces in supporting the adsorption of the fungus. It has been recommended that for metallic Ag, there is interaction of silver species with the sulphydryl groups of respiratory enzymes in the plasma membranes of susceptible pathogens, causing changes in the membrane permeability. SEM showed that *A. niger* hyphae become covered by SiO_2/Ag_2S nanomaterials when the fungus was exposed to the compound. Another hypothesis to explain this mechanism was the binding of Ag^+ to microbial genetic material.

ZnO nanoparticles (70 ± 15 nm) were tested against *B cinerea* at different concentrations (0, 3, 6 and 12 mmol·L^{-1}). The growth of the pathogen was inhibited at concentrations greater than 3 mmol·L^{-1}. The author hypothesized that ZnO nanoparticles affected cellular functions of *B cinerea* causing deformation in pathogen hyphae. Silver nanoparticles (50 nm) at different concentrations (25–100 mg·L^{-1}) were prepared and combined with the fungicide tolclofosmethyl. This combination performed better than individual compounds against *B. cinerea*; in particular, the most significant inhibition for the pathogen was obtained when the ratio between silver nanoparticles to tolclofos-methyl was 100:10 (Sergio Ruffo Roberto et al., 2019).

Many scientists have investigated the development or preparation of nanomaterials coating formulations to extend post-harvest storage period. Regarding the packaging of red table grapes, a chitosan-TiO_2 composite film (70 μm thickness,) was prepared and evaluated against *Escherichia coli, Staphylococcus aureus, Candida albicans* and *A. niger.* The chitosan-TiO_2 film (50–80 nm) was examined as packaging table grapes in order to avoid post- harvest infections and expand the shelf life period. In this study, table grape was preserved for 22 days with composite film compared to two weeks with bare chitosan film before mildew occurred. Nanocomposites of chitosan/ gelatin and silver nanoparticles (25–45 nm) were successfully prepared using the solution casting technique. The grapes wrapped with the nanocomposites hybrid film showed a fresh appearance without putridity, and the surface was kept smooth without any leakage of the grape juice. This nanocomposite led to extending the shelf life period for extra two weeks.

Chitosan nanoparticles (128.3 nm size) were tested against five pathogenic bacteria, including *S. aureus, Listeria monocytogenes, Pseudomonas aeruginosa, Salmonella* spp., and *E. coli*, with minimum inhibition concentration ranging 2–3 $g \cdot L^{-1}$ (Melo et al., 2019). Chitosan nanomaterials efficiently delayed the ripening process and led to reduced mass loss, soluble solids, and sugar contents and increased moisture retention and preservation of the titratable acidity values and sensory characteristics. Also, the lowest decay percentage was recorded in grapes stored under refrigeration as compared to grapes stored at room temperature, and the percentage was much higher than the decay rates found for grapes coated with chitosan nanoparticles. A summary of recent application of different nanomaterials against post-harvest disease of grapes is displayed in Table 18.

Table 18: Recent application of different nanomaterials used as alternative control means to manage post-harvest disease of grapes.

Cultivar/Type	Scientific Name	Nanomaterials	Objective	Size (nm)
Flame Seedless table grape	*Vitis vinifera* L.	Silica nanoparticles (NPs), Copper NPs, Silica-copper NPs, Chitosan NPs	Antifungal	Si (140–150) Cu (25–35) Si-Cu (520–550)
Italia and Benitaka table grapes	*V. vinifera* L.	Chitosan/silica nanocomposite	Antifungal	48
Grapes	*V. vinifera* L.	Nanocapsules/*Thymus vulgaris L.*	Coating/ antifungal	153.9
-	*V. vinifera* L.	SiO_2/Ag_2S nanocomposites	Antifungal	300
-	*V. vinifera* L.	ZnO-nanoparticles	Antifungal	70 ± 15
-	*V. vinifera* L.	Silver-nanoparticles, copper-nanoparticles and silver-copper-nanoparticles	Antifungal	Silver (38) Copper (20)
-	*V. vinifera* L.	Silver-nanoparticles	Antifungal	50
Red grapes	*V. vinifera* L.	Chitosan-TiO_2 composite	Antimicrobial	50–80
Red grapes	*V. vinifera* L.	Chitosan/gelatin and silver nanoparticles	Coating	25–45
Grapes	*Vitis labrusca* L.	Chitosan nanoparticles	Edible coating	128.3

iii. Banana: As a climacteric fruit, it has a relatively short post-harvest period due to several processes that change quality loss after harvest. Coating of banana fruit with chitosan combined with gum arabic at 1% delayed ripening by reducing the respiration rate and improved shelf life period for up to 33 days (Sergio Ruffo Roberto et al., 2019). Banana peel extracts were considered as potential candidate for the synthesis of silver nanoparticles which had a good antibacterial activity against *E. coli, E. aerogenes, Klebsiella* sp. and *Shigella* sp.. For banana fruits, chitosan nanoparticle (102.4–370 nm diameter) was synthesized, characterized and its coating effect on ripening process was investigated. The effectiveness of ajwain (*Trachyspermum ammi*) and neem (*Azadirachta indica*) leaf extracts of copper, silver, nickel and magnesium (68 nm) were tested *in vitro* against the causal agent of banana anthracnose. The results showed that ajwain- magnesium NPs, ajwain-nickel NPs at 0.2% and neem-silver NPs at 0.1% and 0.2% completely inhibited the spore germination of the pathogen. Silver-neem nanoparticles were examined *in vivo* against banana anthracnose disease at different concentrations, and spraying this nanomaterial at 0.2% showed the lowest anthracnose severity (Jagana et al., 2017).

Soybean protein isolate/cinnamaldehyde/ZnO nanocomposite was formulated using: soybean protein isolate, cinnamaldehyde, and ZnO NPs, which are well-known as inorganic nanomaterial with good mechanical properties, barrier capability, and great antimicrobial action. The antifungal efficacy of formulated nanocomposite coating against *A. niger* was 1.25-fold higher than only protein isolate/cinnamaldehyde application, while the pure protein isolate solution alone had no inhibitory effect (Sergio Ruffo Roberto et al., 2019). This behavior indicated that the immediate use of cinnamaldehyde and ZnO nanoparticle produced a synergistic effect for banana coating application. A summary of recent application of different nanomaterials against post-harvest disease of banana is mentioned in Table 19.

Table 19: Recent application of different nanomaterials used as alternative control means to manage post-harvest disease of banana

Group/ Cultivar	Scientific Name	Nanomaterials	Objective	Size (nm)
-	*Musa acuminata* L.	Chitosan/gum arabic	Coating	NM *
Cavendish bananas AAA group	*M. acuminata* L.	Chitosan-nanoparticles	Edible coating	102.4–370
-	*M. acuminata* L.	Ajwain-magnesium nanoparticles, Ajwain-nickel nanoparticles and Silver-neem nanoparticles	Antifungal	68
M. acuminata AAA group	*M. acuminata* L.	Chitosan-nanoparticles	Edible coating	121.2
-	*M. acuminata* L.	Soybean protein isolate/ cinnamaldehyde/ZnO nanoparticles	Antifungal/ film coating	NM *
Nanica	*Musa cavendishii* Lamb.	Chitosan-nanoparticles	Coating film	88.79

NM * = size of the nanomaterials was not mentioned.

iv. Apple: Coating apple with chitosan nanoemulsion significantly reduced mass loss, respiration level, ethylene rate and peroxidase activity as compared to control. Also, chitosan nanoemulsion prolonged the quality and avoided mass loss of apple during storage. In particular, mass loss of control fruit was 5.5% while it was 3.4% and 2.2% for coated fruit with 0.2% and 0.5% nanochitosan, respectively (Sergio Ruffo Roberto et al., 2019). Apple coated with 110 nm-chitosan NPs demonstrated higher antimicrobial action and reduced total microbial contamination levels of a wide range of pathogens evaluated as log colony forming units (CFU) g^{-1}, as compared to other treatments. Apple browning slices coated with normal chitosan or uncoated-control was slightly higher than slices coated with nanochitosan. Fresh-cut apple stored with ZnO NPs packaging had lower fruit decay percentage (21.5%) as compared to control (42.4%). Also, nanopackaging reduced the accumulation of malondialdehyde and decreased polyphenoloxidase and pyrogallol peroxidase activities as compared to control fruit. The primary appearance of apple slices was maintained and the browning index was avoided in nanopackaging 'Fuji' apple fruits stored at 4 °C for 12 days. The nanopackaging was also able to maintain the level of soluble solids and titratable acidity.

Regarding fresh-cut apple, a poly-lactic acid film incorporated with ZnO nanoparticle was prepared as a novel nanopackaging film and tested on the quality attributes of apple (Li et al., 2017). The nanofilm had a higher water vapor permeability and lower oxygen permeability as compared to the pure poly-lactic acid film. Although nanofilms had the highest mass loss at the end of storage compared to the pure poly-lactic acid film, it provided a better preservation of color, firmness, total phenolic content and sensory quality. Concerning the microbial profile ($CFUmL^{-1}$), the combination of ZnO NPs and poly-lactic acid was able to inhibit the growth of pathogens including bacteria, molds and yeasts in fresh-cut apple stored at 4 °C for two weeks.

For 'Red Delicious' apple (*M. domestica* Bork.), nano-calcium and calcium chloride were applied at different concentrations in the field five times, starting 70 days following full-bloom untill one month before apple harvesting. The authors proved that nano-calcium decreased cell wall enzyme activities including polygalacturonase, pectin methylesterase and β-galactosidase (Ranjbar et al., 2018). Fresh-cut 'Red Delicious' apples were coated with tocopherol/nopal mucilage (*Opuntia ficus indica*) nanoemulsion (less 1000 nm) and had a significant impact on pectin methylesterase and polyphenol oxidase activity in the fresh-cut apples. The authors confirmed that nopal mucilage performed as an encapsulant that improves the activity of α-tocopherol. Therefore, color, firmness and mass loss were maintained during refrigerated storage. A summary of recent application of different nanomaterials against post-harvest disease of apple was mentioned in Table 20.

Table 20: Recent application of different nanomaterials used as alternative control means to manage post-harvest disease of apple

Cultivar	Scientific Name	Nanomaterials	Objective	Size (nm)
Golab Kohanz	*Malus domestica* Bork.	Chitosan-nanoemulsion	Coating	≤100
Gala	*M. domestica* Bork.	Chitosan-nanoparticles	Coatings/ antimicrobial	110–300
Fuji	*M domestica* Bork.	Polyvinyl chloride/ZnO nanoparticle	Nanopackaging coating	200–400
Yunnan Zhao Tong	*M. domestica* Bork.	Poly-lactic acid/ZnO nanoparticle	Nanopackaging coating	NM *
Red Delicious	*M. domestica* Bork.	Nano-calcium	Coating	NM *
-	*M. domestica* Bork.	ZnO nanoparticles	Antifungal	70 ± 15
-	*M. domestica* Bork.	ZnO nanoparticles	Antifungal	<50
Red Delicious	*M. domestica* Bork.	Tocopherol/nopal mucilage nanoemulsion	Encapsulant	<1000

NM * = size of the nanomaterials was not mentioned.

v. Mango: Chitosan-silver NPs composite (495–616 nm diameter) was prepared and the silver NPs were distributed in the composite with size 10–15 nm. The composite exhibited higher antifungal effect against the conidial germination of *Colletotrichum gloeosporioides* on mango. In particular, a complete inhibition of spore germination was obtained when chitosan-silver NPs composite was applied at 100 $\mu g{\cdot}mL^{-1}$. Moreover, on detached mango fruit cv. Alphonso, *in vivo* experiments indicated that anthracnose was significantly suppressed by the aforementioned composite. A 45.7% and 71.3% reduction of anthracnose disease was achieved by chitosan-silver nanoparticle composite at 0.5% and 1%, respectively (Sergio Ruffo Roberto et al., 2019).

A nanocomposite edible film containing *Aloe vera* gel, glycerol and ZnO NPs solutions was developed. Mango fruits were coated and stored for nine days at room temperature and quality parameters were evaluated. The overall results showed that increasing the concentration of ZnO NPs led to a reduction of mass loss (%), increasing to titratable acidity, increasing of ascorbic acid content in fruit, maintaining a low soluble solids content, decreasing the pH, increasing thickness and increasing the transmittance and elongation (%) of the formulated films. Copper oxychloride-conjugated silver NPs were synthesized using neem extract and tested for their activity against the causal agent of mango anthracnose (*C gloeosporioides*) by Raghavendra et al., (2019). The resulted nonmaterial had spherical shape and an average particle size of 21–25 nm. The nanomaterials displayed the highest growth inhibition of the pathogen (~187%) as compared to copper oxychloride. The obtained results confirmed that this nanomaterial can be successfully used as alternative control means for controlling anthracnose disease in mango. A summary of recent applications of different nanomaterials against post-harvest disease of mango is shown in Table 21.

Table 21: Recent application of different nanomaterials used as alternative control means to manage post-harvest disease of mango

Cultivar	Scientific Name	Nanomaterials	Objective	Size (nm)
Alphonso	*Magnifera indica* L.	Chitosan-silver nanoparticle composite	Antifungal	495–616
Dasheri	*M. indica* L.	*Aloe vera* gel, glycerol and ZnO nanoparticles	Film coating	NM *
-	*M. indica* L.	Copper oxychloride-conjugated AgNPs	Antifungal	21–25

NM * = size of the nanomaterials was not mentioned.

vi. Tomato : Silver nano-particles (AgNPs) were evaluated for its possible effects on the incidence and severity diseases gray mold for tomato fruits caused by the *Botrytis cinerea.* Treatments were healthy tomato fruits (control), inoculated tomato fruits by *B cinerea*, healthy tomato fruit with silver nanoparticles and combination inoculated tomato fruits with *B. cinerea* and silver nanoparticles) (Salem et al., 2019). The obtained results were, there aren't incidences and severity diseases of gray mold in case of tomato fruits with silver nan particles during forty days and the lowest incidence (0.3) and severity (1.5%) diseases were after thirty days of cold storage in case of treatment tomato fruits were treated with nano–Silver + Inoculated. While, the highest incidence (0.6) and severity (7.5%) diseases percentages were after ten days of cold storage in case of treatment tomato fruits with inoculated by *B cinerea*, comparison with incidence (0.3) and severity (1%) diseases percentages were after twenty days of cold storage in case of un-treatment tomato fruits (control), (c) the highest (TSS) percentage of tomato fruits which treated by nano-particles followed by untreated samples (control). While, the lowest was sample which treated by inoculated + silver nano particles, followed by inoculated samples by *B cinerea* after forty days of cold storage and (d) the highest pH value of tomato fruits which treated by Nano particles, followed by un-treated samples (control), while, the lowest sample which treated by inoculated + silver nano particles, followed by inoculated samples by *B. cinerea* after forty days of cold storage.

To explore alternative control measure of Fusarium wilt, antifungal activities of nickel nanoparticles were investigated for controlling two forma specials of *Fusarium oxysporum, F. oxysporum* f. sp. *lactucae* and *F. oxysporum* f. sp. *lycopersici,* on lettuce and tomato under *in vitro* and greenhouse conditions (Ahmed I. S. Ahmed et al., 2016). Under *in vitro* condition, nickel nanoparticles at the concentration of 100 ppm significantly inhibited the mycelial proliferation and sporulation activities of fungal pathogens. The inhibition of mycelial growth of *F. oxysporum* f. sp. *lactucae* and *F. oxysporum* f. sp. *lycopersici* was 60.23 and 59.77%, respectively. The application of NiNPs as antifungal agents against Fusarium will be better with concentration of 50 ppm. This study suggested that nickel nanoparticles could be used to control Fusarium wilt.

Foliar application of selenium and copper in the form of nanoparticles were evaluated in a tomato crop infested by *Alternaria solani.* The severity of *Alternaria solani,* agronomic variables of the tomato crop, and the changes in the enzymatic and non-enzymatic antioxidant compounds were evaluated (Tomasa Quiterio-Gutiérrez et al., 2019). The joint application of Se and Cu nanoparticles decreases the severity of this pathogen in tomato plants. Moreover, high doses generated an induction of the activity of the enzymes superoxide dismutase, ascorbate peroxidase, glutathione peroxidase (GPX) and phenylalanine ammonia lyase in the leaves, and the enzyme GPX in the fruit. Regarding non-enzymatic compounds in the leaves, chlorophyll a, b, and totals were increased, whereas vitamin C, glutathione, phenols and flavonoids were increased in fruits. The application of nanoparticles generated beneficial effects by increasing the enzymatic and non-enzymatic compounds and decreasing the severity of *Alternaria solani* in tomato plants.

Bacterial canker of tomato caused by *Clavibacter michiganensis* subsp. *michiganensis* (Cmm) is a major limiting factor for tomato production worldwide. Currently there exists no resistant variety of tomato to bacterial canker; only cultural and chemical controls are available. A study by Asma Noshad et al., (2020) used synthesized AgNPs (silver nanoparticles) via a green chemistry route and investigated their bactericidal potential against bacterial canker of tomato (BCT). AgNPs were prepared using mycellial aqueous extract of agriculturally beneficial fungi *Pythium oligandrum.* These results suggested that this process is a strong candidate for industrial scale production of AgNPs. These particles act as an inhibitor and broad spectrum antibacterial agent against cmm, and hence offer a new and eco-friendly alternative in BCT control.

A study was carried out under greenhouse conditions to evaluate the efficiency of silver nanoparticles (AgNPs) as antiviral reagent to induce a systemic acquired resistance (SAR) against Tomato mosaic virus (ToMV) and Potato virus Y (PVY) and suppress their infection on tomato plants (Noha K. El- Dougdoug et al., 2018). Silver nanoparticles (AgNPs) were sprayed 7 days before inoculation with both viruses at 50, 60 and 70 ppm. The results showed that disease severity as well as relative concentration of both viruses was decreased dramatically by the treatment of AgNPs at 50 ppm compared to the other treatments. Examination the clarified sap of both viruses with TEM showed morphological evidenced that AgNPs bind to coat protein virus particles. The induction of SAR against ToMV and PVY was confirmed chemically by several parameters as indicated by increasing photosynthetic pigments, total soluble protein and the activities of peroxidase (POD) and polyphenol oxidase (PPO) compared with infected control.

vii. Brinjal (Eggplant): Two concentrations of Zinc Oxide nanoparticles (ZnO NPs) were applied using three methods i.e., spray, seed treatment and soil inoculation for the management of disease complex of eggplant caused by *Ralstonia solanacearum, Phomopsis vexans* and *Meloidogyne incognita.* The highest increase in plant growth, chlorophyll and carotenoid contents was obtained by spray of NPs followed by seed treatment and soil inoculation (Khan and Siddiqui et al., 2018). Plants sprayed with 200 ppm NPs resulted in higher increase in plant growth, and higher reduction in galling and nematode multiplication; 15 and 8 days early flowering and fruiting were observed in plants sprayed with 200 and 100 ppm NPs, respectively, over seed treatments. Scanning electron microscopy revealed that ZnO NPs had an adverse effect on cell wall/body wall of test pathogens. Wilt and blight indices was three and five when inoculated with single pathogen and inoculated together, respectively. Disease index was reduced to one when plants sprayed with 200 ppm NPs.

An experiment was carried out to test pathological ability of *Rhizoctonia solani* fungus on eggplant and study efficiency of magnesium oxide nanoparticles to control of *Rhizoctonia solani* in the culture media PDA (Amna Mohammed Ali et al., 2018). The results showed that *R. solani* pathological fungus isolation gave high rate infection on eggplant seeds compared with the control sample (without fungus addition) where the rate of infection by the fungus (in germination index) in which the germination ratio without fungus presence was 97.33% and it was with fungus treatment 12%. The results showed also effect of MgO nanoparticles in inhibition of the pathological *R solani* fungus at 1% , 2% and 3% compared with the control sample, and the inhibition rates were 96.33%, 100% and 100% respectively where the MgO nanoparticles affected greatly and reached to 100% in the pathological *R. solani* fungus inhibition on eggplant seeds.

Root-knot nematodes are known to cause significant damage to eggplants. New approaches by green silver nanoparticles (GSN) are used to control plant-parasitic nematode to avoid chemical nematicide hazards. Analyses of the incorporation of different concentrations of nanoparticles on two different algae (*Ulva lactuca* and *Turbinaria turbinata*) were carried ou (Kamal Fouad Abdellatif et al., 2016). Both RAPD and EST molecular markers were used to study the GSN effect on eggplant DNA modification. GSN (17 mg.mL-1) obtained from *U. lactuca* was more effective in reducing second-stage juveniles (J2s) of *M. javanica* (69.44%) population in soil. All treatments improved eggplants growth parameters. Change in DNA profile using of both RAPD and EST markers was noted. GSN (12.75 mg.100 mL-1) were effective on controlling the root-knot nematode (both *T. turbinata* and *U. lactuca* algae), similar to chemical control in eggplants. GSN did not cause any phototoxicity in eggplants under treatment.

viii. Okra: New nano-sulphur synthesized at Indian Agricutural Research Institute (IARI), New delhi, India and three other commercial products namely commercial sulphur (Merck), commercial nano-sulphur (M K Impex, Canada) and Sulphur 80 WP (Corel Insecticide) were evaluated *in vitro* for fungicidal efficacy at 1000 ppm against *Erysiphe cichoracearum* of okra (Gogoi et al., 2013). All the sulphur fungicides significantly reduced the germination of conidia of *E. cichoracearum* as compared to control. Least conidial germination was recorded in IARI nano-sulphur (4.56%) followed by Canadian nanosulphur (14.17%), Merck sulphur (15.53%), sulphur 80 WP (15.97%) and control (23.09%). Non-germinated conidia count was also high in case of IARI nanosulphur followed by Canadian nano-sulphur, Merck sulphur and Sulphur 80WP. Apart from inhibition of conidial germination, cleistothecial appendages were also disrupted in contact with nano-sulphur and the cleistothecia became sterile. Results proved that IARI nano-sulphur is more effective than the commercial formulations and could be applied at lower amount for controlling powdery mildew disease for its better efficacy.

Foliar sprays of three plant resistance inducers, including chitosan (CH), potassium sorbate (PS) ($C_6H_7KO_2$), and potassium bicarbonates (PB) (KHCO3), were used for resistance inducing against *Erysiphe cichoracearum* DC (powdery mildew) infecting okra plants (Soliman et al., 2017). Experiments under green house and field conditions showed that, the powdery mildew disease severity was significantly reduced with all tested treatments of CH, PS, and PB in comparison with untreated control. CH at 0.5% and 0.75% (w/v) plus PS at 1.0% and 2.0% and/or PB at 2.0% or 3.0% recorded as the most effective treatments. Moreover, the highest values of vegetative studies and yield were observed with such treatments. CH and potassium salts treatments reflected many compounds of defense singles which leading to the activation power defense system in okra plant. The highest records of reduction in powdery mildew were accompanied with increasing in total phenolic, protein content and increased the activity of polyphenol oxidase, peroxidase, chitinase, and β-1,3-glucanase in okra plants. Meanwhile, single treatments of CH, PS, and PB at high concentration (0.75%, 2.0%, and/or 3.0%) caused considerable effects. Therefore, application of CH and potassium salts as natural and chemical inducers by foliar methods can be used to control of powdery mildew disease at early stages of growth and led to a maximum fruit yield in okra plants.

ix. Potato: To reduce the excessive reliance on the toxic chemical fungicides, a study by Abdulrhman A. Almadiy and Gomah E. Nenaah (2018) was aimed to isolate the total potato glycoalkaloids (TPAs), and the two steroidal alkaloids α-chaconine and α-solanine from potato. Their structures were characterized using physical and spectroscopic methods including (UV, IR, ^{1}H, ^{13}C--NMR, 2D ^{1}H-^{1}H COSY, HMBC and NOESY). Silver nanoparticles (AgNPs) were prepared from potato alkaloids through a green synthesis approach. Potato alkaloids and their nanoparticles inhibited mycelial growth of the phytopathogenic fungi *Alternaria alternata, Rhizoctonia solani, Botrytis cinerea* and *Fusarium oxysporum* f. sp. *lycopersici* with low minimal inhibitory and minimal fungicidal concentrations. *R. solani* was the most susceptible, while *F. oxysporum* was the most resistant. TPAs were the most fungitoxic (EC50's were 19.8, 22.5, 26.5 and 32.3 μg/ml against *R solani, A. alternata, B. cinerea* and *F. oxysporum* respectively). A mixture of α-solanine and α-chaconine (1:1) showed a marked antifungal activity. AgNPs (size 39.5-80.3 diameter) from alkaloids showed improved fungitoxic activity (EC50's of TPAs nanoparticles ranged between 10.9 and 16.1 μg/ml). Alkaloids exhibited no or a slight phytotoxicity against wheat and radish. Results recommended the potential of using potato alkaloids and their nanoparticles as biorational alternatives to conventional fungicides.

A study investigated the effect of silver and selenium nanoparticles on *Alternaria solani,* the pathogenic fungus causing early blight disease of potato (Abdel-Wahab A. Ismail et al., 2016). Under laboratory conditions, 25 μg/ml concentration of silver nanoparticles completely inhibited *A solani* as compared to Kocide®, fungicide that gave maximum inhibition at 600 μg/ml. The selenium nanoparticles completely inhibited the fungal growth at 800 μg/ml. AgNPs completely inhibited the growth of *A. solani* at low concentrations. Silver nanoparticles might be suitable alternative to chemical fungicides, while, SeNPs can be used as antioxidant for enhancing plant immunity.

A study was conducted to evaluate the effect of antiviral activity of curcumin-milk proteins nanoparticles against potato virus Y (PVY) (Taha et al., 2019). Curcumin-milk proteins nanoparticles were prepared via ionic gelation method. The antiviral activity of the resultant nanoparticles against PVY was evaluated at different concentrations (500, 1000 and 1500 mg/100 mL).

Chlorophyll content as well as the activity of peroxidase (POX) and polyphenol oxidase (PPO) was examined. Curcumin-milk proteins nanoparticles showed inhibitory effect on PVY in a concentration dependent manner. Curcumin-milk proteins nanoparticles displayed a successful tool to control the PVY under green house conditions.

In a study, silver nanoparticles (AgNPs) were prepared by two methods: biologically using *Trichoderma viride* cell free supernatant and chemically with polyvinyl pyrrolidone (PVP) using gamma irradiation. The antimicrobial activity of AgNPs was evaluated in the field against *Alternaria solani* which causes early blight disease of potato. Disease severity, vegetative and biochemical parameters of plant were determined (El-Batal et al., 2016). Application of AgNPs 150 μg/ml + K_2SO_4 (2%) as plant foliar resulted in highly effective inhibitor of disease severity that recorded (8.39 %) compared to the untreated plants (86.17 %) after 75 days of sowing. Also, a significant increase in growth and physiological parameters was recorded. Accumulation of Ag^o in plant tissues reached minimum value. Selenium nanoparticles (SeNPs) 0.5 μg/ml, ascorbic acid 300 μg/ml and K_2SO_4 (2 %) reduced disease severity (9.81 %) significantly. SeNPs 0.5 μg/ ml, AgNPs 150 μg/ml and K_2SO_4 2 %) showed disease severity (12.63%) but all plant parameters were improved including physiological parameters and yield. The application of AgNPs and SeNPs were recommended as plant foliar for controlling plant pathogen and improving plant yield.

A study by Manal A. El-Shazly et al., (2017) was aimed at identifying and molecularly characterizing the isolated PVY infecting potato plants in Egypt, study the efficiency of salicylic acid (SA) at 200 μmol/L, silver nanoparticles (AgNPs) at 0.1μg/μL and mixture of them with the same concentrations to produce resistance against infection with PVY in potato plants and study the effects of these treatments on yield reduction and quality parameters of potato tubers cv. Spunta. To achieve this aim, the virus was isolated from naturally infected potato plants grown in the experimental Farm (Faculty of Agriculture, Cairo University) and biologically purified on both *Chenopodium amaranticolor* and *Chenopodium quinoa* as indicator host plants which showed large and irregular chlorotic local lesions. It was transmitted mechanically and propagated in healthy potato seedlings cv. Spunta. The virus identification was done by direct ELISA technique using specific antiserum and confirmed by reverse transcription polymerase chain reaction (RT-PCR) using two different pairs of primers designed for amplification of two different regions of PVY coat protein gene. All the used compounds could induce resistance to virus infection when applied to the plants as a spray treatment. The most effective treatments were that applied 24h. after inoculation with the virus and treatment with AgNPs at 0.1μg/μL and the mixture of SA and AgNPs at the same concentration applied before 3 and 7 days of virus infection which led to an important decrease in virus concentration (0.280,0.308 and 0.140, respectively) and infection percentage (10, 30 and 10%, respectively). However, weak reduction in virus concentration and percentage of infection was observed when SA was sprayed at the pre-and after viral infection (90,80 and 60%, respectively). Also, all tested treatments gave a significant increase in total number of harvested tubers /plant, total tuber yield, average of plant height (cm), tuber diameter, tuber length, starch and total soluble sugars content compared to healthy plants.

x. Capsicum: Silver nanoparticles are highly reactive as they generate Ag+ ions while metallic silver is relatively unreactive. It was also shown that the nanoparticles efficiently penetrate into microbial cells, which implies lower concentrations of nano-sized silver would be sufficient for microbial control. This would be efficient, especially for some organisms that are less sensitive to antibiotics due to the poor penetration of some antibiotics into cells. The effect of silver nanoparticles wass evaluated against pepper anthracnose (*Colletotrichum* species) under different culture conditions. Silver nanoparticles (WA-PR-WB13R) were applied at various concentrations to determine antifungal activities *in vitro* and in the field (Kabir Lamsal et al., 2011). The application of 100 ppm concentration of silver nanoparticles produced maximum inhibition of the growth of fungal hyphae as well as conidial germination in comparison to the control *in vitro.* In field trials, the inhibition of fungi was significantly high when silver nanoparticles were applied before disease outbreak on the plants. Scanning electron microscopy results indicated that the silver nanoparticles caused a detrimental effect on mycelial growth of *Colletotrichum* species. The results showed that silver nanoparticles treatment suppressed pathogen attack when applied before the disease outbreak on the plant.

A study was aimed to obtain the best formula of hydrolyzed nano chitosan and its potensial in controlling anthracnose (Yadi Suryadi et al., 2017). The hydrolyzed chitosan was prepared using chitinase enzyme extracted from *Burkholderia cepacia* isolate E76. Nano chitosan had an inhibitory activity to *C gloeosporioides in vitro* up to 85.7%. Moreover, it could inhibit 61.2% of *C gloeosporioides* spores germination. It was shown that nano chitosan was also effective to reduce anthracnose disease severity *in vivo* when applied as a preventive measure on chili and papaya fruits. This study could be used as a reference for further fruit coating application using nano chitosan as a promising postharvest biocontrol agent to *C gloeosporioides.*

The potentiality of green synthesized silver nanoparticles (AgNPs) was evaluated to control soft rot disease caused by *Pectobacterium carotovorum* (Pc) in pepper (*Capsicum annuum* L.). Laurel (*Laurus nobilis* L.) leaf extract was used to synthesize silver nanoparticles (Ayisigi et al., 2020). According to *in vivo* tests AgNPs had a protective role against the soft rot disease by limiting disease severity up to 15% on pepper seedlings. Additionally, growth measurements hinted that green synthesized AgNPs have phyto-stimulative effect on growth of pepper seedlings. Green synthesized AgNPs possess an effective and non-toxic solution to control soft rot disease factor *P carotovorum* in pepper cultivation.

Green synthesis of silver nanoparticles was carried out using aqueous leaf extract of *Acalypha indica.* The formation of nanosilver was confirmed by UV-Vis Spectroscopy. The absorbance peak at 450nm indicated the presence of silver nanoparticles in the solution (Rajyasri Ghosh et al., 2018). The antifungal property of silver nanoparticles was evaluated against pathogenic fungus *Colletotrichum capsici* both *in vitro* and in field inoculation experiment. Application of silver nanoparticles produced significant inhibition of growth of fungal mycelia as well as conidial germination in vitro. It has been observed that about 54% and 70% reduction in mycelia growth took place at 100 ppm and 200 ppm concentrations of silver nanoparticles respectively. Results of spore germination test revealed about 97% reduction in germination at 10 ppm concentration of silver nanoparticles. In the field trial experiment significant reduction (46%) in anthracnose disease of bean caused by *C. capsici* was observed when the plants were treated with 200 ppm of nanosilver prior to inoculation. Results of the present study highlighted the eco-friendly approach of plant mediated synthesis of nanosilver and its potential application in the field as an alternative to chemical fungicides for disease management.

xi. Cucumber: Powdery mildew is one of the most devastating diseases in cucurbits. Crop yield can decline as the disease severity increases. The effect of silver nanoparticles were evaluated against powdery mildew under different cultivation conditions *in vitro* and *in vivo.* Silver nanoparticles (WA-CV-WA13B) at various concentrations were applied before and after disease outbreak in plants to determine antifungal activities (Lamsal et al., 2011). In the field tests, the application of 100 ppm silver nanoparticles showed the highest inhibition rate for both before and after the outbreak of disease on cucumbers and pumpkins. Also, the application of 100 ppm silver nanoparticles showed maximum inhibition for the growth of fungal hyphae and conidial germination in *in vivo* tests. Scanning electron microscope results indicated that the silver nanoparticles caused detrimental effects on both mycelial growth and conidial germination.

The effect of bio-agent *Bacillus subtilus*, nano titanium dioxide (TiO_2) and Zinc oxide (ZnO) as well as the combination compared with fungicide (Topas100) was investigated under laboratory and greenhouse conditions against powdery mildew disease caused by *Podosphaera xanthii* (Yaser et al., 2020). Most of treatments were strong to inhibit the spore germination of *P xanthii. B subtilus* and *B. subtilus* + nano-TiO_2 were the most effective followed by Topas100, *B. subtilus* + nano-ZnO, nano-ZnO and Nano-TiO_2 compared with control. Most of treatments significantly reduced the disease symptoms and diseases severity (%) particularly bio-agent, nano-ZnO and combination of *B. subtilus* + nano-ZnO, however, the fungicide treatment showed the best inhibition. Combincd of *B subtilus* and nano-Zinc oxide treatment was the best treatment and promising to be an alternative to fungicide, consequently, decreasing the environmental pollution, cost and harmful effect of fungicide on human health.

xii. Crucifers: The effects of four dose rates 0.27; 0.90; 1.80 and 3.60 kGy/h on the solution of silver (Ag+ 10-2 M, PVP 2%, ethylenglycol 6%) irradiated at 25 kGy were investigated (Pham Thi Le Ha et al., 2013). The results showed that as the dose rates increased, the absorption peak shifted to blue wavelengths and also the particles decreased in size. For field test, nano particles were prepared by irradiation of silver solution at 25 kGy with the dose rate of 3.60 kGy/h. The absorption peaks of the synthesized nanoparticles were obtained at wavelengths of 412 nm and the average diameter of particles were 14 nm. Using two concentrations of 15 and 20 ppm, silver nanoparticles had not affected the growth and development of cabbage but showed antifungal activity against *Plasmodiophora brassicae* cause club root in cabbage. Using nano particles, the clubroot disease index were 9- 10% compared to 5% of nebijin (fungicide), and 12% of control. The yield of cabbage were 55 tons/ha, 63 tons/ha and 70 tons/ha for the control, nanosilver group, and nebijin group, respectively.

Cabbage black rot caused by *Xanthomonas campestris* pv. *campestris* (Xcc) leads to decrease of the production of up to 70% (Hongqiang Dong et al., 2020). In this work, a novel copper-based microbicide was prepared by binding copper with glycine in sodium hydroxide solution (Gly-Cu(OH)2 NPs) and characterized by inductively coupled plasma atomic emission spectroscopy, high-resolution transmission electron microscopy, Fourier transformation infrared spectroscopy, and dynamic light scattering.

The results showed that the prepared Gly-Cu$(OH)_2$ NPs had a mean diameter of 240 nm with copper content more than 25.0% and their antimicrobial efficacies against Xcc were significantly better than Kocide 3000 at 400–800 mg L^{-1} of copper after spraying for 14 days. The phytotoxicity tests under greenhouse conditions showed that Gly-Cu(OH)$_2$ NPs were safer to plants than Kocide 3000 and obviously promoted the plant growth, which led to the increase of fresh weights of Chinese cabbage and tomato seedlings by 6.34% and 3.88% respectively at a concentration of 800 mg L^{-1} of copper. As a novel copper-based microbicide, the Gly-Cu(OH)2 NPs can improve effective utilization of copper-based bactericides and reduce phytotoxicity to plants and would be a potential alternative for sustainable plant disease management.

Silver nanoparticles (SNPs) are widely used as an agent for antifungal treatment. The antimicrobial activity of the nanoparticles varies on the size, shape, and the function of the surface area (Le Thi Anh Tu and Pham Thi Le Ha, 2017). In the liquid culture and soil, SNPs exhibited a controlling effect on *Plasmodiophora brassicae* (Club root disease causal agent) in cabbage. The fungicide activity depended on the SNP concentrations and the medium; no disease symptoms were found at the SNP concentrations of 1.06 ppm in the liquid culture, and 10 and 20 ppm in soil. It was observed that the biomass and height of infected plants were much lower than those of the healthy plants. Comparing to the plants that were grown in the sterilized soil, there was not significant difference between the biomass and the height of plants exposed to pathogen and treated with SNPs.

Green synthesis of silver nanoparticles (AgNPs) was accomplished using different volumes of cauliflower extract and 0.001 M silver nitrate solution at 80°C for 15 min (Alsultani Amjed et al., 2019). The antibacterial activity was tested against Gram-positive and -negative bacteria showed an inhibition zone of 16–27 mm and the antibacterial activity tested for the same bacteria after storage for about 10 months showed an inhibition zone of 6–10 mm. The AgNPs were synthesized biologically using white radish (*Raphanus sativus* var. *aegyptiacus*). The biosynthesis was regularly monitored by UV- Vis spectroscopy (Safaa M Ali et. al., 2015).

The exposure of the snails and soil matrix to AgNPs in a laboratory experiment reduced the activity and the viability of the land snail (20% of AgNPs treated snails died) as well as the frequency of fungal population in the surrounding soil (*Aspergillus, Penicillium, and Trichoderma* were the most common genera. Complete inhibition of *Aspergillus fumigatus, Cochliobolus specifier* and *Fusarium solani*). The synergistic effect of synthesized AgNPs as antifungal was evaluated and clearly revealed that AgNPs can be effectively used against various plant pathogenic fungi. The antifungal mechanism of AgNPs may be due to the fact that the formation of free radicles produced from the nanoparticles could disturb the membrane lipids and then finally spoil the membrane functions. The membrane could be deteriorated by the formation of pits on the surface of the cell wall membrane of microorganisms. The formation of pits on the membrane leads to increase in the permeability and irregular transport that result in the death of the cells. Hence, the green-synthesized silver nanoparticle can be a good source, which is easily produced and extensively useful in agricultural applications.

Pulses: Silver nanoparticles (Ag-NPs) were biosynthesized from aqueous extract of the cyanobacterium, Nostoc sp. PCC7524 (Hamed et al., 2019). *In vitro* assay against the root-knot nematode Meloidogyne javanica showed that Ag-NPs significantly decreased egg hatching of *M. javanica* at different applied concentrations (3, 6, 12, 25, and 50%, v/v). Fifty percent of Ag-NPs induced the highest reduction percent (94.66%). Moreover, Ag-NPs and $AgNO_3$ significantly increased the percentages of larval mortality of the second-stage juveniles (J2) with concentration and time-dependent responses. Ag-NPs or $AgNO_3$ at 2.4 ml/l, 24 h, completely inhibited the growth of J_2 compared to 23% inhibition using aqueous cyanobacterial extract (ACE). *In vivo* effect of Ag-NPs on faba bean-infected plant under greenhouse conditions was *achieved* by treating soil with three different concentrations of 1, 2, and 3 ml/kg soil over two consecutive seasons. Ag-NPs significantly reduced root galling from 39.6 to 78.7% and J_2 population in the soils from 32.2% to 86.7% in the 2018 season and from 21.9 to 78.1% and 40.0 to 81.0% in the 2019 sea*son, resp*ectively. Moreover, 3 ml/kg soil of *Ag-NP* treatment showed statistically comparable effects to that of vydate nematicide but with remarkable enhancement of faba bean growth parameters as compared to those of vydate or $AgNO_3$ *tr*eatments in the two seasons.

Soybeans (*Glycine max*) (V3 stage) were *sprayed* once with nanoparticles (NPs) of AgO, B, CeO, CuO, MnO, MoO_3, SiO, TiO, or ZnO and exposed to *Fusarium virguliforme*, the cause of sudden death syndrome. Up to 80% root rot was observed in greenhouse experiments (Cristian D P Peréz et al., 2020). However, NP CuO, B, MoO_3, or ZnO reduced the root rot severity by 17-25%. Infected roots and shoots had significant changes in B, Mg, P, S, Si, and Zn, but NP treatment restored levels to that of the healthy control. For example, the increased root Mg and Mn contents induced by disease were reversed by NP B and Mn amendments. In vitro assays revealed that the NPs did not inhibit the pathogen. This, along with the restoration of altered nutrient levels in the plant tissue, suggested that modulated plant nutrition increased disease defense. Treatment of seedlings with nanoscale micronutrients may be a new tool in promoting soybean health.

The biofabrication of zinc oxide nanoparticles from aqueous leaf extract of *Melia azedarach* (MaZnO-NPs) was reported through solution combustion method and their novel application in preventing the growth of seed-borne fungal pathogens of soybean (*Cladosporium cladosporioides* and *Fusarium oxysporum*) (Lakshmeesha et al., 2020).The standard blotter method was employed to isolate fungi and was identified through molecular techniques. The synthesized MaZnO- NPs inhibited the growth of *C. cladosporioides* and *F. oxysporum* in a dose dependent manner. Biomass, ergosterol, lipid peroxidation, intracellular reactive oxygen species and membrane integrity determination upon MaZnO- NPs treatment offered significant activities there by confirming the mechanism of action against the test pathogens. Due to the effectiveness of MaZnO-NPs in controlling the growth of *C. cladosporioides* and *F. oxysporum*, the synthesized MaZnO-NPs provides insight towards their potential application in agriculture and food industries.

The isolation and screening of rhizospheric microflora of chickpea and their role in the biosynthesis of silver nanoparticles (AgNPs) was investigated (Kaur Pawan et al., 2018). The antifungal potential of AgNPs for control of wilt disease of chickpea, caused by *Fusarium oxysporum* f. sp. *ciceri* (FOC) was evaluated *in vitro* and *in vivo* pot experiments. The effect of AgNPs on seed germination and soil community was also evaluated. Among microbial strains, two bacteria as *Pseudomonas* sp. and *Achromobacter* sp. and two fungi as *Trichoderma* sp. and *Cephalosporium* sp. were identified for biosynthesis of AgNPs. Pot experiments showed maximum and minimum reduction in wilt incidence over control i.e. 73.33% for AgNPs and copper oxychloride (CuOCl) i.e. 26.67%, respectively. Seeds coated with AgNPs showed high germinability up to 98% and no negative impact was observed on soil community. This study demonstrated the role of AgNPs against the fungal disease of chickpea that can be extended to other diseases of chickpea and other important Indian crops.

Thiamine was incorporated into the chitosan nanoparticles and characterized through FTIR, DLS, SEM, TEM and XRD analyses (Sathiyabama Muthukrishnan et al., 2019). Zeta potential of the synthesized nanoparticles was found to be 37.7 mV. The encapsulation efficiency of chitosan nanoparticle was 90 ± 3%. Application of thiamine loaded chitosan nanoparticle enhanced seed germination and growth of chickpea seedlings when compared to untreated control seeds. Treated seedlings showed enhanced production of indole acetic acid (IAA). Foliar application of synthesized nanoparticle induced defense enzymes in leaves and roots of chickpea plants. Decreased cell death in the chickpea roots of treated plants was observed when compared to control under green house condition. These results showed that the thiamine loaded chitosan nanoparticle can be used as a growth stimulator as well as a defense activator against wilt disease in chickpea.

xiv. Oi Palm: Although fungicides could be the best solution in combating fungal infections in crops, however, the phytotoxic level of fungicides to the crops should be tested first to ensure that it is safe for the crops. Moreover, nanocarrier systems of fungicides could play a significant role in the advancement of crop protection. For this reason, chitosan was chosen in the present study as a nanocarrier for fungicides of hexaconazole and/or dazomet in the development of a new generation of agronanofungicides with a high antifungal potent agent and no phytotoxic effect (Maluin et al., 2020). Hence, the encapsulation of fungicides into the non-toxic biopolymer, chitosan was aims to reduce the phytotoxic level of fungicides. In the present study, the *in vivo* phytotoxicity of chitosan-fungicides nanoparticles on the physiological and vegetative growth of oil palm seedlings was evaluated in comparison to its pure fungicides as well as the conventional fungicides. The results revealed the formation of chitosan-fungicides nanoparticles could reduce the phytotoxic effect on oil palm seedlings compared to their counterparts, pure fungicides. The chitosan-fungicides nanoparticles were seen to greatly reduce the phytotoxic effect compared to the conventional fungicides with the same active ingredient.

xv. Rice: The effect of silver nanoparticles (20-30 nm) against rice leaf blast fungus was evaluated under different cultivation conditions both *in vitro* and *in vivo* (Elamawi Rabab and El-shafey, 2013). Under lab conditions, the application of four concentrations of silver nanoparticles to the culture of *Magnaporthe grisea* showed significant inhibition of both hyphal growth and number of colonies formed in a dose-dependent manner. Number of spores/ml decreased with in all treatment. Under greenhouse conditions, Silver nanoparticles were sprayed in concentrations 0, 25, 50, 100 and 200 ppm on rice seedling leaves at three times (3 hours before inoculation, 1 and 5 days after artificial inoculation with spore suspension). Damaged Leaf Area Percentage (DLA%) indicated that the application of 100 ppm silver nanoparticles was highly efficient before and after inoculation (26.7, 15.3 and 20%), respectively compared to the untreated plants of 80%. Scanning electron microscope results revealed that the silver nanoparticles caused a detrimental effect on mycelial growth.

Synthesis and characterization of Zinc oxide nano-particles (ZnO NPs) and their application as NPs on rice blast fungus (*Pyricularia grisea*) and rice brown spot fungus (*Helminthosporium oryzae*) was carried out *in vitro* under greenhouse conditions (Kalboush et al., 2016). The antifungal effect of ZnO NPs on *P. grisea* and *H. oryzae* was measured as the reduction in the linear growth. ZnO NPs (3, 2 and 1) were the most effective in reducing the linear growth at a concentration of (0.3 and 0.4%) compared to ZnO. ZnO

NPs were used by spraying rice seedlings with the concentrations 0.2 and 0.5%, seven, five and two days before and two days after artificial inoculation with a spore suspension of *P. grisea* and two days after and before artificial inoculation with *H. oryzae* to find out their effects against the infection by both fungi under the greenhouse conditions. Spraying of ZnO NPs five days before inoculation with a spore suspension of *P. grisea* gave the best effect in controlling rice blast disease. On the other hand, spraying of ZnO NPs two days before inoculation with a spore suspension of *H. oryzae* gave the best effect in controlling rice brown spot disease. Chlorophyll content increased with the different treatments compared to the control.

An investigation was carried out to develop an advanced technique or procedure that would allow reliable, reasonably fast and accurate detection of dangerous seed-borne fungal diseases such as *Pyricularia oryzae* in rice (El-Abbasi et al., 2020). Metallic nanoparticles, silver and iron oxide nanoparticles were also involved to enhance DNA extraction which considered an important step in the PCR technology. Silver and Iron oxide nanoparticles were able to enhance the quantity and purity of DNA template for successful PCR assay. Preferably, seed health methods should be sensitive, specific, rapid, robust, inexpensive and simple to be applicable and fit for plant quarantine purposes. Many detection methods occur for different seed-borne pathogens, but, few of the minimum requirements for adequate seed tests were satisfied. Nano based PCR method could be applied as an appropriate test for rapid detection of rice pathogenic fungus *P oryzae.*

Xanthomonas oryzae pv. *oryzae* (Xoo) causes bacterial leaf blight (BLB) disease in rice. Use of biogenic silver nanoparticles (AgNPs) was studied to control BLB in order to cope with the disadvantages of chemical disease control (Ahmed, et al., 2020). Biogenic AgNPs showed substantial antibacterial activity (24.21 ± 1.01 mm) for Xoo. In a pot experiment, AgNPs were found to be effective weapons for BLB by significantly increasing the plant biomass with a decreased cellular concentration of reactive oxygen species and increased concentration of antioxidant enzyme activity.

Copper sulfide nanoparticles (CuS NPs) were synthesized at 1 : 1 and 1 : 4 ratios of Cu and S and their respective antifungal efficacy was evaluated against the pathogenic activity of *Gibberella fujikuroi* (bakanae disease) in rice (Heping Shang et al., 2020). In a 2 d *in vitro* study, CuS (1 : 1) and CuS (1 : 4) NPs at 50 mg L^{-1} decreased *G. fujikuroi* colony-forming units (CFU) by 35.7 and 33%, respectively, compared to controls; commercial CuO NPs caused an 18.7% inhibition. In a greenhouse study, treating with both types of CuS NPs at 50 mg L^{-1} at the seed stage significantly decreased disease incidence on rice by 35.1 and 45.9%, respectively. Comparatively, CuO NPs achieved only 8.1% disease reduction, and the commercial Cu-based pesticide Kocide 3000 had no impact on disease. Foliar-applied CuO NPs and CuS (1 : 1) NPs decreased disease incidence by 30.0 and 32.5%, respectively, which outperformed CuS (1 : 4) NPs (15%) and Kocide 3000 (12.5%). Notably, CuS (1 : 4) NPs also modulated the shoot salicylic acid (SA) and jasmonic acid (JA) production to enhance the plant defense mechanisms against *G. fujikuroi* infection. These findings provided useful information for improving the delivery efficiency of agrichemicals *via* nano-enabled strategies while minimizing their environmental impact, and advance the present understanding of the defense mechanisms triggered by the NPs presence in plants.

Copper bionanoparticles were successfully synthesized using leaf aqueous extract of *Datura innoxia* from copper sulphate. Nanoparticles were characterized with the help of UV–Vis spectroscopy, Field Emission Scanning Electron Microscopy (FESEM), Energy Dispersive X-ray Spectroscopy and Fourier Transform Infrared Spectroscopy (Kala et al., 2016). FESEM analysis showed that the particles were spherical in shape with size ranging from 5 to 15 nm. The antimicrobial activity of copper bionanoparticles revealed that they are effective growth inhibitors against *Xanthomonas oryzae* pv. *oryzae,* the causative organism of bacterial leaf blight of paddy.

A silver nanoparticle (AgNP) formulation was synthesized, characterized, and investigated as a potential nematicide against rice root-knot nematode, *Meloidogyne gramnicola*, on rice (Baronia et al., 2020). A series of lab assays (water and sand screening) and glasshouse experiments (using soilless system, autoclaved soil, and naturally infested soil) were conducted to examine the nematicidal effects of AgNP. The results from lab assays revealed 0.1 μg/ml as the minimum concentration for 100% irreversible nematode mortality after 12 hours in the water screening test. The effective dosage to kill nematodes in field soil assays was determined to be 3 μg/ml, which is lower than the value of 150 μg/ml reported in the literature. No visible adverse effect of AgNP was observed on seed germination or plant growth in all the experiments. The results indicated that AgNP has effective nematicidal activity against *M. graminicola* in rice.

xvi Cotton: The efficiency of four different nanoformulations (F1, F2, F3, and F4) and a former formulation (F0) from the fungus *Talaromyces flavus* were investigated in reducing the incidence of cotton Verticillium wilt in greenhouse conditions (Laleh Naraghi and Maryam Negahban, 2019). Nanoformulations were two types of nanocapsules, one nanoemulsion, and one nanopowder. Results showed decreased percentages of disease severity in treatments of nanoformulations (85-97%) and the former formulation (95%) in comparison to the infected control. The most effective nanoformulations were not significantly different in the disease control efficiency compared with the former formulation.

In addition to decreased disease severity, a ten-day earlier flowering and fruit production were also observed in treatments of nanoformulations and the former formulation than the infected control. In terms of decreasing the disease severity, the nanoformulations presented different results by using two methods of seed impregnation and soil application. No significant differences were found between the two application methods for F2 and F4 nanoformulations. F1 and F3 nanoformulations, however, had better efficiencies in both soil application and seed impregnation methods. The highest disease severity reduction (97%) was recorded in treatments with F2 and F3 nanoformulations in seed impregnation method (1.66) compared to the infected control (66.66). Statistically, disease severity significantly increased in only three F1 treatments with seed impregnation method, and F3 and F4 by soil application compared to the other treatments with no significant differences.

xvii. Maize: Amphiphilic polyanhydride nanoparticles (PAN) were used to enhance efficacy and provide sustained release of several active ingredients in animal systems and have potential for use in crop production. To assess the potential benefits of PANs in seed treatment formulations, two types of experiments were performed with maize seeds (Washington Lauren Alicia, 2017). In the first experiment, seeds were inoculated with *Fusarium graminearum* (seedborne inoculum) and then treated with PANs loaded with varying rates of fludioxonil (FLD) or thiabendazole (TBZ). In the second experiment, seeds were treated with uniform rates of PAN-encapsulated FLD or TBZ and exposed to unfavorable germination environments in the presence of soilborne inoculum. Controls included commercial rates of FLD and TBZ without PAN encapsulation, as well as untreated seeds and noninoculated treatments.

Seedborne inoculum experiments with half rate of TBZ (0.025mg) loaded PANs protected root length similarly to full rate nonencapsulated TBZ, but PAN-encapsulated FLD did not perform as well as nonencapsulated FLD. In soilborne inoculum experiments, disease symptoms were severe and there were few differences among treatments. PAN-encapsulated TBZ performed similarly to non-encapsulated FLD and TBZ in shoot length when emergence was delayed from zero to three weeks. However, PAN-encapsulated fungicides did not increase or prolong efficacy over non-encapsulated formulations.

PAN-formulated seed treatments have the potential to provide tailored release systems for fungicides to protect against soilborne and seedborne diseases under a variety of different environments and potentially using a range of active ingredients. These results indicated that PAN-encapsulation might facilitate use of reduced rates for some active ingredients, but more work is needed to demonstrate their potential to prolong efficacy under delayed emergence conditions.

Sweet potato: A reduction in crop spoilage and an increase in shelf-life is the goal of effective disease managememnt methods. A study by Nivien A. Nafady et al., (2019) was aimed to assess ZnO-nanoparticles (ZnO-NPs) as a safe, new protectant against Rhizopus soft rot of sweet potato. ZnO-NPs had a fungicidal effect against *Rhizopus stolonifer* when used at concentrations above 50 ppm. The results showed that tubers treated with ZnO-NPs exhibited fewer fungal populations (1.2 CFU per segment) than those that did not receive the treatment. Tubers infected with *R. stolonifer* and treated with ZnO-NPs showed no visible decay for up to 15 days, indicating that ZnO-NPs act as a coating layer on tuber surface. The greatest weight loss after 15 days of storage was reported in infected tubers (8.98%), followed by infected tubers treated with ZnO (6.54%) and infected tubers treated with ZnO-NPs (3.79%). The activity of cell-wall degrading enzymes, α-amylase and cellulase, were significantly increased in both infected tubers and those treated with ZnO, compared to the tubers treated with ZnO-NPs. These results confirmed that coating with ZnO-NPs is an effective method of protecting sweet potato tubers from infection, maintaining their quality and increasing their shelf-life for up to 2 months in storage.

General roles of nanotechnology in postharvest management: Antimicrobial packaging of edible food films made with cinnamon or oregano oil, or nanoparticles of zinc, calcium other materials that kill bacteria is being tried (Nivien A. Nafady et al., 2019). Green packaging using nano-fibers made from lobster shells or organic corn (both are antimicrobial and biodegradable) is also a food safety effort. Improved food packaging needs packaging material having strength, barrier properties and stability to heat and cold. These are being achieved using nanocomposite materials. Bayer Polymers have produced a nanocomposite film 'Durethan', It is a film enriched with silicate nanoparticles which reduces the entrance of oxygen and other gases, and preserves moisture, thus preventing food from spoiling. In future, incorporation of silver, magnesium oxide or zinc oxide nanoparticles (which can kill harmful microorganisms) in food or beer packages will save the contamination. Antimicrobial activity can also be imparted through addition of nano-sensors to food packages is also anticipated in the future. McHugh has suggested that these nanosensors could be used to detect chemicals, pathogens and toxins in foods. Radio Frequency Identification (RFID) tags could be incorporated into food packages in the future. These do not require line-of-sight for reading like bar-codes and enable registration of hundreds of tags in a second. Use of nanowheels, nanofibers and nanotubes are being tried to improve the qualities of food packages.

Coatings has proved to be effective in reducing losses in storage as it acts as carriers of useful ingredients like antioxidants, antimicrobial compounds, anti-ripening compounds, color or aroma additives. It has many advantages such as retardation of ripening, reduction of water loss, chilling, mechanical injury and decay (Singh, 2020). It also adds shine to the coated products. Different coating materials can be used like lipids, polysaccharides, proteins or a combination of these materials. Natural antimicrobial compound, Chitosan [Poly b-(1-4) N-acetyl-d-glucosamine] can be used to lower the postharvest decay of horticultural products. It is deacetylated form of chitin which is an eco-friendly means of pre and postharvest diseases control. It has been reported that Chitosan being a natural compound can successfully substitute the synthetic fungicides by improving resistance in tomato against gray mold caused by B. cinerea.

Nanotechnology has been utilized as source of antifungal treatment in many fruits and vegetables. It has produced nanoparticles, zinc oxide using microbial approach, some of which has been used for postharvest storages of banana, carrot, tomato, onion, etc. The introduction of a new generation of packaging coverages (films) has been useful in protecting post-harvest produces. The mixtures of polymers with organic or inorganic additives of certain geometries (spheres, flakes, fibres, particulates) produce polymer composites. With the use of nanoscale fillers, polymer nanocomposites are developed that has become a radical substitute for the conventional polymer composites. The nanocomposites have proved to be successful in expanding the use of biodegradable and edible films Titanium dioxide (TiO_2), semiconductor nanoparticles are focus photocatalysts in photocatalysis because of their physical and chemical stability, ease of availability, low cost and non-toxicity.

The nano-TiO_2 has light catalyzing capability which can oxidize ethylene into water and CO_2. Similarly, silver ions are also important nanoparticles due to their semiconductor photocatalysis, nanocrystallites, photoactivity and antibacterial properties. Nano-Ag has proved to have more effective antibacterial activity compared to $Ag+$ with the property of absorbing and decomposing ethylene. Preservation with Nano-SiOx agents improved the quality of Fuji apples. Nanotechnology aids in improving the quality and strength of packaging material. The packages produced by adding nanoparticles have stronger mechanical and thermal resistance. To ensure food safety, nano-structured materials prevent the entry of bacteria and micro-organisms in food items. Nanobiosensors embedded in packaging can alert the consumer when a food has become non-edible and can be used inplant and insect virus recognition by development and characterization of nanocomposite materials and many other viruses' recognition by using antibody sensor arrays on nanoscale block copolymer patterns.

4.15 Nanoparticles and RNAi in plant disease management

The discovery of the RNAi pathway has heralded new and innovating approaches for the management of pests and pathogens. RNAi, a conserved eukaryotic mechanism, is involved in the growth, development, and host defense against viruses and transposons, that can also be hijacked to target insects, fungi, viruses and weeds (Elizabeth A. Worrall et. al., 2018). Fig. 44 describes the general mechanism of application of dsRNA to target pathogen RNA. In plants, RNAi is triggered by dsRNA, which is processed into small-interfering RNA (siRNA) by Dicer-like (DCL) enzymes. These siRNAs are incorporated into a RNA-induced silencing complex (RISC); siRNAs then direct the RISCs through base pairing to degrade the pathogen RNA, by preventing it from being used as a translation template. Since its discovery, the RNAi pathway has emerged as a powerful tool to combat plant pests and pathogens by genetic modification. However, genetically modified organisms are controversial and highly regulated in most countries; thus, research into new dsRNA delivery methods is being undertaken.

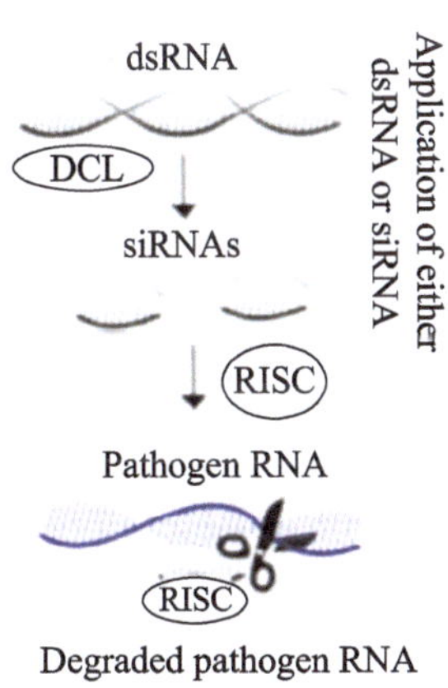

Fig. 44: A brief overview of the RNAi pathway utilized for the application of RNAi-inducing molecules (dsRNA or siRNA) for gene silencing

Topical application of dsRNA has emerged as a highly appealing alternative (Robinson et al., 2014). This method of application faces similar issues to pesticides, such as short longevity due to environmental degradation, and difficulties with site-specific uptake by the targeted pest. To date, nanoparticles acting as carriers of RNAi-inducing molecules have been targeted against viruses, used with honeybees as a model organism for DNA methylation and alternative splicing and targeted against aphids and mosquitoes. Mitter et al. (2017) successfully afforded viral protection against cucumber mosaic virus (CMV) and pepper mild mottle virus (PMMoV) using plants sprayed with dsRNA loaded onto LDH nanoparticles, called BioClay. Significantly, a single spray of BioClay protected plants for 20 days after application on sprayed, and on newly emerged, unsprayed leaves, while plants sprayed with naked dsRNA succumbed to viral infection. Jiang et al. (2014) made the only other report of nanoparticle-facilitated plant uptake of dsRNA through the root tip to induce gene silencing of two endogenous Arabidopsis genes. Each of these reports shows the potential of nanoparticle-facilitated, topical application of RNAi, due to its potential to target a wide range of fugal pathogens, insect pests, and plant viruses with RNAi. Although exogenous application of RNAi-inducing molecules for crop protection holds strong appeal over pesticides, due to its reduced toxicity, topical application of RNAi still faces its own hurdles.

RNAi is a sequence of specific steps for particular protocols, permitting the highly specific targeting sites of individual fungal species. Contamination of foods with mycotoxins is a major concern worldwide, causing severe human and livestock problems. RNAi-based host-induced gene silencing approach represents a robust tool in knock down mycotoxin–related genes in crops (Mohamed A. Mohamed and Kamel A. Abd-Elsalam, 2019). The mechanism of using RNAi in fighting mycotoxicogenic fungi and controlling their toxins was discussed along with the critical role of nanotechnology and the factors influencing the RNAi efficiency. With the noticeable growing number of RNAi products entering the marketing phase, various nanocarriers were suggested to play functional roles in the mechanism of RNAi molecules' delivery in crop plants and stored grains in a way to control mycotoxicogenic fungi and reduce their toxins.

One advantage of using nanoparticles as a delivery system is that they can protect RNAi molecules from enzymatic degradation and immune recognition, also highly efficient in transportation across the plant cell membrane in compared with other conventional carrier. in addition, to prevent excretion if the nanoparticles size and surface coating are appropriate.

Nanoparticles can be classified into two different types when applied as carrier systems in RNAi technology, viz., organic nanoparticles and inorganic substances (Table 22). The RNAi can be combined into organic nanoparticles via chemical bonding or physical embedding.

Table 22: Characteristics of different types of nanoparticles used in functionalized RNAi technology

Type	Common features	Different features	Classification	Nanocarriers
Organic nanoparticles	Low toxicity, good biocompatibility, high securit	Higher thermodynamic stability anddynamic stability	Synthetic polymers: PEI,PLL, PAMAM Natural polymers:Chitosan nanoparticles	Polymeric-based nanoparticles
	Biodegradable,low immunogenicity nanocarrier surfaces should be positively charged	Superior stability, can be sterilized and freeze-dried	Liposomes,micelle emulsions,solid lipid nanoparticles	Lipid-based nanoparticles
Inorganic nanoparticles	High gene transfection efficiency, mass production,low cost	High surface area and large pore volume,strong surface plasmon resonance absorption	Mesoporous silica nanoparticles, carbon nanotubes, quantum dots (DS), gold nanoparticle	

Disease diagnostics, therapeutics and crop protection strategies are fast emerging using nanomaterials preferably nanobiomaterials. It has potential for development of novel nanobiomolecules which offer several advantages over conventional treatment methods. RNA nanoparticles with many unique features are promising candidates in disease treatment (Chaudhary et al., 2018). The miRNAs are involved in many biochemical and developmental pathways and their regulation in plants and animals. These appear to be a powerful tool for controlling various pathological diseases in human, plants and animals, however there are challenges associated with miRNA based nanotechnology. Several advancements made in the field of miRNA therapeutics make it an attractive approach, but a lot more has to be explored in nanotechnology assisted miRNA therapy. The miRNA based technologies can be employed for detection and combating crop diseases as well. Despite these potential advantages, nanobiotechnology applications in the agricultural sector are still in its infancy and have not yet made its mark in comparison with healthcare sector.

RNAi has emerged within the field of applied science as a unique promising and efficient approach for controlling plant diseases. RNAi technology is mainly based on a specific sequence and permits highly specific targeting certain fungal species with no target mistakes. Therefore, most researchers suggested that RNAi technology as a new practical approach considered a unique distinct and preferential solution, compared with chemical fungicides approach that promote the evolution of resistance in targeted and nontargeted.

With the natural occurrence of miRNAs and their role in disease detection and prevention, these small RNAs prove to be a brilliant possibility for use as novel therapeutics. With effective delivery of miRNAs to diseased cells, regulation of specific gene expression and modification of cellular function can be accomplished however successful detection, targeting and delivery of the miRNAs are required. Several distinctive methodologies have been utilised for the delivery of miRNA and different detection procedures are employed using different nanoparticles. However, limitations associated with miRNA delivery can be encountered using RNA nanotechnology, which aims at developing RNA nanoparticles containing a targeting molecule and additionally miRNA (Chaudhary et al., 2018). The self-modelling quality of RNA can be used as a powerful technique to design and structure RNA nanoparticles through the interchange of biological, chemical, physical and computational sciences.

Inspite of difficulties associated with RNA nanotechnology, this emerging field has an extraordinary application in the field of medicine, biotechnology, synthetic biology and nanotechnology, especially in the field of therapeutics where it can be utilized in the treatment of savage diseases like malignancy and viral ailments. Nanotechnology works with smallest possible particles which raises hopes for improving agricultural productivity. It has also been suggested that artificial miRNA poses fewer biosafety or environmental problems when applied to agriculture compared with transgenic breeding. Nanostructured formulation through mechanisms such as targeted delivery or slow controlled release mechanisms and conditional release, could lead to precision release of active ingredients in response to environmental triggers. RNA nanotechnology in plant science holds promise to detect and control various biotic and abiotic stresses through use of miRNA based biomarkers which will revolutionize crop protection. RNA nanotechnology will help to develop new area of research developing biosensors for forecasting and early detection of plant disease.

Future Outlook: High thermodynamic stability and amazing diversity of RNA and its structure and function holds great promise over DNA Nanotechnology. Since it can be chemically modified, it has been efficiently utilized in human and animal therapeutics, however, plant nanotechnology is still an emerging area of research and its applications in plant disease detection and protection is yet to be seen. Nanotechnology assisted strategies will play key role in integrated pest management supplementing conventional methods in view of environmental concerns. Plant transformation has been used in improving disease resistance in crop plants using overexpression of disease resistance genes, RNAi etc. Synthetic miRNAs can be of particular use and provide excellent choice in these approaches. Delivery of synthetic miRNA in plants is a major concern but recently clay nanoparticles have been used in delivering miRNA for viral protection. The making of synthetic RNA is not cost effective at present; the new developments in this field may lead to novel cost effective strategies (Chaudhary et al., 2018). RNA nanoparticles and their use as spray over plants may be advantageous as it is biocompatible, biodegradable and non-toxic making this approach highly flexible and safe.

Nanoparticles synthesized from plants extracts can also be utilized for direct delivery of synthetic miRNA as an efficient method in comparison to Agrobacterium and other direct gene transfer methods in plants. Nanotechnology assisted targeting of miRNA can help in mitigating pests and abiotic stresses under climate change scenario. The miRNAs and transcription factors, being the regulatory molecules, promise to target multigenic traits since these regulatory networks overlap. Rapid and early detection of miRNA responsible for disease in plants is still a challenging task. The miRNA can be detected using nanoparticle tags in diseased state using miRNA assay. The association of electrocatalytic nanoparticles tags to hybridized miRNA molecules generates measurable current upon adding substrate to the solution providing better detection limit over optical miRNA assay. There could be new innovative approaches as well for miRNA detection and their use in therapeutics in humans and animals and crop protection as well in near future.

Bibliography

Abbas Salahi Ardakani, 2013. Toxicity of silver, titanium and silicon nanoparticles on the root-knot nematode, Meloidogyne incognita, and growth parameters of tomato. Nematology, 5: 671–677

Abd El Aty, Abeer A and Moustafa M Zohair, 2020. Green-synthesis and Optimization of an Eco-friendly Nanobiofungicide from Bacillus amyloliquefaciens Mh046937 with Antimicrobial Potential Against Phytopathogens.Environ. Nanotechnol. Monitoring & Management, v. 14, doi: 10.1016/j.enmm.20 20.100309

Abdallah Mohamed Elgorban, Abd El-Rahim Mohamed El-Samawaty, Mohamed Abdallah Yassin, Shaban Rushdy Sayed, Syed Farooq Adil, Khaled Mohamed Elhindi, Marwa Bakri and Mujeeb Khan, 2016. Antifungal silver nanoparticles: synthesis, characterization and biological evaluation. Biotechnol. & Biotechnolog. Equip., 30: 56-62

Abdelmalek, G.A.M. and Salaheldin, T.A.,2016. Silver Nanoparticles as a Potent Fungicide for Citrus Phytopathogenic Fungi. J. Nanomed. Res., 3(5): 00065.

Abd-Elsalam K, A. and Alghuthaymi, M. A., 2015. Nanobiofungicides: are they the Next-Generation of Fungicides? J. Nanotech. Mater. Sci., 2: 38-40.

Abdel-Wahab A. Ismail , Nagwa M. Sidkey , Rawhia A. Arafa , Rasha M. Fathy and Ahmed I. El-Batal, 2016. Evaluation of in vitro Antifungal Activity of Silver and Selenium Nanoparticles against Alternaria solani Caused Early Blight Disease on Potato. British Biotechnol.J.,12: 1-11

Abdulrhman A. Almadiy and Gomah E. Nenaah, 2018 . Ecofriendly Synthesis of Silver Nanoparticles Using Potato Steroidal Alkaloids and their Activity against Phytopathogenic Fungi. Braz. Arch. Biol. Technol. v.61: e18180013 2018

Abdulrhman A. Almadiy and Gomah E. Nenaah, 2018. Ecofriendly Synthesis of Silver Nanoparticles Using Potato Steroidal Alkaloids and Their Activity Against Phytopathogenic Fungi. Braz. Arch. Biol. Technol. vol.61 Curitiba Epub Nov 14, 2018

Agrawal, S. and Rathore, P. 2014. Nanotechnology Pros and cons to agriculture: a review. Int. J. Curr. Microbiol. Appl. Sci., 3: 43 45.

Ahmad, S., Munir, S., Zeb, N., Ullah, A., Khan, B,, Alim J,, Bilalm M,, Omerm M,, Alamzebm M,, Salmanm S,M, and Ali, S., 2019. Green nanotechnology: a review on green synthesis of silver nanoparticles- an ecofriendly approach. Int. J .Nanomed., 14:5087-5107

Ahmed A.A Alsaad, Ahmed A. Hussien and Mowafaq M. Gareeb, 2020. Solid Lipid Nanoparticles (SLN) as a Novel Drug Delivery System: A Theoretical Review. Sys. Rev. Pharm., 11: 259-273.

Ahmed Hammad Nour El-Deen and Bahig Ahmed El-Deeb, 2018. . Effectiveness of Silver Nanoparticles against Root-Knot Nematode, Meloidogyne incognita infecting Tomato under Greenhouse Condition. J. Agri. Sci., 10:148-256

Ahmed, I., Ahmed, S., Dil Raj Yadav and Youn Su Lee, 2016. Applications of Nickel Nanoparticles for

Ahmed, T., Shahid, M., Noman, M., Niazi, M.B.K., Mahmood, F., Manzoor, I., Zhang, Y., Li, B., Yang, Y., Yan, C. and Chen, J., 2020. Silver Nanoparticles Synthesized by Using Bacillus cereus SZT1 Ameliorated the Damage of Bacterial Leaf Blight Pathogen in Rice. Pathogens, 9: 160.

Akther, T. and Hemalatha, S., 2019. Mycosilver Nanoparticles: Synthesis, Characterization and its Efficacy against Plant Pathogenic Fungi. BioNanoSci., 9: 296–301 .

Al Banna, l., Salem, N., Ghrair, A. M. And Habash, S. S., 2018. Impact of silicon carbide nanoparticles on hatching and survival of soil nematodes Caenorhabditis elegans and Meloidogyne incognita. Appl. Eco & Environ.Res., 16: 2651-2662

Alaa, Y., Ghidan. Tawfiq, M. Al Antary, 2019. Applications of Nanotechnology in Agriculture, IntechOpen, DOI: http://dx.doi.org/10.5772/intechopen.88390

Alsultani Amjed , Abdulkadhim Hussein, Jabuk Sura, Fadhel Iman, Hashim Rahmah, Ali Kareen and Alaa Dhay., 2019. Green Synthesis of Silver Nanoparticle by Cauliflower Extract: Characterization and Antibacterial Activity Against Storage. IET Nanobiotechnology, 13 :. 10.1049/iet-nbt.2018.5095.

Aly Derbalah, Mohamed Shenashen, Amany Hamza, Ahmed Mohamed and Sherif El Safty, 2018. Antifungal activity of fabricated mesoporous silica nanoparticles against early blight of tomato. Egyptian J.Basic & Appl.Sci., 5:145-150

Aman Gour and Narendra Kumar Jain, 2019. Advances in green synthesis of nanoparticles. Artificial Cells, Nanomed. & Biotechnol., 47:844-851

Amna Mohammed Ali, Rasha Saad Fawze and Muhannad Mahdi Abd, 2018. Test the Efficiency of Magnesium Oxide Nanoparticles in Rhizoctonia solani Fungus Control in Eggplant. Intl. J. Sci. & Res., 7: 101-104

Anusuya, S. and Sathiyabama, M., 2014. Effect of chitosan on rhizome rot disease of turmeric caused by Pythium aphanidermatum. Hindwai Publishing Corporation. http://dx.doi.org/10.1155/2014/305349.

Ashraf, H., Anjum, T., Riaz, S. and Naseem, S., 2020. Microwave-Assisted Green Synthesis and Characterization of Silver Nanoparticles Using Melia azedarach for the Management of Fusarium Wilt in Tomato. Front. Microbiol., 11:238.

Asma Noshad, Mudassar Iqba ,Crispin Hetherington and Hassan Wahab, 2020. Biogenic AgNPs—A Nano Weapon against Bacterial Canker of Tomato (BCT).Adv. Agri., Volume 2020 |Article ID 9630785

Atef Mohamed Khedr Nassar, Moustafa A Abbassy, Mona A Abdelrasoul and Belal M Soliman, 2018. Effectiveness of Botanical Nanoparticles against Rootknot Nematode (Meloidogyne incognita). J. Indus & Environ. Chem., DOI: 10.4066/2591-7331-C1-002 7th International Conference on Green Chemistry & Technology, June 18-20, 2018 | Dublin, Ireland

Ayisigi, M., Cokislerel, A., Kucukcobanoglu, Y., Yalcin, T. and Aktas, L. Y., 2020. Green synthesized silver nanoparticles for an effective control on soft rot disease pathogen Pectobacterium carotovorum and growth stimulation in pepper. Bulg. J. Agric. Sci., 26:574–584

Baronia, R., Kumar, P., Singh, S.P.and Walia, R.K., 2020. Silver nanoparticles as a potential nematicide against Meloidogyne graminicola. J. Nematol., 52:1-9.

Behiry, S. I., Mohamed, A. A., Younes, H. A., Salem, M. Z. M. and Salem, A. Z. M., 2018. Antigenic and pathogenicity activities of Ralstonia solanacearum race 3 biovar 2 molecularly identified and detected by indirect ELISA using polyclonal antibodies generated in rabbits. Microbial Pathogenesis, vol. 115, pp. 216–221

Bergot, M., Cloppet, E., Pérarnaud, V., Déqué, M., Marçais, B. and Desprez-Loustau, M.L.,2004. Simulation of potential range expansion of oak disease caused by Phytophthora cinnamomi under climate change. Glob. Change Biol., 10:1539–1552

Bhattacharyya Atanu , Duraisamy Prabha, Govindarajan Marimuthu, Buhroo Abdul and Prasad Ram, 2016. Nano-Biofungicides. Emerging Trend in Insect Pest Control. 10.1007/978-3- 319-42990-8_15.

Bhau, B.S., Phukon, P., Barbi Gogoi, Bitupon Borah, J., Baruah, D. K., Sharma and Sawlang Borsingh Wann, 2016. A Novel Tool of Nanotechnology: Nanoparticle Mediated Control of Nematode Infection in Plants. In: Microbial Inoculants in Sustainable Agricultural Productivity. Dhananjaya Pratap Singh, Harikesh Bahadur Singh, Ratna Prabha (Edrs.), pp.pp 253-269, Springer India

Camara, M.C., Campos, E.V.R. and Monteiro, R.A., 2019. Development of stimuli-responsive nano-based pesticides: emerging opportunities for agriculture. J. Nanobiotechnol., 17:100 Campos, E., Oliveira, J. and da Silva, C., 2015. Polymeric and Solid Lipid Nanoparticles for Sustained Release of Carbendazim and Tebuconazole in Agricultural Applications. Sci.Rep., 5: 13809

Canas, J. E., Long, M., Nations, S., Vadan, R., Dai, L., Luo, M., Ambikapathi, R., Lee, E. H. and Olszyk, D., 2008. Effects of functionalized and nonfunctionalized single-walled carbon nanotubes on root elongation of select crop species. Env.& Toxicol.Chem., 27:1922–1931 Chaudhary, V., Jangra, S. and Yadav, N.R., 2018. Nanotechnology based approaches for detection and delivery of microRNA in healthcare and crop protection. .J. Nanobiotechnol., 16: 40 Chauhan, N., Dilbaghi, N. and Gopal, M.,2017. Development of chitosan nanocapsules for the controlled release of hexaconazole. Intl. J.Biol. Macromol., 97:616-624

Chen Juanni, Wu Lintong, Lu Mei, Lu Shasha, Li Ziyan and Ding Wei, 2020. Comparative Study on the Fungicidal Activity of Metallic MgO Nanoparticles and Macroscale MgO against Soilborne Fungal Phytopathogens. Front. in Microbiol., 11: 365

Chen, J. N., Peng, H., Wang, X. P., Shao, F., Yuan, Z. D., and Han, H. Y., 2014. Graphene oxide exhibits broad-spectrum antimicrobial activity against bacterial phytopathogens and fungal conidia by intertwining and membrane perturbation. Nanoscale, 6: 1879–1889

Chowdappa, P. and Gowda, S.K., 2013. Nanotechnology in crop protection: status and scope. Pest Manag. Hortic. Ecosyst., 19:131–151

Cioffi, N., Ditaranto, T. L., Sabbatini, N., Zambonin, L., Tantillo, P. G., Ghibelli, G, D. Alessio, L., Bleve-Zacheo, M. and Traversa, E., 2004. Antifungal activity of polymer-based copper nano-composite coatings. Appl. Physics Lett., 85: 2417-2419

Control of Fusarium Wilt on Lettuce and Tomato. Intl. J. Inno. Res Sci. Engg & Tech., 5: 129- 135

Cristian D P Peréz, Roberto De La Torre Roche, Nubia Zuverza-Mena, Chuanxin Ma, Yu Shen, Jason C White, Edson A Pozza, Adélia A A Pozza and Wade H Elmer, 2020. Metalloid and Metal Oxide Nanoparticles Suppress Sudden Death Syndrome of Soybean. J. Agric. Food Chem., 8:77-87

Debajyoti Ray, Satadru Pramanik, Ranju Prasad Mandal, Sujata Chaudhuri and Swati De, 2015· Sugar-mediated 'green' synthesis of copper nanoparticles with high antifungal activity. Mater. Res. Express, 2: 105002

Derbalah, A. and Elsharkawy, M. A., 2019. New strategy to control Cucumber mosaic virus using fabricated NiO-nanostructures. J. Biotechnol., 306: 134-141

El-Abbasi, I.H., Khalil, A.A. and Awad, H.M., 2020. Nano-diagnostic technique for detection of rice pathogenic fungus Pyricularia oryzae. Indian Phytopathol.,73: 673–682

Elamawi Rabab M. A. and El-shafey, R. A. S., 2013 . Inhibition Effects of Silver Nanoparticles against Rice Blast Disease Caused by Magnaporthe grisea. Egypt. J. Agric. Res., 91: (4) El-Batal, A. I., Nagwa M. Sidkey and Ismail A. A., 2016.. Impact of Silver and Selenium

Nanoparticles Synthesized by Gamma Irradiation and Their Physiological Response on Early Blight Disease of Potato. J. Chem. & Pharmac. Res., 8:934-951

Elsharkawy, M.M and Mousa, K.M., 2015. Induction of systemic resistance against Papaya ring spot virus (PRSV) and its vector Myzus persicae by Penicillium simplicissimum GP17-2 and silica (SiO2) nanopowder. Intl. J. Pest Manag., 61: 353–358

Elwy A. Mohamed, Sherif F. Elsharabasy and Doaa Abdulsamad, 2019. Evaluation of In vitro Nematicidal Efficiency of Copper Nanoparticles against Root-knot Nematode, Meloidogyne incognita. South Asian J. Parasitol., 2: 1-6

Entsar H. Taha, 2016 . Nematicidal Effects of Silver Nanoparticles on Root-knot Nematodes (Meloidogyne incognita) in laboratory and screenhouse. J. Plant Prot. and Path., 7: 333 - 337

Ezzeldin Ibrahim , Muchen Zhang, Yang Zhang, Afsana Hossain, Wen Qiu, Yun Chen, Yanli Wang, Wenge Wu, Guochang Sun and Bin, 2020. Green-Synthesization of Silver Nanoparticles Using Endophytic Bacteria Isolated from Garlic and Its Antifungal Activity against Wheat Fusarium Head Blight Pathogen Fusarium graminearum. Nanomaterials, 10: 219-234

Ezzeldin Ibrahim, Hatem Fouad, Muchen Zhang, Yang Zhang, Wen Qiu Chengqi Yan, Bin Li, Jianchu Mo and Jianping Chen, 2019. Biosynthesis of silver nanoparticles using endophytic bacteria and their role in inhibition of rice pathogenic bacteria and plant growth promotion. RSC Adv., 9: 29293-29299

Farhatun Najat Maluin and Mohd Zobir Hussein, 2020. Chitosan-Based Agronanochemicals as a Sustainable Alternative in Crop Protection. Molecules, 25: 1611-1633.

Farhatun Najat Maluin,, Mohd Zobir Hussein , Nor Azah Yusof,Sharida Fakurazi,Abu Seman Idris,Nur Hailini Zainol Hilmi and Leona Daniela Jeffery Daim, 2020.. Phytotoxicity of chitosan-based agronanofungicides in the vegetative growth of oil palm seedling. https://doi.org/10.1371/journal.pone.0231315

Galdiero, S., Falanga, A., Vitiello, M., Cantisani, M., Marra, V. and Galdiero, M., 2011. Silver nanoparticles as potential antiviral agents. Molecules, 16 : 8894–8918

Ghareeb, R.Y., Alfy, H. and Fahmy, A.A., 2020. Utilization of Cladophora glomerata extract nanoparticles as eco-nematicide and enhancing the defense responses of tomato plants infected by Meloidogyne javanica. Sci. Rep., 10: 19968

Gkanatsiou Ch , Ntalli, N., Menkissoglu-Spiroudi, U. and Dendrinou-Samara, C.,2019. Essential Metal-Based Nanoparticles (Copper/Iron NPs) as Potent Nematicidal Agents against Meloidogyne spp. J. Nanotechnol. Res., 1: 044-058

Gogoi, R., Singh, P.K., Kumar, R., Nair, K.K. and Alam, I., 2013. Suitability of Nano-sulphur for Biorational Management of Powdery mildew of Okra (Abelmoschus esculentus Moench) caused by Erysiphe cichoracearum. J. Pl. Pathol. Microb., 4: 171

González-Melendi, P., Fernández-Pacheco, R., Coronado, M.J., Corredor, E., Testillano, P.S., Risueño, M.C. and Pérez-de-Luque, A.,2008. Nanoparticles as smart treatment-delivery systems in plants: assessment of different techniques of microscopy for their visualization in plant tissues. Ann. Bot., 101:187–195

Gopal, M.A., Gogoi, R.O., Srivastava, C.H., Kumar, R.A., Singh, P.K., Nair, K.K., Yadav, S.A. and Goswami, A.R.,2011. Nanotechnology and its application in plant protection. In: Plant pathology in India: Vision 2030.Indian Phytopathological Society, Indian Agricultural Research Institute, New Delhi, pages 224–232

Gouda H. Attia, Hamad S. Alyami, Mohamed A.A. Orabi, Ahmed H. Gaara and Mohamed A. El Raey, 2020. Antimicrobial Activity of Silver and Zinc Nanoparticles Mediated by Eggplant Green Calyx. Intl. J. Pharmacol., 16: 236-243.

Gregory C. Bernard, Jacob Fitch, Byungjin Min, Naresh Shahi, Marceline Egnin and Inocent Ritte, 2019.Potential Nematicidial Activity of Silver Nanoparticles Against the Root-Knot Nematode (Meloidogyne Incognita). On. J. Complement & Alt. Med., 2:236-242

Guilger-Casagrande, M., Germano-Costa, T. and Pasquoto-Stigliani, T., 2019. Biosynthesis of silver nanoparticles employing Trichoderma harzianum with enzymatic stimulation for the control of Sclerotinia sclerotiorum. Sci. Rep., 9: 14351

Günther Jutz and Alexander Böker,2011. Bionanoparticles as functional macromolecular building blocks – A new class of nanomaterials.Polymer, 52: 211-232,

Hamed, S.M., Hagag, E.S. and El-Raouf, N.A., 2019. Green production of silver nanoparticles, evaluation of their nematicidal activity against Meloidogyne javanica and their impact on growth of faba bean. Beni-Suef. Univ. J. Basic Appl. Sci., 8: 9

Heping Shang, Chuanxin Ma, Chunyang Li, Jason C. White, Tamara Polubesova, Benny Chefetz and Baoshan Xing, 2020. Copper sulfide nanoparticles suppress Gibberella fujikuroi infection in rice (*Oryza sativa* L.) by multiple mechanisms: contact-mortality, nutritional modulation and phytohormone regulation. Env.. Sci-Nano, Issue 9, 2020.

Hongqiang Dong, Renci Xiong, You Liang, Gang Tang, Jiale Yang, Jingyue Tang, Junfan Niu, Yunhao Gao, Zhiyuan Zhou and Yongsong Cao, 2020. Development of glycine-copper(II) hydroxide nanoparticles with improved biosafety for sustainable plant disease management. RSC Adv., 10: 21222-21227

Imada, K., Sakai , S., Kajihara, H., Tanaka, S. and Ito, S., 2016. Magnesium oxide nanoparticles induce systemic resistance in tomato against bacterial wilt disease .Pl.Pathol., 65: 551-560 Irfan Ijaz, Ezaz Gilani, Ammara Nazir and Aysha Bukhari, 2020. Detail review on chemical, physical and green synthesis, classification, characterizations and applications of nanoparticles, Green Chem.Lett.& Rev., 13:223-245

Ismail Ocsoy, Mathews L. Paret, Muserref Arslan Ocsoy, Sanju Kunwar, Tao Chen, Mingxu You and Weihong Tan, 2013. Nanotechnology in Plant Disease Management: DNA-Directed Silver Nanoparticles

Jagana, D., Hegde, Y.R. and Lella, R., 2017. Green nanoparticles: A novel approach for the management of banana anthracnose caused by Colletotrichum musae. Int. J. Curr. Microbiol. Appl. Sci., 6: 1749–1756

Jain, D. and Kothari, S. L., 2014. Green synthesis of silver nanoparticles and their application in plant virus inhibition. J.Mycol.& Pl. Pathol., 44 21-24

Janki N. Thakker, Pranay Dalwadi and Pinakin C. Dhandhukia, 2013. Biosynthesis of Gold Nanoparticles Using Fusarium oxysporum f. sp. cubense JT1, a Plant Pathogenic Fungus. Intl. Scholarly Res. Notices, vol. 2013, Article ID 515091, 5 pages

Jha, A.K., Prasad, K. and Prasad, K.,2009. A green low-cost biosynthesis of Sb2O3 nanoparticles. Biochem. Eng. J., 43:303–306

Jing Cao, Richard H. Guenther, Tim L. Sit, Steven A. Lommel, Charles H. Opperman, and Juanni Chen, Lintong Wu, Mei Lu, Shasha Lu, Ziyan Li and Wei Ding, 2020. Comparative Study on the Fungicidal Activity of Metallic MgO Nanoparticles and Macroscale MgO against Soilborne Fungal Phytopathogens. Front Microbiol., 11: 365

Julie A. Willoughby, 2015. Development of Abamectin Loaded Plant Virus Nanoparticles for Efficacious Plant Parasitic Nematode Control. ACS Appl. Mater. Interfaces, 7: 9546–9553 Kabir Lamsal, Centennial College, Sang Woo Kim, Jin Hee Jung, Yun Seok Kim, Kyong Su Kim and Youn Su Lee, 2011. Application of Silver Nanoparticles for the Control of Colletotrichum Species In Vitro and Pepper Anthracnose Disease in Field. Mycobiology, 39:194-9

Kah, M., Beulke, S. and Tiede, K., 2013. Nano-pesticides: state of knowledge, environmental fate and exposure modelling. Crit. Rev. Environ. Sci. Technol., 43: 1823- 67

Kahru, A. and Dubourguier, H. C., 2010. From ecotoxicology to nanoecotoxicology. Toxicology, 10:105-119

Kala, A., Soosairaj, S., Mathiyazhagan, S. and Raja, P. Green Synthesis of Copper Bionanoparticles to Control the Bacterial Leaf Blight Disease of Rice. Curr. Sci., 110: 2011–2014

Kalboush, Z.A., Hassan, A.A. and Gabr, W.E., 2016. Control of rice blast and brown spot diseases by synthesized zinc oxide nanoparticles. Research Gate, 26: 713-720.

Kamal Fouad Abdellatif, Ragaa Abdelfattah Hamouda and Mostafa Sayed Mostafa El-Ansary, Green Nanoparticles Engineering on Root-knot Nematode Infecting Eggplant plants and Their Effect on Plant DNA Modification, 2016. Iranian J. Biotech., 14: 250-259

Kamel A. Abd-Elsalam and Alexei R. Khokhlov, 2017. Nanoderma: Ecofriendly Nano Biofungicides for Controlling Plant Pathogenic Fungi. Intl.Scholarly and Sci. Res. & Innov., 11(8)

Karousis, N., Tsotsou, G.E., Evangelista, F., Rudolf, P., Ragoussis, N. and Tagmatarchis, N.,2008. Carbon nanotubes decorated with palladium nanoparticles: synthesis, characterization, and catalytic activity. J. Phys. Chem., C112:13463–13469

Kasana, R.C., Panwar, N.R. and Kaul, R.K., 2017. Biosynthesis and effects of copper nanoparticles on plants. Environ. Chem. Lett., 15: 233–240

Kasprowicz, M. J., Gorczyca, A. and Szymocha, A., 2015. Biocontrol Sci. Technol., 25: 873 -887

Katharina Ladewig, Zhi Ping Xu and Gao Qing Max Lu , 2009. Layered double hydroxide nanoparticles in gene and drug delivery . Expert Opin. Drug Deliv., 6:907-22

Kaur Pawan, Thakur Rajesh, Duhan Joginder and Chaudhury Ashok, 2018. Management of wilt disease of chickpea in vivo by silver nanoparticles; biosynthesized by rhizospheric microflora of chickpea (*Cicer arietinum* L.). J. Chem.Technol.& Biotechnol., 93: 10.1002/ jctb.5680.

Keswani, C., 2020. Intellectual Property Issues in Nanotechnology; CRC Press: Cleveland, OH, USA,; ISBN 978-0367482305

Khalid S. Almaary, Shaban R.M.Sayed, Omar H.Abd-Elkader, Turki M.Dawoud and Naglaa F.El Orabi, 2020. Complete green synthesis of silver-nanoparticles applying seed- borne Penicillium duclauxii. Saudi J. Biol. Sci., 27:1333-1339

Khan, M. and Siddiqui, Z.A., 2018. Zinc oxide nanoparticles for the management of Ralstonia solanacearum, Phomopsis vexans and Meloidogyne incognita incited disease complex of eggplant. Indian Phytopathol.,71: 355–364

Khan, M.R. and Rizvi, T.F., 2014. Nanotechnology: Scope and application in plant disease management. Pl.Path.J., 13: 214-231

Khaydarov, R.R., Khaydarov, R.A., Evgrafova, S. and Estrin, Y.,2011. Using silver nanoparticles as an antimicrobial agent. In: Nato science for peace and security series a: chemistry and biology, pp 169–177

Khodakovsky, A., Schr¨oder, P.and Sweldens, W., 2009. Progressive geometry compression, in Siggraph 2000, Computer Graphics Proceedings, pp. 271–278

Kim, S.W., Jung, J.H., Lamsal, K., Kim, Y.S., Min, J.S. and Lee, Y.S., 2012. Antifungal Effects of Silver Nanoparticles (AgNPs) against Various Plant Pathogenic Fungi. Mycobiolog, 40:53-58.

Koka ,J.A., Wani, A.H. and Bhat, M.Y., 2019. Evaluation of antifungal activity of Magnesium oxide (MgO) and Iron oxide (FeO) nanoparticles on rot causing fungi. JDDT [Internet], 9:173-8

Krishnanand Shivanand Iliger, Tariq Ahmad Sofi, Nazir Ahmad Bhat, Farooq Ahmad Ahanger, Jagan Chandra Sekhar, Ahmed Zohier Elhendi, Asma A. Al-Huqail and Faheema Khan,2021. Copper nanoparticles: Green synthesis and managing fruit rot disease of chilli caused by Colletotrichum capsici. Saudi J. Biol. Sci., 28: 1477-1486

Kumar, R., Singh, V.P., maurya, D., Pandey, A.K., 2015. Bionanoparticles: A green nanochemical aaproch.PharmaTutor, 3: 28-35

Kumari, M., Ernest, V. and Mukherjee, A., 2012. In vivo nanotoxicity assays in plant models. Methods Mol. Biol., 926: 399-410.

Kumari, M., Mukherjee, A. and Chandrasekaran, N., 2009. Genotoxicity of silver nanoparticles in Allium cepa. Science of the Total Environment, 407: 5243–5246

Lakshmeesha, T.M., Mahadevamurthy Murali Ansari, Mohammad Udayashankar, Arakere Alzohairy, Mohammad Almatroudi, Ahmad Alomary, Mohammad Asiri, Sarah Ashwini, B.S. Kalagatur, Naveen Nayak and Chandra Niranjana, S.R., 2020. Biofabrication of zinc oxide nanoparticles from Melia azedarach and its potential in controlling soybean seed-borne phytopathogenic fungi. Saudi J. Biol. Sci., 27: 10

Laleh Naraghi and Maryam Negahban, 2019. Biological control of cotton Verticillium wilt by nanoformulations containing Talaromyces flavus. Eurasia J. Biosci., 13: 1177-1185 Lamsal, K., Kim, S.W., Jung, J.H., Kim, Y.S., Kim, K.S. and Lee, Y.S., 2011. Inhibition Effects of Silver Nanoparticles against Powdery Mildews on Cucumber and Pumpkin. Mycobiology, 39:26-32.

Lau, H., Wu, H. and Wee, E., 2017. Specific and Sensitive Isothermal Electrochemical Biosensor for Plant Pathogen DNA Detection with Colloidal Gold Nanoparticles as Probes. Sci Rep., 7: 38896

Le Thi Anh Tu and Pham Thi Le Ha, 2017 . Using γ ray-induced silver nanoparticles on control of clubroot disease of chinese cabbage caused by Plasmodiophora brassicae. Vietnam J. Biotech., 15: 675- 683

Lei Xu, Li-Dong Cao, Feng-Min Li, Xiang-Jing Wang and Qi-Liang Huang, 2014. Utilization of Chitosan-Lactide Copolymer Nanoparticles as Controlled Release Pesticide Carrier for Pyraclostrobin Against Colletotrichum gossypii Southw. J. Dispersion Sci. &Technol., 35: 544-550,

Li, S.j. and Zhu, T.H., 2013. Biochemical response and induced resistance against anthracnose (Colletotrichum camelliae) of camellia (Camellia pitardii) by chitosan oligosaccharide application. Forest Pathol., 43: 67–76

Li, W., Li, L., Cao, Y., Lan, T., Chen, H., and Qin, Y., 2017. Effects of PLA Film Incorporated with ZnO Nanoparticle on the Quality Attributes of Fresh-Cut Apple. Nanomaterials, 7: 207

Liao, Y.Y., Strayer-Scherer, A., White, J.C., 2019. Particle-size dependent bactericidal activity of magnesium oxide against Xanthomonas perforans and bacterial spot of tomato. Sci. Rep., 9: 18530

Lidong Cao, Huirong Zhang and Jiakun Zhang, 2016. Quaternized Chitosan-Capped Mesoporous Silica Nanoparticles as Nanocarriers for Controlled Pesticide Release. Nanomaterials, 6: 126

Lin Cai, Changyun Liu, Guangjin Fan, Chaolong Liu and Xianchao Sun, 2019. Preventing viral disease by ZnONPs through directly deactivating TMV and activating plant immunity in Nicotiana benthamiana. Environ.Sci-Nano, Issue 12

Lin Cai, Juanni Chen, Zhongwei Liu, Hancheng Wang, Huikuan Yang and Wei Ding, 2018. Magnesium Oxide Nanoparticles: Effective Agricultural Antibacterial Agent Against Ralstonia solanacearum. Front. Microbiol., 25 April 2018 https://doi.org/10.3389/fmicb.2018.00790

Luíz Paulo Figueredo Benício, Rejane Alvarenga Silva,Júnia Aparecida Lopes, Denise Eulálio, Rodrigo Morais Menezes dos Santos,Leonardo Angelo de Aquino, Leonardus Vergütz, Roberto Ferreira Novais,Liovando Marciano da Costa,Frederico Garcia Pinto and Jairo Tronto, 2015. Hidróxidos Duplos Lamelares: Nanomateriais Para Aplicações Na Agricultura. Rev. Bras. Ciênc. Solo, vol.39 no.1 Viçosa

Luma S. Al Banna, Nidá M. Salem, Gharam A. Jaleel and Akl M. Awwad, 2020 . Green synthesis of sulfur nanoparticles using Rosmarinus officinalis leaves extract and nematicidal activity against Meloidogyne javanica. Chem. Intl., 6: 137-143.

Madhuri Sharon, Ajoy Kr. Choudhary and Rohit Kumar, 2010. Nanotechnology in agricultural diseases and food safety. J. Phytol., 2: 83–92

Mahendra, R., Shivaji, D., Aniket, G. and Elsalam, K.A.,2012. Strategic nanoparticle-mediated gene transfer in plants and animals—a novel approach. Curr Nanosci., 8:170–179

Mahmoud Abdel-Aziz, 2019. Antifungal Activity of Chitosan Nanoparticles Against Some Plant Pathogenic Fungi In Vitro.Agri.Engg.Intl:CIGR J., 21: 27-32

Maluin, .FN., Hussein, M.Z., Yusof, N.A., Fakurazi, S., Idris, A.S. and Hilmi, N.H.Z., 2020. Phytotoxicity of chitosan-based agronanofungicides in the vegetative growth of oil palm seedling. PLoS ONE 15: e0231315

Maluin, F.N., Hussein, M.Z. and Yusof, N.A., 2019. Preparation of Chitosan-Hexaconazole Nanoparticles as Fungicide Nanodelivery System for Combating Ganoderma Disease in Oil Palm. Molecules, 24:2498.

Manal A. El-Shazly, Yasser A. Attia , Farida F. Kabil , Emad Anis and Mohamed Hazman, 2017. Inhibitory Effects of Salicylic Acid and Silver Nanoparticles on Potato Virus YInfected Potato Plants in Egypt. Middle East J. Agri. Res.,6:835-848

Manish Mathur, Kamna Sharma , Alkesh Tak and Praveen Gehlot, 2018 . Application of Nanotechnology in Plant Pathogen Managements: Knowledge and Gaps . Int.J.Curr.Microbiol. App.Sci ., 7: 4246-4258

Marcela Vargas-Hernandez, Israel Macias-Bobadilla , Ramon Gerardo Guevara-Gonzalez, Enrique Rico-Garcia, Rosalia Virginia Ocampo-Velazquez, Luciano Avila-Juarez and Irineo Torres-Pacheco, 2020. Nanoparticles as Potential Antivirals in Agriculture. Agriculture, 10 : 444-452

Md Nuruzzaman, Jiawei Ren, Yanju Liu, Mohammad Mahmudur Rahman, Ho Kyong Shon and Ravi Naidu, 2020. Hollow Porous Silica Nanosphere with Single Large Pore Opening for Pesticide Loading and Delivery. ACS Appl. Nano Mater., 3: 105–113

Md Nuruzzaman, Jiawei Ren, Yanju Liu, Mohammad Mahmudur Rahman, Ho Kyong Shon Md. Arshad Ali , Temoor Ahmed , Wenge Wu, Afsana Hossain , Rahila Hafeez , Md. Mahidul Islam Masum , Yanli Wang , Qianli An , Guochang Sun and Bin Li, 2020 . Advancements in Plant and Microbe-Based Synthesis of Metallic Nanoparticles and Their Antimicrobial Activity against Plant Pathogens. Nanomaterials,10: 1146

Mehrajud Din Talie, Abdul Hamid Wani, Nusrat Ahmad, Mohd Yaqub Bhat and John Mohd War, 2020. Green synthesis of silver nanoparticles (agnps) using Helvella leucopus pers. and their antimycotic activity against fungi causing fungal rot of apple. Asian J. Pharm. Clin. Res., 13: 161-165

Melo, C.B.N.F., de MendonçaSoares, B.L., Diniz, K.M., Leal, C.F., Canto, D., Flores, M.A.P., da Costa Tavares-Filho, H.J., Galembeck, A., Stamford, M.T.L. and Stamford-Arnaud, T.M., 2018. Effects of fungal chitosan nanoparticles as eco-friendly edible coatings on the quality of post-harvest table grapes. Post-Harvest Biol. Technol., 139: 56–66

Milani, P., França, D., Balieiro, A.G., Faez, R., Milani, P. and França, D., 2017. Polymers and its applications in agriculture. Polímeros, 27:256–66.

Mishra, S., Singh, B.R., Singh, A., Keswani, C., Naqvi, A.H. and Singh, H.B.,2014. Biofabricated Silver Nanoparticles Act as a Strong Fungicide against Bipolaris sorokiniana Causing Spot Blotch Disease in Wheat. PLoS ONE 9(5): e97881.

Mitter, N., Worrall, E. and Robinson, K., 2017. Clay nanosheets for topical delivery of RNAi for sustained protection against plant viruses. Nature Plants, 3: 16207

Mohamed A. Mohamed and Kamel A. Abd-Elsalam, 2019. Nanoparticles and gene silencing for suppression of mycotoxins: what we know and what we should know? In book: Nanomycotoxicology Treating Mycotoxins in the Nano Way (pp.423-448), Publisher: Elsevier Inc.

Momanyi Kerubo Rachael and Rajiv P, 2019. Production and characterization of Green Iron Oxide nanoparticles with antifungal properties against fungal pathogens. DOI: https:// doi. org/10.26452/ijrps.v11i2.2206

Mousa Alghuthaymi, Kamel A. Abd-Elsalam, Priti Paraliker and Mahendra Rai, 2020. Mono and hybrid nanomaterials: Novel strategies to manage postharvest diseases, Multifunctional Hybrid Nanomaterials for Sustainable Agri-Food and Ecosystems, 10.1016/B978-0-12- 821354-4.00013-3, 287-317

Nel A, Xia T, Mädler, L. and Li, N., 2006. Toxic potential of materials at the nanolevel. Science, 311:622-627.

Nguyen, H. M., Hwang, I. C., Park, J. W., Park, H. J., 2012. Enhanced payload and photo-protection for pesticides using nanostructured lipid carriers with corn oil as liquid lipid. J.Microencapsul., 29:596-604.

Nicolaza Pariona, Arturo I. Mtz-Enriquez, D. Sánchez-Rangel, Gloria Carrión, F. Paraguay-Delgado and Greta Rosas-Saito, 2019. Green-synthesized copper nanoparticles as a potential antifungal against plant pathogens. RSC Adv., 9 : 18835-18843

Nicolaza Pariona, Arturo I. Mtz-Enriquez, D. Sánchez-Rangel[ac], Gloria Carrión, F. Paraguay-Delgadoe and Greta Rosas-Saito, 2019. Green-synthesized copper nanoparticles as a potential antifungal against plant pathogens. RSC Adv., 2019, 9, 18835-18843

Nikunj Patel, Purvi Desai, Niti Patel, Anamika Jha and Hemant K Gautam, 2014. Agronanotechnology for Plant Fungal Disease Management: A Review. Int.J.Curr. Microbiol. App.Sci., 3: 71-84

Nivien A. Nafady, Saad A.M. Alamri, Elhagag Ahmed Hassan, Mohamed Hashem, Yasser S. Mostafa and Kamal A.M. Abo-Elyousr, 2019. Application of ZnO-nanoparticles to manage Rhizopus soft rot of sweet potato and prolong shelf-life. Folia Hort., 31: 319-329

Noha, K., El-Dougdoug , A.M., Bondok and El-Dougdoug, K.A., 2018 . Evaluation of Silver Nanoparticles as Antiviral agent against ToMV and PVY in Tomato Plants. Middle East J. Appl. Sci., 8: 100-111

Nugaeva, N., Gfeller, K.Y., Backmann, N., Lang, H.P., Duggelin, M. and Hegner, M., 2005. Micromechanical cantilever array sensors for selective fungal immobilization and fast growth detection. Biosens Bioelectron, 21:849–856

Ocsoy, I., Paret, M.L. and Ocsoy, M.A., 2013. Nanotechnology in plant disease management: DNA-directed silver nanoparticles on graphene oxide as an antibacterial against Xanthomonas perforans. ACS Nano., 7:8972-8980.

Olivija Plohl, Sašo Gyergyek and Lidija Fras Zemljič, 2020. Mesoporous silica nanoparticles modified with N-rich polymer as a potentially environmentally-friendly delivery system for pesticides, Microporous & Mesoporous Mat., 310: 1387-1811

on Graphene Oxide as an Antibacterial against Xanthomonas perforans. ACS Nano, 7 : 8972–8980

Othman-Al, M. R., Abd EL-Aziza, A. R. M., Mahmoud, M. A., Eifan, S. A., Shikh-EL, M. S. and Majrashi,M., 2014. Application of Silver nano-particles as Antifungal and antiaflatoxin B1 Produced by Aspergillus Flavus. Digest J. Nano-mat.& Bio str., 9: 151 157.

Pankaj K. Tyagi, Rashmi Mishral, Fahad Khan , Dhriti Guptal, Deepak Gola, 2020 . Antifungal Effects of Silver Nanoparticles Against Various Plant Pathogenic Fungi and its Safety Evaluation on Drosophila melanogaster. Bioninterfase Res. in Appl.Chem., 10 :6587 - 6596 Park, H. J., Kim, S. H., Kim, H. J. and Choi, S. H., 2006. A New Composition of nanosized silicasilver for control of various plant diseases. J. Pl. Pathol., 22: 295-302

Park, M., Lee, C.I. and Seo, Y.J., 2010. Hybridization of the natural antibiotic, cinnamic acid, with layered double hydroxides (LDH) as green pesticide. Environ. Sci.and Pollution Res. Intl., 17:203-209

Parul Sharma, Adikshita Sharma, Monica Sharma, Nikhil Bhalla, Pedro Estrela, Aditya Jain, Preeti Thakur, and Atul Thakur, 2017. Nanomaterial Fungicides: In Vitro and In Vivo Antimycotic Activity of Cobalt and Nickel Nanoferrites on Phytopathogenic Fungi. Global Challenges, 1: 1700041

Peralta-videa, J., Huang, Y., Parsons, J., Zhao, L., Lopez-Moreno, M., Hernandez-Viezcas, J. and Gardea-Torresdey, J., 2016. Plant-based green synthesis of metallic nanoparticles: Scientific curiosity or a realistic alternative to chemical synthesis? Nanotechnol. Environ. Eng., 1: 4

Pham Thi Le Ha, Nguyen Tan Man, Nguyen Duy Hang, Le Hai, Tran Thi Tam, Pham Thi Sam, Le Huu Tu, Tran Thu Hong, Tran Thi Thuy and Nguyen Tuong Ly Lan, 2013. Field Test of Capability to Prevent Cabbage Clubroot Disease Caused By Plasmodiophora brassicae of silver nanoparticles synthesized by gamma radiation. The Annual Report for 2013, VINATOM. The 10th National Conference on Nuclear Science and Technology, Vung Tau, 8/2013 (in Vietnamese).

Pragati Jamdagni, Jogender Singh Rana, Poonam Khatri and Kiran Nehra, 2018. Comparative account of antifungal activity of green and chemically synthesized Zinc Oxide nanoparticles in combination with agricultural fungicides. Int. J. Nano Dimens., 9 : 198-208

Pragati Jamdagni, J S Rana and Poonam Khatri, 2018. Comparative study of antifungal effect of green and chemically synthesised silver nanoparticles in combination with carbendazim, mancozeb, and thiram. IET Nanobiotechnol. 12:1102-1107

Pulugu Sai, Vijay Kumar and Ankita Garkoti, 2020. Applications of Nanotechnology in Plant Disease Management. Intl. J. Sci. and Res., 9 : 1506-1508

Qian, K., Shi, T. and Tang, T., 2011. Preparation and characterization of nano-sized calcium carbonate as controlled release pesticide carrier for validamycin against Rhizoctonia solani. Microchim Acta, 173: 51–57

Raghavan Charudattan, 2019. Delivery of nematicides to control plant parasitic nematodes by using Tobacco mild green mosaic tobamovirus as a nanocarrier. SBIR-STTR, Americas Seed Fund, Department of Agriculture, USDA

Raghavendra, S.N., Raghu, H.S., Divyashree, K. and Rajeshwara, A.N., 2019. Antifungal efficiency of copper oxychloride-conjugated silver nanoparticles against Colletotrichum gloeosporioides which causes anthracnose disease. Asian J. Pharm. Clin. Res., 12: 230–233 Rahul, S., Chandrashekhar, P. and Hemant, B., 2014. Nematicidal activity of microbial pigment from Serratia marcescens. Natural Product Res.,28 : 1399-1404

Rai, M. and Ingle, A., 2012. Role of nanotechnology in agriculture with special reference to management of insect pests. Appl. Microbiol. Biotechnol., 94: 287- 293.

Rajyasri Ghosh, Sushmita Dutta and Shampa Bhattacharyya, 2018. Green Synthesis of Silver Nanoparticle by Acalypha indica and its Antifungal Effect against Phytopathogen Colletotrichum capsici. Acta Sci. Agric.,2: 27-31.

Ranjbar, S., Rahemi, M. and Ramezanian, A., 2018. Comparison of nano-calcium and calcium chloride spray on post-harvest quality and cell wall enzymes activity in apple cv. Red Delicious. Sci. Hortic. 240: 57–64

Rawhia A. Arafa and Rasha M. Fathy, 2016. Evaluation of in vitro Antifungal Activity of Silver and Selenium Nanoparticles against Alternaria solani Caused Early Blight Disease on Potato. British Biotechnol. J., 12:1-11

Rishu Sharma, Sujaya Dewanjee and Chittaranjan Kole, 2016. Utilization of Nanoparticles for Plant Protection. In : Plant Nanotechnology . C. Kole et al. (Eds.),, Springer International Publishing Switzerland, 305-327.

Robinson, K.E., Worrall, E.A. and Mitter, N., 2014. Double stranded RNA expression and its topical application for non-transgenic resistance to plant viruses. J. Plant Biochem. Biotechnol., 23: 231–237

Safaa M. Ali, Naeima M. H. Yousef and Nivien A. Nafady, 2015. Application of Biosynthesized Silver Nanoparticles for the Control of Land Snail Eobania vermiculata and Some Plant Pathogenic Fungi. J. Nanomat., vol. 2015, Article ID 218904, 10 pages .

Salem, E.H., Nawito, M.A.S.and El-Raouf Ahmed, A.E.A., 2019. Effect of Silver Nano-Particles on Gray Mold of Tomato Fruits. J. Nanotechnol.Res., 1 :: 108-118.

Sánchez-López, E., Gomes, D., Esteruelas, G., Bonilla, L., Lopez-Machado, A.L., Galindo, R., Cano, A., Espina, M., Ettcheto, M. and Camins, A., 2020. Metal-Based Nanoparticles as Antimicrobial Agents: An Overview. Nanomaterials,10: 292

Sardella, D., Gatt, R. and Valdramidis, V.P., 2017. Physiological effects and mode of action of ZnO nanoparticles against post-harvest fungal contaminants. Food Res. Int. 101: 274–279. Sarlak, N., Taherifar, A. and Salehi, F., 2014. Synthesis of nanopesticides by encapsulating pesticide nanoparticles using functionalized carbon nanotubes and application of new nanocomposite for plant disease treatment. J. Agric. Food Chem., 62 :4833-8

Sathiyabama Muthukrishnan, Indhumathi Murugan and Muthukumar Selvaraj, 2019. Chitosan nanoparticles loaded with thiamine stimulate growth and enhances protection against wilt disease in Chickpea. Carbohydrate Polymers, 212: 169-177,

Sergio Ruffo Roberto, Khamis Youssef, Ayat Farghily Hashim and Antonio Ippolito, 2019. Bioprocessing strategies for costefective large-scale biogenic synthesis of nano-MgO from endophytic Streptomyces coelicolor strain E72 as an anti-multidrugresistant pathogens agent. Sci. Rep., 8: 3820

Sharma Rishu, Dewanjee Sujaya and Kole Chittaranjan, 2016. Utilization of Nanoparticles for Plant Protection. P.3-24. In : Plant Nanotechnology Principles and Practices Nanotechnology for Crop Improvement. 305-327pp

Shuang Wang, Li-Yan Chang , Yong-Jun Wang, Qi Wang , Ching-Hong Yang and Ru-Hong Mei, 2009.. Nanoparticles aject the survival of bacteria on leaf surfaces. FEMS Microbiol Ecol, 68 : 182–191

Siddiqi, K.and Husen, A. 2017. Plant Response to Engineered Metal Oxide Nanoparticles. Nanoscale. Res. Lett., 12: 49

Siddiqi, K.S. and Husen, A., 2016. Fabrication of Metal Nanoparticles from Fungi and Metal Salts: Scope and Application. Nanoscale Res. Lett., 11: 98

Siddiqi, K.S. and Husen, A., 2020. Current status of plant metabolite-based fabrication of copper/copper oxide nanoparticles and their applications: a review. Biomater Res., 24: 11 Singh, 2020. Nanotechnology: Advanced Technique in Postharvest Management of Horticultural Crops. Research Today, 2: 456-458

Singh, J., Dutta, T. and Kim, KH., 2018. 'Green' synthesis of metals and their oxide nanoparticles: applications for environmental remediation. J.Nanobiotechnol ., 16 : 84

Soliman, M.H. and El-Mohamedy, R.S.R.,2017. Induction of Defense-Related Physiological and Antioxidant Enzyme Response against Powdery Mildew Disease in Okra (Abelmoschus esculentus L.) Plant by Using Chitosan and Potassium Salts. Mycobiol., 45:409-420.

Susana C. De la Rosa-García, Pablo Martínez-Torres, Sergio Gómez-Cornelio, Mario Alberto Corral-Aguado, Patricia Quintana, Nikte M. Gómez-Ortíz, "Antifungal Activity of ZnO and MgO Nanomaterials and Their Mixtures against Colletotrichum gloeosporioides Strains from Tropical Fruit", J. Nanomaterials, vol. 2018, Article ID 3498527, 9 pages

Taha, S.H., El-Sherbiny, I.M. and Salem, A.S.,2019.. Antiviral Activity of Curcumin Loaded Milk Proteins Nanoparticles on Potato Virus Y. Pakistan J.Biol. Sci., PJBS. 22:614-622

Thakur, R.K. and Shirkot, P.,2017. Potential of Biogold Nanoparticles to Control Plant Pathogenic Nematodes. J. Bioanal. Biomed., 9: 220-222

Theint Theint Win, Sikandar Khan and PengCheng Fu, 2020. "Fungus- (Alternaria sp.) Mediated Silver Nanoparticles Synthesis, Characterization, and Screening of Antifungal Activity against Some Phytopathogens", J. Nanotechnol., vol. 2020, Article ID 8828878, 9 pages. Tomasa Quiterio-Gutiérrez , Hortensia Ortega-Ortiz , Gregorio Cadenas-Pliego , Alma Delia Hernández-Fuentes , Alberto Sandoval-Rangel , Adalberto Benavides-Mendoza ,

Marcelino Cabrera-de la Fuente and Antonio Juárez-Maldonado, 2019. The Application of Selenium and Copper Nanoparticles Modifies the Biochemical Responses of Tomato Plants under Stress by Alternaria solani. Int. J. Mol. Sci., 20: 1950-1956

Wade Elmer, Chuanxin Ma and Jason White, 2018. Nanoparticles for plant disease management. Curr. Opinion Environ Sci. & Health, 6: 66-70

Wang, Yan & Cui, Haixin & Sun, Changjiao & Zhao, Xiang & Cui, Bo. (2014). Construction and evaluation of controlled-release delivery system of Abamectin using porous silica nanoparticles as carriers. Nanoscale Res. Lett., 9: 2490. 10.1186/1556-276X-9-655.

Wani A. H. and Shah M. A., 2012 . A unique and profound effect of MgO and ZnO nanoparticles on some plant pathogenic fungi. J.Appl. Pharmaceutical Sci., 2: 40-44

Washington Lauren Alicia, 2017. "Utilization of polyanhydride nanoparticle encapsulated fungicide seed treatments against seedborne and soilborne Fusarium graminearum on maize". Graduate Theses and Dissertations., Iowa State University. 15639.

Xia, Z.K., Ma, Q.H., Li, S.Y., Zhang, D.Q., Cong, L., Tian, Y.L. and Yang, R.Y.,2016. The antifungal effect of silver nanoparticles on Trichosporon asahii. J. Microbiol. Immunol. Infect., 49: 182–188

Xing, K., Zhu, X., Peng, X. and Qin, S., 2015. Chitosan antimicrobial and eliciting properties for pest control in agriculture: A review. Agron. Sustain. Dev., 35: 569–588.

Xiuping Wang, Zilin Zhou and Fangfang Chen, 2017. Surface Modification of Carbon Nanotubes with an Enhanced Antifungal Activity for the Control of Plant Fungal Pathogen. Materials, 10: 1375

Yadi Suryadi, Tri Puji Priyatno , I Made Samudra , Dwiningsih Susilowati , Tuti Septi Sriharyani and Syaefudin, 2017. Control of Anthracnose Disease (Colletotrichum gloeosporioides)

Using Nano Chitosan Hydrolyzed by Chitinase Derived from Burkholderia cepacia Isolate E76. J. AgroBiogen., 13:111–122

Yao, K.S., Li, S.J., Tzeng, K.C., Cheng, T.C. and Chang, C.Y., 2009. Fluorescence silica nanoprobe as a biomarker for rapid detection of plant pathogens. Adv. Mater. Res., 79:513– 516

Yaser M. Hafez, Kotb A. Attia, Said Kamel, Salman F. Alamery, Safaa El-Gendy, Abdullah A. Al- Doss, Farid Mehiar, Abdelhalim I. Ghazy, Eid I. Ibrahim and Khaled A.A. Abdelaal,2020. Bacillus subtilis as a bio-agent combined with nano molecules can control powdery mildew disease through histochemical and physiobiochemical changes in cucumber plants. Physiol.& Mol. Pl. Pathol., 111: 101489, ISSN 0885-5765,

Yifen Shang, Md. Kamrul Hasan, Golam Jalal Ahammed, Mengqi Li, Hanqin Yin, Jie Zhou, 2019. Applications of Nanotechnology in Plant Growth and Crop Protection: A Review. Molecules, 24: 2558

Youssef, K., de Oliveira, A.G., Tischer, C.A., Hussain, I. and Roberto, S.R., 2019. Synergistic effect of a novel chitosan/silica nanocomposites-based formulation against gray mold of table grapes and its possible mode of action. Int. J. Biol. Macromol., 141: 247–258

Yuan Ge, John H. Priester, Laurie C. Van De Werfhorst, Sharon L. Walker, Roger M. Nisbet, Youn-Joo An, Joshua P. Schimel, Jorge L. Gardea-Torresdey and Patricia A. Holden, 2014. Soybean Plants Modify Metal Oxide Nanoparticle Effects on Soil Bacterial Communities Environ. Sci. Technol., 48: 13489–13496

Yugandhar, P., Vasavi, T. and Uma Maheswari Devi, P., 2017. Bioinspired green synthesis of copper oxide nanoparticles from Syzygium alternifolium (Wt.) Walp: characterization and evaluation of its synergistic antimicrobial and anticancer activity. Appl Nanosci., 7, 417– 427

Zhang, L., Wu, L., Si, Y. and Shu, K., 2018. Size-dependent cytotoxicity of silver nanoparticles to Azotobacter vinelandii: Growth inhibition, cell injury, oxidative stress and internalization. PLoS ONE, 13, e0209020.

Zhao, X., Cui, H.X., Wang, Y., Sun, C.J., Cui, B. and Zeng, Z.H., 2017. Development Strategies and Prospects of Nano-based Smart Pesticide Formulation. J. Agric. Food Chem., doi:10.1021/acs.jafc.7b02004.

Zhinan Fu , Kai Chen , Li Li , Fang Zhao ID , Yan Wang , Mingwei Wang , Yue Shen , ID , Haixin Cui , Dianhua Liu and Xuhong Guo ,2019. Spherical and Spindle-Like Abamectin-Loaded Nanoparticles by Flash Nanoprecipitation for Southern Root-Knot Nematode Control: Preparation and Characterization. Nanomaterials 8: 449; doi:10.3390/ nano8060449

5

Nanotechnology and Pesticides

With the development of improved systems for monitoring environmental conditions and delivering pesticides appropriately, nanotechnology can further improve our understanding of the biology of different crops and thus potentially enhance yields. It can offer routes to plant disease diagnostics, insect-pest management and efficient pesticides utilization. Nanostructure catalysts will help in increasing the efficiency of pesticides allowing lower doses to be used. The environmental problems caused by overuse of agrochemicals have attracted lot of attention of scientists in the recent years due to high toxicity and non-biodegradability of pesticides and lack of scientific pesticide formulations. Techniques at the nano-scale are being applied in an attempt to enable the targeted delivery or increased toxicity of pesticide applications. This includes insertion of nano-scale active ingredients into pesticides. The specific properties of these nano-scale materials, such as their ability to dissolve in water more effectively than the existing products or their increased stability, are designed to maximize the effectiveness of these pesticides. There are different types of nano formulations that can be used for agrochemicals like nano emulsions, nano encapsulations, nano suspensions and others. Nano-emulsions are water or oil-based, contain uniform suspensions of pesticidal nanoparticles, have many potential applications as disease and pest prevention measures. They exhibit greater stability and increased coating of leaves and uptake through plant cell walls, as a result of low surface tension (Shaon Kumar Das et al., 2018). The main advantages are solubilisation of hydrophobic pesticides, no precipitation or creaming (therefore, no need for constant mixing), increased stability and improved uptake.

5.1 Nanopesticides

Current status and future possibilities: At present, there is no generally accepted definition of the term "nanopesticide" which is currently being used differently in legislation, science, public and business. One proposal is "A nanoparticide is a plant protection product that contains components in the nanometer size range (up to 1000 nm) and has new properties that are related to the small size of their components." So, a size-based definition is not enough for nanopesticides, unlike other nanomaterials. Generally, pesticides always consist of several substances. Similar to a drug, the actual active ingredient is mixed with other accompanying compounds. Those allow the active ingredient to be easy to apply to the plant, to evenly distribute and to remain stable after use. In most of the proposed applications for nanopesticides, the nanomaterial itself is not the active agent, but acts as an auxillary compound to stabilize the active agent or enhance its controlled release (e.g. as a nanocapsule). Nano-copper and nano-silver are examples of nanomaterials as active agents.

In order to protect the active ingredient from the adverse environmental conditions and to promote persistence, a nanotechnology approach, namely "nano-encapsulation" can be used to improve the insecticidal value (Manjunatha et al., 2019). Nano encapsulation comprises nano-sized particles of the active ingredients being sealed by a thin-walled sac or shell (protective coating). Nanoencapsulation of insecticides, fungicides or nematicides will help in producing a formulation which offers effective control of pests while preventing accumulation of residues in soil. In order to protect the active ingredient from degradation and to increase persistence, a nanotechnology approach of "controlled release of the active ingredient" may be used to improve effectiveness of the formulation that may greatly decrease amount of pesticide input and associated environmental hazards.

Nano-pesticides will reduce the rate of application because the quantity of product actually being effective is at least 10-15 times smaller than that applied with classical formulations, hence a much smaller than the normal amount could be required to have much better and prolonged management. Several pesticide manufacturers are developing pesticides encapsulated in nanoparticles. These pesticides may be time released or released upon the occurrence of an environmental trigger (for example, temperature, humidity, light). Clay nanotubes (halloysite) have been developed as carriers of pesticides at low cost, for extended release and better contact with plants, and they will reduce the amount of pesticides by 70-80%, thereby reducing the cost of pesticide with minimum impact on water streams.

Nanoparticles tagged to agrochemicals or other substances could reduce the damage to other plant tissues and the amount of chemicals released into the environment. These materials would release pesticides or fertilizers at a specific time and targeted location (Srilatha, 2011). Nano particles (NPs) use their optical scattering properties for imaging and diagnostics, and their photo thermal properties for various types of therapies. The situation was improved by using active molecular targeting with cell-specific molecules (peptides, antibodies) attached to NPs and coupling to cognate receptors at the membranes of specific target (diseased) cells.

Application of agricultural fertilizers, pesticides, antibiotics, and nutrients is typically by spray or drench application to soil or plants, or through feed or injection systems to animals. Delivery of pesticides or medicines is either provided as "preventative" treatment, or is provided once the disease causing organism has multiplied and symptoms are evident in the plant . In this context, nanotechnologies offer a great opportunity to develop new products against pests. Nanotechnology improves their performance and acceptability by increasing effectiveness, safety, patient adherence, as well as ultimately reducing health care costs. Nanoscale devices are envisioned that would have the capability to detect and treat an infection, nutrient deficiency, or other health problem, long before symptoms were evident at the macro-scale. This type of treatment could be targeted to the area affected with a greater awareness of the hazards associated with the use of synthetic organic pesticides, there has been an urgent need to explore suitable alternative products for pest control. The broad application of molecular biology revolutionized the field of diagnostics

Current status: Nanocapsules have shown controlled release and slow degradation properties of active ingredient (AI), making them more efficient in controlling plant disease and pest. Different type of polysaccharide materials have been incorporated in synthesis i.e chitosan, poly ethylene glycol (PEG), starch, cellulose and polyester substance (Hemraj Chhipa, 2017). Bhan et al. (2014) synthesized solid lipid and polymeric nano-capsules loaded with carbendazim and tebuconazole for fungicide applications. To minimize the harmful effect of herbicides, Poly (epsilon-caprolactone) (PCL) nanocapsule has been developed for controlled release of atrazine and found enhanced herbicidal activity in comparison to commercial formulation of atrazine.

Nanoemulsions (NEs) were also developed to improve solubility and spreading capacity of pesticide by dispersion into two liquid phases. The advantages of NEs were found in greater spreadability, wettability, and superior mechanical stability in comparison to normal emulsion. These characteristic of NEs found helpful in less degradation and volatilization of active ingredient (AI) and improve their bioavailability for long time period. Still, nanoemulsion of different fatty acid methyl esters, organosilicones, alkyl glucosides and Tebuconazole (TBZ) were formulated to improve their efficacy against different pest. Further, pheromone based nanogels were also synthesised using methyl eugenol to control fruit pest. Chitosan nanogel was used with copper to enhance antifungal activity against *Fusarium graminearum* and found synergistic effected of chitosan and copper to inhibit the fungal pathogen.

Nanoform of copper (Cu) showed four order higher activity against bacterial blight on pomegranate at 10000 times less concentration of recommended Cu.

Titanium dioxide nanoparticles were also explored for bacterial spot disease of tomato and rose in pristine form or doped with zinc and silver. Nano-formulation were also developed to improve efficiency, stability and reduction of effective pesticide concentration, like nanoformulation of pyridalyl insecticide coated liposome, neem oil, garlic essential oil, Artemisia arborescens essential oil, imidacloprid, thiamethoxam, carbofuran, thiram and β-cyfluthrin, carbofuran, mancozeb, atrazine and simazine, oilcore silica-shell nanocapsule for fipronil and lansiumamide B.

Further possibilities: Recent studies have shown that nano-pesticides have the capability to reduce toxic impact of chemical pesticide and provide target specific control of crop pest, can be helpful into development of intelligent nano systems for minimization of adverse problem to agriculture like environmental imbalance, food security and food productivity (Hemraj Chhipa, 2017). These nano system have shown great capability of controlled release pattern of active ingredient (AI) make them more efficient for long time period usability that can be solve eutrophication and residual pesticide accumulation problem. In addition, nanopesticide showed improved solubility and stabilities of active ingredient for effective control of pests.

a. Use of green chemistry and environmental sustainability principals in nanopesticide development to maximize their efficiency

b. Process development for up scaling of nanopesticide for commercial level

c. Comparison of nano-formulation activity with preexisting commercial product at field level to determine practical utility

d. Environmental impact assessment of nanopesticide to determine toxicity level

e. Improvement in regulation for nanomaterial application in agriculture.

f. Development of smart nanopesticide will provide many solution to the agro-chemical industry i.e. solubility of active ingredient, stability, controlled release and targeted delivery of active ingredient but still lots of research is required to understand the fate of nanopesticide in environment.

Two potential scenarios that the development of nanoagrochemicals might follow in the future can be considered (Kah,2015). In the first, developments continue along the current path and nanoagrochemicals are likely to become, or at least be perceived as, the next emerging category of contaminants associated with agricultural practices. Alternatively, nanotechnology could become a potential source of emerging solutions to mitigate contamination by pesticides and fertilizers. This second scenario can only be achieved by rapid changes by industry, researchers, and regulatory agencies, following for instance, some of the directions suggested below.

- Available analyses of nanoagrochemicals typically consider all products discussed in literature, many of which represent an unacceptable risk, without being relevant for agricultural applications (e.g., ingredients that are extremely persistent or whose efficacy is marginal). Increasing collaborations between disciplines that are involved at all stages of the development and evaluation of agrochemicals (e.g., formulation, plant, material, and environmental scientists) will support the development of products fitting within the multiple constraints of the agrochemical sector and that are likely to bring an added value relative to existing products.
- The requirements of the latest EU directive regarding a better evaluation of formulations are typically viewed as additional constraints. Current approaches to chemical regulatory assessment often consist of applying incremental safety factors to account for the increasing level of uncertainty. Alternatively, new science-based tools should be developed to assess and fully exploit what the science of formulation has to offer, based on the risks and benefits over the entire life cycle of the products.
- With increasing regulatory pressure and risk of stigmatization, incentives are needed to promote innovation that may lead to the development of more intelligent solutions for plant protection and nutrition. Promotion of more collaboration across sectors (e.g., research, industry, and regulators) and integration of social science and law will ensure public/consumer acceptance and the development of suitable legal frameworks.
- Establishing a size threshold to distinguish "nano" from "non-nano" is of limited value for agrochemicals. Moving to a broader concept of nano- enabled technology and building on the experience from other sectors (e.g., food science, nanometrology) will be more valuable to support the development of more sustainable agrochemicals.

5.2 Benefits of nanopesticides

Often, the active ingredients in pesticides are poorly soluble in water, which makes their use difficult. Nanomaterials can increase the water solubility of the active ingredients as accompanying substances. In this way, the release and distribution can be better controlled and the active ingredient can be better protected against premature degradation. This not only improves the uptake and effectiveness of the pest control agent, it also significantly reduces the amount of pesticide used. Furthermore, alternative active ingredients and products are intended to be less harmful to the environment and possibly reduce the development of resistance of pests to pesticides. At present, however, there is little knowledge on the gain in effectiveness of nanopesticides compared to conventional pesticide.

In recent years, the utilization of nanotechnology to create novel formulations has shown great potential for diminishing the indiscriminate use of pesticides and providing environmentally safer alternatives. Smart nano-based pesticides are designed to efficiently delivery sufficient amounts of active ingredients in response to biotic and/or abiotic stressors that act as triggers, employing targeted and controlled release mechanisms. The current status of stimuli-responsive release systems with potential to be used in agriculture, highlighting the challenges and drawbacks that need to be overcome in order to accelerate the global commercialization of smart nanopesticides were discussed (Camara et al, 2019).

Nano-emulsions can be used for hydrophilic and hydrophobic pesticides, but are largely being developed for those that are poorly water soluble. Nano-emulsion particles are thermodynamically driven to fuse with lipid containing organisms. Nano-encapsulation techniques enable greater control over the circumstances in which encapsulated pesticides will be released. Pesticides could be released quickly or slowly depending on need and under specific conditions, such as moisture and temperature. Such nano-encapsulation techniques not only provide in-built pesticides for crops but also in built switches to control the release and subsequent availability of pesticides. Nano-encapsulated pesticides meet these demands in that they enable smaller quantities of the pesticides to be used effectively over a given period of time interval and in that their design enables them to resist the severe environmental processes that act to eliminate conventionally applied pesticides, *i.e.* leaching evaporation and photolytic, hydrolytic and microbial degradation.

Nanosuspensions consist of the pure poorly water-soluble pesticide without any matrix material suspended in dispersion. It is submicron colloidal dispersion of pure particles of pesticide stabilized by surfactants. By formulating nanosuspensions problems associated with delivery of poorly water soluble pesticide and lipid-soluble pesticides can be solved. Preparing nanosuspensions is preferred for the compounds that are insoluble in water (but are soluble in oil) with high log P value. Conventionally, the pesticides that are insoluble in water but soluble in oil phase system are formulated in liposome, emulsion systems but these lipidic formulation approaches are not applicable to all pesticides. In these cases nanosuspensions are preferred. In case of pesticides that are insoluble in both water and in organic media instead of using lipidic systems nanosuspensions are used as a formulation approach. Agrochemicals companies are reducing the particle size of existing chemical emulsions to the nanoscale, or are encapsulating active ingredients in nanocapsules designed to break open in certain conditions in an alkaline conditions in an insects stomach (Shaon Kumar Das et al., 2018). Similar to the nanocapsules and nanoemulsions being developed for the food and packaging sectors, the smaller size of nanoparticles and emulsions used in agrochemicals is intended to make them more potent.

Leading agrochemicals companies are engaged in nanotech research. Agrochemical companies, currently retail a number of chemicals with emulsions that contain nanoparticles, e.g., Primo MAXX Plant Growth Regulator, Banner MAXX Fungicide, Apron MAXX RFC seed treatment and Cruise MAXX Beans. Primo MAXX ® plant growth regulator, if applied prior to the onset of stress such as heat, drought, disease or traffic can strengthen the physical structure and allow it to withstand ongoing stresses throughout the growing season.

Environmental exposure varies on the basis of condition such as the way in which materials are handled in the workplace, how nanomaterials partition to various phases (e.g., water and air), mobility of nanomaterials in each of these phases, their persistence, and the magnitude of the sources. What is also needed is an evaluation of the expected quantities and concentrations of NP in environmental systems. To date nothing is known about this issue, neither from an analytical point of view (e.g., actual measurements of NP in the environment) nor with respect to theoretical or modeling studies. As a starting point to risk assessment, exploring the sources and environmental pathways helps to identify relevant applications and situations where a subject deserving protection may face exposure to nanotechnology. Although several approaches have been undertaken to develop nanoformulations of agrochemicals around the world, but research on molecular mechanism of action of nano pesticides in insects, biosafety and molecular interaction with plant, soil and environment is scanty. When the nano formulations are applied as foliar spray or in the soil, the carrier and pesticide interact with the soil, insect, plant and atmosphere. It is not completely clear how these nano encapsulated chemicals are degraded in the soil and environment.

Nanopesticides are one of a new strategies being used to address the problems of non-nanopesticides. Nanopesticides cover a wide variety of products, some of which are already on the market. They cannot be considered as a single entity; rather such nanoformulations combine several surfactants, polymers (organic), and metal nanoparticles (inorganic) in the nanometer size range. The lack of water solubility is one of the limiting factors in the development of crop-protecting agents. Microencapsulation has been used as a versatile tool for hydrophobic pesticides, enhancing their dispersion in aqueous media and allowing a controlled release of the active compound. Polymers often used in the nanoparticle production have been reported. Nanocopper particles suspended in water have been used since at least 1931, in a product known as Bouisol as fungicide in the growing of grapes and fruit trees (Sekhon, 2014).

In the research and development stage, nanosized agrochemicals or nanoagrochemicals are mostly nano-reformulations of existing pesticides and fungicides. Nanoformulations are generally expected to increase the apparent solubility of poorly soluble active ingredients, to release the active ingredient in a slow/targeted manner, and/or to protect against premature degradation. Nanopesticides offer a way to both control delivery of pesticide and achieve greater effects with lower chemical dose. Agrochemical companies are reducing the particle size of existing chemical emulsions to the nanoscale, or are encapsulating active ingredients in nanocapsules designed to split open, for example, in response to sunlight, heat, or the alkaline conditions in an insect's stomach.

The smaller size of nanoparticles and emulsions used in agrochemicals is intended to make them more potent. Nanocapsules can enable effective penetration of herbicides through cuticles and tissues, allowing slow and constant release of the active substances. Viral capsids can be altered by mutagenesis to achieve different configurations and deliver specific nucleic acids, enzymes, or antimicrobial peptides acting against the parasites. Silver nanoparticles at 100 mg/kg inhibited mycelia growth and conidial germination on cucurbits and pumpkins against powdery mildew. Silver nanoparticles have received significant attention as a pesticide for agricultural applications. The potential of nanomaterials in insect pest management as modern approaches of nanotechnology has been reported.

Nanosilica has been successfully employed to control a range of agricultural insect/pest and ectoparasites in animals (Sekhon, 2014). Such nanoparticles get absorbed into cuticular lipids (used by insects to prevent death from desiccation) by physisorption and cause insect death by physical means when applied on leaves and stem surfaces. Antifungal activities of polymer-based copper nanocomposites against pathogenic fungi and silica–silver nanoparticles against *Botrytis cinerea, Rhizoctonia solani, Calllectotrichum gloeosporioides, Bipolaris sorokiniana*, and *Magnaporthe grisea* have been reported. Copper nanoparticles in soda lime glass powder showed efficient antimicrobial activity against gram-positive and gram-negative bacteria and fungi. A novel photodegradable insecticide involving nanoparticles has been reported.

Recent advances in nanoscale engineering have created a new class of particulate bionanotechnology that uses biomimicry to better integrate adjuvant and antigen. These pathogen-like particles can come from a variety of sources, ranging from fully synthetic platforms to biologically derived, self-assembling systems. By employing molecularly engineered targeting and stimulation of key immune cells, recent studies utilizing pathogen-like particles as vaccine delivery platforms have shown great promise against high-impact, unsolved vaccine targets ranging from bacterial and viral pathogens to cancer and addiction. Controlling pests and investigation of the plant pathology field on the basis of nanobiotechnology is challenging.

5.3 Categories of nanopesticides

The research in nanotechnology has led to the development of different nanoformulation which can be applied in crop protection viz., nano-insecticides, nanoherbicides, nano-fungicides and nano-nematicide. Nano-pesticides are formulated according to their intended purpose as formulations improving solubility, slow release of active ingredients, prevent degradation etc. For achieving these purposes, modifications in the chemical nature carrier molecule have been modified and classified as organic polymer-based formulations, lipid-based formulations, nanosized metals and metal oxides, clay based nanomaterials etc. There are many types of nanopesticides. The nanomaterial is usually used as a carrier for the actual active ingredient in emulsions, capsules, gels, fibers and particles. Nanoparticles can also act alone as nanopesticides (e.g. Copper) (Table 23) (Christoph Steinbach, 2019).

In addition to the above formulations, solid nanoparticles can also be used as nanopesticides. The inert dusts, such as silica, alumina, and clays cause damage to the wax coating on the insect cuticle through both sorption and abrasion. This physical damage can cause to lose water and thus resulting in dehydration to the insect (Rajna et al., 2019). Nano-silica particle can be suggested as an eligible candidate among solid nanoparticles due to its additional benefit to aid in increasing tolerance to abiotic and biotic stresses by the plant. However, high dosages have to be applied for a better result which can adversely affect the grain properties. Nanostructures alumina was also reported to have pesticide properties. Moreover, nano-clays developed from montmorillonite were also shown to have pesticidal function but low toxicity.

Table 23: Types of nanopesticides

Type	Relation to ′nano′	Benefits
Nanoemulsions	The active ingredient is located in nanoscale oil droplets that float in an aqueous solution.	Improved distribution and increased uptake by organisms.
Nanocapsule	The active ingredient is packed in nanoscale capsules.	Increased uptake and thus increased effectiveness against pests.
Nanogel	The active ingredient is distributed in a gel, which consists of nanoscale building blocks.	Easier handling, decreased evaporation of active ingredient.
Nanofibres	The active ingredient is incorporated in a nanoscaled fibre.	Improved distribution, slow release of active ingredient.
Liposomes	Liposomes are in nano sized particles.	Slow-release of active ingredient.
Lipid nanoparticles	Nanoparticles composed of lipids that carry the active ingredient.	Slow release of the active ingredient, protection from degradation.
Inorganic nanoparticles	Inorganic nanoparticles as active ingredient.	Improved distribution, high effectiveness of active ingredient against pests.
Inorganic nanoparticles associated with organic active ingredient	Inorganic nanoparticles act as carriers of organic active ingredient.	Enables slow release of the active ingredient.

5.4 Risks of nanopesticides for the environment

Nanomaterials in nanopesticides are deliberately released into the environment due to the nature of their application. Therefore, benefits and risks to humans and the environment must be considered carefully. The EU Biocidal Products Regulation states that every manufacturer must prove that a pesticide is safe before admitting a product. Since 2013, nanomaterials have also been explicitly included in the regulation. Despite the many ideas on how nanomaterials can be used to improve conventional pesticides, there are currently no nanopesticide products on the market. On the one hand, the development and approval of a pesticide is very tedious and expensive, it is currently unclear how great the savings by using a nanopesticide compared to a conventional pesticide really is (Christoph Steinbach, 2019). The current understanding is not yet sufficient to reliably assess the benefits and risks of nanopesticides. Further studies on the fate and risks of nanopesticides in the environment are thus needed.

In the last few years, application of nanotechnology in agriculture has created new opportunities for developing nanosized agrochemicals that have the potential to improve efficiency, enhance stability, prolong the effective duration and at the same time reduce environmental loads (Xiaoxia Zhang et al., 2019). Nanosized agrochemicals can be classified into two kinds: nanosized active ingredient (AI) and molecular AI-nanocarrier complexes, wherein AI is the key component for agrochemicals. Among nanosized agrochemicals available in the market, copper(II) hydroxide ($Cu(OH)_2$) nanopesticide (i.e., nanosized $Cu(OH)_2$ as the AI) is being used instead of conventional copper(II) sulfate pesticide ($CuSO_4$), which would cause an increase of annual input amounts of $Cu(OH)_2$ nanopesticide. Considering the special physicochemical properties of nanosized AI, it is of utmost importance to assess the impacts of $Cu(OH)_2$ nanopesticide on the environment prior to its widespread utilization. $Cu(OH)_2$ nanopesticides and organic insecticides are continuously applied to soil at a temporal interval, while knowledge about the impact of $Cu(OH)_2$ nanopesticides on organic insecticides degradation is currently scarce, resulting in poorly comprehensive evaluation of the potential environmental risks of $Cu(OH)_2$ nanopesticides.

Degradation of organic pollutants in soil is generally affected by bacterial communities. To date, effects of metal-based nanomaterials like Ag-, ZnO, CuO-, CeO_2- and TiO_2-NPs on soil bacterial communities have been extensively reported.

5.5 Nanoparticle-enabled Smart Delivery Options

Nanotechnology is considered as one of the key technologies in the twenty-first century that promises to advance traditional agricustural practices and offer sustainable development by improving the management and conservation tactics with reduced waste of agricultural inputs (Yifen Shang et al., 2019). The delivery systems of agrochemicals and organic molecules including transport of DNA molecules or oligonucleotides into the plant cells are important aspects of sustainable agricultural production as well as precision farming. In conventional methods, agrochemicals are generally applied to crops by spraying and/or broadcasting. As a result, a very low number of agrochemicals reaches the target sites of crops, which is much below the minimum effective concentration required for successful plant growth. The losses are due to leaching of chemicals, degradation by photolysis, hydrolysis and also by microbial degradation. For instance, in the case of the application of fertilizer, more emphasis should be made of the bioavailability of nutrients due to the chelation effects of soil, degradation by microorganisms, evaporation, over-application, hydrolysis, and run-off problems.

In the case of pesticide applications, the efficacy enhancement with spray drift management is to be focused on. In order to ensure eco-friendly agricultural practices, the recent advancement of nanotechnology-based synthesis of slow or controlled release fertilizers, pesticides and herbicides has, therefore, received an extra attention in agriculture farming. With the passage of time, nanotechnology has gradually moved from the lab-based experimental trials to practical applications. The aim of controlled delivery techniques is to release measured amount of necessary and sufficient quantities of agrochemicals over a period of time and to obtain the full biological competency with minimizing the loss and harmful effects (Yifen Shang et al., 2019). Nanoparticles offer the advantages of effective delivery of agrochemical due to their large surface area, easy attachment and fast mass transfer. For these reasons, micronic or submicronic particles are incorporated into the agrochemicals through several mechanisms, such as capsulation, absorption, surface ionic or weak bond attachments and entrapment into the nano-matrix of active ingredients. Apart from the agricultural applications, the convergence of nanotechnology with biotechnology also provides opportunities as new tools of molecular transporter to modify genes and even produce new organisms. For instance, nanobiotechnologies implicate nanoparticles, nanocapsules, and nanofibres to carry foreign DNA and the chemicals that facilitate to modify the target genes. During the delivery of genetic materials, viral gene delivery vectors face numerous challenges, such as limited host range, limited size of inserted genetic material, transportation across the cell membrane and also trafficking problem of the nucleus (Yifen Shang et al., 2019). In contrast, the recent breakthroughs in nanobiotechnology provide greater opportunities to the researchers to completely replace the genetic material of one species by another. In genetic engineering, silicon dioxide nanoparticles have been devised to deliver DNA fragments/sequences to the target species, such as tobacco and corn plants without any undesirable side effects. In addition, NP-assisted delivery system is also used to develop insect resistant novel crop varieties.

The application of nanotechnology for the controlled delivery of pesticides, herbicides and fertilisers provides new approaches to improve safety of products with increasing the efficiency of production and decreased pollution (Mehrazar, et al., 2015). The development of nanodevices such as smart delivery systems to target specific sites, also nanocarriers for chemical controlled release are currently important aspects in novel agriculture.

Nanoscale carriers can be utilized for the efficient delivery of fertilizers, pesticides, herbicides, plant growth regulators, etc. The mechanisms involved in the efficient delivery, better storage and controlled release include: encapsulation and entrapment, polymers and dendrimers, surface ionic and weak bond attachments among others (Allah Ditta, 2012). These mechanisms help improve stability against degradation in the environment and ultimately reduce the amount to be applied, which reduces chemical runoff and alleviates environmental problems. These carriers can be designed in such a way that they can anchor the plant roots to the surrounding soil structure and organic matter. This can only be possible through the understanding of molecular and conformational mechanisms between the delivery nanoscale structure and targeted structures and matter in soil. These advances will help in slowing the uptake of active ingredients, thereby reducing the amount of inputs to be used and also the waste produced.

Many nanoscale carriers, including encapsulation and entrapment, polymers and dendrimers, surface ionic and weak bond attachments and other mechanisms may be used to store, protect, deliver, and release by control of intended payloads in crop production processes. One of the advantages of nanoscale delivery vehicles in agronomic applications is its improved stability of the payloads against degradation in the environment, thereby increasing its effectiveness while reducing the amount applied. Environmental consideration including precision farming can reduce pollution to a minimum. The advancement in nanotechnology- enabled gene sequencing is expected to introduce rapid and cost effective capability within a decade (Hongda Chen and Rickey Yada, 2011).

An interesting and fascinating area of nanoparticles is 'smart' or targeted drug delivery in the biological system. This is being vigorously pursued in cancer treatment (Kukowska-Latallo et al., 2005). Use of nanoparticles in 'smart' delivery was postulated as 'magic bullets' more than a century ago by P. Ehrlich (Himmelweit, 1960). In order to develop smart treatment-delivery system in plant, Gonza′lez-Melendi et al. (2008) worked with *Cucurbita pepo*, which were treated with carbon-coated Fe nanoparticles in vitro. The magnetic core consisting of Fe nanoparticles allow themselves to be guided to a place of interest in the body (affected part) of an organism using small magnets that create a magnetic field. The carbon coating provides biocompatibility and acts as a surface for adsorption where various types of molecules of interest (drug/DNA/chemical/enzyme) can be adsorbed. Gonza′lez-Melendi et al. (2008) was the first to report the penetration and transport of nanoparticles inside whole plant. These results indicate the possibility and potential of nanoparticles in delivery of substances inhibitory to various plant pathogens.

Technologies such as encapsulation and controlled release methods have revolutionised the use of pesticides and herbicides. Many companies make formulations which contain nanoparticles within the 100-250 nm size range that are able to dissolve in water more effectively than existing ones (thus increasing their activity). Other companies employ suspensions of nanoscale particles (nanoemulsions), which can be either water or oil-based and contain uniform suspensions of pesticidal or herbicidal nanoparticles in the range of 200- 400 nm. These can be easily incorporated in various media such as gels, creams, liquids etc, and have multiple applications for preventative measures, treatment or preservation of the harvested product (Tiju Joseph and Mark Morrison, 2006).

Nanotechnology enables companies to manipulate the properties of the outer shell of a capsule in order to control the release of the substance to be delivered. 'Controlled release' strategy is highly prized in medicine since it allows drugs absorbed more slowly, at a specific location in the body or at the say-so of an external trigger. Pesticides containing nano-scale active ingredients are already in the market and many of the world's leading agrochemical firms are conducting research on the development of new nano-scale formulations of pesticides. Nanoparticles comprising a crop protection agent involve an active ingredient whose ideal particle size is between 10 and 150 nm (Chinnamuthu and Murugesa Boopathi, 2009). The advantage of nano-formulation is that the pesticide dissolves more easily in water; it is more stable and the killing-capacity of the chemical (herbicide, insecticide or fungicide) is optimized.

Nanopesticides cover a wide variety of products, some of which are already on the market. They cannot be considered as a single entity; rather such nanoformulations combine several surfactants, polymers (organic), and metal nanoparticles (inorganic) in the nanometer size range. The lack of water solubility is one of the limiting factors in the development of crop-protecting agents. Microencapsulation has been used as a versatile tool for hydrophobic pesticides, enhancing their dispersion in aqueous media and allowing a controlled release of the active compound. Controlled-release (CR) formulations of carbofuran are in market. Nanomaterials including polymeric nanoparticles, iron oxide nanoparticles, gold nanoparticles, and silver ions have been exploited as pesticides. Researchers have reported various aspects of nanoparticle formulation, characterization, effect of their characteristics, and their applications in management of plant diseases (Bhupinder singh Sekhon, 2014). Nanoencapsulation helps slow release of a chemical to the particular host for insect pest control through release mechanisms that include dissolution, biodegradation, diffusion, and osmotic pressure with specific pH. particles suspended in water have been used since at least 1931, in a product known as Bouisol as fungicide in the growing of grapes and fruit trees.

In the research and development stage, nanosized agrochemicals or nanoagrochemicals are mostly nano-reformulations of existing pesticides and fungicides. Nanoformulations are generally expected to increase the apparent solubility of poorly soluble active ingredients, to release the active ingredient in a slow/targeted manner, and/or to protect against premature degradation. Nanopesticides offer a way to both control delivery of pesticide and achieve greater effects with lower chemical dose. Many companies make formulations that contain nanoparticles within the 100–250 nm size range that are able to dissolve in water more effectively than existing ones, thus increasing their activity (Kumar et al., 2010).

As a potential "green" nanocarrier of pesticides, novel amphiphilic self-assembled octahydrogenated retinoic acid-glycol chitosan (OR-GC) nanoparticles have been synthesized and characterized, which can effectively load the poorly water-soluble pesticide AZA (about 74%). Compared with free AZA in organic solvents, the AZA-OR-GC nanoparticles showed favorable inhibition of cell proliferation and sustained drug release performance in Sl-1 cells *in vitro* (Wei Lu et al., 2013).

Nanotechnology- enabled delivery of pesticides: Nanoscale carriers can be utilized for the efficient delivery of fertilizers, pesticides, herbicides, plant growth regulators, etc. The mechanisms involved in the efficient delivery, better storage and controlled release include: encapsulation and entrapment, polymers and dendrimers, surface ionic and weak bond attachments among others (Allah Ditta, 2012). These mechanisms help improve stability against degradation in the environment and ultimately reduce the amount to be applied, which reduces chemical runoff and alleviates environmental problems. These carriers can be designed in such a way that they can anchor the plant roots to the surrounding soil structure and organic matter. This can only be possible through the understanding of molecular and conformational mechanisms between the delivery nanoscale structure and targeted structures and matter in soil. These advances will help in slowing the uptake of active ingredients, thereby reducing the amount of inputs to be used and also the waste produced.

Many nanoscale carriers, including encapsulation and entrapment, polymers and dendrimers, surface ionic and weak bond attachments and other mechanisms may be used to store, protect, deliver, and release by control of intended payloads in crop production processes. One of the advantages of nanoscale delivery vehicles in agronomic applications is its improved stability of the payloads against degradation in the environment, thereby increasing its effectiveness while reducing the amount applied. This reduction helps address agricultural chemicals run-off and alleviate the environmental consequence. The nanoscale delivery vehicles may be designed to "anchor" to plant roots or the surrounding soil structures and organic matter if molecular or conformational affinity between the delivery nanoscale structure and targeted structures and matters in soil could be utilized. Controlled release mechanisms allow the active ingredients to be slowly taken up, hence, avoiding temporal overdose, reducing the amount of agricultural chemicals used, and minimizing the input and waste. Environmental consideration including precision farming can reduce pollution to a minimum. The advancement in nanotechnology- enabled gene sequencing is expected to introduce rapid and cost effective capability within a decade (Hongda Chen and Rickey Yada 2011).

5.6 Nanoformulations

Nanoparticles can be used in the preparation of new formulations like pesticides, insecticides, insect repellents, pheromones and fertilizers (Rashmi Urkude, 2019.). Nanomaterial offer a wider specific surface area crucial for the sustainable development of agriculture systems The potential uses and benefits of nanotechnology in all stages of production, processing, storing, packaging and transport of agricultural products are enormous especially the recent advancement of nanotechnology-based synthesis of slow or controlled release pesticides is reported to be efficient over other pesticide delivery vehicles. One of the key aspects of precision farming is the delivery of agrochemicals to plants using control delivery systems. The aim of controlled delivery techniques is to release measured amount of necessary and sufficient quantities of agrochemicals over a period of time and to obtain the full biological competency with minimizing the loss and harmful effects.

Nano-pesticides are formulated according to their intended purpose as formulations improving solubility, slow release of active ingredients, prevent degradation etc. (Rajna et al., 2019). For achieving these purposes, modifications in the chemical nature carrier molecule have been modified and classified as organic polymer-based formulations, lipid-based formulations, nanosized metals and metal oxides, clay based nanomaterials etc. Some foremost nano-formulations are mentioned below.

i. **Nano-emulsions:** Generally an oil-in-water (O/W) emulsion is more common as a nano-emulsion where, active ingredient of the chemical is dispersed as nanosized droplets in water, with surfactant molecules confined at the pesticide-water interface. Nano-emulsions get further classified based on the quantity and type of surfactants, as thermodynamically stable and kinetically stable. If the pesticide is partially soluble in the aqueous phase and spontaneous formation of a stable emulsion happens when surfactant, pesticide, and water components are brought together, that is a thermodynamically stable nano-emulsion. The insolubility of the active ingredient make the pesticide and surfactant to initially form a two-phase system and thus, a continuous shearing make them to mix together and pesticides droplets in the nano-emulsion will remain dispersed for an extended period of time and so are considered to be kinetically stable. Eg: Oil in water nanoemulsion of neem oil has been developed for insect management using Tween 20 as the surfactant.

ii. **Nano-suspension:** Nano-suspensions, also termed as nano-dispersions are formulated by dispersing the pesticide as solid nanosized particles in aqueous media. In nanodispersions, the surfactant molecules get confined at the particle surface where polar portions extending into the aqueous solution and the non-polar portions associating with the solid pesticide. Eg: Aqueous dispersions of nano-permethin, novaluron and β cypermethrin have been developed by researchers.

iii. **Polymer based nano-particles:** Polymer-based pesticide nanocarriers are majorly deployed in the slow and controlled release of active ingredients to the target site. Moreover, they can serve to improve dispersion in aqueous media and also as a protective reservoir. Nano-encapsulation, nano-spheres, nano-gels, nano-fibers, etc are some of them falling in this category.

iv. **Nano-encapsulation:** Nano-capsules or nano-encapsulation are heterogeneous reservoir type structure containing an inner central cavity which confines the hydrophobic or hydrophilic active ingredient, surrounded by a polymer coating or membrane. The active ingredient in neem-azadiractin formulation can be protected through this formulation. Eg: Controlled-release nano-formulation of the neonicotinoid insecticide i.e., acetamipirid and imidacloprid have been developed.

v. **Nanospheres:** These are homogeneous vesicular structures, in which the bioactive ingredient is uniformly dispersed throughout the polymer matrix. Eg: Polymer stabilized bifenthrin nanoparticles are developed as nanospheres.

vi. **Nanogels:** These are also known also hydrogel nanopartciles. These are formulated by cross linking of polymeric particles having hydrophilic groups, thus absorb higher quantities of water. Chitosan nanogel is an example for this.

vii. **Nano-fibres:** Nano-fibres are developed through electrospinning, thermal induced phase separation. Researchers have developed electospun nano-fibers loaded with the chemical, (Z)-9-dodecenyl acetate, an ingredient of pheromone which get embedded in the polymer matrix for the management of many lepidopteran insect pests.

Nanoencapsulation: Nanoencapsulation is defined as the technology of packaging nanoparticles of solid, liquid, or gas, also known as the core or active, within a secondary material, named as the matrix or shell, to form nanocapsules (Fig. 45) (Cano-Sarabia and Maspoch,2012). The core contains the active ingredient (drugs), while the shell isolates and protects the core from the surrounding environment. This protection can be permanent or temporal, in which case the core is generally released by diffusion or in response to a trigger, such as shear, pH, or enzyme action, thus enabling their controlled and timed delivery to a targeted site.

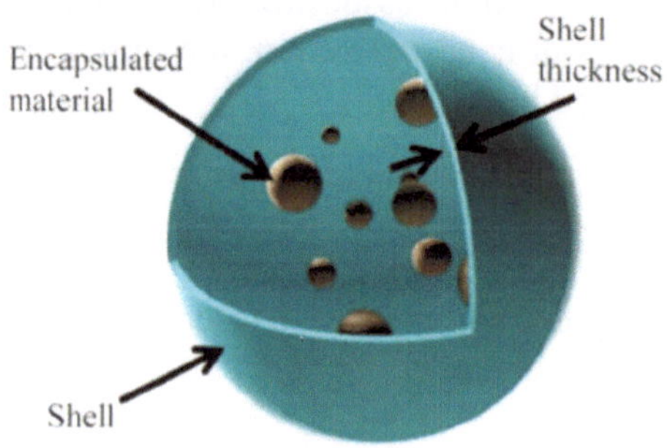

Fig. 45: Nanoencapsulation

Three general classes of microencapsulation exist, *viz.*, nanoencapsulation 30-200 nm, microencapsulation 0.1-1000 μm and macroencapsulation >1 mm. Microencapsulated pesticides are mixed with water and sprayed in the same manner as other sprayable formulations. After spraying, the capsule wall breaksdown and slowly releases the active ingredient. Breakdown of the microencapsulated materials to release the pesticide sometimes depends on weather conditions. Under certain conditions, the microencapsulated materials may break down more slowly than expected. This could leave higher residues of pesticide active ingredient in treated areas beyond normal restricted-entry or harvest intervals with the potential to injure fieldworkers. For this reason, regulations require long restricted-entry intervals for some microencapsulated formulations. The cost of microencapsulation can vary considerably and is largely dependent upon the technique employed. Some techniques require the use of specialized equipment, whilst others do not. Some techniques used expensive process chemicals, whilst others use very cheap ingredients. Processes which involve heat are always more expensive than those which do not. Removal of the continuous phase to produce a "dry" product will require an extra processing step adding to the cost. Some products, particularly those of high value or low volume, are better able to absorb such an increased cost. To economically encapsulate high volume products or those which produce a low profit margin it is necessary to employ one of the cheaper techniques if this is possible for the application in question.

Pesticides via encapsulation is a more sophisticated approach to formulating nanoscale pesticides involves encapsulation -packaging the nano-scale active ingredient within a kind of tiny 'envelope' or 'shell'. Both food ingredients and agrochemicals in microencapsulated form have been on the market for several decades. According to industry, reformulation of pesticides in microcapsules has triggered 'revolutionary changes', including the ability to control under what conditions the active ingredient is released. According to the agrochemical industry, re-formulating pesticides in microcapsules can also extend patent protection, increase solubility, reduce the contact of active ingredients with agricultural workers and may have environmental advantages such as reducing run-off rates.

Nanotechnology operates at the same scale as a virus or disease-infecting particle, and thus holds the potential for very early detection and eradication. Smart treatment delivery systems are envisioned for biology and bioactive systems such as drugs, pesticides, nutrients, probiotics, nutraceuticals and implantable cell bioreactors (Chinnamuthu and Murugesa Boopathi, 2009). Experts in agriculture and nanoscience identified some of the important areas in the field of agriculture that have enormous potential for the application of nanotechnology. Some of the products like nano pesticides have arrived already in the market, while many others are under developing stage and it may take many years before they are commercialized. These applications are largely intended to address some of the limitations and challenges like problem weed management, slow release fertilizers, conditional release of pesticides and herbicides, precise micro-management of soils, the more efficient and targeted use of inputs and new toxin formulations for pest control .

Microencapsulation-like nanoencapsulation is used to develop the quality of products of desired chemicals delivery to the target biological process. Recently, few chemical companies openly promote nanoscale pesticides for sale as "microencapsulated pesticides." Some products from Syngenta (Switzerland) such as Karate ZEON, Subdue MAXX, Ospray's Chyclla, Penncap-M, and microencapsulated pesticides from BASF may fighting fit for nanoscale (Ram Prasad et al., 2018). Syngenta also markets in the Australia some products such as the Primo MAXX, Banner MAXX, Subdue MAXX, etc. Despite they are known as microemulsions in the market, however, they are really nanoscale emulsions. It confirms very thin interface between the term of microemulsion and nanoemulsion. This technique is commonly used for formulations of organic NPs containing active agrochemicals or substances of interest.

Nano-based viral diagnostics, including multiplexed diagnostic kit development, have taken momentum in order to detect the exact strain of virus and stage of application. Detection and utilization of biomarkers that accurately indicate disease stages with differential protein production in both healthy and diseased states lead to the identification of the development of several proteins during the infection cycle. These nano-based diagnostic kits not only increase the speed of detection but also increase the power of the detection. These nano and micro-formulations are being developed and patented by agribusiness and food corporations such as Monsanto, Syngenta and Kraft (Chinnamuthu and Murugesa Boopathi, 2009).

Pesticides containing nano-scale active ingredients are already in the market, and many of the world's leading agrochemical firms are conducting research on the development of new nano-scale formulations of pesticides. For example, BASF of Germany, the world's fourth ranking agrochemical corporation, recognizes nanotech's potential usefulness in the formulation of pesticides. BASF is conducting basic research and has applied for a patent on a pesticide formulation, "Nanoparticles Comprising a Crop Protection Agent," that involves an active ingredient whose ideal particle size is between 10 and 150 nm. The advantage of nano-formulation is that the pesticide dissolves more easily in water; it is more stable and the killing-capacity of the chemical (herbicide, insecticide or fungicide) is optimized. Bayer Crop Science of Germany has also applied for a patent on agrochemicals in the form of an emulsion in which the active ingredient is made up of nanoscale droplets in the range of 10-400 nm. The company refers to the invention as a "microemulsion concentrate" with advantages such as reduced application rate, "a more rapid and reliable activity" and "extended long-term activity".

Like Bayer Crop Science, Syngenta claims that its Primo MAXX Plant Growth Regulator and Banner MAXX fungicide are extremely small particle size of about 100 nm (or 0.1 micron) prevents spray tank filters from clogging, and the chemicals mix so completely in water that they neither settle in the spray tank not separate from water for up to one year. Additionally the fungicides will be absorbed into the plant's system and cannot be washed off by rain or irrigation.

Encapsulating control nanotechnology enables companies to manipulate the properties of the outer shell of a capsule in order to control the release of the substance to be delivered. 'Controlled release' strategies are highly prized in medicine since they can allow drugs to be absorbed more slowly, at a specific location in the body or at the say-so of an external trigger (Anonymous, 2004). With potential applications across the food chain (in pesticides, vaccines, veterinary medicine and nutritionally-enhanced food), these nano- and micro-formulations are being developed and patented by agribusiness and food corporations such as Monsanto, Syngenta and Kraft.

Examples of Nano and Microcapsule designs

Slow release: The capsule releases its payload slowly over a longer period of time (e.g., for slow delivery of a substance in the body).

Quick-release: The capsule shell breaks upon contact with a surface (e.g. when pesticide hits a leaf) Specific release: The shell is designed to break open when a molecular receptor binds to a specific chemical (e.g., upon encountering a tumour or protein in the body).

Moisture release : The shell breaks down and releases contents in the presence of water (e.g., in soil) 48 Heat-release: The shell releases ingredients only when the environment warms above a certain temperature.

pH release: Nanocapsule breaks up only in specific acid or alkaline environment (e.g., in the stomach or inside a cell).

Ultrasound release: The capsule is ruptured by an external ultrasound frequency.

Magnetic release: A magnetic particle in the capsule ruptures the shell when exposed to a magnetic field.

DNA nanocapsule: The capsule smuggles a short strand of foreign DNA into a living cell which, once released, hijacks cell machinery to express a specific protein (used for DNA vaccines).

Advantages of encapsulation: Longer-lasting biological activity , less soil binding for better control of pests in soil, reduces worker exposure, improves safety by removing flammable solvents, reduces damage to crops, less pesticide lost by evaporation, less effect on other species, reduced environmental impact, prevents degradation of active ingredients by sunlight and makes concentrated pesticide safe and easy to handle by growers.

Concerns raised by encapsulation: Both biological activity and environmental/ worker exposure can be longer-lasting; beneficial insects and soil life may be affected, could nano-scale pesticides be taken up by plants and smuggled into the food chain?, pesticides can be more easily aerosolized as a powder or droplets – therefore inhale-able, and perhaps a greater threat to human health and safety, could pesticides formulated as nanocapsules or nano-scale droplets exhibit different toxicity and enter the body and affect wildlife through new exposure routes, for example, across skin, potential for use as a bioweapons delivery vehicle., what other external triggers might affect the release of the active ingredient (e.g., chemical binding, heat or break down of the capsule)?, microcapsules are similar in size to pollen and may poison bees and/or be taken back to the hives and incorporated in honey (Anonymous, 2004). Because of their size, "microencapsulated insecticides are considered more toxic to honey bees than any formulation so far developed."55 Will nanocapsules be more lethal?, It is not known how 'unexploded' nanocapsules will behave in the human gut if ingested with food.

Implications of encapsulation for Nanobioweaponry: Nanocapsules and microcapsules make an ideal vehicle for delivering chemical and biological weapons because they can carry substances intended to harm humans as easily as they can carry substances intended to kill weeds and pests. By virtue of their smallsize, DNA nanocapsules may be able to enter the body undetected by the immune system and then become activated by the cells' own mechanisms to produce toxic compounds. The increased bioavailability and stability of nano- encapsulated substances in the environment may offer advantages to the Gene Giants, but the same features could make them extremely potent vehicles for biological warfare. In addition, because of their increased bioavailability only a small quantity of the chemical is needed. When programmed for external triggers such as ultrasound or magnetic frequencies, activation can be controlled remotely, suggesting a number of grim scenarios. Could agrochemical/ seed corporations remotely activate triggers to cause crop failure if the farmer infringes the company's patent or fails to follow prescribed production practices? What if nanocapsules containing a potent compound are added to a regional water supply by a foreign aggressor or terrorist group?

5.7 Fungicide Nanoencapsulation

i. Propiconazole: Using nanoencapsulation in fungicide delivery has created many opportunities for safe application of for example, nanoencapsulation of fungicide to obtain agrochemical delivery systems, involving the formation of fungicide -loaded particles or –entrapped particles having a diameter within the nanorange (Felipe Barrera-Méndez et al., 2019). According to the definition of nanoparticle, this size range should be between 1 to 100 nm in at least one dimension. Nevertheless, the term "nano" for encapsulated pesticides may be referred to a particle size bigger than 100 nm. This may be based on the efficacy and applications of their novel small-sized particulate and not only in their size. The large surface area obtained increases the affinity to the target species and reduces the pesticide required for pest control. Nanocarrier materials, such as biodegradable polymers, also protect the active ingredient from premature degradation and allow them to be released in a controlled way . Polylactic acid (PLA) and Poly(lactic-co-glycolic) acid (PLGA) are two polymers that have been used as nanocarriers for delivery systems, since their biodegradation by- products are non-toxic . Previous studies suggest that the biodegradation in soil of these polymers could take up to several months , leaving the molecular diffusion as the dominant mechanism of fungicide liberation to the medium. An attempt was made to obtain a nanoencapsulated propiconazole delivery system to reduce costs in the treatment of the Fusarium dieback disease (Fusarium euwallaceae) on avocado.

The nanoencapsulation of pesticides in biodegradable polymers confers several advantages to conventional agrochemicals, such as protection against losses due volatilization and degradation of the active ingredient, as well as the augment of water dispersion, allowing for their application on crops without requiring the use of organic solvents that could harm the user and the environment (Felipe Barrera-Méndez et al., 2019). These characteristics could enhance the productivity, reducing both costs and environmental pollution. In this work the propiconazole fungicide, forming part of a commercial formulation as well as in its pure state, was encapsulated using as carriers the biodegradable polymer poly lactic acid (PLA) and the biodegradable co-polymer poly (lactic- co-glycolic) acid, all with the aim of generate systems of controlled release of fungicide to augment the efficiency of the treatments of the Fusarium dieback disease.

The most efficient system obtained presented nanospheres of 146.28 nm and an encapsulation efficiency over 42%. The antifungal activity tests showed that the use of this nanoencapsulated fungicide system enhances the growth inhibition percentage in 5%, obtaining a formulation that presents good dispersion in water without the need of organic emulsifier agents. The release profile test showed that while the commercial formulation dialyzed 50 and 90% of the propiconazole in approximately 5 and 30 hours, respectively, it took the nanoencapsulated propiconazole almost twice as much time to dialyze the same percentage of load, effectively prolonging the availability of the fungicide without diminishing the antifungal activity.

ii. Fluazinam : Two kinds of poly (lactic-co-glycolic acid)(PLGA) nanoparticles (NPs) were prepared with polyvinyl alcohol (PVA) and sodium dodecyl sulfate (SDS) surfactants (Qi Wang et al., 2019). These were used to encapsulate the fungicide fluazinam (Flu) against *Rhizoctonia solani* using the Shirasu Porous Glass (SPG) membrane emulsification method. Both nanoparticles had uniform spherical shapes with average diameters of 314.13 nm (SDS) and 612.80 nm (PVA). The slow-release microspheres had excellent sustained- release properties, resistance to UV degradation, storage stability, leaf surface coverage and antifungal efficacy compared to the commercial formulation.

iii. Nanoencapsulation to enhance the antifungal activity of essential oils (EOs): Essential oils represent promising agents for reducing food decay and development of pathogen microorganisms. The overall aim of the present study was to encapsulate Iranian thyme (*Thymus daenensis* L) and dill (*Anethum graveolens* L.) EOs in copper nanoparticles (NPs) and to evaluate their antifungal activity against the phytopathogen, *Colletotrichum nymphaeae* (Weria Weisany et al., 2019). Encapsulation of thyme and dill EOs with copper NPs resulted effective in reducing the mycelium growth even by 90% after 9 days treatment. Furthermore a strong inhibitory effect on conidia germination of *C. nymphaeae* was observed. SEM observations indicated that *C. nymphaeae* treated with EOs, copper NPs and encapsulated EOs, showed deformation in mycelial growth and hyphae twisting. In summary, these results showed that nano-encapsulation of EOs with NPs, using safe materials, increased the perspective of their effectiveness as new plant disease management strategies.

iv. Nano formulation of Tricyclazole: A novel nano formulation of Tricyclazole, a systemic fungicide was developed to improve the quality of rice production by polymeric encapsulation. Polypropylene glycol was used as encapsulating agent. The size distribution of polypropylene encapsulated Tricyclazole was obtained by pade laplace dispersion and was noted at 50-60nm.The characterization was carried by using Fourier Transform infra-red analysis and Electron Microscope analyses. The anti-fungal and anti-bacterial assay was carried and the comparison with commercial pesticide Nano encapsulated Tricyclazole showed a better bio-efficacy

Preparation of nano- tricyclazole: Tricyclazole was fully grounded in a mortar. 200 ml of water solution of pesticide sample (1gm of grounded pesticide was dissolved in 200 ml of water) and 200ml of Polypropylene glycol were mixed in an ultrasonic bath for 40 minutes. This process would help to disperse the pesticide particles in polypropylene glycol. The sonicated solution was subjected to continuous stirring for 8 hours at 1500rpm.

Instrumentation: Particle size distribution was obtained by Dynamic light scattering (DLS) VASCO, CORDOUAN. Fourier transform infrared (FTIR) (Bruker) spectroscopy was used to characterize and identify the organic molecule. Surface topology was determined through scanning electron microscopy (Zeiss Evol8). Morphology and structure of nano Tricyclazole was studied using Transmission electron microscopy (Jeol JEM 2100, Tokyo, Japan) capable of point to point resolution.

Application of nano-tricyclazole on Rhizoctonia solani: Rhizoctonia Solani is responsible for several plant diseases including Brown patch, sheath blight of rice, belly rot of cucumber, red rot of sugarcane. In one litre of distilled water 39 grams of potatoe Dextrose Agar (Himedia-MO96) was suspended, sterilized at 121 °C for 15 minutes. The sterilized potato Dextrose Agar solution was allowed to cool in water both at 55 – 60 °C. The nano-Tricyclazole stock solution was mixed with potato Dextrose Agar solution. Control was made with only distilled water. The mixtures of nano Tricyclazole and potato Dextrose Agar solutions were poured directly into petridishes and allowed to cool at room temperature. Freshly made fungal suspension of *Rhizoctonia Solani* was inoculated aerobically at 35 °C. In the petri dish containing nano Tricyclazole the growth of *R.solani* was inhibited upto a maximum extent where as control containing only distilled water shows growth of *R. solani.*

v. Nanoencapsulation of *Zataria multiflora* essential oil: A study was undertaken to investigate the nanoencapsulation of *Z. multiflora* essential oil (ZEO) in chitosan nanoparticles (CSNPs) in order to enhance antifungal activity and stability of the oils against one isolate of *Botrytis cinerea*, the causal agent of gray mould disease (Mohammadi Ali et al., 2015). ZEO was encapsulated by an ionic gelation technique into CSNPs with an average size of 125–175nm as observed by transmission electron microscopy (TEM). From UV-vis spectrophotometry results, the drug encapsulation and loading efficiency of ZEO decreased from 45.24% to 3.26% and from 9.05% to 5.22%, respectively, upon increasing initial ZEO content from 0.25 to 1g/g chitosan. *In vitro* release studies also demonstrated a controlled and sustained release of ZEO for 40days. The superior performance of ZEO when encapsulated by CSNPs under both in vitro and in vivo conditions in comparison with unmodified ZEO against *B. cinerea* was revealed. The *in vivo* experiment also showed that the encapsulated oils at 1500ppm concentration significantly decreased both disease severity and incidence of Botrytis-inoculated strawberries during 7days of storage at 4°C followed by 2–3 more days at 20°C. These findings revealed the promising role of CSNPs as a controlled release system for EOs in order to enhance antifungal activities.

Industrial relevance: Application of plant essential oil (EOs) treatment at pre- or postharvest stage has been considered as an alternative treatment to the use.

vi. Chitosan nanoparticles encapsulated with *Cymbopogon martinii* essential oil : The *Cymbopogon martinii* essential oil (CMEO) induced the death of fungi through elevating intracellular reactive oxygen species and lipid peroxidation, and depleting ergosterol content. Regrettably, essential oils are highly volatile and become unstable and lose their biological features on exposure to light, heat, pH, moisture, and oxygen. To overcome these issues, chitosan encapsulated CMEO nanoparticles (Ce-CMEO-NPs) were prepared (Naveen K. Kalagatur et al., 2018). The synthesized Ce-CMEO-NPs have spherical morphology with Zeta potential of 39.3–37.2 mV and their corresponding size was found in range of 455–480 nm. The Fourier transform infrared analysis confirmed that bio-active constituents of CMEO were well stabilized due to chitosan conjugation and successfully formed Ce-CMEO-NPs. The *in vitro* release assay observed that the release of CMEO is stabilized due to the complex formation with chitosan and thereby, increases the lifetime antifungal activity of CMEO by gradual release of antifungal constituents of Ce-CMEO-NPs. In conclusion, antifungal and antimycotoxin activities of CMEO and Ce-CMEO-NPs against *F. graminearum* were assessed in maize grains under laboratory conditions over a storage period of 28 days. Interestingly, Ce-CMEO-NPs were presented efficient and enhanced antifungal and antimycotoxin activities related to CMEO, and it could be due to perseverance of antifungal activity by controlled release of antifungal constituents from Ce-CMEO-NPs. The study concluded that Ce-CMEO-NPs could be highly appropriate as mycobiocides in safeguarding the agricultural commodities during storage period in agricultural and food industries.

vii. Nanoencapsulated methyl salicylate: In view of the suspected negative impact of synthetic fungicides to the human health, nutritional quality and non-targeted organisms, the use of plant-based antifungal agents has gained considerable interest to the agri-food industries (Anupam Kujur et al., 2019). The aim of this study was to explore the antifungal and aflatoxin B1 (AFB1) inhibitory activity of chitosan (low molecular weight) encapsulated methyl salicylate. The nanoencapsulation of methyl salicylate (Ne-MS) has been characterized by SEM, FTIR, and XRD analysis. The encapsulation efficiency and loading capacity ofNe-MS ranged between 32–34% and 5–7% respectively. The minimum inhibitory concentration of Ne-MS (1.00 μL/mL) against the growth and aflatoxin B1 production by *Aspergillus flavus* was found to be lower than the free MS (1.50 μL/mL). Mode of action studies demonstrated that the Ne-MS cause a significant decrease in the ergosterol content, leakage of vital ions (Ca2+, Mg2+, and K+), utilization of different carbon source by the *A. flavus*. Further, the docking result showed ver1 and omt A gene of AFB1 biosynthesis are the possible molecular site of action of methyl salicylate. The *in situ* study revealed that Ne-MS had no significant negative impact on the organoleptic properties of the food system (maize) which strengthen its potential as a biorational alternative of synthetic fungicides.

viii. Chitosan nanocapsules: Chitosan nanoparticles and chitosan nanocapsules incorporated with lime essential oil were synthesized by nanoprecipitation and nanoencapsulation methods, respectively. The samples were characterized by transmission electron microscopy (TEM) and Fourier Transform Infrared Spectroscopy (FTIR), and Z potential was measured (Sotelo-Boyás, et al., 2017). Also, particle size distribution was analyzed by dynamic light scattering (DLS) and the antibacterial activity was studied. According to TEM, the average size of nanocapsules was higher than for nanoparticles. When lime essential oil was incorporated, the particle size increased. Lime essential oil incorporation was evidenced by FTIR. Chitosan nanocapsules showed higher Z potential value compared to chitosan nanoparticles. The antibacterial activity was tested against four food-borne pathogens, being higher for nanoparticles than for nanocapsules. The highest antibacterial activity was observed for chitosan nanoparticles incorporated with lime essential oil applied against *Shigella dysenteriae*, attaining an inhibition halo (IH) value of 3.5 cm for 40 μL of minimum inhibitory volume (MIV). The novelty of incorporating lime essential oil into chitosan nanoparticles and nanocapsules and the study of their enhancing effect on antibacterial activity were shown.

Nanoencapsulation,the hydrophobic partitioning and toxicity of an active ingredient : Given the costs associated with designing novel active ingredients, new formulations focus on the use of other ingredients to modify existing formulations. Nanosized encapsulated pesticides offer a variety of enhanced features including controlled release and improved efficacy (Matthew Slattery et al., 2019). Despite the presence of nanosized capsules in current-use pesticide formulations, the analytical and toxicological implications of encapsulation are uncertain. To explore this issue quantitatively, the capsules of a commercially available encapsulated insecticide formulation (γ-cyhalothrin active ingredient) was fractioned into two size ranges: a large fraction (LF), with an average hydrodynamic diameter (HDD) of 758 nm, and a small fraction (SF), with an average HDD of 449 nm. A novel extraction method was developed demonstrating a time-dependent inhibition of γ-cyhalothrin from capsules for up to 48 h.

Innovations in pesticide formulation are important for manufacturers, who benefit from the economic opportunity associated with pesticide sales, as well as other stakeholders concerned with environmental impacts caused by pesticide use. The advantages of encapsulation technologies should not be understated, and combining these features with the recent developments in nanoenabled products is becoming a fruitful area of research. As these investigations continue, it would be mutually beneficial for risk assessors and pesticide manufacturers to minimize the barriers of proprietary technology so that studies like this can result in more effective products that reduce environmental impacts. With the litany of nanoscale capsule designs available, identifying and describing their effects on active ingredients is an important pathway towards innovative pesticide products. In future studies, attention to the rates of release and desorption of active ingredients from nanoparticle carriers, and subsequent incorporation into risk assessment frameworks, will be a valuable contribution to predicting the fate of nanoenabled pesticides. Besides helping to bolster the pesticide market, keeping pace with the evolution of encapsulated pesticides is vital for regulatory agencies. These modifications can alter the risk profile of pesticide products by disrupting foundational assumptions about chemical behavior, such as the reliance upon a KOW value in active ingredient transport. Given concerns surrounding surface water pesticide contamination, recognizing these nuances may lead to more accurate environmental modeling and highlight scenarios in which other ingredients deserve unique attention.

Solid nanoparticles as nanopesticides: In addition to the above formulations, solid nanoparticles can also be used as nanopesticides (Rajna et al., 2019). The inert dusts, such as silica, alumina, and clays cause damage to the wax coating on the insect cuticle through both sorption and abrasion. This physical damage can cause to lose water and thus resulting in dehydration to the insect. Nano-silica particle can be suggested as an eligible candidate among solid nanoparticles due to its additional benefit to aid in increasing tolerance to abiotic and biotic stresses by the plant. However, high dosages have to be applied for a better result which can adversely affect the grain properties. Nanostructures alumina was also reported to have pesticide properties. Moreover, nano-clays developed from montmorillonite were also shown to have pesticidal function but low toxicity. Among metals, silver, titanium oxide and copper are most preferred as nanoparticles.

The bactericidal and viricidal activity of silver nanoparticles makes them favourable by nanotechnology researchers. The low toxicity, inherent charge, larger surface area and crystallographic structure increase its preference. The use of titanium dioxide to crops has proved effective antimicrobial and antifungal activity. Nano-copper formulations can cause cell wall damage of bacterial cells and found effective against pomegranate bacterial blight at very low concentrations. Cell wall damage was observed in nano-copper treated bacterial cell. From past one decade, considerable research has been made in the field of nanotechnology. Yet, it is only recently we are beginning to recognize how nanoparticles can impact on our lives and lead to both positive and negative results. Research in nanopesticide development is being taken up in a faster rate but, this subject matter has not reached the public awareness or regulatory authorities so far. By and large, innovation all the time results in both benefits and problems for human and environmental health. The current level of knowledge regarding environmental fate does not allow us to a fair assessment of the advantages and disadvantages that will result from the use of nanopesticides.

Nanoencapsulation for smart delivery of pesticides: Nanoencapsulation is another part of nanotechnology that offer variety of desirable features that include reduction in human exposure to active ingredients, controlled release, longer residual concentrations, elimination of organic solvents and increased efficacy (Rashmi Urkude, 2019). Encapsulation formulations have revolutionized the application of pesticides due to the development of nanotechnology in insect pest management as it reduces the doses compared to the conventional pesticide to get maximum effect with more target oriented action of the pesticides. In this method the core material (pesticide) is coated *i.e* encapsulated by capsulation material or shell and the size of the pesticide reduces up to the nano size.

Nanoencapsulation is another part of nanotechnology that offer variety of desirable features that include reduction in human exposure to active ingredients, controlled release, longer residual concentrations, elimination of organic solvents and increased efficacy. Encapsulation formulations have revolutionized the application of pesticides due to the development of nanotechnology in pest management as it reduces the doses compared to the conventional pesticide to get maximum effect with more target oriented action of the pesticides. In this method the core material (pesticide) is coated *i.e.*, encapsulated by capsulation material or shell and the size of the pesticide reduces up to the nano size. In nanoencapsulation, pesticide is slowly and efficiently released for insect pest control. In this process individual particles or droplets of solid or liquid core material are surrounded or covered with a continuous film of polymeric material and pesticide is released only in the targeted environment like in specific pH (e.g., in the stomach or inside a cell), specific temperature, moisture, external ultrasound frequency or in the occurrence of explicit compounds (Rashmi Urkude, 2019). Some encapsulation is done in such way that it gets absorbed in the surface of the plant and facilitates protract release which lasts for longer time compared to conventional pesticides which is washed away in the rain thus prevent frequent use of pesticides.

The incorporation of nanotechnology as a means for nanopesticides is in the early stage of development. The main idea behind this incorporation is to lower the indiscriminate use of conventional pesticides to be in line with safe environmental applications. Nanoencapsulated pesticides can provide controlled release kinetics, while efficiently enhancing permeability, stability, and solubility. Nanoencapsulation can enhance the pest-control efficiency over extended durations by preventing the premature degradation of active ingredients (AIs) under harsh environmental conditions. A review by Sandeep Kumar et al., (2019) critically assessed the significant role of nanotechnology for encapsulation of AIs for pesticides. The smart delivery of pesticides is essential to reduce the dosage of AIs with enhanced efficacy and to overcome pesticide loss (e.g., due to leaching and evaporation). The future trends of pesticide nanoformulations including nanomaterials as AIs and nanoemulsions of biopesticides were also explored.

Packaging insecticides in nanocapsules may make them more toxic: Environmental toxicologist Stacey Harper and her team found that a common insecticide in its "capsule suspension" formulation - with molecules of the active ingredient encapsulated in tiny, inert plastic pellets - was more toxic than the same amount of active ingredient delivered straight up in water. Their study is published in the journal Nanomaterials ("Pesticide Encapsulation at the Nanoscale Drives Changes to the Hydrophobic Partitioning and Toxicity of an Active Ingredient"). They Studied a commercial pyrethroid-type insecticide with an encapsulated active ingredient, gamma-cyhalothrin. The insecticide is primarily used in the home and garden for ants, bed bugs, ticks and other insects. The capsules encasing the product's active ingredient range from micron-sized (a human hair is 40-75 microns thick), to nanometer-sized, a thousand times smaller.

Chemical manufacturers have offered encapsulated formulations of pesticides for more than 50 years, because encapsulation is thought to improve the product's dispersal and durability. It was said that currently, the only testing that's done after the final formulation are hazards like corrosivity and flammability, but not toxicity. They have found that encapsulation makes a difference in toxicity and that it is size-dependent.

5.8 Nano-pesticides v/s Conventional pesticides

- Nanotechnology offers a tool for developing novel formulations of eco-friendly pesticides as majority of nano-pesticide formulations are highly target specific.
- Generally, targeted delivery and controlled release of nano-pesticides can improve pesticide utilization and reduce residue and pollution. For example, Nano-microcapsule formulations have slow release and

protection performance because they have been prepared using light-sensitive, thermo-sensitive, humidity-sensitive enzyme-sensitive and soil pH-sensitive high polymer materials to deliver pesticides.

- Nano-pesticide formulations improve adhesion of droplets on plant surface (reduces drift losses) which intern improves the dispersion and bio-activity of active ingredient (a.i.) of pesticide molecules. Therefore Nano-pesticides will have high efficacy compared to the conventional pesticide formulations (i.e., DDust, G-Granule, P-Pellet, EC-Emulsifiable Concentrate, WP-Wettable Powder, WDG-Water Dispersible Granule etc.) and due to their small size, improvable pesticide droplet ductility wettability and target adsorption when sprayed in fields has made these nano-pesticides provide efficient and environmental friendly advantages.
- Nano-pesticides are extraordinary means for setting up an eco-friendly and sustainable agriculture system because it reduces the overall chemical usage, decreases the toxic residues and enhances the overall crop protection.

The limitations of nano-pesticides

- The risk that nano-particles (nano-pesticides) may pose to human and environment health is not yet fully understood.
- Nano-pesticides may also create new kinds of contamination of soils and waterways since nano-pesticides are apparently much more persistent and have higher degrees of toxicity when compared to their traditional counterparts.
- Hence, a better understanding of the fate and effect of nano-pesticides after their application is required. It is a good thing that all necessary safety precautions are taken before deciding to go ahead and use new technologies on a large scale.

Beneficial improvements of nano-based pesticide formulations: Nanotechnology created novel nano-based pesticide formulations i.e., Nanopesticides. Nanopesticides can be prepared using emulsion, polymer, lipid,ceramics, proteins and metals (Rashmi Urkude, 2019). The common methods for preparation of nanopesticide are by using oil in water emulsion system (microemulsion and nanoemulsion) and the conversion of emulsion to organic nanoparticles by milling, solvent evaporation, coacervation and precipitation techniques. Nano-pesticides can also be developed by directly processing into nanoparticles (nanosized pesticides) or by loading pesticides with nanocarriers in delivery systems. In nano-carrier systems, pesticides are loaded in following manner, encapsulation inside the nanoparticulate polymeric shell, absorption onto the nanoparticle surface, attachment on the nanoparticle core via ligands or entrapment within the polymeric matrix. A variety of nanoformulation types have been developed, including nanoemulsions, nanocapsules, nanospheres, nanosuspensions, solid lipid nanoparticles, mesoporous nanoparticles and nanoclays.

Nanoparticles offer the advantages of effective delivery of agrochemical due to their large surface area, easy attachment and fast mass transfer. Nanoparticles release agrochemicals to plants using control delivery systems compared to conventional pesticide formulation where more than 90% of pesticides run off into the environment and reside in agricultural products in the process of application. The control delivery system facilitate the sustained release of the active ingredient at the target site for an extended period of time thus increasing targeted delivery efficiency of pesticide into action targets such as plants, insects and pathogen, increasing solubility and dispersion for fat-soluble chemicals in aqueous solution, reducing pesticides application and treatment frequency by extended lasting validity period enhancing the bioefficacy, reducing the chemical input to plants, solving the problem of nontarget toxicity and improving chemical stability for lightsensitive compounds by restricting photo-degradation. The sustained release of pesticide protects biodiversity in ecosystem.

5.9 Nanotechnology in pesticides removal from aqueous solutions

There are a broad range of physicochemical properties that make nanomaterial specific candidate and reactive media for pesticides removal. Nanomaterials can also be functionalized with various chemical groups to increase their efficiency for removal of desired target compounds. In recent years, water pollution and pesticide residues in the food chain have become a serious environmental and health hazard problem. Therefore, an efficient technology is essential for complete mineralization of pesticides to non-toxic forms. Nanotechnology offers many potential benefits to improve existing environmental technologies using new materials with effective performance, resulting in less consumption of energy and materials. The aim of this review is to compile and study current publications regarding pesticides removal by nanotechnology. A study by Taghizade Firozjaee et al., (2018) discusses the applications, advantages and limitations various nanotechnology processes for removal of pesticides. An overview of recent advances in nanotechnologies for removal of pesticides in three main capabilities: adsorption, filtration and degradation were provided.

Pesticides residues which have been detected in various natural waters in many countries are of special concern because of their persistence in the aquatic environment and potential adverse health effects. It is an emerging problem in developing countries, and there is an essential need for efficient technologies. Research indicated that there is a significant potential for pesticides removal by the use of the different processes of nanotechnology. In this study, three main processes categories for pesticides removal including adsorption, filtration and degradation were discussed. The researchers applied nanomaterials either individual or composite.

In many cases, the multitude of parameters has been considered for nanomaterial chemical reactivity enhancement and its use in aqueous solution for removal of pesticide: shape- controlled synthesis procedures, increase in specific surface area, increase in reactivity and incorporation of nanoparticle on support structures. Overall, different processes of nanotechnology have similarities to natural processes including energy efficiency, use of smaller quantities of materials, conversion efficiency, etc. Although it needs to be studied further about to large scale application of nanotechnology process in water and wastewater treatment to eliminate of pesticide and other pollutants associated with the investigation on potential risks of nanomaterials for environmental and human health.

i. Removal of organochlorine pesticides: Silver oxide nanoparticles embedded chitosan beads were studied to remove pesticides from water. A simple and an effective approach was used in regard to preparing AgO nanoparticles on the surface of Chitosan (Rahmanifar et al., 2014). The chitosan–AgO nanoparticles (CS–AgONPs) composite was characterized by infrared spectroscopy (FTIR), X-ray diffraction, and scanning electron microscopy. The CS–AgONPs composite beads were optimized to remove maximum permethrin as the model pesticide, with amount of sorbent, agitating time, initial concentration of pesticide, and pH parameters. In optimum conditions, room temperature, pH 7, and the CS–AgONPs beads could recover 99 % pesticides of permethrin solution (0.1 mg L-1) which were characterized by using UV spectrophotometer at 272 nm. In comparison with the pure chitosan, the removal capacity of CS–AgONPs beads has been increased to 49 %. The CS–AgONPs composite beads have high capacity as an adsorbent which could explore a new biocompatible and eco-friendly strategy for pesticide removal and appears to be the new promising material in water treatment application.

ii. Removal of organophosphorus pesticides from water: Adsorption of pesticides is one of the most commonly used strategies for this task. Tamara Lazarević-Pašt et al., (2018) analyzed the adsorption of two organophosphorus pesticides, dimethoate (DMT) and chlorpyrifos (CPF), on graphene-based materials. The adsorption was found to be very sensitive to the structure of the adsorbents used. In particular, aliphatic DMT was found to prefer hydrophilic oxidized graphene surfaces. The CPF molecule, which contains an aromatic moiety, prefers adsorption on the surface of a graphene basal plane with high structural order and preserved π electron system. The toxicity of pesticide solutions is reduced after adsorption, suggesting that there is no oxidation of DMT and CPF to more toxic oxo forms. It was emphasized that the combination of structural properties of adsorbents and adsorbates defines the adsorption of organophosphorus pesticides on graphene-based materials, while the specific surface area of adsorbents is not the major factor.

5.10 Advances in targeted pesticides with environmentally responsive controlled release by nanotechnology

Using nanotechnology to design and prepare targeted pesticides with environmentally responsive controlled release via compound and chemical modifications has also shown great potential in creating novel formulations. A review by Bingna Huang et .al., (2019), pay special attention to intelligent pesticides with precise controlled release modes that can respond to micro-ecological environment changes such as light-sensitivity, thermo-sensitivity, humidity sensitivity, soil pH, and enzyme activity. Moreover, establishing intelligent and controlled pesticide release technologies using nanomaterials are reported. These technologies could increase pesticide-loading, improve the dispersibility and stability of active ingredients, and promote target ability.

Problems with traditional pesticide formulations have motivated a new research direction in precise controlled release in recent years. Controlled pesticide release systems are used on farmlands by selecting a suitable and timely administration route to precisely regulate the target pest. This approach aims to minimize crops' demand for pesticides to gradually achieve more effective, safe pesticide usage via intelligent design that slows and controls pesticide release (Bingna Huang et .al., 2019). To date, studies on the controlled release of pesticides show the system only allows for a simple, slow release rather than a responsive release based on environmental conditions. Thus, if we can achieve precise pesticide release according to environmental light, temperature, soil pH, humidity, and enzyme changes, it may be possible to greatly improve the use of pesticides by reducing waste and pollution. In the past, some pesticides such as pyraclostrobin have been shown to exert a positive effect on crops affected by rice blast fungus. This chemical can induce physiological changes in crops and improve their resistance to the disease, increasing yields considerably. However, this pesticide is highly toxic to aquatic organisms, which has limited its rice crop application (Guan et al., 2015).

Nano-microcapsules or nanospheres have been prepared using light-sensitive, thermo-sensitive, humidity-sensitive, and enzyme- and soil pH-sensitive high polymer materials to deliver pesticides. These nanostructured materials are prepared via processes such as adsorption, coupling, encapsulation, and embedding. Such formulations can protect pesticides' active ingredients and enhance stability, control the release of core materials, reduce or obscure odors, decrease volatility, and isolate components. In recent years, the application of nanomaterials and technologies in the field of pesticides has made considerable strides. In particular, targeted delivery and controlled release of nanomaterials can improve pesticide utilization and reduce residue and pollution. Pesticide nanocapsule formulations have slow release and protection performance and, due to their small size, improvable pesticide droplet ductility, wettability, and target adsorption when spraying fields; these methods provide efficient and environmentally friendly advantages.

While progress has been made in research on pesticide nanocapsule formulations, preparation technology warrants further study. Popular pesticide microcapsule preparation methods are shown in table 24 (Bingna Huang et. al., 2019). Nanostructured materials can also respond to external stimuli such as temperature, light, humidity and pH to achieve an environmental response and controlled release of pesticides. Therefore, intelligent microcapsules that can respond to the external environment have attracted considerable attention. Although environmentally- responsive nano-drugs have been rapidly developed and applied in the medical field, the application of these materials to pesticides has only recently begun.This paper comprehensively summarizes progress in the applications of polymer materials sensitive to light, temperature, humidity, soil pH, enzymes, and other factors to enhance pesticide release.

Table 24: Main pesticide microcapsule preparation methods

Method	Preparation Process	Example
Interfacial polymerization	Two reactive monomers are dissolved in two different solvents. When one solvent is dispersed in another solvent, the two monomers undergo a polycondensation reaction at the phase interface to form microcapsules	Natural pyrethrin nanocapsules
In situ polymerization	Two or more water-soluble monomers are polymerized to form a water-insoluble polymer and are deposited on the surface of the core material for coating	S-ethyl dipropylthiocarbamate, acetochlor, atrazine, methotrexate

Emulsion polymerization	A solvent-insoluble monomer is dispersed in a solvent to form a uniform emulsion via mechanical agitation, high-speed shearing, and vigorous shaking with a surfactant (an emulsifier). Then, the polymerization reaction is initiated to form the polymer to achieve core material encapsulation	Abamectin nanocapsule suspension, natural suspension, natural
Membrane emulsification	The dispersion phase enters the continuous phase through a shirasu porous glass membrane under inert gas pressure, and the continuous phase breaks into droplets on the membrane surface by the shear forces of the SPG membrane and droplets.	Chlorantraniliprole nanocapsules, avermectin nanocapsules
Solvent evaporation	The wall material and core material are dispersed in the organic phase, added to the solution immiscible with the wall material, and the wall material is precipitated to form the microcapsule by heating and evaporating the solvent	Spinosad nanocapsules
Nano-precipitation	The interfacial interaction between solvent and non-solvent disperses the polymer and drug from the oil phase into the aqueous phase. This material can quickly wrap the drug and obtain nanocapsules through precipitation	Pyrazole azoxystrobin nanocapsules, azoxystrobin microcapsules
Double coacervation	Two oppositely charged water-soluble polymers form a wall around the water-insoluble pesticide active ingredient, which is a spontaneous liquid- to-liquid separation caused by electrostatic interactions.	Azoxystrobin microcapsules

5.11 Nanoremediation- Remediation of pesticides using nanomaterials

Nano remediation is the use of nanoparticles for environmental remediation It is being explored to treat ground water, wastewater, soil, sediment, or other contaminated environmental materials (Crane and Scott, 2015). Nanoremediation is an emerging industry; by 2009, nano remediation technologies had been documented in at least 44 cleanup sites around the world, predominantly in the United States. In Europe, nano remediation is being investigated by the EC funded NanoRem Project. A report produced by the NanoRem consortium has identified around 70 nano remediation projects worldwide at pilot or full scale. During nano remediation, a nanoparticle agent must be brought into contact with the target contaminant under conditions that allow a detoxifying or immobilizing reaction. This process typically involves a pump-and-treat process or *in situ* application.

Environmental remediation relies mainly on using various technologies (e.g., adsorption, absorption, chemical reactions, photocatalysis, and filtration) for the removal of contaminants from different environmental media (e.g., soil, water, and air). The enhanced properties and effectiveness of nanotechnology-based materials makes them particularly suitable for such processes given that they have a high surface area-to-volume ratio, which often results in higher reactivity. A review by Fernanda D. Guerra et al., (2018) provides an overview of three main categories of nanomaterials (inorganic, carbon-based, and polymeric-based materials) used for environmental remediation. The use of these nanomaterials for the remediation of different environmental contaminants, such as heavy metals, dyes, chlorinated organic compounds, organophosphorus compounds, volatile organic compounds, and halogenated herbicides was reviewed. Various recent examples were extensively highlighted focusing on the materials and their applications.

Some nano remediation methods, particularly the use of nano zero-valent iron for groundwater cleanup, have been deployed at full-scale cleanup sites. Other methods remain in research phases.

Materials used: A variety of compounds, including some that are used as macro- sized particles for remediation, are being studied for use in nanoremediation (Anonymous, 2014). These materials include zero-valent metals like zero- valent iron, calcium carbonate, carbon-based compounds such as graphene or carbon nanotubes, and metal oxides such as titanium dioxide and iron oxide.

Challenges: When using *in situ* remediation the reactive products must be considered for two reasons. One reason is that a reactive product might be more harmful or mobile than the parent compound. Another reason is that the products can affect the effectiveness and/or cost of remediation. TCE (trichloroethylene), under reducing conditions by nanoiron, may sequentially dechlorinate to DCE (dichloroethene) and VC (vinyl chloride). VC is known to be more harmful than TCE, meaning this process would be undesirable (Lowry, 2017).

Nanoparticles also react with non-target compounds. Bare nanoparticles tend to clump together and also react rapidly with soil, sediment, or other material in ground water. For *in situ* remediation, this action inhibits the particles from dispersing in the contaminated area, reducing their effectiveness for remediation. Coatings or other treatment may allow nanoparticles to disperse farther and potentially reach a greater portion of the contaminated zone. Coatings for nZVI include surfactants, polyelectrolyte coatings, emulsification layers, and protective shells made from silica or carbon. Such designs may also affect the nanoparticles' ability to react with contaminants, their uptake by organisms, and their toxicity. A continuing area of research involves the potential for nanoparticles used for remediation to disperse widely and harm wildlife, plants, or people (Wiesner Mark et al., 2007). In some cases, bioremediation may be used deliberately at the same site or with the same material as nanoremediation. Ongoing research is investigating how nanoparticles may interact with simultaneous biological remediation.

Principles of nanobioremediation technologies: Every year around 10 million tons of toxic chemical compounds are released by industry (Edgar Vázquez-Núñez et al., 2020). After release, these compounds may further react to form chemicals, for instance, polychlorinated dibenzo-p-dioxins or polychlorinated dibenzofurans, which are by-products of certain chemical processes involving chlorine.

There is high variability in the physical and chemical properties of these chemical compounds, and their cytotoxicity and multiple interactions with biotic and abiotic environmental factors, i.e., microorganisms, plants, animals, water, minerals, organic matter, wind, etc., have complicated the successful implementation of remediation technologies. The combined use of NMs and NPs with biotechnologies could offer a step-change in remediation capabilities, avoiding process intermediates, and increasing the speed of degradation. Besides physical and chemical technologies to remediate polluted sites, biological treatments have become relevant due to their low cost and wide range of applications. Bioremediation includes biosorption, bioaccumulation, biotransformation, and biological stabilization, among others. These technologies use plants and some microorganisms including bacteria and fungus, as well as combinations of them.

During recent years, NMs have been integrated with biological processes to accelerate and promote the removal of toxic compounds from the environment. The term nanobioremediation was used for processes where NPs and microorganisms or plants are used to remove contaminants. These types of practices were named according to the nature of the organism utilized for the remediation of contaminants. Thus, they were more specific and named the techniques as phyto-nanoremediation, microbial nanoremediation, and zoo-nanoremediation.

In any case, since bioremediation uses living organisms to remediate contaminated environments, a proper interaction between nanoparticles (NPs) and living organisms is essential. In this context, some aspects are of paramount importance. For example, it is known that nanotoxicity, NPs size, and nanonutrition may affect the living organisms and this in turn may affect the whole bioremediation process. The physical and chemical interactions between NMs, biota and contaminants depend on a variety of parameters including NMs size and shape, surface coating, chemical nature of the NMs and contaminant, type of organism, media, pH, and temperature, among others. Given the number of potential parameters influencing such interactions, this phenomenon becomes complex. For example, pH media as well as temperature play an essential role in the proper development of living organisms. These parameters in turn, may influence the stability of the NMs as well as that of the contaminant. It was proved that Au NPs were stable in MilliO water and a buffer; however, this stability was lost at pH of 4, 7, 8 and 10. Different synthetic methods influenced the thermal stability of Cu NPs were also proved. To the best of our knowledge, no comprehensive studies exist in the literature regarding the influence of the parameters shown above on the nanobioremediation of contaminants. Proper experimental designs should be applied to determine, for example, to what extent temperature and pH affect the synergistic effect of NMs and living organisms for the remediation of contaminants.

International markets and regulations of nanotechnologies applied in bioremediation: Nanotechnologies used in bioremediation processes are expected to drive the technological evolution for the improvement of the environmental quality in developed and emerging countries (Edgar Vázquez-Núñez et al., 2020). A significant amount of research has been performed to determine the mechanisms of decontamination and remediation. As it has been discussed previously, the biosafety related to the use and application of nanomaterials is a high concern due to the lack of knowledge and validated protocols to measure the impact of these materials on human health, loss of biodiversity and bioaccumulation and transport of NMs in trophic chains. Diverse international institutions such as USEPA, European Observatory for Nanomaterials (EON), the OECD Working Party on Manufactured Nanomaterials (WPMN) and ISO Technical Committee TC 229 "Nanotechnologies" have established international cooperation in order to increase the application of available regulations.

Furthermore, nanotechnology and bioremediation world markets are expected to continue growing and developing new niches to improve not only environmental aspects but also human lifestyle. The global nanotechnology market is expected to exceed US$125 billion by 2024; this sector continues to have an impact on other markets, fundamentally in electronics, energy, biomedical, cosmetics, defense, energy and agriculture and, according to the US-EPA it is estimated that bioremediation and phytoremediation technologies could have an annual growth of US$1.5 billion per year. The development of new methodologies and incorporation of nanotechnologies will expand new opportunities for treating sewage, lakes, rivers, and ponds, among others, creating new consumers and strengthening world trade.

An intensive use of pesticides in agriculture has caused serious health and environmental problems. Various adsorbents and photocatalysts have been used for the remediation of pesticides from aqueous environment. In recent years, nanomaterials viz. nanocomposites and nano-biocomposites are playing a promising role towards the removal of pesticides. A review by Sahithya and Nilanjana Das (2015) focused the updated information on pesticide removal using nanocomposites and nano-biocomposites through adsorption and photocatalytic degradation. Leaching and persistence of chemical pesticides are the main reasons for the organic pollution in the environment. Industrial wastewaters are also the significant sources of the pesticide pollution in aquatic system which is a great threat to general public health due to their high toxicity. Several approaches such as oxidation, fluid-extraction, biodegradation, ozonation, chlorination and membrane separation were reported for the removal of pesticides from aquatic environment. However, these techniques do not lead to the complete removal of pesticide and further treatments are needed to convert the pesticides into harmless products such as CO_2 and H_2O_4.

With the recent developments in the field of nanotechnology, the correlation of material properties with filler size has become a point of great interest. Nanocomposites based on metallic nanoparticles have witnessed an exponential growth in the field of nanotechnology. They have also emerged as suitable alternatives to overcome the limitations of microcomposites and monolithics. The unique properties derived from the nanocomposites can be exploited in various fields of science and technology. To overcome the disadvantages associated with biopolymers, a unique approach of incorporation the nanofillers into biopolymer matrix was introduced to improve their mechanical, barrier properties and thermal stability. These 'nano-biocomposites' have received increasing attention as potential adsorbents towards reducing environmental pollution which incorporates the advantages of both biopolymers and nanoparticles.

The rise of global population has resulted in an impressive growth in food production, which has been attributed to the overly increased use of agrochemicals such as pesticides and fertilizers. However, unthoughtful use of agrochemicals also leads to deterioration of environmental quality and pose severe threats to terrestrial and aquatic ecosystems. Even though persistent and harmful chemicals have been phased out and replaced by less persistent and biodegradable chemicals, legacy residues from the past and recent residues are still a painful challenge to solve. In this chapter the advances of nanotechnology and their application for remediations of agrochemicals are discussed. Depending on the decontamination process, these techniques have been classified into two categories: reactive and adsorptive treatments.

Lienda Handojo et al., (2020) reviewed the utilization of various nanomaterials including metal- and metal-based silica, carbonaceous, and polymeric nanomaterials for remediation techniques. Besides, ecotoxicological concerns associated to the application of nanomaterials in the environment was also included.

5.12 Nanomaterials as photocatalysts

Semiconducting metal oxide nanostructures play an important role in developing smart materials that are well efficient for sensing and simultaneously destroy harmful chemical contaminants from our environment. A review by Abdullah (2016) highlighted some recent advances of nano-science in the area of environmental hazardous contaminant detection by sensing followed by the remediation; focused on water remediation especially through photo-catalysis. In addition, the state-of the- art research activities involved in importance of metal oxide and their various applications were also discussed. The compilation consists of three main sections vis a vis semiconducting metal oxides and applications, Photocatalysis: fundamentals, processes and mechanism and sensors: processes, types and their respective applications to monitor levels of environmentally important parameters. Also an inclusion of a section on future trends which discusses some of the new approaches being used to improve the selectivity and sensitivity of metal oxide semiconductors. This review concluded the perspectives and outlook on the future developments in the metal oxide nanostructure research area as well as summarizes a comprehensive compilation of the work done in order to address the challenges followed by prevailing achievements till date.

Nanomaterials including silver, titanium dioxide, zinc oxide were used as photocatalysts for the heterogeneous degradation of pesticides (Sahithya and Nilanjana Das, 2015).

i. **Nano-composites:** Nanocomposites are prepared by reinforcing the nano sized inclusion (nanofiller) into the material which aims to improve the optical, catalytic, mechanical, magnetic, thermal, barrier and other properties. Depending on the type of nanofiller used, the nanocomposites could show the drastic modifications in their properties. Nanofillers are generally classified into three types based on their geometry and aspect ratio: Layered silicates (e.g. clays), spherical (e.g. silica) and acicular (e.g. CNT). At present, layered silicates are most widely studied due to the ease of availability, versatility and eco-friendly nature.

ii. **Nano-biocomposites:** Nano-biocomposites are the new class of composite materials where nanofillers are incorporated into biopolymer matrix using various techniques. The incorporation of appropriate nanoparticles into biopolymers will provide special properties which can be used for wide range of both conventional and emerging applications. The hydrophilic nature of biopolymers makes them potential adsorbents for remediation of pollutants from aqueous environment. Use of biopolymers is an innovating answer to replace the conventional and durable synthetic polymers. The reported performances of nano- composites on pesticide removal are quite encouraging. Since industries are concerned with sustainable developments, and the production cost of biopolymers is decreasing, strong developments of biopolymers-based nano-biocomposites is needed in near future. Extensive research should be carried out to promote the large scale use of nano-biocomposites for the effective removal of pesticides from wastewater

iii. **Titanium dioxide (TiO_2):** In recent years, a light-mediated Advanced Oxidation Process (AOPs) has been adopted to resolve pesticide residue issues in the field. Among the current available semiconductors, titanium dioxide (TiO_2) is one of the most promising photocatalysts. In a study Xiangying Liu et. al., (2019), investigated the photocatalytic degradation of profenofos and triazophos residues in Chinese cabbage, *Brassica chinensis*, using a Cerium-doped nano semiconductor TiO_2 (TiO_2/Ce) under the field conditions. The results showed that the degradation efficiency of these organophosphate pesticides in *B chinensis* was significantly enhanced in the presence of TiO_2/Ce. Specifically, the reactive oxygen species (ROS) contents were significantly increased in *B. chinensis* with TiO_2/Ce treatment, accelerating the degradation of profenofos and triazophos. Ultra- performance liquid chromatography–mass spectroscopy (UPLC-MS) analysis detected 4-bromo-2-chlorophenol and 1-phenyl-3-hydroxy-1,2,4-triazole, the major photodegradation byproducts of profenofos and triazophos, respectively. To better understand the relationship between photodegradation and the molecular structure of these organophosphate pesticides, the spatial configuration, the bond length and Mulliken atomic charge using quantum chemistry were investigated. Ab initio analysis suggested that the bonds connected by P atom of profenofos/triazophos were the initiation cleavage site for photocatalytic degradation in *B. chinensis*

Titanium dioxide (TiO_2) is the most widely accepted photocatalyst because it is non-toxic, stable to photocorrosion, low cost and can potentially work using sunlight rather than artificial sources of light. When titanium dioxide is illuminated by UV radiation, the absorption of photons of energy is then equal to or greater than its band gap width (Termtanun Mutsee, 2013). This artefact leads to the formation of conduction-band electrons and valence-band holes on the surface of TiO_2, which yield hydroxyl radicals, the primary oxidising species needed for the photocatalytic degradation of pollutants. Supercritical water hydrothermal synthesis (ScWHS) is one of novel approaches for nanoparticle manufacture which involves the mixing of an aqueous metal salt stream with a supercritical water stream to produce nano-sized metal oxide particles. The engineering design for the mixing of these two fluids is critical and a novel nozzle reactor has been developed at the University of Nottingham that can produce high quality particles with an excellent control over particle size and particle size distribution.

By application of this technique, titanium dioxide in nanoparticle form, was produced and used for the photocatalytic treatment of wastewater. In this thesis, the photocatalytic degradation of the three pesticides isoproturon, simazine and propazine was measured using three different types of reactors: thin film fixed bed reactor (TFFBR), a stirred reactor, and a fluidised bed photoreactor. Various conditions were used: for example without UVC and UVA illumination, with commercial TiO_2, with the synthesised TiO_2, without any photocatalyst, low concentration, high concentration, and at different TiO_2 concentrations. The optimum TiO_2 concentration for the treatment of the three chosen pesticides was equal to 5 g litre-1. The efficiency in decreasing the pesticides concentration of the synthesised TiO_2 (from ScWHS technique) with all three types of reactors was lower than that of the commercial titanium dioxide (P-25 TiO_2). The fluidised bed reactor appeared to give the highest performance amongst three reactors.

iv. Carbendazim: The carbendazim is a widely used benzimidazole fungicide. It is of major concern due to its suspected hormone disrupting effects. The degradation of carbendazim considered as a "priority hazard substance" by the

Water Framework Directive of the European Commission was investigated by UV/solar photocatalytic process using P25 TiO_2 and LR grade TiO_2 (Taranjeet Kaur, et al., 2015). A batch-type photoreactor was employed and the influence of catalyst loading, initial concentration, area/volume ratio, pH of the solution and light conditions were studied. Aeroxide P25 shows better degradation efficiency than LR grade TiO_2 in both UV and sunlight conditions for optimized parameters. Degradation of fungicide was negligible with TiO_2 under UV without catalyst. However, no adsorption took place with catalyst in dark. Also, the addition of oxidant (H_2O_2) had no appreciable increase in the rate of degradation for both the catalysts. Under UV light with TiO_2 carbendazim has shown substantial degradation at different time of irradiation. The optimized parameters under UV for the degradation of carbendazim are catalyst loading 1 g L^{-1}, area/volume ratio 0.919 cm^2ml^{-1}, intensity 30 W m^{-2}, pH 6.5 and around 85% mineralization of carbendazim was achieved.

Iron (Fe) and Silica (Si) doped TiO_2 nanoparticles were synthesized by surface impregnation method. The concentration of dopants, Fe (1%, 2%, 3% and 4%) and Si (3%, 5%, 7% and 9%) were varied for the preparation of doped TiO_2 which were characterized by XRD, FESEM and TEM analysis (Taranjeet Kaur et al., 2016). The TEM measurement shows that the particle size is in the range of 25- 35 nm for Fe and Si doped TiO_2. With the optimum dopant concentration i.e. 2% for Fe and 5% for Si, doped TiO_2 showed nearly 98% degradation of fungicide carbendazim under sunlight. Optical measurements indicated that the absorption band edge of Fe and Si dopants were shifted towards visible wavelength. Band gap of doped and undoped TiO_2 nanoparticles was measured by UV-Vis spectroscopy. Fe and Si doped TiO_2 degrades fungicide much faster than the undoped TiO_2 at optimum catalyst loading of 1 g L.-1

Result signifies that doped TiO_2 (Fe, Si) have remarkable efficiency to degrade complex fungicide into smaller molecules, in the presence of sunlight

v. Thiram: Jingzhe Xue et al, (2014) reported a residue-free green nanotechnology which synergistically enhance the pesticides efficiency and successively eliminate its residue. A composite antifungal system was built by a simple pre-treating and assembling procedure for investigating synergy. Investigations showed 0.25 g/L ZnO nanoparticles (NPs) with 0.01 g/L thiram could inhibit the fungal growth in a synergistic mode. More importantly, the 0.25 g/L ZnO NPs completely degraded 0.01 g/L thiram under simulated sunlight irradiation within 6 hours. It was demonstrated that the formation of ZnO-thiram antifungal system, electrostatic adsorption of ZnO NPs to fungi cells and the cellular internalization of ZnO-thiram composites played important roles in synergy. Oxidative stress test indicated ZnO-induced oxidative damage was enhanced by thiram that finally resulted in synergistic antifungal effect. By reducing the pesticides usage, this nanotechnology could control the plant disease economically, more significantly, the following photocatalytic degradation of pesticide greatly benefit the human social by avoiding negative influence of pesticide residue on public health and environment.

vi. Benalaxyl: Studies on the dissipation behavior of benalaxyl beneath direct sunlight by using Fe doped TiO_2 nanoparticles ($FeTiO_2$) as catalystwas taken up (Rao et al., 2016). $FeTiO_2$ nanoparticles are synthesized and characterized by way of Scanning Electron Microscopy (SEM) combined with energy Dispersive X-ray evaluation (EDX) and Fourier Transform Infrared Spectroscopy (FT-IR). The photo catalytic studies were conducted by the 1 l of Milli-Q water, pH 4.0, 7.0 and 9.0 buffer were spiked with 1 mL of 1000 mg/L stock solution of benalaxyl 35% WP to get 1 μg/ml concentration of pesticide active in water. Three sets of such samples in triplicates were prepared and sonicated for 10 min to get homogeneous concentration of pesticide active in water and labeled as S1, S2, and S3. Two sets of samples (S1 and S2) were added $FeTiO_2$ nanoparticles to get 0.06 g/L content of photo catalyst. S1 and S3 sets of samples were exposed to sun light. S2 set of samples were kept in an oven at 40°C in dark. The Milli-Q water spiked with stock solution of pesticide and without addition of $FeTiO_2$ nanoparticles (S3) were also exposed to sun light for the measurement of photolytic degradation of pesticide in water. The $FeTiO_2$ added Milli-Q water without spiking of pesticides (S4) was also maintained as untreated control for the identification of absence of pesticide. The collected samples were quantified using a validated HPLC-UV method. Kinetic parameters such as rate constant (k), DT50 and DT90 were calculated using the dissipation data of benalaxyl.

vii. Tricyclazole: Studied the dissipation behavior of tricyclazole under direct sunlight was studied using Fe doped TiO_2 nanoparticles ($FeTiO_2$) as catalyst (Tentu Nageswara Rao et al., 2016). $FeTiO_2$ nanoparticles are synthesized and characterized by scanning electron microscopy (SEM) combined with energy Dispersive X-ray analysis (EDX) and Fourier transform Infrared Spectroscopy (FT- IR). The photocatalytic studies were conducted by the 1 litre of milli-Q water, pH 4.0, 7.0 and 9.0 buffer were spiked with 1 mL of 1000 mg/L stock solution of tricyclazole 75 % WP to get 1μg/mL concentration of pesticide active in water. Three sets of such samples in triplicates were prepared and sonicated for 10 minutes to get homogeneous concentration of pesticide active in water and labeled as S1, S2, and S3. Two sets of samples (S1 and S2) were added $FeTiO_2$ nanoparticles to get 0.06g/L content of photocatalyst. S1 and S3 sets of samples were exposed to sun light. S2 set of samples were kept in an oven at 40°C in dark. The milli-Q water spiked with stock solution of pesticide and without addition of $FeTiO_2$ nanoparticles (S3) were also exposed to sun light for the measurement of photolytic degradation of pesticide in water. The $FeTiO_2$ added Milli-Q water without spiking of pesticides (S4) were also maintained as untreated control for the identification of absence of pesticide. The collected samples were quantified using a validated HPLC-UV method. Kinetic parameters such as rate constant (k), DT50 and DT90 were calculated using the dissipation data of tricyclazole.

5.13 Nanomaterials in pesticide-based plant protection

The assistance of nanotechnology in plant protection products has exponentially increased to achieve higher crop production. In general conventional crop protection practices involve applications of large-scale and over-dose of fungicides, herbicides and insecticides. Among the applied pesticides, more than 90% are either lost in the environment or unable to reach the target sites essential for effective pest control (Yifen Shang et al., 2019). This not only increases the expenses of crop production, but also causes the depletion of environmental systems. It is to be noted that the presence of active ingredients in minimum effective concentration of a formulation at the target sites is essential for assurance of better protection of plants from pest invasion and subsequent crop loss. In this connection, the development of new plant protection formulation has long been a very cognitive field of agricultural research. One such technology is nanoformulation or encapsulation ofpesticides which has revolutionized the plant protection sector. Nanoformulation of pesticides contains a very tiny number of particles that act as active ingredients of pesticides, whereas other engineered nano-structures also have useful pesticidal properties (Haq et al., 2019). The nanoencapsulation of pesticides is the coating of active ingredients of pesticides with another material of various sizes at nano-range where encapsulated materials are referred to as internal phase of the core material (pesticides) and capsulation materials are known as external phase, *i.e.*, the coating nanomaterials.

Nanoformulations or encapsulations of pesticides facilitate the persistence or controlled release of active ingredients in root zones or inside plants without compromising effectiveness. On the other hand, conventional formulations of pesticides or herbicides not only limit water solubility of pesticides, but also injure other organisms, leading to increased resistance to target organisms. In contrast, nanoformulations assist to overcome the above-mentioned limitations. Nanoformulations of pesticides boost crop yields by increasing pesticide efficacy through regulating transport potential of pesticide. They found that nanoformulations combining polymeric nanocapsules and the pyrethroid bifenthrin (nCAP4-BIF) display increased elution with time and enhanced transport potential even upon the addition of fertilizer in loamy sand soil saturated with artificial porewater containing Ca^{2+} and Mg^{2+} cations. This means that nCAP4 could be a promising delivery vehicle of pesticides like pyrethroid in plant protection. This could possibly due to the increased potential of dispersion and wettability of nanoformulations that reduce organic solvent runoff and unwanted pesticidal movement. Moreover, nanomaterials in pesticide formulation show some useful properties like increased stiffness, permeability, thermal stability, solubility, crystallinity and also biodegradability essential for sustainable agro-environmental system. More importantly, the timely and controlled release of active ingredients reduce the total amount of pesticides required for pest and disease control, an important feature of integrated pest management (IPM) (Yifen Shang et al., 2019).

Furthermore, sustainable agriculture demands minimum use of agrochemicals, so as to protect environmental depletion and others non-target species. Moreover, the minimum use of pesticides reduces the cost involved in crop production. It has been estimated that on a global scale annual agricultural crop losses account for 2000 billion US dollars due to plant diseases, insect pests and weeds, and only in United States for pathogen control efforts simply through fungicide applications cost exceeds $600 million. Fascinatingly, in such circumstances, the use of NPs is being reported as efficient alternative to directly suppress pathogen infection and activity, leading to the increased crop growth and yield. For example, halloysites a types of clay nano tubes used as cost-effective carriers of pesticides in agriculture. These nano tubes not only show extended release period of active ingredients (AI), but also provide assurance of better contact with minimum environmental effect. Nano silica is one such example that is hydrophobic in nature and can absorb into the cuticle layer of the insects upon contact, leading to ultimate death of insects (Nair et al., 2010).

In addition, many investigations have provided evidence that nanoformulations of pesticides facilitate the widening of plant-based systemic acquired resistance (SAR) against pests. For examples, silica nanosphere formulations can increase ability of pesticides to penetrate through the plant and reach the cell sap, thereby exerting systemic effect to control chewing or sucking type insects like aphid. Such types of hollow formulation also protect pesticides from photodegradation due to direct exposure to sun rays (Panpatte et al., 2016). It has also been observed that nanoformulations alter non-systemic behavior of pesticides. The nonsystemic behavior of ferbam can alter and increase the penetrating potential into tea leaves when formulated with metallic NPs (AuNPs). Such kinds of findings obviously will provide a new frontier of development of pesticide formulations in order to acquire plant-based systemic resistance. However, further study is needed to unravel the behaviors and fate of pesticides and their interactions with biomacromolecules present in living system or environment.

The potential application of engineered nanomaterials in agriculture is also ascertained in disease and weed management Inorganic NPs, such as ZnO, Cu, SiO_2, TiO_2, CaO, MgO, MnO and AgNPs play important roles in various arena of plant protection including microbial activity and bacterial diseases (Yifen Shang et al., 2019). For example, ZnO nanoparticles have recently been shown to provide effective growth control of *Fusarium graminearum*, *Penicillium expansum*, *Alternaria alternata*, *F. oxysporum*, *Rhizopus stolonifer*, *Mucor plumbeus* and *A. flavus* as well as *Pseudomonas aeruginosa* (Dwivedi et al., 2016). Nano-Cu application was found to be more effective against *Phytophthora infestans* compared to currently available non-nano Cu formulations in tomato. Besides, Si, and TiO_2 have been found promising to suppress crop diseases directly, through antimicrobial activity. MONPs inhibit the development of fungal conidia and conidiophores which cause ultimate death of fungal hyphae. Likewise, weeds are also considered as serious threat to global agricultural production as they compete with crops for their nutrient, water and light. However, the application engineered nanomaterials containing herbicides provide the solution in eco-friendly manner. The nanomaterials in pesticides, fungicides and herbicides have a tremendous scope in sustainable agricultural development.

5.14 Nano-pesticides and Greener Agriculture

Research is urgently needed to evaluate the risks and benefits of nano-pesticides to human and environmental health. Melanie Kah and Thilo Hofmann from the Department of Environmental Geosciences of the University of Vienna recently performed an extensive analysis of this emerging field of research. The results were published June 6th in the internationally recognised journal "Critical Reviews in Environmental Science and Technology". The study presents the current scientific state of art on nano-pesticides and identifies direction priorities for future research (Melanie Kah et al., 2012). Nanotechnology has developed tremendously in the past decade and was able to create many new materials with a vast range of potential applications. Some of those innovative materials are promising to reduce environmental pollution. For instance, carbon nanotubes and metal nano-particles are great candidate materials for cleaning polluted water and soils.

However, the risk that nano-particles may pose to human and environment health is not yet fully understood. The precautionary principle therefore suggests keeping environmental release of nano-particles minimal until their fate and toxicity is better understood. "A good understanding of nano-materials is essential to evaluate whether the benefits overcome potential new risks", explains Thilo Hofmann, dean elected at the Faculty of Geosciences, Geography and Astronomy of the University of Vienna.

Among numerous proposed applications, nanotechnology has the potential to revolutionize agricultural practices and food systems. Research has been extremely active over the past few years to develop new pesticides products based on nanotechnology. Nano-pesticide research is emerging at high speed at the agrochemical labs, however, this topic has not reached public awareness or state authorities so far, nor are any products available at the marked. Since those nano-pesticides have new or enhanced properties, this will change in near future and will inevitably result in both new risks and new benefits to human and environmental health. Nano-pesticides encompass a great variety of products, some of which are already on the market. The application of nano-pesticides would be the only intentional diffuse input of large quantities of engineered nano-particles into the environment. Innovation always results in both drawbacks and benefits for human and environmental health. Nano-pesticides may reduce environmental contamination through the reduction in pesticide application rates and reduced losses. However, nano-pesticides may also create new kinds of contamination of soils and waterways due to enhanced transport, longer persistence and higher toxicity.

The current level of knowledge does not allow a fair assessment of the advantages and disadvantages that will result from the use of nano-pesticides. As a prerequisite for such assessment, a better understanding of the fate and effect of nano-pesticides after their application is required. The suitability of current regulations should also be analyzed so that refinements can be implemented if needed. Research on nano-pesticides is therefore a priority for preserving the quality of both the food chain and the environment.

5.15 Ecological risk assessment of nano-enabled pesticides

Nanotechnology is emerging as a highly attractive tool for the formulation and delivery of pesticide active ingredients, as well as enhancing and offering more effective and environmentally friendlier crop protection products (Walter de Gruyter, 2017). A number of formulation types have been suggested, including emulsions (*e.g.* nanoemulsions), nanocapsules (*e.g.* with polymers), and products containing pristine engineered nanoparticles (such as metals, metal oxides and nanoclays). Some nano-enabled pesticides are in the pipeline and are expected to be presented for registration to the regulatory agencies. For example, US EPA recently registered a nano-enabled pesticide, AZteroid, which is based on Vive Crop Protection's Allosperse (a polymer-based delivery system). The increasing interest in the nano-enabled pesticides raises questions as to how to assess the environmental risk of these products for regulatory purposes. Therefore, there is a need to develop a set of guiding principles to facilitate a harmonized approach for the environmental risk assessment (ERA) of nano-pesticides.

In response to the above, an IUPAC project was conceded on "Guiding principles to facilitate a harmonized ecological risk assessment framework for nano-pesticides in the environment". The project has laid down some guiding principles and suggested an approach for consideration by regulatory agencies and industry. The project found that the existing tiered risk assessment approach, such as that used in the European Union, remains a useful framework for the risk assessment of nano-pesticides. In many situations, existing approaches can be used with some modifications and adaptations to cover nano-pesticides. In this regard, a crucial decision making step relates to the durability of a nano-enabled pesticide product and the fate of the product in soil. This criteria allows an assessment of whether a new nano-enabled product could be treated as a conventional pesticide, or if it needs to be treated differently, as a product that demonstrates an altered efficacy or fate or toxicity profile. For example, a nano-formulation that is short-lived, *i.e.,* it releases the active ingredient (a.i.) relatively rapidly, the product can be treated as a conventional product. For relatively durable formulations, potential alterations in fate and behavior resulting from the nanoformulation can be accounted for, with some adaptions to existing approaches.

Plant protection products containing nanomaterials that alter the functionality or risk profile of active ingredients (nano-enabled pesticides) promise many benefits over conventional pesticide products (Glen W. Walker et al., 2018). These benefits may include improved formulation characteristics, easier application, better targeting of pest species, increased efficacy, lower application rates, and enhanced environmental safety. After many years of research and development, nano-enabled pesticides are starting to make their way into the market. The introduction of this technology raises a number of issues for regulators, including how does the ecological risk assessment of nano-enabled pesticide products differ from that of conventional plant protection products.

There is growing interest in more deliberate applications of nanotechnology in the development of new plant protection products. Dendrimer technology is being applied to a number of active ingredients (a.i.) to develop nano-enabled pesticides with enhanced efficacy, and some products are in an advanced stage of development.

Effect on pesticide degradation, earthworms and uptake by plants: An increasing number of nanoformulated pesticides (nanopesticides) have been developed in recent years with the aim to improve pesticide efficiencies and reduce their impact on the environment and human health. However, knowledge about their environmental fate and effects is still very limited (Dana Fojtová et al., 2019).

Risk assessment frameworks for nano-enabled pesticides: The risks associated with nanotechnologies will also form a complex risk landscape rather than a homogenous set of risks. The emphasis on what kind of risks is key to consider will depend on the perspective of the particular organisation involved in nanotechnologies (Christoph Lauterwasser, 2016). To name just a few business risks involved with marketing of nanotechnology enabled products; risks related to the protection of intellectual property; political risks regarding the impact on the economical development of countries and regions ; risks regarding privacy when miniature sensors become ubiquitous; environmental risks from the release of nanoparticles into the environment; safety risks from nanoparticles for workers and consumers; futuristic risks like human enhancement and self replications of nano machines.

The catch-all term "nanotechnology" is not sufficiently precise for risk governance and risk management purposes. From a risk-control point of view it will be necessary to systematically identify those critical issues, which should be looked at in more detail. This risk identification process is a task for all parties involved and it should remain a dynamic process which always takes into account new scientific, technological, societal and legal trends. The risk and safety discussion related to free nanoparticles will concern only a fraction of the applications of nanotechnologies. In most applications nanoparticles will be embedded in the final product and therefore not come into direct contact with workers, consumers or the environment. They are unlikely to raise concerns because of their immobilisation. Exceptions are possible when the products or materials within which nanoparticles are enclosed are discarded, burned or otherwise destroyed.

Ideally, case-by-case risk assessment of nanomaterials should be performed in order to take the unique properties of specific nanomaterials into consideration (Anonymous, 2012). Traditional chemical risk assessment is based on the notion that the chemical identity governs the risk of a chemical. However, for nanomaterials a number of other characteristics may be relevant for assessing human health risks, and it is known that a large number of nanoparticle characteristics may influence the overall hazard. On the one hand, this argues in favour of a case-by-case risk assessment approach because this is the only manner in which one can obtain a univocal determination of the risk of a given nanoparticle and its specific properties. Adopting a case-by-case risk assessment, changes in the specific properties (size, surface area, etc.) can lead to totally different risks, requiring separate estimation.

On the other hand, the number of potential combinations of various material properties makes case-by-case risk assessment of nanomaterials so demanding as to be impracticable, unless the key specific properties driving the critical outcome of interest are well known. It has for instance been suggested that there are up to 50,000 potential combinations of single-walled carbon nanotubes (SWCNTs), depending on structural types, length, manufacturing and purification processes, and surface coatings. Each one of these SWCNTs has different chemical, physical and biological properties that may determine their overall hazard. Not all of these SWCNT varieties are expected to be of commercial relevance, but there are numerous other kinds of nanoparticles, such as fullerenes, quantum dots, metals and metal oxide nanoparticles, resulting in effectively countless types of nanomaterials, which may pose different types of risk.

Thus, identifying patterns and similarities between various sets and families of nanomaterials, which can be addressed collectively by applying grouping and read-across with regards to their toxicological profile and possible health implications, would be extremely beneficial. It is also a practical necessity if risk governance of nanotechnology is to be pursued. This requires reporting of case-by-case specific material properties and their implication for toxicological and environmental behaviour, to allow the recognition, description and evaluation of such families of nanomaterials and/or applications.

Risk governance: Due to data gaps, scientific uncertainty and ambiguity, traditional risk analysis, where regulation and risk management measures are based foremost on an independent and complete scientific risk assessment, may tend to lead to "paralysis by analysis" (Anonymous, 2012). Too much time can be spent waiting for the completion of risk assessment procedures and too little focus is put on implementing measures to prevent or reduce possible risks. Thus, many review exercises concur about the need for approaches that build on risk assessment but are better designed to handle complexity. The Risk Governance Framework developed by the International Risk Governance Council (IRGC), provides an example of such an approach for nanomaterial risk analysis, as it combines aspects of the traditional risk assessment framework with societal and ethical characteristics. This framework was considered particularly relevant for nanotechnology due to the uncertainties surrounding the potential health and environmental effects, and hence the need to involve a wide range of stakeholders when making decisions regarding nanotechnology's use and implementation.

Risk communication: Communication is a key element of this framework as one cannot choose "not to communicate" (Grobe, 2012). Lack of communication and understanding about the science, application and regulation of nanotechnology among all stakeholders may have negative effects on societal impressions and political/regulatory decision making. Hence, communication plays a central role as it is of major importance throughout the entire value chain and the product's life cycle. Good communication serves a number of purposes. First of all, it should facilitate understanding among stakeholders, for example about the rationale behind the risk appraisal and risk management phases. Second, communication should help stakeholders make informed choices about risk when they are themselves involved in risk-related decision-making. From a more holistic perspective, the heavy focus on communication between risk assessors and managers, between scientists and policy-makers, as well as civil society, is intended to "…foster tolerance for conflicting viewpoints and provides the basis for their resolution, and creates trust in the institutional means for assessing and managing risk and related concerns".

Commercialization of nanotechnologies entails a regulatory requirement for understanding their environmental, health and safety (EHS) risks (Danail Hristozov et al., 2016). Today we face challenges to assess these risks, which emerge from uncertainties around the interactions of manufactured nanomaterials (MNs) with humans and the environment. In order to reduce these uncertainties, it is necessary to generate sound scientific data on hazard and exposure by means of relevant frameworks and tools. The development of such approaches to facilitate the risk assessment (RA) of MNs has become a dynamic area of research. The analysis concluded that none of the reviewed frameworks were able to fulfill all evaluation criteria.

Many of the existing modelling tools are designed to provide screeninglevel assessments rather than to support regulatory RA and risk management. Nevertheless, there is a tendency towards developing more quantitative, higher-tier models, capable of incorporating uncertainty into their analyses. There is also a trend towards developing validated experimental protocols for material identification and hazard testing, reproducible across laboratories. These tools could enable a shift from a costly case-by-case RA of MNs towards a targeted, flexible and efficient process, based on grouping and read-across strategies and compliant with the 3R (Replacement, Reduction, Refinement) principles. In order to facilitate this process, it is important to transform the current efforts on developing databases and computational models into creating an integrated data and tools infrastructure to support the risk assessment and management of MNs.

Ecological risk assessment of pesticide a.i. and formulated products is a highly developed process, which forms an integral part of pesticide regulatory frameworks (Glen W. Walker et al., 2018). The issues associated with assessing the ecological risks of nanomaterials within the conventional risk-assessment paradigm used for chemicals and chemical products have been explored by the Organisation for Economic Co-operation and Development (OECD) Working Party on Manufactured Nanomaterials. This work concluded that the existing risk-assessment paradigm used for chemicals can be adapted to the risk assessment of nanomaterials. Problem formulation is a key initial phase in the ecological risk assessment of chemicals. This phase of assessment is intended to define the nature of the problem to be addressed by risk assessment and to develop a plan or strategy by which the risks can be subsequently characterized.

The consistent and structured application of problem formulation principles to the risk assessment of nano-enabled pesticide products will be essential in formulating sound recommendations to risk managers and decision makers in regulatory agencies. It will also help identify the most relevant assessment information, including hazard and exposure data, and available risk reduction measures at an early stage in the risk assessment process. These are acknowledged to be key factors influencing the conduct of effective risk assessments.

Given the development of nanotechnology within numerous scientific disciplines, it is likely that nanoscale products have been and will be used for agricultural, vector, and urban pest control prior to a complete evaluation of exposure and risk (Stone David et al., 2010). Significant differences may exist between nanotechnology-based pesticides (NBPs) and conventional pesticides, primarily due to size and surface characteristics. These differences may result in changes in bioavailability, sensitivity, dosimetry, and pharmacokinetics. This paper considers the role of exposure assessment in the regulation of NBPs. While the existing regulatory infrastructure for pesticides is well established, several issues specific to NBP exposure are discussed, including disclosures of nanoparticle characteristics in product formulations, additional uncertainty factors for NBPs with inadequate data, route-specific approaches for assessing exposure, testing with the commercial form of NBPs, initiation of a health surveillance program and development of educational programs.

Exposure risk: Nanomaterial exposure measurement is essential for hazard characterization and risk assessment. Migration of nanomaterial or its degraded products in non-nano form (its type and quantity) from agri produce or via animals for food production or from food contact materials (like packaging) should be considered in exposure measurement and hazard characterization. Specific testing protocols for analysis of migrated products are required. Food sampling, variability in composite sampling and concentration variations between samples are critical sampling issues in exposure evaluation. The decision framework defined by OECD (ENV/JM/MONO (2019)12) may be followed for inclusion of physico-chemical parameters for exposure and fate assessment of nanofertilizer, nanopesticide and NMs in food. Guidance on risk assessment of the application of nanoscience and nanotechnologies in the food and feed chain (EFSA, 2018) may also be followed.

Uncertainty analysis: Analysing possibility of uncertainty in assessing the above-mentioned assessments. Some of the possible reasons for uncertainty in assessing nanomaterials include non-availability of standard methods for physico-chemical characterisation of various ENM, structures and associated properties, sample preparation procedures and calibration of the analytical equipment dictate characterization, accuracy Differences in the physical principles applied by various measurement techniques, aggregation/agglomeration behaviour of nanomaterials and other factors such as dilutions or dispersions, vary with their interaction with various environmental factors. Reduced information can be provided when no exposure to nanomaterials is confirmed by data indicating no migration from food contact materials or when complete degradation/dissolution is demonstrated with no absorption of engineered nanomaterials as such.

5.16 Stimuli-responsive nano-based pesticides

The utilization of nanotechnology to create novel formulations has shown great potential for diminishing the indiscriminate use of pesticides and providing environmentally safer alternatives. Smart nano-based pesticides are designed to efficiently delivery sufficient amounts of active ingredients in response to biotic and/or abiotic stressors that act as triggers, employing targeted and controlled release mechanisms (Camara, et al., 2019). This review discussed the current status of stimuli-responsive release systems with potential to be used in agriculture, highlighting the challenges and drawbacks that need to be overcome in order to accelerate the global commercialization of smart nanopesticides.

The advantages of controlled release for the delivery of agrochemicals include reductions of phytotoxicity, leaching losses, volatilization, drift, and soil degradation, as well as increased safety during application. However, the release of active agents from loaded particles mainly occurs by means of passive diffusion, capsule erosion, or osmotic pressure, resulting in poor control of pesticide release. Hence, stimuli-responsive systems are promising candidates for improving controlled release properties, promoting site-specific and smart release of pesticides in response to biotic or abiotic stimuli (Liang et al., 2017). Abiotic and biotic factors enabling the use of stimuli-responsive nanomaterials for site-specific release of active substances, in order to increase plant resistance are represented in fig. 46. Biotic (plant pathogens, insects, and weeds) and abiotic (temperature, drought, flooding, and salinity) stressors are responsible for reducing global crop productivity by about 50%. Therefore, prospecting new responsive carriers that respond to environmental stimuli such as pH, temperature, redox conditions, light, and enzymes could be an important strategy for improving the resistance of crops to the stresses caused by these environmental factors.

The use of stimuli-sensitive carriers in nanoformulations can enable the release of different compounds in an effective manner and at defined times, hence increasing bioefficacy, while reducing side effects, the dosage required, and the number of applications. The use of stimuli-responsive particles is a promising technology that is still in an early stage of development and clearly needs further exploitation to extend its use in agroindustry. Therefore, the aim of this review is to discuss the current state of development of stimuli-responsive carrier systems, where the mechanisms of release of active ingredients are activated by different triggers, in response to biotic and abiotic factors, as well as to highlight the perspectives and commercial challenges for the use of smart delivery systems in the agricultural sector.

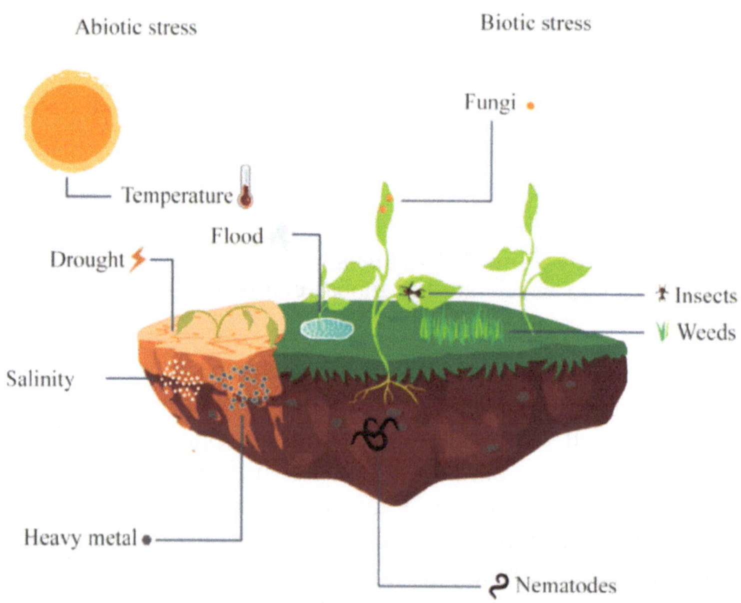

Fig. 46: Abiotic and biotic factors enabling the use of stimuli-responsive nanomaterials

Stimuli-responsive systems in Agriculture: The development of stimuli-responsive smart delivery systems represents a new contribution to the goal of achieving sustainable agricultural practices. (Camara, et al., 2019) Stimuli-responsive nanoformulations can release their active ingredients in response to environmental triggers and biological demands more precisely, by means of targeted delivery or controlled release mechanisms. A variety of factors act to restrict crop growth and production, including biotic stress caused by insects, pathogenic diseases, nematodes, and weed competitors, as well as abiotic factors such as extreme temperatures, poor soils, unavailability or poor uptake of nutrients, metal-contaminated soils, drought, flooding, and high salinity. Therefore, stimuli-responsive materials have been studied for the release of agrochemical compounds in response to pH, temperature, redox conditions, enzymes, and light.

Stimuli-responsive nanocarriers for control of phytopathogens

i. pH-responsive systems : Systems that are pH-responsive are also being studied for the control of diseases caused by phytopathogens. An example is a system based on poly(succinate), which is a biodegradable hydrophilic biopolymer that is liable to hydrolysis at alkaline pH (Wu et al., 2019). A poly (succinate) nanocarrier system was developed intended for the control of diseases affecting plant phloem. The pH of most plant tissues is acidic, while that of the vascular phloem tissue is alkaline. Therefore, this type of nanocarrier system could be used against pathogens such as *Citrus huanglongbing* (known as greening), which specifically infects the phloem, leading to death of the plant.

Systems responsive to pH could also be used to prevent the contamination of fruits by pathogens, which can reduce yields by up to 30%. Contamination of fruits by fungi results in lowering of pH, due to the production of enzymes such as esterases and pectinases. Fruit ripening may also be accompanied by increasingly acidic pH. Nanocarrier systems can be developed to control the growth of contaminant organisms in food, as well as to avoid early maturation. For example, silica nanoparticles coated with chitosan were developed for delivery of the fungicide prochloraz. In the absence of the esterase enzyme and at pH 7, only 18.9% of the fungicide was released for 30 days. However, decrease of the pH and the presence of the enzyme stimulated the system, with 82% release obtained during the same period. At acidic pH, chitosan becomes protonated, exposing the pores of the silica nanoparticles and consequently enabling release of the active agent. A system of graphene oxide nanoparticles was described for the release of salicylaldehyde, where the binding of salicylaldehyde to the graphene oxide nanoparticles was mediated by the hydrazone molecule. At acidic pH, 50% of the active agent was released after 50 h, compared to only 15% at neutral pH. The release was due to the hydrolysis of hydrazone at acidic pH, releasing the active agent and reducing precocious ripening.

ii. Redox-responsive systems: A redox-responsive system was developed for the plant growth regulator salicylic acid, which is a hormone that regulates plant defense mechanisms in the presence of biotrophic pathogens (Yi et al.,

2015). Silica nanoparticles were functionalized with thiol groups, followed by coating with decanediol, forming specific disulfide bonds that could be cleaved in the presence of glutathione. In *in vitro* release assays, there was an increase in glutathione levels, which increased the release of salicylic acid. Plant assays, using Arabidopsis as a biological model, evaluated expression of the PR-1 defense gene that is mediated by salicylic acid. Increase of the glutathione concentration in the plants resulted in the nanocarrier system starting to release the salicylic acid, which induced high expression of the PR-1 gene that can increase the resistance of plants against pests.

Temperature-responsive systems: Silica nanoparticles containing the fungicide thymol were developed, with functionalization using –OH, –NH2, and –COOH groups (Mattos et al., 2018). When the temperature was increased, the nanoparticles functionalized with amino groups showed higher release of the fungicide, compared to the nanoparticles functionalized with hydroxyl and carboxyl groups.

iii. Enzyme-responsive systems: Pectinases from phytopathogenic fungi have been investigated for triggering the release of antibiotics in order to control plant diseases. Liu et al. (2015) synthesized kasugamycin conjugated with modified pectin by amide bonds and evaluated its triggered release performance in the presence of *Pseudomonas syringae* pv. *lachrymans* A sustained-release profile was observed for kasugamycin-pectin, which was controlled by the pectin degradation enzymes produced by *P. syringae*. The release rate of kasugamycin increased with time, indicating that greater quantities of pectinase were released during microbial growth, resulting in faster release of the active compound.

The preparation of enzyme-responsive kasugamycin microcapsules was described, using mesoporous silica conjugated with pectin by addition of disulfide bonds. Dual-responsive kasugamycin microcapsules were obtained, with protective effects against degradation by temperature and light. The delivery system was triggered by either pectinase or gluthatione stimuli, with the release of kasugamycin reaching 88.13% after 30 h in the presence of both stimulants, while values of 68.87% and 77.74% were obtained in the presence of pectinase or gluthatione alone, respectively. The antimicrobial efficacy of the kasugamycin microcapsules against *Erwinia carotovora* was lower during the first few days, subsequently increasing over time due to continuing production of the triggers by the microorganism. It was concluded that the product could remain effective during approximately 3 weeks, indicating its great potential for use in agricultural applications.

iv. For nematode management: pH-responsive systems can be used for nematodes that damage agricultural crops. For example, Liang et al., (2018), developed a nanoparticle system consisting of γ-polyglutamic acid and chitosan for the transport of avermectin, used as a nematicide. The system showed an initial release of 20%, independent of pH, while subsequently there was release of 69.5% at pH 8.5, compared to 60.4% at pH 7 and 57.5% at pH 5.5. The nanocarrier system employed showed high stability at pH 5.5, due to the electrostatic interactions between chitosan and the carboxyl groups of γ-polyglutamic acid. As the pH increased, this interaction diminished, hence increasing the release of the nematicide. In agricultural systems, pH variations occur in soils, in different plant organs, during physiological processes of the plant, and in the maturation of fruits, as well as due to the presence of pathogens or agricultural pests Therefore, nanocarrier systems able to release active agents in response to pH changes represent a major strategy for the development of new nanopesticides, since the release kinetics is modulated according to pH changes.

A hotspot of agricultural nanotechnology is the development of stimuli-responsive nanoformulations able to maintain the stability of the active ingredient under environmental conditions, deliver the active ingredient to the target, decrease its dispersal in the environment, and prolong its biological activity. Currently, although stimuli-responsive systems are well developed in the area of medicine, their applications in agriculture remain limited. The pesticide field also requires continued systematic research for the development of improved environmentally responsive, targeted, controlled-release pesticide formulations. Effective systems that are responsive to internal bio-stimulation present a great challenge in this respect.. In order to ensure the safe use of nanopesticides, it is essential to develop nanoformulations that are based on green nanotechnology and that offer low cost, simple procedures, and controlled-release features. In conclusion, the use of smart delivery nanopesticides is highly promising as an effective tool for sustainable agricultural development.

5.17 Fate of the nano-enabled pesticide formulations in soil

The fate and behavior of nano-enabled pesticides in the environment are likely to be dependent upon the functional characteristics of the carrier and the durability of the a.i. carrier complex (Glen W. Walker et al., 2018). Both characteristics should be considered in problem formulation of nano-enabled pesticides because the spatial and temporal nature of exposure to non-target organisms could change significantly when compared to conventional pesticide formulations. Durability is a measure of how long a pesticide nanocarrier complex maintains its integrity after application in the field. Durability is likely to be dependent upon the exposure conditions. For example, a nano-enabled pesticide may release the a.i. at different rates in soil depending upon factors such as soil moisture or soil pH. These will be important considerations for the risk assessment. The development of nanopesticides has recently received an increased level of attention. However, there are very few data about the environmental fate of these new products, and it is not known whether nanoformulations can be evaluated within the current pesticide regulatory framework (Melanie Kah, et al., 2016). Sorption and degradation parameters of the insecticide bifenthrin were measured in two soils for the pure active ingredient, three nanoformulations, and a commercially available formulation. In most cases, fate parameters derived for the nanopesticides were significantly different from those derived for the pure active ingredient (factors of up to 10 for sorption and 1.8 for degradation), but discrepancies were not easy to relate to the characteristics of the nanocarriers. In some cases, differences were also observed between the commercial formulation and the pure active ingredient (factors of up to 1.4 for sorption and 1.7 for degradation). In the regulatory context, the common assumption that formulations do not influence the environmental fate of pesticide active ingredients after application seems therefore not always adequate. In the absence of direct measurement, an inverse modeling approach was successfully applied to evaluate the durability of the formulations in soil (release half-life ranged between 11 and 74 days). Predicted groundwater concentrations very much depended on the modeling approach adopted but overall suggest that the nanoformulations studied could reduce losses to groundwater.

Fojtová Dana et al., (2019) compared the soil fate and bioaccumulation of four model nanopesticides (chlorpyrifos and tebuconazole loaded on polymeric and lipid nanocarriers) relative to the conventional formulations and pure active ingredients (all added at 0.5 mg kg^{-1}) in microcosms containing earthworms Eisenia fetida and lettuce, *Lactuca sativa* in two soils (LUFA 2.1 and 2.4) over a period of four months. Overall, the microcosm results demonstrated that nanoformulations can significantly alter the fate and bioavailability of pesticides. However, generalisations were difficult to make as the impacts depended on the type of nanocarrier, pesticide, soil used and time scale. We believe that our study contributes towards the critical mass of case studies needed to enable a judging of the benefits versus risks of nanopesticides.

i. Effect on soil health : Globally, there is a lot of research being conducted in the field of application of nanoparticles in agriculture. In recent time nano-based materials like nanoparticles and nanoformulations, nano-based fertilizers, nano-pesticides and insecticides, nano fungicides and other plant disease control formulations are available in the market (Zoya Javed et al.,2019). Though this technology offers many advantages, yet there are some vices which is causing the concern among the researchers and practitioners. The unregulated exposure of the nanoparticles to the soil is expected to cause adverse effects on the soil microbiota and in some cases negatively impact the important adaphic factors like soil infertility and toxicity. The studies conducted so far have shown diverse responses of nanoparticles by soil fungi and soil bacteria. The enzymatic activities of the microbes have been reported to be affected when the stress is induced from different sources. In some studies, silver nanoparticles have been found to decrease FDA- hydrolysis in soil. More research still needs to be done on addressing the nanoparticle induced metal-stress by different microbes. A review by Zoya Javed et al., (2019) provided the comprehensive account of side effects of the nanoparticles on soil fertility and their effect on the different properties of soil and plant growth along with the research gaps in the field of soil application of nanoparticles and way forward to introduce an ecofriendly application of nanoparticles.

ii. Effect of nanoparticles on pH of soil : Soil pH is a significant factor that directly relates to soil fertility and health. It is an indication of acidic or alkalinity of the soil. Soil pH affects the availability of the nutrients for plants. It is proved that the nature of acidity or basicity of soil depends upon the composition of the soil. In general optimum pH range for plants is 5-7. It is observed that the pH of the soil may be influenced by the accumulation of the different type of nanoparticles like Zn, Ag, Au, Cu etc. It was found that the pH of the soil also influences the toxicity of nanoparticles on the microorganisms and nematodes. The effect of organic matter and pH of the soil on the toxicity of ZnO nanoparticles on *Folsomia candida* . It was found greater toxicity the effect of Zn nanoparticles then dissolved Zn on the bacterial community . Very recently found that copper oxide nanoparticles may able to change the pH of paddy soil. Copper oxide nanoparticles cause an increase in the pH of soil which ultimately affects soil property. Uptake of Silver nanoparticles accumulated in soil by insects may also be influenced by the pH of the soil.

Soil pH seems to be the principal factor that governs concentrations of soluble metals. Zn solubility correlates negatively with pH. Moreover, clay content in calcareous soil was higher than in acidic soil, which increase Zn retention in soil and hence, decreases Zn availability to microorganisms, as clays are believed to adsorb zinc ions across both ion exchange and specific adsorption. Zinc availability has also been connected to soil organic matter content. In this study, the soil organic matter content effect of content on Zn availability to microorganisms should be considered insignificant because values were very low and similar in both soils (Zoya Javed et al., 2019). The impacts of nanomaterials are firmly subject to the change forms they experience in nature. AgNM are oxidized in nature, and the release of silver particles might be in charge of the lethal effect. Both the oxidation of AgNMs and the release of particles are reliant on the pH of the soil.

This concurs with previous findings about demonstrating that the toxicity of AgNMs towards earthworms is related with the particle content in the soil pore water. The impact of ZnO nanomaterials is additionally caused for the most part by the released of particles, which proposes the outcomes ought to be similar with those watched for AgNMs. In like manner, the impact of ZnO nanomaterials on the proliferation of *Folsomia candida* demonstrated that the poisonous quality of metallic nanomaterials was reliant on soil pH, and again that the danger expanded in progressively acidic soils. The impact of Zn danger on the multiplication of *Eisenia fetida* is affected by pH for all types of Zn, reflecting the impact of pH on Zn disintegration .

iii. Effect of nanoparticles on soil organic carbon: The effect of the natural carbon content on AgNM toxicity was studied since this parameter may impact the impact of nanomaterials (Zoya Javed et al.,2019). The natural carbon content differed between 0.93% and 3.85%, however there was no connection between the natural carbon substance of the five soils we tried and the lethality of either AgNM or silver nitrate. There are signs in the writing that natural issue can change the impacts of nanomaterials, particularly those which discharge particles. For instance, the poisonous quality of AgNMs towards the freshwater zooplankton *Ceriodaphnia dubia* and *Daphnia magna* declined altogether as the broke down natural carbon content expanded. AgNMs are less versatile in soils with a high substance of natural issue. in light of the fact that natural materials, for example, sewage slop connect with AgNMs, with sorption rates more prominent than 90% for uncoated particles and lower rates for functionalized particles. Concentrates with biofilm networks and unadulterated societies have additionally appeared natural issue, for example, exopolysaccharides and humic acids can secure microorganisms and diminish the poisonous quality of AgNMs. For instance, the nearness of Suwannee River humic acids made it difficult to decide an impact on the development of *Pseudomonas fluorescens.*

In any case, these discoveries are for the most part from investigations in which the impact of natural carbon was tried on amphibian life forms though comparable information for soil life forms is increasingly constrained. It is already demonstrated that AgNMs and the particles discharged by them communicate with natural issue added to soil, diminishing the long haul harmful consequences for the aggregate of microbial exercises bringing about nitrate gathering. The proposed mechanisms for the restraint of AgNM disintegration incorporate surface adsorption onto common natural issue (which thus forestalls AgNM oxidation and the arrival of silver particles) and the decreasing action of humic/fulvic acids which changes over oxidation items once more into silver metal .

iv. Effect of nanoparticles on microbial diversity: As indicated by the examination TiO_2 and CuO nanoparticles diminish the microbial biomass of the form soil and action of the chemical, influenced on their structures in overflowed paddy soil. Comparable outcomes were acquired biometal who contemplated the impact of ZnO, TiO_2, CeO_2, and Fe_3O_4 nanoparticles on enzymatic activities of soil (invertase, urease, catalase, and phosphatase) and bacterial networks of the saline-soluble base and black soils. The outcomes demonstrated an effect on soil chemical exercises and changes in soil bacterial network and danger on natural nitrogen obsession. Results gotten have demonstrated that a high convergence of Fe_3O_4 NPs altogether diminished the substance of microscopic organisms in the soil. Zinc oxide and CeO_2 NPs influenced the plate checks of *Azotobacter*, P-solubilizing, and-solubilizing microscopic organisms and restrain enzymatic exercises (Zoya Javed et al.,2019).

Titanium dioxide NPs diminished the plenitude of utilitarian soil microscopic organisms and enzymatic activities and hindering impact on microbial action, occurrence, and diversity. The biogenic Au NPs up to the grouping of 33 mg kg^{-1} don't influence the soil properties and their processes and can be delegated not destructive. Be that as it may, antibacterial exercises (development of a zone of inhibition) were seen on clinical disconnects (*Bacillus subtilis, Escherichia coli, Klebsiella pneumonia, Pseudomonas aeruginosa, Salmonella typhi, and Staphylococcus aureus*) treated with ZnO NPs incorporated by organic and concoction techniques. Copper ions, released from the Cu NPs, were considered as a major cause of lethality to both the pathogenic and the beneficial bacteria. It was investigated the effect of CuO and AgNPs on leaf microbial decomposition showing that exposure to these NPs led to a decrease in leaf decomposition rate (Table 25).

Table 25: Effect of nanoparticles on soil microbial community and enzymes

Nanomaterials	Major effect
Au	Hinderance in enzymatic activities of soil
TiO_2	The loss in diversity of the bacterial community
Cu	Decreases C, N Biomass
Ag	Loss of microbial community, hindrance in enzymatic activities
Mixture of Cu and Fe	Loss of microbial community
Mixture of Co, Fe, Ni and Ag	Not much effect on the soil community
Ag	β-Glucosidase, acid phosphatase, dehydrogenase, Urease
C_{60}	eucine-aminopeptidase, β-cellobiohydrolase, acid phosphatase, β-Glucosidase, chitinase, xylosidase

It was observed that TiO_2, at high concentrations, can disturb the symbiosis relationship between plant and bacteria and have a negative effect on the growth of clover. In natural soil, the mobility of the nanoparticles was weak and no increase in the uptake of titanium dioxide by the plants was found.

v. On nematode community: The mobility of MWCNT in soils was also weak; the experiments showed a concentration-independent uptake of MWCNT by the plants and a reduction in the number of red clover blossoms in the case of high concentrations. Both TiO_2 and MWCNT in very high concentrations resulted in a change in the composition of the nematode communities that interact with the plants (Table 26) (Zoya Javed et al.,2019).

Table 26: Effect of nanoparticles on soil nematodes community

Test Nanoparticles	Test species
MWNT, SWNT	*Eisenia foetida*
Tio_2, ZnO	*Eisenia foetida*
Al_2O_3	*Eisenia foetida*
Pt	*Caenorhabditis elegans*
Ag	*Caenorhabditis elegans*
ZnO	*Caenorhabditis elegans*
CeO_2, TiO_2	*Caenorhabditis elegans*

Nanoparticles mobility and toxicity in soil: Nanoparticles movement in soil is not affected by gravitational settling but is guided by Brownian motion into the soil pores. In particular, soils are composed of micropores (network of humic materials and soil particles) and macropores; single NPs can enter into the micropores and if they are sorbed on mobile colloids, the mobility is highly enhanced, while aggregates of NPs remain in the macropores, and if sorbed on non-mobile particles their mobility is inhibited (Stefania Mura et al., 2013). The attachment between NPs and soil molecules depends on the shape of NPs and the collector and on the different properties that change the environment surrounding the particles. Therefore, soil conditions can enhance or inhibite NP mobility.

Soils overlying buried gold (Au) and platinum (Pt) deposits are often Au/Pt-anomalous, widening exploration footprints. This may be a result of the dispersion of (bio) geochemically formed Au/Pt nanoparticles and their sorption/deposition by solid soil phases (Reith Frank and Cornelis Geert,2017). However, the influence of soil properties on nanoparticle dispersion, sorption and partitioning are little understood.

Engineered oxide nanoparticles have emerged as a highly promising tool for soil remediation. However, their behaviour under complex environmental exposure (*i.e.* in natural soils) remains poorly understood and a relevant issue to complete their risk assessments (Maya Al-Sid-Cheikh et al., 2019).

The widespread use of silver nanoparticles (AgNPs) as a bactericide will ultimately result in their increased concentration in soils. (Nadir Saleeb et al., 2019). The likely mobility, toxicity, and plant uptake of Ag applied to soil was determined, as either AgNPs or Ag^+. Given the widespread interest in AgNPs, there is limited knowledge on the behavior of Ag^+ in soil and this should be the subject of future research.

The exposure pathways and environmental impacts of titanium dioxide nanoparticles (TiO_2 NPs) released into soils could be significantly influenced by their stability and transport behaviors (Zhang et al., 2020). They investigated the stability and transport of TiO_2 NPs in three variable-charge soils and to determine the key factors controlling these behaviors. It was concluded that the stability of TiO_2 NPs in three variable-charge soil suspensions was dependent on the SOM. However, the mobility of TiO_2 NPs in soils was not directly related to their stability in the soil suspensions. The difference in amorphous iron oxide content could induce the disparity in mobility of TiO_2 NPs in soils.

The mobility of synthetic and virus-based model nanopesticides were tested by combining soil column experiments with computational modeling (Chariou et al., 2019). It was found that the tobacco mild green mosaic virus and cowpea mosaic virus penetrate soil to a depth of at least 30 cm, and could therefore deliver nematicides to the rhizosphere, whereas the Physalis mosaic virus remains in the first 4 cm of soil and would be more useful for the delivery of herbicides. It was confirmed that plant viruses are superior to synthetic mesoporous silica nanoparticles and poly(lactic-co-glycolic acid) for the delivery and controlled release of pesticides, and could be developed as the next generation of pesticide delivery systems.

5.18 Nanoparticle toxicity in soil

In recent years, there has been a strong interest in the use of NPs in soils for different applications, considering NPs as environmentally benign materials, as reported in different papers (Stefania Mura et al., 2013). However, researchers are developing new ways to monitor possible NP toxicity and to evaluate the benefits or risks to organisms of interest in soil science, such as bacteria, fungi, plants and other living beings (fish, rats, etc.). In particular, it was found that TiO_2, SiO_2, ZnO NPs suspended in water for *Bacillus subtilis* and *Escherichia. coli* are toxic for Gram positive and negative, and in the first ecotoxicological test for the earthworm *Eisenia fetida*, AgNPs produced reproductive failure and an avoidance mechanism of this pathogen in soils contaminated by this type of NPs. Negative effects of NPs were also demonstrated on soil bacterial communities that were altered by ZnO NPs, reducing the microbial biomass and diversity.

Furthermore, soil enzyme activities were influenced by treatments with ZnO NPs, with an especially large decrease in dehydrogenase activity in response to Zn^{2+}. With this treatment, also biomass and root length decreased, showing a phytotoxicity due to immobilisation and aggregation of NPs in the soil. Toxicity of ZnO NPs was also demonstrated on ryegrass and in corn where it inhibited germination. Other experiments carried out in plants showed that aluminium oxide NPs and rare earth oxide NPs can be toxic to a variety of crops (carrots, cabbage, soybeans, corn, cucumbers) as inhibitors of root elongation. NPs can also be absorbed from soil into plants, as demonstrated in a study on gold NPs. The effects that these new materials can have on a living organism once absorbed are far from understood and the difficult question is "are NPs taken up into plants toxic when consumed by animals and humans?"

Assessing the environmental risk requires a clear understanding on NP mobility, bioavailability, reactivity, ecotoxicity and persistency. Currently, not much is known about biodegradation of nanomaterials or whether biodegradation products may be toxic. Nations across the worlds are now trying to put global regulation policy on nanomaterials. Sometimes, they are reported to be more toxic than the corresponding ion and micromaterials. Therefore, safety research of nanoparticles has huge implications on a national economics. Woo-Mi Lee et al., (2010) evaluated and analyzed the research trend of ecotoxicity of nanoparticles in soil environment. Test species included terrestrial plants, earthworms and soil nematode. Soil enzyme activities were also discussed. It was observed that the results of nanotoxicity studies were affected by many factors such as physicochemical properties, size, dispersion method and test medium of nanoparticle, which should be considered when conducting toxicity researches. In particular, more researches on the effect of physicochemical properties and fate of nanoparticles on toxicity effect should be conducted consistently.

Nanoparticles are being utilized for the welfare of mankind; but at the same time, the unprecedented use and uncontrolled release of nanomaterials into the environment poses enormous threat to living organisms. Silver nanoparticles (AgNPs) are used in several industries and its continuous release may hamper many physiological and biochemical processes in the living organisms including autotrophs and heterotrophs (Durgesh K. Tripathi et al., 2017). The present review gives a concentric know-how of the effects of AgNPs on the lower and higher autotrophic plants as well as on heterotrophic microbes so as to have better understanding of the differences in effects among these two groups. It also focuses on the mechanism of uptake, translocation, accumulation in the plants and microbes, and resulting toxicity as well as tolerance mechanisms by which these microorganisms are able to survive and reduce the effects of AgNPs.

Increasing exposure to engineered inorganic nanoparticles takes actually place in both terrestric and aquatic ecosystems worldwide. Although harmful effects of AgNP on the soil bacterial community is known, information about the impact of the factors functionalization, concentration, exposure time, and soil texture on the AgNP effect expression are still rare (Grün et al., 2019) .

5.19 Nanomaterials in water

A 2004 US-EPA report estimated that it will take 30-35 years and cost up to $250 billion to clean up the nation's hazardous waste sites. EPA anticipates that these high costs will provide an incentive to develop and implement clean-up strategies and technologies that will result in better, cheaper, and faster site cleanups (Stefania Mura et al., 2013). Nanoremediation has the potential not only to reduce the overall cost of cleaning up large-scale contaminated sites, but it can also reduce clean-up time, eliminate the need for treatment and disposal of contaminated dredged soil, reduce some contaminant concentrations to near zero, and, what is more, can be carried out in situ. In situ nanoremediation methods involve the application of reactive nanomaterials for transformation and detoxification of pollutants *in situ* or below ground. No groundwater is pumped out for above-ground treatment, and no soil is transported to other places for treatment and disposal. Because of the high cost and lengthy operating periods for pump-and-treat remedies, in situ groundwater treatment technologies are increasing.

In addition to groundwater remediation, nanotechnology holds promise in reducing the presence of non-aqueous phase liquids. Recently, a material utilizing nano-sized oxides (mostly calcium) was used in situ to clean up heating oil spills from underground oil tanks. Preliminary results from this redox-based technology suggest faster, cheaper outcomes, and, ultimately, lower overall contaminant levels compared to previous remediation methods. Although the technology is likely a beneficial replacement of current practices for site remediation, potential risks are poorly understood. The factors and processes affecting ecotoxicity are complex, and knowledge of the potential impacts of manufactured NPs in the environment and on human health is still limited.

Nanotechnologies can be applied to water for different purposes as transport of organic or inorganic nutrients in nanocapsules or NPs; modification of concentration and bioavailability of dissolved species; remediation of contaminated water and aquifers; detection of contaminants [pesticides, pathogens, polychlorinated biphenyls (PCBs), pollutants]. In particular, iron nanopowders can be used as effective nanoremediation tools for cleaning up contaminated soil and water. In fact, iron catalyses the oxidation and breakdown of organic contaminants (trichloroethene, carbon tetrachloride, dioxins, and PCBs) to simpler carbon compounds much less toxic. Furthermore, it is extremely effective for binding and removing arsenic from groundwater, something that affects the water supply of millions of people in the developing world, and for which there is no effective existing solution.

Among some of the applications would be nanocomposites of dioxide of titanium, which in presence of ultraviolet (UV)-visible and solar light have bactericide and antiviral properties. Other application would be the hypercatalysis with palladium covered by gold NPs that would eliminate. A variety of nanomaterials is in various stages of research and development, each possessing unique functionalities that is potentially applicable to the remediation of industrial effluents, groundwater, surface water and drinking water. Nanoscale iron particles and their derivatives offer more alternatives to many remediation technologies.

The stability and aggregation behaviour of CeO_2 nanoparticles (NPs) was investigated to predict their fate in the agricultural environment (Ekta Tiwari, et al., 2020). The results showed that critical coagulation concentration (CCC) values obtained for CeO_2 NPs, *i.e.* 26.5 mM and 7.9 mM for NaCl and $CaCl_2$ respectively, drastically decreased to 16.2 mM and 1.87 mM in the presence of bentonite clay colloids, which may lead to their deposition within the soil matrix. However, the presence of bicarbonate ions (0.1–2 mM) along with DOM (1–20 mg L^{-1}) may result in their stabilization and co-transport of CeO_2 NPs with clay in water bodies having low ionic strength. It was also observed that the negative charge of a bentonite clay suspension was completely reversed with an increase in CeO_2 concentration by 37.5 times.

5.20 Nano-eco-toxicity effects on the environment

Over recent years, there has been an increased awareness of the potential risks associated with manufactured nanomaterials. Legally, manufactured nanomaterials are covered by the definition of substances as mentioned in the Registration, Evaluation, Authorisation and Restriction of Chemicals (REACH) legislation (Stefania Mura et al., 2013). Risks associated with substances have to be evaluated. There have been several recent reviews of the emerging health issues concerning products of nanotechnologies. Eco-toxic effects on environmental species have been demonstrated; aquatic species have been the most studied. One of the major problems in testing eco-toxicological fate and effects is the absence of consistent and broadly applicable information on how nanomaterials are to be suspended in the various exposure media used in testing. Particular attention needs to be paid to exposure media, how materials are mixed with the media, and the question of realistic exposures. A reduction in size to the nanoscale level results in an enormous increase in surface to volume ratio, so relatively more molecules of the chemical are present on the surface, thus increasing the intrinsic toxicity. This may be one of the reasons why NPs are generally more toxic than larger particles of the same insoluble material when compared on a mass dose base. The eco-toxicity is likely to be altered by environmental factors that influence the colloid behaviour of particles; these include pH, ionic strength, divalent ions such as Ca^{2+}, and the presence of organic matter.

Studies so far have collected information on the direct relevance to risk assessment, such as lethal concentration estimates, as well as fundamental research on possible mechanisms of toxicity, sub-lethal effects, and uptake processes. Existing risk assessment methods are generally applicable to nanomaterials, but specific aspects need to be developed. They include methods for both estimating exposure and identifying hazards. The highest potential risks come from free, insoluble NPs, either dispersed in a liquid or as dust. The most important properties of a nanomaterial to be characterised for risk assessment include size and distribution of free particles and fibres/rods/tubes. These may be produced during the manufacture, use (including wear), and/or disposal/recycling of the nanoproduct; specific surface area; stability in relevant media (including the ability to aggregate and disaggregate); surface adsorption properties; water solubility; chemical reactivity.

The genotoxic effects of conventional particles are driven by two mechanisms: direct and indirect (mediated by inflammatory processes). One of the main limitations in the risk assessment of nanomaterials is the general lack of high- quality exposure and dosimetry data both for humans and for the environment. One of the issues is the difficulty in determining the presence of nanomaterials, and measuring them correctly on a routine basis in various substrates. In contrast to the other exposure routes, analytical instruments are generally available to determine exposure (size distribution of mass and number) for air-borne nanomaterials. NPs may act via either of these pathways since they can cause inflammation and can also enter cells and cause oxidative stress. There is some evidence that their small size may allow NPs to penetrate into sub-cellular compartments such as the mitochondria and nucleus. The presence of nanomaterials in mitochondria and the nucleus can result in oxidative stress-mediated genotoxicity, and/or direct interaction with DNA. For some manufactured nanomaterials, genotoxic activity has been reported, mainly associated to reactive oxygen species (ROS) generation, while for others contrasting results were obtained. Besides oxidative stress, additional mechanisms of genotoxicity that may be specific for nanomaterials also need to be considered, such as possible mechanical interferences during cell division, and other sources of genotoxic effects (i.e. metal release by nanomaterials).

The high presence of NPs in the environment can lead to toxic effects, and recent studies seem to suggest that their mechanism can be summarised as, NPs can generate reactive oxygen species (redox activity) and pose oxidative stress to organisms; NPs can be adsorbed on cell membrane, disturbing its permeation properties, puncturing cell membrane, and interfering with physiological activities) NPs could retain electrons and disturb electron transfer in organisms (phosphorylation and energy transfer); NPs could interact with proteins and disturb the transfer of biosignals or gene information (Stefania Mura et al., 2013).

Although these seem to be the causes of toxicity, there are also some controversial results. In fact, nano TiO_2 stimulates the growth of spinach when applied to the seeds or sprayed onto the leaves through the increased activity of several enzymes. It also promotes nitrate adsorption and enhances the efficiency of transforming inorganic nitrogen to organic nitrogen but, at the same time, toxicity was reported for algae and daphnids. Several authors showed that these results may be due to a lack of standard procedures of toxicity experiments for NPs. Therefore, experimental procedures such as NP preparation, selection of organism species, environmental conditions and toxicity testing need to be standardized.The safety of nanomaterials is of high priority. Whereas toxicity focuses on human beings and aims at protecting individuals, ecotoxicity looks at various trophic organism levels and intend to protect populations and ecosystems. Ecotoxicity includes natural uptake mechanisms and the influence of environmental factors on bioavailability and thereby on toxicity.

i. Ecotoxicity of Fullerenes : C_{60} fullerenes have been among the first engineered nanoparticles to be investigated with respect to ecotoxicity, starting with the pioneer work of Eva Oberdörster and colleagues. In Oberdörster's first report on fullerene toxicity, she showed that low concentrations of C_{60} (0.5 mg/L) could cause oxidative damage (lipid peroxidation in the brain) and enzyme changes (glutathione reduction in gills) in fish (largemouth bass, Micropterus salmoides). C_{60} ecotoxicity is influenced by mode of C_{60} preparation (Sachindri Rana and Kalaichelvan, 2013). Bacterial tests of the toxicity of C60 fullerenes have shown reduced growth and respiration using *Escherichia coli* and *Bacillus subtilis*. Other studies have shown that bacterial membrane composition and fluidity are sensitive cellular characteristics that are affected by low nC_{60} concentrations. *Pseudomonas putida* decreased its levels of unsaturated fatty acids and increased the proportions of cyclopropane fatty acids in the presence of nC_{60}, possibly to protect the bacterial membrane from oxidative stress. *B. subtilis* responded to a low dose of nC_{60} (0.01 mg/L) by significantly increasing the levels of branched fatty acids, and to a high, growth-inhibiting concentration of nC_{60} (0.75 mg/L) by increasing synthesis of monounsaturated fatty acid. Changes in community structure were indicative of growth impediment for certain species of the indigenous microbial community and demonstrated that even in the presence of a highly complexing medium like soil ecotoxicological effects of nC_{60} may be encountered. The highest concentration of C_{60} was introduced in a dry state without any prior suspension in water. The absence of any ecotoxicological effects could indicate that the bioavailability of C_{60} in the latter experiment was lower in spite of higher total concentrations.

Ecotoxicity of Carbon nanotubes: Unprocessed single-walled carbon nanotubes (SWCNT) and double-walled carbon nanotubes were investigated for ecotoxicity to zebrafish (*Danio rerio*) under different salinity conditions (Sachindri Rana and Kalaichelvan, 2013). SWCNTs induced a significant hatching delay in Zebrafish embryos at concentrations greater than 120 mg/L. Double-walled carbon nanotubes also induced a hatching delay at concentrations of greater than 240 mg/L, while carbon black did not affect hatching. Embryonic development was not affected at SWCNT concentrations of up to 360 mg/L. In an experiment with Rainbow Trout being exposed to dispersed SWCNT prepared using the surfactant sodium dodecyl sulphate (SDS) and sonication, a dose-dependent rise was observed in ventilation rate and gill pathologies (oedema, altered mucocytes, and hyperplasia) at concentrations between 0.1 and 0.5 mg/L. SWCNTs precipitated on the gill mucus, and the authors concluded that SWCNTs can act as a respiratory toxicant in trout. Copepods ingested purified SWCNTs, but this had no significant effects on mortality, development, or reproduction. Also, exposure to multiple-walled carbon nanotubes (MWCNTs) at concentration higher than 1 mg/L induced a dose-dependent growth inhibition of the cells.

Ecotoxicity of Metal nanoparticles : The toxicity of a metal is influenced by several factors like solubility, binding specificity to a biological site, and so forth. The toxic effect or heavy metal poisoning is defined as, any functional or morphologic change in the body produced by an ingested, injected, inhaled, or absorbed drug and chemical or biological agent (Sachindri Rana and Kalaichelvan, 2013). Metal nanoparticles also exhibit antibacterial activities. Metal nanoparticles exert cytotoxicity depending on the charge at membrane surface. Gram positive cells are less prone to nanotoxic effects due to the presence of thicker peptidoglycan layer compared to Gram negative cells. Nanotoxicity may be attributed to electrostatic interaction between nanoparticles with membrane and their accumulation in cytoplasm.

Ecotoxicity of nanocomposites: The promise of nanocomposites lies in their multifunctionality, the possibility of realizing unique combinations of properties unachievable with traditional materials (Sachindri Rana and Kalaichelvan, 2013). Depending on the nature of the nanophase and the matrix, a wide variety of nanocomposites can be prepared. The idea of using polymer-metal nanocomposites can be advantageous from two different points of view. Firstly, the development of polymer-stabilized metal nanoparticles is considered to be one of the most promising solutions to the issue of nanoparticles stability, by preventing their selfaggregation. Secondly, the use of immobilized nanoparticles reduces the chances of their appearance in the environment. In addition, the incorporation of metal nanoparticles into polymeric matrices can endow the polymer with distinctive properties. The properties of metal-polymer nanocomposites are not determined only by the properties of the metal nanoparticles. The formation of metal nanoparticles within the polymer matrices may strongly modify the polymer morphology, for example, due to the appearance of nanoporosity, which enhances the rate of mass transfer inside the nanocomposites as well as some other structural parameters which are of great importance in their practical applications.

Titanium dioxide nanoparticles seem to have a low toxicity to terrestrial organisms, though few studies are published in this area. TiO_2 used in sunscreens are nanocomposites where TiO_2 has been coated with magnesium, silica, or alumina, as well as amphiphilic organics like polydimethyl siloxane (PDMS), and these coatings are modified by ageing. No bioaccumulation of TiO_2 nanocomposites was observed in the experiment carried out in the anecic earthworm Lumbricus terrestris. The stabilization and immobilization of silver nanoparticles in different matrices (nanocomposites) has recently gained great attention from scientists and technologists for two main reasons. (i) Immobilization in the matrix can improve the safety of the material (due to the reasonable doubt of their human toxicity). (ii) The immobilization improves the handling of metal nanoparticles and simplifies their final application.

ii. Ecotoxicity of Oxide nanoparticles: The ecotoxicity of TiO_2 (APS 330 nm), SiO_2 (APS 205 nm), and ZnO (APS 480 nm) nanoparticles to Gram-positive (*Bacillus subtilis*) and Gram-negative (*Escherichia coli*) bacteria in water suspensions containing citrate and low PO4 concentrations was investigated (Sachindri Rana and Kalaichelvan, 2013.). Antibacterial activity generally increased from SiO_2 to TiO_2 to ZnO, and *B. subtilis* was the most sensitive to such effects. Concentration effects were observed for all nanoparticles up to the highest concentrations (5000 mg/L), except for ZnO which showed high and constant inhibition of *B subtilis* at all concentrations from 10 mg/L. Bacterial growth inhibition was also observed under dark conditions where ROS production was expected to be low. Enhanced toxicity induced by preilluminating TiO_2 particles shows that the photocatalytic activity of nanoparticles can last for a period of time.

Magnetic nanoparticles are of great interest for researchers from a wide range of disciplines, including catalysis, biotechnology/biomedicine, and environmental science and technology. Regarding the materials safety, it is important to notice that due to their low level of toxicity and their good magnetic properties, the use of ferrites is very convenient for biological applications. Core/shell metal nanoparticles of this type can be easily prepared by using IMS technique and the resulting nanocomposites can be applied in catalysis or water treatment. In catalysis, for instance, due to the magnetic properties of metal nanoparticles, the nanocomposites can be easily recovered and reused in sequential catalytic cycles, which is particularly important for metal nanoparticles containing platinum group metals.

Phytotoxicity of nanoparticles has been demonstrated for Zn and ZnO as inhibition of seed germination and root growth after 2 h exposure to nanoparticle suspensions in deionized water. Five types of nanoparticles (multiwalled carbon nanotubes, aluminum, alumina, zinc, and zinc oxide) and six plant species (radish, rape, ryegrass, lettuce, corn, and cucumber) were screened. Fifty percent inhibition of root growth (the most sensitive parameter) was observed for nano-Zn and nano-ZnO at approximately 50 mg/L for radish, and about 20 mg/L for rape and ryegrass whereas other nanoparticle-plant combinations showed weaker inhibition.

Results have shown that pure alumina particles significantly reduce the root elongation in all plant species, thus potentially slowing the plants' growth. Since alumina nanoparticles can be released in the atmosphere by exhaust systems, they can also mix with other particles present in the air. Thus the researchers repeated the experiment by loading alumina nanoparticles with phenanthrene (Phen), which is a major constituent of polycyclic aromatic hydrocarbons in the atmosphere and can be absorbed to particular matter in the air. When nanoparticles were loaded with Phen, their phytotoxicity significantly decreased showing no adverse effects on plants' roots. However, the scientists argue that in the field, it is possible that Phen associated to nanoparticles may decompose under UV radiations, thus even the Phen loaded particles in the natural environment may impose adverse effects on plant growth.

iii. Ecotoxicity of other nanoparticles: Quantum dots (q-dots) is a diverse group of substances with different composition, size,and coatings. Some q-dots have a strong cytotoxic effect but data on ecotoxicity are lacking (Sachindri Rana and Kalaichelvan, 2013). The low amounts necessary for detection of q-dots in medical imaging and stringent policies for recovery of medical waste in hospitals indicate that loss of q-dots to the environment will be very low. Dendrimers are repeatedly branched organic molecules occasionally used to form hollow cages for transport of drug or other therapeutic agents in nanomedicine. There has been some indication that dendrimers display cytotoxicity and that cationic dendrimers were more cytotoxic and hemolytic than anionic or PEGylated dendrimers as cationic molecules in general can destabilize cell membranes, resulting in cell lysis.

5.21 Nanoethics

Nanoethics or the study of nanotechnology's ethical and social implications, is an emerging field of study that concerns with the study of ethical and social implications of nanoscale science and technology. With these implications of Nanotechnologies, there has always been the need of regulation concerned with the associated risks. Nanoethics focus on these public and policy issues related to the Nanotechnology research and development. Nanoethics are the ethics of nanotechnology. If one believes nanotechnology is just straightforward applied chemistry and nothing more, then nanoethics becomes the ethics of chemistry at best. Or, if one believes nanotechnology refers only to fanciful mechanisms that in principle cannot exist, then the value of nanoethics is dubious.

i. Ethical issues and analysis: The rapid development of nanoscience and nanotechnology is not simply a technological initiative, but has social aspects as well. While fueled by scientific ingenuity, it is also motivated by political pressures, competition for new international markets, venture capital ambitions, and competing conceptualizations of the public good. There is a sense of urgency that because of potential dangers (such as freely migrating carbon nanotubes penetrating plant, animal, and human cells, or uncontrollable self-assemblers) science must learn how to respond effectively and proactively to avert any consequential and irreversible social and environmental harms (Anonymous, 2020). The use of nanotechnology to design improved surveillance systems raises the issue of the privacy rights of individuals. The potential of nanotechnology to produce powerful and precise new weapons calls into question the purposes of advanced and redefined forms of military combat and intervention. Miniaturization and hybridization of commonly used electronic devices tests the assumption that faster and cheaper is equal to better, and demands examination of how market imperatives could supercede other social goods and respected human values.

ii. Policy responses : Matters to be resolved include how government is to be held accountable for funding stipulations that influence actual nanoscience research, timeline and reporting of results, the ethics of basic research questions that grantees study, and the technologies they are asked to develop. Provisions for access to education and technical training in this new field are also a matter of public policy. Who will pay for and provide the specialized retraining needed for teachers, or for the equipment, facilities and supplies needed for the schools? How will society assure democratic inclusion and full public access in this fast moving new initiative?

There are multiple questions to be considered and new policies to be debated regarding who will receive the benefits of nanotechnology developments, and at what cost and to whom. Ownership, power, and control issues regarding devices and processes that are fundamentally invisible to the human eye present interesting ethical challenges both legally and socially. Some political rhetoric uses the language of competition, describing the international climate of nanoscience initiatives as a race. The very notion of a race raises the questions of why science is in such a hurry and to what end. The issues of who will win this race, and how world powers will implement and control the applications of nanotechnology have not as yet been effectively examined. Public policy must also respond to the potential for private individuals to gain access to the raw materials of nanotechnology, such as carbon nanotubes, or eventually, assemblers. Who, then, will oversee or control the use individuals make of those materials, such as for the building of experimental devices or weapons of mass destruction? To protect society from possible harm, external controls may have to be put in place to regulate and govern the types of nanotechnology that corporations can develop (Anonymous, 2020). Moral responsibility dictates that corporations adhere to rigorous self regulation, abide by widely adopted rules, principles and codes, such as those proposed by the Foresight Institute, and/or become involved in public policy, citizen review groups, and the like.

Public policy must also address the management of nano-related toxicity, release and control of nano-scaled, self-replicating artifacts, subtleties of nano-scaled surveillance mechanisms, inequities in access to power, and other unpredictable nano-related implications for society.

Nanotechnology is a part of our nation's future. Clearly this research could potentially be of great benefit to society through applications in agriculture and food systems. However, any new technology carries an ethical responsibility for its wise application and the recognition that there are potential unforeseen risks that may come with the enormous potential benefit (Stefania Mura et al., 2013). One of the most important aspects of nanotechnology from an ethical point of view is the fact that nanomaterials show novel characteristics compared to the same material at a larger scale. In recent years, several authors have argued that the desirability of novel technologies should be assessed early, when they are still emerging. Such an ethical assessment of emerging technologies is by definition focused on an elusive object. Usually the estimated results, expectations, and benefits of the technology are taken as a starting point. This jeopardises the traditional risk-assessment approach that has, in recent years, represented the most natural option in terms of environmental and other applied ethics, focusing on safety, environment and human health. Consequentialist nanoethics seem to move in a circle.

There are many argumentative schemes in nanoethics. Consequentialist argumentative schemes are fuelled by two general perspectives on technology, viz., the optimistic view, claiming that technological progress is basically beneficial and the pessimistic view of technology, seen as inherently risky. Both perspectives come to terms with uncertainty in their own way. The optimistic view assumes that, in the long term, the new nanotechnology will be shown to be as beneficial as almost all other technology in human history, the presumption being that no technology is inherently morally bad, and that it is only the way it is used that can be drawn into question (Anonymous, 2020). The pessimistic view assumes on its side that we should mistrust progress, technology, and artificial, unnatural things and practices.

Along with consequentialist argumentative schemes, different non-consequentialist schemes have emerged in nanotechnology, although they seem to be less popular and merely add to consequentialist schemes, for example, deontological arguments, justice arguments, good life arguments, etc.. Such argumentative schemes have actually faced the problem of uncertainty and ignorance by trying to scaling down the problem of risks and toxicity, and focusing on different ethically relevant aspects which are prior to or independent of the consequences. However, non-consequentialist arguments, being a minority, reveal that the issue of consequences and risks in nanoethics cannot be satisfactorily bypassed. It is hard to accept that the right thing to do is to ban nanoresearch, also because we know that, if we do, someone elsewhere in the world will continue anyway. Consequently, one of the most reasonable objectives for future nanoethics should be to balance our goal of avoiding risks with the need for further research into nanotechnology in order to develop knowledge about its potential consequences, risks and toxicity.

In a nut shell, through the tools now available, extensions of human hands and eyes such as the atomic force and atomic probe microscopes, allow scientists to observe and manipulate atoms directly, move them, rearrange them, and reconfigure them. The resultant potential, to create atomically built hybrids of synthetic, mechanical, and biological components and turn them into novel devices, suggests that society is embarking on an incredibly powerful, tremendously exciting, but possibly dangerous undertaking (Anonymous, 2020). The development of nanotechnology could mean fundamental and beneficial changes to our relationships with the physical world, as human beings gain greater power to manipulate their bodies and environment. Where might such awesome abilities lead? What will happen when nanoscience and nanotechnology advance enough to achieve the results aimed for by scientists? What society does with this new knowledge may determine the changing substance of the physical, social, cultural, economic, moral, and perhaps even spiritual lives of humankind. Are people fully cognizant of and fully prepared to accept and adapt to those changes? Are science and the public in general proceeding with conscientious commitment? The ethical challenges are as daunting as the technical ones.

5.22 Global Nanopesticides Market

The global Nanopesticide market is valued at US$ xx million in 2020 is expected to reach US$ xx million by the end of 2026, growing at a CAGR of xx% during 2021-2026. This report focuses on Nanopesticide volume and value at the global level, regional level and company level. From a global perspective, this report represents overall Nanopesticide market size by analysing historical data and future prospect. Regionally, this report focuses on several key regions: North America, Europe, China and Japan etc.(Sha, 2020). The use of nanopesticides suggests a significant reduction in the impact of the active substances of plant protection products on the environment and an improvement in quality and quantity of the yields (Anonymous, 2020 a). Although, much of the research in nanopesticides is in the developmental stage, the future of the product is linked to the balance between their advantages in terms of increasing efficiency and reducing the use and impact of plant protection products, combined with their disadvantages related to contamination of nanoparticles by people and the environment.

i. Market size and forecast : The nanopesticides market is observing vibrant growth on account of growing demand for food crop production around the globe, in order to achieve the target indicator 2.1.1 of the Sustainable Development Goals (SDG) framework, and mitigate the demand gap (Anonymous, 2020a). According to the statistics of the United Nations Development Program (UNDP), the number of undernourished people reached 821 million in 2017. Furthermore, growing utilization of numerous types of pesticides globally is also estimated to boost the growth of the nanopesticides market. According to the Food and Agricultural Organization of the United Nations (FAO), pesticides use doubled globally from 2 million tonnes in 1990 to 4 million tonnes in 2017. The market is anticipated to record robust CAGR throughout the forecast period, i.e. 2019-2027. The nanopesticides market is segmented by utility into food crop and industrial crop, out of which, food crop segment is anticipated to hold largest market share on the back of growing need for cultivation of food crops, so as to mitigate the demand supply gap around the globe. By application, the nanopesticides market is segmented into production, protection, harvesting and packaging, out of which, harvesting segment is anticipated to gain significant market share owing to the growing need for controlling global food loss index (GFLI) occurred due to issues faced in the supply chain, i.e. from producer to retail level. According to FAO, around 40% of losses occur at post-harvest and processing levels of developing countries and more than 40% losses happen at retail and consumer levels in industrialized countries.

ii. Growing exports of fresh fruits and vegetables: There is a rising demand among consumers for the consumption of fresh foods and vegetables around the globe. The demand also suffices the need for seasoned and unseasoned products to be made available around the globe throughout the year (Anonymous, 2020a). Agro-economic countries are therefore focusing on exporting fresh fruits and vegetables to mitigate the demand gap, which is raising the demand for the use of nanopesticides in packaging of these fresh foods to minimize loss in the supply chain and maintain the freshness in the products. According to the Agricultural and Processed Food Products Export Development Authority (APEDA), Government of India (GOI), exports of fresh fruits and vegetables grew from 0.67 billion USD in 2010-11 to 1.30 billion USD in 2018-19. Also, the export value have already reached 0.21 billion USD until the month of May for 2019-20.

iii. Market segmentation: In-depth analysis of the nanopesticide market includes the following segments:

By type: Nanoemulsions, Nanocapsules, Nanogels, Nanofibers and Nanoparticles

By application; Production, protection, harvesting and packaging

By target organism; Algicides, Avicides, Bactericides, Fungicides, Herbicides, Insecticides, Miticides, Molluscicides, Nematicides, Rodenticides, Slimicides and Virucides,

By utility: Food Crop and industrial Crop

By end-user: Farmer, pest control agents and others

By region: On the basis of regional analysis, the nanopesticides market is segmented into five major regions including North America, Europe, Asia Pacific, Latin America and Middle East & Africa region. Asia-Pacific is expected to hold largest market share on the back of extensive use of pesticides in agricultural products in the region. Furthermore, the region is known for being the largest producer of many food and cash crops, for instance, China and India are known for being the largest producers of cotton. Moreover, higher demand for food grains, limited availability of arable land, elevating exports, growth in horticulture and floriculture, and others are some of the factors raising the demand for nanopesticides among producers in the region.

The nanopesticide market is further classified on the basis of region as follows:

- North America (U.S. & Canada) Market size, Y-O-Y growth & Opportunity Analysis
- Latin America (Brazil, Mexico, Argentina, Rest of Latin America) Market size, Y-O-Y growth & Opportunity Analysis
- Europe (U.K., Germany, France, Italy, Spain, Hungary, Belgium, Netherlands & Luxembourg, NORDIC, Poland, Turkey, Russia, Rest of Europe) Market size, Y-O-Y growth & Opportunity Analysis
- Asia-Pacific (China, India, Japan, South Korea, Indonesia, Malaysia, Australia, New Zealand, Rest of Asia-Pacific) Market size, Y-O-Y growth & Opportunity Analysis.

- Middle East and Africa (Israel, GCC (Saudi Arabia, UAE, Bahrain, Kuwait, Qatar, Oman), North Africa, South Africa, Rest of Middle East and Africa) Market size, Y-O-Y growth & Opportunity Analysis.

iv. Market segment analysis : The research report includes specific segments by Type and by Application. This study provides information about the sales and revenue during the historic and forecasted period of 2015 to 2026. Understanding the segments helps in identifying the importance of different factors that aid the market growth.

Segments by type, the nanopesticide market is segmented into Nanoemulsions, Nanocapsules, Nanogels, Nanofibers and Nanoparticles.

Segments by application: Production, Protection, Harvesting and Packaging The key regions covered in the Nanopesticide market report : North America (U.S., Canada), Europe (Germany, France, U.K., Italy, Russia), Asia-Pacific (China, Japan, South Korea, India, Australia, Taiwan, Indonesia, Thailand, Malaysia, Philippines, Vietnam), Latin America (Mexico, Brazil, Argentina) and Middle East & Africa (Turkey, Saudi Arabia, U.A.E.).

i. Regional analysis: The Nanopesticide market is analysed and market size information is provided by regions (countries). The report includes country-wise and region-wise market size for the period 2015-2026. It also includes market size and forecast by Type and by Application segment in terms of sales and revenue for the period 2015-2026.

The key regions covered in the Nanopesticide market report: North America, U.S., Canada, Europe, Germany, France, U.K., Italy, Russia, Asia-Pacific, China, Japan, South Korea, India, Australia, Taiwan, Indonesia, Thailand, Malaysia, Philippines, Vietnam, Latin America, Mexico, Brazil Argentina, Middle East & Africa, Turkey, Saudi Arabia, and U.A.E.

ii. Competitive analysis: This section of the report identifies various key manufacturers of the market. It helps the reader understand the strategies and collaborations that players are focusing on combat competition in the market. The comprehensive report provides a significant microscopic look at the market. The reader can identify the footprints of the manufacturers by knowing about the global revenue of manufacturers, the global price of manufacturers, and sales by manufacturers during the forecast period of 2015 to 2019.

The major players in global Nanopesticide market: Marrone Bio Innovations, Bioworks, Valent Biosciences, Andermatt Biocontrol, Stockton, Bayer, Camson Bio Technologies and Corteva.

5.23 Global Nanopesticide Market Forecast (2020-2026)

The market research report is a very useful resource that can help manufacturers, stakeholders, decision-makers, and other market participants to become familiar with every factor impacting the growth of the global Nanopesticide market. The analysts authoring the report have closely studied key strategies adopted by top players of the global Nanopesticide market. The report includes SWOT, and other market analyses to provide a clear and deep understanding of important aspects of the global Nanopesticide market (Anonymous, 2020 b). Readers of the report can become informed about current and future trends of the global Nanopesticide market and how they will impact market growth during the forecast period. The global market size is accurately estimated in terms of revenue as well as volume for each year of the forecast period. The report offers reliable price analysis by region, manufacturer, each type segment, and global price analysis for the period 2015-2020. Furthermore, it provides price analysis by each type segment and manufacturer for the period 2015-2025.

Geographic segmentation: The report offers exhaustive assessment of different region-wise Nanopesticide markets such as North America, Europe, China and Japan, etc. Key regions covered in the report are U.S., Canada, Germany, France, U.K., Italy, Russia, China, Japan, South Korea, India, Australia, Taiwan, Indonesia, Thailand, Malaysia, Philippines, Vietnam, Mexico, Brazil, Turkey, Saudi Arabia, U.A.E. For the period 2015-2025, the report provides country-wise revenue and volume sales analysis and region-wise revenue and volume analysis of the global Nanopesticide market. For the period 2015-2020, it provides sales (consumption) analysis and forecast of different regional markets by each application as well as type segment in terms of volume.

Manufacturers: The report includes comprehensive company profiling of leading and emerging companies competing in the global Nanopesticide market. It provides a detailed list of players operating at a global level. The players have been listed according to the type of products they offer in the global Nanopesticide market and other factors. As part of the company profiling, the analysts authoring the report has provided the market entry year of each player considered for the research study. The report also offers extensive price, volume sales, and revenue analysis by the manufacturer at the global level for the period 2015-2020.

By type and application segments: The report includes a detailed analysis of leading and type and application segments of the global Nanopesticide market. All of the segments covered in the report are broadly analyzed based on some deciding factors. The segmental analysis section of the report offers revenue sales analysis and forecast of the global Nanopesticide market by each type segment for the period 2015-2025. It also offers volume sales (consumption) analysis and forecast of the global Nanopesticide market by each type segment for the same period. Also, it provides volume sales (consumption) analysis and forecast of the global Nanopesticide market by each application segment for the same period.

This report includes the following manufacturers: Marrone Bio Innovations, Bioworks, Valent Biosciences, Andermatt Biocontrol, Stockton, Bayer, Camson Bio, Technologies, Corteva.

Market segment by type; Nanoemulsions, Nanocapsules, Nanogels, Nanofibers, Nanoparticles.

Market segment by application: Production, Protection, Harvesting, Packaging, Research Methodology.

To compile the detailed study of the global Nanopesticide market, a robust research methodology has been adopted that aids in determining the key insights and also evaluates the growth prospects of the Nanopesticide market. Our Research analysts have conducted in-depth primary and secondary research to obtain crucial insights into the Nanopesticide market. To carry out secondary research, the analysts have collected the information through company annual reports, journals, company press releases, and paid databases that were referred to gain and identify better opportunities in the global market.

5.24 Future of nanopesticides

The most urgent needs for research related to environmental impact of nanoparticles is to establish the degree of environmental mobility and bioavailability. These parameters will decide whether nanoparticles can be taken up and cause harm to various organisms (including plants). This is a pre-requisite for ecological damage as well as effects on public health through entry into drinking water and the human food chain. Research on environmental impact of engineered nanoparticles has started and will face major methodological obstacles regarding detection, characterization, and tracing, as well as a dilemma on which nanoparticles to be examined (Sachindri Rana and Kalaichelvan, 2013). The first is due to the small size of nanoparticles and the complexity of the environments where it should be studied (water, sediments, soils, and ecosystems), and the latter to the multitude of engineered nanoparticles that exist and their derivatives (e.g., surface modifications of pristine materials).

To study ecotoxicity of nanoparticles, criteria to be considered include the selection of appropriate terrestrial/aquatic organisms and end points for judging nanoparticles ecotoxicity (or judging fitness for purpose of existing methods), establishing procedures for realistic exposure of organisms to nanoparticles during ecotoxicity measurements, establishing dose-response relationships and toxico-kinetics, including translocation, excretion dynamics, acute versus chronic toxicity, and toxicity mechanisms, (as for basic toxicological studies above). To study environmental behaviour of nanoparticles, the following should be considered, i.e., mobility of nanoparticles in soils, sediments and waste, adsorption/desorption behavior in relation to organic, mineral, and biological components of soil, sediments, and water.

Recent studies have shown that nano-pesticides have the capability to reduce toxic impact of chemical pesticide and provide target specific control of crop pest, can be helpful into development of intelligent nano systems for minimization of adverse problem toagriculture like environmental imbalance, food security and food productivity (Hemraj Chhipa, 2017). These nano system have shown great capability of controlled release pattern of active ingredient make them more efficient for long time period usability that can be solve eutrophication and residual pesticide accumulation problem. In addition, nanopesticide showed improved solubility and stabilities of active ingredient for effective control of pest. Still, there is need to improve the techniques for significant contribution in agricultural practices, some aspects were identified like use of green chemistry and environmental sustainability principals in nanopesticide development to maximize their efficiency; Process development for up scaling of nanopesticide for commercial level; Comparison of nano-formulation activity with preexisting commercial product at field level to determine practical utility; Environmental impact assessment of nanopesticide to determine toxicity level; Improvement in regulation for nanomaterial application in agriculture; Development of smart nanopesticide will provide many solution to the agro-chemical industry i.e. solubility of active ingredient, stability, controlled release and targeted delivery of active ingredient but still lots of research is required to understand the fate of nanopesticide in environment.

Two potential scenarios that the development of nanoagrochemicals might follow in the future were discussed (Melanie Kah, 2015). In the first, developments continue along the current path and nanoagrochemicals are likely to become, or at least be perceived as, the next emerging category of contaminants associated with agricultural practices. Alternatively, nanotechnology could become a potential source of emerging solutions to mitigate contamination by pesticides and fertilizers. This second scenario can only be achieved by rapid changes by industry, researchers, and regulatory agencies, following for instance, some of the directions suggested below.

- Available analyses of nanoagrochemicals typically consider all products discussed in literature, many of which represent an unacceptable risk, without being relevant for agricultural applications (e.g., ingredients that are extremely persistent or whose efficacy is marginal). Increasing collaborations between disciplines that are involved at all stages of the development and evaluation of agrochemicals (e.g., formulation, plant, material, and environmental scientists) will support the development of products fitting within the multiple constraints of the agrochemical sector and that are likely to bring an added value relative to existing products.
- The requirements of the latest EU directive regarding a better evaluation of formulations are typically viewed as additional constraints. Current approaches to chemical regulatory assessment often consist of applying incremental safety factors to account for the increasing level of uncertainty. Alternatively, new science-based tools should be developed to assess and fully exploit what the science of formulation has to offer, based on the risks and benefits over the entire life cycle of the products
- With increasing regulatory pressure and risk of stigmatization, incentives are needed to promote innovation that may lead to the development of more intelligent solutions for plant protection and nutrition. Promotion of more collaboration across sectors (e.g., research, industry, and regulators) and integration of social science and law will ensure public/consumer acceptance and the development of suitable legal frameworks.
- Establishing a size threshold to distinguish "nano" from "non-nano" is of limited value for agrochemicals. Moving to a broader concept of nano- enabled technology and building on the experience from other sectors (e.g., food science, nanometrology) will be more valuable to support the development of more sustainable agrochemicals.

Nanotechnology provides a much better effective way of environment detection, sensing and bioremediation. It can enhance agricultural productivity by using nanoporous zeolotes (for controlled release and efficient amount of water, fertilizer etc.), nanocapsules (for delivering of herbicide, vector and management of pests), nanosensors (for detecting aquatic toxins and pests) , nanoscale biopolymers (proteins and carbohydrates) based nanoparticles with few properties such as low impact on human health and the environment may be used in disinfection and recycling of heavy metals, nanostructured metals can be explored in decomposition of harmful organics at room temperature, smart particles can be useful in effective environmental monitoring and purification processes (Manjunatha et al., 2019).

Nanotechnology will transform agricultural practices including advanced pest management in future. Over the next 20 years, green revolution would be hastened by means of nanosciences. Nanomaterials would be beneficial in the development and formulation of next generation pesticides, insecticides and insect repellents. Thus, nanotechnology is considered as the best possible solution of the problems present in food and agriculture sector.

Bibliography

Abdullah, M.M., 2016Role of Nanomaterials and their Applications as Photo-catalyst and Senors: A Review. Nano Res Appl., 2:1.

Anonymous, 2004. The Impact of Nano-Scale Technologies on Food and Agriculture, Down on The Farm, ETC Group, Canada

Anonymous, 2012. Nanotechnology and human health: Scientific evidence and risk governance. Report of the WHO expert meeting 10–11 December 2012, Bonn, Germany. Copenhagen, WHO Regional Office for Europe, 2013.

Anonymous, 2014. U.S. EPA (2012-11-14). "Nanotechnologies for environmental cleanup". Retrieved 2014-07-29.

Anonymous, 2020 b. Global Nanopesticide Market By Manufacturers, Regions, Type, Application, Forecast 2020-2026. Chemical & Material category. American Research, 110 pp

Anonymous, 2020. Nanoethics. Encyclopaedia.com

Anonymous, 2020a. Chemicals and Advanced Materials, Global Nanopesticides Market, Report ID: 1932

Anupam Kujur, Amrita Yadav, Akshay Kumar, Prem Pratap Singh and Bhanu Prakash, 2019. Nanoencapsulated methyl salicylate as a biorational alternative of synthetic antifungal and aflatoxin B1 suppressive agents. Environ. Sci. and Pollu. Res., 26:18440–18450

Bhan S, Mohan L, Srivastava, C.N.,2014. Relative larvicidal potentiality of nano-encapsulated Temephos and Imidacloprid against Culex quinquefasciatus. J. Asia-Pacific Entomol.,17: 787-791.

Bingna Huang, Feifei Chen, Yue Shen, Kun Qian, Yan Wang, Changjiao Sun, Xiang Zhao, Bo Cui, Fei Gao, Zhanghua Zeng and Haixin Cui, 2019. Advances in Targeted Pesticides with Environmentally Responsive Controlled Release by Nanotechnology. Nanomaterials Basel, 8: 102

Camara, M C., Campos, E.V.R. and Monteiro, R.A. 2019. Development of stimuli-responsive nano-based pesticides: emerging opportunities for agriculture. J. Nanobiotechnol., 17: 100 Cano-Sarabia, M. and Maspoch, D.,2012. Nanoencapsulation. In: Encyclopedia of Nanotechnology. Bhushan B. (eds). Springer, Dordrecht. https://doi.org/10.1007/978-90-481-9751-4_50

Chariou, P.L., Dogan, A.B. and Welsh, A.G., 2019. Soil mobility of synthetic and virus-based model nanopesticides. Nat. Nanotechnol., 14: 712–718

Christoph Lauterwasser, 2016. Opportunities and risks of Nanotechnologies. OECD The OECD International Futures Programme, Allianz Center for Technology , France

Christoph Steinbach, 2019. DECHEMA e.V.Theodor-Heuss-Allee 25, 60486 Frankfurt A.M., Germany

Crane, R. A. and Scott, T. B., 2015. Nanoscale zero-valent iron: Future prospects for an emerging water treatment technology. J. of Hazardous Materials, Nanotechnologies for the Treatment of Water, Air and Soil. 211–212: 112–125

Dana Fojtová , Jana Vašíčková , Renato Grillo , Zuzana Bílková , Zdeněk Šimek , Natália Neuwirthová , Melanie Kah and Jakub Hofman , 2019. Nanoformulations can significantly affect pesticide degradation and uptake by earthworms and plants. Environ. Chem., 16 : 470-481

Danail Hristozov, Stefania Gottardo, Elena Semenzin, Agnes Oomen, Peter Bos, Willie Peijnenburg, Martie van Tongeren, Bernd Nowack, Neil Hunt, Andrea Brunelli, Janeck J. Scott-Fordsmand , Lang Tran and Antonio Marcomini, 2016 . Frameworks and tools for risk assessment of manufactured nanomaterials. Environ. Intl., 30 : 1-18

Durgesh K. Tripathi, Ashutosh Tripathi Shweta, Swati Singh, Yashwant Singh, Kanchan Vishwakarma, Gaurav Yadav, Shivesh Sharma, Vivek K. Singh, Rohit K. Mishra, R. G. Upadhyay, Nawal K. Dubey, Yonghoon Lee and Devendra K. Chauhan, 2017. Uptake, Accumulation and Toxicity of Silver Nanoparticle in Autotrophic Plants, and Heterotrophic Microbes: A Concentric Review. Front. Microbiol., 26 , doi: 10.3389/fmicb.2017.00007

Dwivedi, S., Saquib, Q., Al-Khedhairy, A.A. and Musarrat, J., 2016. Understanding the role of nanomaterials in agriculture. In: Microbial Inoculants in Sustainable Agricultural Productivity. Singh D.P., Singh H.B., Prabha R., (Edrs.), Springer; New Delhi, India: 2016. Edgar Vázquez-Núñez, Carlos Eduardo Molina-Guerrero, Julián Mario Peña-Castro, Fabián Fernández-Luqueño and Ma. Guadalupe de la Rosa-Álvarez, 2020. Use of Nanotechnology for the Bioremediation of Contaminants: A Review. Processes, 8 : 826

Ekta Tiwari, Mithu Mondal, Nisha Singh, Nitin Khandelwal, Fazel Abdolahpur Monikh and Gopala Krishna Darbha, 2020. Effect of the irrigation water type and other environmental parameters on CeO2 nanopesticide–clay colloid interactions.. Environ Sci.- Provesses & Impact, Issul 1, 2020.

Felipe Barrera-Méndez, Diter Miranda-Sánchez , Diana Sánchez-Rangel, Israel Bonilla-Landa, Benjamín Rodríguez-Haas, Juan Luis Monribot-Villanueva and José Luis Olivares- Romero, 2019. Propiconazole nanoencapsulation in biodegradable polymers to obtain pesticide-controlled delivery systems. J. Mex. Chem. Soc., 63 : 50-60

Fernanda D. Guerra, Mohamed F. Attia, Daniel C. Whitehead ID and Frank Alexis, 2018. Nanotechnology for Environmental Remediation: Materials and Applications. Molecules, 23 : 1760

Fojtová Dana, Vašíčková Jana, Grillo Renato, Bílková Zuzana, Šimek Zdeněk, Neuwirthová Natália, Kah Melanie and Hofman Jakub, 2019. Nanoformulations can significantly affect pesticide degradation and uptake by earthworms and plants. Environ. Chem., 16: 470-481. Glen W. Walker, Rai S. Kookana, Natalie E. Smith, Melanie Kah, Casey L. Doolette, Philip T. Reeves, Wess Lovell, Darren J. Anderson, Terence W. Turney and Divina A. Navarro, 2018 . Ecological Risk Assessment of Nano-enabled Pesticides: A Perspective on Problem Formulation. J. Agric. Food Chem., 66 : 6480−6486

Grobe, A., 2012. The IRGC Risk Governance Framework: Applications in food and cosmetics. WHO Workshop on Nanotechnology and Human Health: Scientific Evidence and Risk Governance. Bonn, Germany, 10–11 December 2012.

Grün, A., Manz, W. and Kohl, Y.L. , 2019. Impact of silver nanoparticles (AgNP) on soil microbial community depending on functionalization, concentration, exposure time, and soil texture. Environ. Sci. Eur., 31: 15

Guan, L., Zhang, P., Wang, X.K., Ren, Y.P., Guo, B.B. and Liu, F., 2015. Photodegradation of Pyraclostrobin in Water Environment and Microencapsulation Effect on Its Photostability. J. Agro-Environ. Sci., 34:1493–1497.

Haq, I.U. and Ijaz, S., 2019. Use of Metallic Nanoparticles and Nanoformulations as Nanofungicides for Sustainable Disease Management in Plants. In:. Nanobiotechnology in Bioformulations. Prasad R., Kumar V., Kumar M., Choudhary D., (Edrs.), Springer; Cham, Switzerland: 2019. pp. 289–316.

Harper, 2019. Packaging insecticides in nanocapsules may make them more toxic. Nanowerk News. Oregon State University

Hemraj Chhipa, 2017. Nanopesticide: Current Status and Future Possibilities. Agri Res & Tech: Open Access J., 5: 1-4

Jingzhe Xue, Zhihui Luo, Ping Li, Yaping Ding, Yi Cui and Qingsheng Wu., 2014. A residue-free green synergistic antifungal nanotechnology for pesticide thiram by ZnO nanoparticles. Sci Rep., 4: 5408.

Kumar, R,, Sharon, M. and Choudhary, A.K., 2010. Nanotechnology in agricultural diseases and food safety. J. Phytol., 2:83–92

Liang, W., Yu, A., Wang, G., Zheng, F., Jia, J. and Xu, H., 2018. Chitosan-based nanoparticles of avermectin to control pine wood nematodes. Int. J. Biol. Macromol., 112:258–63

Liang, Y., Guo, M., Fan, C., Dong, H., Ding, G. and Zhang, W., 2017. Development of novel urease-responsive pendimethalin microcapsules using silica-IPTS-PEI as controlled release carrier materials. ACS Sustain Chem. Eng.,5 :4802–4810.

Lienda Handojo, Natasha Agustin Ikhsan, Rino R.Mukti and Antonius Indarto, 2020. Nanomaterials for remediations of agrochemicals. In : Agrochemicals Detection, Treatment and Remediation-Pesticides and Chemical Fertilizers, 535-567 pp

Liu, Y., Sun, Y., Ding, G., Geng, Q., Zhu, J. and Guo, M., 2015. Synthesis, characterization, and application of microbe-triggered controlled-release kasugamycin-pectin conjugate. J. Agric. Food Chem., 63:4263–8.

Lowry, G. V.,2007. Nanomaterials for groundwater remediation. In: Environmental Nanotechnology. Wiesner, M.R.; Bottero, J. (Eds.), The McGraw-Hill Companies, New York, NY, pp. 297-336.

Manjunatha, R.L., Dhananjay Naik and Usharani, K.V/. 2019. Nanotechnology application in agriculture: A review. J. Pharmacognosy and Phytochemistry, 8: 1073-1083

Matthew Slattery, Bryan Harper and Stacey Harper, 2019. Pesticide Encapsulation at the Nanoscale Drives Changes to the Hydrophobic Partitioning and Toxicity of an Active Ingredient. Nanomaterials, 9 : 2-11

Mattos, B.D., Tardy, B.L., Pezhman, M., Kämäräinen, T., Linder, M. and Schreiner, W.H., 2018. Controlled biocide release from hierarchically-structured biogenic silica: surface chemistry to tune release rate and responsiveness. Sci. Rep., 8:5555

Maya Al-Sid-Cheikh, Mathieu Pédrot, Aline Dia, Mélanie Davranche Laurent Jeanneau, Patrice Petitjean, Martine Bouhnik-Le Coz, Marc-André Cormier and Fabien Grasse, 2019. Trace element and organic matter mobility impacted by Fe3O4-nanoparticle surface coating within wetland soil. Environ. Sci-Nano, 10: 22-35

Mehrazar, Elham & Rahaie, Mahdi & Rahaiee, Somayeh, 2015. Application of nanoparticles for pesticides, herbicides, fertilisers and animals feed management. Intl.l J.Nanopart., 8: 1. 10.1504/IJNP.2015.070339.

Melanie Kah, 2015. Nanopesticides and Nanofertilizers: Emerging Contaminants or Opportunities for Risk Mitigation. Front. Chem., 3: 1-6

Melanie Kah, Anne-Kathrin Weniger and Thilo Hofmann, 2016. Impacts of (Nano) formulations on the Fate of an Insecticide in Soil and Consequences for Environmental Exposure Assessment. Environ. Sci. Technol., 50:10960–10967

Melanie Kah, Sabine Beulke, Karen Tiede and Thilo Hofmann, 2012. .Nano-pesticides: state of knowledge, environmental fate and exposure modeling: Critical Reviews of Environmental Science and Technology

Mohammadi Ali, Hashemi Maryam and Hosseini Seyed Masoud, 2015. Nanoencapsulation of Zataria multiflora essential oil preparation and characterization with enhanced antifungal activity for controlling Botrytis cinerea, the causal agent of gray mould disease. Innov. Food Sci.& Emerging Technol., ISSN: 1466-8564

Nadir Saleeb, Ravi Gooneratne, Jo Cavanagh,Craig Bunt, A. K. M., Mofasser Hossain,Sally Gaw and Brett Robinson, 2019. The Mobility of Silver Nanoparticles and Silver Ions in the Soil-Plant System. J Environ.Qty., 48: 1835-1841

Nair, R., Varghese, S.H., Nair, B.G., Maekawa, T., Yoshida, Y. and Kumar, D.S., 2010.

Nanoparticulate material delivery to plants. Plant Sci., 179:154–163

Nam, S., Il Kwak, J. and An, Y., 2018. Quantification of silver nanoparticle toxicity to algae in soil via photosynthetic and flow-cytometric analyses. Sci. Rep., 8: 292

Naveen K. Kalagatur, Oriparambil S. Nirmal Ghosh, Naveen Sundararaj and Venkataramana Mudili, 2018. Antifungal Activity of Chitosan Nanoparticles Encapsulated with Cymbopogon martinii Essential Oil on Plant Pathogenic Fungi Fusarium graminearum. Front Pharmacol. ; 9: 610.

Panpatte, D.G., Jhala, Y.K., Shelat, H.N. and Vyas, R.V., 2016. Microbial Inoculants in Sustainable Agricultural Productivity. Springer; New Delhi, India: 2016. Nanoparticles: The next generation technology for sustainable agriculture; pp. 289–300.

Phenny Mwaanga, 2018. Risks, uncertainties and ethics of nanotechnology in agriculture. New Visions in Plant Science, Özge Çelik, IntechOpen, DOI: 10.5772/intechopen.76590.

Qi Wang, Mengfei Shen, Wenjing Li, Wanwan Li and Fang Zhang,2019. Controlled-release of fluazinam from biodegradable PLGA-based microspheres, J. Environ. Sci. and Health, Part B, 54:: 810-816,

Rahmanifar, B. and Moradi Dehaghi, S., 2014. Removal of organochlorine pesticides by chitosan loaded with silver oxide nanoparticles from water. Clean Techn Environ Policy, 16, 1781– 1786 .

Rajna S., Paschapur A. U. and Raghavendra K. V., 2019. Nanopesticides: Its Scope and Utility in Pest Management. Indian Farmer, 6: 17-21

Ram Prasad, Atanu Bhattacharyya and Quang D. Nguyen, 2017. Nanotechnology in Sustainable Agriculture: Recent Developments, Challenges, and Perspectives. Front Microbiol., 8: 1014

Rao, T.N., Patrudu, .T.B. and Rao, M.V.B., 2016. Degradation Study of Benalaxyl Fungicide Residues in Different pH Waters Using Ferrous Doped Titanium Dioxide Nanoparticles. Anal. Chem. Ind. J.,16:108.

Rashmi Urkude, 2019. Application of Nanotechnology in Insect Pest Management. Int. Res. J. of Science & Engineering, 7 : 151-156.

Reith Frank and Cornelis Geert, 2017. Effect of soil properties on gold- and platinum nanoparticle mobility. Chem. Geol., 466. 10.1016/j.chemgeo.2017.06.033.

Sachindri Rana and Kalaichelvan, P. T., 2013. Ecotoxicity of Nanoparticles. Hindawi Publishing Corporation, ISRN Toxicology, Volume 2013, Article ID 574648, 11 pages

Sahithya, K .and Nilanjana Das, 2015. Remediation of Pesticides using Nanomaterials : An overview. Int.J. ChemTech Res. 8 : 86-91.

Sandeep Kumar, Monika Nehra, Neeraj Dilbaghi, Giovanna Marrazza, Ashraf Aly Hassan and Ki Hyun Kim, 2019. Nano-based smart pesticide formulations: Emerging opportunities for agriculture. J. Controlled Release, 294 : 131-153

Sekhon, B., 2014. Nanotechnology in agri-food production: an overview. Dovepress, 7: 31-53 Sha, F., 2020. Global Nanopesticide Market: Facts & Figures by Region and Country Wise Analysis in a Latest 2020 Research Report. Internet shots, P 95

Shaon Kumar Das, R. K., Avasthe, R. Gopi , Aniruddha Roy and Matber Singh, 2018. Nanotechnology in pesticide formulation: A new era in plant protection. Publication from the ICAR Research complex for NEH Region, Gangtok and Meghalaya, India

Sotelo-Boyás, M. E., Correa-Pacheco, Z. N., Bautista-Baños, S., & Corona-Rangel, M. L.,2017. Physicochemical characterization of chitosan nanoparticles and nanocapsules incorporated with lime essential oil and their antibacterial activity against food-borne pathogens. LWT - Food Science and Technology, 15-20.

Srilatha, 2011. Nanotechnology in Agriculture. J. Nanomedic. Nanotechnol., 2:2-7

Stefania Mura, Giovanna Seddaiu, Fabio Bacchini, Pier Paolo Roggero and Gian Franco Greppi, 2013. Advances of Nanotechnology in Agro-Environmental Studies. Italian J. Agron., DOI: 10.4081/ija.2013.e18

Stone David, Harper Bryan Lynch Iseult Dawson, Kenneth and Harper Stacey., 2010. Exposure Assessment: Recommendations for Nanotechnology-Based Pesticides. Intl. J. Occupational & Environ. Health, 16 : 467-474.

Taghizade Firozjaee , T., Mehrdadi, N. , Baghdadi. M. and Nabi Bidhendi, G. R. , 2018. Application of Nanotechnology in Pesticides Removal from Aqueous Solutions - A review. Int. J. Nanosci. Nanotechnol., 14 : 43-56

Tamara Lazarević-Pašti, Vladan Anićijević, Miloš Baljozović, Dragana Vasić Anićijević, Sanjin Gutić, Vesna Vasić, Natalia V. Skorodumova[f] and Igor A. Pašti, 2018. The impact of the structure of graphene-based materials on the removal of organophosphorus pesticides from water. Environ. Sci.-Nano, Issue 6.

Taranjeet Kaur, Abhishek Sraw, Wanchoo, R.K. and Amrit Pal Toor, 2016. Visible – Light Induced Photocatalytic Degradation of Fungicide with Fe and Si Doped TiO2 Nanoparticles. Materialstoday-Proc.. 3 : 354-361

Taranjeet Kaur, Amrit Pal Toor and Ravinder Kumar Wanchoo, 2015. Parametric study on degradation of fungicide carbendazim in dilute aqueous solutions using nano TiO2, Desalination and Water Treatment, 54 : 122-131

Tentu Nageswara Rao , Benarji Patrudu, T., Basaveswara Rao, M. V., Karri Apparao and Krishna Rao, N., 2016. Photo Catalytic Applications of Ferrous Doped Titanium Dioxide Nanoparticles In Dissipation Kinetics of Tricyclazole Fungicide Residues In Different pH Waters Under Direct Sun Light. IOSR J. Environ. Sci., Toxicol., and Food Technol., (IOSR-JESTFT), 10 : 109-115

Termtanun Mutsee, 2013. Photocatalytic degradation of pesticides using TiO2 nanoparticles. Silpakorn University

Venugopala, N.V.S. and Sainadhb, N.V.S., 2018. Synthesis and characterization of Nano-tricyclazole-A systemic fungicide. Materialstoday: Proceedings. , 5: 16807-16813

Walter de Gruyter , 2017. Environmental Fate and Risks of Nano-enabled Pesticides. Chem. Intl., 39: 24-25

Weria Weisany, Saadi Samadi, Jahanshir Amini, Somaieh Hossaini, Shima Yousef and Filippo Maggi, 2019. Enhancement of the antifungal activity of thyme and dill essential oils against Colletotrichum nymphaeae by nano-encapsulation with copper NPs. Industrial Crops and Products, 132: 213-225

Wiesner Mark, R., Greg V. Lowry, Pedro Alvarez, Dianysios Dionysiou and Pratim Biswas, 2007. Assessing the Risks of Manufactured Nanomaterials. Environ. Sci. & Technol., 40 : 4336–434

Woo-Mi Lee, Shin Woong Kim, Jin Il Kwak, Sun-Hwa Nam, Yu-Jin Shin and Youn-Joo An, 2010. Research Trends of Ecotoxicity of Nanoparticles in Soil Environment. Toxicol Res., 26: 253–259

Wu, C., Zhu, Y., Wu, T., Wang, L., Yuan, Y. and Chen, J., 2019. Enhanced functional properties of biopolymer film incorporated with curcurmin-loaded mesoporous silica nanoparticles for food packaging. Food Chem.,288:139–45.

Xiangying Liu, You Zhan, Zhongqin Zhang, Lang Pan , Lifeng Hu, Kailin Liu, Xuguo Zhou and Lianyang Bai, 2019 . Photocatalytic Degradation of Profenofos and Triazophos

Residues in the Chinese Cabbage, Brassica chinensis, Using Ce-Doped TiO2. Catalysts, 9 : 294

Xiaoxia Zhang, Zhenlan Xu, Mansha Wu, Xiaoting Qian, Daohui Lin, Hangjun Zhang, Juan Tang, Tao Zeng, Weijun Yao, Juliane Filser, Lingxiangyu Li, Virender K. Sharma,2019. Potential environmental risks of nanopesticides: Application of Cu(OH)2 nanopesticides to soil mitigates the degradation of neonicotinoid thiacloprid. Env. Intl., 129: 42-50

Yi, Z., Hussain, H.I., Feng, C., Sun, D., She, F. and Rookes, J.E., 2015. Functionalized mesoporous silica nanoparticles with redox-responsive short-chain gatekeepers for agrochemical delivery. ACS Appl. Mater. Interfaces, 7:9937–9946.

Yifen Shang, Md. Kamrul Hasan, Golam Jalal Ahammed, Mengqi Li, Hanqin Yin and Jie Zhou, 2019. Applications of Nanotechnology in Plant Growth and Crop Protection: A Review. Molecules, 24: 2558

Zhang, R., Tu, C. and Zhang, H., 2020. Stability and transport of titanium dioxide nanoparticles in three variable-charge soils. J. Soils Sediments, 20: 1395–1403

Zoya Javed, Kavya Dashora, Mansi Mishra, Vinayak D Fasake and Ayushi Srivastva, 2019. Effect of accumulation of nanoparticles in soil health- a concern on future. Front Nanosci. Nanotech., 5: DOI: 10.15761/FNN.1000182

6

Nanotechnology in Pesticides Delivery

Biocompatible, biodegradable, intelligent and responsive materials are currently an emerging area of interest in the field of efficient, safe, and green pesticide formulation. Using nanotechnology to design and prepare targeted pesticides with environmentally responsive controlled release via compound and chemical modifications has also shown great potential in creating novel formulations. Special attention was paid to intelligent pesticides with precise controlled release modes that can respond to micro-ecological environment changes such as light-sensitivity, thermo-sensitivity, humidity sensitivity, soil pH and enzyme activity (Huang et al., 2018). Establishing intelligent and controlled pesticide release technologies using nanomaterials were reported. These technologies could increase pesticide-loading, improve the dispensability and stability of active ingredients, and promote target ability.

Problems with traditional pesticide formulations have motivated a new research direction in precise controlled release in recent years. Controlled pesticide release systems are used on farmlands by selecting a suitable and timely administration route to precisely regulate the target pest. This approach aims to minimize crops' demand for pesticides to gradually achieve more effective, safe pesticide usage via intelligent design that slows and controls pesticide release. To date, studies on the controlled release of pesticides show the system only allows for a simple, slow release rather than a responsive release based on environmental conditions. Thus, if we can achieve precise pesticide release according to environmental light, temperature, soil pH, humidity, and enzyme changes, it may be possible to greatly improve the use of pesticides by reducing waste and pollution.

Earlier, some pesticides such as pyraclostrobin have been shown to exert a positive effect on crops affected by rice blast fungus. This chemical can induce physiological changes in crops and improve their resistance to the disease, increasing yields considerably. However, this pesticide is highly toxic to aquatic organisms, which has limited its rice crop application. Seltima, a product of BASF, was introduced in 2016 to address pyraclostrobin's toxicity. Seltima's microcapsule preparation permits slow release; it is released in a directed manner onto rice leaves based on the pesticide's humidity sensitivity. Thus, pyraclostrobin can be safely applied to rice and other crops without harming aquatic organisms. The introduction of this product has led to further studies on environmentally responsive pesticide formulations. For example, the temperature-responsive pyraclostrobin microcapsules prepared by Wang et al. (2017) offer strong temperature response characteristics and controlled-release functions. They embedded abamectin in hollow porous SiO_2 nanospheres, which improved the pesticide's action time and stability while controlling its release rate to increase efficiency.

In the future, nanoscale devices with novel properties could be used to make agricultural systems "smart". For example, devices could be used to identify plant health issues before these become visible to the farmer. Such devices may be capable of responding to different situations by taking appropriate remedial action. If not, they will alert the farmer to the problem. In this way, smart devices will act as both a preventive and an early warning system. Such devices could be used to deliver chemicals in a controlled and targeted manner in the same way as nanomedicine has implications for drug delivery in humans (Tiju Joseph and Mark Morrison, 2006).

Nanotechnology enables companies to manipulate the properties of the outer shell of a capsule in order to control the release of the substance to be delivered. 'Controlled release' strategies are highly prized in medicine since they can allow drugs to be absorbed more slowly, at a specific location in the body or at the say-so of an external trigger (Manjunatha et al., 2019). Pesticides encapsulated in nanomaterials for controlled release, stabilization of biopesticides with nanomaterials, slow release of nanomaterial assisted fertilizers, biofertilizers and micronutrients for efficient use and field applications of agrochemicals, nanomaterials assisted delivery of genetic material for crop improvement, with potential applications across the food chain in pesticides, vaccines, veterinary medicine and nutritionally-enhanced food). Examples of nano and microcapsule designs include the following.

1. **Slow release:** The capsule releases its payload slowly over a longer period of time (e.g., for slow delivery of a substance in the body).
2. **Quick release:** The capsule shell breaks upon contact with a surface (e.g., when pesticide hits a leaf)
3. **Specific release:** The shell is designed to break open when a molecular receptor binds to a specific chemical (e.g., upon encountering a tumour or protein in the body)

4. **Moisture release:** The shell breaks down and releases contents in the presence of water (e.g., in soil)
5. **Heat release:** The shell releases ingredients only whenthe environment warms above a certain temperature
6. **pH release:** Nanocapsule breaks up only in specific acidor alkaline environment (e.g., in the stomach or inside a cell)
7. **Ultrasound release**: The capsule is ruptured by an external ultrasound frequency
8. **Magnetic release**: A magnetic particle in the capsule ruptures the shell when exposed to a magnetic field
9. **DNA nanocapsule:** The capsule smuggles a short strandof foreign DNA into a living cell which, once released, hijacks cell machinery to express a specific protein (used for DNA vaccincs)

Smart release in response to environmental agents has a promising future for revolutionizing agricultural production. The development of nanomaterials and related technologies has provided new ways of creating intelligent nano-pesticides. Nanomaterials feature unique surface properties, a small size, and quantum size effects. Nano-microcapsules or nanospheres have been prepared using light-sensitive, thermo-sensitive, humidity-sensitive, and enzyme- and soil pH-sensitive high polymer materials to deliver pesticides. These nanostructured materials are prepared via processes such as adsorption, coupling, encapsulation, and embedding. Such formulations can protect pesticides' active ingredients and enhance stability, control the release of core materials, reduce or obscure odors, decrease volatility, and isolate components. In recent years, the application of nanomaterials and technologies in the field of pesticides has made considerable strides. In particular, targeted delivery and controlled release ofnanomaterials can improve pesticide utilization and reduce residue and pollution. Pesticide nanocapsule formulations have slow release and protection performance and, due to their small size, improvable pesticide droplet ductility, wettability, and target adsorption when spraying fields; these methods provide efficient and environmentally friendly advantages (Khot et al., 2012). While progress has been made in research on pesticide nanocapsule formulations, preparation technology warrants further study. Popular pesticide microcapsule preparation methods are shown in table 27.

Table 27: Major pesticide microcapsule preparation methods

Method	Preparation Process	Example
Interfacial polymerization	Two reactive monomers are dissolved in two different solvents. When one solvent is dispersed in another solvent, the two monomers undergo a polycondensation reaction at the phase interface to form microcapsules	Natural pyrethrin nanocapsules
In situ polymerization	Two or more water-soluble monomers are polymerized to form a water-insoluble polymer and are deposited on the surface of the core material for coating	S-ethyl dipropylthiocarbamate, acetochlor, atrazine, methotrexate
Emulsion polymerization	A solvent-insoluble monomer is dispersed in a solvent to form a uniform emulsion via mechanical agitation, high-speed shearing, and vigorous shaking with a surfactant (an emulsifier). Then, the polymerization reaction is initiated to form the polymer to achieve core material encapsulation	Abamectin nanocapsule suspension, natural pyrethrin nanocapsules
Membrane emulsification	The dispersion phase enters the continuous phase through a shirasu porous glass membrane under inert gas pressure, and the continuous phase breaks into droplets on the membrane surface by the shear forces of the SPG membrane and droplets.	Chlorantraniliprole nanocapsules, avermectin nanocapsules
Solvent evaporation	The wall material and core material are dispersed in the organic phase, added to the solution immiscible with the wall material, and the wall material is precipitated to form the microcapsule by heating and evaporating the solvent	Spinosad nanocapsules
Nano-precipitation	The interfacial interaction between solvent and non-solvent disperses the polymer and drug from the oil phase into the aqueous phase. This material can quickly wrap the drug and obtain nanocapsules through precipitation	Pyrazole azoxystrobin nanocapsules, azoxystrobin microcapsules
Double coacervation	Two oppositely charged water-soluble polymers form a wall around the water-insoluble pesticide active ingredient, which is a spontaneous liquid-to-liquid separation caused by electrostatic interactions	Azoxystrobin microcapsules

Nanostructured materials can also respond to external stimuli such as temperature, light, humidity, and pH to achieve an environmental response and controlled release of pesticides. Therefore, intelligent microcapsules that can respond to the external environment have attracted considerable attention. Although environmentally-responsive nano-drugs have been rapidly developed and applied in the medical field, the application of these materials to pesticides has only recently begun. Bingna Huang et al., (2018) summarized the progress in the applications of polymer materials sensitive to light, temperature, humidity, soil pH, enzymes, and other factors to enhance pesticide release.

The advances in the development of nano-delivery systems using nature-derived polymers for agri-food applicationswere discussed (Kaarunya Sampathkumar et al., 2020). This review briefs different types of nature-derived polymers currently being used in various applications in the agri-food industry with a special mention on microbial extracellular polymeric materials. The major applications of nano-delivery systems in the food sector, such as food fortification and food preservation, as well as in the agricultural sector for controlled release of agrochemicals using nature-derived polymers are discussed. The review also deals with a perspective on the safety and public perception of nano-enabled foods with a concluding remark on future directions of incorporating nano-delivery systems for agri-food purposes.

The efficacy of conventional pesticide application methods is limited due to loss by leaching, evaporation, and degradation. As much as 90% of the applied pesticide is estimated to be lost, which not only increases the cost of peat management but also leads to environmental pollution. Smart delivery systems using bio-degradable polymers offer several advantages viz., improved specificity, ease and safety in handling, and minimum ecological side effects (Rameash, 2014). Nanoscale delivery system offer additional benefits like lower application rates and enhanced bioavailability owing to their huge surface area/volume ratio. In this perspective, a study has been undertaken to produce nanoparticles incorporated with pesticides as a smart delivery system. Imidacloprid, an insecticide and tebuconazole, a fungicide has been incorporated into poly (lactic-co-glycolic acid) [PLGA] biopolymer (Mw= 50.000 - 75.000 g mol-1) and nanoparticles were produced by electro spraying process. The parameters of the electro spraying were optimised for getting uniform nanoparticles. PLGA @ 6.4 wt % in the acetonitrile incorporated with imidacloprid (10 mg mL-1) and tebuconazole (0.4 mg mL-1) produced uniform nanoparticles with the following parameters of electro spraying viz., needle size 25 G; feed rate of infusion pump @ 30 μl/min; voltage 15 kV and distance between electrodes 10 cm. The morphology of the biodegradable nanoparticles produced through electro spraying was observed by field emission scanning electron microscopy (FESEM).The size of particles ranged from 80 nm to 120 nm in diameter. The resultant smart delivery system of biodegradable nanoparticles was utilised as a successful seed treatment formulation in cotton and wheat.

6.1 Nano-delivery systems

Ideally, nano delivery systems should fulfill certain technical and economical requirements. Table 28 presents a summary of the characteristics of an ideal nano-carrier for biomedical and agricultural purposes (Pablo Vega-Vásquez et al., 2020). First, the materials used as carriers should not trigger any adverse response in the recipient organism. Also, not only the matrix material should be biocompatible, but its degradation products. Second, the mechanical properties of the polymer must provide prolonged protection to its cargo allowing chemical stability over time. Third, the scalability of the fabrication process should be technologically feasible and economically viable. Accordingly, the processes employed for the elaboration of nano-carriers should yield consistent results in a batch to batch basis, in terms of size, polydispersity, encapsulation efficiency, and stability. Finally, the materials to act as nano-carriers should be carefully selected since they not only must meet the technical criteria to address mandatory regulations prior to being commercialized but they also must display good performance in terms of cost/benefit and eco-compatibility.

Table 28: Characteristics of an ideal nano-carrier for agricultural purposes.

Fabrication conditions	Encapsulation properties	Release profile
• Mild conditions	• Stable	• Controlled
• Scalable	• No early cargo release/leakage	• Targeted
• Low-cost	• Non-toxic	• Stimuli sensitive (pH, light, temperature)
• Reproducible	• Biodegradable	
• Low batch-to-batch variability	• Eco-compatible Water soluble	

Table 29 presents a summary of the advantages and disadvantages of drug delivery nano-carriers with potential use in agriculture (Pablo Vega-Vásquez et al., 2020). In general, drug nano-encapsulation depends on the physicochemical nature of the encapsulation matrix, the cargo, and the method to carry out the process. However, regardless of the encapsulation matrix and cargo nature; or the method used to fabricate the drug-loaded nano-carriers, a plethora of reports confirm that some processes to elaborate them have the potential to be standardized since their reproducibility is fairly consistent.

Table 29: Summary of advantages and disadvantages of drug delivery nano-carriers with potential use in agriculture

Type of nano-carrier	Advantages	Disadvantages
Mesoporous silicon-based materials	Stable structure	Inorganic
	Tunable and uniform pore size	Non-biodegradable
	Controlled release of cargo	Potential cell lysis caused by silanol groups interacting with membrane lipids
Solid lipidic nanoparticles (SLNs)	Improves solubility in water of hydrophobic cargo	Low load capacity
	Hydrophilic cargo possible	Low Encapsulation efficiency
	Relatively inexpensive production	High water content in dispersions (70–99.9%)
	Biocompatible/biodegradable	Premature cargo release during storage
	Feasible production scaling-up	
Nano-emulsions	- Highly stable to gravitational separation and aggregation - Improves solubility in water of hydrophobic cargo - Biocompatible/biodegradable - Relatively inexpensive production - Suitable for incorporating lipophilic cargo - Increase efficacy of antimicrobial agents	- High amounts of surfactant needed to achieve oil droplets of nanometric sizes
Dendrimers	Functionalization of peripheral groups determines solubilization and enables targeted delivery of cargo	Cytotoxicity reported on cationic dendrimers
	Suitable for incorporating lipophilic or lipophobic cargo	Toxicity correlated with the number of surface amine groups
	PAMAM dendrimers are reported to be relatively resistant to hydrolysis	Pharmacokinetics, biodistribution, biodegradation, and chronic toxicity of PAMAM dendrimers are not yet clearly understood
Nanocrystals	Carrier-free (i.e., they are almost 100% drug)	Difficult to control morphology and crystallinity of final product
	Improves bioavailability of water-insoluble compounds Improves drug adhesiveness to surface cell membranes	Highly time/money/energy demanding. Need for large amounts of organic solvents (Bottom-up approach)
	Enhances particle stability in suspension	Residual presence of surfactants, solvents or stabilizers (top-down approach)

	Increase drug dissolution velocity	Specialized equipment is needed
Hydrogels	Complete bio and eco-compatibility Relatively inexpensive production Easy to fabricate	Batch to batch variation due to the heterogeneity of the polymer Fine tuning formulation required to achieve stable particles

6.2 Nanoparticles and Nano-carriers

i. Metallic Nanoparticles : Due to their chemical nature, metallic nanoparticles such as gold and silver display enhanced physicochemical properties when presented as nanometric particles (Pablo Vega-Vásquez et al., 2020). Taking advantage of these properties, major efforts on research has focused on the development of devices, predominantly in the biomedical field, for detection and treatment. Chemical sensors are one of the most prominent biomedical applications of metallic nanoparticles. For instance, gold nanoparticles conjugated with specific oligonucleotides can sense complementary deoxyribonucleic acid (DNA) strands, detectable by color changes. Furthermore, gold nanoparticles can be readily functionalized with antibodies and oligonucleotides, enzymes. These hybrid nanostructures are also active elements of a number of biosensor assays to detect gene products in plants, drug and gene delivery systems. Although metallic nanoparticles are widely used in detection, these have limited applications as delivery systems.

ii. Mesoporous Silicon-based Nano-Carriers (MPSNPs): Silicon-based mesoporous materials belong to the group of inorganic nano-carriers widely used as drug delivery systems. This approach takes advantage of the highly stable porous surface of silicon mesoporous materials to fill with bioactive cargo. Ideally, loaded pores are capped and the cargo is released intracellularly. One of the main advantages of MPSNPs is their stability, which confers the ability to cope with physical stress such as temperature and pH variations in their surrounding environment. Moreover, their tunable and uniform pore size (3–50 nm) allows them not only to load relatively high amounts of drug cargo due to their high surface area and large pore volume but to selectively functionalize candidate molecules onto its surface (Pablo Vega-Vásquez et al., 2020) (Fig. 47). The bioactive cargo can be loaded into the porous spaces via passive adsorption or active anchoring. Stimuli responsive caps can be design to prevent early cargo release and detach from its pore allowing controlled release. Targeted cargo delivery can be performed by attachment of targeting agents onto previously functionalized particle surface. Different synthesis protocols to obtain fine-tuned large-pore mesoporous nano-carriers and their suitability in the delivery of proteins, enzymes, antibodies, and nucleic acids were explored.

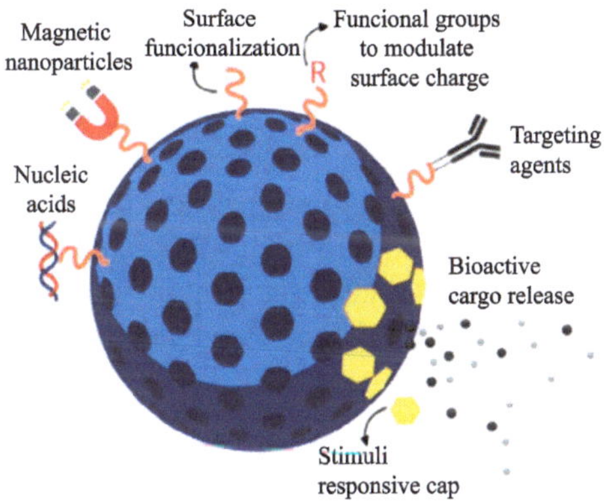

Fig. 47: Mesoporous silicon-based nano-carriers

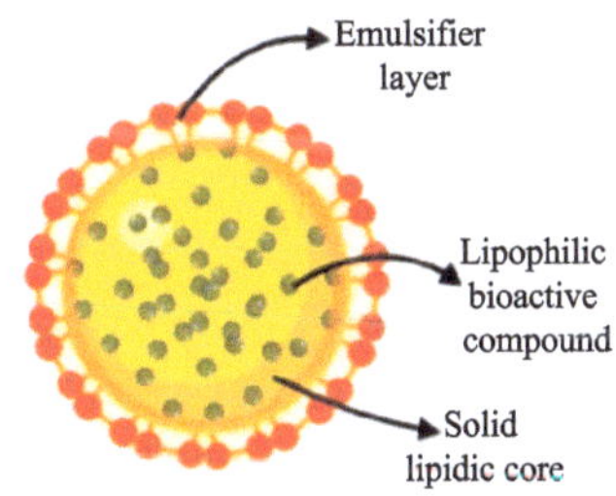

Fig. 48: Solid lipidic nanoparticle

The extended use of silicon-based mesoporous nano-carriers in clinical applications has been delayed due to the lack of pharmacokinetic-pharmacodynamic studies concerning biodistribution, clearance, therapeutic efficacy, and safety are important parameters that need further attention in the quest of providing competent porous nanoparticles For instance, it has been demonstrated that mesoporous silica nanoparticles is not completely hemo-compatible; such phenomena have been attributed to the surface density of silanol groups interacting with the surface of phospholipids or the red blood cell membranes resulting in hemolysis. One of the potential drawbacks of the use of MPSNPs in agriculture is its non-biodegradability and lack of data on bioaccumulation to meet regulatory standards. However, due to their intrinsic physico-chemical properties, the scope of use of MPSNPs include a wide range of applications such as water decontamination through adsorption of radioactive pollutants , separation of dyes, catalysis, delivery of agrochemicals, chromatography to mention a few.

iii. Solid Lipid Nanoparticles (SLN): Solid lipid nanoparticles (SLN) (Fig. 48) are spherical nanoparticles, which makes these ideal candidates for the encapsulation of lipophilic bioactive compounds. The main advantage of SLN relies on their relatively low fabrication cost with the potential for scaling-up of production. However, potential disadvantages for its use in agriculture include poor cargo loading capacity and early cargo expulsion after polymorphic transition during storage (Pablo Vega-Vásquez et al., 2020). During SLN fabrication, a lipophilic bioactive cargo is dissolved in a liquid hot lipid matrix. Under proper formulation and operational conditions, nanoparticles are formed assisted by an emulsifier as the lipidic core solidifies at room temperature.

SLN have been successfully implemented in a wide range of applications. In the biomedical field, for instance, it has been used to increase both the solubility of several poorly soluble drugs just to mention some. In the cosmetic industry, they have been used to encapsulate UV blockers such as 3,4,5-trimethoxybenzoylchitin (TMBC), 2-hydroxy-4-methoxybenzophenone and vitamin E for use as sunscreen. In the food industry, SLNs have been used to encapsulate antioxidant molecules such as ferulic acid and tocopherol, natural antimicrobial compounds, and hydrophobic flavoring agent.

iv. Nano-Capsules : Nano-capsules are nano-vesicular systems in which drugs are enclosed in an inner cavity created by a unique polymeric membrane (Fig. 49). Nano-encapsulation enhances drug delivery and efficacy, but the different methods used for the preparation of nano-capsules frequently produce dispersions with low drug loading. This is a serious disadvantage when the aim is to obtain therapeutic concentrations (Pablo Vega-Vásquez et al., 2020). Bioactive cargo is encapsulated into a core-shell polymeric matrix. Polymer surface can be functionalized and decorated with targeting agents enabling targeted delivery. Similar to NLPs, the application of nano-capsules also extends from the pharmaceutical sector for the encapsulation and delivery of drugs, to the food industry and agriculture, as well as application in cosmetics and personal care in the form of cosmeceuticals.

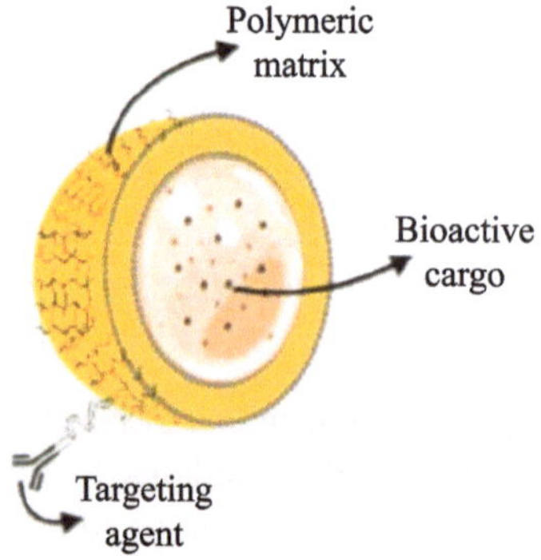

Fig. 49: Core shell nano-capsules for drug delivery

Drug loaded nano-capsules are especially useful for skincare and dermatological treatments because of their enhanced bioavailability in dermal cells. Ebselen (Eb) is an example of a repurposed drug with poor aqueous solubility which requires a sophisticated delivery system such as nano-encapsulation for topical application as a promising, safe and complementary alternative to the treatment of cutaneous candidiasis . Examples of commercially available cosmeceutical products are “Hydra flash bronzer” a facial skin moisturizer, “Soleil soft-touch anti-wrinkle sunscreen," "Soleil instant cooling sun" and "primordiale optimum lip" produced by Lancôme®. These products claim to contain nano-capsules of vitamin E and antioxidant agents as active ingredient. A more comprehensive list of readily available cosmeceuticals products containing nano-capsules and SLN is available in the literature.

iv. Micelles, Liposomes, and Nano-Emulsions : Micelles are spontaneously self-arranged spherical aggregates made of surfactant molecules. Liposomes are spherical vesicles with at least one lipid bilayer, and nano-emulsions are surfactant-assisted homogeneous suspensions of nano-sized droplets of a dispersed phase in a continuous phase. They all display spherical shape and facilitated a controlled release of cargo. Besides their inherent biocompatibility, their surface can be modified and functionalized for conjugation with targeting moieties which enable targeting to specific sites, improving efficacy and potency. In general, liposomes (Fig.50) are used to encapsulate water-soluble compounds because they are comprised of a lipid bilayer separating an aqueous internal compartment from the bulk aqueous phase. Oil in water (O/W) nano-emulsions are used to encapsulate hydrophobic compounds. In contrast, polymeric micelles (Fig. 50A) are used to encapsulate both hydrophobic and hydrophilic compounds depending on the design (Pablo Vega-Vásquez et al., 2020). Block copolymers have a hydrophilic and a lipophilic block. Block-copolymers can easily reach NP size higher than 20 nm and close the liposomes.

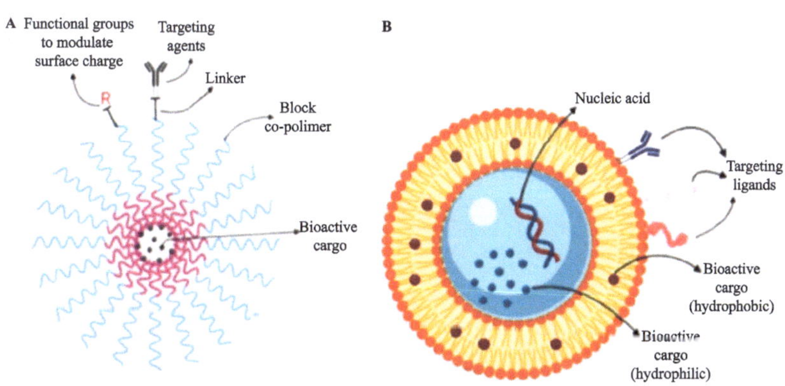

Fig. 50: Polymeric micelles and liposomes. **(A)** Schematic representation of a polymeric micelles composed by a co-block polymer (red and blue wavy lines). The core shell is formed encapsulating the bioactive cargo inside. The surface can be functionalized with linker molecules and further decorated with targeting ligands to enable targeted delivery. **(B)** Depiction of a liposome containing hydrophilic cargo in its core a hydrophobic cargo allocated in the bilayer. Surface functionalization can be achieved by anchoring of targeting ligands such as antibodies, proteins and aptamers

The applications of micelles, liposomes, and nano-emulsions include the encapsulation of poorly water-soluble bioactive molecules to be further incorporated into aqueous products. For instance, for biomedical purposes, a plethora of different types of drug-loaded nano-emulsions is available including oral, topical, intranasal and ocular administration. In the food industry, several types of different nano-emulsions have also been used as carriers of natural occurring, but poorly soluble flavors, colors, preservatives and antioxidant agents. Increased attention has been focused on the nano-emulsification of essential oils because it has been proven that when presented on a nanometric scale, their antimicrobial activity is enhanced. Moreover, its long-term stability is also enhanced. In a recent work D-limonene was used to prevent the formation of biofilms on *E. coli* O157:H7 at sub-lethal doses, by blocking the quorum sensing mediated autoinducer-2 (AI-2) communication and curli-related gene expression

v. Dendrimers: Dendrimer structures are comprised of three components (Fig. 51) a focal core, dendrons, and cavities formed between dendrons. Some of the desirable characteristics of dendrimers are their uniform molecular weight and their three-dimensional structure with peripheral groups that determine solubility, making them relatively easy to design upon specific demands. Further, their smaller hydrodynamic volume and lower molecular volume compared with linear polymers of similar molecular weight. Bioactive cargo can be encapsulated into cavities. Dendron ends can be functionalized allowing targeting ligand attachment, fluorophore molecules, nucleic acids among other molecules of interest.

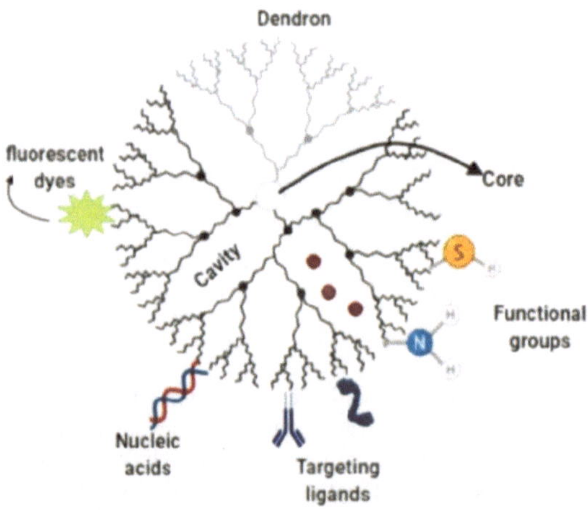

Fig. 51: Dendrimer structure and functionalization; Schematic representation of a Dendron comprising dendron units branching out of a focal core interspaced by cavities

Exposed terminal groups in dendrimeric particles mostly control their chemical interactions with the molecular environment. Their properties such as nanometer size range, ease of preparation and functionalization, also their multiple copies of surface groups displaying stability, make them an attractive system for drug delivery (Pablo Vega-Vásquez et al., 2020). However, despite their initial popularity in drug delivery, at present, serious concerns exist on the cytotoxicity of cationic dendrimers, which has led to further investigation of alternatives to overcome this issue. The toxicity of dendrimers mainly comes from the high cationic charge density in the periphery, where charges interact with the biological cell membrane and then result in membrane disruption.

vi. Nanocrystals: Nanocrystals are another nanotechnological approach to deliver poorly soluble drugs. In contrast to the prior mentioned drug delivery system platforms, nanocrystals have several unique traits. Drug delivery nanocrystals are carrier-free colloidal delivery systems (i.e., they are almost 100% drug). Thus, drug nanocrystals possess the merits of improving the oral bioavailability of water-insoluble compounds, reducing administered dose, avoiding abnormal absorption thus minimizing utilization of large excipients, increasing dissolution velocity, increasing adhesiveness to surface cell membranes, and increasing particle stability. Conventionally, drug nanocrystals can be produced whether from a top-down or a bottom-up approach. The demand for energy, time and money is high for top-down approaches such as milling or high-pressure homogenization. For instance, high-pressure methods require specialized equipment able to deliver up to 1,700 bar for over 100 homogenization cycles, and the milling method requires hours if not days to achieve the desired particle size, depending on the drug properties (Lu et al., 2016). Moreover, the grinding process may contaminate or denature labile drugs which may lead to unexpected side effects on the recipient patient. Further concerns exist on the potential loss of bioactivity and molecular integrity due to severe thermogenesis derived from the milling process. Other disadvantages for the top-down methods are: The lack of complete control of the morphology and crystallinity of the final product; particle aggregation/ agglomeration issues; losses of the product due to drug adherence to equipment surfaces and residual presence of surfactants, solvents or stabilizers.

vii. Nanogels: Nanogels are hydrophilic cross-linked networks forming polymer chains that absorb substantial amounts of aqueous solutions. Due to their conformational tridimensional structure, hydrogels are capable of imbibing bioactive molecules solubilized in water or aqueous fluids. The presence of chemical crosslinks (tie-points or junctions) or physical crosslinks, such as entanglements or crystallites, are responsible for their characteristic conformational structure and size, which can be fine-tuned via chemical control of the formulation and the process to obtain the hydrogel nanoparticles. The main advantage of hydrogels, when used as drug delivery systems, is their complete biocompatibility due to their high content of water. On the other hand, one of the major drawbacks of these types of particles is the batch-to-batch variation due to the heterogeneity of the polymer itself, such as the case for chitosan-based drug delivery systems.

Coacervation or ionic gelation method is one of the most common processes carried out to produce this type of nanoparticles because it is easy to implement and requires un-expensive materials. In general, the process involves the mixture of two aqueous phases, where one of which is the polymer and the other is the dissolved cross-linker. It is common to use an oil/water emulsion as one of the aqueous phases containing the bioactive hydrophobic molecule or drug of interest to be encapsulated within the forming capsule. (Pablo Vega-Vásquez et al., 2020). The method is relatively easy to perform since it does not require sophisticated equipment, which is imperative for the scaling up. However, the final characteristics of the produced nanoparticles, such as size, polydispersity, and stability, are highly sensitive to changes in the fabrication conditions, such as pH, ionic strength, stirring speed, addition rate, and type and concentration of polymers and cross-linkcrs. Chitosan-based nano-carriers are of special interest in agriculture. However, the literature regarding both, nanoparticle formation via ionic gelation and cargo release profile, is overwhelmingly inconsistent.

6.3 Nano-delivery systems based on nature-derived polymers

Nanomaterials based on nature-derived biopolymers are gaining significant attention for their potential applications in both food and agriculture sectors owing to their advantageous properties as well as sustainability by minimizing ecological impacts. The current limitation encountered by inorganic nanomaterials like TiO_2, ZnO, and SiO_2 and organic nanomaterials like carbon nanostructures is the lack of robust information on their toxicity as most of the toxicity studies conducted are *in vitro* rather than *in vivo* (Kaarunya Sampathkumar et al., 2020). Some of the *in vivo* studies conducted are also short-term studies, investigating the effects of high-dose exposure. A comprehensive study on the chronic exposure of such nanomaterials would be more representative of realistic conditions. This, combined with the lack of proper regulation and legislation, has now shifted the focus to nature-derived polymeric nanoparticles owing to their biocompatible and low or negligible toxic properties. Bottom of Form Nature-derived polymeric nanoparticles can be utilized in varying applications including nanoherbicides, nanodetectors, and nanofertilizers to resolve conventional challenges of agriculture including environmental contaminations and human health concerns. These engineered nanoproducts help enhance food production and nutritional value by improving the quality of pesticides, fertilizers, and growth regulators. For instance, nanocarriers are employed to carry and deliver pesticides in a more controlled and slow release profile to achieve "precision farming," which targets only crop productions without affecting the water and soil. The main advantage of nanoencapsulation of pesticides is that encapsulation greatly reduces the amount of pesticides used, by solubilizing them and providing a controlled release of these chemicals at the target site. This in turn checks the development of resistant pest species that arise from overexposure and contamination of soil and water bodies with excess chemicals.

Nano complexation is an encapsulation technique using cyclodextrins, which are cyclic derivatives of starch containing a hydrophilic exterior and a hydrophobic interior, forming excellent complexes with molecules that have low solubility in aqueous media. Campos et al. have exploited this property of cyclodextrin to form inclusion complexes with two plant-derived compounds, carvacrol and linalool, that have insecticidal and repellent properties. The cyclodextrins were then used to functionalize chitosan before forming chitosan nanoparticles to facilitate easier application to crops. Given the high volatility of the natural compounds, nano complexation improved the lifetime of the compounds when exposed to the environment. Another work from the same group uses zein, a nature-derived protein, to form nanoencapsulates of plant-derived insecticides, namely, citronella oil, geraniol, eugenol, and cinnamaldehyde. The use of such botanical extracts helps to check development of resistance in insects to commonly used insecticides as well. Encapsulation was shown to help prevent the degradation of the compounds, lower the IC_{50}, and also provide prolonged release of compounds.

Biopolymeric nanocarrier made up of natural polymers, chitosan, and pectin was developed by Sandhya et al. to provide a sustained and controlled release of encapsulated carbendazim with good bio-efficacy and inhibition against fungi such as *Fusarium oxysporus* and *Aspergillus parasiticus*. Moreover, the phytotoxicity assessment showed that carbendazim-loaded polymeric nanoformulation provides better germination and root growth for the seeds of *Zea mays*, and *Cucumis sativa*. Encapsulation in chitosan alginate systems has shown to decrease the toxicity by 50% and improve the efficiency of two herbicides, imazapic and imazapyr, used to combat weeds in maize and peanut fields. Similar studies have also been done previously to demonstrate the ability of nanoencapsulation in chitosan nanoparticles to reduce the toxicity of the herbicide, Paraquat, by four folds, while preserving its activity. Nanocapsules prepared by the polyelectrolyte complexation of chitosan and alginate were used to retard the fast release of the water-soluble insecticide acetamiprid. The carrier was able to slow down the release of the chemical, providing a controlled release of the chemical for 36 h (Kumar, et al., 2015).

A nanoscale platform that targets and delivers nanomaterials with biochemicals to plant photosynthetic organelles (chloroplasts) was demonstrated using a guiding peptide recognition motif (Santana et al.,2020). Quantum dot (QD) fluorescence emission in a low background window allows confocal microscopy imaging and quantitative detection by elemental analysis in plant cells and organelles. QD functionalization with β-cyclodextrin molecular baskets enables loading and delivery of diverse chemicals and nanoparticle coating with a rationally designed and conserved guiding peptide targets their delivery to chloroplasts. Peptide biorecognition provides high delivery efficiency and specificity of QD with chemical cargoes to chloroplasts in plant cells in vivo (74.6 ± 10.8%) and more specific tunable changes of chloroplast redox function than chemicals alone. Targeted delivery of nanomaterials with chemical cargoes guided by biorecognition motifs has a broad range of nanotechnology applications in plant biology and bioengineering, nanoparticle-plant interactions, and nano-enabled agriculture.

6.4 Novel adhesive pesticide nano-delivery system

Developing new eco-friendly pesticides is the foundation for preventing major biological disasters and ensuring food security. However, it has been estimated that 70% to 90% of the applied pesticides are either lost in the air or run- off, which has caused many adverse effects, such as pest resistance, risk to humans and non-target organisms and environmental contamination. In the recent decade, the development of nanopesticide formulations has shown a great potential to improve the performance of pesticides by constructing nano-delivery systems. Moreover, in terms of the crop foliage microstructure, the surface of nanoparticles can be easily modified by affinity groups to improve adhesion and decrease the loss from crop foliage (Changjiao SUN, 2019).

New formulations of avermectin (Av) were developed. In order to decrease the organic solvent pollution and prevent premature degradation of avermectin, stable and controlled release of pesticide, formulations with high affinity for crop foliage and long retention time on crop foliage were constructed to increase the effective utilization rate of pesticides and minimize loss to the environment. The efficacy of the nanoformulations was evaluated as well on pests and non-target species. First, surfactant-modified mesoporous activated carbon (MAC) was employed to absorb Av in order to improve its photo-stability and allow for sustained release of avermectin. Sodium dodecyl sulfate (SDS) modified MAC had excellent absorption of avermectin, and the absorption could be represented by the Langmuir isotherm model. The Av-MAC-SDS delivery system significantly improved sustained release of avermectin and also effectively inhibited the photo-degradation of avermectins (Changjiao SUN, 2019). Then, biocompatibility and biodegradable polylactic acid (PLA) was employed as the carrier material.

Tannic acid (TA), a bioadhesive natural molecule, was applied to modify abamectin nano-delivery systems to enhance retention time on foliage. The nanoparticles showed excellent continuous sustained release and photo-stability. Compared with unmodified nanopesticides, the retention rate of modified nanoparticles on the foliage was remarkably enhanced by more than 50% and indoor toxicity with dipping method against Myzus persicac was also increased. Finally, insecticidal activity of PLA-based nano-formulated abamectin was examined on the pea aphid, Acyrthosiphon pisum and the aphid predator Adalia bipunctata. A Potter Precision Laboratory Spray Tower was used to conduct direct spray laboratory bioassays. A comparable insecticidal effect of TA modified nanoformulation was observed compared to commercial emulsifiable concentrate (EC) against the aphid. And the nanoformulations had lower stomach toxicity on non-target lady beetles. These results are expected to be beneficial to develop novel leaf-adhesive nanopesticides with high retention time and bioavailability.

A nanocapsule is composed by a shell that contains an active compound, like an agrochemical product for the protection of the plant against pests or diseases. The shell can be constituted by different elements, such as polymers, lipids, viral capsids or nanoclays. Its main function is to protect the active compound until it is released, but it can also improve the solubility and the penetration of the compound into the plant tissues (Alejandro Perez-de-Luque, 2014). Depending on the specific characteristics of the shell, the active compound can be released slowly and gradually, or completely after the shell opening is triggered by certain circumstances (e.g. pH changes or enzymatic degradation). Nanoparticles have a solid core or a matrix that can be composed by different materials (such as metals or polymers) and is surrounded by linkers and biomolecules. Due to the small size, the ratio between surface area and volume is increased in the nanomaterials (compared with bulk forms), improving the biochemical reactivity and conferring unusual and valuable physical properties (e.g. superparamagnetism).

An example of application of nanocapsules for plant protection delivery is the use of nanodisks for delivering amphotericin B, an important antimicotic. The nanodisk is a matrix composed by a bilayer of phospholipids containing the molecules of amphotericin inside.. This structure protects amphotericin molecules against the degradation by external agents (e.g. pH or light) while improving its solubilisation. While medical applications are intended to protect or cure one individual at a time, plant protection in agriculture is a massive treatment for thousands of plants. Therefore, it is important that the active compound is applied in relatively small doses to cover large plant surfaces. The characteristics of the product must thus be designed taking into account these wide treatments, and also if the mechanism of action is systemic or by contact. An important aspect concerning plant protection products involves the way in which they are absorbed by the plant and their translocation within the plant tissues and organs. The formulation of the product varies if the absorption of the active ingredient takes place through the leaves or the roots. Absorption through the roots could be easier due to their biological role for nutrient assimilation, but the advantage of getting absorption through leaf tissues is that they are more easily available for field treatments. However, the knowledge about how these mechanisms work with nanomaterials is still to be improved. Magnetic carbon-coated nanoparticles can be tracked and used to analyse nanomaterials behaviour moving all along the plant structure. They are easily detectable and their magnetic properties allow their accumulation in tissues by way of using magnetic fields.

The use of nanocapsules and nanoparticles for plant protection products offers important advantages. Since active compounds are protected in capsules, they are not degraded by external agents or the crop plant itself, and are not involuntarily dispersed into the soil, allowing the use of a reduced amount of active compounds for plant treatments and consequently causing a lower environmental impact. Other environmental benefits of using nanodevices, e.g. nanoclays, derive from the potential reduction of leaching and further water contamination. Additionally, thanks to the protective capsule, it is possible to use also labile chemical products for plant protection, which can be less harmful to the environment and are currently not employed in agriculture because of their quick degradation. Finally, nanoparticles linked with biomolecules with specific affinity (e.g. antibodies or aptamers) guarantee the selectivity and specificity of targets.

Delivery systems in agriculture are important for application of pesticides and fertilizers as well as during genetic material mediated plant improvement. Application systems for pesticides need to focus on efficacy enhancement and spray drift management while fertilizers face problems of bioavailability due to soil chelation, over-application and runoffs. A viable alternative for these problems is provided by controlled delivery systems for the pesticide and fertilizer application. Controlled delivery technique aims towards measured release of necessary and sufficient amounts of agrochemicals over aperiod of time, to obtain the fullest biological efficacy and to minimize the harmful effects. For this purpose micronic and sub-micronic particles were explored as agrochemical delivery vehicles (Manjunatha et al., 2019).

In comparison to micronic particles (1000 nm), nano particles (100 nm) offered the advantage of effective loading due to the larger surface area, easy attachment and fast mass transfer. Controlled release of the active ingredient is achieved due to the slow release characteristics of the nanomaterial, bonding of the ingredients to the material and the environmental conditions. In case of genetic material, delivery systems face challenges such as limited host range, transportation across cell membrane and trafficking to the nucleus. Nevertheless, the use of NPs assisted delivery for genetic material to develop insect resistant plant varieties is being studied. For instance, DNA-coated gold NPs are used as bullets in 'gene gun' system, for bombardment of plant cells and tissues to achieve gene transfer.

Infield, application of NPs for the delivery of pesticides and biopesticides face several challenges such as multiple environmental perturbations, large areas under spray coverage and finally cost effectiveness. In the usual spraying regime the whole crop is sprayed with the chemical for the ease of application involving a high volume, low value preparation. Whereas nanomaterial based preparations are expected to involve a low volume, high value applications. Such controlled nanoparticulate delivery systems will require a targeted delivery approach focused using the knowledge of the life-cycle and the behaviour of the pathogen or pest.

6.5 Encapsulation of Active Ingredients

A review was organized to critically assess the significant role of nanotechnology for encapsulation of active ingredients (AIs) for pesticides (Sandeep Kumara et al., 2019). The smart delivery of pesticides is essential to reduce the dosage of AIs with enhanced efficacy and to overcome pesticide loss (e.g., due to leaching and evaporation). The future trends of pesticide nanoformulations including nanomaterials as AIs and nanoemulsions of biopesticides are also explored. This review offers a valuable guide for establishing regulatory frameworks related to field applications of these nano-based pesticides in the near future.

Nanocarriers or nanoencapsulation: For integrated pest management, the potential uses of nanotechnology lie in the controlled delivery of active ingredients (AIs) with enhanced activity at low drug concentration and efficient monitoring of pesticide interactions with the environment. Nanocarriers are highly useful for the protection of pesticides against harsh environmental conditions (like high temperature and radiation), while considerably increasing their chemical stability. These also increase the dispersion and wettability of pesticides, thereby lowering the risk of organic solvent runoff. The other interesting properties of pesticidal nanoformulations can be seen in large surface area, thermal stability, biodegradable nature, and increased affinity to the target pest species. Smart delivery systems have been developed for agricultural purposes to regulate the release of AIs at the target site in a designated manner. These systems are also effective to monitor the effects of pesticides on people, animals, insects, plants, soil, and the environment. These systems can be operated for individual goals or a combination of different ones such as time-controlled release, spatially-targeted release, and self or remotely-regulated delivery to avoid the biological barriers in successful targeting.

The main functions of nanocarrier or nanoencapsulation include, to protect the AIs before their target-specific release, to improve the solubility and penetration of the AIs into the targeted plant tissues and to modify or control the release functions of AIs into the surrounding environment. All these factors are mainly dependent on the specific characteristics of the encapsulation material. The encapsulation of AIs improves their mode of action with reduced toxicity (Maruyama et al., 2016). The controlledrelease formulations are highly desirable to maximize the utilization efficiency of pesticides. For instance, a stimuli-responsive nanoformulation can intelligently respond to the stimuli produced by the pest and/ or the surrounding environment that ultimately triggers the release of the AIs to control the pests effectively. In particular, nanoparticles can achieve a high rate of solubility and large coverage of plant/ leaf surfaces due to their high surface area.

The nano-encapsulation materials are efficient delivery system of pesticides for their controlled release as prolonged release of AIs is desirable for plants. These nanomaterials can be designed to specify the controlled release of AIs in terms of slow release, quick release, specific release, moisture release, heat release, pH release, ultrasound release, magnetic release, and DNA-based release profile. The design of encapsulation material is significantly important not only in defining the release profile of AIs but also to protect the AIs from surrounding environment conditions, the remarkable photostability of pesticidal nanoformulations is desirable to prolong their effect on target pests (Sandeep Kumara et al., 2019). Beyond these properties, the nanoformulations help in better and improved functioning of pesticides.

The properties of several nanomaterials for the encapsulation of AIs are discussed in detail including polymeric nanomaterials, lipid nanoparticles, nanoclays, metal organic frameworks (MOFs) and other greener nanoformulations (Sandeep Kumara et al., 2019).

i. Polymer-based encapsulation : Polymeric nanomaterials or nanocomposites are most commonly used for the encapsulation of AIs, mainly due to their environmentfriendly and biodegradable nature. These nanomaterials do not result in the production of any harmful by-products. The development of several synthetic and natural polymers (such as polyethylene glycol [PEG], poly-ε-caprolactone [PCL], cellulose, chitosan, and alginate-gelatin) have been reported for encapsulation of AIs . Different polymeric nanoformulations have been reported to offer significant reductions in the consumption of AIs while maintaining nontoxic behavior, low cost, and biodegradable properties. The use of functional nano-dispensers (i.e., PLA) was reported for encapsulation of imidacloprid with approximately 200 times lower concentration than commercial formulations against phytopathogen vectors.

Polymeric nanoformulations can provide efficient utilization of pesticides while reducing the detrimental effects on the environment and human health. These effects are mainly due to the presence of organic solvents and surfactants in pesticide formulations. Therefore, research efforts have been directed toward maximizing the efficiency of pesticides along with minimizing the detrimental effects of appurtenants. The drug-release profile is strongly dependent on the type of polymeric nanoformulations, which include granule-based, water-based, microcapsule, and microemulsion formulations (Sandeep Kumara et al., 2019). The nanocapsules offered pH-dependent release behavior for acetamiprid with reduced side effects. Moreover, the selection of a particular surfactant strongly influenced the size and drug-dispersity of the polymeric nanoformulation that ultimately affected their drug-entrapment efficiency. A slow drug-release profile is desirable to improve pesticide efficiency and to help prolong the controlled effects of the drug in the surrounding environment.

Poly (N-isopropylacrylamide) (PNIPAm)-capped polydopamine (PDA) microspheres were reported that exhibited a high drug-loading capacity of imidacloprid and excellent NIR-light and temperature-sensitive release profiles (Xu et al. 2017). The drug-release rate is strongly dependent on the temperature and shows negligible drug release at 15 °C, ~20.5% drug release within 5h at 25 °C and 64.3% drug release within 5 h at 40 °C. This temperature-dependent drug-release profile is beneficial in identifying the best surrounding environmental conditions for field application of these nanoformulations. Under NIR light irradiation, the cumulative drug release was 15.7% after 0.5h than 5.2% without light. The drug release rates should be slow to maintain long-term toxicity of the pesticide nanoformulations to the targeted pest.

There are different types of parameters for polymeric nanoformulations that influence the composition of pesticides including solubility and particle size stability. These parameters can be crucial in determining the performance of poorly water-soluble pesticides that need solvents and/or cosolvents for nanoformulations (Sandeep Kumara et al., 2019). Nanogels are promising nanocarriers due to their high stability, even in the caseof a diluted medium. However, the amounts of cross-linker and nanofiller strongly affected the swelling behavior of the hydrogels.

Micelles are very important in the controlled delivery of pesticides or other bioactive substances. In micelles, hydrophobic cores act as reservoirs for several bioactive agents, while the hydrophilic nature of the outer shell helps in steric stability, aqueous solubility, and inactivation of AI in the delivery process. Among the various stimuli for drug release, the photoresponsive properties of micelles are of significant interest and can affect spatial and temporal factors for the controlled release of AIs in response to external stimuli (e.g., light). The photo-controlled sustained release behavior of diuron-loaded 2-nitrobenzyl succinate-carboxymethyl chitosan (NBS-CMCS)was reported. Research efforts have been focused on the selection of particular nanofillers in the bio-sourced polymer matrix to combine advanced properties (e.g., slow drug-release profile with water resistance and good mechanical properties). A montmorillonite (MMT) reinforced wheat gluten matrix was fabricated for encapsulation of ethofumesate. The nanoclays were potentially useful in modulating the transfer properties of AIs in the polymer(Sandeep Kumara et al., 2019). Exploration of interaction between nanofillers and the polymer matrix with pesticides can further help control the drug release pattern more efficiently.

ii. Lipid nanomaterial-based encapsulation: Lipid nanomaterials have been reported as promising colloidal nanocarriers for controlled drug-delivery systems due to their beneficial properties such as good physiochemical storage stability, non-toxic nature, high loading capacity, feasibility, and target-oriented release profile (Sandeep Kumara et al., 2019). Among the different types of lipid nanocarriers, nanostructured lipid carriers (NLC), solid lipid nanocarriers (SLC), and nanoemulsions are the most commonly used nanoformulations. The NLCs have already been reported for nutraceutical delivery due to their encapsulation stability and safety for food systems. However, only a few studies have been reported on their use as pesticide nanocarriers. Lipid nanocarriers have been developed to overcome the photodegradation of pesticides without involving the use of any UV absorbers.

In comparison to other drug delivery systems (e.g., polymeric nanoparticles, emulsions, and liposomes), SLNs have several benefits including: the ability to incorporate both hydrophobic and hydrophilic AIs, reduced chemical degradation of AIs, and feasible large-scale production. The physical state of the lipid matrix of nanocarriers has a significant effect on their penetration into plant roots and their further transportation to the stem to target pests entering the plant system. Three different lipid nanocarrier formulations were fabricated including lipid-based nanoemulsions, NLC, and SLN through a combination of hot homogenization and sonication methods with corn oil (as liquid lipid) and beeswax (as solid lipid). The nanoemulsions penetrated faster into the roots, taking only 1 day as compared to NLC (i.e., 3 days) and SLN (i.e., 6 days). These approaches were effective in prolonging the protection time against pests by minimizing the chances of pesticide loss under diverse environmental conditions. However, it is desirable to further explore the interactions between lipid nanocarriers and the stem cells of the plant. In fact, the SLN did not exhibit any phytotoxic effects during germination and seeding development in the case of *Zea mays, Pisum satium*, and *Brassica rapa*. In contrast, such patterns were no longer evident in the case of polymeric chitosan/tripolyphosphate nanoparticles. The SLN loading capacity of AIs generally depends on several factors including the physiochemical structure of the solid lipid matrix, the polymeric state of the lipid materials, and the solubility of AIs in the liquid lipid.

The modern trends in lipid nanocarriers can be explained at least partially by the diversity of their sources, especially renewable vegetable resources, e.g., raspberry seed oil and rice bran oil. These nanocarriers offered several attractive properties including: minimal side effects, self-antioxidative properties, and ability for the simultaneous co-encapsulation and release of two AIs, similar to polymeric nanocarriers. The fabrication of NLC has been reported using Beeswax, propolis wax, and pomegranate seed oil. These NLCs offered excellent physical stability and low polydispersity with an average nanoparticle size of 71 –350 nm. The stability of AIs strongly depends on the resistance of the nanocarriers (both liquid and solid lipid) against di ffusion of agents that degrade the environment. Some simulative studies confirmed the e ffects of several factors, including the sizes of the nanoparticles, domain size, and solid domain fraction on the di ffusion of degradation agents into the nanoformulations (Nelson et al., 2017). However, the interaction of the lipid nanocarriers with chemical pesticides demands further toxicological studies to accurately assess the toxic profile of these nanoformulations.

iii. Clay nanomaterial-based encapsulation: Clay-based nanoformulations have been developed to promote the adsorption and slow release of neutral and hydrophobic AIs. Nanoclays are considered to be economically viable and biocompatible materials that can o ffer great opportunities in this field due to their potential role as nanocarriers in materials science and agricultural applications. The modification of clay nanomaterials has been conducted to provide them with beneficial properties (e.g., large surface area, charge density, hydrophilic nature, and porosity). The modification of clay with organic cations helped to enhance their affinity for adsorption of hydrophobic AIs such as alachlor, types of lipid-based nanocarriers for pesticide encapsulation, ochlor, metolachlor and metolachlor norflurazon.

Positively charged mesoporous silica nanoparticles via a post-grafting method was developed to load anionic pesticide, e.g., 2,4-D sodium salt. The charged nature of the silica nanoparticles offered significant improvement in pesticide loading (21.7%) in comparison to pristine forms (1.5%). Further, polymer-based clay nanocomposites have been developed to achieve controlled release of the anionic AIs. The use of biopolymer chitosan (or Fe3+ cations) was reported for modification of a natural smectite to encapsulate imazamox. These nanocomplexes exhibited much slower release of this herbicide (from 27 to 75%) into water in the first 10 min than that of commercial imazamox (86%). The total soil-leaching losses were also reduced by 15% with these nanoformulations.

Photo-thermal instability can cause adverse effects to achieve maximum bioactivity at the targeted site. Fan et al. (2017) developed pectin-crosslinked silica microcapsules via particular disulfide bonds. These silica microcapsules offered an enhanced loading efficiency (20% w/w) for kasugamycin as well as improved photo-thermal stability. These microcapsules offered a dual-triggered drug release mechanism between pectinase and glutathione, as described by the Riger-Peppas model. The enzymatic responsive mesoporous silica nanoparticles offered excellent loading ability (42% w/w) for the pesticide chlorantraniliprole.

Major parameters: There are several parameters that can affect the intercalation of AIs inside clay such as spacing and the presence of water molecules between interlayers and exchangeable cations (Sandeep Kumara et al., 2019). Beyond chemical compounds, the clay-based nanocarriers are also useful for controlled release of bioactive compounds due to their attractive properties (e.g., dimensional and thermo-chemical stability). Recent work has focused on the development of renewable and inexpensive silica nanostructures that are very dependent on plants (silica-accumulating or Equisetaceae plants). On a different topic, the interaction of silica nanostructures with biocides (i.e., plant extracts) is still an area for further investigation. The use of biogenic silica nanostructures was reported for efficient and controlled release of neem oil. The covalent bonding of neem extract with silica offered enhanced stability of the nanoformulation in water, good thermal stability, high antioxidant capacity in comparison to natural neem bark extracts, and a slow drug release profile resulting in 100% mortality of *Acromyrmex crassispinus* ants. However, the uncontrolled release profile (including release rate and duration) of silica nanoparticles limited their use in agricultural applications.

iv. Stimuli-responsive microcapsules: Enormous efforts have been aimed at developing stimuli-responsive (e.g., pH, enzymes, temperature, redox, etc.) silica microcapsules to obtain controlled and smart AI release profiles. Of various types of stimuli, enzymes are very attractive due to their accuracy, specificity, and efficiency for several substances. Urease-responsive delivery of pendimethalin was fabricated using isocyanate-functionalized silica cross-linked microcapsules. These microcapsules showed longer duration herbicidal activity against weeds, i.e., *Amaranthus retroflexus* L. (RPW) and *Echinochloa cruss-galli* L. (BYG). The outer shell layer of pendimethalin microcapsules was beneficial to prevent the initial burst release of drugs to obtain a sustained drug-release profile (Sandeep Kumara et al., 2019). Despite these advancements in nanocarrier applications, the presence of crystalline impurities is still a major concern in clay nanomaterials. The development of efficient synthesis procedures and suitable surface modification techniques are thus required to improve their application in pesticide delivery.

v. Metal organic framework-based encapsulation (MOFs): In recent years, MOFs have gained significant attention due to their novel structure-property relations. They have been reported as efficient nanocarriers for targeted-drug delivery due to their efficient drug adsorption in the pore structures. They can be used to overcome several limitations of conventional therapeutic agents (e.g., non-selective biodistribution and poor solubility) that can cause cardiotoxicity effects and damage healthy tissue. The MOFs exhibit several novel properties, including large surface area, high pore volumes, tunable pore size, effective surface chemistry, high thermal stability, and multiple topologies. Beyond MOFs, there are several other porous matrices including adsorption resin, activated clay, activated alumina and activated carbon. These formulations have prolonged effective lifetimes of AIs. However, the non-biodegradable nature of these materials is still a major issue to be resolved . MOFs (especially green MOFs like Ca- and Fe-based MOFs) have been introduced as an alternative solution for encapsulation of AIs to suppress their volatility. These green MOFs can be easily decomposed in water to its components (Ca and Fe) that function as nutrients in soil, ultimately reducing the environmental pollution.

MOFs are synthesized from metal clusters or ions that work as coordination centers interconnected by organic ligands. MOFs can offer multiple active sites of different strengths for delivery of active species. They have excellent ion exchange capabilities (Rozek and Dincă, 2014), high adsorption capacity and excellent electronic properties. The size of entrapped species should be larger than the pores of the matrix to ensure the effective encapsulation of AIs in a porous matrix as well as to avoid the leaching of AIs through pores.

There are two different synthetic approaches to be followed to support the conditions of encapsulation, including assembling of AIs within the pores of the MOFs and assembling the hollow MOFs around the AIs. Both types of encapsulation methods depend upon the type of AIs to be encapsulated and the chemistry of the support. The use of MOFs can be further explored for green applications in terms of either the use of eco-friendly metal ions or the delivery of natural components to resolve the issue of toxicity. Porous structures (430 and 160 m2 /g) loaded with the fumigant cis-1,3-dichloropropene based on Ca2+ ions and lactate with good drug-release kinetics were reported. These sorption-based formulations offered enhanced release kinetics, i.e., 100 times slower release in MOF-1201 than in MOF-1203. However, further research efforts are needed to explore MOF-based encapsulation in terms of drug-loading capacity, toxicity to non-target plants, and the amount of residue in the final product.

vi. Greener encapsulations: Soils are facing several new environmental stressors in the form of nanomaterials (e.g. TiO_2). These emerging pollutants are rapidly increasing in our ecosystems and agricultural fields. The exposure of TiO_2 nanoparticles to an agricultural soil was tested to evaluate the response of the microbial community over 90 days. This emergence of nanoparticles had strong negative impacts on both the populations of microorganisms that oxidize ammonia and the activities of nitrification enzymes. The negative impact of nanoparticles requires further research to utilize these pollutants for modification of soil health and ecosystem function. Thus, the use of non-toxic materials (e.g., starch, chitin, and nanoclays) should be considered to eliminate such risks due to their biocompatible and biodegradable nature (Sandeep Kumara et al., 2019). In this respect, numerous research efforts have been made to develop starch-based nanocontainers to facilitate the delivery of crop protection molecules, nutrients, and biostimulants into plant-tissues. The potential application of starch-based nanocontainers has also been realized to protect plants against harmful pathogens.

It is essential to promote greener production of these nanoformulations. For instance, starch nanoparticles are usually produced by the acid-hydrolysis of starch granules; this process suffers from several drawbacks such as low yield, long processing time, and environmentally malignant residues (e.g., toxic wastes). To resolve these issues, research efforts should be directed to the development of eco-friendly approaches (like ultrasound without any additional chemical reagents). Surface functionalization of the as produced nanoparticles is a practical option to chemically and/or enzymatically modulate their physicochemical and biological properties for appropriate entrapment of AIs.

Microorganisms as nanocarriers: Besides the biodegradable polymers, microorganisms (like cyanobacteria) have also been reported as nanocarriers due to their large abundance, biocompatible nature, and microalgae surface having heterogeneous functional groups (e.g., carboxylic, hydroxyl, phosphoric, and amine groups) that can support the pesticide loading (Giessen and Silver, 2016). The several species of algae and cyanobacteria have been reported for adsorption and removal of heavy metals ions and/or organic compounds. These microorganisms are environment-friendly and can be used for the effective fixation of nitrogen in soil, especially in rice fields. Cyanobacteria are also found to influence the physicochemical and biological properties of soil and the soil-water interface. The inoculation of cyanobacterial in soil was indeed found to improve the soil stability due to excretion of lipids and polysaccharides. However, the beneficial effects of these microorganisms have not been studied as much as pesticides for the purpose of suppressing weeds. The improved performance of soil inoculated with cyanobacteria as nanocarriers was reported for stimuli-controlled delivery of avermectin with higher photostability (against UV radiations) than that of the free avermectin. The considerable binding affinity of avermectin to cyanobacteria was found in isopropanol. The use of cyanobacteria prolonged the release of avermectin in ethanol/water (4:1, v/v) mixture which can be further controlled by coating of carbopol resin on avaermectin-loaded cyanobacteria. The carbonol resin offers more hydrogen binding sites for avermectin, therefore, thicker carbonol resin layer creates longer distance for avermectin to travel in order to diffuse into solvents. The overall process offers slow release rates with low release amount of the pesticide.

As an alternative to synthetic nanoparticles, there are several plant-based viral nanomaterials that occur naturally and can be utilized for packaging and delivery of AIs. The use of tobacco mild green mosaic virus (TMGMV) was also reported to deliver plant parasitic nematodes. These plant viral nanoparticles are economically viable, robust in harsh environmental conditions, efficient, provide excellent soil motility, and have a biodegradable nature. The use of plant pathogenic bacteria (e.g., *Agrobacterium tumefaciens, Rhodococcus fascians, Pseudomonas syringae, A. rhizogenes* and *Streptomyces*) was reviewd to enhance plant productivity and yield (Tarkowski and Vereecke, 2014). As such, greener nanoformulations for the application of pesticides will offer significant opportunities in near future due to the various benefits explained above.

vii. Others: The nanocarriers for pesticides should possess good pesticide- and plant-binding properties to resist drift loss. They should have good degradation potential for the pesticides after a certain threshold time for pest control. Different nanocomposites have been pursued for pesticide delivery to reduce non-target accumulation (Sandeep Kumara et al., 2019). TiO_2-based nanocomposites have been reported as nanocarriers with the enhanced potential for pesticide degradation by UV-aided photocatalytic activity. Likewise, the coating of *Bacillus thuringiensis* (Bt) Cry11Aa toxins with $Mg(OH)_2$ nanoparticles significantly affected their bioactivity and anti-UV ability. $Mg(OH)_2$ nanoparticles worked as coating clothes to protect the Cry protein to achieve slower degradation rates (from ~64% to ~17%) with increased toxicity against *Culex quinquefasciatus*. MWCNT-graft-poly(citric acid) (MWCNT-gPCA) nanocomposites were reported for zineb and mancozeb encapsulation that resulted in superior toxicity on *Alternaria alternata* (maximum effect after 10 days) (sandeep Kumara et al., 2019). The encapsulation of dichlorvos and chlorpyrifos using starchsilver nanoparticles showed excellent encapsulation efficiencies of 95% and 98%, respectively. These nanoformulations provided the controlled release of insecticides and offered antimicrobial activity for silver nanoparticles.

Graphene oxide (GO): The use of graphene oxide was also explored for the spontaneous adsorption of chlorpyrifos, endosulfan, and malathion. GO turned out to be highly favorable for hydrophobic interactions, hydrogen bonding, π- π interactions, and the presence of electron donors (like S, P, and N). The potential of GO was further investigated for controlled release of non-covalently bound hydrophobic AIs via any of the stimuli effects including the photothermal effect, hydrophobic medium, and pH change . When employed for the delivery of AIs, the nanocomposites of rGO-Cu_2-xSe exhibited a drug loading capacity of 40% (w/w). The nanocomposite serves as a reservoir of AIs in the leaf, and has been used to treat the pests without any drift losses, offering an enhanced larval mortality of approximately 35%. The GO nanocomposites can also provide sustained degradation of pesticide after a significant time period. That is, 50% AIs remained after 6 days when considering the 8 h average solar irradiation time per day. The ultrasonic-assisted surface modification of porous carbon (derived from natural waste, i.e., silkworm excrement) was reported as a nanocarrier. The surface treatment of porous carbon with low valence states of metal ions (i.e., Lewis acid metal ions) offers the opportunity for an enhancement in the total release amount of thiamethoxam and its release rate through weakening of the interaction between the nanocarrier and the hard- base pesticide.

Many research efforts have been made to utilize layered metal hydroxides for drug delivery applications with specific magnetic and catalytic properties. In particular, zinc hydroxide nitrate (ZHN) is one of the most studied nanomaterials for biomedical and agricultural applications due to its highly biocompatible nature and easy surface functionalization. Beyond anionic pesticides, the intercalation of charge neutral and poorly watersoluble AIs is highly dependent on the hydrophobic nature and accessibility of anionic surfactant. The release rate and equilibrium release amounts of chlorpyrifos were found to be strongly dependent on the structure of the ZHN nanocomposite and the release medium.

Expansion of nanocarriers: The benefits of nanomaterials as carriers can be further expanded by adding tracking ability beyond the primary goal (e.g., controlled release of pesticide). In this respect, fluorescent photoresponsive nanocarriers have attracted considerable attention for pesticide nanoformulations (Sandeep Kumara et al., 2019). Fluorescent photoresponsive nanoparticles can help assess the morphological changes induced by the pesticide inside the plant system using appropriate imaging techniques. Ultimately, such photoresponsive nanocarriers can lead to improvement in the pesticide activity of AIs along with a drug-release profile for efficient photoregulation. The application of pesticide nanoformulations based on fluorescent organic perylene-3-ylmethanol nanoparticles for controlled release of 2,4-dichlorophenoxyacetic acid was reported. The presence of functional groups on the nanoparticle surface may improve the adhesive property and retention time of the nanoformulation resulting in 72.3% retention rate of bioinspired mussel avermectin nanoparticles on weakly hydrophilic cucumber-foliage.

Some recent advances in the rapid expansion of supercritical solutions (RESS) processing technique wre discussed (Yaping Sun et al., 2005). Also includes the simple but significant modification to the traditional RESS by using a liquid solvent or solution at the receiving end of the supercritical solution expansion, or the rapid expansion of a supercritical solution into a liquid solvent (RESOLV), and applications of the technique to the preparation of nanoparticles.

6.6 Prospects of nano-delivery system technology

Based on the data collected from the literature, it is expected at least two main positive impacts of the extended, prolonged and improved use of nano-delivery technology translocated into the food production chain. The first is related to the technical aspects of pesticide usage (Pablo Vega-Vasquez, 2020). Similar to the role they play in the medical field, nano-delivery systems can increase the controlled-release properties of the pesticide, increased solubility of active ingredients, protection against premature degradation and increase the stability of active ingredients. Another advantage is that non-target surrounding or distant flora and fauna will be less affected as a result of reducing exposure to toxic chemical compounds. In addition, the technical constraints concerning the massive production of nano-carriers for use in agriculture should be correlated with the economical boundaries which limit the production costs and configures the potential revenues for producers. Additional studies are required to assess, not only the fate of nano-encapsulation materials and payloads, and the resulting physical-chemical and biological performance, but the long-term environmental risks and economic viability.

To conclude, it is clear that there is an immense need to develop methods or technologies that allow us to cope with the contrasting challenges of the food supply chain. For instance, the toxicity threshold of materials used in the delivery system is species-dependent and responses to these are driven by a series of factors including not only the nanomaterial itself but the environmental and physiological conditions on which they are applied. Another noteworthy factor is the broader impact of the delivery system to the environment, while in the medical systems, it is localized to the individual receiving the treatment. Impacts on plant growth, and therefore on product yield and food quality, have been reported. However, several gaps exist in understanding the dynamics of interactions between plants and engineered nanomaterials (ENMs). Given the lack of experimental standardization and the divergent responses, even within similar plant species, it is challenging to foresee the challenges on the use of ENMs in plants (Zuverza-Mena et al., 2016). Finally, there is an imperative need to standardize and validate protocols to assess the positive and negative impact of nano-carriers in an experimental setting, and scale-up of testing can yet be another challenge. Most of the currently available information stems from experiments under controlled conditions, making it difficult to predict the real potential of functional prototypes. Research efforts could focus on controlled release, particle stability, and environmental fate and toxicity to make this a fully-embedded technology.

More than 90% of pesticides run off into the environment and residue in agricultural products in process of application, due to the disadvantages of conventional pesticide formulation such as use of harmful solvent, poor dispersion, dust drift, etc. In recent years, using nanotechnology to create novel formulations has shown great potential in improving the efficacy and safety of pesticides. The development of nano-based pesticide formulation aims at precise release of necessary and sufficient amounts of their active ingredients, in responding to environmental triggers and biological demands through controlled release mechanisms (Zhao Xiang et al., 2017). several scientific issues and strategies regarding development of nano-based pesticide formulations were discussed viz., Construction of water-based dispersion pesticide nanoformulations, mechanism on leaf-targeted deposition and dose transfer of pesticide nano-delivery system; mechanism on increased bioavailability of nano-based pesticide formulation and impacts of nanoformulations on natural degradation and bio-safety of pesticide residues.

Nano-delivery systems have been designed to deliver effective components to targeted sites within a plant to provide potential solutions to some devastating crop diseases which cannot be effectively managed with conventional methods. However, nano-enabled agriculture is still in its infancy and its applications are mostly theoretical. Therefore, more research is needed to develop biodegradable, cost-effective and safe nanomaterials for future application. Moreover, systematic studies are crucial to safeguard our food production system, while making efforts to raise public awareness of nanotechnology (Xiaoping Xin et al., 2020). To date, a large variety of nanomaterials have been synthesised as nanocarriers including the usage of metallic NPs, CNTs, silica NPs, polymeric NPs and liposomes as nanocarriers in agriculture.

6.7 Release and mechanisms of nano-delivery systems

i. Loading: The mechanisms of active fertilisers or pesticide molecules loading onto NPs include adsorption, encapsulation, covalent bonding and entrapment. For example, CNTs could adsorb a range of organic contaminants from soil, such as organophosphorus pesticides, organochlorine pesticides, pharmaceuticals and personal care products, and pyrethroid insecticides. Given their hydrophilic exterior and hydrophobic interior, polymeric NPs have been used to encapsulate lipid soluble drug molecules for targeted nanodelivery to improve stability, bioavailability and controlled release of the encapsulated drugs (Xiaoping Xin et al., 2020) Covalent binding of Ag NPs onto cotton fabric through carboxymethyl chitosan was reported as an effective method for obtaining antimicrobial functions and outstanding laundering durability. Active molecules may also be entrapped inside NPs to protect host molecules from unexpected degradation. For example, mesoporous silica nanoparticles (MSN) have demonstrated potential for delivering prochloraz in cucumber and fenoxanil in rice

Release : In nano-delivery systems, effective components can be control-released in response to external stimuli, including physical, chemical and/ or biological stimuli. Accomplished by the above stimuli, NPs can be used to precisely release active ingredients to targeted sites with maximum drug efficacy, while minimising potential damage to healthy cells. For example, magnetic NPs have been used to deliver drugs to cancerous tissues with the aid of small magnets in cancer therapy. pH-responsive polymeric NPs have been widely used to deliver drugs directly to tumour sites where an acidic microenvironment triggers the release. Similarly, sugar and pH dual-responsive nanocarriers have been synthesised by using mesoporous silica-coated Fe_3O_4 magnetic NPs for cargo delivery. Combinations of NPs and enzyme systems could unload their cargo in response to a specific enzymatic stimulus, especially tumour-associated enzymes that have been harnessed in drug delivery.

Layer by layer biodegradable nanodelivery systems: Nanoparticulation of solid and hollow nanoparticles using edible proteins and carbohydrates from plant and animal origin was reviewed and shared new research conducted in our laboratory on fabrication of biopolymer based nanodelivery systems (Hazal Turasan et al., 2020). The properties of nanoparticles fabricated were compared using desolvation, coacervation, and layer by layer assembly. The parameters affecting the morphological and physical states of the nanodelivery systems as well as their abilities to encapsulate bioactive compounds, and the mechanisms of the nanoparticulation process were discussed in detail. Methods that need to be used with each nanoparticulation technique along with key characterization data obtained were discussed. The effect of varying a wide range of parameters on particle size, size distribution, morphology, effect of solvents and non solvents, effect of surface charge of polyelectrolytes, effect of fabrication conditions were presented.

Biopolymer based nanodelivery systems have great potential to be used in the nutraceutical, food, pharmaceutical industries as well as agriculture. The biodegradable and edible nature of these nanodelivery systems make them safe for incorporating into food and nutraceutical materials, because, the biopolymers used in their fabrication are biocompatible polymers of food origin that are generally recognized as safe. Therefore, use of edible polymer based nanoparticles would be safer and less toxic and consequently better for all these industries. For example, in the food industry, antioxidant compounds can be encapsulated which would then be incorporated into packaging materials and that would help maintain freshness and good quality and would prevent oxidative deterioration of foods to increase shelf life.

Pesticides and insecticides can also be encapsulated in these nanodelivery systems for agricultural applications. Because they are biodegradable and originated from natural biopolymers, their impact on the environment and the quality of the soil would be minimized compared to current applications that use some burdensome chemicals. Biopolymer based nanodelivery systems would also be of high interest for the pharmaceutical industry. Through the efficacy and extra stability gained with encapsulation, drugs and bioactives would be better protected from the acidic environment of the stomach and be released more controllably in the intestines. Further functionalization of nanodelivery systems with recognition elements, such as antibodies or aptamers, would enable more specifically targeted delivery.

Controlled-release delivery system of Abamectin

Photolysis and poor solubility in water of Abamectin are key issues to be addressed, which causes low bioavailability and residual pollution. A novel hydrophilic delivery system was developed through loading Abamectin with porous silica nanoparticles (Abam-PSNs) in order to improve the chemical stability, dispersity and the controlled release of Abamectin (Yan Wang et al., 2014). These results suggested that Abam-PSNs can significantly improve the performance of controllable release, photostability, and water solubility of Abamectin by changing the porous structure of silica nanoparticles, which is favorable to improve the bioavailability and reduce the residues of pesticides.

Controlled release behaviors of Abam-PSNs: Compared with pure Abamectin, initial burst release of Abam-PSNs is not so obvious. As expected, Abam-PSNs exhibit slower release rates due to the porous structure on the surface which is favorable to prolong validity period and reduce leaching loss of Abamectin. The release rate of Abam-PSNs samples was relatively fast at initial stage and then gradually became slow with increased time, as the Abamectin adsorbed on the surface of Abam-PSNs was easier to release than that loaded within the carriers. For the sample with less internally porous structure (Abam-PSNs-1), Abamectin concentrated on the surface nanopores of silica carrier, which make it easier to release pesticide with the fastest release rate than other Abam-PSNs samples. Whereas, for the sample with richer porous structure (AbamPSNs-3, Abam-PSNs-4), the Abamectin were mainly loaded on the internal porous structure of silica carriers, resulting in longer release time due to relatively slow and smooth release. For the sample of Abam-PSNs-2, the irregular release rate of Abamectin may be due to the mutual interference between fast release on the surface and slow release resulting from decreased surface area which leads to the decreased diffusion gradients and driving force of Abamectin. The results indicated that Abamectin loaded by Abam-PSNs can slowly and controllably be released by adjusting the porous structure of PSNs carriers, resulting in the validity of Abamectin in the target lasting for a longer time. The properties could not only increase the efficiency of pesticide applications and decrease the spraying dosage but also reduce residues and improve the environmental protection.

6.8 Nanodelivery for the Aquaculture Development

Aquaculture has been globally recognized as the fastest growing food production sector which plays a major role in meeting the increasing demand for animal protein requirement. A consensus is growing that a dramatic increase in aquaculture is needed to supply future aquatic food needs (Md. Aklakur et al., 2016). However, there are sustained problems with the aquaculture like disease outbreaks, chemical pollution, the environmental destruction, and inefficient feed utilization. These altogether raise question mark on sustainability of aquaculture. In spite of the several strategy adopted on national and international level, as improved laboratory facilities, diagnostic expertise, and control and therapeutic strategies in order to handle disease outbreaks more effectively. Aquaculture industry is under uncertainty and the progress has not matched that of the rapidly developing aquaculture sector.

In order to control disease prevalence and ensure better health of system and sustainable production, the sector demand more technical innovation for the drug use, disease treatment, water quality management, production of tailored fish for suiting better health, productivity drive by epigenetic and nutrigenomic interaction, better breeding success by efficient delivery of maturation and spawning inducing agent, nutraceutical delivery for rapid growth promotion and culture time reduction, successful use of autotransgenic, and effective vaccine. Nanotechnology has a tremendous potential to revolutionize agriculture and allied fields including aquaculture and fisheries. For these multiple purposes effort, importance of nanotechnology and nanodelivery of drugs, vaccine, nutraceutical, inducing hormones, and growth-promoting anabolics open tremendous opportunity.

6.9 Nanodelivery systems based on encapsulation technology

Nano-encapsulation in the form of micelles, liposomes or biopolymer-based carrier systems has been used to develop delivery systems for additives and supplements for use in food and beverage products. Nano-encapsulation is the technological extension of microencapsulation, which has been used by the industry for food ingredients and additives for many years (Anonymous, 2010). Nano-encapsulation offers benefits that are similar to, but better than, those of microencapsulation, in terms of preserving the ingredients and additives during processing and storage, masking unpleasant tastes and flavours, controlling the release of additives, better dispersion of water-insoluble food ingredients and additives, as well as improved uptake of the encapsulated nutrients and supplements. The modified optical characteristics of nanocarriers mean that they can be used in a wide range of products, such as clear beverages. The improved uptake and bioavailability alone has opened up a vast area of applications in food products that incorporate nanosized vitamins, nutraceuticals, antimicrobials, antioxidants, etc. After food packaging, nanoencapsulation is currently the largest area of nanotechnology application in the food sectors, and a growing number of products based on nanocarrier technology are already available on the market.

The concept of nanodelivery systems seems to have originated from research on the targeted delivery of drugs and therapeutics. While it can offer many benefits to the consumer from increased absorption, uptake and improved bioavailability of nutrients and supplements, it also has the potential to alter the distribution of the substances in the body.

Biopolymer-based nanoparticles help local delivery of fertilizers and pesticides without polluting soil and air. The immense impact of encapsulation technologies in the food industry is apparent with the global food encapsulation market projected to be $54.846 billion by 2024 (Kaarunya Sampathkumar et al., 2020). Nature-derived polymers are also being studied for pesticide encapsulation. Seltima is an example of a commercial product using polymers for encapsulation of pesticides. Launched by BASF to protect rice plants, the product uses a humidity responsive controlled release technology to release pesticides selectively to the leaves of rice while avoiding release of the chemicals into the water in paddy fields thereby greatly reducing the contamination of water sources (Huang et al., 2018).

Nanoencapsulation is more prominently applied to pesticide encapsulation than fertilizers. With regards to nanotechnology in fertilizer applications, inorganic nanoparticles of zinc (Zn), copper (Cu), Fe, cerium (Ce), and titanium (Ti) are studied extensively both in laboratory and field conditions when compared with encapsulation of these nutrients into nanoparticulate systems. Chitosan is one of the nature-derived polymers that has many implications in agriculture. With its inherent growth enhancement and antimicrobial properties, chitosan could be exploited in agriculture, either stand alone or as an encapsulation matrix for nutrients (Kaarunya Sampathkumar et al., 2020). Owing to the greater dispersibility of chitosan nanomaterials, it is preferred to be used in the nano from rather than in the macro form. Heba et al. have demonstrated the ability of chitosan nanoparticles loaded with nutrients N, P, and K, to accelerate plant growth, thereby shortening the life cycle of wheat plants they were tested on from 170 to 130 days. Barley field tests on using chitosan nanoparticles by itself have demonstrated their ability in improving grain yield through soil application to barley plants under drought stress.

The main advantage of nanoencapsulation of pesticides is that encapsulation greatly reduces the amount of pesticides used, by solubilizing them and providing a controlled release of these chemicals at the target site. This in turn checks the development of resistant pest species that arise from overexposure and contamination of soil and water bodies with excess chemicals. Nano complexation is an encapsulation technique using cyclodextrins, which are cyclic derivatives of starch containing a hydrophilic exterior and a hydrophobic interior, forming excellent complexes with molecules that have low solubility in aqueous media. This property of cyclodextrin was exploied to form inclusion complexes with two plant-derived compounds, carvacrol and linalool, that have insecticidal and repellent properties. The cyclodextrins were then used to functionalize chitosan before forming chitosan nanoparticles to facilitate easier application to crops. Given the high volatility of the natural compounds, nano complexation improved the lifetime of the compounds when exposed to the environment. Biopolymeric nanocarrier made up of natural polymers, chitosan, and pectin was developed by Sandhya et al. to provide a sustained and controlled release of encapsulated carbendazim with good bio-efficacy and inhibition against fungi such as *Fusarium oxysporus* and *Aspergillus parasiticus*. Moreover, the phytotoxicity assessment showed that carbendazim- loaded polymeric nanoformulation provides better germination and root growth for the seeds of *Zea mays*, and *Cucumis sativa* (Sandhya et al., 2017). Chitosan nanoparticles have also been shown to boost the inherent immunity of plants thereby inducing disease resistance in them. Maize plants treated with Cu-chitosan nanoparticles were able to resist Curvularia leaf spot disease by boosting its antioxidant activities and defense enzymes.

Bibliography

Alejandro Perez-de-Luque, 2014.Nanomaterials as smart delivery systems for disease and pest control in plants . In: Proc.of a workshop on "Nanotechnology for the Agricultural Sector: From Research to the Field. Luxembourg: Publications Office of the European Union

Anonymous, 2010. FAO/WHO Expert meeting on the application of nanotechnologies in the food and agriculture sectors: potential food safety implications Meeting report. Food and Agriculture Organization of the United Nations and World Health Organization, Rome 2010, 129 pp.

Bakht Ramin Shah and Jan Mraz, 2020 . Advances in nanotechnology for sustainable aquaculture and fisheries. Rev. Aquaculture, 12 : 925-942

Bingna Huang, Feifei Chen, Yue Shen, Kun Qian, Yan Wang, Changjiao Sun, Xiang Zhao, Bo Cui, Fei Gao, Zhanghua Zeng and Haixin Cui, 2018. Advances in Targeted Pesticides with Environmentally Responsive Controlled Release by Nanotechnology. Nanomaterials (Basel), 8: 102.

Changjiao SUN, 2019. Development and efficacy evaluation of novel adhesive pesticide nano-delivery systems. (Ph.D. Dissertation in English). Gembloux, Belgium, Gembloux Agro-Bio Tech, Université de Liège. 130 pages

Fan, C., Guo, M., Liang, Y. , Dong, H., Ding, G., Zhang, W., Tang, G., Yang, J., Kong, D. and Cao, Y., 2017. Pectin-conjugated silica microcapsules as dual-responsive carriers for increasing the stability and antimicrobial efficacy of kasugamycin, Carbohydr. Polym. 172 : 322–331

Fasil Dawit Moges, Paritosh Patel, S.K.S. Parashar, Biswadeep Das,2020. Mechanistic insights into diverse nano-based strategies for aquaculture enhancement: A holistic review. Aquaculture, 519: 734-770

Giessen, T.W. and Silver, P.A., 2016. Encapsulation as a strategy for the design of biological compartmentalization, J. Mol. Biol., 428: 916–927

Hazal Turasan, Jose Bonilla , Fatih Bozkurt , Luis Maldonado, Xingfei Li , Tuncay Yilmaz, Rohollah Sadeghi and Jozef Kokini, 2020. Comparison of the fabrication methods, formation dynamics, structure, and delivery performance of solid nanoparticles and hollow layer by layer edible/biodegradable nanodelivery systems. J.Food Process Engg., 43: e13413

Huang, F., Chen, Y., Shen, K., Qian, Y., Wang, C., Sun, X., Zhao, B., Cui, F., Gao, Z., Zeng, H. and Cui, 2018. Advances in targeted pesticides with environmentally responsive controlled release by nanotechnology.Nanomaterials, 8 : 102

Kaarunya Sampathkumar, Kei XianTan and Say Chye Joachim Loo, 2020. Developing Nano-Delivery Systems for Agriculture and Food Applications with Nature-Derived Polymers. iScience, 23 101055

Khot, L.R., Sankaran, S., Maja, J.M., Ehsani, R. and Schuster, E.W., 2012. Applications of Nanomaterials in Agricultural Production and Crop Protection: A review. Crop Prot., 35:64–70.

Kumar, S., Chauhan, N., Gopal, M. , Kumar, R. and Dilbaghi, N., 2015. Development and evaluation of alginate–chitosan nanocapsules for controlled release of acetamiprid. Int. J. Biol. Macromol., 81 : 631-637

Lu, Y., Li, Y. and Wu, W., 2016. Injected nanocrystals for targeted drug delivery. Acta Pharm Sin B., 6:106-113

Md. Aklakur, Mohd Asharf Rather and Neeraj Kumar, 2016. Nanodelivery: An Emerging Avenue for Nutraceuticals and Drug Delivery. Crit. Rev. in Food Sci. & Nutr.,. 56: 2352-2361

Manjunatha, R.L ., Dhananjay Naik and KV Usharani, 2019. Nanotechnology application in agriculture: A review. J.Pharmacognosy and Phytochem., 8: 1073-1083

Maruyama, C.R., Guilger, M., Pascoli, M. , Bileshy-José, N., Abhilash, P.C., Fraceto, L.F. and de Lima, R., 2016. Nanoparticles based on chitosan as carriers for the combined herbicides imazapic and imazapyr. Sci. Rep., 6 :19768

Nelson, D.D., Pan, Y., Tikekar, R.V., Dan, N. and Nitin, N., 2017. Compound stability in nano particles: the effect of solid phase fraction on diffusion of degradation agents into nanostructured lipid carriers. Langmuir 33 :14115–14122

Pablo Vega-Vásquez, Nathan S. Mosier and Joseph Irudayaraj, 2020. Nanoscale Drug Delivery Systems: From Medicine to Agriculture. Front Bioeng Biotechnol., 8: 79.

Rameash, K., 2014. Biodegradable nanoparticles as carrier in smart delivery of pesticides. 3rd International Conference on Agriculture & Horticulture, October 27-29, 2014 Hyderabad International Convention Centre, India

Rozek, C.K. and Dincă, M., 2014. Cation exchange at the secondary building units of metalorganic frameworks, Chem. Soc. Rev., 43: 5456–5467

Sandeep Kumara, Monika Nehraa, Neeraj Dilbaghia , Giovanna Marrazzac, Ashraf Aly Hassane and Ki-Hyun Kim, 2019. Nano-based smart pesticide formulations: Emerging opportunities for agriculture . J. Controlled Release, 294: 131-153

Sandhya, S., Kumar, D. and Kumar, Dilbaghi, N., 2017. Preparation, characterization, and bio-efficacy evaluation of controlled release carbendazim-loaded polymeric nanoparticles. Environ. Sci. Pollut. Res. Int., 24: 926-937

Santana, I., Wu, H. and Hu, P., 2020. Targeted delivery of nanomaterials with chemical cargoes in plants enabled by a biorecognition motif. Nat Commun., 11: 2045.

Tarkowski, P. and Vereecke, D., 2014. Threats and opportunities of plant pathogenic bacteria. Biotechnol. Adv., 32: 215–229

Tiju Joseph and Mark Morrison , 2006. Nanotechnology in Agriculture and Food- Nanoforum Report. European Nanotechnology Gateway

Wang, N., Qi, L., Wang, Y. and Li, X.G., 2017. Preparation and Performance of Thermo-sensitive Pyraclostrobin Microcapsules. Chin. J. Pestic. Sci., 19:381–387.

Xiaoping Xin, Jonathan D. Judy, B Brent . Sumerlin and Zhenli He, 2020. Nano-enabled agriculture: From nanoparticles to smart nanodelivery systems. Environ. Chem., 17: 413–425

Xu, X., Bai, B., Wang, H.and Suo, Y., 2017. A near-infrared and temperature-responsive pesticide release platform through core-shell polydopamine@PNIPAm nanocomposites. ACS Appl. Mater. Interfaces, 9 : 6424–6432.

Yan Wang, Haixin Cui, Changjiao Sun, Xiang Zhao and Bo Cui , 2014. Construction and evaluation of controlled-release delivery system of Abamectin using porous silica nanoparticles as carriers . Nanoscale Res.Lett., 9:655

Yaping Sun, Meziani Mohammed, Pathak Pankaj and Qu Liangwe, 2005. Polymeric Nanoparticles from Rapid Expansion of Supercritical Fluid Solution. Chemistry (Weinheim an der Bergstrasse, Germany). 11: 1366-1373.

Zhao Xiang, Cui Haixin, Wang Yan, Sun Changjiao, Cui Bo and Zeng, Zhanghua.,2017. Development Strategies and Prospects of Nano-based Smart Pesticide Formulation. J.Agri.& Food Chem., 66: 10.1021/acs.jafc.7b02004.

Zuverza-Mena, N., Martínez-Fernández, D., Du, W., Hernandez-Viezcas, J. A., Bonilla-Bird, N. and López-Moreno, M. L., 2016. Exposure of engineered nanomaterials to plants: insights into the physiological and biochemical responses-a review. Pl. Physiol. Biochem., 110: 236–264

7

Regulations and Risks in Nanotechnology

Nanoscience and nanotechnologies are set to transform the global industrial landscape, but the debate on how to regulate environmental, health and safety risks is lagging behind technological innovation. Current regulatory efforts are primarily focused on the national and regional level, while the international dimensions of nanotechnology governance are still poorly understood and rarely feature on the international agenda. With the ongoing globalization of nanosciences and the rapid expansion of international trade in nanomaterials, however, demand for international coordination and harmonization of regulatory approaches is set to increase. However, uncertainty about nanotechnology risk poses a profound dilemma for regulators and policy-makers. Uncertainty both creates demand for and stands in the way of greater international cooperation and harmonization of regulatory approaches. Falkner Robert and Jaspers Nico (2012) reviewed the emerging debate on nanotechnology risk and regulatory approaches, investigates the current state of international cooperation and outlines the critical contribution that a global governance approach can make to the safe development of nanotechnologies.

Nanotechnology risk is a relatively new and hitherto largely neglected, concern in the study of global environmental politics. Most of the debate on its environmental and health risks has been confined to scientific and regulatory expert circles. Academic research on nanotechnology regulation has so far concentrated on the domestic legal, political and social dimensions, within the subfields of science and technology studies, applied ethics, regulation studies and law. By contrast, the international dimensions of nanotechnology regulation have received relatively little attention and are only slowly coming into focus. Because of the ongoing controversy on the implications of nanotechnology, there is significant debate concerning whether nanotechnology or nanotechnology-based products merit special government regulation. This mainly relates to when to assess new substances prior to their release into the market, community and environment. Regulatory bodies such as the United States Environmental Protection Agency and the Food and Drug

Administration in the U.S. or the Health and Consumer Protection Directorate of the European Commission have started dealing with the potential risks posed by nanoparticles. Regulation of nanotechnology will require a definition of the size, in which particles and processes are recognized as operating at the nano-scale. Regulatory frameworks for chemicals tend to be triggered by mass thresholds (Anonymous, 2004).

7.1 Legal and Ethical Issues

Nanotechnology is a vast field which includes a range of technologies at the nano scale, such as pharmaceuticals, biotechnology, genomics, neuroscience, robotics and information technologies. Nanotechnology is the latest technological innovation in global debates on risk regulation and international cooperation (Agnieszka Baran, 2016). Regulatory bodies have started dealing with the potential risks posed by nanoparticles. Since 2004 the UE has been developing a regulatory policy to tighten control and to improve regulatory adequacy and knowledge of nanotechnology risks. Currently, specific provisions on nanomaterials have been introduced for biocides, cosmetics, food additives, food labelling and materials in contact with foodstuff. The main problems are public trust, potential risks, issues of environmental impact, transparency of information, responsible nanosciences and nanotechnologies research.

The primarily raised question in international literature is the one whether nanotechnology should be included in the legal framework or whether the so called soft law mechanisms can be used? Currently, in relation to nanomaterials, within the European Union we are dealing with both the legal acts with binding force (regulations, directives), as well as non-binding acts such as recommendations concerning the fair conduct of scientific research or the application of a uniform definition of nanomaterial. Since nanotechnology has been within the interests of the European legislator for almost twenty years, it is also the subject of discussions regarding the legal doctrine. There has been a different approach to the matter of regulation of nanotechnology.

Some authors see the possibility of influencing the strengthening of the safety of use of nanomaterials, through non-judicial means. Due to the dynamics of change in this area, some people see an urgent need for the use of soft law. Soft law includes resolutions, guidelines, explanations, messages, programs, plans, mostly issued by the authorities involved in the legislative processes, the result of which is a generally applicable legislative act. Construction of soft law instruments is intended to constitute a foundation for the future hard legal regulations. Also important are the voluntary programs aimed at gathering information useful for the design of legal instruments.

According to the position of EU bodies, some authors suggest the strengthening of the existing forms of cooperation in the field of nanotechnology and cheering for States to build internal legal regulations (Agnieszka Baran, 2016). Relevant legal regulations in the European Union do not relate directly to nanotechnology, which is why new tools should be created in order to prevent the possible harmful effects of the use of nanomaterials. The view regarding the possibility of the future creation of a division of law of a collective nature is not often presented as it may take place only under the condition of attainment of full knowledge regarding the properties of nanomaterials. The most important problems in the sphere of nanotechnologies which should be soon resolved have been indicated. It is a question of applying a uniform definition of the nanomaterial, development of metrology tools in relation to nanomaterials, development of safety tests and risk assessment methodologies.

Some authors are unanimous about the fact that the regulation of nanotechnology at the international level is a big challenge because of the different ways of the use of nanomaterials. It seems, however, that in the future comprehensive regulations in the field of nanotechnology will be created. It may be significantly affected by the creation of the best practices for dealing with nanomaterials, which will be applied at the international level. Nanotechnologies constitute an interdisciplinary subject, which translates into a very broad spectrum of potential applications. Nowadays nanotechnologies encompass most areas of the sphere of technology. In the literature, there appears the concept of nano-ethics, which is defined as moral reflection on the development and application of nanotechnology, or the manipulation of matter at the molecular level.

Ethical aspects: The second half of the twentieth century saw the rapid development of new technologies. The XXI century is the century of biotechnology, genetic engineering and nanotechnology. The development of biotechnology has raised a lot of controversy, despite the regulation of this sector to a large extent, there is still no consistent law regarding, for example genetic research. Due to intensive scientific research and the creation of new solutions, the EU legislator has introduced patent protection for biotechnological inventions. Although nanotechnology inventions do not arouse such controversy as biotechnology, perhaps it is worth considering whether similar solutions in the patent law for nanotechnology should be introduced in the future. In literature, one may come across the statement that these two areas, contrary to appearances, have a lot in common. Nanotechnology is based on techniques and methods of manipulating matter on the nanoscale, but it makes no difference whether it is living or inanimate matter. For this reason, it is possible to combine the essential elements of both disciplines in nanobiotechnology (Balcerzak, 2013, p. 155). To some extent, it is possible to use the legislation concerning biotechnology. Public acceptance of new technologies is a key aspect of their development. At the moment in Europe, we are seeing the controversy surrounding the nerve implants, which give new opportunities to those suffering from the Parkinson's disease or patients who require the restoration of sight. The concerns relate to the possible use of implants for purposes unrelated to treatment. The European Commission considered to take into account the expectations and concerns of the society (Agnieszka Baran, 2016). Nanotechnology must not only be safe and useful, but also have to gain the support of public opinion. The public must be aware of the benefits, potential risks and necessary measures related to the use of nanotechnology. Public debate should also be encouraged in order to help people in forming an independent opinion. A major role in this process should be played by scientists, explaining the principles and applications of nanotechnology to the public.

Social acceptance and elimination of concerns about nanomaterials will have a positive impact on the future and the development of nanotechnology. The social dialogue should involve the representatives of the civil society and scientists, as well as other stakeholders.

7.2 Legal Regulation in the field of Nanotechnology

In Second Implementation Report for the years 2007-2009 it was found that nanotechnology offers significant potential to improve the quality of life and the competitiveness of industry in Europe (Agnieszka Baran, 20167.2). Its development and use should not be delayed, unbalanced or left to chance. At the time, an initial review of the legal framework in the field of nanotechnology was performed to examine whether new regulatory action was required to cover the risks in relation to nanomaterials. Preliminary results showed that the existing provisions in principle included the health issues and environmental impact. The regulatory authorities of the Member States were commissioned to assess the national legislation and identify gaps in it. Even then a preventive approach in the field of nanotechnology was recommended, due to the absence of complete knowledge about nanomaterials, it was recommended to use the existing legal mechanisms in relation to thresholds, the authorization of substances and ingredients, qualification of hazardous waste, reinforcing conformity assessment procedures, restrictions on the introduction of chemical substances and preparations to the market and their use.

In 2009, the European Parliament (EP) passed a report on regulatory aspects of nanomaterials, taking also into account the Commission Communication of 17 June 2008 entitled "Regulatory aspects of nanomaterials". The report shows that the European Commission sees the benefits of nanotechnology development but at the same time is aware of the risks arising from this development for humans and the environment. The EC confirmed the absence of complete knowledge about the potential risks posed by nanomaterials, and the evidence relating to the risks caused by certain nanomaterials and the overall lack of methods for the proper assessment of the risks associated with raising concerns about nanomaterials. In the context of many doubts concerning the use of nanomaterials, there appears an urgent need to include this sphere into the legal framework. Therefore, the European Commission, starting from 2008, has been reviewing the existing regulations concerning the use of nanomaterials, and has indicated the actions that ought to be taken in the future.

The European Commission, in the second regulatory review dedicated to nanomaterials, emphasizes the need to improve EU law in order to ensure the safe use of nanomaterials. More and more is known about the hazardous properties of nanomaterials. They do not succumb to categorization, and this justifies the need to carry out assessments of the risk associated with particular applications. The European Commission, stresses that an individual approach should be applied to risk assessment, using strategies based on the information about the potential risks in respect of exposure or hazard. In recent years we have seen undertaking legislative work to regulate the use of nanomaterials by a majority of Member States. To ensure safety it is advisable to compile a register of nanomaterials and products containing nanomaterials, such a register will facilitate the supervision of entities introducing nanomaterials to the market and ensure the transparency of data on products for buyers. At the moment such an obligation is relevant for cosmetic products (Jurewicz, 2014).

In the international arena, activities are also undertaken to regulate nanotechnology. The potential risks arising from the development of this area are seen by many states. In the United States in 2008, the US National Research Council issued a report calling for greater regulation of nanotechnology. Australia and Canada are also rather active on nanoregulation. Both have important programmes on Environmental Health and Safety (EHS) research and have published in-depth reviews of their regulations to assess eventual limits when dealing with nanotechnology. Even though no specific laws have been set up, the adoption of a precautionary approach principle, when dealing with nanotechnology application, is envisaged in both countries.

In Japan, China, Korea and Taiwan, which are deeply involved in nanotechnology, there are also, at different levels, important research initiatives dealing with EHS issues such as risk assessment and risk management of nanomaterials and nano-related products. They do participate in the worldwide debate on nanoregulation but no specific initiatives on the matter have been taken so far in these countries. The actions to regulate nanotechnology focus on the national level, and at the international level initiatives of the nature of joint research programs are observed.

7.3 Nanotechnology and regulation: Global and Indian Perspective

The scale of nanotechnology's operation has raised concerns over nanotechnology risk and safety aspects with respect to environment and health safety (EHS) as well as its ethical, legal and social implications (ELSI) (Amit Kumar, 2014). The World over, countries are vying to establish a regulatory framework to address these concerns. The initiatives in this regard comprise enacting new regulations or amending existing regulations/laws for nanotechnology; enforcing safety guidelines for researchers/workers in the laboratories in universities/research centres or companies; and establishing a global standard.

The European Union has also been active in the aspect of addressing these concerns. In a non-binding resolution, the European Parliament asked for tighter controls on nanotechnologies, in particular in the case of chemicals and materials, cosmetics, foods, occupational health and worker safety, environmental safety and waste. The most important among the steps taken is the initiative to revise REACH (Registration, Evaluation, Authorisation and Restriction of Chemicals). REACH regulates the use of chemical substances in Europe. In March 2008, the European Chemicals Agency (ECHA) established the Competent Authorities Sub Group on Nanomaterials (CASG Nano) for this purpose. The Sub Group published two technical guidelines and launched three projects in 2010 devoted to the application of REACH to nanomaterials, as well as the application of the related CLP regulation (classification, labelling and packaging of substances and mixtures) (Amit Kumar, 2014.). Regarding chemicals, the way nanomaterials are considered in REACH will strongly influence regulatory actions at national level, in particular in countries such as France, Germany, Austria, Belgium and Italy whcih are considering introducing notification and registration mechanisms for nanomaterials.

US: In the US, the Environmental Protection Agency (EPA), under the Toxic Substances Control Act (TSCA), has been mandated to regulate nanomaterials. A task force on nanotechnologies has been active within the Food and Drug Administration (FDA) since 2007. The general approach to nanoregulation of FDA is that the existing regulation adequately covers nano forms of substances, though a careful review is generally devoted to nanotech products (Mantovani et al. 2011).

Canada and Australia: Environmental Health and Safety (EHS) and regulatory issues are receiving increasing resources within their national strategies for nanotechnologies, and the need to adopt a precautionary approach is explicitly stated. There is a growing involvement of authorities in different sectors that are working to develop regulatory, product-specific guidance documents for nanomaterials (Amit Kumar, 2014.). The lack of a definition of nanomaterials is considered one of the key gaps to enable regulatory actions.

Japan: The Ministry of Economy, Trade and Industry (METI), since 2008, has supported a voluntary gathering initiative related to risk management of nanomaterials at industry level. Research reports on occupational health and safety (OHS) issues of nanomaterials have been published by the National Institute of Occupational Safety and Health Japan (JNIOSH).

The Republic of Korea: It is developing the National Nano-safety Strategic Plan (2011-2015). Moreover, a specific Risk Management Platform Technology for Nanoproducts is also at development stage and aims at providing a certification system for nano-related products. Several research programmes on EHS and ELSI are also in progress.

Taiwan: Within the framework of their Strategic Plan for Responsible Nanotechnology (2009-2014), the Nanomark Certification System has been active since 2004. This is a voluntary reporting and certification scheme that aims to increase public confidence in nanotechnology products.

Thailand: Nanosafety is among the priorities of the national policy on nanotechnologies. A strategic lan on safety and ethical issues is expected to be proposed to the government by the National Nanotechnology Center. This will include plans for the creation of an industrial standards certification for nanotechnologies related products (called NanoQ).

India: In India, the issue of nanotechnology risk regulation has been raised in terms of the concerns related to human health safety, environmental pollution, toxicity and towards general societal impacts relating to labour dislocation and livelihood disruption (Amit Kumar, 2014). Regarding the regulatory aspects many researchers are of the view that the government is very lax in establishing a regulatory body to foresee the safe development and commercialisation of nanotechnology products (Jayanthi et al. 2012). Few others have also raised the issue of labelling of nano-based products to enable people make an informed choice. In fact, on the sidelines of an International Conference on Nano Science and Technology (ICONSAT) in 2010 itself, Prof. CNR Rao, Nano Mission Council Chairman, said that 'as Indian industries are coming with various nanotechnological products including water filters, biomedical products, chemicals, cosmetics and paints, we are in the process of forming a regulatory body for the nanotechnology to regulate the products for the safety and benefit of the society.

The International Standards Organisation (ISO) Technical Committee (TC) 229, in conjunction with many national standards bodies, has been directing activities on nanotechnologies standards since 2004. The ISO TC 229 is organised into four working groups that focus on issues that are crucial for the development of an effective regulation for nanotechnologies-related products, viz. terminology and nomenclature; measurements and characterisation; health, safety, and environment; and materials specification (Amit Kumar, 2014). The Bureau of Indian Standards (BIS) is an active participating member of ISO TC/229 Committee which is presently engaged in devising Nanotechnology Standards. Till date, about 35 such standards have been published and some are in review stage.

Regarding safety of workers/researchers in laboratories, a number of countries require manufacturers of industrial chemicals, pharmaceuticals, veterinary drugs, pesticides, cosmetic products, food products, feed additives, etc., to establish through data that use of these products does not pose any hazards to human health and the environment. Non-hazardous nature needs to be established through studies and data, which is to be examined by the regulatory authorities of the concerned countries.

Various existing Indian laws/regulations that need relevant and timely interventions/changes to encompass nanotechnology related consequences can be classified into the following categories.

Production and Marketing: Drugs and Cosmetics Act, 1940, National Pharmacovigilance Protocol, Medical Devices Regulation Bill, and Insecticides Act, 1968.

Occupational Health and Safety: Factories Act, 1948 and OHS under other legislation.

Environmental Risk Management: Pollution control laws, Environment Protection Act, and Public Liability Insurance Act.

Waste Disposal: Factories Act, Hazardous Material (Management, Handling and Transboundary Movement) Rules 2007, BioMedical Waste (Management and Handling) Rules 1998, Municipal Solid Wastes Rules 2000.

As of now, there has been no effort made to streamline these acts/rules in order to address peculiarities that nanotechnology carries with itself. However recently, the DST has made efforts made towards establishing of a regulatory system. The Nano Mission has organised national dialogues to promote R&D in development of standards for nanotechnology and for laying down a National Regulatory Framework Road-Map for Nanotechnology (NRFR-Nanotech). It is aptly to quote here that so far there are no nano-specific rules or regulations in any government department or ministry, be it Environment Ministry or Health Ministry or Labor Ministry or Law Ministry. The deliberations have been going on to whether incorporate nano-specific concerns within the existing rules and regulations or create new ones exclusively dealing with nanotechnology.

According to the NISTADS Report (2012), it is argued that "The multifaceted dimensions and implications of nanotechnology do not fit into the compartments delineated by the present regulatory framework in India. An effective risk governance system is urgently required both because of the inadequate picture of present nanotechnology regulatory scene and because of the perplexities presented by technological advancements." According to another researcher, "While leading nanotechnology nations are debating on the best strategy to ward off risk, Indian government has so far adopted a "learn by doing" approach in nanotechnology development. However, it is high time India's science and technology research bodies showed some seriousness towards nanotechnology risk governance." (Jayanthi et al., 2012). On the experience of debates on GM, nuclear plant and nanomaterials in India, Chaturvedi (2013) very rightly observes that 'it is high time that we, as a society, develop institutional mechanism for technological assessments where, along with the safety aspects, knowledge streams from the social sciences are brought in to assess social and economic implications.

A study of the regulatory regimes for health and food will provide a more holistic perspective on the manner in which nano-applications in food and health will be regulated in India. Effective national regulation is not only fundamental to ensuring responsible development of nano-technology but can also act as a foundational experience to developing international regulatory frameworks that could benefit from national regulatory experiments (Chowdhury Nupur, 2006). From the outset, certain caveats need to be made. First, it is important to differentiate between regulation of technology and regulation of applications of that technology. Regulation of a technology per se is not only difficult but, as many regulatory experts have suggested, is also not desirable since technology developments often results in redundant regulatory categories that fail to fulfill regulatory objectives. Nevertheless, in the context of emerging technology, one has witnessed the development of general regulatory principles (vis-à-vis uncertainty and risk regulation) that may have implications for the regulation of nanotechnologies. Thus, regulatory experience garnered in the biotechnology sector (faced within comparable risks) could have an influence on the architecture and design of regulatory structure and the choice of regulatory tools applied in the regulation of nanotechnologies.

7.4 Risk, Uncertainty and Precaution in Global Environmental Governance

Scientific uncertainty in risk governance is one of the major challenges facing global environmental politics. Risk assessment and to a lesser extent risk management are often portrayed as being determined solely by scientific criteria and findings (Falkner Robert and Jaspers Nico, 2012). In reality, however, both processes, which are about determining whether certain risks exist and deciding on how to deal with them, cannot be separated from political decision-making. Even though scientists play a central role in creating knowledge about potential harm to humans and the environment, their scientific judgments alone cannot suffice as the basis for risk assessment and management. This is most clearly the case with regard to risk management decisions, which involve decisions on the level of risk that is socially acceptable. It also concerns risk assessment itself, particularly with regard to the definition of harm, identification of causal links and establishment of thresholds as triggers for regulatory action.

The distinction between uncertainty and risk is analytically important in that it demarcates the realm of calculable and controllable risk from the murkier field of uncertain knowledge about risk. In an ideal scenario, scientific research would provide regulators with sufficiently precise knowledge to remove the veil of uncertainty from decision-making. To a certain extent, this is the ideal that informs legal interpretations of risk assessment and management in the context of World Trade Organization (WTO) rules. In this view, regulatory decisions that interfere with international trade are to be based on reliable scientific evidence. Where such evidence is missing, further research is needed to seek to remove remaining uncertainties, even if full scientific certainty can never be achieved.

As recent international trade disputes have shown, uncertainty is a pervasive phenomenon in environmental and health regulation. It may exist with regard to the selection of appropriate risk categories and models; it may arise from unreliable measurement and testing techniques; and it may afflict the underlying causal models used in risk assessment. Scientific uncertainty is particularly pronounced in emerging technologies, such as biotechnologies, nanotechnologies or synthetic biology (Falkner Robert and Jaspers Nico, 2012). Scientists and regulators often have to operate without a comprehensive understanding of what types of risk need addressing, what testing methodologies need to be applied or developed in order to assess those risks, and what exposure paths are to be taken into account during the life-cycle of particular products.

The inherently political nature of technology risk is one important reason why leading industrialized countries have only partially succeeded in harmonizing international rules on risk regulation (Falkner Robert and Jaspers Nico, 2012). Some limited progress has been made in the WTO, e.g. on animal and food safety standards and safety-related technical standards (Agreements on the Application of Sanitary and Phytosanitary Measures (SPS) and Technical Barriers to Trade (TBT), but deeper international harmonization has proved elusive. In fact, national differences in risk regulation have repeatedly spilled over into commercial and political conflicts. The global conflict over trade in genetically modified organisms (GMOs), which led to acrimonious negotiations over the Cartagena Protocol on Biosafety and a WTO dispute over the EU's temporary ban on GMO imports, is but one high profile case in a long line of such disputes.

While the US has repeatedly insisted that regulatory trade restrictions should be based on "sound science" in line with WTO law, the EU has pushed for the global expansion of precautionary standards and a re-balancing of the relationship between WTO rules and environmental policies in favor of the latter (Falkner Robert and Jaspers Nico, 2012). Subsequent treaties, such as the 2000 Cartagena Protocol on Biosafety, have gone one step further and have included references to precaution not only in the preamble but also in the operational parts of the treaty (Graff Laurence., 2002).

Countries will continue to use precaution selectively and on an ad hoc basis, but the danger exists that these divisions will lead to further international regulatory polarization and conflicts (Wiener Jonathan 2007). Such profound differences in interpretation pose a serious challenge to the creation of comprehensive and effective global governance for emerging technology risk. They suggest that regulatory harmonization will prove difficult where scientific uncertainty prevails and the economic stakes involved are high. At a time when the need for global risk governance is constantly growing as technological innovation and economic globalization continue unabated, a gap is emerging between the demand for and supply of global policy solutions. This is increasingly the case for nanotechnologies, where uncertainties about risk simultaneously require and impede the establishment of a global governance structures.

7.5 Opportunities, Risks and Regulatory Challenges

Over the last decade, nanosciences have emerged from relative obscurity to occupy a central place in the scientific and to some extent, public imagination. At the nanoscale, common materials often have different physico-chemical properties than in bulk-form: gold changes color, aluminum becomes explosive, silicon turns from being an insulator to being a conductor, silver becomes a highly effective antimicrobial, and carbon becomes extremely strong and stiff (Falkner Robert and Jaspers Nico, 2012).

The future commercial value of nano-enabled products is difficult to predict, with some estimates as high as several trillion US dollars over the next five to ten years. In 2007, the materials and manufacturing sector alone accounted for $97 billion in nanoenabled products.24 Investment in nanotechnology research and development has risen rapidly, with cumulative public spending reaching almost $50 billion worldwide in 2009 (Cientifica, 2009). As more and more nano-products are being marketed, scientists and regulatory experts have voiced growing concerns about the safety of some nanomaterials. These concerns are focused on the miniscule size of nanomaterials and their unique physico-chemical characteristics. For example, some nanomaterials may enter the human body through mucous membranes or the skin and migrate via the bloodstream to vital organs including the brain. Some nanomaterials can also enter cells, interact with their molecular structure, and have cytotoxic or genotoxic effects (Monteiro-Riviere Nancy, 2005). Traditional protective equipment such as gloves or masks may not provide sufficient protection against accidental inhalation or absorption through the skin. Changes to the physico-chemical properties, especially when they lead to increased toxicity, may produce unexpected negative effects in the human body or the environment, particularly so as nanomaterials may change as a consequence of their interaction with living systems (ICON, 2008).

Governments in industrialized countries have poured large sums of public money into nanosciences in the hope of nurturing the next industrial revolution (Falkner Robert and Jaspers Nico, 2012). But balancing scientific freedom, technological innovation and an adequate level of environmental, health and safety protection can prove to be an elusive goal.

Regulatory systems, thus face challenges on a number of fronts. They must adapt to deal with novel and uncertain risks. They must expand scientific capacity against the background of a fiscally constrained environment and competition for scientific talent between public and private sectors. They must also anticipate rapidly changing scientific and technological systems, and develop globally oriented information-gathering systems to cope with the ongoing globalization of nanotechnology. These challenges are of such a fundamental nature that they go beyond the capacity of individual states and require international coordination and cooperation.

National and Regional Regulatory Approaches: Convergence or Divergence?

The nanotechnology safety debate has gathered momentum since the early 2000s and has come to focus more systematically on the appropriateness of existing regulatory policies. A series of scientific reports have highlighted persistent scientific uncertainties and knowledge gaps regarding environmental and health impacts, most notably the 2004 report on nanosciences and nanotechnologies by the UK's Royal Society and Royal Academy of Engineering (Falkner Robert and Jaspers Nico, 2012). Other studies have also highlighted the scientific uncertainty and regulatory challenges surrounding nanotechnology. Regulatory authorities themselves have acknowledged some of the findings of the aforementioned reports but conclude that, by and large, existing frameworks provide appropriate regulatory coverage for nanotechnology risks.

Basic regulatory approaches: The key challenge in the current debates on regulation of nanomaterials originates from a conflict of two different regulatory approaches. One position can be - in a way stylized – summarised as strongly precautionary-oriented, putting nanomaterials under general suspicion because of their new properties and the limited knowledge about their (potential) environmental, health and safety implications. In this approach, nanomaterials are usually defined rather broad and a number of strong measures are proposed to supervise and control the entire life cycle of nanomaterials or products containing nanomaterials or being manufactured using nanotechnologies (Anonymous, 2011). Given the considerable broadness of the definitions of nanomaterials and nanotechnologies, a large number of both natural and artificial materials and products as well as various technological processes will be affected by this regulation. Important questions to be discussed in connection with this approach are: Do the regulatory agencies and other affected parties have sufficient resources to implement and enforce this regulation? What are the implications of this approach on existing and future social practices, technological innovation and economic development? Are there mechanisms to "release" nanomaterials from that regulatory regime, assumed they were proven to be "safe"? And how "safe" is safe enough to justify this decision?

One more regulatory approach is closely linked to evidence from toxicological, ecotoxicological and biological research. Its proponents argue that particularly (or solely) those nanomaterials should be regulated that give rise to concerns regarding their EHS implications, either because toxicological research has shown that a hazard exists or because the physico-chemical properties of the nanomaterial allow to predict a certain hazard potential (e.g. when the nanomaterials exist in free form, are known to be insoluble, biopersistent, etc.).

7.6 International Regulatory Coordination and the Global Governance Gap

In contrast to the rapid pace with which the commercialization of nanotechnologies has proceeded, international governance initiatives have so far been slow off the mark and limited in scope. Profound scientific and regulatory uncertainties have held back efforts to advance global governance approaches for this emerging technology (Falkner Robert and Jaspers Nico, 2012). Regulators remain focused on developing and implementing national regulations.As the global dimensions of nanotechnology are becoming apparent, however, international standardization, coordination and harmonization have slowly emerged on the international agenda.

The EU and the US have taken a lead in international coordination efforts, working primarily through the Organization for Economic Cooperation and Development (OECD) and the International Organization for Standardization (ISO) as well as through bilateral links. Other international organizations are only beginning to develop nanotechnology-specific work programs. Furthermore, a small number of private initiatives have sprung up to promote information gathering, standardization and international debate about regulatory challenges. However, most of these are still at an early stage and involve only low levels of international coordination.

The international institutionalization of research and regulatory coordination began in 2005. Although an early proposal by the European Commission to establish an international "code of good conduct" in research as a "global agreement on base principles for the responsible development of nanotechnology" was met with skepticism, notably from the US, both Europe and America have repeatedly expressed their commitment to work together in other international forums (European Commission, 2004).

Aside from the OECD, ISO has taken a lead in developing internationally harmonized terminology and standards. Its technical committee on nanotechnologies (TC 229), which was created in 2005, has published 11 international standards to date and continues to work on various issues relating to the standardization of terminology, definitions, toxicity testing and environmental studies protocols, measurement techniques, calibration procedures, and reference materials. The growing push to internationalize the safety and regulatory debate is also evident from the number of public-private and private initiatives that provide forums of debate and informal coordination mechanisms such as the International Council on Nanotechnology (ICON), the International Alliance for NanoEHS Harmonization and the aforementioned International Risk Governance Council.

Genuine multilateral nanotechnology governance initiatives that are based on UN processes are few and far between. The United Nations Industrial Development Organization's International Centre for Science and High Technology (ICS UNIDO), for instance, has hosted dialogues that seek to promote regional networking between scientists, industrialists and policy-makers. And the WHO's Intergovernmental Forum on Chemical Safety (IFCS) has only recently begun to work on nanotechnology-related safety issues. The WHO's Dakar Statement on Nanotechnology and Manufactured Nanomaterials was adopted in 2008 and calls for more international cooperation in information sharing and risk assessment (IFCS, 2008).

Of perhaps greater significance for future nanotechnology governance is the adoption of the Strategic Approach to International Chemicals Management (SAICM) in 2006 by UNEP's International Conference on Chemicals Management (ICCM). SAICM emerged out of the collaboration of various international organizations and stakeholders, including the WHO's IFCS, and supports the goal adopted at the 2002 World Summit on Sustainable.

To sum up, as this brief survey of international initiatives has shown, the internationalization of the regulatory debate in nanotechnologies is now well underway, but remains limited in its scope and degree of international participation. International governance of nanotechnology risk as it exists today is mostly limited to scientific and technical standardization and coordination efforts by the leading nanotechnology countries in the OECD, with a few other international forums having emerged to consider broader policy challenges. No deeper structures for global governance of nanotechnology have been created despite the rapid globalization of nanotechnologies.

Uncertainty about the nature and scope of nanotechnology related risk is one of the factors that stands in the way of a broader political consensus on how to create global governance structures. At the same time, however, leading nanotechnology countries have engaged in some forms of international collaboration on regulatory policy.

Role of Global Governance: As an analysis shows, a global governance gap has emerged that is likely to grow in the coming years unless governments and other stakeholders step up current coordination and cooperation efforts (Falkner Robert and Jaspers Nico, 2012). While recent regulatory and policy reviews have highlighted the need for greater international cooperation, proposals for filling the global governance gap vary considerably, ranging from the use of soft law approaches and building sector-specific governance capacity to the creation of an international framework convention. At this point, the shape of any future global governance architecture is uncertain.

Governments face a dilemma with regard to the global governance of nanotechnology. For the profound uncertainty that characterizes risk regulation cuts both ways, at the same time requiring and impeding closer international cooperation. On the one hand, reducing underlying uncertainty, in terms of the scientific understanding of risks, regulatory coverage and institutional capacity, is of utmost importance, and global governance approaches can make a significant contribution in this respect. On the other hand, the very uncertainty that pervades risk assessment and management militates against the early creation of appropriate global governance structures. Faced with an uncertain cognitive environment, governments will want to develop national regulatory approaches first before committing to international efforts. They are likely to be hesitant to undertake potentially costly steps towards a global regulatory regime for what are as yet uncertain risks. In other words, while early international cooperation may be desirable and mutually beneficial, the associated transaction costs are likely to be prohibitive.

Where the coordination problem and the interests and capacities of key players are uncertain, international cooperation is bound to be more tentative and exploratory in nature (Abbott, Sylvester and Marchant 2010). It is thus more realistic to expect global governance functions to be built in a flexible, step-by-step, approach. Three areas of international cooperation provide opportunities for developing the first building blocks of a global governance architecture for nanotechnology. They are concerned with improving the cognitive environment of risk assessment, encouraging greater coordination of risk management approaches, and supporting capacity building in developing countries to strengthen their regulatory systems and promote their involvement in international processes.

Improving the Cognitive Environment: One of the most urgent tasks for global governance approaches is to build a sound and reliable cognitive basis for identifying and assessing environmental and health risks throughout the life- cycle of nanomaterials (Falkner Robert and Jaspers Nico, 2012). This relates, in first instance, to technical and scientific standardization for risk assessment. It also includes the internationally coordinated promotion of research into environmental and health risks and the creation of more reliable knowledge on the presence of nanomaterials in international markets. Technical and scientific standardization is progressing already, albeit slowly.

The OECD has recently created an internet- based database of current safety research on nanotechnologies, and is also hosting a series of research projects on the potential hazards of selected nanomaterials. What is missing, however, is a more concerted effort at defining an internationally coordinated research strategy, which would pull together the different strands of the current research landscape and promote coherence and strategic vision. A further area of uncertainty that plagues current regulatory approaches and that requires international cooperation concerns the state of commercialization of nanomaterials. While a vast number of commercial applications of nanotechnologies have emerged in recent years, regulators are still uncertain about the extent to which nanomaterials have penetrated international trade. Discussions about creating a market register for nanomaterials are currently underway in various national settings (OECD 2010). But given the rapid globalization of nanotechnologies, international cooperation on developing a register would provide regulators with a reliable basis for determining pathways of nanomaterials and their potential risk to humans and the environment.

Promoting convergence of risk assessment and risk management: A further, and arguably more demanding, step towards global nanotechnology governance would be the creation of internationally agreed principles and rules for risk assessment and management. The international community has sought to create such internationally harmonized rules in a number of areas of environmental and health risk, though with varying degrees of success. For example, among the more successful and widely accepted agreements is the WTO's SPS Agreement on food safety and animal and plant health, while the Cartagena Protocol on Biosafety remains contested by most GMO-exporting countries. No such international regime exists for nanotechnologies, however. Bilateral talks between regulatory agencies and multilateral discussions at the OECD eschew the bigger question of international harmonization. They are aimed at dialogue and information exchange in risk assessment and tend to exclude the more politically charged dimensions of risk management. Yet, given the global nature of nanotechnologies, demand for international coordination, and eventually harmonization, is likely to grow soon.

As leading nanotechnology countries apply existing regulatory frameworks and begin to take regulatory decisions, differences in national laws and regulatory culture may create an uneven regulatory field for the increasingly global market in nanotechnology applications (Jaspers Nico, 2010). In the transatlantic context, for example, newly created requirements in the EU for labeling nanomaterials in food and cosmetics products go well beyond similar provisions in US law, creating in fact the first nanotechnology-specific labeling system.81 Just as in other regulatory policy areas, such differences often end up souring international trade relations and may become a source of international commercial and political conflict. An early and proactive approach to promoting convergence in regulatory approaches to nanotechnologies would go some way towards preventing such conflicts.

7.7 Capacity-building and engagement of developing countries

As in previous technological revolutions, leading industrialized countries are at the forefront of developing regulatory policies for nanotechnology. They are also leading international standardization and coordination efforts, centered mainly on the ISO and OECD. By contrast, the vast majority of developing countries have been absent from international nanotechnology debates. A few emerging economies that engage in nanotechnology research and production (e.g. China, Thailand) have been involved in the OECD working parties in an ad hoc fashion (Falkner Robert and Jaspers Nico, 2012). Overall, however, participation by developing countries in the global regulatory debate ranges from weak to nonexistent. The current North-South divide in global nanotechnology governance poses a number of challenges. Already, a number of emerging economies are investing heavily in nanotechnology R&D and some of them are emerging as major sites of nanomaterials production (e.g. China). Even those developing countries that are not engaged in research and commercialization will become importers of nanomaterials and nano-enabled products in what is already a global business. Developing countries and emerging economies will soon be faced with similar regulatory challenges that industrialized countries are addressing today, yet few of them are actively engaged in international processes.

As in other areas of technology governance, one of the primary tasks in global governance will be to promote capacity building in developing countries. In order effectively to engage in global regulatory debates and carry out risk assessment and management at home, developing countries will need to build up scientific and regulatory capacity, by training toxicologists and regulatory experts, establishing testing facilities, creating international scientific networks and promoting information exchange, and developing safety guidelines and implementation guidelines for regulatory frameworks (Raines, 2003). The specific needs will vary according to the level of economic development and state of investment in nanosciences and nanotechnologies, but international support will be of critical importance to fill the capacity gaps that are fast emerging in many countries.

Finally, a global governance approach also promises a higher degree of legitimacy for emerging norms and rules for nanotechnology regulation. In the past, global governance initiatives have suffered from legitimacy deficits due to the weak involvement of developing countries. Developing countries have on numerous occasions been marginalized in international rule-making and only weakly represented in private standardization process. Given that global nanotechnology debates are only just starting up, the international community still stands a chance of creating a more inclusive approach to developing global governance in this area.

Conclusion: Nanotechnologies are set to transform the global industrial landscape. The speed with which this change is occurring is breathtaking. Only a few decades ago, the manipulation of matter at the molecular level belonged to the realm of scientific speculation. It has now become a commercial reality. Nanotechnologies also introduce profound uncertainties that require international responses while at the same time impeding the creation of global nanotechnology governance. As in other emerging technologies, global regulatory debates and governance initiatives are lagging behind nanotechnology innovation (Falkner Robert and Jaspers Nico, 2012). The international dimensions of nanotechnology regulation are still poorly understood and rarely feature on the international agenda. Leading nanotechnology countries in the OECD have started to promote international coordination, and UNEP and the WHO have begun to debate nanotechnology as an emerging concern. Yet, the global debate on how to address nanotechnology and its associated risks has barely begun. As more and more nanomaterials and nanotechnology-enabled products enter international trade, a global governance gap is emerging with regard to environmental, health and safety regulation.

In the last five years, the first international coordination mechanisms have been created, primarily through the OECD and ISO. The scope of their efforts is limited, however, as is their authority to develop a more broadly-based global governance framework. Several analysts and regulatory reviews have called for increased international cooperation, standard-setting and capacity-building.

Developing nanotechnologies in a responsible manner will require governments to work together to reduce uncertainties and promote coordination and cooperation at an early stage. Persistent uncertainty both creates demand for and stands in the way of greater international cooperation and harmonization of regulatory approaches. Against the background of this profound dilemma of international risk regulation under conditions of uncertainty, global nanotechnology governance is likely to emerge only slowly and in incremental fashion. Three areas were recognized in which international approaches could and should be taken forward, as building blocks for an emerging global governance architecture. These are first, improvements in the cognitive environment for regulatory approaches, second, the promotion of convergence in the principles and rules for risk assessment and management, and third, capacitybuilding and greater involvement of developing countries in international processes. By developing global governance in an incremental way, governments can respond to emerging nanotechnology risks in a more coordinated and informed yet flexible manner.

7.8 Criteria for a legal definition

The process towards the development of a harmonised legal/regulatory definition of nanomaterials should be intensified (Anonymous, 2011). Following four arguments might be helpful to assist this process.

- Legal definitions of nanomaterials have to describe the object of regulation sufficiently precise to be clear to all parties affected by it. They have to consider practices of production and application of nanomaterials as well as to be enforceable by the responsible authorities.
- A legal definition of nanomaterials incorporates not only scientific and technological knowledge (and its respective uncertainties), but also includes the results of policy choices and political decisions. It therefore should be science-based but does not necessarily have to be identical to scientific definition(s) of the same term.
- The breath of the legal definition has to be matched with both the regulated artefact and the regulatory goals. A legal definition of nanomaterials has to take into account that these may occur in nature including in a number of natural products that are consumed by humans, could be incidentally produced as results of various human activities, or can be intentionally manufactured. This situation results in different hazard assessments, diverse exposure scenarios and various starting points for regulatory intervention, depending on the aims of the regulation. Meaningful regulation is limited to human activities; therefore a legal definition of nanomaterials should focus on manufactured nanomaterials.
- A legal definition of nanomaterials based on "new" properties occurring at the nanoscale might be difficult to achieve. Therefore, a size range in which the most size-dependent properties appear could serve as an appropriate, albeit imperfect, heuristic. Although any choice of a size range would be imperfect with respect to certain regulatory goals, since there are no direct, material-independent relations between size and "nanoscale properties", a size range from 1 nm to a value not above 100 nm might cover many configurations of materials that give reasons for regulatory concern. For various reasons, an upper size limit cannot directly be derived from scientific results but would be the result of a balancing of goals and interests and therefore should be subject to political decisions.

7.9 Risks posed by nanomaterials

Effective regulation of nanotechnology-enabled products requires an understanding and consideration of their unique properties. In reality, nanoparticle application in the various industries has outpaced the research that is needed to determine which of their characteristics might pose unique hazards. A better understanding of the link between nanomaterial properties and biological behaviour in humans and the environment is also needed to guide risk assessment paradigms (Anonymous, 2014). Manufactured nanomaterials are found in many consumer products on the market today, including sunscreens, cosmetics and clothing. The public want to know if these products are safe; some people contend that the risks posed by nanomaterials are unknown and there have been calls nationally for mandatory labelling and a register of nanomaterials as a fundamental right to know.

The safety of nanomaterials is an emotive and controversial matter that often features in the popular press. The three cases that follow are typical of those reported. This first case involves safety concerns relating to nanoparticle-based sunscreens. Media reports assert that nanoparticles incorporated in some UV filters penetrate the skin, enter the bloodstream, and cause possibly adverse effects. Scientists contend there is no definitive evidence for nanoparticles in sunscreens penetrating the skin. Rather the nanoparticle 'marker ions' detected in the blood and urine of human trial subjects who have applied sunscreens containing nanoparticles result from the dissolution and ionization of nanoparticles on the skin,so it's the ions and not nanoparticles per se that have been absorbed percutaneously.

Nanoscale silver (nanosilver) has been the subject of numerous media reports. In addition the National Institute for Public Health and the Environment (RIVM) located at Bilthoven, the Netherlands, conducted a hypothetical registration of nanoscale silver according to the EU Registration, Evaluation, Authorisation and Restriction of Chemicals (REACH) legislation in 2009. Most products containing nanosilver are used for their antibacterial properties; in the home to keep surfaces clean or to reduce odours, as a coating in medicinal applications, such as artificial joints and pacemakers, and in ceramic filters for water purification. There are also concerns that overusing nanosilver could lead to antimicrobial resistance. Nanosilver highlights the importance of life- cycle considerations, including production, transport, storage, use and disposal or recycling, when regulating nanomaterials.

A third case focuses on media reports about research published by Poland et al (2008), who reported that injecting carbon nanotubes (CNTs) into the abdominal cavities of mice resulted in asbestos-like lesions (Anonymous,

2014). The study demonstrates that CNTs conform to a structure-activity relationship based onaspect ratio, to which asbestos and other macro- and micron-sized pathogenic fibres conform. However, while the findings confirm that fibres with high aspect ratios can result in long-term harm, this behaviour is not restricted to nanofibres but applies to all fibres, regardless of size.

Size matters: Size affects the potential risks posed by nanomaterials in at least three ways. First, materials at the nanoscale dimension are of similar size to natural biomolecules. For example, the diameter of a DNA double helix is about 2 nm while the diameter of bucky balls is about 5 nm (Klaine et al, 2012). The similarity in size translates into an increased potential for manufactured materials to interfere with biological processes. This could include the behaviour of cell membranes, biochemical pathways in cells, or even the genetic code itself. Second, the surface areas (relative to volume) of nanomaterials are orders of magnitude larger than those of non-nanomaterials. A consequence of a large surface-tovolume ratio is that the properties of the surface molecules dominate. It therefore follows that surface chemistry is an important determinant of nanomaterial hazard. Third, nanoparticles may translocate to locations in the human body or the environment not accessible to their conventional counterparts (Anonymous, 2014). All three of these size-related nanomaterial behaviours pose potential risks to human health and the environment, and are addressed in registration applications and assessed by Government regulators.

A complete and accurate characterization of manufactured nanomaterials is required in order to fully understand both the benefits and the potential toxicity of nanoparticles in biological systems. When a full characterization is not possible and it becomes necessary to priorities the parameters for characterization, the criteria are to be considered, viz., the context within which a material is being evaluated, the importance of measuring a specific parameter within that context and the feasibility of measuring the parameter within a specific context. The OECD Working Party on Manufactured Nanomaterials has recently published a Report of the OECD Expert Meeting on the physical chemical properties of manufactured nanomaterials and test guidelines, No. 41 (OECD, 2014). The report provides guidance for assessing the aggregation and agglomeration of nanomaterials, and determining the size, surface area, porosity and surface reactivity of nanoparticles.

Traditional test methods need to be applicable to the nanomaterial under consideration. For example, whether a material is soluble, insoluble, or partially soluble may affect the suitability of a traditional toxicity test. If a traditional toxicity testing method cannot be satisfactorily modified to be fit for purpose, new methods may be needed. In addition, caution is needed if extrapolating the results of in vitro studies to an in vivo assessment since a direct translation seldom applies. It is also important to keep in mind that in vitro tests are most useful in providing information on mechanistic processes and in clarifying mechanisms and modes of action suggested by studies in whole animals.

Biokinetics, which deals with the absorption, distribution, metabolism and excretion (ADME) of substances in the body, determines the internal exposure of organs to potentially toxic substances. In addition to a sound knowledge of the biokinetics of a nanomaterial being necessary, the possibility of nanoparticles having a 'Trojan horse' effect by acting as carriers of contaminants (Anonymous, 2014) must also be considered. The impact of nanomaterials on the environment has been the subject of intense and ongoing research. The impact is more pronounced with non-biodegradable nanoparticles that persist and accumulate in the environment. The current knowledge base highlights the need to track the full lifecycle of manufactured nanomaterials and adopt 'safe by design' concepts for reducing, or even preventing altogether, the detrimental impact of nanomaterials on the environment. Two examples of 'safe by design' concepts applied to nanoparticles are those designed to dissolve very slowly, and those coated with inert compounds.

Nanotechnology-enabled products used in food-producing animals and crops need to be evaluated for their potential to leave residues in food. A recent expert meeting noted that the current risk assessment approaches used by FAO/WHO and Codex for residues in food are suitable for manufactured nanomaterials, but any additional safety concerns arising from the characteristic properties of nanomaterials would need to be addressed. It is also noted in reference to Codex standards that: 'neither the specifications nor the ADI for food additives that have been evaluated in other forms are intended to apply to nanoparticulate materials'.

Potential environmental risks : The general categories of information required to carry out a risk assessment for nanomaterials include identity information physicochemical properties, industrial and consumer uses and environmental releases, environmental fate, absorption, distribution, metabolism and excretion (ADME) and the potential toxicity of nanomaterials (Anonymous, 2014).

7.10 Risk assessment of nanomaterials

The steps in the risk assessment paradigm are the same as for other chemicals and include exposure assessment, effects assessment (hazard identification and classification), and risk characterization.

In the case of nanotechnologies, the potential number of nanomaterials, which may each have different properties, uses and exposure pathways, requires that a range of important issues be considered or addressed to enhance their risk assessment. These issues relate principally to the physicochemical properties of nanomaterials and possible exposure pathways. The problem formulation stage is a problem-scoping exercise. It identifies the relevant sources and targets of suspected harm and guides the remainder of the assessment. It may provide an opportunity to consider the physical and chemical properties of the nanomaterial upstream to reduce or avoid downstream risks. For example, if the problem formulation stage suggested a 'safe by design' approach for environmental sustainability reasons, the manufacturing process might be modified accordingly.

Elements of risk assessment: The exposure assessment analyses the physicochemical properties of the nanomaterial and possible exposure pathways whereas the effects assessment (hazard identification and classification) evaluates ecotoxicology information. The elements of the risk assessment at these stages include the behaviour of nanomaterials in various media, the persistence of nanomaterials (both chemical and physical persistence) transportation/distribution, Predicted Environmental Concentrations (PECs), transformation products and impurities, bioaccumulation and effects/Predicted No Effect Concentration (PNEC). Risk characterisation draws together the various lines of evidence from the assessment of the nanomaterial. A weight-of-evidence approach is used to determine the potential for harmful effects.

Limitations of the risk assessment of nanomaterials: Both positions, in different ways, have to deal with profound limitations of the risk assessment of nanomaterials. The methodology for the assessment of chemicals risks including, but not limited to nanomaterials – applied in most countries consists of four parts: hazard identification, hazard assessment (including dose-response relationships), exposure assessment, and risk characterization (Anonymous, 2011). Each of these four elements holds a number of limitations that are not easy to overcome.

The majority of nanotoxicological work done contributed to the field of hazard identification, attempting to reveal the toxicity of MPNs in respect its type and characteristics. The current knowledge suggests that inhalation is the main portal of entry of MPNs into the body. Epidemiological studies about MPNs are not available therefore studies of ambient ultrafine particle (< 100 nm) toxicology are taken into consideration to study human adverse health effects by nanoparticles. Various studies showed that inhaled MPN size-dependently deposit in different regions of the lungs. It was demonstrated that to a certain amount MPNs can be removed by clearance mechanisms (especially in bronchia) and/or immune system (especially in alveoli) of the lungs. Thereby, it seems that these mechanisms are less effective by a decreasing particle size. If insoluble particles are deposited in a certain area of the lung they will undergo clearance mechanisms or will be accumulated in particular areas where they even may pass membrane barriers and enter individual cells causing biological or toxicological effects. At high doses, certain MPNs (e.g. fibre like carbon nanotubes or nanosilver particles) led to pathologic conditions and can cause toxic effects.

The assumption that the move to the nanoscale implicates not only novel material properties but also entails novel environmental and health risks was confirmed on a scientific basis. However, the relevance of the data from the various in vivo and in vitro studies is still unclear. Thus, the available data provide a basis for further investigations by providing knowledge about fate and behaviour (ADME-profiles) as well as the toxicity including underlying mechanism – however only for certain MPNs. It was shown that the shape of certain MPNs as well as their purity is important for toxicity, e.g. carbon nanotubes, seem to be more toxic if trace impurities of iron or solvents were present.

Beside, toxicity testing faces some intrinsic limitations; some of them can be overcome in future, others won't. As mentioned above, there is evidence that some manufactured particulate nanomaterials may be hazardous to human health, depending on their characteristics. A problem repeatedly discussed in this context is that so-called "no-effect studies", i.e. nanotoxicological studies that have "failed" to show effects of MPN on various endpoints, to a large extent remain unpublished. The reasons for that are manifold and span from methodological challenges to limited opportunities and incentives for publication due to the scientific system. Then again, no-effects studies are a valuable repository for hazard characterization and its limited accessibility could be seen as a waste of scientific resources.

Hence, risk characterization that builds on hazard and exposure assessment is at this time (and most probably in short- and medium-term) not feasible or certainly not scientifically reasonable and only preliminary.

7.11 Challenges for risk assessment and risk governance

The risk assessment methodology as a whole is inadequate to timely inform political decisions regarding the regulation of nanomaterials, at least in the short- to medium-term (Anonymous, 2011). In the light of the various knowledge gaps, it would need enormous efforts to perform valid and broadly accepted risk assessments for specified nanomaterials. Whether these materials are considered "reasonably safe" or "of high concern", both claims will remain unproven for many years. Moreover, its role and validity as justifications for regulatory strategies of these claims will be contested.

Regulation under uncertainty raises the fundamental political question of how policy makers should regulate risk in the face of limited scientific evidence. In this context, it is of particular importance that regulations represent not only a restriction for companies, but can also serve as a guideline for strategic decision and legal certainty. Lawmakers on national and European level are dealing already with the implementation of nanospecific aspects in an incremental case by case approach. These activities imply a wide range of provisions and instruments, depending on the application and life cycle stage and different levels of attention and risk assessment. While a nanospecific framework does not exist, the adaption of existing regulations is an ongoing process, concerning the scope and the threshold limits as well as adequate nanospecific assessment procedures. REACH seems to provide a powerful framework to regulate nanomaterials, but there are open gaps and problems. Other policies concerning nanomaterial aspects are mentioned in this project, mainly the regulation on cosmetic products and the currently discussed amendments on the Novel Foods Regulation, the proposal for a Biocidal Products Regulation as well as the Medical Devices Directive. Besides these mandatory provisions, also voluntary measures based on an increased self-responsibility of producers are important. They are somehow temporary actions in between the establishment of a firmer scientific evidence for specific policy decisions. Advantages and problems of voluntary registers and code of conducts are discussed in the light of governance, regulation and control of nanomaterials.

Another question still under debate is whether existing legislation can be – or should be – adapted to MPN or whether a new regulatory framework for nanomaterials should be developed. Most scholars and practitioners in regulatory law as well as most political decision makers prefer a so-called incremental approach. They favour to adapt the existing legal framework to enable nanotechnology regulation and amending it in order to deal with the unintended implications of this technology. This approach has a number of challenges, limitations and potential gaps since existing legislation is not designed to accommodate some specific aspects of nanomaterials or nanotechnologies as follows, which need to be addressed as soon as possible for the incremental approach to be successful and to go along with a responsible development and use of nanomaterials and nanotechnology.

- Developing a legal definition for nanomaterials;
- Integration of nanomaterials into the REACH classifications and procedures, including the development of suitable guidance documents;
- Being able to identify and address the relevant adverse effects of the production, use and disposal of nanomaterials and nanoproducts;
- Enabling appropriate integration of nanospecific aspects into existing pieces of legislation for sectors, applications, products, or substances;
- Covering borderline products (like medical devices or nanomedicinal products) that cross different classic regulatory contexts and for whom regulators have additional uncertainties for the regulatory coverage of emerging nanomaterials risks;
- Finding adequate regulatory instruments;
- Enforcing compliance with existing and emerging regulation.

Some scholars as well as some stakeholders argue that the limitations of the incremental approach are so serious that an entirely new regulatory framework for nanomaterials is needed. But many voices do not further conceptualize this idea. Therefore an exploratory process towards the development of a new regulatory framework for nanomaterials should be encouraged that also tests its feasibility and discusses its advantages and disadvantages compared to the current incremental approach. This discussion could become more urgent since various technology vision documents forecast the development of future-generation nanomaterials, including active nanomaterials with overlapping aspects of information technology, biotechnology and cognitive science.

Although these trends are difficult to foresee, regulators will have to monitor these developments and therefore need both the (scientific and budgetary) resources and the regulatory instruments for being able to answer with flexible responses.

7.12 Preliminary considerations for risk characterization

The first step toward the environmental risk assessment of a nanoparticle, like AgVet product is to determine whether or not a special nanomaterial risk assessment must be applied (Anonymous, 2014). Aside from the size range (1- 100 nm) that initially separates nano AgVet products from traditional products, another important consideration for preliminary assessment would be nanomaterial stability, does the nanomaterial remain in the nano-scale in the formulation and when delivered? Nanomaterial dissolution and aggregation are two processes that would directly affect nanomaterial size and can significantly affect the nanomaterial activity. Dissolution refers to the release of individual ions and/or component molecules that are soluble in water. Homoaggregation refers to the aggregation/attachment of two similar particles forming larger units. Homoaggregation of nanoparticles would be important in areas such as formulation stability and ecotoxicity testing. However, heteroaggregation of nanoparticles with natural colloids is likely to control their fate and ecotoxicity in aquatic and terrestrial environments.

Environmental fate and transport: Nanomaterials that are released from products are likely to go through the different environmental processes, viz.,: physical and chemical transformation during use, retention/partitioning between soil and water, transport and remobilisation from soil to water, abiotic/biotic degradation/transformation of nanoparticles in soil and water, uptake by terrestrial and aquatic organisms. While these processes are not unique, because of their dynamic nature the responses of nanomaterials to these processes may differ from those of traditional chemicals (Anonymous, 2014).

Assessment of ecotoxicological effects

Physico-chemical properties and nanomaterial behavior: The physical and chemical properties of nanomaterials are important factors influencing their fate and ecotoxicity in aquatic and terrestrial environments (Anonymous, 2014). The properties of nanomaterials recognised as relevant to toxicological testing have been compiled by a number of organisations, including the Organisation for Economic Co-operation and Development (OECD) Working Party on Manufactured Nanomaterials (WPMN) and the International Organization for Standardization.

i. Size : The size of nanoparticles will significantly affect their dissolution, aggregation, surface area and reactivity. Nanomaterials with smaller size/dimensions could potentially have enhanced mobility through soil pores, reactivity (ie increased surface area), biological uptake via passive transport and toxicity. Kim et al found in zebra fish embryos that 20 nm Ag nanoparticles were more toxic than 110 nm Ag nanoparticles (Kim et al. 2013). Ag nanoparticles coated with polyvinylpyrrolidone (PVP) that remain unagglomerated in suspension were also observed to be more toxic than Ag nanoparticles coated with citrate, despite similar sizes of the core nanoparticles. Soil retention of Ag, CeO_2, and ZnO nanoparticles has been shown to be dependent on particle size with significantly higher numbers of nanoparticles retained compared to micron-sized (bulk) particles.

ii. Morphology/shape/surface area: The morphology/shape of nanomaterials is recognized as an important property that can significantly affect the surface area and reactivity of the nanomaterials (Batley et al. 2012). Higher surface area-to-volume ratios suggest an increased number of available sites for interaction that may increase adsorption and enhance reactivity, cellular uptake and toxicity. The effect of surface area on the photoactivity of TiO_2 particles revelaed that those with a larger specific surface area induce higher cytotoxicity (UV absent) and phototoxicity (UVactivated) to cells after 24 h incubation (Xiong et al. 2013). A large number of hydroxyl radicals were detected from TiO_2 particles with larger surface areas. It was suggested this was due to their increased interaction with biomolecules.

In the case of inert and porous materials used in controlled-release applications, their high pore volumes could enable sorption and facilitate transport of other contaminants that enter and are retained in their pores, especially once the active ingredient (ie pesticide) has been delivered.

iii. Solubility: 'Solubility' is often loosely applied in nanomaterial studies, especially in relation to carbon-based nanomaterials, often meaning 'forming stabilised suspensions' rather than truly dissolving (Anonymous, 2014). Hydrophobic nanomaterials are virtually insoluble, for example fullerene solubility has been calculated as 10-18 mol L-1 and require the addition of organic solvents to form stabilised dispersions. Interactions of hydrophobic nanomaterial with different aqueous/soil components, such as dissolved organic matter, may result in a variety of nanoparticle transformations, including partial/complete degradation, chemical transformation, and surface modification that can promote nanomaterial solubility/dispersibility. Hydrophobic fullerenes (C60) exhibit increased dispersion in the presence of dissolved organic matter (possibly due to changes in surface chemistry via hydrophobic or pi-pi interactions) and increased solubility through transformation into more water soluble poly-oxygenated/hydroxylated C60 derivatives (oxygenation and hydroxylation via photoactivation or ozonation).

Most metal-based nanoparticles are hydrophilic with solubility dependent on size, shape, composition and reactivity. In many studies solubility is not measured, despite the soluble ionic metal fraction being the most toxic to aquatic and terrestrial biota. Nanoparticulate ZnO rapidly dissolve to produce 6 mg L-1 of dissolved Zn within 6 h and 16 mg L-1 in 72 h in a buffered pH 7.5 soft water, in excess of the 5 mg Zn L-1 that would be toxic to most aquatic biota. By contrast, nanoparticulate CeO2 has a very low solubility (ng L-1), so the effects of nanoparticle versus micron-sized particle toxicity could be readily investigated (in the absence of the confounding influence of dissolution). Higher toxicity to algae exposed to nanoparticulate CeO_2 was noticed compared to its micron-sized equivalent.

Aggregation and agglomeration: Aggregation (homo- and hetero-) and agglomeration are important properties that determine the fate and toxicity of nanomaterials in the environment (Batley et al. 2012). Aggregation results when the repulsive forces that keep colliding particles apart are weakened (or when the attractive forces are enhanced). This process is heavily influenced by the properties of the suspension such as pH, ionic strength, and the presence of other colloids. Because nanoparticles can aggregate with each other (homoaggregation) and with other natural particles (heteroaggregation), the main route for the removal of most nanoparticles in the environment will be through aggregation, followed by sedimentation. Aquatic and terrestrial environments contain a wide range of natural particles, including colloidal clays, iron and manganese hydrous oxides, and dissolved organic matter (comprising fulvic and humic acids) and fibrillar colloids (exo-polymers) that are exudates from algae and other microorganisms (the exudates are largely polysaccharides and some proteins). Heteroaggregation of artificial nanomaterials with natural colloids is therefore likely to control the fate of most nanoparticles (Anonymous, 2014). Despite this, most studies have focused on investigations of homoaggregation. The aggregation process may change the overall size of the nanomaterials, surface area and reactivity that will affect their mobility and ecotoxicity in the environment.

Nanoparticle (20 nm) aggregation in river water containing natural organic matter (NOM) followed first-order kinetics towards a residual concentration of the nanoparticles in the water phase (Quik et al. 2012).

The uptake sites in/on organisms such as fish, invertebrates, plants and microorganisms consist of cell surfaces (eg gill epithelial cells and root epidermis cells) that contain polysaccharides, proteins, and other ligands with an overall slight negative charge. Nanomaterials with a positive charge can be attracted through electrostatic attractions to negatively charged cell surfaces that can result in shear forces causing either particle aggregation or higher concentrations of the nanoparticles at the surface of the organism. In contrast, negatively charged nanoparticles are electrostatically repelled from the cell surfaces/membrane and are less likely to be taken up.

iv. Surface chemistry: The surface chemistry of a nanomaterial has implications for its dissolution, aggregation, reactivity and toxicity. In AgVet products, the nanomaterial surface could be expected to be coated/modified with multiple organic/inorganic coating combinations in the different formulations (Anonymous, 2014). These modifications are typically done for the following reasons: to passivate the surface of the nanoparticles (remove free electrons and dangling bonds), control the size of the nanoparticles, render the nanoparticles dispersible in a particular solvent/formulation, and, in some cases, to improve specificity (ie plant uptake via biologically recognisable ligands). Nanomaterials with hydrophobic coatings can render the nanoparticles dispersible in hydrophobic formulations, while hydrophilic (charged) coatings make nanomaterial dispersible in hydrophilic formulations. In the similar way, the hydrophobicity/hydrophilicity of the nanoparticle surface can influence its overall soil retention, where transport and mobility within soil may be promoted by more hydrophilic (typically charged) surface chemistries (Kirschling et al., 2011).

The influence of surface chemistry on uptake and toxicity has been demonstrated in a number of studies. The effect of surface hydrophobicity was studied on toxicity, such as cytotoxicity and consequently DNA damage, using gold (Au) nanoparticles (2 nm core) featuring quaternary ammonium functionality with varied hydrophobic alkyl chain (Chompoosor et al. 2010). In this study, the group found increasing hydrophobicity on the surface of the nanoparticles resulted in higher cytotoxicity with concomitant ROS production. Carboxyl-modified (negatively charged) cadmium selenide/zinc sulfide quantum dots (QDs) were taken up to a greater extent by Daphnia magna than either the amino-terminated (positively charged) or polyethylene glycol (uncharged) QDs (Anonymous, 2014).

v. Concentration and composition: The concentration and composition of manufactured nanomaterials (and their impurities) can affect their fate (eg dissolution, adsorption and aggregation) and ecotoxicity in environments. Xia et al found changing the composition of ZnO nanoparticles by doping with iron (1-10 wt%) decreased ZnO nanoparticle dissolution and reduced toxicity to zebra fish embryos (Anonymous, 2014). Quantum dots, such as CdS, CdSe, Cd/Te and CdSe/ZnS, contain toxic components (eg cadmium and zinc) that can adversely affect organisms and biological processes. The observed toxicity may be directly related to the nanoparticles or release of the toxic component ions, such as Cd2+. Quantum dots containing Cd are more toxic than those containing only Zn due to the higher toxicity of Cd2+. Metallic impurities in CNTs (resulting from synthesis) may lead to toxicity which is unrelated to the CNT.

Nanomaterials have been shown to display dose-response relationships to aquatic and terrestrial organisms (Handy et al. 2012). However, ecotoxicity testing of nanomaterials can be complicated and often difficult to interpret due to changes in the actual dose (dissolution, aggregation/agglomeration, precipitation, adsorption, and complexation, for example, onto gill tissue surfaces) during the test period. The retention (Kr) of CeO_2 nanoparticles onto solid phases in soils was significantly increased with the rate of CeO+nanoparticle addition (Cornelis et al., 2011). It was suggested higher CeO+ nanoparticle concentrations may have increased the collision efficiency and thus the fraction of larger aggregates.

vi. Reactivity: Evaluating nanomaterial reactivities, catalytic activity, redox potential, radical formation potential is important, particularly when considering potential ecotoxicity (ie nanomaterials facilitating the formation of toxic ROS) (Anonymous, 2014). Reactivity is closely associated to other fundamental properties such as size, surface area, composition, and crystalline structure. For example, TiO_2, in its anatase form, is known for its photocatalytic activity (TiO_2 nanoparticles used to enhance pesticide degradation, while it is photostable in rutile form (ie TiO_2 nanoparticles used to protect a system from photodegradation. When doped with different metals (ie Ce-doped TiO_2 used as a fungicide), the photocatalytic response of TiO_2-anatase is changed significantly, shifting from UV to the visible region. When surface coated with other materials (ie $Al(OH)_3$ or SiO_2 as in sunscreen formulations, surface reactions, such as generation of ROS, are diminished. Nanomaterials known to produce these toxic species would then be of environmental concern, especially if fate and transport data suggest them to be unintentionally active and available to non-target organisms.

The ability of nanomaterials to enhance penetration or bioavailability and toxicity is among the main drivers for some of the emerging nano-formulations of AgVet chemicals (Anonymous, 2014). Therefore, it is crucial that such properties of nanoformulations are given adequate consideration in effects assessments on non-target organisms.

Methods of assessing the ecotoxicological impact of engineered nanoparticles may require considerable adaptation and careful control during the tests. The challenges that ecotoxicologists face during effect assessment of nano products are significant. Ecotoxicological test systems consists of characterization of formulated product and associated nanoparticles, test organisms, organs and endpoints, exposure conditions and their environmental relevance, use of reference materials and positive and negative controls.

7.13 State of the art in environmental risks

Studying ecotoxic effects, environmental conditions have to be either simulated at the laboratory level or real life conditions have to be investigated within the environment. Since MPNs are hard to measure and almost untraceable within the environment real life investigations cannot be performed (Anonymous, 2011). Therefore, so called marker organisms are used to test ecotoxic effects under laboratory conditions. These aquatic organisms are often vulnerable to noxes and can be used to measure different biological endpoints. Factors that characterize ecotoxicity are similar to those that are identified for human toxicity, like biopersistence, chemical or biological reactivity, chemical composition and especially surface functionalisation. Little is known about the influence of the environment on the MPNs, since for example the metal speciation may change due to changed redox-conditions or salt content. Also the degradability and the accumulation of MPNs can change the biological effectiveness of the MPNs; therefore information is fundamental to determine the environmental hazardousness of MPNs. In general, since only a few ecotoxicological studies are available, it is not possible to draw any common conclusion on the ecotoxicological effects by MPNs. Also no clear pattern on species sensitivity, suitability as a model organism for nano-ecotoxicity testing or relevance of endpoints is seen (ENRHES, 2010).

There is evidence for potential adverse effects of MPNs in the environment. For example, it seems that MPNs are passing through the food chain from smaller to larger organisms. Within the food chain MPNs can potentially harm aquatic organisms like fishes and invertebrates and, already at low concentrations, microbes, earthworms and (crop) plants. Since MPNs have been shown to be capable to harm microbes, it is discussed that MPNs can have an impact on functional ecosystems which is dependent on an intact micro flora. Much attention has been drawn on aquatic ecosystems. All investigated groups of manufactured nanoparticles (fullerenes, CNTs, metal and metal oxide nanoparticles) have shown to be toxic to aquatic organisms such as zebrafish, daphnia, algae, invertebrates and rainbow trout.

There is an ongoing controversial discussion whether MPNs are responsible for the observed effects or rather the toxicity testing methods including the used chemicals (e.g. solvents) are inducing them. For example, C60-solvent interactions and solvent degradation products may increase the C60 ecotoxicity. Metal and metal oxide nanoparticles, especially silver and copper nanoparticles showing toxic effects in high doses, seem to release continuously ions reaching toxic levels in small areas (e.g. cell). Moreover, some studies point out that these MPNs are not degradable and therefore of high concern. It has to be noticed that the behaviour of the MPNs within the environment is highly complex, since it can be influenced by several factors like changes in pH, temperature, UV-radiation, the presence of other reactive substances, etc. (Anonymous, 2011).

Need for a pragmatic approach: For nano-metal oxides or nano clays, the nanoparticles are expected to be persistent. However, the changes in their surface characteristics (either during product formulation or in the environment) may have implications for their fate and potential effects (Anonymous, 2014). For the nano-active ingredient complex involving complex formulations, information on the durability of a product, ie the fate of the product in soil or water, may become crucial in making decisions as to what material (conventional active ingredient, nano-active ingredient complex or free nanoparticles) should be tested and analysed in aquatic/terrestrial fate studies and in aquatic/terrestrial ecotoxicity studies. For example, if the polymer cage or capsule does not persist in the environment, and readily releases the active ingredient without releasing any persistent nanoparticles, the conventional risk assessment approach may be appropriate. Conversely, if the formulation results in the release of a significant number of nanoparticles, such as TiO_2 catalyst, the product has to be treated differently and assessed for risk as a nanomaterial.

Special considerations for fate and effect assessment: There are some major challenges that nanoparticles present during fate and effect assessment and they require a very different approach to that for conventional nano chemicals. The key considerations for fate and effect studies include the following (Anonymous, 2014).

- The fate and effects of nanoparticles are likely to be governed by particle chemistry and their interactions, heteroaggregation for example, rather than the traditional molecular interactions with the environmental media and organism surfaces used for solutes.
- The aggregation chemistry and ecotoxicity of nanoparticles may be related to particle and surface properties such as size, shape, surface area and surface charge.
- Environmental conditions such as pH, salinity, water-hardness and the presence of NOM may modify the physico-chemistry of nanoparticles and consequently their fate and effects.
- Conventional measures of fate and bioaccumulation potential such as sorption coefficient (Koc) or octanol:water partition coefficient (Kow) may not be directly relevant to nanoparticles. New measures such as attachment efficiency and retention coefficient (Kr) may be more relevant. However, these new indices are still under development and require further testing and standardisation.
- A critical weakness in current ecotoxicological studies in the literature is the lack of characterisation of nanoparticles during the test exposure time. Due to continuously changing properties (aggregation, deposition)

with time, the exposure conditions must be controlled for a proper assessment of ecotoxicological effects.

Occupational and environmental exposures to a limited number of engineered nanomaterials have been reported. Uncertainties in health and environmental effects associated with exposure to engineered nanomaterials raise questions about potential risks from such exposures. EPA's mission and mandates call for an understanding of the health and environmental implications of intentionally produced nanomaterials. A challenge in evaluating risk associated with the manufacture and use of nanomaterials is the diversity and complexity of the types of materials available and being developed, as well as the seemingly limitless potential uses of these materials. A risk assessment is the evaluation of scientific information on the hazardous properties of environmental agents, the dose-response relationship, and the extent of exposure of humans or environmental receptors to those agents (EPA, 2007).

A number of authors have reviewed characterization, fate, and toxicological information for nanomaterials and proposed research strategies for safety evaluation of nanomaterials. A general framework was proposed for risk assessment of nanomaterials which identified nanomaterial characteristics, such as particle size, structure/properties, coating, and particle behavior, that are expected be important in developing nanomaterial risk assessments. Other governments have also undertaken efforts to identify research needs for nanomaterial risk assessment. The European Union's Scientific Committee on Emerging and Newly Identified Health Risks has also overviewed existing data on nanomaterials, data gaps, and issues to be considered in conducting risk assessments on nanomaterials.

Chemical identification and characterization of nanomaterials: The identification and characterization of chemical substances and materials is an important first step in assessing their risk. Understanding the physical and chemical properties in particular is necessary in the evaluation of hazard (both toxicological and ecological) and exposure (all routes) (EPA, 2007). Chemical properties that are important in the characterization of discrete chemical substances include, but are not limited to, composition, structure, molecular weight, melting point, boiling point, vapor pressure, octanol-water partition coefficient, water solubility, reactivity, and stability. In addition, information on a substance's manufacture and formulation is important in understanding purity, product variability, performance, and use. The diversity and complexity of nanomaterials makes chemical identification and characterization not only more important but also more difficult.

A broader spectrum of properties will be needed to sufficiently characterize a given nanomaterial for the purposes of evaluating hazard and assessing risk. Chemical properties such as those listed above may be important for some nanomaterials, but other properties such as particle size and size distribution, surface/volume ratio, shape, electronicproperties, surface characteristics, state of dispersion/agglomeration and conductivity are also expected to be important for the majority of nanoparticles.

A given nanomaterial can be produced in many cases by several different processes yielding several derivatives of the same material. For example, single-walled carbon nanotubes can be produced by several different processes that can generate products with different physical-chemical properties (e.g., size, shape, composition) and potentially different ecological and toxicological properties. It is not clear whether existing physical-chemical property test methods are adequate for sufficiently characterizing various nanomaterials in order to evaluate their hazard and exposure and assess their risk. It is clear that chemical properties such as boiling point and vapor pressure are insufficient. Alternative methods for measuring properties of nanomaterials may need to be developed both quickly and cost effectively.

Because of the current state of development of chemical identification and characterization, current chemical representation and nomenclature conventions may not be adequate for some nanomaterials. Nomenclature conventions are important to eliminate ambiguity when communicating differences between nanomaterials and bulk materials and in reporting for regulatory purposes. EPA's OPPT is participating in new and ongoing workgroup/panel deliberations with the American National Standards Institute (ANSI), the American Society for Testing and Materials (ASTM), and the International Organization for Standardization (ISO) concerning the development of terminology and chemical nomenclature for nanosized substances, and will also continue with its own nomenclature discussions with the Chemical Abstracts Service (CAS).

7.14 Environmental detection and analysis of nanomaterials

The challenge in detecting nanomaterials in the environment is compounded not only by the extremely small size of the particles, but also by their unique physical structure and physicochemical characteristics. The varying of physical and chemical properties can significantly impact the extraction and analytical techniques that can be used for the analysis of a specific nanomaterial. As noted above, the chemical properties of particles at the nanometer size can significantly differ from the chemical properties of larger particles consisting of the same chemical composition. Independent of the challenges brought on by the intrinsic chemical and physical characteristics of nanomaterials, the interactions of nanomaterials with and in the environment, including agglomeration, also provide significant analytical challenges (EPA, 2007). Some nanomaterials are being developed with chemical surface treatments that maintain nanoparticle properties in various environments. These surface treatments can also complicate the detection and analysis of nanomaterials.

In characterizing an environmental sample for intentionally produced nanomaterials, one must be able to distinguish between the nanoparticles of interest and other ultra-fine particles, such as nanoscale particles in the atmosphere generated from coal combustion or forest fires, or nanoscale particles in aquatic environments derived from soil runoff, sewage treatment, or sediment resuspension. Information used to help characterize nanomaterials includes particle size, morphology, surface area and chemical composition.

The level of effort needed and costs to perform analysis for nanomaterials will depend on which environmental compartment samples are being taken from, as well as the type of desired analytical information. The analysis of nanomaterials from an air matrix requires significantly less (if any) "sample" preparation than samples taken from a soil matrix where it is necessary to employ greater efforts for sample extraction and/or particle isolation. Analytical costs also depend on the degree of information being acquired.

The information available from the bulk analysis of nanomaterials from environmental samples has limitations when one is trying to identify a specific nanomaterial. As stated previously, nanoscale particles generated by natural and other anthropogenic sources cannot be separated from nanomaterials of interest using sampling methodologies based upon particle size. During analysis, detected signals generated by nanoscale particles that are not of interest can mask or augment the signals of nanomaterials of interest, resulting in inadequate or erroneous data.

Even given all the challenges presented in analyzing for specific nanomaterials of interest, methods and technologies are available that have demonstrated success. For aerosols, multi-stage impactor samplers are available commercially that can separate and collect nanoparticle size fractions for subsequent analysis (EPA, 2007). These technologies provide nanoparticle fraction separation based upon the aerodynamic mobility properties of the particles. Aerodynamic mobility-based instruments include micro-orifice uniform deposit impactors (MOUDIs), and electrical low-pressure impactors (ELPIs). There are also aerosol fractionation and collection technologies based upon the electrodynamic mobility of particles. These technologies use the mobility properties of charged nanoparticles in an electrical field to obtain particle size fractionation and collection. Instruments employing this technology include differential mobility analyzers (DMAs) and scanning mobility particle sizers (SMPSs).

Available technologies for the size fractionation and collection of nanoparticle fractions in liquid mediums include size-exclusion chromatography, ultrafiltration and field flow fractionation (Powers et al., 2006). On-line particle size analysis in liquid mediums can be done using various techniques including dynamic light scattering (DLS) to obtain a particle size distribution and inductively-coupled mass spectrometry (ICP-MS), a technique that provides chemical characterization information. For more definitive analytical data, single-particle analytical techniques can be employed. Single-particle laser microprobe mass spectrometry (LAMMS) can provide chemical composition data on single particles from a collected fraction (McMurry, 2000). Electron microscopy techniques [e.g., transmission electron microscopy (TEM), scanning electron microscopy (SEM)] can provide particle size, morphological and chemical composition information on collected single nanoparticles in a vacuum environment.

7.15 Limitations and difficulties of risk assessment

The major disadvantage in many methods used for characterization and detection of MPNs is that the equipment is developed for research use, requiring large, expensive and immobile instruments, extensive sample preparation and often laborious conditions (e.g. ultra-high vacuum) as well as highly trained personal. These circumstances increase the errors in quantity and quality analysis especially since sampling techniques are often inadequate, prevent in situ (within the living cell/organism) analysis and the sample might be not representative and representing just a snapshot of a particle concentration at a certain location (Anonymous, 2011). Moreover, analyses of MPNs in complex media can induce changes in physico-chemical properties of the MPNs which are often different from pure or neutral media used in research and development . Unfortunately, none of the available techniques is capable to provide all necessary information about the properties of a certain MPN. Further, biological samples are often limited in size and thus only to a small degree of the different properties can be determined. And most available characterization techniques destruct the material itself; therefore samples cannot be analyzed twice or by more than one technique. Another limitation of the available methods is the incapability to distinguish between naturally occurring (e.g. minerals, protein complexes, combustion products, etc.), unintentionally produced nanoparticles (e.g. soot, diesel or welding fumes) and MPNs.

To differentiate between background contamination and MPNs electron microscopically methods are needed. Thus, the development of new techniques and instruments to assess exposure of MPNs in air, water, soil and biological probes are required. New techniques should be able to deal with complex samples, minimizing sample alterations, avoiding artefacts and providing as much information as possible. Moreover, new instruments should be inexpensive, portable and equipped with smart sensors that for example can combine information on various aspects of exposure and hazard (Tiede et al. 2008). The development of such techniques and instruments is quite time costly but already present in ongoing research.

Many exposure related studies are published on occupational scenarios while much fewer studies are published on environmental and consumer exposure. Thus, the available studies are useful in mapping the "exposure landscape"; however, a clear numerical estimation of exposure is still impossible. Research is usually hypothesis based activity whereby experiments are highly specifically designed for individual studies and thus not necessarily for the purpose of risk assessment. Hence, results are often not comparable since the used conditions are differing from study to study (e.g. different cell types/ animals, testing conditions, handling, etc). Furthermore, studies are often performed without standardized techniques (e.g. guidelines) and therefore the reproducibility of results is often not possible.

Additionally, experiments are often performed without positive and/or negative controls and never in a blinded manner. Whereby it is a matter of fact that latter can unintentionally influence the results. There is an ongoing debate on high dose *in vitro* or *in vivo* studies. There are various arguments for high dose studies, e.g. that they are valuable for hazard identification, proof of principle and are applicable for high throughput assays. However, these studies are little contributing for risk assessment. In ecotoxicological studies it is difficult to simulate real environmental scenarios, since the dose is quite unknown, and the extrapolation of data very limited. Another important difficulty is the missing definition or concept for dose/concentration of MPNs.

Surface functionalisation which are detected in most real-life applications of MPNs (e.g. titanium dioxide in sunscreens should be functionalized to reduce photocatalytic activity) are not addressed by many studies. As mentioned above, the use of solvents in case of non-dispersing nanoparticles (e.g. fullerenes) in aqueous media is problematic due to various reasons, e.g. it may produce testing artefacts (Henry et al. 2007). However, these facts are well known within the scientific community and addressed in many publications.

Limitations: One of the central challenges of risk assessment is the large and rapidly growing number of manufactured particulate nanomaterials to be tested for biological activity along with the limited human financial and time resources (Anonymous, 2011). To assess the risk of MPNs, two components have to be identified: hazard and exposure. Regarding hazard it is possible to make some general statements that represent current sound knowledge: The assumption that the move to the nanoscale implicates not only novel material properties but also entails novel environmental and health risks was confirmed on a scientific basis. It seem to be clear, that certain MPNs (e.g. fibre like carbon nanotubes or nanosilver particles) induce at high doses pathologic conditions and can cause toxic effects. The various in vivo and in vitro studies have shown that several types of MPNs cause multiple types of biological effects, even though the relevance of these data is unclear. Thus, the available data provide a basis for further investigations by providing knowledge about fate and behaviour (ADME-profiles) as well as the toxicity including underlying mechanism, however only for certain MPNs. It was shown that the shape of certain MPNs as well as their purity is important for toxicity, e.g. carbon nanotubes seem to be more toxic if trace impurities of iron or solvents were present.

However, there are intrinsic limitations of detection and toxicity testing methods and very little knowledge is available about the exposure to NPMs; summarizing these limitations in brief:

- Information about acute (high) or chronic (low) exposure of the general public to MPNs are not available. Also data about occupational exposure is insufficient.
- Lack of appropriate instruments to detect MPNs in gas and liquid phase being appropriate for on-line measurements of multiple MPN properties.
- Toxicity is determined by a variety of MPN properties (size, shape, surface, etc.) therefore case by case studies are necessary. Plus, it is not clear which metric is the most meaningful for toxicity.
- A vast number of in vivo and in vitro studies have been published showing that certain MPNs are causing several types of biological effects; however, the biological relevance is unclear. This limits our ability to characterise hazards adequately.
- Dose-dependent studies (dose-response correlations) using good controls and reproducible standardised experiments are missing.

The missing exposure data prevent risk assessment of MPNs since this requires two components namely hazard AND exposure. Another aspect of the recent development in nanotoxicology is those leading toxicologists propose a critical view on conceptions of earlier work based on high dose studies along with the assumption of high translocation rates. Similarly, one should note that certain MPNs have also beneficial properties. Though, a study even suggested that C60 fullerenes might prevent toxicity by being antioxidants. It is thus hardly surprising that an increasing number of medical applications of MPNs exist.

Addressing further developmental needs to overcome the present hindrance of risk assessment the following may be identified.

- Further research activity is needed especially in assessing the exposure of the general public and the environment. Therefore, novel instruments and techniques need to be developed or existing ones need to be refined to meet requirements (i) to detect MPNs in complex and small samples especially in biological setting, (ii) to allow discrimination of naturally occurring/background nanoparticles and MPNs, (iii) inexpensive, (iv) portable and (v) allow measurement of multiple properties at the same time.
- The standardization of methods is needed including (i) reliable and validated procedures for characterization and measurement of dose and exposure, (ii) a broad characterization of the used MPN and (iii) the use of an agreed dose units which can be used in hazard and exposure assessments.
- Until now, most of the experiments conducted are not standardised since academic research does not primarily aim to produce data for risk assessment purposes. Hence, one should think of implementation of rewards for or even mandatory usage of harmonization tools into e.g. financial support of research. A standardised format for reporting the results of toxicological studies focusing on physicochemical characterization might be helpful.
- Moreover, an (inter)national coordination of research and development activities (e.g. Road map, master plan) along with the above mentioned standardization of methods would be beneficial in respect to expedite the progress and increase comparability for risk assessment. This should be accompanied with significantly increased.
- The publication of no-effect studies is important and recommended.
- In this context the proposal to create a third-party body, similar to the Cochrane model6 but comprised of stakeholders from all groups,

to review and provide authoritative interpretations of research in nanotoxicology is very interesting. As well as the one to develop a multidisciplinary community of practice and informationsharing forum for researchers and all interested stakeholders (Balbus et al. 2007).

- One should promote the multidisciplinary training of young scientist in the nanotechnology field.

7.16 Safety of engineered nanoparticles

With the production of engineered nanoparticles we are confronted with a new class of materials that have novel properties compared to bulk material. Information describing the health risk of engineered nanoparticles is only evolving and many questions are still open. The uncertainties involved have to be seen against the background of the ever more demanding public views regarding the safety of products from new technologies and an increasing potential for exposure as the quantity and types of nanoparticles used in society grow. From animal experiments and analogies to studies on incidentally produced ultra fine particles (such as from burning of fuels) it is possible that at least some nanoparticles are hazardous for the human body and that the exposure to these nanoparticles should be avoided or at least minimized.

One promising path to prevent potential health hazards proposed by a number of scientists is to make the particles biodegradable. Particles that are degradable either by water or by lysis with enzymes will greatly reduce the risks involved because they do not persist in the body. It is self-evident that the slower the particles are cleared (high persistence), the higher the tissue burden can be. With a short bio-durability, long-term effects can be minimised or even excluded. Biocompatability can serve as one major engineering parameter for nanomaterials in the future (Christoph Lauterwasser, 2005). There is a broad variety of nanoparticles being investigated and a high number of parameters that influence the functionality of these nanoparticles as well as their interaction with the human body. It is expected that it will take several years until some major critical risk assessment issues regarding hazards and exposure can be answered. Research needs related to nanoparticles include the following.

- Ecotoxicity of nanoparticles for cases of inadvented contact by children, adults and susceptible individuals,
- Biomonitoring studies of specific biological impact of nanoparticles, according to their most likely routes of exposure,
- Standards development and (certified) reference materials,

- Sensors for detecting nanoparticles and assessing exposure, both stationary and portable with a focus on low-cost, highly specific sensors giving a real time response to environment an health relevant properties,
- A strategic concept for a comprehensive risk assessment.

The regulatory risk assessment of chemicals will in the future be carried out under the New European Union Chemicals Policy (Registration, Evaluation and Authorisation of Chemicals). In its current state REACH will be concerned only with the safe use of bulk materials, and will not take account of size. Additionally, under REACH production mass limits are planned (limited evaluation procedure for production of less than 100 tons, no registration for production volume below 1 ton). Many nanoparticles that are produced in low volume will therefore generally not be accounted for in the REACH framework. Further examinations of this regulatory situation might be required on the basis of new scientific evidence related to manufactured nanoparticles.

It will be a challenge for industry, legislators and risk assessors to fill all relevant knowledge gaps and to construct a set of high throughput and low cost tests for nanoparticles. In that way a risk assessment of each nanoparticle should be established before the larger quantities are manufactured. This process is unlikely to be fast enough without active steering and support by governments and the EU. The uncertainties involved, especially long term, will have to be addressed by all organisations involved in the process of the introduction of nanotechnologies. It will be a necessary for the various industries involved to perform life-cycle assessments and demonstrate the safety aspects to a broader public. Successful communication will depend on a general feeling of trust towards nanotechnologies. This in turn needs an open dialogue and interaction involving all different stakeholders, including a high degree of transparency regarding scientific results.

7.17 Risk management approach

A more reflexive, incremental, and cooperative risk management approach is required, which not only will help manage emerging risks from nanotechnology applications, but will also create a new risk management model for managing future emerging technologies. As a management process, risk management is used to identify and avoid the potential cost, schedule, and performance/ technical risks to a system, take a proactive and structured approach to manage negative outcomes, respond to them if they occur, and identify potential opportunities that may be hidden in the situation.

Risk awareness: There is a much uncertainty about emerging risks associated with nanotechnologies. It will take years for studies about exposure routes, the effects on human health and the environment to reach conclusive results (Christoph Lauterwasser, 2005). While it is still too early to make conclusive statements, our own risk management will need constantly to "put its feelers out." The first step in our risk management toolbox is to create an awareness of the risks and an understanding of the hazards. The first step is to determine how underwriters and risk engineers should deal with critical issues such as direct exposure to nanoparticles or their release into the environment.

Risk identification and risk evaluation: The next step is to identify and evaluate the risks in a continuing process that considers scientific, technical, legal and regulatory factors. Public opinion is also important because that influences many political, industrial, and legal decisions. Yet close scrutiny of the individual risk, often with support from risk-consulting units, is a prerequisite to taking economically sustainable decisions. Often more important than the underlying basic risk, the risk management practices in these companies will range from very basic to highly sophisticated. The last point is of key importance: even if a certain technology creates new risks, the quality of risk management practices will determine whether these hazards will actually materialise. This is an area where the insurer can contribute through his experience with a multitude of clients from various industries.

Risk handling and accumulation control: Given the knowledge gap about risks outlined above, it is a matter of debate whether one main criterion of insurability, namely the assessability of risks with respect to probability and severity, is or can be fulfilled. The combination of limited evidence about the hazards and potential latency claims again warrants close monitoring of the risk.

On the other hand, it seems neither feasible nor appropriate to start a debate about a general exclusion of nanotechnologies from the insurance coverage today. There are several arguments for this as follows.

- Nanotechnologies cover a very broad field with far from uniform risk characteristics,
- The terminology used in nanotechnologies is very broad. No uniform language or set of definitions exist,
- The sheer variety of nanotechnologies and their applications across broad sections of industrial segments mean a positive diversification effect for insurance portfolios,
- The exposure of the general population to nanotechnologies is still comparatively low

But all this does not rule out specific applications such as the use of nanoparticles in environmental remediation from being subject to more intense risk analysis and possibly even being excluded from cover. A major challenge with nanotechnologies as well as with other emerging risks will not only be selection and pricing, but also adequate risk capital allocation. A "wait-and-see" attitude is not appropriate, especially because of the existence of latent risks, which can take many years to show.

7.18 Positive effects on human health and the environment

A fair assessment of the risks of any new technology must also consider positive contributions to increased safety. The basic innovations that come from nanotechnologies have the potential to contribute to human health and environmental safety in many ways. They have the potential to contribute to solve urgent issues like the provision of clean drinking water or more efficient energy conversion and energy storage. The potential of nanotechnologies regarding economic benefits, the potential to create jobs, wealth and well being is very high. At the moment, public awareness about nanotechnology is limited. What happens over the next few years will determine how the public comes to view it. A transparent discussion of benefits and risks will help people reach a considered, balanced view. This will enable a greater public acceptance, which, in turn, will enable society as a whole to profit from these fundamental technological developments while, at the same time, the risks are kept under control.

Improved diagnosis: There are quite a few technological developments that promise enhanced diagnostic possibilities, new ways to monitor patients, new ways to treat diseases like cancer and to reduce side effects. For example,

- Nanoparticles can be used as carriers for targeted drug delivery. Their ability to penetrate certain protective membranes in the body, such as the bloodbrain barrier, can be beneficial for many drugs. This could open the way for new drugs from active substances that have not been able to pass clinical trials due to less precise delivery mechanisms,
- Allows for the simultaneous measurement of multiple parameters (Rickerby and Morrison, 2007).
- Nanosensors and lab-on-a-chip-technologies will foster early recognition and identification of diseases and can be used for continuous monitoring of patients with chronic diseases,
- New therapeutic methods for the treatment of cancer with the help of nanoparticles are investigated.

- Nanoparticles can be beneficial in catalytic and remediation application. For example, metal oxide nanocatalysts are being developed for the prevention of pollution due to industrial emissions (e.g. DeNOx catalysts could be applied for the removal of nitrogen oxides from fossil fuel power plant emission gases) and the photo-catalytic properties of titanium dioxide nanoparticles can be exploited to create self-cleaning surfaces that reduce existing pollution.

The use of nanomaterials and nanoparticles can also lead to significant savings in resources and efficiency increases in manufacturing and energy related applications. For example, energy related applications include nanostructured electrode materials for improving the performance of lithium ion batteries and nanoporous silicon and titanium dioxide in advanced photovoltaic cells.

Ultrasensitive detection of substances will have implications for safety in many other areas such as industrial medicine, environmental medicine and food safety.

Water quality improvement: With the help of nanotechnology, water quality can be improved. Some of the nanomaterials that can be used for remediation of water are carbon nanotubes (CNTs), zeolites, nanoparticles of zero valent iron (ZVI), silver nanoparticles, etc (Ray Unmesha, 2019). Other nanomaterials like zinc oxide (ZnO), titanium dioxide (TiO_2), tungsten oxide, serve as a photocatalyst. These photocatalysts can oxidize organic pollutants into harmless materials. TiO_2 is the most preferred material as it has high photostability, high photoconductivity, easily available, inexpensive and non-toxic. Silver nanoparticles have antimicrobial effect. Also, many polymeric nanoparticles are being used for wastewater treatment..

Another new technology is known as nanofiltration which can be used in water treatment in homes, offices, and industries. Molybdenum disulphide (MoS_2) nonporous membrane is used for energy efficient desalination of water which filters five times more than the conventional ones. To clean oil spills in the water bodies, a nanofabric paper towel has been developed which are woven from tiny wires of potassium manganese oxide that can absorb oil 20 times its weight. Thus, nanotechnology provides a solution to clean the contaminated water and prevent new pollution.

Clearing toxic gases: With the help of nanotechnology, toxic gases in the air can be cleaned. But first, we have to detect the pollutants at the molecular level using precise sensors. A sensor called nanocontact sensor has been developed which can detect the heavy metal ions and radioactive elements. These sensors have a small size, are inexpensive and are easy to use on-site. Currently, single-walled nanotubes (SWNTs) are being used for the detection of NO_2 and NH_3 gases. Also, SWNTs sensor can accomplish high sensing activity at room temperature when compared to the conventional sensors which work at 200 to 600°C. Cantilever sensors have been developed to sense VOCs, heavy metals and pesticides. A mixture of CNTs with gold particles helps adsorb toxic gases like NOx, SO_2 and CO_2. Another porous nanomaterial manganese oxide has better adsorption of toxic gases due to its large surface area.

By detecting pollutants by specific sensors, we can help protect the sustainability of human health and the environment. Thus, nanotechnology provides us with a new approach to cut down the waste production, reduce the emission of greenhouse gases and discharge of hazardous chemicals in water bodies.

Nanotechnology increases the strengths of many materials and devices, as well as enhances efficiencies of monitoring devices, remediation of environmental pollution, and renewable energy production. In general, nanotechnology devices consume less energy, reduce material wastes, and help in monitoring. Nanotechnology can also be used to reduce and prevent the toxicity of nanoparticles in environment more efficiently. Following are certain areas of manufacturing currently benefiting from the development of nanotechnology.

Energy consumption: Nanotechnology is finding application in traditional energy sources and is greatly enhancing alternative energy approaches to help meet the world's increasing energy demands. Many scientists are looking into ways to develop clean, affordable, and renewable energy sources, along with means to reduce energy consumption and lessen toxicity burdens on the environment: Nanotechnology is improving the efficiency of fuel production from raw petroleum materials through better catalysis. It is also enabling reduced fuel consumption in vehicles and power plants through higher-efficiency combustion and decreased friction. Nanotechnology is also being applied to oil and gas extraction through, for example, the use of nanotechnology-enabled gas lift valves in offshore operations or the use of nanoparticles to detect microscopic down-well oil pipeline fractures. Researchers are investigating carbon nanotube "scrubbers" and membranes to separate carbon dioxide from power plant exhaust.

The use of graphene into a coating material results in the need for only one layer, which does not require a multifunctional film coating. Two applications for a graphene based coating are to apply it to a blade used in wind turbines or on the body of an airplane. It saves the weight increasing efficiency.

Cost saving on materials: An alternative energy method such as hybrid automobiles will decrease the price by novel developments in nanotechnology.

Less waste on raw materials: Large sample testing will be done on a smaller scale and simultaneously use of raw materials will become more efficiency.

Nanoscale chemical reagents (or catalysts) increase the reaction rate and other efficiency of chemical reactions.

Environmental Remediation: Utilizing advanced nanotechnology, a detector was made to detect a nuclear leak faster and more accurate at the Fukushima Daiichi Nuclear Power Plant, which is one of the best radiation detector in Washington and can sense the faintest amount of radiation.

Nanotechnology could help meet the need for affordable, clean drinking water through rapid, low-cost detection and treatment of impurities in water.

- Engineers have developed a thin film membrane with nanopores for energy-efficient desalination. This molybdenum disulphide (MoS2) membrane filtered two to five times more water than current conventional filters.
- Nanoparticles are being developed to clean industrial water pollutants in ground water through chemical reactions that render the pollutants harmless. This process would cost less than methods that require pumping the water out of the ground for treatment.
- Researchers have developed a nanofabric "paper towel" woven from tiny wires of potassium manganese oxide that can absorb 20 times its weight in oil for cleanup applications. Researchers have also placed magnetic water-repellent nanoparticles in oil spills and used magnets to mechanically remove the oil from the water.
- Many airplane cabin and other types of air filters are nanotechnology-based filters that allow "mechanical filtration," in which the fiber material creates nanoscale pores that trap particles larger than the size of the pores. The filters also may contain charcoal layers that remove odors.
- Nanotechnology-enabled sensors and solutions are now able to detect and identify chemical or biological agents in the air and soil with much higher sensitivity than ever before. Researchers are investigating particles such as self-assembled monolayers on mesoporous supports (SAMMS™), dendrimers, and carbon nanotubes to determine how to apply their unique chemical and physical properties for various kinds of toxic site remediation. Another sensor has been developed by NASA as a smartphone extension that firefighters can use to monitor air quality around fires.

Biological applications: Developing ultra-small probes on planetary surfaces for agricultural applications and control of soil, air, and water contamination.

Biomedical applications: This includes the medical diagnostic and treatments.

Positive effects on environment : Nanotechnology offers potential economic, societal and environment benefits. Nanotechnology also has the potential to help reduce the human footprint on the environment by providing solutions for energy consumption, pollution, and green gas emissions. (Zhang et al., 2011). Nanotechnology offers the potential for significant environmental benefits, including cleaner, more efficient industrial processes; improved ability to detect and eliminate pollution by improving air, water, and soil quality ; high precision manufacturing by reducing amount of waste ;clean abundant power via more efficient solar cells ; removal of greenhouse gases and other pollutants from the atmosphere ; decreased need for large industrial plantsand remediating environmental damages.

The nanoscale products that utilize graphene in an industrial use or research can benefit the environment in following ways.

- Graphene based nanocomposites reduce the weight of airplanes by substituting traditional metals and composites, and the consequence of the weight saving results in a reduction of a thousand tons of gasoline
- Graphene thin films or graphene buckypapers can be substituted in place of metal meshes around the fuselage of airplane used to prevent the direct and indirect effects of lightning strikes
- The eminent properties of graphene increases the efficiency of advanced renewable energy processes, such as reducing the weight of a wind turbine blades and increasing the energy converse efficiency.

Bibliography

Agnieszka Baran, 2016. Nanotechnology: legal and ethical issues. Econ.& Management , 8: 47-54

Aitken, R.J., Creely, K.S. and Tran, C.L. 2004. Nanoparticles: An Occupational Hygiene Review. Research Report 274. Prepared by the Institute of Occupational Medicine for the Health and Safety Executive, North Riccarton, Edinburgh, England.

Amit Kumar, 2014. Nanotechnology Development in India An Overview. RIS (Research & Information System for Developing Countries) Discussion Papers. 40 pp.

Anonymous, 2004. Royal Society and Royal Academy of Engineering (2004). "Nanoscience and nanotechnologies: opportunities and uncertainties". Retrieved 2008-05-18

Anonymous, 2011. NanoSafety – Risk Governance of Manufactured Nanoparticles. European Technology Assessment Group, Karlsruhe and Vienna

Anonymous, 2014. Regulatory Considerations for Nanopesticides and Veterinary Nanomedicines- A Draft APVMA Report. Commonwealth of Australia 2014

Balbus, J.M., Maynard, A.D., Colvin, V.L., Castranova, V., Daston, G.P., Denison, R.A., Dreher, K.L., Goering, P.L., Goldberg, A.M., Kulinowski, K.M., Monteiro-Riviere, N.A., Oberdörster, G., Omenn, G.S., Pinkerton, K.E., Ramos, K.S., Rest, K.M., Sass, J.B., Silbergeld, E.K and, Wong, B.A., 2007. Meeting report: hazard assessment for nanoparticles--report from an interdisciplinary workshop. Environ Health Perspect, 115:1654-9.

Balcerzak, M., 2013. Zagadnienia nanotechnologii w prawie, Czy nanotechnologia może czerpać z doświadczeń biotechnologii? [The issues of nanotechnologies in law, Is nanotechnology can draw on the expertise of biotechnology?]. In: Innowacje i komercjalizacja w biotechnologii [Innovation and commercialization in biotechnology], D. M. Trzmielak (Ed.), Poznań – Łódź, Poland: Uniwerstytet Łódzki

Batley, G.E., Kirby, J.K. and McLaughlin, M.J.,2012. Fate and risks of nanomaterials in aquatic and terrestrial environments. Accounts Chem. Res., 46:854-862

Chompoosor, A., Saha, K. and Ghosh, P.S., 2010. The role of surface functionality on acute cytotoxicity, ROS generation and DNA damage by cationic gold nanoparticles. Small, 6:2246-2249.

Chowdhury Nupur, 2006. Nanotechnology Regulation in India: A Framework for Exploring the Risks and Opportunities. Conference: International Studies Association Annual Convention, ResearchGate

Cientifica, 2009. Nanotechnology Takes a Deep Breath...and Prepares to Save the World! Cientifica Report.

European Commission, 2008. Regulatory Aspects of Nanomaterials. Communication from the Commission to the European Parliament, the Council and the European Economic and Social Committee, COM(2008) 366 final.

Falkner Robert and Jaspers Nico, 2012. Regulating nanotechnologies: risk, uncertainty and the global governance gap. Global Enviro. Politics, 12 : 30-55

Graff, 2002. The Precautionary Principle. In : The Cartagena Protocol on Biosafety: Reconciling Trade in Biotechnology with Environment and Development? Christoph Bail, Robert Falkner and Helen Marquard (Edrs.), 410-422. London: Earthscan

Handy ,R.D., Cornelis, G. and Fernandes, T., 2012. Ecotoxicity test methods for engineered nanomaterials: Practical experiences and recommendations from the bench. Environ. Toxicol. & Chem., 31:15-31

Henry, T.B., Menn, F.M., Fleming, J.T., Wilgus, J., Compton. R.N. and Sayler, G.S. 2007. Attributing effects of aqueous C60 nano-aggregates to tetrahydrofuran decomposition products in larval zebrafish by assessment of gene expression. Environ Health Perspect, 115:1059-65.

Ibrahim, R.K., Hayyan, M. and AlSaadi, M.A., 2016. Environmental application of nanotechnology: air, soil, and water. Environ. Sci. Pollut. Res., 23, 13754–13788

ICON, 2008. Towards Predicting Nano-Biointeractions ICON Report 4.

IFCS, 2008. Final Report, Sixth Session of the Intergovernmental Forum on Chemical Safety. IFCS/FORUM-VI/07w

Jaspers Nico, 2010. Nanotechnology: How to Avoid International Trade Conflicts. European J. Risk Regul.,2: 167-173.

Jayanthi, A.P., Beumer, K., Bhati, M. and Bhattacharya, S., 2012. "Nanotechnology: 'Risk Governance' in India". Econ. & Political Weekly, Vol. XLVII, 4: 34-40.

Jurewicz, M.,2014. Nanotechnologia, Regulacje prawne, Legislacja Unii Europejskiej [Nanotechnology, Regulations, European Union Legislation]. Warszawa, Poland: Difin.

Khan, M.R. and Akram, M., 2020. Nanoparticles and Their Fate in Soil Ecosystem. In: Biogenic Nano-Particles and their Use in Agro-ecosystems. Ghorbanpour M., Bhargava P., Varma A., Choudhary D. (Edrs). Springer, Singapore. https://doi.org/10.1007/978-981-15-2985-6_13

Kim, K.T., Truon,g L., Wehmas, L. and Tanguay, R.L.,2013. Silver nanoparticle toxicity in the embryonic zebrafish is governed by particle dispersion and ionic environment. Nanotechnology, 24: doi:Artn 115101

Kirschling, T.L., Golas, P.L. and Unrine, J.M., 2011. Microbial bioavailability of covalently bound polymer coatings on model engineered nanomaterials. Environ. Sci. & Technol., 45:5253-5259.

Klaine, S.J., Koelmans, A.A. and Horne, N., 2012. Paradigms to assess the environmental impact of manufactured nanomaterials. Environ. Toxicol. & Chem., 31:3-14.

Luther, W., 2004. Technological Analysis, Industrial Application of Nanomaterials - Chances and Risks. Future Technologies Division, VDI Technologiezentrum GmbH, Düsseldorf, Germany

Mantovani, E., Porcari, A., Morrison, M.D. and Geertsma. R.E., 2011. Developments in Nanotechnologies Regulation and Standards . Report of the Observatory Nano.

McMurry, P. H., 2000. A Review of Atmospheric Aerosol Measurements. Atmospheric Environ., 34:1959-1999

Monteiro-Riviere Nancy, Robert J Nemanich, Alfred O Inman, Yunyu Y Wang, Jim E Riviere., 2005. Multi-Walled Carbon Nanotube Interactions with Human Epidermal Keratinocytes. Toxicol. Lett., 155 : 377–384.

Nistads, 2012. Knowledge Creation and Innovation in Nanotechnology: Contemporary and Emerging Scenario in India. National Institute of Science, Technology and Development Studies. New Delhi.

OECD, 2004. Nanotechnology: Emerging safety issues? ENV/JM (2004)32, quoted in Small Sizes That Matter: Opportunities and Risks of Nanotechnologies. Allianz Report in co-operation with the OECD International Futures Programme. Available at http://www.oecd.org/dataoecd/32/1/44108334.pdf, accessed 11 January 2011

OECD, 2010. Current Developments/Activities on the Safety of Manufactured Nanomaterials. Tour de Table at the 7th Meeting of the Working Party on Manufactured Nanomaterials. Paris, France, 7-9 July 2010 (ENV/JM/MONO(2010)42). Paris: OECD.

Powers, K.W., Brown, S.C, Krishna, V.B., Wasdo, S.C., Moudgil, B.M. and Robert, S.M., 2006. Research Strategies for Safety Evaluation of Nanomaterials, Part VI: Characterization of Nanoscale Particles for Toxicological Evaluation. Toxicol. Sci., 90: 296-303.

Quik, J.T.K., Stuart, M.C., Wouterse, M., Peijnenburg, W., Hendriks, A.J. and van de Meent, D.,2012. Natural colloids are the dominant factor in the sedimentation of nanoparticles. Environ. Toxicol. & Chem., 31: 1019-1022.

Raines Susan S., 2003. Perceptions of Legitimacy and Efficacy in International Environmental Management Standards: The Impact of the Participation Gap. Global Environ. Politics, 3: 47-73.

Ray Unmesha, 2019. The Environmental Impact of Nanotechnology. AZoNano. Retrieved on November 11, 2020

Rickerby, D.G. and Morrison, M.,2007. Nanotechnology and the environment: A European perspective. Sci. & Technol. Adv.Mat., 8: 19-24

Tiede, K., Boxall, A.B., Tear, S.P., Lewis, J., David, H. and Hassellov, M., 2008. Detection and characterization of engineered nanoparticles in food and the environment. Food Addit. Contam. Part A Chem. Anal Control Expo. Risk Assess, 25:795-821.

U.S. Environmental Protection Agency. 2004. Office of Research and Development. Air Quality Criteria for Particulate Matter. Report Number EPA/600/P-99/002a,bF. October.

Wiener Jonathan, B., 2007. Precaution. In :The Oxford Handbook of International Environmental Law, Daniel Bodansky, Jutta Brunnée and Ellen Hey (Edrs.), 597-612. Oxford: Oxford University Press.

Wiesner, M.R., Lowry, G.V., Alvarez, P., Dionysiou, D. and Bisawas, P., 2006. Assessing the Risks of Manufactured Nanomaterials. Environ. Sci. Tech., 40:4336-4345

Zhang, B., Misak , H., Dhanasekaran , P.S., Kalla D. and Asmatulu, R., 2011. Environmental Impacts of Nanotechnology and Its Products. Proc. of the 2011 Midwest Section Conference of the American Society for Engineering Education Chaturvedi, Sachin. 2013. "Where Society Meets Science". The Indian Express, 3 September

Zoya Javed, Kavya Dashora, Mansi Mishra, Vinayak D Fasake and Ayushi Srivastva, 2019. Effect of accumulation of nanoparticles in soil health- a concern on future. Front. Nanosci. Nanotech., 5: DOI: 10.15761/FNN.1000182

8

Nanotoxicity

Nanotoxicity is the study of the toxicity of nanomaterials that focuses on how the size and unique properties of nanoparticles influences their impact on plant helath, human health and the environment. Plants are essential fundamental components of all ecosystems and the interaction between nanoparticles (NPs) and plants is an indispensable aspect of the risk assessment. NP phytotoxicity, an important precondition to promote the application of nanotechnology and to avoid the potential ecological risks was reviewed (Jie Yang et al., 2017).

The large-scale and unrestricted use of NPs has led researchers to consider the problems, challenges and consequences of their environmental impact (Gottschalk et al. (2015). the concentration of NPs in the environment is much lower than the toxic concentration. The potential health and environmental effects of NPs need to be thoroughly evaluated before they are widely commercialized. When NPs enter the soil through agricultural application, atmospheric deposition, rain erosion, surface runoff or other pathways, the NPs will accumulate in the soil as time goes because of their weak migration ability in soil. Exposure modeling also indicated that the concentrations of NPs in soil are higher than those in water or air, implying that soils might be the main source of NPs released into the environment.

8.1 Potential harmful effects of nanoparticles

Interaction of nanoparticles with living organisms: Nanoparticles, can have the same dimensions as biological molecules such as proteins. In living systems, they may immediately adsorb onto their surface some of the large molecules they encounter as they enter the tissues and fluids of the body. This ability of nanoparticles to have molecules "sticking" to their surface depends on the surface characteristics of the particles and can be relevant for drug delivery uses. Indeed, it is possible to deliver a drug directly to a specific cell in the body by designing the surface of a nanoparticle so that it adsorbs specifically onto the surface of the target cell (Scenihr, 2006). But the interaction with living systems is also affected by the dimensions of the nanoparticles. For instance, nanoparticles not bigger than a few nanometres may reach well inside biomolecules, which is not possible for larger nanoparticles. Nanoparticles may cross cell membranes. It has been reported that inhaled nanoparticles can reach the blood and may reach other target sites such as the liver, heart or blood cells.

Key factors in the interaction with living structures include nanoparticle dose, the ability of nanoparticles to spread within the body, as well as their solubility. Some nanoparticles dissolve easily and their effects on living organisms are the same as the effects of the chemical they are made of. However, other nanoparticles do not degrade or dissolve readily. Instead, they may accumulate in biological systems and persist for a long time, which makes such nanoparticles of particular concern. There remain many unknown details about the interaction of nanoparticles and biological systems and more information on the response of living organisms to the presence of nanoparticles of varying size, shape, chemical composition and surface characteristics is needed to understand and categorize the toxicity of nanoparticles.

Characteristics of nanoparticles relevant for health effects: Studies specifically dealing with the toxicity of nanoparticles have only appeared recently and are still scarce. Most of the information available comes from studies on inhaled nanoparticles and from pharmaceutical studies in which nanomaterials are used, among other things to improve drug delivery (Scenihr, 2006).

The characteristics of nanoparticles that are relevant for health effects are as follows.

Size: In addition to being able to cross cell membranes, reach the blood and various organs because of their very small size, nanoparticles of any material have a much greater surface to volume ratio (i.e. the surface area compared to the volume) than larger particles of that same material. Therefore, relatively more molecules of the chemical are present on the surface. This may be one of the reasons why nanoparticles are generally more toxic than larger particles of the same composition.

Chemical composition and surface characteristics: The toxicityof nanoparticles depends on their chemical composition, but also on the composition of any chemicals adsorbed onto their surfaces. However, the surfaces of nanoparticles can be modified to make them less harmful to health.

Shape: Although there is little definitive evidence, the health effects of nanoparticles are likely to depend also on their shape. A significant example is nanotubes, which may be of a few nanometres in diameter but with a length that could be several micrometres. A recent study showed a high toxicity of carbon nanotubes which seemed to produce harmful effects by an entirely new mechanism, different from the normal model of toxic dusts.

Inhaled nanoparticles affect health: Particulate matter present in air pollution, especially from traffic emissions, is known to affect human health, although it is not clear exactly how. Epidemiological studies on ambient air pollution have not proved conclusively that nanoparticles are more harmful than larger particles, but these studies may not be well suited to demonstrate such differences. Inhaled particulate matter can be deposited throughout the human respiratory tract, and an important fraction of inhaled nanoparticles deposit in the lungs. Nanoparticles can potentially move from the lungs to other organs such as the brain, the liver, the spleen and possibly the foetus in pregnant women. Data on these pathways is extremely limited but the actual number of particles that move from one organ to another can be considerable, depending on exposure time. Even within the nanoscale, size is important and small nanoparticles have been shown to be more able to reach secondary organs than larger ones. Another potential route of inhaled nanoparticles within the body is the olfactory nerve; nanoparticles may cross the mucous membrane inside the nose and then reach the brain through the olfactory nerve (Scenihr, 2006). Out of three human studies, only one showed a passage of inhaled nanoparticles into the bloodstream. Materials which by themselves are not very harmful could be toxic if they are inhaled in the form of nanoparticles.

The effects of inhaled nanoparticles in the body may include lung inflammation and heart problems. Studies in humans show that breathing in diesel soot causes a general inflammatory response and alters the system that regulates the involuntary functions in the cardiovascular system, such as control of heart rate. The pulmonary injury and inflammation resulting from the inhalation of nanosize urban particulate matter appears to be due to the oxidative stress that these particles cause in the cells.

Health implications of nanoparticles used as drug carriers: Nanoparticles can be used for drug delivery purposes, either as the drug itself or as the drug carrier. The product can be administered orally, applied onto the skin, or injected. The objective of drug delivery with nanoparticles is either to get more of the drug to the target cells or to reduce the harmful effects of the free drug on other organs, or both. Nanoparticles used in this way have to circulate long distances evading the protection mechanisms of the body. To achieve this, nanoparticles are conceived to stick to cell membranes, get inside specific cells in the body or in tumours, and pass through cells. The surfaces of nanoparticles are sometimes also modified to avoid being recognized and eliminated by the immune system. With dermal administration, it was found that particle size was less important than the total charge in terms of permeation through the skin (Scenihr, 2006). For instance, only negatively charged particles were found to overcome the skin barrier and only when concentration of charge was high enough.

Nanoparticles may be used effectively to deliver genes to cells, to treat cancer, as well as in vaccination. The use of nanoparticles as drug carriers may reduce the toxicity of the incorporated drug but it is sometimes difficult to distinguish the toxicity of the drug from that of the nanoparticle. Toxicity of gold nanoparticles, for instance, has been shown at high concentrations. In addition, nanoparticles trapped in the liver can affect the function of this organ. Nanoparticles have the potential to cross the blood brain barrier, which makes them extremely useful as a way to deliver drugs directly to the brain. On the other hand, this is also a major drawback because nanoparticles used to carry drugs may be toxic to the brain.

Assessment of harmful effects of nanoparticles: Traditionally, doses are measured in terms of mass because the harmful effects of any substance depend on the mass of the substance to which the individual is exposed. However, for nanoparticles it is more reasonable to measure doses also in terms of number of particles and their surface area because these parameters further determine the interactions of nanoparticles with biological systems. Several hypotheses were proposed for the adverse health effects of nanoparticles as part of ambient air pollution. These hypotheses address nanoparticle characteristics, their distribution, and their effects on organ systems, including effects on immune and inflammatory systems.

However, some of these hypotheses may be of limited or no relevance for engineered nanoparticles. For instance, the adhesion of toxic substances onto the surface of nanoparticles may be of less relevance for production and handling facilities of large volumes of engineered nanoparticles compared to the particles in ambient air (Scenihr, 2006). In addition, drawing conclusions from tests on healthy animal models may be unsuitable as some of the effects of nanoparticles may only be a risk for susceptible organisms and predisposed individuals, but not to healthy people. For instance, age, respiratory tract problems and other pollutants can modify the pulmonary inflammation and oxidative stress induced by nanoparticles.

Because of the specific characteristics of nanoparticles, conventional toxicity tests may not be enough to detect all their possible harmful effects. Therefore, a series of specific tests was proposed to assess the toxicity of nanoparticles used in drug delivery systems. One mechanism of toxicity of nanoparticles is likely to be the induction of oxidative stress in cells and organs. Testing for interaction of nanoparticles with proteins and various cell types should be considered as part of the toxicological evaluation. With the exception of airborne particles delivered to the lung, information on the behaviour of nanoparticles in the body including distribution, accumulation, metabolism, and organ specific toxicity is still minimal.

Effects of nanoparticles on the environment: There are almost no publications on the effects of engineered nanoparticles on animals and plants in the environment. However, a number studies have examined the uptake and effects of nanoparticles at a cellular level to evaluate their impact on humans; it can reasonably be assumed that the conclusions of these studies may be extrapolated to other species, but more research is needed to confirm this assumption (Scenihr, 2006). Moreover, careful examination and interpretation of existing data and careful planning of new research is required to establish the true impact of nanoparticles on the environment, and the differences with larger, conventional forms of the substances. Persistent insoluble nanoparticles may cause problems in the environment that are much greater than those revealed by human health assessments.

8.2 Effects of nanoparticles on plants

8.2.1 Phytotoxicity

When adverse reactions are reported in plants, these may correspond to slower growth, increased oxidative stress, chromosomal abnormalities, decreased photosynthetic rate, disturbances in water transport and the water status of the plant, decreases in the concentration of growth hormones such as IAA, metabolic disorders or necrosis, extensive changes in the transcriptional profile of many genes ot increased susceptibility to natural toxins such as As (América Berenice Morales-Díaz et al., 2017). It is important to note that depending on the experimental situation, the same NPs or NMs may indicate conflicting results: at times toxic, and at other times without apparent changes or even with positive effects. An example of this was described by researchers who applied 20 and 50 mg l–1 of NPs biomanufactured from Ag to Pennisetum glaucum, which initially improved germination but later had a negative impact on seedling growth. A similar effect was observed in ryegrass, barley and flax upon the application of NPs of Fe and Ag.

When determining NP toxicity in plants, it is necessary to consider the fact that their response will vary according to the element or chemical elements that constitute the NPs, their concentration and aggregation state , the different metabolic abilities of the plant species, the environmental conditions of their surroundings that normally extensively modify the properties of the NPs, and the exposure time.

Considering the effect of the different NP coating types is important when estimating their environmental impact because some surface coatings will increase NP bioavailability and toxicity. The pH and presence of complexing agents such as citrate or ascorbate in the water or soil solution (where part of the root exudates form) and bacterial or root exopolysaccharides may significantly modify NP toxicity. This occurs because these factors increase the stability of NPs and decrease their aggregation, which increases the opportunity to contribute potentially toxic ions even from impurities (not expected but present) in supposedly pure NMs. At the same time, other environmental factors such as temperature can change the aggregation speed, which is why different interaction scenarios between various abiotic and biotic factors can multiply extensively (América Berenice Morales-Díaz et al., 2017). NM toxicity can be direct by absorption or adsorption by the plant or indirect through the release of ions from the decomposition of NMs or by the induction of amplified responses of other environmental toxins. The toxic effect in plants is due per se to NPs from Fe_2O_3, CO_3O_4, Cr_2O_3 and NiO, and CuO toxicity is due to both the NPs and the released ions, compared with only released ions for ZnO (Wang et al., 2016) . For bacteria, the toxic effects of NPs from CuO, ZnO, NiO and Sb_2O_3 have been reported to be dependent per se on the NPs rather than the ionic form of the elements. In another rstudy that included algae, bacteria, and protozoans, it was found that toxicity from CuO and ZnO (EC50 with bacteria > protozoa).

NP and NM toxicity caused to soil microorganisms can occur directly or indirectly by modifications in the radical exudates of plants. This can result in changes in the total populations and/or profile or biodiversity of microbial symbionts. This in turn can modify the response capacity or adaptive capacity of the plants or other organisms faced with environmental factors or stress. A problem that arises in studies conducted in soils is the lack of information about the interactions between NPs and the organic and inorganic abiotic soil components. Some studies assume that these interactions do not generate a response to NPs in microorganisms and plants, but in reality, there is little information on the effect of these interactions. Ideally, long-term experiments using several types of soil should be conducted.

Positive responses of soil bacteria have been reported in some cases, as was the case when 1 or 10 mg kg^{-1} of NPs7 of FeO were applied to the soil, although the application of NPs of Ag (0.1 to 10 mg kg^{-1}) had a negative effect. The indirect impact on the soil microorganisms may partially explain some mixed positive and negative results on the plants caused by NPs applied to the soil. Another factor to be considered for determining NP and NM toxicity is that many studies have been conducted under controlled conditions (in Petri dishes and growth chambers), with low irradiance, temperature control, and absence of stress and interactions with different physico-chemical and biotic factors commonly found in situations of extensive agricultural production or under natural conditions. When plants are subjected to less favorable conditions, it is possible that the presence of NPs and NMs (for example TiO_2, CeO_2 and $Cu(OH)_2$) may interact negatively with abiotic factors that cause stress. The toxicity of the NPs and consequently, plant responses to the NPs are modified according to their environmental context, for example, the plant's own concentration of NPs, the dynamic equilibrium resulting from the release and aggregation of ions, irradiance, the presence of other metal ions, pH, the oxidation-reduction potential (ORP), and the amount of organic material in the soil. Biotic factors also modify the response to nanomaterials, for example, CuO and ZnO NPs interact with crop root exudates in sand and induce plant toxicity, with toxicity being an inverse function of NP size.

The types of NM toxicity reported vary depending on the types of NMs studied, their shape, diameter and reactivity, the experimental design used, the plant species and the stages of the plant life cycle when the materials were applied (Canivet et al., 2015) In some cases, exposure to NMs does not appear to induce harm, but the effect can be different for different plant species in the same experiment because they may differ in their absorptive capacity of the NMs or in their ability to metabolize them. In addition, an experimental model may be used, such as for seed germination and seedling growth, which indicated that toxicity did not occur even at high NP concentrations, similar to those used for nanoremediation process, i.e., higher than 30 000 mg l^{-1} of NPs of Fe. Although the above-mentioned authors found no adverse effects on germination and seedling growth, these concentrations would certainly be toxic to these same plants over a longer time period.

The results of many studies on NP and NM plant toxicity are extremely valuable from the nanoscience point of view because they illustrate the response of different plant organisms to NPs and NMs (América Berenice Morales-Díaz et al., 2017) However, considering that the behavior of complex systems is modified to include new components, at different temporal and spatial scales, the results obtained in laboratories or microcosms cannot be completely extrapolated to conditions of commercial agricultural production systems and their surrounding natural ecosystems. The findings of these studies should be considered carefully before concluding that certain types of NPs or NMs exert an adverse, neutral or beneficial effect at the ecosystem scale.

Usage of nano-technological products has spread nanoparticles into the environment during its manufacture, usage or disposal through water, air, and soil. The unintentional spread of nanoparticles has accelerated a robust debate among the scientific community and have drawn attention towards the potential impact of nanotoxicity in the environment. The physiochemical properties and reactivity of nanoparticles differ not only between nanoparticle with different chemical composition but also among identical nanoparticles with different shape, size, surface properties, and crystalline structure (Syeda Hafsa Ali et al., 2019). Phytotoxicity occurs as nanoparticles are uptaken, translocated, or localized in a plant. Consequently, affecting germination rate, physiological processes that disrupt cell integrity at the molecular level and causes detrimental effects on plant growth and development of various crops. This toxicity produces Reactive Oxygen Species (ROS) that causes DNA damage and lipid peroxidation. However, these free radicals are actively scavenged by antioxidant enzymes to repair the damage and help the plant to withstand the stress. However, the continuous increase of nanoparticles can permanently damage the plant thereby reducing its ability to withstand. Therefore, cost-effective strategies are required to overcome the risk of nanoparticles.

As primary producers, plants are key for any community to function as they are responsible for converting solar energy into organic matter that can be used by other trophic groups. Plants serve as a potential pathway for the transportation of NPs. Through the food chain, NPs can be accumulated in high trophic-level consumers. Organisms in the ecosystem could suffer from oxidative stress induced by NPs. In recent years, research in this area has been focused on the interaction between plants and NPs, and the effects of NPs on ecology, the food chain and human health; evaluating the pros and cons of NPs requires interdisciplinary knowledge.

(Abbreviations: $^1O_2$: singlet oxygen; AA: ascorbate; ABA: abscisic acid; APX: ascorbate peroxidase; B-Ag NPs: biosynthesized silver nanoparticles; Bt: Bacillus thuringiensis; CAT: catalase; CDs: carbon dots; Cu/Zn-SOD: copper/ Zinc SOD; DHA: dehydroascorbate; DHAR: dehydroascorbate reductase; GA: gibberellic acid: GA-Ag NPs: coated silver nanoparticles; GPOX: guaiacol peroxidase; GPX: glutathione peroxidase; GR: glutathione reductase; GSH: glutathione; GSSG: oxidized glutathione; GST: glutathione S-transferases; •OH: hydroxyl radical; HA: high amylose; IAA: indole-3-acetic acid; LA: low amylose; MA: medium amylose; MDA: malondialdehyde; MDHAR: monodehydroascorbate reductase; NPs: nanoparticles; O·–2O2·–: superoxide radical; POD: promoting peroxidase; ROS: reactive oxygen species; SOD: superoxide dismutase; t-ZR: trans-zeatin-riboside).

Most studies showed that NPs could produce toxic effects above a certain concentration. In contrast, some studies have reported that NPs might have a positive effect on plants. Mixed nano-TiO_2 and nano-SiO_2 was introduced into soybean, causing an increase in nitrate reductase activity; this treatment hastened plant germination and increased growth by enhancing the absorption and utilization of water and fertilizer by the crop. Some studies showed that TiO_2 NPs promoted nitrogen metabolism and photosynthesis, thereby improving spinach growth when provided at a suitable concentration. Carbon dots (CDs) at a concentration of 0–1.0 mg/mL played a role in promoting mung bean growth. The results showed that CDs could enhance the ability of mung bean to absorb and utilize nutrients. Therefore, it was concluded that CDs induced a physiological response in the mung bean plant and had a positive effect on the growth of mung bean. The unique optical properties of the CDs allowed their transport by means of the apoplastic paths from the roots to the stems and leaves by the vascular system (Li et al. 2016). These studies have increased our understanding of NPs in phytotoxicology. However, toxicity mechanisms have not yet been completely elucidated for most NPs, and little is known about the potential effects of plants and NPs on their subsequent fate in the food chain.

Phytotoxicity of any nanoparticle is largely influenced by its shape, size chemical composition, and coating material composition. Sometimes, the phytotoxicity of nanoparticles may be as a result of the toxicity of substances which were used for its preparation (Ruttkay-Nedecky, et al., 2017). Further phytotoxicity may depend on the environment and on the physical and chemical nature of the plant species. The nanoparticles may have potentiating or inhibitory effects on plant growth in different developmental stages. Some nanoparticles are taken up by plant roots and transported to the aboveground parts of the plant through the vascular system, depending on the composition shape, size of nanoparticle, and anatomy of the plant. Some nanoparticles remain adhered to the plant roots. In the discussed studies, sometimes nanoparticles have not been properly characterized and/or their composition vs. their shapes have not been considered, which is one the biggest obstacles need to be overcome for further planning of nanoparticles-plant research. Despite the fact that a lot of knowledge has been acquired through many previous studies, many questions still remain unanswered such as the fate and behavior of nanoparticles in plant systems, or the role of surface area or activity of nanoparticles on phytotoxicity, and the role of plant cell walls in the internalization of nanoparticles.

8.2.2 Other effects on plant

Plant physiological indices: The main plant physiological indices of the toxic effects of NPs include the germination percentage, root elongation, biomass and leaf number. NPs can have substantial negative effects, such as reduction in seed germination and suppression of plant elongation, and can even cause plant death. Several previous plant nanotoxicity studies have determined the inhibition of plant species such as soybean, maize, wheat (Triticum aestivum), ryegrass and barley by exposure to NPs (MWCNTs, single-wall carbon nanotubes, ZnO NPs, Ag NPs and Fe NPs); several aspects of plant growth were affected including seed germination, shoot length, biomass and gene expression (Jie Yang et al., 2017). Growth inhibition was observed when

i. *Bacillus thuringiensis* (Bt)-transgenic cotton was exposed to SiO_2 NPs: When wheat plants were grown in a sand matrix, CuO NPs inhibited their growth and changed the structure of the roots. CuO NPs significantly reduced the fresh weights and root length of Arabidopsis seedlings, and the germination rate and biomass of rice seeds.

Rare earth oxide NPs (CeO_2, La_2O_3, Gd_2O_3 and Yb_2O_3) had a detrimental effect on plant growth in radish, tomato, rape, lettuce, wheat, cabbage, cucumber and corn when added to roots at high concentrations. TiO_2 NPs increased the content of total chlorophyll and catalase (CAT) and decreased ascorbate peroxidase (APX) content in leaves.

Plant hormones: Plant hormones are active organic materials that are produced by plant metabolism, which can regulate physiological responses during plant growth and mediate responses to challenges (Santner et al. 2009). The content and activity of plant hormones are considered an important index of toxicity in plants. CeO_2 NPs had no significant effect on indole-3-acetic acid (IAA), abscisic acid (ABA) and gibberellic acid (GA) in the leaves of Bt-transgenic and conventional cotton compared with the control group. When conventional cotton was exposed to 500 mg/L CeO_2 NPs, the content of trans-zeatin-riboside (t-ZR) in the leaves decreased by 25% when compared with the control group alone. The IAA and ABA content in the roots of transgenic and non-transgenic rice was increased in response to γFe_2O_3. NPs had a significant influence on the production of plant hormones. Hao et al. (2016) demonstrated that decreases in phytohormone concentrations were evident in rice seedlings upon exposure to carbon nanotubes. Because Ag ions inhibited the production of ethylene, the interaction between IAA and ethylene would be substantially weakened.

ii. Crop quality: Previous studies of hydroponic plants have shown that the accumulation of NPs in the environment could profoundly change the soil-based food crop quality and yield.Protein was insensitive to stimulation by Ag NPs when compared with carbohydrates, with protein content increased only at high Ag NP concentrations (100 mg/L). ZnO increased the starch and protein content and decreased the micronutrient (Cu and Mo) concentrations in cucumber. Rice treated with $nCeO_2$ had less Fe, S, prolamin, glutelin, lauric and valeric acids, and starch than the control, and the antioxidant properties of the treated rice were weakened. CeO_2 NPs changed the content of amino acids, fatty acids, nonreducing sugars and phenolics in plants. Le Van et al. (2016) also documented the effects of NPs on plant nutrient content.

8.2.3 Factors affecting nanoparticle phytotoxicity

In combination with recent studies and our own knowledge, we have concluded that the main factors influencing the effects of NPs on plants are the characteristics of the nanomaterials themselves (concentration/size/category/stability), the plant seed (size/species), the plant growth medium, the plant growth stage and the NP coating material. The influence of various factors is discussed below.

i. Effects of different nanomaterials on NP phytotoxicity: Treatment of tomato with Ag NPs resulted in a reduction in biomass and root length. TiO_2 NPs could significantly improve the germination rate of seeds; however, bulk TiO_2 can have inhibitory effects on seed germination (Jie Yang et al.,2017).

Because of the high reactivity of small materials, NPs usually show greater toxicity than the same material of larger size. However, other studies have shown that phytotoxicity increases with particle size. All Ag NP treatment groups had no effect on the growth of castor, but in the treatments where the Ag was used in its bulk form, inhibition was observed. On the fifth day of NP exposure, synthesized silver NP treatments had higher Ag accumulation than biosynthesized silver NP (B-Ag NP) treatments (10 and 100 mg/L) in water hyacinth; at the high concentration (100 mg/L), B-Ag NPs improved plant growth.

ii. Effects of different seeds on NP phytotoxicity : To some extent, growth and antioxidant defense responses differ from species to species. Some studies have speculated that the differences in the toxicity of NPs might be related to the differences in seed size, the single leaf and the xylem structure of double-leaf plants (Lee et al. 2008). Three rice varieties (high (HA), medium (MA) and low (LA) amylose) were cultivated in soil amended with CeO2 NPs; a comparison between the treatments showed that relative to the control, the Ce content in HA was not significantly different. Treatment with CeO_2 NPs enhanced the Ce content in MA and LA treatments. Transgenic seeds are likely to have different responses when compared with conventional seeds. When exposed to ZnO NPs, leaf surface area was statistically reduced for cultivar Zhu Liang You 06 (Oryza sativa) when compared with cultivar Qian You No. 1 (Oryza sativa) under control conditions for both cultivars.

iii. Effects of different growth media on NP phytotoxicity: NPs must be studied in the specific environments in which they occur, as their characteristics differ widely depending on the environmental conditions (Song et al. 2013). When the biomass of radishes was compared in silty loam (2.21% ± 0.04% soil organic matter) and loamy sand (11.87% ± 0.56% soil organic matter), the former had significantly higher root biomass and the presence of 1000 mg/kg CeO_2 NPs made no substantial difference. The Ce accumulation in fine roots and in the storage root showed that growth in loamy sand was higher than that in silty loam. Kidney bean plants exposed to CeO_2 NPs for 52 days in a low organic matter soil usually possessed higher Ce concentration than the same tissues collected from an organic matter-enriched soil. Likewise, different levels of phytotoxicity caused by CeO_2 NPs were found in lettuce seedlings incubated with three kinds of media, including agar, potting mix soil and sand. Ag NP toxicity decreased with increasing clay content and pH, butseemed to be unaffected by the organic carbon content of the soil.

iv. Effect of different growth stages on NP phytotoxicity: Photosynthetic parameters changed throughout the whole growth period in cucumber; seedling leaf size decreased when compared with the control when treated with 200 mg/L of CeO_2 NPs and CuO NPs, but when mature leaves were measured, there were no significant differences among the treatments. At 15 days, the

CAT and APX activities in the roots showed a sharp decrease when compared with the treatments at day 7, except for the control treatment. Total soluble protein content in roots showed the opposite results, with the exception of the 500 mg/L CeO_2 NP treatment.

v. Effect of different coating material on NP phytotoxicity: Coating not only alters the function of NPs, but also changes its effects on soil–plant systems. The types of coating also influence the toxicity and solubility of NPs. The suspension of Ag NPs led to the increased germination rate of 5 of 11 species, whereas exposure to coated silver NPs (GA-Ag NPs) only significantly contributed to the germination rate of one species. This was because of the presence of a co-ligand in the soil, which caused a preferential reduction in the bioavailability and toxicity of Ag^+. The toxicity of GA-Ag NPs was longer lasting than that of $AgNO_3$. No significant differences in the biomass production rate (fresh weight) of rice were found among plants treated with Fe-Co-filled carbon nanotubes (FeCo-CNTs) compared with the non-stressed plants; MWCNTs and Fe-CNTs significantly decreased the biomass production rate at 30 and 50 mg/L (Hao et al. 2016).

8.3 Mechanisms of toxicity and detoxicity in plants induced by nanoparticles

With regard to the effects of NPs on plants, recent reports have revealed ambiguous results. Many reports found NPs beneficial effects, whereas, some reports that show harmful effects (Jie Yang et al., 2017). Plant species, NP properties (size, shape, types, structure and defects, surface coating, etc.) and culture media may all potentially cause the divergent results.

8.3.1 Phytotoxicity mechanism in plants induced by nanoparticles

Reactive oxygen species (ROS), in general, contain both free radicals such as hydroxyl radical (•OH) and superoxide radical (O·−2), and non-radical molecules such as singlet oxygen ([1]O) and hydrogen peroxide (H−O). ROS is 22; 2 the product of aerobic metabolism in an ordinary plant, which acts as signaling molecules, whereas excess ROS would cause kinds of adverse effects called oxidative stress, which emerges when the ROS level exceeds the defense mechanisms (It will be described in detail in the next section), and is able to pose threat to cells by inducing DNA damage, protein oxidation, electrolyte leakage, lipid peroxidation and membrane damage, finally causing cell death. Published studies on metal and metal-based NP phytotoxicity suggest that NPs can induce oxidative stress in many plant species (Zhang et al. 2017). To clarify ZnO NP phytotoxicity to ryegrass, it was proposed that the particle- dependent ROS formation and lipid peroxidation occurred on the surface of cellular membranes. Likewise, due to a high correlation between the decrease in viable cells and ROS generation at concentrations of Ag NPs up to 10 mg/ mL over a week of treatment period, the phytotoxicity of Ag NPs was caused by ROS generation.

ROS can lead to oxidative damages in nuclear, chloroplastic and mitochondrial DNA. DNA molecule is a cell's genetic material and any damage to the DNA would cause changes in the encoded proteins, which may result in malfunctions and/or complete inactivation of the encoded proteins. Enhanced DNA degradation has been observed in plants exposed to kinds of environmental stresses such as metal-based NPs and carbon nanomaterials. High concentrations of CeO_2 NPs had negative effects on DNA of *G. max*. Similarly, *Allium cepa* roots suffered various negative effects such as total chromosomal aberrations and increased mitotic index when exposed to bismuth oxide NPs. Several studies have also investigated the potential genotoxic effects of TiO_2 NPs on various species of plants.

A study on *Nicotiana tabacum* and *A. cepa* has found that genotoxic effects occur with exposure to different concentrations of TiO_2 NPs (100 nm). The production of ROS is based on the physicochemical properties of NPs as well as the test species. Various determinants, such as size and shape, solubility and particle dissolution, metal ions released from metal and metal oxide NPs, biotransformation of NPs, light and so forth, may cause the ROS generation and phytotoxicity (Rui et al. 2015).

The toxicity of three types of CeO_2 NPs were compared to different kinds of Lactuca genus plants. Compared to the control, 7 nm CeO_2 caused a significant increase in the MDA level, which means the membrane damage happened in the root cells. However, there was no obvious discrepancy in the MDA level among control, 25 nm CeO_2 and its bulk counterpart treatment. Other reports have studied the effects of size and shape of NPs on various plant species as well. Plant responses to CeO_2 exposure varied with particle sizes and the growth stages of plants.

Compared to high-solubility NPs such as ZnO NPs and Ag NPs, CeO_2 NPs are generally regarded as stable regardless of the biological or environmental systems involved, which were always used as a model material to study the phytotoxicity mechanisms of NPs. Moreover, recent studies indicated that CeO_2 NPs can be biotransformed to $CePO_4$ and $Ce(CH_3COO)_3$ in several plant species (Ma et al. 2015).

Furthermore, phosphates play a vital role in the transformation process of CeO_2 NPs in plants. It is well known that phosphates widely exist in the environment and are used as a basic component of many culture media for toxicity testing. A recent study has also confirmed that the phytotoxicity to romaine lettuce of CeO_2 NPs was determined by phosphates in a sand culture.

The phytotoxicity of ZnO NPs was studied on *Lolium perenne* (ryegrass), and found that concentrated ZnO NPs in the rhizosphere solution and absorbed on the root surface caused potential effects on the plant's growth. Results also revealed that the concentration of released Zn^{2+} ions was far below the toxic threshold to the plant and, herein, the phytotoxicity effect cannot be attributed to ions dissolution. Toxicity by dissolution of meta-based NPs was also excluded from the main mechanisms in some other investigations. Recently a study was conducted on the toxicogenomic effects of CuO NPs on *Arabidopsis thaliana* by using microarray analysis. The results also agreed with previous opinions that the released Cu^{2+} ions from CuO NPs contributed to partial toxicity during CuO NPs exposure (Tang et al. 2016).

Compared with Ag^+ ions, Ag NPs showed higher toxicity on the root elongation of *A. thaliana*, even though seeds germination was not affected. Results also found that seedlings adsorbed very few amounts of Ag^+, which could not affect chloroplast structure compared to Ag NPs. Ag NPs could change the balance of oxidation and antioxidant systems and further affect the homeostasis of water and other small molecules within the plant body.

8.3.2 Detoxification mechanism in plants

Certain environmental conditions, such as heavy metal, salt stress, nutrient deficiency and diverse metal-based NPs exposure, clearly affect the level of cellular ROS, for example, H_2O_2 and O·–2 $^1O_2$ and •OH. ROS is a signaling oring molecule, which plays a vital role in the subtle equilibrium between the levels of ROS generation and scavenging (Jie Yang et al., 2017). ROS have been involved in intercellular signaling cascades as second messengers which regulate various plant responses in plant cells such as gravitropism programmed cell death, stomatal closure and so forth. They also regulate the activities of numerous components by signaling, including transcription factors protein phosphatases and protein kinases. However, excess ROS could cause oxidative stress in plants under NPs exposure. To protect themselves against these toxic oxygen intermediates, herein, plant cells and its organelles such as mitochondria, peroxisomes and chloroplast make use of antioxidant defense systems to scavenge excess ROS production. Antioxidant defense systems of plants contain both nonenzymatic antioxidants including thiols, glutathione (GSH), phenolics, ascorbate (AA) and so forth, and enzymatic components such as CAT, SOD, guaiacol peroxidase (GPOX), APX, glutathione reductase (GR), monodehydroascorbate reductase, dehydroascorbate reductase (DHAR) glutathione S-transferases (GST) and glutathione peroxidase (GPX).

8.3.3 Nonenzymatic agents of the antioxidant defense system

Thiols and AA are the most significant low molecular weight antioxidants. AA is a crucial antioxidant that can resist oxidative stress induced by the promoted level of ROS generation. Owing to the intrinsic ability to donate electrons of AA in diverse nonenzymatic and enzymatic reactions, it can scavenge O·–2 and •OH and regenerate α-tocopherol from the tocopheroxyl radical to protect membranes directly. It also acts as the first barrier of defense against the potential negative external oxidants.

GSH, one of the fundamental metabolites (a nonprotein thiol with low molecular weight) in plants, plays a key role in intercellular antioxidant defense against ROS-induced oxidative stress. In plant tissues, GSH is commonly found in all cell compartments, including chloroplasts, mitochondria, vacuole, cytosol, as well as in peroxisomes and endoplasmic reticulum ((Jie Yang et al., 2017).

Due to possessing antioxidant properties, phenolic compounds can scavenge ROS species, chelate transition metal ions and inhibit lipid peroxidation by means of trapping the lipid alkoxyl (–OCH3) polyphenols. Carotenoids, in particular, are the member of lipophilic antioxidants and can detoxify many forms of ROS (Jie Yang et al., 2017). Tocopherols, containing α-, β-, γ- and δ-tocopherol, represent a class of lipophilic antioxidants that participate in the scavenging of lipid peroxidation radicals, oxygen-free radicals and ^1O. By chemically reacting with O_2 and physically quenching in chloroplasts, tocopherols can protect lipids and other membrane components, herein protecting the membrane structure and PS II.

Enzymatic antioxidants of the antioxidant defense system: SOD, the most effective intracellular metalloenzyme, plays a fundamental role in the antioxidant defense system against oxidative stress by catalyzing highly toxic ROS (O·–2O2·–) to less toxic H_2O_2 and O_2. It contains three isozymes including iron SOD (Fe-SOD), manganese SOD (Mn-SOD) and copper/zinc SOD (Cu/Zn-SOD) (Jie Yang et al., 2017), which are ubiquitous not only in aerobic organisms but also in subcellular compartments to prevent against the toxic effects of enhanced levels of ROS production. Several reports have recorded that the promotion of SOD is related to the increase in tolerance of plants against environmental stress, such as NP toxicity, which means that SOD could act as an indirect selection criterion for researching oxidative stress.

The SOD activity in rice was elevated under 250 mg/L CeO_2 NPs exposure, whereas at a higher concentration treatment there was no obvious change in enzyme activity. Once H_2O_2 is overproduced either from alternative abiotic stress or from SOD defense system induction, to avoid oxidative stress such as lipid peroxidation, CAT, APX and oxidized GSH contain the main antioxidant enzyme pathway that would be activated to detoxify H_2O_2.

CAT, among all of the antioxidant enzymes, is the enzyme that was first discovered and characterized and plays an indispensable role in ROS detoxification under stress. It is widely found in tetrameric heme-containing enzymes and converts H_2O_2 into O_2 and H_2O directly. APX, regarded as the most significant ROS scavenger to protect higher plants, avoids oxidative stress and plays a central role in regulating the AA–GSH cycle.

APX can reduce the level of H_2O_2, which was formed by SOD converting it to H_2O (Rico et al. (2015). Owing to DHAR existence, regenerated AA was used by APX for the diminution of H_2O_2 again. Likewise, various antioxidant enzymes can be activated by numerous NPs, for example, Fe_3O_4 NPs, Co_3O_4 NPs and CeO_2 NPs induced CAT; Au NPs, MnO_2 NPs, Fe_3O_4 NPs, CuO NPs, Co_3O_4 NPs and CeO_2 NPs induced GPX and fullerene; and Pt NPs and CeO_2 NPs induced SOD on *Phaseolus vulgaris*, showed that phytotoxicity in plants caused by CeO_2 NPs exposure disturbed the antioxidant defense system.

The defense mechanisms of *A. thaliana* was studied and found that the activities of CAT, SOD, POD and APX were significant ameliorated under CeO_2 NPs exposure, while the elevations were only found for POD and SOD under In_2O_3 NPs exposure(Jie Yang et al., 2017). Similarly, an article on the effects of Nd_2O_3 NPs and its bulk counterparts in pumpkin found that not only pumpkin growth was inhibited, but also the K, Ca, Mg and S levels in pumpkin tissues were decreased under 100 mg/L Nd_2O_3 NPs and bulk exposure. They also reported that the enzymatic activities of SOD and POD elevated, whereas that of CAT and APX reduced, which could be due to other pathways scavenging H_2O_2 or their biosynthesis was inhibited under overproduced H_2O_2.

8.4 Assay-dependent phytotoxicity of nanoparticles to plants

The effects of five nanomaterials (multiwalled carbon nanotubes [MWCNTs], Ag, Cu, ZnO, Si) and their corresponding bulk counterparts on seed germination, root elongation, and biomass of Cucurbita pepo (zucchini) were investigated (Dimitrios Stampoulis et al., 2009) The plants were grown in hydroponic solutions amended with nanoparticles or bulk material suspensions at 1000 mg/L. Seed germination was unaffected by any of the treatments, but Cu nanoparticles reduced emerging root length by 77% and 64% relative to unamended controls and seeds exposed to bulk Cu powder, respectively. During a 15-day hydroponic trial, the biomass of plants exposed to MWCNTs and Ag nanoparticles was reduced by 60% and 75%, respectively, as compared to control plants and corresponding bulk carbon and Ag powder solutions.

Although bulk Cu powder reduced biomass by 69%, Cu nanoparticle exposure resulted in 90% reduction relative to control plants. Both Ag and Cu ion controls (1–1000 mg/L) and supernatant from centrifuged nanoparticle solutions (1000 mg/L) indicate that half the observed phytotoxicity is from the elemental nanoparticles themselves. The biomass and transpiration volume of zucchini exposed to Ag nanoparticles or bulk powder at 0–1000 mg/mL for 17 days was measured. Exposure to Ag nanoparticles at 500 and 100 mg/L resulted in 57% and 41% decreases in plant biomass and transpiration, respectively, as compared to controls or to plants exposed to bulk Ag. On average, zucchini shoots exposed to Ag nanoparticles contained 4.7 greater Ag concentration than did the plants from the corresponding bulk solutions. These findings demonstrate that standard phytotoxicity tests such as germination and root elongation may not be sensitive enough or appropriate when evaluating nanoparticle toxicity to terrestrial plant species.

Phytotoxicity of nanomaterial products based on essential metals: Ruttkay-Nedecky et al., (2017) summarized the latest findings on the phytotoxicity of nanomaterial products based on essential metals used in plant protection. The metal nanoparticles (NPs) synthesized from essential metals belong to the most commonly manufactured types of nanomaterials since they have unique physical and chemical properties and are used in agricultural and biotechnological applications, which are discussed. The paper discusses the interactions of nanomaterials and vascular plants, which are the subject of intensive research because plants closely interact with soil, water, and atmosphere; they are also part of the food chain. Regarding the accumulation of NPs in the plant body, their quantification and localization is still very unclear and further research in this area is necessary.

Not all plants treated with nanoparticles exhibit toxic effects; substantially more studies showed positive or no consequential effects on plants. Nanotoxicity mechanisms remain unknown, however, it can be assumed to be closely related to the chemical composition, chemical structure, size, and surface area of nanoparticles. The presence of nanoparticles on the root surface can change the surface chemistry of the roots and consequently affect the uptake of nutrients into the plant root, thus, these have to be taken into consideration too. Generally, the toxicity of nanoparticles is attributable to two different steps , viz., chemical toxicity based on chemical composition, for example the release of (toxic) ions, and stress stimuli caused by surface, size, or shape of the particles.

Future research needs: Scientists have made some consensus on the environmental behavior and ecological effects of nanomaterials; but there are still a lot of controversies and problems that need to be further studied (Jie Yang et al., 2017).

i. Further understanding of the environmental fate and mechanism of NPs: It is not enough to rely on the current detection methods for phytotoxicity only; how to correctly use the positive effect and eliminate the negative effect of NPs is the research motivation and goal of researchers. On the one hand, the dose- and time-dependent negative effects of NPs on plant growth observed in experimental conditions should lead to a more careful use in crop management. On the other hand, the phytotoxic mechanism of NPs must be thoroughly understood before being applied in the field environment under the toxic concentration; an effective method to reduce the toxicity also needs to be developed.

ii. Change the exposure time, concentration and mode in the experiment: Unfortunately, very few reports covered the life cycle of plants treated with NPs to make a conclusive assessment of their long-term risks and benefits. Lin et al. (2009) further found that soluble organic matter treated C_{70} and NWCNTs can reduce the seed setting rate, and even affect the next generation. So it is necessary to strengthen the research on the toxicity and absorption of NPs to plants in a long-term dose exposure test. Similarly, increasing the diversity of material exposure is also important, such as changing short-term exposure with high concentration to long-term exposure with low concentration.

iii. Make the experiment condition more close to the real field environment: These limited data, as show above, are only based on laboratory or greenhouse studies. Due to the results of the single element in the laboratory and the properties of NMs are greatly influenced by environmental media, the results of field trails and laboratory studies are often contradictory. Therefore, it is needed to strengthen the toxicity and absorption of NPs in soil or simulated soil (e.g. quartz sand). In addition, the study on the effects of NPs on population and ecological structure should be strengthened in the future.

iv. Analyze problems using multidisciplinary knowledge objectively: The plant germination rate decreased, but the amount of root elongation had been promoted (Jie Yang et al., 2017). In this case, how to evaluate the effects of NPs on plants? It is done by assuming that under moderate concentrations the root elongation could promote plant growth and increase yield, so as to make up for the loss caused by the germination rate and even exceed, resulting in an increase in the total production, which would use the knowledge of economics. Furthermore, for the analysis of the crops quality, food safety or food nutrition knowledge is needed as support.

v. Develop a series of safety evaluation and toxicological risk assessment standards: At present, the standard research methods and evaluation indexes of the toxicity of nanomaterials are relatively simple. The analysis of morphological, anatomical, physiological and biochemical indexes cannot fully reflect the plant toxicity to NPs. NPs have unique physical and chemical properties, so the obtained scientific knowledge is still needed further inspection. A series of safety evaluation and toxicological risk assessment standards must be formulated, including exposure route, safe exposure dose, etc. The purpose of this work is to provide a scientific basis for the application of NPs.

vi. Others: As to the accumulation of NPs in plants, the problem should also be analyzed objectively. For example, if NPs accumulate in the tomato leaves, not the edible part, then the NPs will not enter the human body through the consumption of the fruit, and it will not affect our health. At this point, we should focus on the study of NPs in the environment and how to recycle them and reduce their release. The research on cucumber, pumpkin, radish, lettuce and other edible plants should be based on the same theory. When the study of NPs in agricultural soils has matured to a certain extent, the study realms should be expanded, such as saline land, tropics droughts, floods and some other extreme conditions, in order to observe whether the special properties of NPs in extreme environments will produce special reactions.

8.5 Methods for testing the nano-phytotoxicity

There are no specific test guidelines for nanotoxicity, hence EPA48 or OECD49 directives from the US for chemical testing are currently used(Ruttkay-Nedecky et al., 2017). Phytotoxicity tests generally use plants recommended by these guidelines. These are mostly species of crops, and include both monocotyledonae and dicotyledonae. Species that are recommended most are bean (*Phaseolus vulgaris*), cabbage (*Brassica oleracea*), carrot (*Daucus carota* subsp. *sativus*), cucumber (*Cucumis sativus*), lettuce (*Lactuca sativa*), maize (*Zea mays* subsp. *mays*), oat (*Avena sativa*), onion (*Allium cepa*), radish (*Raphanus sativus*), rice (*Oryza sativa*), ryegrass (*Lolium perenne* L.), soybean (*Glycine max*), tomato (*Solanum lycopersicum*), and wheat (*Triticum aestivum*). Recently, research model species such as the well characterized thale cress (*Arabidopsis thaliana*) were also included.

Phytotoxicity tests are carried out in two stages of plant development:, first, during germination, when the germination percentage is measured, where the seeds must be exposed to the test solution for the duration of germination (preferably at least 4 days) and second, during seedling growth, in which root/shoot elongation and dry weight are frequently used variables to assess the effects of plant exposure to harmful substances. The aforementioned protocols have been applying for testing the effects of nanoparticles in water, wastewater, sediment, and slurry. For phytotoxicity testing different media for the growth of plants are used. The simplest medium is water. Other applications include soft gels or agars, which better represent the soil, and finally soil itself is also often used. Nanoparticles tend to adsorb to soil matrix and aggregate in the natural environment which reduces their mobility and bioavailability. Also, the study of the interactions of nanoparticles with plants in alternative substrates does not take into account the potential interaction of nanoparticles with soil and the associated water phase. For example, nanoparticles in soil can influence the growth of soil bacteria, which may then indirectly affect the plant growth.

Nanoscale zerovalent iron (Fe^0) and bimetallic Ni^0–Fe^0 nanoparticles have emerged as effective redox media for the detoxification of organic and inorganic pollutants in aqueous solutions. These nanomaterials (10–100 nm) have larger surface areas and reactivity than bulk Fe^0 particles. Mn deficiency has been widely reported all over the world, especially in soils with higher pH (>6.0) or in calcareous, sandy, peat, or muck soils. Therefore, Mn fertilization is very important to improve agronomic production Manganese nanoparticles (MnNPs) have been also proposed as a suitable alternative to commercially used manganese salts $MnSO_4$ (MS) for nanobiotechnology based crop management studies.

8.5.1 Nanomaterials based on Zinc

Zinc nanoparticles (NPs) are spherical or polished metal particles with a high specific surface area. The applications of zinc nanocrystals include antimicrobial, antibiotic and antifungal agents, which are a part of painting buildings, dressing materials, nanofibers, plastics, and textiles. Moreover, ZnO NPs are widely used in personal care products such as sunscreens, cosmetics, textiles, lipsticks, and hair dyes.

On the basis of thermodynamics, ZnO NPs should dissolve faster and to a greater extent than bulk ZnO particles (equivalent spherical diameter >100 nm). These novel solubility features of ZnO NPs might be exploited to improve the efficiency of Zn fertilizers. In this field, coated monoammonium phosphate granules show greater Zn solubility and faster dissolution rates in sand columns compared to coated urea granules, which may be related to pH differences in the solution surrounding the fertilizer granules.

Besides supplementation by Zn, ZnO NPs, synthesized by soil fungi in a concentration 10 mg L^{-1}, has been shown to enhance the mobilization of native phosphorus in the mung bean (*Vigna radiata*) rhizosphere. The analyses made by authors showed that they synthesized ZnO NPs with average diameter as 22.4 nm as they claimed to have stable nanoparticles due to in situ corona formation by fungal extracellular protein used in the synthesis procedure. Zn acts as a cofactor for phosphorus-solubilizing enzymes such as phosphatase and phytase, and ZnO NPs increased their activity. The level of resultant phosphorus uptake in *V. radiata* was increased by 10.8%. In addition, biosynthesized ZnO NPs also improved plant plienology such as stem height, root volume, and biochemical indicators such as leaf protein and chlorophyll contents. Not only the effect but also the way of application was considered by authors as they choose foliar application on 2-week-old mung bean plants. The observed positive effects of ZnO NPs on plants are summarized in Table 30.

Table 30: The observed positive effects of ZnO NPs on plants

Plant	Particle size (nm)	Particle concentration	Comment	Observed effect
Vigna radiata	22.4 ± 1.8	10 mg L^{-1}	NPs were synthesized by soil fungi	Increase in stem height and root length. Increase in phosphorus mobilizing enzymes and phosphorus uptake by 10.8%
Zea mays, Glycine max, Cajanas cajan, Abelmoschus esculentus	<100	25 or 50 mg Zn g^{-1} seed	Seeds were coated with ZnO NPs	Improved germination Coated seeds 93–100% Uncoated seeds 80%
Triticum aestivum	30	100 mM Zn	Wheat plants were inoculated with *Fusarium graminearum*	Reduction in number of CFU of F. graminearum

Ecotoxicity studies on ZnO NPs are most abundant in bacteria and are relatively lacking in other species. These studies suggest relative high acute toxicity of ZnO NPs (in the low mg L^{-1} levels) to environmental species, although this toxicity is highly dependent on test species, physicochemical properties of the material, and test methods.

The impact of different zinc oxide (ZnO) NPs on green pea plants (*Pisum sativum* L.) were studied (Ruttkay-Nedecky et al., 2017). Although all treated plants showed higher tissue Zn content than controls, those exposed to Al_2O_3/ZnO NPs at 1000 mg kg^{-1} had greater Zn concentration in roots and seeds, compared to bulk Zn and the other NPs treatments. In leaves, Al O /ZnO NPs at 250 mg kg^{-1} significantly increased chlorophyll-α and carotenoid concentrations relative to the bulk, ionic, and other NPs treatments.

ZnO NPs at concentrations of 2000 mg L^{-1} have been shown to inhibit root elongation (50.45% for maize and 66.75% for rice) of two crop plants(Ruttkay-Nedecky et al., 2017). Both the production of free hydroxyl groups and the Zn bioaccumulation in roots or shoots resulted in toxicity of ZnO NPs to Chinese cabbage seedlings.

The effect of exposure to 100 mg L^{-1} ZnO NPs on gene expression in *A. thaliana* roots was studied using microarrays. The genes induced by ZnO NPs include mainly ontology groups annotated as stress responsive, including both abiotic (oxidative, salt, water deprivation) and biotic (wounding and defense to pathogens) stimuli. The down-regulated genes upon ZnO NPs exposure were involved in cell organization and biogenesis, including translation, nucleosome assembly, and microtubule based process(Ruttkay-Nedecky et al., 2017). Soybean plants (*G. max*) were grown through the seed production stage in soil amended with ZnO NPs (0, 50, 100 or 500 mg kg^{-1}).

The phytotoxicity and oxidative stress by ZnO NPs in cucumber (*Cucumis sativus*) was studied (Ruttkay-Nedecky et al., 2017). Bioaccumulation of ZnO NPs was estimated in plants, reactive oxygen species enzyme [superoxide dismutase (SOD), catalase (CAT), and peroxidase (POD)] activities in plant root tissue, and observed ZnO NPs with transmission electron microscopy. They found that the seedling biomass significantly decreased to 35% of that of control at 1000 mg L^{-1} of ZnO NPs. These results indicate that ZnO NPs altered both phytotoxicity and oxidative stress in plant assays.

The effects of zinc oxide NPs (ZnO NPs) were tested on the root growth, root apical meristem mitosis, and mitotic aberrations of garlic (*Allium sativum* L.) were investigated. ZnO NPs caused a concentration-dependent inhibition of root length. An overview of the phytotoxicity of nanoparticles based on ZnO is shown in Table 34.

Table 34: Phytotoxicity of nanoparticles based on ZnO

Plant	Type of nanoparticle, particle size (nm)	Particle concentration	Comment	Observed effect
Glycine max	ZnO NPs <50 nm	500 ppm	Effect of ZnO NPs on soybean seedlings was studied	Decrease in root growth (length and weight), loss of root cell viability, accumulation of superoxide and decrease in leaf weight, down regulation of oxidative cascade related genes
Pisum sativum	Bare ZnO NPs 10 nm, Al_2O_3/ZnONPs 15 nm, KH550/ZnO NPs 20 nm	250, 1000 mg L^{-1} of soil	Pea plants were grown for 65 days in soil amended with three types of ZnO NPs	Al_2O_3/ZnO NPs at 250 mg kg^{-1} significantly increased chlorophyll-α and carotenoid concentrations. Al_2O_3/ZnO NPs at 1000 mg kg^{-1} significantly increased sucrose concentration of green peas
Zea mays, Oryza sativa	ZnO NPs <50 nm	500, 1000, 2000 mg L^{-1}	Seed germination was investigated	ZnO NPs inhibited root elongation at 2000 mg L^{-1} (50.45% for maize and 66.75% for rice) of two crop plants
Brassica pekinensis	Spheric ZnO NPs 30 nm, spheric ZnO NPs 50 nm, columnar ZnO NPs 90 nm, hexagon rod-like ZnO NPs 150 nm	1, 5, 10, 20, 40, 80 mg L^{-1}	There were no significant differences in observed effects between different NPs	ZnO NPs inhibited the root and shoot elongation of Chinese cabbage seedlings. The highest inhibition of root elongation at 80 mg L^{-1} was observed
Pisum sativum	ZnO NPs <50 nm	250, 500, 750, 1000 mg L^{-1}	No impact on germination	ZnO NPs (500–1000 mg L^{-1}) significantly inhibited root elongation
Vigna unguiculata	ZnO NPs <100 nm	25 mg L^{-1}	More pronounced effects were observed with $ZnCl_2$ than with ZnO NPs	Significant decrease in biomass production of roots and leaves observed in solution culture, but not observed in soil culture

Plant	Type of nanoparticle, particle size (nm)	Particle concentration	Comment	Observed effect
Arabidopsis thaliana	ZnO NPs <100 nm	100 mg L^{-1}	Effect of ZnO NPs on gene expression in plant roots were studied	Induction of stress responsive genes, down regulation of genes involved in cell organization and biogenesis
(Glycine max)	ZnO NPs 10 nm	50, 100, and 500 mg kg^{-1} of soil	ZnO NPs were added to the soil	Zn bioaccumulated in all tissues and especially in the leaves
(Fagopyrum esculentum)	ZnO NPs <50 nm	10, 100, and 1000 mg kg^{-1} of soil	ZnO NPs were added to the soil and growth of plant seedlings were observed	Inhibition of shoot growth
Triticum aestivum	ZnO NPs <100 nm	500 mg kg^{-1} sand	ZnO NPs were added to the sand	Reduced root growth, increased lipid peroxidation and oxidized glutathione in roots. Bioaccumulation of Zn and decreased chlorophyll content in the shoots
Cucumis sativus	ZnO NPs 50 nm	10, 50, 100, 500, 1000 mg L^{-1}	Hydroponic experiments	Decrease in seedling biomass. ZnO NPs adhered to the root cell wall, and some of them were observed in the root cells
Allium sativum	ZnO NPs 3–5 nm	10, 20, 30, 40, 50 mg L^{-1}	Hydroponic experiments	Concentration-dependent inhibition of root length, observed mitotic aberrations
Zea mays	ZnO NPs 370–410 nm	20 mg L^{-1}	ZnO NPs were added to the sandy loam soil or to the water	ZnO NPs aggregates penetrated the root epidermis and cortex. Some of the NPs aggregates were also present in xylem vessels

8.5.2 Nanomaterials based on copper

Cu/CuO NPs are used in optoelectronics, catalysis, solar cells, as semiconductors, as they are also used as pigments, and fungicides (Ruttkay-Nedecky et al., 2017). Copper as fungicide is especially used in vineyards and in organic farming. The ability of copper ions to prevent spore germination of fungi has been known for a long time, but to achieve this effect it is necessary to apply a large amount of copper (500–1500g ha^{-1}). It is believed that the stimulatory effects of CuNPs are related to the induction of antioxidant activity. The positive effect of CuNPs on *S. lycopersicum* is shown in Table 31 in the same way as in the case of Zn NPS.

Table 31: The observed positive effects of CuNPs on *Solanum lycopersicum*

Plant	Particle size (nm)	Particle concentration	Comment	Observed effect
Solanum lycopersicum	<100	15, 30, 60, 150 mg L^{-1}	CuNPs were adsorbed on chitosan hydrogels	Application of chitosan hydrogels with CuNPs was favorable to tomato growth and quality

CuNPs toxicity mechanisms have been extensively studied in animal/human systems. In plants, toxicity of Cu and Cu ions was thoroughly investigatedbut not so Cu/CuO NPs phytotoxicity (Ruttkay-Nedecky et al., 2017). The response of cucumber plants in hydroponic culture investigated at early development to two concentrations of CuNPs (10 and 20 mg L^{-1}). Results showed that CuNPs interferes with the uptake of a number of micro- and macronutrients such as Na, P, S, Mo, Zn, and Fe. Metabolomics data revealed that CuNPs at both levels triggered significant metabolic changes in cucumber leaves and root exudates. The root exudate metabolic changes revealed an active defense mechanism against CuNPs stress: up-regulation of amino acids to sequester/exclude Cu/CuNPs, down regulation of citric acid to reduce the mobilization of Cu ions, ascorbic acid up-regulation to combat reactive oxygen species, and up-regulation of phenolic compounds to improve antioxidant system. It also observed a decrease in root length, reduction of root biomass, and bioaccumulation of Cu mainly in roots.

A study was conducted to assess the effects of laboratory-prepared CuNPs in low concentrations (<50 ppm) on the germination of lettuce (*L. sativa*) seeds in a water medium. The data showed that CuO NPs were slightly more toxic than Cu^{2+} ions and a reduction of seed germination and root elongation was observed. In the next study 18-day-old hydroponically grown lettuce (*L. sativa*) seedlings were treated for 15 days with core–shell Cu/CuO NPs at two concentrations (10 and 20 mg L^{-1}). The results showed that Cu^{2+} ions or Cu/CuO NPs reduced water content, root length, and dry biomass of the lettuce plants. ICP-OES results revealedthat Cu/CuO NPs treatments produced significant accumulations of Cu in roots compared to the Cu^{2+} ions. In roots, all Cu treatments increased CAT activity but decreased APX activity. In addition, relative to the control, nano-Cu/CuO altered the nutritional quality of the lettuce, since the treated plants had significantly more Cu, Al, and S, but less Mn, P, Ca, and Mg (Ruttkay-Nedecky et al., 2017). In another study Atha et al. reported that copper oxide nanoparticles induced DNA damage in agricultural and grassland plants. Significant accumulation of oxidatively modified, mutagenic DNA lesions (7,8-dihydro-8-oxoguanine; 2,6-diamino-4-hydroxy-5-formamidopyrimidine; 4,6-diamino-5-formamidopyrimidine) and strong plant growth inhibition was observed for radish (*R. sativus*), perennial ryegrass (*L. perenne*), and annual ryegrass (*Lolium rigidum*) under controlled laboratory conditions.

The phytotoxicity and accumulation of copper oxide (CuO) NPs *toElsholtzia splendens* (a Cu-tolerant plant) was taken up under hydroponic conditions. The Cu content in the shoots treated with 1000 mg L^{-1} CuO NPs was much higher than those exposed to the comparable 0.5 mg L^{-1} soluble Cu and CuO bulk particles. CuO NPs-like deposits were found in the root cells and leaf cells. Cu K-edge X-ray absorption near-edge structure analysis further revealed that the accumulated Cu species existed predominantly as CuO NPs in the plant tissues. All these results suggested that CuO NPs can be absorbed by the roots and translocated to the shoots in *E. splendens*. Phytotoxicity of CuO NPs was assessed in two crop plants, maize (*Z. mays*) and rice (*O. sativa*). The results showed that seed germination was not affected by CuO NPs at any of the investigated concentrations. However, at the concentration of 2000 mg L^{-1}, the root elongation was significantly inhibited by CuO NPs (95.73% for maize and 97.28% for rice), and the shoot length of maize was reduced by 30.98%. An overview of phytotoxicity of nanoparticles based on Cu/CuO is shown in Table 35.

Table 35: Phytotoxicity of nanoparticles based on Cu/CuO

Plant	Type of nanoparticle, particle size (nm)	Particle concentration	Comment	Observed effect
Cucumis sativus	CuNPs 40 nm	10, 20 mg L^{-1}	Analysis of plants and root exudates	Decrease in root length, reduction of root biomass, bioaccumulation mainly in roots, a little in stems
Lactuca sativa	CuO NPs 5–15 nm	0.02, 0.04, 0.4, 4, 8mg L^{-1}	5-day seed germination test	Reduction of seed germination and root elongation
Lactuca sativa	Core–shell NPs Cu/CuO 13/9 nm	10, 20 mg L^{-1}	15-days treatment of hydroponically grown lettuce	Reduction of water content, root length, and dry biomass of the plant, alteration of the nutritional quality of lettuce
Raphanus sativus, Lolium perenne, Lolium rigidum	CuO NPs 6 nm, CuO bulk particles 200 nm	10, 100, 500, 1000 mg L^{-1}	The seeds were allowed to germinate for 6 days	Oxidative damage to plant DNA, inhibition of seedling growth (root and shoot growth)
Elsholtzia splendens	CuO NPs 34–52 nm, CuO bulk particles 1000 nm	100, 200, 500, 1000, 2000 mg L^{-1}	Cu—tolerant plant, the seeds were allowed to germinate for 5 days, hydroponic experiments	No effect on seed germination, reduction of root length, accumulation of CuO NPs in root and leaf cells
Zea mays, Oryza sativa	CuO NPs 40–80 nm	500, 1000, 2000 mg L^{-1}	Seed germination was investigated	CuO NPs inhibited root elongation at 2000 mg L^{-1} (95.73% for maize and 97.28% for rice) of two crop plants and reduced shoot length of maize by 30.98%

8.5.3 Nanomaterials based on Iron

Iron nanoparticles (INPs) represent a new generation of environmental remediation technologies that could provide cost-effective solutions to some of the most challenging environmental issues. Because of large surface areas and high surface reactivity, INPs have found their main application in remediation. This method is relatively cheap and uses both free (soil application, where INPs can penetrate ground water) and into matrix fixed nanoparticles (cleaning water or air) (Ruttkay-Nedecky et al., 2017). In the greatest extent INPs are used to decompose substances such as chlorinated hydrocarbons (e.g. trichloroethylene), organochlorine pesticides, and polychlorinated biphenyls. Besides decomposing, INPs can be further applied to bind, for example, to a significant pollutant, arsenic ions. Materials composed of nanoscaled iron particles exhibit high absorbency and a second advantage is their response to external magnetic fields by which they can be, even with bound arsenic compounds, removed. The mentioned procedure can also be used for other metals such as mercury or lead.

In agriculture, Fe_2O_3 NPs may be used instead of Fe fertilizers. The effects of the Fe_2O_3 NPs and a chelated-Fe fertilizer (ethylenediaminetetraacetic acid-Fe; EDTA-Fe) on the growth and development of peanut (*Arachis hypogaea*), a crop that is very sensitive to Fe deficiency, were studied in a pot experiment(Ruttkay-Nedecky et al., 2017). The results showed that Fe_2O_3 NPs increased root length, plant height, biomass, and soil plant analysis development (SPAD) values of peanut plants. The Fe_2O_3 NPs promoted the growth of peanuts by regulating phytohormone contents and antioxidant enzyme activity. The Fe contents in peanut plants with Fe_2O_3 NPs and EDTA-Fe treatments were higher than the control group. The next study was conducted to examine the effect FeNPs (prepared by reduction with a gum kondagogu) on the growth of a mung bean (*V. radiata*). The radical length and biomass was increased in seeds exposed to FeNPs in comparison with the ions. In the following study, the uptake of iron oxide (Fe_2O_3) nanoparticles by spinach (*Spinacea oleracea*) via hydroponics was demonstrated and its effects on the growth rate and productivity of the spinach plant were examined. The experimental studies such as plant growth (stem and root length) and biomass analysis revealed a dose and time dependent increase due to the uptake of Fe_2O_3. An overview of the positive effects of INPs and iron oxide nanoparticles (IONPs) on plants is shown in Table 32.

Table 32: The observed positive effects of iron/iron oxide NPs on plants

Plant	Particle size (nm)	Particle concentration	Comment	Observed effect
Arachis hypogaea	γ-Fe2O3, 20 nm	2, 10, 50, 250, 1000 mg kg^{-1} of soil	Fe_2O_3 NPs were applied into soil and compared with a chelated-Fe fertilizer	Fe_2O_3 NPs increased root length, plant height, biomass, and SPAD values of peanut plants. Fe O NPs adsorbed onto sandy soil and improved the availability of Fe to the plants. Fe_2O_3 NPs can replace traditional Fe fertilizers in the cultivation of peanut plants
Vigna radiata	FeNPs 2–6 nm +0.2% gum, +0.4% gum	1 mM Fe^{2+}ions	Natural biopolymer gum kondagogu as reducing and capping agent was used	The radical length and biomass was increased in seeds exposed to Fe NPs in comparison to Fe^{2+} ions. The α-amylase activity was increased in the seeds exposed to Fe NPs
Spinacea oleracea	α-Fe2O3 50 nm	100, 150, 200 mg kg^{-1} of soil	Experiments were performed in a solid hydroponic medium consisting of sawdust and coco peat and adequate amounts of water	Positive effects on spinach plant due to uptake of Fe_2O_3 nanoparticles such as increase in stem and root lengths, biomass production and magnetic properties were observed
Lactuca sativa	Core–shell NPs Fe/ Fe O 13/9 nm	10, 20 mg L^{-1}	15-days treatment of hydroponically grown lettuce	The nano-Fe/Fe O at 10 and 20 mg L^{-1} and $FeSO_4 \cdot 7H_2O$ at 10 mg L^{-1} did not affect lettuce growth and chlorophyll content
Vigna unguiculata	<100 nm	25, 500 mg L^{-1}	The elements were applied 56 and 72 days after sowing over the leaves, and data was collected after day 85	Iron had significant effect on yield, leaf Fe content, stem Mg content, plasma membrane stability, and chlorophyll content, probably as a result of more efficient photosynthesis

Glycine max	γ-Fe_2O_3 (IONPS) and citrate coated IONPs 6 nm	500, 1000 mg L^{-1}	Plant physiological performance was assessed after foliar and soil IONPs fertilization	IONPs produced a significant positive effect on root elongation. IONPs-Cit significantly enhanced photosynthetic parameters when sprayed foliarly. More pronounced positive effects of IONPs via foliar application than by soil treatment was observed
Glycine max	Superparamagnetic iron oxide NPs (SPIONs) 8–12 nm	200, 400, 1000 and 2000 mg L^{-1}	Seed germination of soybean exposed to SPIONs was investigated	SPIONs, which were entered and translocated in the soybean, increased chlorophyll levels, with no trace of toxicity

Most of the available studies on the phytotoxicity of iron nanomaterials have focused mainly on their advantages while relatively few have examined the mechanisms of phytotoxicity, uptake, translocation, and bioaccumulation (Ruttkay-Nedecky et al., 2017). If water uptake by the roots could be affected by the adsorption of γ-Fe2O3 nanoparticles (50, 100 mg L^{-1}) was studied on the root surface of *Helianthus annuus.* The main effect was related to the reduction of the root hydraulic conductivity (Lo) and the nutrient uptake. The concentrations of the macronutrients Ca, K, Mg, and S in the shoot were reduced relative to the control plants, which resulted in lower contents of chlorophyll pigments.

NPs were not only less toxic than their ionic counterparts but also significantly stimulated the growth of root elongation by 12–26% in a concentration range (5–20 mg L^{-1}). Conversely, at 50 mg L^{-1} root elongation was inhibited by 12%. (Ruttkay- Nedecky et al., 2017).

The activities of antioxidant enzymes were significantly higher as a factor of low γ-Fe_2O_3 NP treatment concentration and significantly lower at high NPs concentrations, but only among transgenic plants. An overview of phytotoxicity of nanoparticles based on iron oxides is shown in Table 36.

In phytotoxicity studies on nanoparticles, the most urgent need is to build a connection between the characteristics of nanoparticles (surface area, particle size, surface tension) and phytotoxicity. Equally important is the need to understand the role of plant species and composition of the nanoparticles phytotoxicity. Finally, most studies on phytotoxicity and uptake of nanoparticles plants were performed in a hydroponic setup. Hydroponic studies do not reflect the interaction of nanoparticles with soil and soil microorganisms. It can be concluded that the nanoparticles prepared from essential heavy metals and their oxides have proven to be suitable for use in the agriculture. The least phytotoxic of these appear to be nanoparticles made of iron oxides and manganese oxides.

Table 36: Phytotoxicity of nanoparticles based on iron oxides

Plant	Type of nanoparticle, particle size (nm)	Particle concentration	Comment	Observed effect
Helianthus annuus	γ-Fe_2O_3 20–100 nm	50, 100 mg L^{-1}	Effect on the root functionality was investigated	The treatment with 50 mg L^{-1} FeNPs significantly reduced the root hydraulic conductivity (Lo) by up to 26% at 100 mg L^{-1} FeNPs, but it had no effect on plant biomass production, on shoot or root elongation, and it did not induce oxidative stress in the plant
Solanum lycopersicum	nZVI < 50nm γ-Fe O 20–100 nm	50, 100 mg L^{-1}	Effect on the root functionality was investigated	The treatment with 100 mg L^{-1} of Fe O NPs inhibited 40% of the root hydraulic conductivity (Lo), with nZVI no effect on Lo was observed
Lactuca sativa	FeOx NPs <50 nm	1, 5, 10, 20, 50 mg L^{-1}	A 5-day seed germination test was used to test how different FeOx NPs affected the plant growth in comparison with their respective ionic or solid counterparts	FeOx NPs significantly enhanced root elongation of lettuce seedlings by 12%–26%, indicating that FeOx NPs could be used as an Fe fertilizer as well at low application rates (5–20 mg L^{-1}). At a concentration of 50 mg L^{-1}, FeOx NPs decreased root elongation of lettuce seedlings by 20%
Oryza sativa	γ-Fe2O3 7–13 nm	2, 20, 200 mg L–1	A 7-day seed germination test was used	Root phytohormone inhibition abscisic acid (ABA) and indole-3-acetic acid (IAA) was positively correlated with Fe2O3 NPs concentrations, indicating that Fe2O3 had a significant influence on the production of these hormones

8.5.4 Nanomaterials based on Manganese

Manganese (Mn) is an essential micronutrient for growth regulation and the development of plants. It plays a pivotal role in oxygenic photosynthesis both directly and indirectly. The major drawbacks associated with Mn deficiency are plant nutritional disorders. To circumvent this nutritional disorder of plants, nanoparticle mediated crop management has of late found potential applications (Ruttkay-Nedecky et al., 2017). The effect of manganese nanoparticles (MnNPs) on nitrogen uptake in mung bean plants (*V. radiata*) was investigated. MnNPs were modulated to affect the assimilatory process by enhancing the net flux of nitrogen assimilation through NR-NiR and GS-GOGAT pathways.

MnNP-treated chloroplasts showed greater photophosphorylation, oxygen evolution with respect to control, and MnSO4-treated chloroplasts as determined by biophysical and biochemical techniques. Positive effects on root and shoot elongation was observed. MnNP-treated plants did not trigger oxidative stress.the effects of laboratory-prepared MnOx NPs was investigated on the germination of lettuce (*L. sativa*) seeds in a water medium(Ruttkay-Nedecky et al., 2017). MnOx NPs only slightly reduced the germination percentage from 84% (control) to 63% even at a high concentration of 50 mg L^{-1} and was not significantly different from that of the control. Furthermore, MnOx NPs specifically improved the growth of lettuce seedlings by enhancing root elongation. An overview of the positive effects of MnNPs and manganese oxide nanoparticles (MnOx NPs) in plants is shown in Table 33.

Table 33: The observed positive effects of MnNPs on plants

Plant	Particle size (nm)	Particle concentration	Comment	Observed effect
Vigna radiata	MnNPs	50, 100, 500, 1000 mg L^{-1}	Leaf and root enzyme extract was analyzed for use as nanofertilizer	Nitrogen uptake, its assimilation, and metabolism was increased after MnNPs soil application
Vigna radiata	MnNPs	50, 100, 500, 1000 mg L^{-1}	Leaf and root enzyme extract was analyzed. Chloroplasts from leaves were isolated and analyzed for their level of photophosphorylation and oxygen evolution	MnNP-treated chloroplasts showed greater photophosphorylation, oxygen evolution with respect to control and $MnSO_4$-treated chloroplasts. Positive effects on root and shoot elongation was observed

Plant	Particle size (nm)	Particle concentration	Comment	Observed effect
Lactuca sativa	MnOx NPs 5–15 nm	0.25, 0.5, 5, 10, mg L^{-1}	Overall, the data suggests that MnOx NPs can be used as an Mn fertilizer (better than their soluble or bulk solid counterparts) for crop growth improvement	MnOx NPs specifically improved the growth of lettuce seedlings by enhancing root elongation

A good understanding of the mechanisms of the nanoparticle phytotoxicity is important for the targeted application of nanoparticles. Essential metal NPs can cause phytotoxicity via the dissolution and release of higher concentration of essential ions such as Zn^{2+} and Cu^{2+} or the production of excess reactive oxygen species (ROS) through redox cycling and the Fe^{2+}-mediated Fenton reaction (Ruttkay-Nedecky et al., 2017).

8.5.5 Metal oxide nanoparticles

All metal oxide nanoparticles influence the growth and development of plants. They generally enhance or reduce seed germination, shoot/root growth, biomass production and physiological and biochemical activities (Siddiqi et al., 2017). Some plant species have not shown any physiological change, although significant variations in antioxidant enzyme activity and upregulation of heat shock protein have been observed. Plants have evolved antioxidant defence mechanism which involves enzymatic as well as non-enzymatic components to prevent oxidative damage and enhance plant resistance to metal oxide toxicity. The exact mechanism of plant defence against the toxicity of nanomaterials has not been fully explored. The absorption and translocation of metal oxide nanoparticles in different parts of the plant depend on their bioavailability, concentration, solubility and exposure time. Further, these nanoparticles may reach other organisms, animals and humans through food chain which may alter the entire biodiversity.

The plants grown in presence of nanoparticles may absorb and translocate them in different tissues. The increased activities of enzymes and that of HSP70 are due to the induced reaction against CeO_2 nanoparticle(Siddiqi et al., 2017). CeO_2 nanoparticle- plant-root interaction and translocation in hydroponically grown wheat and pumpkin plant for 8 days (17–100 nm) at 100 mg/L in the absence and presence of fulvic acid and gum arabic have been reported. The above plants did not exhibit any reduction in root growth. Plant response against some metal oxide nanoparticles is summarized in Table 37.

Table 37: Plant response to some metal oxide nanoparticles

Nanoparticle	Size (nm)	Plant	Concentration	Plant response
CeO_2	7	Soybean	0, 500, 1000, 2000, 4000 mg/L	Genotoxicity recoded at 2000 and 4000 mg/L concentration; a new band in the roots' RAPD profile was observed
	7	Alfalfa, corn, cucumber, tomato	0, 500, 1000, 2000, 4000 mg/L	In corn, tomato and cucumber seed germination was reduced at 2000 mg/L; promoted root elongation for corn and cucumber; reduced root growth of alfalfa and tomato
	8.0 ± 1.0	Coriander	125 mg/kg	Increased shoot, root length and biomass; increased ascorbate peroxidase activity in roots and catalase activity in shoots
	<8.0 ± 1.0	Rice	0, 62.50, 125, 250, 500 mg/L	Reduced H_2O_2 generation in shoots and roots; increased electrolyte leakage and lipid peroxidation in shoots
	8 ± 1	Corn	0, 400, 800 mg/kg	No impact on chlorophyll contents and gas exchange
	8 ± 1	Barley	0, 125, 250, 500 mg/kg	Increased the plant height, chlorophyll contents, biomass, reduced spike production; increased Ca, K, Zn, Mg, Cu, Al, Fe, P and S in grains
	8 ± 1	Wheat	0, 100, 400 mg/kg	Changes in microstructure of leaf cells, swollen chloroplasts, squeezed nuclei, bent and loosely arranged thylakoids; decreased chlorophyll contents and exhibits variation in protein content
	10 ± 3.2	*Bacillus thuringiensis* transgenic cotton	0, 100, 500 mg/L	Swollen and destructed chloroplasts, reduced Zn, Mg, Fe and P levels in xylem sap of cotton
	50–105	Tomato	20 mg/kg	Increased Ca, K, Mg, P in roots; Ca, Mg in stems; decreased Na contents stems; K, Na, P and S in leaves
	8 ± 1	Wheat	0, 125, 250, 500 mg/L	Changes the amounts S and Mn in grains, amino acid composition and linolenic acid contents
ZnO	8	Soybean	0, 500, 1000, 2000, 4000 mg/L	No change in germination; genotoxicity recoded at 4000 mg/L concentration; a new band in the roots' RAPD profile was observed
	10	Soybean	0–500 mg/kg	Reduced Fe at all treatments; Mg and K were decreased at 5 mg Zn/kg treatment

Nanoparticle	Size (nm)	Plant	Concentration	Plant response
	<50	Soybean	500 mg/kg	Reduced roots and shoots; had smaller surface area and volume; no seed formation
	20	Radish, rape, ryegrass, lettuce, corn, cucumber	2000 mg/L	Reduced root growth and elongation
	<10	Zucchini	1000 mg/L	Reduced biomass (78–90%)
	10	Cucumber	400–800 mg/kg	No impact on growth, gas exchange or chlorophyll contents
	90	Corn	800 mg/kg	Reduced growth and inhibition of arbuscular mycorrhizal
	10	Alfalfa	250, 500, 750 mg/kg	Reduced root biomass (80%)
	44.4	Arabidopsis	400, 2000, 4000 mg/L	Reduced seed germination, root elongation and number of leaves
	<100	Arabidopsis	100 mg/L	Reduced biomass (81.4%), seed germination, 660 up-regulated genes and 826 down-regulated genes
	<50	Garden pea	100–1000 mg/L	No impact on germination; root length, stem length, leaf surface area, transpiration and root nodulation was affected
	1.2–6.8	Clusterbean	10 mg/L	Increased biomass (27.1%), shoot length, root length, root area, chlorophyll content and total soluble leaf protein
	25	Tomato	0–1000 mg/L	Plant height was increased (24%) at 250 mg ZnO/Kg; increased root length in foliar sprayed plants with 250 mg ZnO/L; concentrations above 250 mg ZnO/kg affected root length in both methods of application
	<100	Wheat	50 mg/kg	Reduced biomass
	<100	Wheat	500 mg/kg	Reduced root growth, increased reactive oxygen species production
CuO	<50	Arabidopsis	0, 0.5, 1, 2, 5, 10, 20, 50, 100 mg/L	Reduced biomass, root growth retardation, increased reactive oxygen species production
	<50	Indian mustard	0, 20, 50, 100, 200, 400, 500 mg/L	Reduced shoot and root growth

Nanoparticle	Size (nm)	Plant	Concentration	Plant response
	10–50	Mung bean	0, 20, 50, 100, 200, 500 mg/L	Reduced biomass and root length at all concentrations; reduced chlorophyll content above 100 mg/L; no changes in carotenoid content; increased H_2O_2 and lipid peroxidation; increased reactive oxygen species production with increase in concentration; modulations in gene expression
	<50	Wheat	500 mg/kg	Inhibition in root and shoot growth; produced oxidative stress possibly due to Cu released from nanoparticles, Cu bioaccumulates
	<50	Squash	0, 100, 500 mg/L	Reduced growth and transpiration (60–70%)
	<100	Radish, grasses	10, 100, 500, 1000 mg/L	Growth inhibition; DNA damage
TiO_2/ inorganic bentonite clay	30/1–60	Maize	300, 1000 mg/L	Inhibited hydraulic conductivity, leaf growth and transpiration
Activated carbon-based TiO_2	30–50	Tomato	0–500 mg/L	Improved germination, reduced germination time
	30–50	Mung bean	0–500 mg/L	Improved germination, reduced germination time
TiO_2	–	Soybean	0, 0.01, 0.03, 0.05%	Increased height (0.05%) and dry weight
	<100	Wheat	~91 mg/kg	Reduced biomass, nanoparticles found mostly stick on surface of roots
	<25	Tobacco	0, 0.1, 1, 2.5, 5%	Reduced biomass, inhibited germination and root length; upregulation of alcohol dehydrogenase and ascorbate peroxidase
	4–6	Spinach	0.25%	Improved growth; increased glutamate dehydrogenase, glutamine synthetase and glutamic piruvic transaminase activity
	7–40	Chickpea	2–10 mg/kg	Reduction in electrolyte leakage and malondialdehyde content at 5 mg/kg treatment

Nanoparticle	Size (nm)	Plant	Concentration	Plant response
	6.22	*Ulmus elongata*	0.1–0.4%	Increased Cu accumulation in leaves; reduced net photosynthetic rate; increased carbohydrates and lipids
	27 ± 4	Cucumber	0, 250, 500, 750 mg/kg	Enhanced catalase; activity in leaves; enhanced P and K availability in fruit
Fe3O4	20	Pumpkin	500 mg/L	No toxic effect; nanoparticles are translocated throughout the plant tissues, detected in stem and leaves, accumulated on the surface of root
	7	Cucumber, lettuce	62, 100, 116 mg/L	Low to zero toxicity on germination
	6	Lettuce, radish, cucumber, spinach, tomato, leek, peppers	0.67 mg/mL	Reduced germination
	25	Ryegrass, pumpkin	30, 100 and 500 mg/L	Increased root elongation; no uptake; block of aquaporins; oxidative stress
Fe2O3	20–100	Sunflower	50, 100 mg/L	No uptake and translocation; reduced root hydraulic conductivity
	22–67	Arabidopsis	4 mg/kg	Reduced biomass and chlorophyll contents
	–	Soybean	0, 0.25, 0.5, 0.75, 1.0g/L	Increased leaf and pod dry weight; increased grain yield (48%)
	246	Lettuce, radish, cucumber	1000 mg/L	Found to be adsorb on the surface of seed
Al2O3	13	Maize, cucumber, carrots, cabbage	2000 mg/L	Reduced root growth
	–	Corn	2000 mg/L	Reduced root length
	–	Tobacco	0, 0.1, 0.5, 1%	Increased root length, biomass; decreased leaf count; the seedlings significantly decreased; 1% Al_2O_3 exposure has shown extreme increase in microRNA expression

Barley treated with CeO_2 nanoparticles (250 mg/kg) enhanced methionine, aspartic acid, threonine, arginine and linolenic acid contents in the grain. It is clear that the moderate concentration (125–250 mg/kg) of CeO_2 nanoparticles is highly beneficial while the higher doses (500 mg/kg) are toxic to barley(Siddiqi et al., 2017).

The phytotoxicity of CuO nanoparticles in a 1:1 mixture of CuO and ZnO nanoparticles in plants colonized by *Pseudomonas chlororaphis* in a sand matrix has been investigated by Dimkpa et al. (2015). Bean root growth was inhibited and shoot was elongated by CuO nanoparticles. The CuO nanoparticles were found to release copper to the shoot where maximum copper loading was noted with the minimum dose of 100 mg/kg CuO nanoparticles and lower level of copper with higher doses of 250 and 500 mg/kg. It has been observed that anatase TiO_2 nanoparticle of diameter lower than 140 nm is accumulated in wheat roots. They are translocated to leaves if their diameter is smaller than 36 nm even if it is below the detection limit. TiO_2 nanoparticles neither dissolve nor their crystal phase undergoes any change during translocation (Siddiqi et al., 2017). The exposure of wheat plant to 14 and 22 nm TiO_2 nanoparticles causes enhancement in root elongation without influencing seed germination, vegetative growth, photosynthesis or redox reaction.

Nanomaterials of the same metal with different structural motifs have different effects on plants, although there is no distinction in their chemical behaviour. Of the three crystalline structures of TiO_2 nanoparticles (anatase, rutile and brookite), anatase exhibits the highest catalytic activity (Yin et al., 2005) and can inhibit the growth of many microbes such as algae, fungi and bacteria. It promotes carotene and chlorophyll synthesis in cucumber. The smaller nanoparticles penetrate into plant cells or microbial cells but larger ones adhere to cell wall causing agglomeration. Effect of Al_2O_3 on the growth and development of *Nicotiana tabacum* and role of microRNA has been investigated (Burklew et al., 2012). It has been observed that as the concentration of Al_2O_3 nanoparticles increases from 0.1 to 1.0%, the root length, average biomass and the leaf count of each tobacco seedlings decrease.

8.6 Disruptive mechanisms of nanoparticles

Nanoparticles are used in agriculture to enhance seed germination and plant growth and to protect crops from biotic stresses, such as insects, fungi and bacteria. However, given the unique properties of nanoparticles (NPs), living organisms in the ecosystem could suffer from oxidative stress induced by NPs (Nhan et al., 2015).

8.6.1 Disruption of the uptake of nutrient elements

A common conclusion from recent studies was that the nanotoxicity might depend on either metal speciation or the plant species. Exposure to 2000 mg·L^{-1} NPs from the lanthanide series, such as lanthanum oxide (La_2O_3) gadolinium oxide (Gd_2O_3) and ytterbium oxide (Yb_2O_3), greatly inhibits root growth in radish, tomato, rape, lettuce, wheat, cabbage and cucumber plants.

A concentration of 200 mg·L^{-1} of nano-sized Zn (zinc, 35 nm) and ZnO (zinc oxide, 20 nm) inhibited germination in ryegrass and corn, respectively. Cerium oxide (CeO) NPs disrupted the uptake of nutrient elements in Bt-transgenic cotton compared with its parental conventional cotton plant. Thus, the phytotoxic mechanism of nanomaterials must bc thoroughly understood before application in fields.

A study by Nhan et al., (2015) was focused on determining the phytotoxic mechanism of CeO_2 nanoparticles (NPs): destroying chloroplasts and vascular bundles and altering absorption of nutrients on conventional and Bt-transgenic cottons. Experiments were designed with three concentrations of CeO_2 NPs including: 0, 100 and 500 mg·L^{-1} and each treatment with three replications. Results indicate that absorbed CeO_2 nanoparticles significantly reduced the Zn, Mg, Fe and P levels in xylem sap compared with the control group and decreased indole-3-acetic acid (IAA) and abscisic acid (ABA) concentrations in the roots of conventional cotton.

Transgenic plants offer numerous advantages to producers, consumers and the environment. Indeed, engineered plants are highly tolerant to pesticides and herbicides, demonstrate resistance towards pathogens caused by microorganisms and exhibit increased production yields and enhanced biofuel content, which are considered novel and sustainable energy sources Bt-transgenic cotton contains one foreign gene derived from the bacterium Bacillus thuringiensis that encodes the Cry toxin, which effectively kills bollworms and therefore controls yield losses.

The various concentrations of CeO_2 NPs were tested on the growth of cucumber, alfalfa, tomato and corn seedlings and reported that the maximum growth (140%) was recorded at a concentration of 2000 mg/l and minimum growth (50%) at a concentration of 500 mg/l compared to control treatments. (Raul Cuero, 2016). A few the uptake, storage and differential toxicity of CeO_2 Exposure of CeO_2 NPs as aerosol or suspension on corn leaves revealed that about 50 µg of cerium per gram of leaves was either absorbed or incorporated into maize leaves but there is no evidence of translocation of cerium into the newly developed leaves. These results clearly indicated that the absorption and the translocation ability of CeO_2 NPs are dependent on the plant species.

An ultra-small TiO_2 complex with Alizarin red S nano-conjugates were directly up taken and translocated the nanomaterials into the seedlings of *A. thaliana*. It was concluded that root of the experimental plants released mucilaginous substances, which may form a pectin hydrogel capsule around the root and playing a significant role in the inhibition or facilitation of the TiO_2 complexes with Alizarin red S (Raul Cuero, 2016). The experiment conducted on rye-grass confirmed that the adsorption and accumulation of the ZnO NPs to the root surfaces.

The Al uptake was observed in red kidney beans when exposed Al NPs and their results showed that the Al concentration in the red kidney beans were not significantly different compared to untreated control treatments. The uptake of Ag NPs was studied in zucchini compared to bulk (control) treatments and stated that the silver concentration in shoots was higher, when the plants were exposed Ag NPs compared to bulk and concluded that release of silver ions from Ag NPs is accounted for an increase in the concentration of silver in shoots.

The presence of engineered nanoparticles (ENMs) could interfere with nutrient acquisition and transport, subsequently causing nutritional imbalance and compromised plant health. For example, Ag NPs exposure to wild type Crambe abyssinica caused significant decrease in Fe accumulation in the plant tissues (Chuanxin Ma et al., 2015).

In spite of the observed negative impacts of certain ENMs on nutrient levels in plants, other ENMs have potential as nano-enabled fertilizers and/or growth stimulants for the purposes of enhanced crop performance. Numerous examples of positive effects of ENMs on plant germination, biomass, and crop yield have been reported (Liu and Lal, 2015). Not only could amendments of nano-scale nutrient elements promote plant growth, but also, due to their antimicrobial property, these NPs could inhibit activities of both bacterial and fungal pathogens, slowing disease course and increasing yield.

8.6.2 Inhibition of plant growth and development

Usage of nano-technological products has spread nanoparticles into the environment during its manufacture, usage or disposal through water, air, and soil. The unintentional spread of nanoparticles hasaccelerated a robust debate among the scientific community and have drawn attention towards the potential impact of nanotoxicity in the environment (Syeda Hafsa Ali et al., 2019). The physiochemical properties and reactivity of nanoparticles differ not only between nanoparticle with different chemical composition but also among identical nanoparticles with different shape, size, surface properties, and crystalline structure. Phytotoxicity occurs as nanoparticles are uptaken, translocated, or localized in a plant. Consequently, affecting germination rate, physiological processes that disrupt cell integrity at the molecular level and causes detrimental effects on plant growth and development of various crops. This toxicity produces Reactive Oxygen Species (ROS) that causes DNA damage and lipid peroxidation. However, these free radicals are actively scavenged by antioxidant enzymes to repair the damage and help the plant to withstand the stress. However, the continuous increase of nanoparticles can permanently damage the plant thereby reducing its ability to withstand. Therefore, cost-effective strategies are required to overcome the risk of nanoparticles.

The use of lanthanum oxide nanoparticles (La_2O_3 NPs) in life products have increased dramatically in the past decades, which are inevitable released into natural environment. In a study, Yinglin Liu et al., (2018) determined the phytotoxicity of La_2O_3 NPs to maize grown in one-fourth strength Hoagland solution. The toxicity of colloidal solutions of metal nanoparticles (Cu, Mn) and their oxides, obtained in the absence and presence of a stabilizer, was examined and compared with the use of the standard test system of *Allium cepa* L. (YevheniiaKonotop et al., 2019). It was established that solutions of stabilized metal nanoparticles were at given concentration toxic to *Allium cepa* L. according to the integral index of roots growth, however, were not cytotoxic.

Effects of four metal oxide nanoparticles, aluminum oxide (nAl_2O_3),silicon dioxide ($nSiO_2$), magnetite (nFe_3O_4), and zinc oxide (nZnO), on the development of *Arabidopsis thaliana* (Mouse ear cress) were studied (Chang Woo Lee et al., 2010). Overall, this study shows that direct exposure to nanoparticles significantly contributed to phytotoxicity and underscores the need for eco-responsible disposal of wastes and sludge containing metal oxide nanoparticles.

8.6.3 Disruption at the morphological level

Silver nanoparticles (AgNPs) are one of the most commonly used nanomaterials in various fields, especially in the agricultural sector. Plants are the basic component of the ecosystem and the most important source of food for mankind; therefore, understanding the impacts of AgNPs on plant growth and development is crucial for the evaluation of potential environmental risks on food safety and human health imposed by AgNPs. A review by An Yan and Zhong Chen (2019) summarized uptake, translocation, and accumulation of AgNPs in plants, and exemplifies the phytotoxicity of AgNPs on plants at morphological, physiological, cellular, and molecular levels. It also focused on the current understanding of phytotoxicity mechanisms via which AgNPs exert their toxicity on plants. In addition, the tolerance mechanisms underlying survival strategy that plants adopt to cope with adverse effects of AgNPs were discussed.

After exposure to AgNPs, significant changes in the morphology of plants were observed. Growth potential, seed germination, biomass, and leaf surface area are the commonly used parameters for assessing the phytotoxicity of AgNPs in plants. It was demonstrated that AgNP exposure could inhibit seed germination and root growth, and reduce biomass and leaf area. Jiang et al. found that AgNPs significantly decreased plant biomass, inhibited shoot growth, and resulted in root abscission in *Spirodela polyrrhiza.* Exposure to higher concentrations (from 5 to 20 mg/L) of AgNPs resulted in reduction of the biomass in Arabidopsis. AgNPs reduced the length of shoots and roots of wheat in a dose-dependent manner in wheat. AgNPs significantly reduced root elongation, and shoot and root fresh weights in rice. AgNPs (>100 mg/L) inhibited seed germination and reduced biomass in zucchini (*Cucurbita pepo*). Similar results regarding the toxicity on seed germination, biomass accumulation, and root and shoot growth by AgNPs were reported in other studies involving various plant species, including *Arabidopsis, Brassica* nigra, Lemna, *Phaseolus radiatus* and *Sorghum bicolor*, *Lolium multiflorum*, rice, wheat, *Lupinus termis* L., and so on. A summary of compiled descriptions of the effects of AgNPs in plants is shown in table 38.

Table 38: Summary of studies on phytotoxicity of silver nanoparticles (AgNPs) in plants

Size (Diameter in nm)	Concentration	Species	Impacts
25–70; 7.5–25.0	10, 20, 40, 50 ppm	Wheat	Caused various types of chromosomal aberrations
5–10	0, 0.1, 0.3, 0.5 mg/L	Lupinus termis	Reduction in shoot and root elongation, shoot and root fresh weights, total chlorophyll, and total protein contents; Decreased sugar contents and caused significant foliar proline accumulation;
37.4 ± 13.4 (AgNP- Ba); 29.0 ± 6.0 (AgNP-PVPb); 21.5 ± 4.2 (AgNP-Citrate)	5, 10 µg/mL	Bryophyte (Physcomitrella patens)	Inhibited the growth of the protonema; Changed the thylakoid and chlorophyll contents
79.0 ± 8.0	0.05–2 mg/L	Lemna minor	Caused decays on growth rate and fronds per colony; Induced oxidative stress
3.1–8.7	20, 200, 2000 mg/kg	Wheat	Caused lower biomass, shorter plant height, and lower grain weight;\ Decreases in the contents of micronutrients (Fe, Cu, and Zn); Decreased the contents of arginine and histidine
17.2 ± 0.3	0, 1, 10 and 30 mg/L (soybean); 0, 0.1, 0.5, 1 mg/L (rice)	Soybean; Rice	Significantly reduced plant biomass; Increased the malondialdehyde and H2O2 contents of leaves
12.9 ± 9.1	0.01, 0.05, 0.1, 0.5, 1 mg/L	*Capsicum annuum*	Decreased plant height and biomass; Causued a significant increase in total cytokinins in the leaves
20	1000, 3000 µM	*Pisum sativum*	Declined growth, photosynthetic pigments, and chlorophyll fluorescence; Inhibited activities of glutathione reductase (GR) and dehydroascorbate reductase (DHAR).
61.2 ± 33.9 (AgNP2- 5, 50, 75, 100 µM Citrate); 9.4 ± 1.3 (AgNP-PVP); 5.6 ± 2.1 (AgNP-CTAcB)		*Allium cepa*	Caused oxidative stress; Led to strong reduction of the root growth
20	5, 10, 20 mg/L	*Allium cepa*	Induced various chromosomal aberrations in both mitotic and meiotic cells

Size (Diameter in nm)	Concentration	Species	Impacts
200–800	1 mg/L	*Trigonella foenum-graecum*	Enhancement in plant growth and diosgenin synthesis
20	10–150 mg/L	*Arabidopsis thaliana*	Inhibited root gravitropism; Reduced auxin accumulation in root tips; Down regulated expression of auxin receptor-related genes
47	1, 3 mM	Mustard (*Brassica* sp.)	Declined growth of Brassica seedlings; Induced oxidative stress
35, 73	10 mg/L	Cucumber (*Cucumis sativus*); Wheat	Teduced growth; Upregulation of genes involved in the ethylene signalling pathway
5–50	800 μg/kg	*Vicia faba* L.	Declined germination; Decreased shoot and root length; Retarded the process of nodulation; Caused early senescence of root nodules
20	0, 2, 10, 20 mg/L	Potato	Total reactive oxygen species (ROS) and superoxide anions were increased; Significant increases in the activities ofsuperoxide dismutase (SOD), catalase, ascorbate peroxidase, and glutathione reductase (GR); Higher ion leakage and cell death
35–40	50, 75 mg/L	*Triticum aestivum; Vigna sinensis; Brassica juncea*	50 ppm treatment promoted growth and increased root nodulation in Improved shoot parameters at 75 ppm in Brassica
2	0, 125, 250, 500 mg/L	*Raphanus sativus*	Water content was reduced; Root and shoot lengths were reduced at 500 mg/L treatment; Significantly less Ca, Mg, B, Cu, Mn, and Zn
41	100–5000 mg/L	*Arabidopsis thaliana*	Reduced root length, leaf expansion and photosynthetic efficiency; Induced ROS accumulation; Induced Ca2+ in cytoplasm, inhibited plasma membrane K+ efflux and Ca2+ influx currents
100	50–100 μM	*Arabidopsis thaliana*	Accumulated more amino acids

Size (Diameter in nm)	Concentration	Species	Impacts
10	1, 2, 5, 8, 10 mg/L	*Wolffia globosa*	Caused oxidative damage, higher malondialdehyde (MDA) content and an upregulation of SOD activity; Decreased contents of chlorophyll a, carotenoids and soluble protein
20	5 mg/L	*Arabidopsis thaliana*	111 genes were unique in AgNPs and enriched in three biological functions: response to fungal infection, anion transport, and cell wall/ plasma membrane related.
10, 20, 40, 80	0.2 μg/L	*Arabidopsis thaliana*	Inhibition of root hair development; Repressed transcriptional responses to microbial pathogens, resulting in increased bacterial colonization
60–100 (Ag2S-NPs); 15–20 (AgNPs)	0–20mg/L (Ag2S- NPs); 0–1.6 mg/L (Ag-NPs)	Cowpea (*Vigna unguiculata* L. Walp.); Wheat (*Triticum aestivum* L.)	Ag2S-NPs reduced growth by up to 52%; Ag accumulated as Ag2S in the root and shoot tissues after exposed to Ag2S-NPs
20	75–300 μg/L	*Arabidopsis thaliana*	Prolonged vegetative and shortened reproductive growth; Decreased germination rates of offspring
6, 20	0.5, 5, 10 mg/L	*Spirodela polyrhiza*	Dose dependent increase in levels of ROS, SOD, peroxidase, and the antioxidant glutathione content; Chloroplasts accumulated starch grains and had reduced intergranal thylakoids.
20	0, 0.2, 0.5, 1 mg/L	*Arabidopsis thaliana*	Significantly reduced total chlorophyll and increased anthocyanin content; Increased lipid peroxidation; a dose-dependent increase in ROS production; Significant upregulated the expression of sulfur assimilation, glutathione biosynthesis, glutathione S-transferase, and glutathione reductase genes one reductase genes

Size (Diameter in nm)	Concentration	Species	Impacts
20	0, 0.2, 0.5, 1 mg/L	*Oryza sativa* L.	Significant reduction in root elongation, shoot and root fresh weights, total chlorophyll, and carotenoids contents; Caused significant increase in H2O2 formation and lipid peroxidation in shoots and roots, increased foliar proline accumulation, and decreased sugar contents; Caused a dose dependent increase in ROS generation; Changes in mitochondrial membrane potential in the roots of
8, 45, 47	2–100 μM	*Arabidopsis thaliana*	Induced root growth promotion (RGP) and Cu/Zn superoxide dismutase (CSD2) accumulation; Inhibited ethylene (ET) perception and could interfere with ET biosynthesis
20,30–60, 70–120, 150	0.1, 1, 10, 100, mg/L	*Oryza sativa* L.	Seed germination and seedling growth were decreased
20, 40, 80	67–535 μg/L	*Arabidopsis thaliana*	Inhibited seedling root elongation; AgNPs were apoplastically transported in the cell wall and aggregated at plasmodesmata
20	5–25 mg/L	*Arabidopsis thaliana*	Upregulation of stress related genes, downregulation of pathogen and hormonal stimuli genes; Oxidative stress
10	0.2, 0.5, 3 mg/L	*Arabidopsis thaliana*	Root growth inhibition; Disrupted the thylakoid membrane structure and decreased chlorophyll content; Caused alteration of transcription of antioxidant and aquaporin related genes
11 ± 0.7 (Citrate)	0.05, 0.1, 1, 18.3, 36.7, 73.4 mg/L	*Zea mays Brassica oleracea*	Cell erosion in maize root apical meristem
18.34	0.30–60 mg/L	*Oryza sativa* L.	Damage the cell morphology and its structural features; Total soluble carbohydrates significantly declined; Caused production of the ROS and local root tissue death
10	0.5, 1.5, 2.5, 3.5, 5 mg/kg	*Triticum aestivum*	Teduced the length of shoots and roots; Caused oxidative stress in roots; Induced expression of a metallothionein gene involved in detoxification

Size (Diameter in nm)	Concentration	Species	Impacts
10–15	0, 100, 1000 mg/L	Tomatoes (*Lycopersicon esculentum*)	Significant decreases in root growth; Decreased chlorophyll contents and Higher SOD activity; Less fruit productivity,
5, 10, 25	0.01–100 mg/L	*Arabidopsis thaliana*; poplars	Stimulatory effect on root elongation, fresh weight, and evapotranspiration at sublethal concentrations; Toxicity increased with decreasing AgNPs size
20 (AgNP-PVP) 6 (AgNP-GA[d])	1, 10, 40 mg/L	Eleven species of wetland plants	40 mg/L AgNPs-GA exposure significantly reduced the germination rate of three species and enhanced the germination rate of one species.
<100	250, 750 mg/L	*Cucurbita pepo*	Reduction in plant biomass and transpiration
5–25	0, 5, 10, 20, 40 mg/L	*Phaseolus radiates; Sorghum bicolor*	Inhibition of plant growth
<100	0, 100, 500 mg/L	Cucurbita pepo	Decreased rate of transpiration
60	12.5, 25, 50, mg/L	10Vicia faba	Increased the number of chromosomal aberrations and micronuclei, and decreased the mitotic index
190–1100	0, 25, 50, 100, 400 mg/L	*Brassica juncea*	Increase in root length and increase in vigor index; Improved photosynthetic quantum efficiency and higher chlorophyll contents; Induced the activities of antioxidant enzymes, resulting in reduced reactive oxygen species levels
20, 100	5 μg/L	*Lemna minor* L.	Inhibition of plant growth
25	50, 500, 1000 mg/L	*Oryza sativa*	Broken the cell wall and damaged the vacuoles of root cells 24
55	0–80 mg/L	*Allium cepa*	Induced cell death and DNA damage through generation of ROS
<100	100 ppm	*Allium cepa*	Disturbed mitosis, reduction in mitotic index, declined metaphase, sticky chromosome, disintegration and breakdown of cell wall
20	100 mg/L	*Green asparagus*	Higher ascorbic acid and total chlorophyll contents

[a]AgNP-B: AgNPs without surface coating; [b] AgNP-PVP: polyvinylpyrrolidone-coated silver nanoparticles; [c]AgNP-CTAB: cetyltrimethylammonium bromide-coated silver nanoparticles; [d]AgNP-GA: gum arabic-coated silver nanoparticles

8.6.4 Disruption at the physiological level

Phytotoxicity of AgNPs to plants at the physiological level is predicted by reduction of chlorophyll and nutrient uptake, decline of transpiration rate, and alteration of hormone (An Yan and Zhong Chen , 2019). AgNPs can disrupt the synthesis of chlorophyll in leaves and, thus, affect the photosynthetic system of the plants. Qian et al. showed that AgNPs could accumulate in Arabidopsis leaves, further disrupt the thylakoid membrane structure, and decrease chlorophyll content, leading to the inhibition of plant growth. Nair and Chung reported that, after exposure to AgNPs for one week, total chlorophyll and carotenoids contents were decreased significantly in rice (*Oryza sativa* L.) seedlings. AgNPs could accumulate in mustard (*Brassica* sp.) seedlings and caused severe inhibition in photosynthesis. A recent study showed that AgNP exposure changed the thylakoid in Physcomitrella patens, and AgNPs decreased the chlorophyll b content and disturbed the balance of some essential elements in the leafy gametophytes (Liang et al., 2018). In *Lupinus termis* L. seedlings, after exposure to AgNPs for ten days, the shoot and root elongation and fresh weights, total chlorophyll, and total protein contents were significantly reduced. In *Cucurbita pepo*, the rate of transpiration was remarkably reduced after AgNP exposure.

In addition, AgNPs can affect the fluidity and permeability of the membrane and, consequently, influence water and nutrient uptake. AgNP exposure on radish sprout caused a decrease in water content in a dose-dependent manner; the nutrient content (Ca, Mg, B, Cu, Mn, and Zn) was also significantly reduced, suggesting that AgNPs may affect plant growth by changing water and nutrient content. It was reported that AgNPs also affect plant hormones. The root gravitropism of Arabidopsis seedling was inhibited by exposure to AgNPs in a dose-dependent manner. Further analysis indicated that AgNPs reduced auxin accumulation, while gene expression analysis suggested that auxin receptor-related genes were downregulated upon AgNP exposure.

The effect of nanoparticles on the morphological and physiological changes in plants, their uptake and accumulation in the various biological systems and their consequences on the cellular and genetic toxicity in plants were investigated. The nanoparticles synthesized by biological means are advantageous compared to chemical or microbial synthesis because of their nontoxic, rapid and cost-effective method (Raul Cuero, 2016). The nanoparticles synthesized from the plant materials are referred as biogenic or phytosynthesized nanoparticles. The phytosynthesized nanoparticles have distinct assets *viz,* greater surface area and stability, uniform shape and size compared to physically or chemically synthesized methods. The phytosynthesised nanomaterials are the major concern for the researchers, whose are still trying to optimize the nanomaterials to be used as nano-fertilizers for agriculturally important crop plants in respect to the guarantee of their safe use.

Various nanomaterials used in modern agricultural practices have both positive and negative impacts on the cultivated crop plants. The effect of very low concentration of alumina nanoparticles was obsereved on the carrot, cabbage, corn, cucumber and soybean. They concluded that alumina nanoparticles have shown phytotoxic effect on all the tested plants. The exposure to higher concentrations of aluminium, alumina, multi-walled carbon nanotube, zinc and zinc oxide nanoparticles on root development and seed germination has a phytotoxic effect on the tested plant species. However, the effect of the phytotoxicity of functionalized and non-functionalized single-walled carbon nanotubes on the root growth and development of cabbage, carrot, cucumber, lettuce, onion, and tomato has been investigated. They concluded that amongst all the various tested crop plants the phytotoxicity vary greatly in between the functionalized and non-functionalized single-walled carbon nanotubes, which may also lead into an adverse effect of the root growth and development.

The effects of nanoparticles on the physiological changes of plants are summarized in table and the effects of nanoparticles on the physiological changes of plants are tabularized in table 39.

Table 39: Effects of nanoparticles on the morphological changes in plants

Types of NPs	Concentration	Morphological changes
TiO_2 ZnO	10g/kg 5g/kg	The biomass of wheat are reduced by the application of both nanoparticles under field conditions. The TiO_2 NPs were adherent to cell walls of plants and retained in the soil for long periods. However, the ZnO NPs dissolved in the soil, and enhance the zinc uptake by plants.
Fe3O4, TiO_2, Carbon	100-5000 μg/ml	Cucumber seed germination was inhibited by all the nanoparticles but less inhibition of seed germination was recorded for carbon nanoparticles compared to other tested nanoparticles.
Ag	10-100 ppb	It is evident from the results that the various concentrations of Ag NPs have not significantly affected the seed germination of *B. monnieri.*
ZnO	10-50 mg/l	ZnO NPs caused a concentration dependent inhibition in root length of garlic. The total percentage of abnormal cells increased with the increase of ZnO NPs concentration and the prolongation of treatment time.
Ag	40 mg/l	Broken epidermis and root cap were observed in Lolium multiforum.
Fe_3O_4, TiO_2	50-500 mg/l	Root morphology showed that initiation in the formation of root hairs on tomato in hydroponic condition.

Ag	10 mg/l	Seedlings treated with different concentrations of PVP coated Ag NPs or $AgNO_3$ for 5 days showed an increase in root length of *Eruca sativa.*
ZnO	10 mg/l	Foliar spray of ZnO NPs significantly increased the plant biomass, shoot length, root length and root area on cluster bean.
MWCNT	20-2000 mg/l	Results showed that root and shoot length of red spinach, lettuce, and cucumbers were significantly reduced by the exposure of multi-walled carbon nanotubes only at higher concentrations (1000 mg/l and 2000 mg/l) after 15 days inoculation. Similar toxic effects occurred regarding cell death and electrolyte leakage. Moreover, red spinach and lettuce were most sensitive to MWNTs, followed by rice and cucumber.
Ag	10 mg/l	Results showed that the exposure of wheat seedlings to Ag NPs adversely affected the seedling growth and induced morphological modifications in root tip cells.
Effect of nanoparticles on the physiological changes in plants:		
TiO_2	10-2000 mg/l	Exposure of *Lemna minor* to TiO_2 NPs increased the activities of various enzymes (POD, SOD, and CAT) below concentration of 200 mg/l because of eliminating accumulated reactive oxygen species in plant cells. The
Al_2O_3	10-1000 mg/l	The activity of SOD and CAT were increased with the treatments of Al_2O_3 NPs at a concentration of 200 and 500 mg/l.
ZnO	10 mg/l	Treatment of cluster bean with foliar sprays of ZnO NPs caused a significant increase in chlorophyll content (276.2 %), total soluble leaf protein (27.1%), acid phosphatase (73.5%), alkaline phosphatase (48.7%), and phytase (72.4%) over control.
Fe_2O_3	20 mg/l	Increase in root activity, activity of catalase, peroxidase, superoxide dismutase, chlorophyll, malondialdehyde contents, ferric reductase activity, the root apoplastic iron content were recorded by the translocation of the significant amount of Fe_2O_3 NPs suspended in a liquid medium to various tissues of plants.
SNP ZnO	500-4,000 ppm	Application of various concentrations of both sulfur and zinc oxide nanoparticles in significantly increase the total lipids, proteins, amino acids, thiol and chlorophyll contents compared to untreated control but no significance difference has been observed among the treatments with various concentrations for both nanoparticles.
TiO_2	200 mg/l	Application of TiO_2 NPs had a noticeable effect on chlorophyll a and b and carotenoid contents on Mentha.

As roots are the major target of AgNPs toxic effects, several authors investigated root morphology of various plant species treated with AgNPs of different sizes, concentrations and coatings.In several studies the impact of AgNPs on the seedlings roots was analysed in the organelles at the ultrastructural level by transmission electron microscopy (TEM). Major changes were observed in plastids, vacuole and endoplasmic reticulum (ER) of root cap, meristems and differentiating cells (Mirta Tkalec et al., 2019). The number of the amyloplasts and the size of the smooth endoplasmic reticulum (sER) in the root cap columella cells of rocket seedlings were reduced after both AgNP-PVP- and $AgNO_3$-treatments . However, only AgNPs induced morphological modifications of sER in the region of cell elongation and differentiation of root samples; in particular, an extensive swelling was observed. In wheat seedlings exposed to AgNP-PVP , no starch grains were noticed in the root cap columella amyloplasts and in the plastids of meristematic cells. However, there are some investigations showing that observed morphological effects of AgNP exposure can be a result of the immediate uptake of AgNPs by root cells as discussed in part uptake and accumulation.

Beside changes in root cells, AgNPs also induce morphological modifications in the leaves of seedlings of various plant species. Morphological modifications and ultrastructural changes of roots, stems and leaves were also observed in the fully developed stage of various plant species. In A. cepa, the highest applied AgNP concentration of 100 ppm induced complete disintegration of cell walls for most of the cells. the root tip cells of tobacco adult plants were highly vacuolated after 7 days of exposure to both 100μM AgNP-citrate and $AgNO_3$. Furthermore, TEM study showed that after both types of treatments only nuclei could be observed within the root cells, due to large vacuoles; however, exposure to $AgNO_3$ also resulted with partly destroyed root cells, while nuclei were highly damaged (Cvjetko et al., 2017).

8.6.5 Cytotoxicity and genotoxicity

AgNPs can also cause toxicity at the cellular and molecular level in plants. Many studies showed that the inhibition of plant growth after AgNP exposure is accompanied with alteration of cell structure and cell division. *Lolium multiflorum* seedlings failed to develop root hair, and the cortical cells were highly vacuolated and collapsed, while the epidermis and root cap were also damaged after exposure to 40 mg/L AgNPs. AgNPs could reduce the size of the vacuole and lead to the reduction of cell turgidity and cell size in maize and cabbage. After AgNPs enter the cell of *Brassica campestris*; vacuoles and cell wall integrity were damaged, and other organelles might also be affected. AgNPs with a concentration of to 60 μg/mL could penetrate the cell wall, and damage the cell morphology and its structure in rice (Mirzajani et al., 2013). AgNP exposure in *Allium cepa* significantly decreased the mitotic index and impaired cell division, resulting in chromatin bridge, stickiness, disturbed metaphase, multiple chromosomal breaks, and cell disintegration. AgNP treatment significantly increased the chromosomal aberrations and micronuclei, and decreased the mitotic index (MI) in root tip cells of broad bean (*Vicia faba* L.), suggesting that cell cycle and mitosis in root tip cells was disrupted by AgNPs. The root tip cells of wheat could readily internalize the AgNPs. After AgNP internalization, the root tip cells exhibited various types of chromosomal aberrations, such as incorrect orientation at metaphase, chromosomal breakage, spindle dysfunction, fragmentation, unequal separation, and distributed and lagging chromosomes, which seriously interfered with cell function.

The toxicity of multi-walled carbon nanotube suspension of rice cells were studied and reported that multi-walled carbon nanotubes at low concentrations cause cell death and at high concentration cause cell mortality (Raul Cuero, 2016). The concentration and size of single-wall carbon nanotubes cause the toxicity in the rice and *Arabidopsis* protoplast cells and the breakdown of genetic material in *Arabidopsis* cell directly correlated with the genetic toxicity of the single walled carbon nanotubes. The cellular toxicity of CeO_2 and ZnO NPs was studied on soybean seedlings and detected the genetic toxicity of both NPs by RAPD assay. It was elaborated that the interaction of genetic material with the zinc ions is the main reason behind the increase in the toxicity level. TiO_2 NPs were genotoxic and cytotoxic due to the generation of superoxide radicals during lipid peroxidation.

Titanium dioxide nanoparticles (TiO_2NPs) are widely used and may impact the environment. Thus, this study used a high concentration of TiO_2NP (1000 mg/L) to verify the defense mechanisms triggered by a plant system, an indicator of toxicity (Ronaldo dos Santos Filho et al., 2019). Nucleolar alterations and plant defense responses (*ie*, increased lytic vacuoles, oil bodies and NP phase change) were observed in meristematic cells exposed to TiO_2NP at 1000 mg/L. In summary, TiO_2NPs can damage the genetic material of plants; however, plants displayed defense mechanisms against the deleterious effects of these NPs. In addition, *A. cepa* was found to be a suitable test system to evaluate the cyto- and genotoxicity of NPs.

Studies were conducted on the toxicity of graphene on different cell types and genetic material. Various studies include cytotoxicity factor of GFNs (Singh, 2016). In biological microenvironment, biomolecules bind onto nanoparticles forming corona and endow nanoparticles a new biological identity. The so-called "protein corona" formed in serum medium decreased the cellular uptake of GO, thus significantly mitigating its potential cytotoxicity. Molecular dynamic simulations also revealed that the adsorbed bovine serum albumin in effect weakened the interaction between the phospholipids and graphene surface due to reduction of the available surface area and an unfavorable steric effect, thus significantly reducing the graphene penetration and lipid bilayer damaging.

8.6.6 Induced Oxidative Stress

Induction of oxidative stress studies published so far suggest that exposure to metal NPs can induce increased generation of reactive oxygen species (ROS) (Yang et al., 2017) which can react with proteins, lipids and DNA molecules, resulting in a number of metabolic disorders, destruction of cell membranes and, in consequence, cell death. Production of superoxide radicals O_2 and hydrogen peroxide (H_2O_2), common ROS was studied and observed their increased levels in *A. cepa* roots after 2 h-exposure to AgNPs (37 nm, 0, 5, 10, 20, 40, 80mg L1).

The study in which adult tobacco plants were treated with 61 nm AgNP-citrate (25, 50, 75, 100 and 500 µM) no changes in ROS accumulation were recorded in comparison to the control, thus suggesting that the response to AgNP-imposed stress might also be dependent on the plant developmental stage. Still, in the majority of presented studies. However, at this point it cannot be stated with certainty if AgNPs induce generation of ROS directly or indirectly through Ag+ ions (Mirta Tkalec et al., 2019). When generation of ROS exceeds the capacity of the cellular antioxidant defence system, oxidative stress occurs, which may lead to inactivation and damage of membrane lipids, proteins and DNA molecule.

In several studies, comparison between plants exposed to AgNPs with those exposed to the same concentration of $AgNO_3$ was performed. In some of them $AgNO_3$-exposure imposed greater stress than AgNP-exposure, which confirms that Ag+ is generally more toxic for plants than Ag nanoparticles and that the AgNP-toxicity probably does not depend solely on the Ag+ ions dissociated from nanoparticles. However in some cases the results are not unambiguous. Results obtained on oxidative damage to lipids, proteins, and DNA molecule as well as on changes in the activity of plant antioxidant enzymes in plants treated with AgNPs suggest that oxidative stress could have an important role in the phytotoxicity of AgNPs.

The main mechanism underlying the phytotoxicity of AgNPs is the production of excess reactive oxygen species (ROS) induced by AgNPs, resulting in oxidative stress in plant cells (An Yan and Zhong Chen, 2019). A number of studies demonstrated that ROS production is significantly elevated in plants after exposure to AgNPs.The production of excess ROS induced by AgNP exposure can subsequently lead to oxidative stress, cause peroxidation of polyunsaturated fatty acids (known as lipid peroxidation), and damage the cell membrane permeability and alter cell structure, directly damaging protein and DNA, resulting in potential cell death and growth inhibition in plants.

AgNP-P(phyto-synthesizedfromsilvernitrateAgNO3)orAgNP-S(commercial AgNPs from Sigma–Aldrich) application in *Allium cepa* significantly increased the generation of superoxide (O2•–) and H_2O_2; they also induced cell death to different extents in a dose-dependent fashion, following an order of AgNP-S > AgNP-P at doses ≥20 mg/L. Moreover, AgNP-P significantly decreased the mitotic index. Comet assay suggested that DNA damage was significantly enhanced after AgNP-P and AgNP-S treatments in a dose-dependent manner, whereby AgNP-S (threshold dose ≥ 10 mg/L) is more genotoxic than AgNP-P (threshold dose ≥ 20 mg/L) (Panda et al., 2011). AgNPs could accumulate in Arabidopsis leaves and change the transcription of antioxidant and aquaporin genes, suggesting that AgNPs can change the balance between oxidant and antioxidant systems. Lipid peroxidation increased significantly after exposure to 0.2, 0.5, and 1 mg/L AgNPs in Arabidopsis.

8.6.7 Silver-specific toxicity

It was shown that AgNPs can leach ionic silver (Ag^+) into the surroundings through the oxidation of zero-valent Ag (An Yan and Zhong Chen, 2019). During AgNP uptake and translocation, Ag^+ is released from AgNPs, resulting in oxidative stress through the generation of ROS and disturbing cell function, causing phytotoxicity by binding to cell components and modifying their activities. The ion release kinetics of AgNPs was analyzed in the pollen culture medium, and found that AgNPs rapidly dissolved into ions and reached a maximum of 11.8 wt.% ion release. The released Ag^+ caused a fivefold increase in H_2O_2 production over controls; moreover, the released Ag^+ damaged pollen membranes and inhibited germination to a greater extent than the AgNPs themselves, suggesting that Ag^+ may excert its impacts mostly through chemical or physicochemical interactions with nucleic acids to induce DNA damage. A gene expression study by microarray in Arabidopsis compared gene expression profiles between AgNP and silver ion (Ag^+) treatments and found a significant overlap between genes differentially expressed in the two treatments, suggesting a similarity between plants' responses to AgNPs and Ag^+. Actually, when AgNPs oxidize in water, they can make bonds with anions and transform into the characteristics of heavy metals, which is more hazardous. It was demonstrated that the conversion of AgNPs to a complex of anion or heavy metal could cause toxic effects on various living organisms.

Ag^+ can also affect photosynthesis through competitive substitution of Cu^+ in plastocyanin (Pc). Pc is a soluble copper-binding protein found in the thylakoid lumen of the chloroplast.

Although the phytotoxicity of AgNPs was associated with the impact of dissolved Ag^+ on plants, the phytotoxicity effect could not be explained solely by the activity of the released Ag^+ ions. In some case, AgNPs can be even more toxic than free Ag^+ ions even at the same concentrations of Ag^+ (An Yan and Zhong Chen, 2019). In another study, AgNP exposure to *Cucurbita pepo* caused more reduction in biomass and transpiration when compared with bulk Ag. These studies suggest that free Ag^+ ions contribute only partially to the phytotoxicity of AgNPs, while the intrinsic properties of AgNPs are critical for the phytotoxicity of AgNPs.Indeed, the physical interactions between AgNPs and plant cell-transport pathways can influence the phytotoxicity of AgNPs. Uptake of AgNPs into plant tissue may cause inhibition of apoplastic trafficking by clogging of pores and barriers in the cell wall or the nano-sized plasmodesmata, thereby effectively inhibit the apoplastic flow of water and nutrients.

A number of studies suggested that the effect and phytotoxicity of AgNPs are closely associated with the nature of the interactions between plants and AgNPs, which are determined by the intrinsic properties of AgNPs. These physical and chemical properties of AgNPs, including size, shape, exposure concentration, surface coating, Ag form, and aggregation state, greatly influence the effect of AgNPs on different aspects of plant morphology, physiology, and biochemistry.Among these properties, the size of AgNPs is critical for the phytotoxicity of AgNPs. The smaller-sized AgNPs have a larger surface area to mass ratio, which allows better interference with cell membrane function by directly reacting with the membrane. Meanwhile, a higher proportion of the atoms of the particle on the surface can affect the interfacial reactivity and the ability to pass through physiological barriers. It was shown that smaller AgNPs could accumulate to higher levels in plants and be more toxic than their bulk particles. Moreover, smaller-sized AgNPs had a greater impact on root browning. Smaller AgNPs (5 and 10 nm) accumulated to higher levles in poplar tissues than the larger 25-nm AgNPs when applied within the particle subinhibitory concentration range; both Arabidopsis and poplar showed susceptibility to the toxic effects of AgNPs, and this susceptibility increased with decreasing AgNP size.

Although these studies demonstrated that smaller AgNPs cause more phytotoxicity than larger AgNPs, this correlation between AgNP size and phytotoxicity is not always true for every AgNP exposure scenario. Further analysis found that 20-nm AgNPs were trapped in the roots rather than transported to the leaves, thereby causing less phytotoxicity on seedling growth than 150-nm AgNPs (Thuesombat et al., 2014)

The phytotoxicity of AgNPs is positively correlated with the concentration of AgNPs during exposure. AgNPs can only cause negative effects on plants when applied with a concentration above a certain threshold. AgNPs were unable to change cell morphology or structure of rice root when present in low concentrations (30 µg/mL), whereas, with an increased concentration of 60 µg/mL, AgNPs not only penetrated the cell wall, but also destroyed the cell morphology and the structural features.

The increasing release of silver (Ag) nanoparticles (NPs) into the environment highlights the importance of exploring the interactions between Ag NPs and plants, which are the basis of most ecosystems (Di Cui, et al., 2014). In this study, two plant species, cucumber and wheat were exposed to Ag NPs and Ag+ (added as AgNO3) at the germination and vegetative growth stages. Above certain concentrations, Ag NPs and Ag+ were toxic to the two plants. However, stimulatory effects were observed on root elongation for the cucumbers that were exposed to Ag NPs at concentrations below 200 mg/L, and Ag+ at concentrations below 5 mg/L. The two plants were more susceptible to the toxicity of Ag NPs at the vegetative growth stage than the germination stage. Cysteine, a strong chelating ligand of Ag+, could completely eliminate the effects of Ag NPs on cucumber and wheat, suggesting that the phytotoxicity of Ag NPs was possibly caused by the release of Ag+.

8.6.8 Delayed germination

Graphene family nanomaterials (GFNs) have experienced significant development in recent years and have been used in many fields. Despite the benefits, they bring to society and the economy, their potential for posing environmental and health risks should also be considered. The increasing release of GFNs into the ecosystem is one of the key environmental problems that humanity is facing. Although most of these nanoparticles are present at low concentrations, many of them raise considerable toxicological concerns, particularly regarding their accumulation in plants and the consequent toxicity introduced at the bottom of the food chain. Qinghai Wang et al., (2019) reviewed the recent progress in the study of toxicity caused by GFNs to plants, as well as its influencing factors. The phytotoxicity of GFNs is mainly manifested as a delay in seed germination and a severe loss of morphology of the plant seedling (Fig.52).

The potential mechanisms of phytotoxicity were summarized. Key mechanisms include physical effects (shading effect, mechanical injury, and physical blockage) and physiological and biochemical effects (enhancement of reactive oxygen species (ROS), generation and inhibition of antioxidant enzyme activities, metabolic disturbances, and inhibition of photosynthesis by reducing the biosynthesis of chlorophyll). In the future, it is necessary to establish a widely accepted phytotoxicity evaluation system for safe manufacture and use of GFNs.

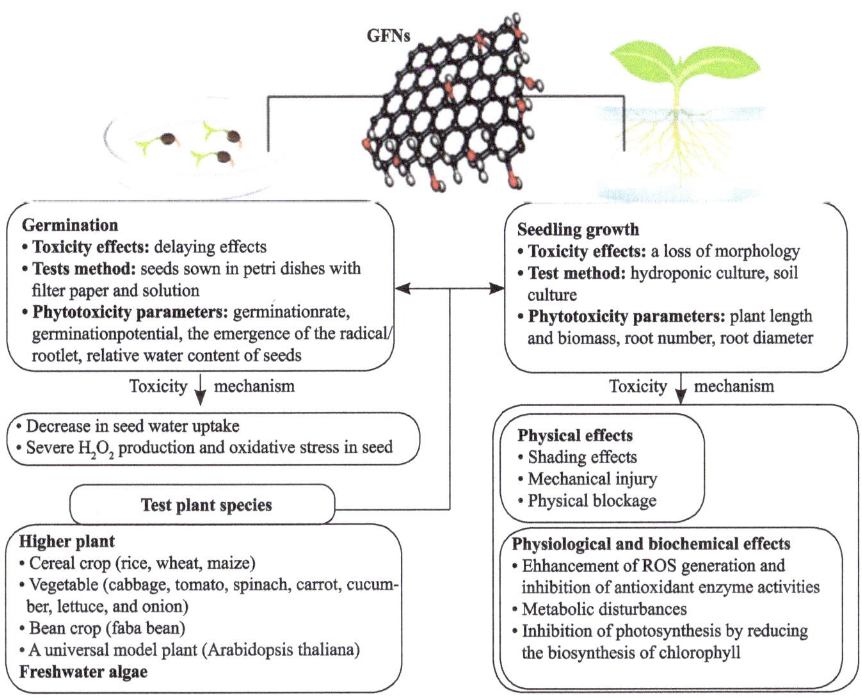

Fig. 52: Schematic diagram of the disruption /phytotoxicity and mechanisms of GFNs

The potential effects of GFNs depend on many factors, such as their physicochemical, the exposure concentration and time, and the plant species, which deserve further attention. Phytotoxicity is an important consideration to understand the potential environmental impacts of manufactured nanoparticles, which ultimately accumulated in soil. In a study, Sarvendra Kumaret al., (2016) assessed the effects of two metal oxide nanoparticles (NPs), nano-ZnO, nano-Fe2O3 and one carbon based (fullerene) NPs, on seed germination, root and shoot elongation of wheat, rice, green gram and cucumber plants. These parameters were quantified following exposure to each type of NPs at different concentrations, from 10 mg/L to 1000 mg/L. Inhibition of root growth varied greatly among NPs and plant types. Among NPs, nano-ZnO found to be most phytotoxic, followed by nano-Fe_2O_3 and fullerene.

Reduction in root and shoot growth found to be a more sensitive parameter than germination percentage. Results also indicated that NPs at low concentrations can be useful for seedling growth.

The biological effects of nanoparticles (NPs) based on seed germination and root elongation tests were assessed (Wu et al., 2012). Lettuce, radish and cucumber seeds were incubated with various metal oxide NPs (CuO, NiO, TiO_2, Fe_2O_3, Co_3O_4), of which only CuO and NiO showed deleterious impacts on the activities of all three seeds. Metal oxide NPs tended to adsorb on seed surfaces in the aqueous medium and released metal ions near the seeds.

Therefore, metal oxide NPs had higher phytotoxicity than free metal ions of the equivalent concentrations. Further, the surface area-to-volume ratio of seeds may also affect NPs phytotoxicity, whereby small seeds (i.e., lettuce) were the most sensitive to CuO and NiO NPs in our experiments.

Studies were conducted on the phytotoxicity of TiO_2 nanoparticles (NPs) compared with bulk TiO_2 particles B on: germination parameters, vegetative traits, roots viability, Biomass of seedling and photosynthetic pigments of Amber 33 variety of Rice (*Oryza sativa*) *in vitro* (Raghad et al., 2015).

There were inductions in seed germination percentage using different concentrations of B particles, while it decreased when the seeds exposed to NPs at concentrations (10, 1 and 0.01) mg/ml compared with control. There were differences between the effects of NPs compared with B particles on germination percentage. Other concentrations reduced the germination rate. All concentrations of NPs reduced mean germination time.

The toxicity of various metal oxide ENPs (Al_2O_3, CuO, Fe_3O_4, MnO, TiO_2, and ZnO), including nanowires were tested together with their bulk counter particles and soluble metal salts, on germinating seeds of *Sinapis alba* L. Fe_3O_4, TiO_2, MnO_2, and Al_2O_3 ENPs did not negatively affected seed germination at any tested concentrations (Landa et al., 2016). However, CuO and ZnO ENPs showed a dose-dependent inhibition of germination. Metal ions were more toxic than metal oxide particles at corresponding concentrations. The highest toxicity was exhibited by Cu, followed by Zn, Fe, Al, and Mn ions. A comparison of ENPswith bulk materials did not reveal significantly higher ENP toxicity. Similarly, nanowires showed effects similar to other nanoparticles and bulk materials. The results indicated that the nanosize or shape of particles did not play a crucial role, whereas metal ions released into cultivation media and accumulated in seedlings contributed significantly to the phytotoxicity of metal oxides.

Effects of five types of nanoparticles (multi-walled carbon nanotube, aluminum, alumina, zinc, and zinc oxide) on seed germination and root growth of six higher plant species (radish, rape, ryegrass, lettuce, corn, and cucumber) were investigated (Daohui Lin, 2017). Seed germination was not affected except for the inhibition of nanoscale zinc (nano-Zn) on ryegrass and zinc oxide (nano- ZnO) on corn at 2000 mg/L. Inhibition on root growth varied greatly among nanoparticles and plants. Suspensions of 2000 mg/L nano-Zn or nano-ZnO practically terminated root elongation of the tested plant species. Fifty percent inhibitory concentrations (IC50) of nano-Zn and nano-ZnO were estimated to be near 50mg/L for radish, and about 20mg/L for rape and ryegrass. The inhibition occurred during the seed incubation process rather than seed soaking stage. These results are significant in terms of use and disposal of engineered nanoparticles.

Because of the growing trend of ZnO nanoparticles (nZnO) which is one of the most widely used nanoparticles being released into the environment, it has attracted the attention for more studies to be done on the effects of this nanoparticle on organisms (Ehsan Borzouyan Dastjerdi et al., 2016). This study was carried out to investigate the phytotoxicity effect of nZnO on groundnut seedlings in Murashige and Skoog (MS) medium. The results of this study showed that radicle and plumule length of groundnut seedlings were affected by nZnO exposure, in a way that length of radicles in 50 mg.L-1 nZnO and higher concentrations was significantly lower than that of control treatment and the shortest plumule length was observed in 2000 mg.L-1 nZnO concentration treatment. Both the radicle and plumule wet weight were also decreased as the nanoparticle concentration was increased. However, despite the decreasing in radicle and plumule dry weight with increasing in nZnO concentration, this increase was not significant.

Potential environmental impacts of engineered nanoparticles (ENPs) can be understood taking into consideration phytotoxicity. Libralato et al.,(2016) reported on the effects of ionic (FeCl3), micro- and nano-sized zerovalent iron (nZVI) about the development of three macrophytes, *Lepidium sativum, Sinapis alba* and *Sorghum saccharatum*. Four toxicity indicators (seed germination, seedling elongation, germination index and biomass) were assessed following exposure to each iron concentration interval: 1.29–1570 mg/L ($FeCl_3$), 1.71–10.78 mg/L (micro-sized iron) and 4.81–33,560 mg/L (nano-iron). Exposure effects were also observed by optical and transmission electron microscopy.

Results showed that no significant phytotoxicity effects could be detected for both micro- and nano-sized zerovalent irons, including field nanoremediation concentrations. Biostimulation effects such as an increased seedling length and biomass production were detected at the highest exposure concentrations. Ionic iron showed slight toxicity effects only at 1570 mg/L and, therefore, no median effect concentrations were determined.

The impact of AgNPs on seed germination and growth, morphology and ultrastructure of plants depends on various AgNP characteristic such as size, chemical composition, surface structure and oxidative dissolution (Mirta Tkaleca et al., 2019). Their influence also seems to depend on the plant species and growth stage of plant, the type of substrate (soil or different nutrient media) the concentrations of nanoparticles involved and the manner of the application (foliar or soil). After AgNP accumulation in plants, they generally start degrading the quality of plants with generally negative impact on the root growth of germinating seedlings and the fresh biomass of the plant through the reduction in root elongation and biomass. Since roots are the first target tissue to confront with AgNP solution, toxic symptoms appear more in roots rather than in shoots, but AgNPs also induce morphological modifications in the stem and leaves. The main cell targets affected by AgNP toxicity are chloroplasts, mitochondria and nucleus, although there is still an open debate is the toxicity caused by AgNPs themselves or released $Ag+$.

8.6.9 Physical effects

The main physical mechanisms for the phytotoxicity of GFNs include the shading effect, mechanical injury, and physical blockage. Both shading effect and mechanical injury were observed (Zhao et al, 2017). They found that the dispersed and darkened GO reduced light transmittance, thus decreasing the available light required to support plant growth, resulting in approximately 16% of growth inhibition; and more interestingly, direct penetration into algal cells by graphene materials was discovered for the first time. The physical blockage is closely related to the size of GFNs. If nanoparticle diameter is larger than the diameter of root cell wall pores, particles will accumulate at the root surfaces and form surface layers, thus decreasing the hydraulic conductivity and uptake of nutrients.

Gold nanoparticles (AuNPs) in particular, have attracted significant attention due to their broad applications. However, most previous toxicity studies have been heavily focused on animal and human cell lines, with limited information available for the AuNPs plant interactions. A study by Ma et al., (2016) aimed to elucidate the impact of surface charge on the phytotoxicity and plant uptake of AuNPs by *Phaseolus vulgaris* (bush bean), an important food crop. The results showed that 5 mg/L of AuNPs exerted inconsequential impact on the physiological processes of plants, regardless of the surface property. However, AuNPs displayed significant surface charge-dependent effects on important biochemical parameters such as the concentrations of hydrogen peroxide in plant roots and the activities of several enzymatic antioxidants. Deposition of AuNPs on plant root surface was observed for all types of AuNPs. However, surface charge affected the internalization of AuNPs in plant root cells and their subcellular localization. Overall, the study revealed that the physical and chemical properties AuNPs play important roles in their interactions with agricultural crops and consideration of the unique properties of nanoparticles is important in assessing their food safety concerns.

8.6.10 Physiological and biochemical effects

GFNs enhance the generation of reactive oxygen species (ROS) and inhibit antioxidant enzyme activities, resulting in oxidative stress, which has been recognized as one of the most important mechanisms in growth-limiting effects on plants (Chen et al., 2017). GO also caused metabolic disturbances linked to key biological processes, such as inhibiting carbohydrate and amino acid metabolisms, and increasing the ratios of unsaturated to saturated fatty acids, changing the flux of nitrogen metabolism.It is well-known that photosynthesis is critical for plant survival and growth. Graphene significantly inhibited the biosynthesis of chlorophyll and decreased chlorophyll content in plants, leading to impaired photosynthesis and reduced growth. Glyconic acid and aconitic acid were upregulated by graphene, and these metabolites were negatively correlated with the biosynthesis of chlorophyll.

The potential phytotoxicity of silver nanoparticles (AgNPs) on tobacco plants was investigated and compared it with the effects of the same AgNO3 concentrations (Cvjetko et al., 2017). Accumulation of silver in roots and leaves was equally efficient after both AgNP and $AgNO_3$ treatment, with predominant Ag levels found in the roots.Exposure to AgNPs did not result in elevated values of oxidative stress parameters either in roots or in leaves, while $AgNO_3$ induced oxidative stress in both plant tissues.

Recent works have demonstrated that nanomaterials (NMs) can penetrate, translocate and accumulate in plants. However, studies about the effects of the NMs on plants are still limited because most investigations are carried out in the initial stage of plant development (Falco et al., 2019). The study aimed to evaluate and characterize the photochemical efficiency of photosystem II (PSII) of broad bean (*Vicia faba*) leaves when subjected to silver nanoparticles (AgNPs) with diameters of 20, 51, and 73 nm as well as to micrometer-size

Ag particles (AgBulk). The results revealed that AgNPs induce a decrease in the photochemical efficiency of photosystem II (PSII) and an increase in the non-photochemical quenching. It was concluded that AgNPs may negatively affect the photosynthesis process when accumulated in the leaves and that the NPs themselves were mainly responsible since negligible Ag+ release was detected.

The effects of various concentrations of zinc and zinc oxide nanoparticles (nZn, nZnO) were evaluated in tomato and wheat (Rayhaneh Amooaghaie et al., 2017). Results showed that at lower concentrations, nZn and nZnO augmented seed germination and growth parameters, whereas with higher concentrations, the nanoparticles reduced these traits. Zn concentrations corresponding to Zn dissolved (3–23 mg $Zn{\cdot}L^{-1}$) from nanoparticles (NPs) did not significantly affect the germination indices in either species. The results suggested that the toxicity of NPs could be due to the particle itself, or from the Zn^{2+} ions dissolved from NPs. Moreover, nanotoxicity, like other stresses, caused oxidative stress in both plants, and the differences in proline accumulation and the antioxidant enzyme activities of leaves, especially APX activity, at least in part, explained the higher sensitivity of tomato to NPs than wheat.

8.6.11 On Photosynthesis

Photosynthetic process is known to be very sensitive to stress caused by adverse environmental conditions, such as high light, drought, salinity, heat and heavy metals. Chlorophyll as a major photosynthetic pigment is responsible for capturing light energy and its content can be used as an indicator of plant photosynthetic capacity (Mirta Tkalec et al., 2019). A decrease in chlorophyll content as a parameter of AgNP phytotoxicity has been noticed in various organisms including marine and freshwater microalgae aquatic plants, crop plants as well as a model plant *Arabidopsis.*

Several authors have evaluated AgNP phytotoxicity by measuring different fluorescence parameters, most often maximum quantum yield (Fv/Fm) of photosystem II (PSII), which corresponds to the efficiency by which an absorbed photon will be trapped by PSII reaction centres and found that PSII efficiency in various algae and plants was significantly reduced after AgNP exposure (Qian et al., 2013). The change of fluorescence parameters under a lightadapted state, which can provide information about the efficiency of photochemical reactions and/or heat dissipation of chlorophyll excitation energy, has also been analysed. More recently, a measurement of rapid fluorescence induction curves with high resolution and the parameters of the so-called JIP-test have been used to assess the behaviour of various components of photosynthetic apparatus in plants and algae exposed to AgNPs.

Regarding investigations on crop plants, in seedlings of wheat and sunflower treatments with AgNPs (uncoated, 20–30 nm, 10–100 ppm) for 24 h, did not cause any significant alteration in PSII efficiency ,while in lettuce 7-day foliar exposure to uncoated 40 nm AgNPs (1, 10 or 100μg AgNPs gl FW) did not exhibit toxic effects on photosynthetic pigment content (Larue et al., 2014). Contrary, in rice seedlings after 1-week exposure to AgNPs, a significant reduction in total chlorophyll and carotenoids contents was observed .

Majority of literature data on AgNP toxicity in various species showed that AgNP-induced decrease in the chlorophyll content was in correlation with the reduced chlorophyll fluorescence yield. Several studies showed that the negative effect of AgNP-exposure on photosynthetic parameters can be correlated with the increase in ROS generation (Jiang et al., 2017). Exposure to AgNPs has a detrimental impact on the structure and function of the photosynthetic apparatus. Certain divergence in the AgNP-imposed effect observed can be attributed to differences in the type of coating, size, time of exposure or plant species used in experiment as has already been established for toxic effects on other parameters like growth, germination, etc.

8.7 Ecotoxicity of nanoparticles in soil environment

Safety research of nanoparticles has huge implications on a national economics. Test species include terrestrial plants, earthworms, and soil nematode. Soil enzyme activities were also discussed (Woo-Mi Lee,et al., 2010), who found that the results of nanotoxicity studies were affected by many factors such as physicochemical properties, size, dispersion method and test medium of nanoparticle, which should be considered when conducting toxicity researches. In particular, more researches on the effect of physico chemical properties and fate of nanoparticles on toxicity effect should be conducted consistently.

Nanotoxicity in soil environment: Aquatic toxicity studies usually focused on carbon based nanoparticles such as fullerene (C60), single-wall carbon nanotubes (SWNTs) , and multi-wall carbon nanotubes (MWNTs) while soil toxicity studies focused on metal (oxide) based nanoparticles. Tables 40 to 43 lists the major nanotoxicological studies on terrestrial plants, earthworms, soil nematodes, soil enzyme activities (Woo-Mi Lee et al., 2010).

Table 40: Nanotoxicity studies for terrestrial plants

Test NP	Test species	Note	Country
Al_2O_3	Maize, Cucumber, Soybean, Wild cabbage, Carrot	Phytotoxicity of Al nanoparticles loaded with and without phenanthrene	USA
MWNT, Al, ZnO, Zn, Al_2O_3	Rape, Radish, Ryegrass, Lettuce, Naize, Cucumber	Phytotoxicity of five species nanoparticles	China & USA
TiO_2	Spinach	N2 fixation in plant by TiO_2	China
SWNTs	Cabbage, Carrot, Cucumber, Onion, Tomato , Lettuce	Phytotoxicity of functionalized and nonfunctionalized SWNT on plant	USA
Fe_3O_4	Pumpkin, Lima bean	Uptake, translocation and accumulation of Fe_2MO_3	USA
Carbon coated- Fe	Pumpkin	NP transformaton in plant	Spain
Pd	Barely	Bioaccumlation and growth by Pb	Germany
ZnO	Rye grass	Uptake and toxicity of ZnO	China & USA
Cu	Mung bean, Wheat	Phytotoxicity of CuNP using plant agar test	Korea
Al	California red kidney bean, Rye grass	Effect of AlNP to plant	USA
Ag	Onion	Cytotoxic and genotoxic impact of plant	India
Ag, MWCNT, Cu, ZnO, Si	Pumpkin	NP and BP effect to seed germination, root elongation, and biomass	USA
Au, Ag, Fe_3O_4	Lettuce , Cucumber	Several nanoparticles toxicity effect to plant and microbial	Spain
Si, Pd, Au, Cu	Lettuce	4 metal nanoparticles toxicity effect to seed germination	USA
SWNT	Tobacco	SWNT transports in plant cells	China
MWNT, C70	Rice	Uptake and bioaccumulation of CNT in plant	USA
MWNT	Rice cell	Effect of MWNT to rice cell	Japan
MWCT	Tomato seeds	Effect of MWNT to tomato seed germination and growth rate	USA
MWNT, TiO_2, CeO_2	Wheat	Proposed investigation method of NP in plants cell	UK
TiO_2	Willow tree	Acute toxicity test of TiO_2	Denmark

Test NP	Test species	Note	Country
TiO_2	Maize	Effect of TiO_2 on root water transport	Israel
TiO_2	Arabidopsis	Uptake and distribution of TiO_2in plant	USA
Al_2O_3, SiO_2, Fe_3O_4, ZnO	Arabidopsis	4 metal NP and bulk materials toxicity to plant	USA
La_2O_3, Gd2O_3, CeO_2, Yb_2O_3	Rape, Radish, Wheat, Lettuce, Cabbage, Tomato, Cucumber	Two rare earth oxide nanoparticles effect to plant	China
ZnO, CeO_2	Soybean	Random amplified polymorphic DNA (RAPD) assay	Puerto Rico
Pb	Kiwi	Pb toxicity to kiwifruit pollen	Italy
Fe_3O_4	Lettuce, Cucmber	Physiological effect of magnetic NP on plant	USA & China

Table 41: Nanotoxicity studies for earthworms

Test NP	Test species	Note	Country
MWNT, SWNT	*Eisenia foetida*	CNT toxicity and bioaccumulation to earthworm	USA
DWNT, C_{60}	*Eisenia veneta*	CNT toxicity to earthworm	Danmark
TiO_2, ZnO	*Eisenia foetida*	Acute and chronic toxicity test of ZnO and TiO_2 nanoparticles	USA
Al2O3	*Eisenia foetida*	Effect of NP and BP of aluminum in earthworm	USA
TiO_2, ZnO	*Eisenia foetida*	Toxicity of TiO_2 and ZnO to earthworm	China

Table 42: Nanotoxicity studies for soil nematodes

Test NP	Test species	Note	Country
Pt	*Caenorhabditis elegans*	Toxicity and antioxidant on C. elegans to PtNP	Japan
Ag	*Caenorhabditis elegans*	Gene expression analysis	Korea
CeO_2, TiO_2	*Caenorhabditis elegans*	Gene expression, growth, mortality of C. elegans exposed to CeO_2 and TiO_2	Korea
ZnO	*Caenorhabditis elegans*	Toxicity of ZnO in C. elegans	USA
ZnO, Al_2O_3, TiO_2	*Caenorhabditis elegans*	Toxicity of ZnO, Al_2O_3, and TiO_2 NP and BP in C. elegans	USA

Table 43: Nanotoxicity studies for soil enzyme activities

Test NP	Enzyme eucine-	Note	Country
C_{60}	aminopeptidase, β-cellobiohydrolase, acid phosphatase, β-Glucosidase, chitinase, xylosidase	Impact of C60 on soil microbial	USA
Ag	β-Glucosidase, acid phosphatase, dehydrogenase, Urease	Effect of AgNP to soil microbial activity	German

8.7.1 Tolerance mechanisms

Phytotoxicity of AgNPs is highly associated with oxidative stress, which is caused by the production of excess amounts of ROS after AgNP exposure. To avoid the detrimental effects of ROS, a set of antioxidant defense mechanisms are activated in plant cells (An Yan and Zhong Chen, 2019). The defense mechanism involves the activities of enzymatic antioxidants such as superoxide dismutase (SOD), catalase (CAT), ascorbate peroxidase (APX), guaiacol peroxidase (GPX), dehydroascorbate reductase (DHAR), and glutathione reductase (GR). As different types of ROS have different modes of action and exhibit different effects on cellular organelles of plant cells, they can be balanced or removed by specific antioxidant enzymes. For example, there are three types of SOD in plant cells, including Fe-SOD, Mn-SOD, and Cu-Zn-SOD, and they can rapidly convert highly toxic ROS (O•–) to less toxic species (H_2O_2). CAT can convert H_2O_2 to H_2O and O_2. APX is able to convert H_2O_2 to H_2O via ascorbate oxidation into monodehydroascorbate (MDA) and dehydroascorbate (DHA), both of which can be recycled to produce more ascorbate via the catalysis of MDA reductase (MADR) and DHAR.

Upon exposure to AgNPs, activities of these enzymatic antioxidants are elevated in plant cells to protect the cells from oxidative stress. For example, Zou et al. observed obvious oxidative damage to *Wolffia globosa* when the plants were exposed to 10 mg/L AgNPs. Meanwhile, the SOD activity was increased by 2.52 times, suggesting that the ROS-scavenging mechanism was activated. Similarly, elevated SOD activity was also observed after AgNP exposure in tomatoes. Enhancement of peroxidase and catalase activity was also observed in *Bacopa monnieri* (Linn.) after AgNP treatment. Jiang et al. found that the catalase activity in cells of Spirodela polyrhiza was significantly increased. Moreover, the SOD and peroxidase activity, and the antioxidant glutathione content were increased in a dose-dependent manner after exposure to 6-nm AgNPs (Jiang, et al., 2014). In addition, Bagherzadeh Homaee and Ehsanpour examined the effects of AgNPs on potato (*Solanum tuberosum* L.) and observed that the activities of SOD, CAT, APX, and GR were all significantly increased in AgNP-treated plantlets.

Non-enzymatic antioxidants, such as anthocyanin, ascorbate, glutathione, and thiols, also contribute to the antioxidant defense mechanisms (Tripathi et al., 2017). Anthocyanin is a kind of pigment that is implicated in tolerance to various biotic or abiotic stresses, such as herbivores and pathogens, drought, cold, ultraviolet (UV) radiation, and heavy metals. Anthocyanin commonly serves as a non-enzymatic antioxidant to scavenge free radicals and chelate metals under stress conditions. It was reported that anthocyanin accumulation was significantly induced in the spherical AgNP-treated Arabidopsis seedlings and was dose-dependent. Similarly, anthocyanin accumulation was also significantly increased after exposure to higher concentrations of AgNPs in turnip. In addition, other antioxidants such as ascorbic acid, carotenoids, and proline are also implicated in antioxidant defense responses of plants to AgNPs.

At the molecular level, the expression changes of genes that are associated with the response to AgNPs may underlie the antioxidant defense mechanisms of plants in response to AgNPs. The transcription of a gene encoding metallothionein (MT), which is a cysteine-rich protein was involved in detoxification by metal ion sequestration, and found that the expression of this gene was highly induced after AgNP treatment in wheat (Dimkpa et al., 2013).

To conclude, phytotoxicity of any nanoparticle is largely influencd by its shape, size, chemical composition, and coating material composition. Sometimes, the phytotoxicity of nanoparticles may be as a result of the toxicity of substances, which were used for its preparation. Further, phytotoxicity may depend on the environment and on the physical and chemical nature of the plant species (Branislav Ruttkay Nedecky et al., 2017). Some nanoparticles are taken up by plant roots and transported to the aboveground parts of the plant through the vascular system, dependingon the composition, shape, size of nanoparticle, and anatomy of the plant. Some nanoparticles remain adhered to the plant roots.

Some metal based particles were both beneficial and toxic, but the right reason for these misleading findings lies in very high doses used together with a number of artifacts and misinterpretations especially regarding description of nanoparticles uptake (Branislav Ruttkay Nedecky et al., 2017). Despite the fact that a lot of knowledge has been acquired through many previous studies, many questions still remain unanswered such as the fate and behavior of nanoparticles in plant systems, or the role of surface area or activity of nanoparticles on phytotoxicity and the role of plant cell walls in the internalization of nanoparticles. Finally, it can be concluded that the nanoparticles prepared from essential heavy metals and their oxides have proven to be suitable for use in the agriculture. The least phytotoxic of these appear to be nanoparticles made of iron oxides and manganese oxides.

8.8 Fate and behavior of nanoparticles in the environment

The fate and behavior, metrology, transformations, bioavailability, toxicity mechanisms, and environmental impacts, with a focus on terrestrial and aquatic systems were discussed by Jamie R. Lead et al.,(2018). The findings were that despite substantial developments, critical gaps remain, in large part due to the lack of analytical, modeling, and field capabilities, and also due to the breadth and complexity of the area ; a key knowledge gap is the lack of data on environmental concentrations and dosimetry generally; substantial evidence shows that there are nanospecific effects (different from the effects of both ions and larger particles) on the environment in terms of fate, bioavailability, and toxicity, but this is not consistent for all NMs, species, and relevant processes; a paradigm is emerging that NMs are less toxic than equivalent dissolved materials but more toxic than the corresponding bulk materials; and translation of incompletely understood science into regulation and policy continues to be challenging.

Biologically mediated processes include biodegradation and biomodification, most likely microbially mediated. These are encapsulated in a conceptual process model. Although it is clear that transformations will depend on the nature of the NMs and the environmental conditions, the complexity and variability of both these factors make understanding and prediction extremely challenging (Ibrahim *et al.,* 2016). The general structure of the materialflow model is represented in figure 53. The model's principle is to track engineered NM flows throughout the entire life cycle: engineered NM production; incorporation into products; engineered NM release from products during use; transport and fate of engineered NM between and within sewage treatment plants, waste incineration plants, landfill, and recycling processes (technosphere); transfer from technosphere to air, soil, water, and sediments (ecosphere); and transport within environmental compartments. The amounts of engineered NM in the compartments provide the basis for calculating the overall environmental concentrations of engineered NM.

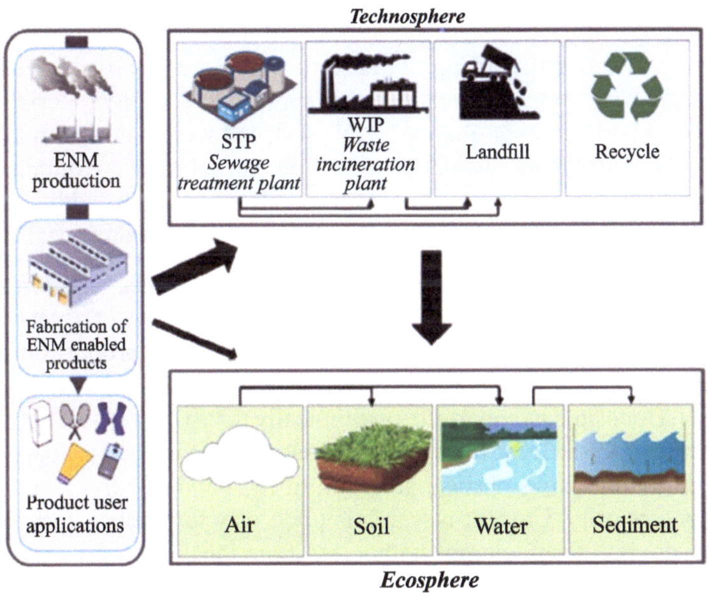

Fig. 53: General structure of the material-flow model. NM = nanomaterial; ROS = reactive oxygen species; NOM = natural organic matter; NP = nanoparticle.

As more products containing nanomaterials are developed, there is greater potential for environmental exposure. The fundamental properties concerning the environmental fate of nanomaterials are not well understood (European Commission, 2004), as there are few available studies on the environmental fate of nanomaterials.

The fate of nanomaterials in the atmosphere, in soils and in water is briefed below.

i. Fate of nanomaterials in air: Nanoparticles containing metals in their structure constitute an increasingly large group of substances present in the air. They come from both natural and anthropogenic sources. The wider and wider use of such particles means that increasingly complicated structures are present in the air we breathe. These particles are in the ultrafine size range (<100 nm), readily deposit in the lower areas of the respiratory tract and can be translocated directly into human body.

Several processes and factors influence the fate of airborne particles in addition to their initial dimensional and chemical characteristics: the length of time the particles remain airborne, the nature of their interaction with other airborne particles or molecules, and the distance that they may travel prior to deposition. The processes important to understanding the potential atmospheric transport of particles are diffusion, agglomeration, wet and dry deposition, and gravitational settling. These processes are relatively well understood for ultrafine particles and may be applicable to nanomaterials as well. However, in some cases, intentionally produced nanomaterials may behave quite differently from incidental ultrafine particles, for example, nanoparticles that are surface coated to prevent agglomeration. In addition, there may be differences between freshly generated and aged nanomaterials.

With respect to the length of time particles remain airborne, particles with aerodynamic diameters in the nanoscale range may follow the laws of gaseous diffusion when released to air. The rate of diffusion is inversely proportional to particle diameter, while the rate of gravitational settling is proportional to particle diameter. Deposited nanoparticles are typically not easily resuspended in the air or re-aerosolized. Because physical particle size is a critical property of nanomaterials, maintaining particle size during the handling and use of nanomaterials is a priority. Current research is underway to produce carbon nanotubes that do not form clumps either by functionalizing the tubes themselves, or by treatment with a coating or dispersing agent.

Many nanosized particles are reported to be photoactive (Colvin, 2003), but their susceptibility to photodegradation in the atmosphere has not been studied. The human body is an excellent medic for itself. The lungs exhale a good proportion of micron-sized particles (1,000 times bigger than nanoparticles), and can expel a certain quantity of nanoparticles too. But long exposures to high concentrations of nanoparticles bring a number of adverse health effects such as lung problems, and cardiovascular disease.

There are currently no air quality regulations in any part of the world to control exposure to nanoparticles, for numerous reasons.

Knowledge on the toxicological character of nanoparticles from biofuels is limited. There are substantial efforts to control public exposure through regulating air quality. The science surrounding their behaviour, measurement, dispersion, health and environmental impact has improved appreciably over the past decade (Anonymous, 2013). With the notable recent advancement of measuring instruments it is now possible to measure atmospheric particles down to 1nm. But there are still number of technical hurdles to be crossed before any nationwide regulatory framework can be put into practice.

Release of nanoparticles into the air : So far most published measurement campaigns concentrated on assessing workers' exposure by measuring or calculating ENM concentrations in air at the workplace whilst there has been little study of the release of engineered nanomaterials to the environmental air compartment during the product life cycle (John et al., 2017). Indeed, field measurements of ENMs in the environment are scarce and often measurements of releases of ENMs to air from industrial plants are done by the industry under strict confidential conditions. The following section aims at detailing the current knowledge of the various release scenarios that can lead to emission of ENMs to ambient air.

ENM life-cycle : Each step of the ENM life cycle from production, processing and use to the end-of-life may lead to emissions into ambient air. The emissions can be continuous, e.g., resulting from the evacuation of polluted air flows from production, transformation or waste treatment plants, the emission from the secondary material being processed (e.g., release from various matrixes following mechanical, thermal, chemical or combined stresses) either during the production of ENMs or during their use. ENMs can also be released accidentally, resulting from a sudden uncontrolled loss of containment, a fire or an explosion (John et al., 2017). An overall estimate of ENM release to different compartments considers that currently, around 63–91% of the ten most produced ENMs are disposed in landfills, whereas releases into water are around 0.4–7% of total global production.

ENM transformation during the airborne phase: While a lot of research has been done on transport and alterations of ENMs in aquatic media during the last years, information on ENM transformations within the atmosphere is still quite scarce. However, information and studies about the emission and behavior of unintentionally produced UFP are readily available, which can be used as starting point for the ENM assessment. While engineered nanomaterials consist of primary particles, UFP can be released as primary particles (e.g., soot) or can be formed in the atmosphere as secondary particles from gaseous precursors (e.g., formation of $(NH_4)_2SO_4$). Modest differences between the characteristics and behavior of atmospheric ENMs and UFP have been observed, originating from the higher ENM uniformity in shape, size and functionality (Kumar et al., 2010).

Visibility impairment is generally caused by a build–up of suspended particles in the atmosphere. It increases with relative humidity and atmospheric pressure and decreases with temperature and wind speed (Tsai, 2005).

There is a natural background of NPs (ultrafine particles-smaller than 0.1 gm) in the atmosphere and, despite little work has been done on ecological system, it is known that, when compared to anthropogenic sources and potential released manufactured NPs, the natural background present in the air is considerably low (Remédios et al., 2012). The key species in the atmosphere are sulfuric acid,nitric acid, and organic gases. Some conditions (e.g., temperature, residence time, dilution rate, dilution ratio, and concentration of carbon particles) are necessary for production of these secondary NPs.

There is a natural background of NPs in the atmosphere and, despite little work has been done on ecological system, it is known that, when compared to anthropogenic sources and potential released manufactured NPs, the natural background present in the air is considerably low (Remédios et al., 2012). NPs have short residence time. However they attach to accumulation mode particles and remain in the atmosphere as long as the accumulation mode particles participate in the atmospheric chemistry and physics. Diesel NPs are one of the concerning NPs released into the atmosphere. These NPs contribute to global warming since they absorb light and also endanger human health. When ultrafine particles are directly released from the source, they are called primary particles; volatile NPs (secondary NPs) composed by nucleation mode may be formed in the atmosphere by photochemical processes.

ii. *Fate of nanomaterials in water*

Factors that control the fate of nanomaterials in water include aqueous dispersibility or solubility, nanomaterial interactions between the natural and anthropogenic chemicals in the system, and biological and abiotic processes. Particles in the upper layers of aquatic environments, in water droplets in the atmosphere, and on soil surfaces are exposed to sunlight. Light-induced photoreactions are often critical in determining the environmental fate of chemical substances. These reactions may change the chemical and physical properties of nanomaterials and therefore alter their behavior in aquatic environments (Reginald Davey, 2020). Fate of nanomaterials in aqueous environments is controlled by aqueous solubility or dispersability, interactions between the nanomaterial and natural and anthropogenic chemicals in the system, and biological and abiotic processes. Waterborne nanoparticles generally settle more slowly than larger particles of the same material. However, due to their high surface-area-tomass ratios, nanosized particles have the potential to sorb to soil and sediment particles. Where these soil and sediment particles are subject to sedimentation, the sorbed nanoparticles can be more readily removed from the water column.

Certain organic and metallic nanomaterials may possibly be transformed under anaerobic conditions, such as in aquatic (benthic) sediments. Nanoparticle photochemistry is being studied with respect to its possible application in water treatment. Processes that control transport and removal of nanoparticles in water and wastewater are being studied to understand nanoparticle fate.

Zinc oxide nanostructures synthesized by a vapor-solid process. The fate of nanosized particles in wastewater treatment plants is not well characterized. Wastewater may be subjected to many different types of treatment, including physical, chemical and biological processes, depending on the characteristics of the wastewater, whether the plant is a publicly owned treatment work or onsite industrial facility, etc. Broadly speaking, nanosized particles are most likely to be affected by sorption processes (for example in primary clarifiers) and chemical reaction.

Nanomaterials that escape sorption in primary treatment may be removed from wastewater after biological treatment via settling in the secondary clarifier. Industrial products and wastes tend to end up in waterways, increasing the possibility of NPs contamination, although almost no current data on behavior and fate of NPs in aquatic systems is available (Remédios et al., 2012) Knowing NPs size, surface charges, and chemical reactivity, it is possible to predict some potential biological effects in the aqueous environment, such as generation of ROS, phototoxicity and photocatalytic activity, and interaction with soluble pollutants , but due to the aquatic system complexity this physical behavior is hard to predict.

Oceans exhibit changes in physicochemical characteristics with depth (e.g., temperature) that may influence aggregation and colloid chemistry. Analogously to freshwater aggregates of NPs may sink into the ocean floor, as well as accumulate at the interface between cold and warm currents or be recycled by biota. This may threaten pelagic species since they feed at these zones, while deposition in the sediment may present a risk of exposure to benthic species. NPs may also accumulate on the surface microlayers of the oceans, presenting a route of aerosol exposure risk to marine birds and mammals, as well as the organisms living on the surface microlayer (Remédios et al., 2012).

iii. *Fate of nanomaterials in soil*

Soil and sediments are the ultimate sinks of NPs and, whether directly or indirectly released (e.g., via sewage treatment plants, aerial deposition, or waste handling), NPs will end up in soil; this system may present the most significant exposure avenue for assessing environmental risk. Being soil one of the main sinks, it raises concern about the entry of NPs into food webs and human access to contaminated agriculture (Remédios et al., 2012). In terrestrial ecosystems, natural NPs are present (soil colloids), such as organic matter, clays, iron oxides, and other minerals that play an important role in biogeochemical processes. Therefore, the introduction of manufactured NPs in this ecosystem may change the profile of the natural NPs and interfere with soil development (pedogenesis) or behavior (e.g., crusting and dispersion). The fate of NPs in terrestrial ecosystems is difficult to elucidate, and there is few data that report transport of NPs through soils. NPs are made to have different properties than their similar bulk and macroscopic material, and these differences may change fate and effects of NPs in terrestrial environment. The high surface area, reactivity, size, aggregation, and absorbance are important determinants of their adhesion to soil, and thus their mobility and transport.

Bioavailability of NPs also depends on specific properties of each NP, in addition to composition of the soil matrix, porosity, hydraulic conductivity, groundwater gradient and flow velocity, and geochemical properties (e.g., dissolved oxygen, pH, ionic strength, and concentrations of nitrate, nitrite, and sulfate).

The NPs of Ag, CuO, and ZnO showed both toxicity on pathogenic bacteria (e.g., *Escherichia coli* and *Staphylococcus aureus*) and beneficial effects on microbes, as *Pseudomonas putida*,which has bioremediation potential and is a strong root colonizer. These NPs showed higher toxicity when compared to their equivalent bulk materials (Hardman, 2006). Therefore, being terrestrial systems the larger and most reactive sink for NPs, more research and standardized testing protocols are needed for fate/behavior and food web accumulation of NPs.

Nanoparticles (NPs) of ZnO and CuO are used in an array of different commercial products. NPs could be introduced into soils by contamination after production from distribution and use, especially from their end-of-life fate. Purposeful introduction to soils may occur in formulations as pesticides, because of their antimicrobial properties, and as fertilisers, to increase the loading of the essential metals Zn and Cu into plant tissues (Anne Andersonet. al., 2017). These findings indicated that phytotoxicity of these metal oxide NPs in soils will vary with the properties of the soils. Consequently soil properties will influence how the NPs could be used as fertilisers.

Interactions between nanomaterials and soil contaminants: Nanomaterials have been demonstrated to alter the partitioning behavior of chemicals between environmental compartments and between the environment and living organisms. Furthermore, several nanomaterials are reactive toward chemicals in the environment, generate reactive species, or catalyze reactions of other chemicals. These properties are currently under study for use in waste remediation operations. It should be noted that the potential also exists for nanomaterials to effect unforseen changes, if released to the environment in large quantities.

Two types of effects under study for possible exploitation are sorption and reaction. The high surface area of nanosized particles provides enhanced ability to sorb both organic and inorganic chemicals from environmental matrices compared to conventional forms of the same materials. No single overall effect can be described for the sorption of chemicals to nanomaterials based on their size or chemical makeup alone. In air, aerosolized nanoparticles can adsorb gaseous or particulate pollutants. In soil or sediments, nanomaterials might increase the bioavailability of pollutants, thereby increasing the pollutant=s availability for biodegradation. Depending on the conditions, nanosized carbon such as C60 or nanotubes could either enhance or inhibit the mobility of organic pollutants (Cheng et al., 2004). Stable colloids of hydrophobic nanomaterials in an aqueous environment could provide a hydrophobic microenvironment that suspends hydrophobic contaminants and retards their rate of deposition onto soils and sediments.

Several studies investigating the sorption of organic pollutants and metals in air, soil, and water to nanosized materials have recently been reported in the literature. The sorption of naphthalene to C60 from aqueous solution was compared to sorption to activated carbon. The investigators observed a correlation between the surface area of the particles and the amount of naphthalene adsorbed from solution. In other studies, nanoparticles made of an amphiphilic polymer have been shown to mobilize phenanthrolene from contaminated sandy soil and increase its bioavailability (Tungittiplakorn et al., 2005). It has been reported that magnetite crystals adsorb arsenic and chromium (CrVI) from water, suggesting potential purification techniques for metal-laden drinking water.

Nanoscale materials are typically more reactive than larger particles of the same material. This is true especially for metals and certain metal oxides. In the environment, nanomaterials have the potential to react with a variety of chemicals; their increased or novel reactivity coupled with their sorptive properties allows for accelerated removal of chemicals from the environment. Many groups are currently investigating the use of nanomaterials for the destruction of persistent pollutants in the environment. Nanoscale iron particles have been demonstrated to be effective in the in situ remediation of soil contaminated with tetrachloroethylene. Nanosized clusters of C60 have been shown to generate reactive oxygen species in water under UV and polychromatic light (EPA, 2007). Similar colloids have been reported to degrade organic contaminants and act as bacteriocides (Boyd et al., 2005). Fullerol ($C60(OH)_{24}$) has also been demonstrated to produce reactive oxygen species under similar conditions.

8.9 Nanoparticles and Potential Risk in Human Health

8.9.1 Human exposures

As the use of nanomaterials in society increases, it is reasonable to assume that their presence in environmental media will increase proportionately, with consequences for human and environmental exposure. Potential human exposures to nanomaterials, or mixtures of nanomaterials, include workers exposed during the production, use, recycling and disposal of nanomaterials, general population exposure from releases to the environment as a result of the production, use, recycling and disposal in the workplace, and direct general population exposure during the use of commercially available products containing nanomaterials (EPA, 2007). This section identifies potential sources, pathways, and routes of exposure, discusses potential means for mitigating or minimizing worker exposure, describes potential tools and models that may be used to estimate exposures, and identifies potential data sources for these models.

a. **Exposure to nanomaterials:** The exposure paradigm accounts for a series of events beginning from when external mechanisms (e.g., releases or handling of chemicals) make a chemical available for absorption or other mode of entry at the outer body boundary to when the chemical or its metabolite is delivered to the target organ. Between outer body contact with the chemical and delivery to the target organ, a chemical is absorbed and distributed. Depending on the nature of the chemical and the route of exposure, the chemical may be metabolized.

b. **Populations and sources of exposure:** The potential for intentionally produced nanomaterials to be released into the environment or used in quantities that raise human exposure concerns are numerous given their predicted widespread applications in products.

c. **Occupational exposure:** Workers may be exposed to nanoscale materials during manufacturing/synthesis of the nanoscale materials, during formulation or end use of products containing the nanoscale material, or during disposal or recycling of the products containing the nanoscale materials. Because higher concentrations and amounts of nanoscale materials and higher frequencies and exposures are more likely in workplace settings, occupational exposures warrant particular attention. While there are several potential exposure sources for each manufacturing process, packaging, transfer, and cleaning operations may provide the greatest potential for airborne levels of nanomaterials and resultant occupational exposures. "The risk of particle release during production seems to be low, because most production processes take place in closed systems with appropriate filtering systems. Contamination and exposure to workers is more likely to happen during handling and bagging of the material and also during cleaning operations" (Luther, 2004).

During the formulation of the nanomaterials into products (e.g., coatings and composite materials), workplace releases and exposures may be most likely to occur during the transfer/unloading of nanoscale material from shipping containers and during cleaning of process equipment and vessels. During the use of some of these products in workplace settings, releases of and exposures to nanoscale material are highly dependent upon the application. For example, workers who manually apply spray coatings often have higher levels of occupational exposure. Alternately, components of composites are usually bound in the composite matrix, and workers handling the composites would generally have lower levels of occupational exposure. Exposure could also occur during product machining (e.g., cutting, drilling and grinding), repair, destruction and recycling (National Institute for Occupational Health and Safety NIOSH, 2005). It has issued additional documents on nanotechnology and workplace safety and associated research needs.

Release and general population exposure: It may occur from environmental releases from the production and use of nanomaterials and from direct use of products containing nanomaterials. During the production of nanomaterials, there are several potential sources for environmental releases including the evacuation of production chambers, filter residues, losses during spray drying, emissions from filter or scrubber break-through, and wastes from equipment cleaning and product handling. No data have been identified quantifying the releases of nanomaterials from industrial processes or of the fate of nanomaterials after release into the environment (EPA, 2007). However, due to the small size of nanomaterials, they will likely stay airborne for a substantially longer time than other types of particulate. The most likely pathway for general population exposure from releases from industrial processes is direct inhalation of materials released into the air during manufacturing (UK Royal Society, 2004). Releases from industrial or transportation accidents, natural disasters, or malevolent activity such as a terrorist attack may also lead to exposure of workers or the general public.

Nanoscale materials have potential applications in many consumer products resulting in potential general population exposure. Electronics, medicine, cosmetics, chemistry, and catalysis are potential beneficiaries of nanotechnology. Widespread exposure via direct contact with products from these sectors is expected.

Exposure Routes : Much remains to be scientifically demonstrated about the mechanisms by which human exposure to nanomaterials can occur. Intentionally produced nanomaterials share a number of characteristics, such as size and dimensions, with other substances (e.g., ultrafine particles) for which a large body of information exists on how they access the human body to cause toxicity. The data from these other substances focus primarily on inhalation as the route of exposure. However, as the range of applications of nanomaterials expands, other routes of exposure, such as dermal, ocular, and oral, may also be found to be significant in humans.

Inhalation exposure: A UK Health and Safety Executive reference suggests that aerosol science would be applicable to airborne nanoparticle behavior. Aerosol behavior is primarily affected by particle size and the forces of inertia, gravity, and diffusion. Other factors affecting nanoparticle airborne concentrations are agglomeration, deposition, and re-suspension. (UK Health and Safety Executive, 2004) All of these issues, which are discussed in more detail in the reference, are relevant for understanding, predicting, and controlling airborne concentrations of nanomaterials. One reference study was found to have investigated issues involved with aerosol release of a single-walled carbon nanotube (SWCNT) material. This study noted that while laboratory studies indicate that sufficient agitation can release fine particles into the air, aerosol concentrations of SWCNT generated while handling unrefined material in the field at the work loads and rates observed were very low on a mass basis (Maynard et al., 2004). The study suggests that more research will be needed in this area.

Ingestion exposure: Information on exposure to nanoscale environmental particles via oral exposure is lacking. In addition to traditional ingestion of food, food additives, medicines and dietary supplements, dust and soil (particularly in the case of children), ingestion of inhaled particles can also occur (such as through the activities of the mucocilliary escalator). The quantity ingested is anticipated to be relatively small in terms of mass.

Dermal exposure: Dermal exposure to nanomaterials has received much attention, perhaps due to concerns with occupational exposure and the introduction of nanomaterials such as nanosized titanium dioxide into cosmetic and drug products. One reference study was found to have investigated issues involved with potential dermal exposure to a SWCNT material. The study suggests that more research will be needed in this area. This study noted that airborne particles of SWCNT may contribute to potential dermal exposure along with surface deposits due to material handling. Surface deposits on gloves were estimated to be between 0.2 mg and 6 mg per hand (Maynard et al., 2004).

There is an ongoing debate over the potential for penetration through "healthy/ intact" versus damaged skin. Physiological characteristics of the skin were discussed that may permit the absorption of nanosized materials. In particular the review highlights a conceivable route for the absorption of nanoparticles as being through interstices formed by stacking and layering of the calloused cells of the top layer of skin. Movement through these interstices will subsequently lead to the skin beneath, from which substances can be absorbed into the blood stream. Nanomaterials also have a greater risk of being absorbed through the skin than macro-sized particles. Reports of toxicity to human epidermal keratinocytes in culture following exposure to carbon nanotubes have been made. A significant amount of intradermally injected nanoscale quantum dots were found to disperse into the surrounding viable subcutis and to draining lymph nodes via subcutaneous lymphatics. It has recently been reported that quantum dots with different physicochemical properties (size, shapes, coatings) penetrated the stratum corneum and localized within the epidermal and dermal layers of intact porcine skin within a maximum 24 hours of exposure. Drug delivery studies using model wax nanoparticles have provided evidence that nanoparticle surface charge alters blood-brain barrier integrity and permeability.

Ocular exposure : Ocular exposure to nanomaterials has received little attention. However, the potential for ocular exposure to nanomaterials from deposition of airborne powders or mists in the eyes or from splashing of liquids must also be considered.

d. Exposure Mitigation: Approaches exist to mitigate exposure to fine and ultrafine particulates. Some approaches such as engineering controls are applicable to the work place and may mitigate environmental releases while others such as personal protective equipment (PPE) are primarily applicable to the workplace. NIOSH suggests considering the range of control technologies and personal protective equipment demonstrated to be effective with other fine and ultrafine particles. In the hierarchy of exposure reduction methods, engineering controls are preferred over PPE.

Engineering controls: Engineering controls, and particularly those used for aerosol control, should generally be effective for controlling exposures to airborne nanoscale materials. Depending on particle size, nanoparticles may diffuse rapidly and readily find leakage paths in engineering control systems in which containment is not complete. However, a welldesigned exhaust ventilation system with a high efficiency particulate air (HEPA) filter should effectively remove nanoparticles. As with all filters, the filter must be properly seated to prevent nanoparticles from bypassing the filter, decreasing the filter efficiency. Aitken et al. (2004) recommends that engineering controls (e.g., enclosures, local exhaust ventilation, fume hoods) used to control exposure to nanoparticles need to be of similar quality and specification as those typically used for gases. However, the report also notes that no research has been identified evaluating the effectiveness of engineering controls for nanoparticles.

Efficient ultrafine particle control devices (e.g., soft x-ray enhanced electrostatic precipitation systems) may have applicability to nanoparticles control. HEPA filters may be effective, and validation of their effectiveness is currently being studied. Magnetic filter systems in welding processes have proven effective in capturing magnetic oxides and the use of nanostructured sorbents in smelter exhausts to prepare ferroelectric materials may also have applicability.

Personal Protective Equipment (PPE): Properly fitted respirators with a HEPA filter may be effective at removing nanomaterials. Contrary to intuition, fibrous filters trap smaller and larger particles more effectively than mid-sized particles. Small particles (80 nm and < 2000 nm) can remain suspended in air for the longest time. NIOSH certifies particulate respirators by challenging them with sodium chloride (NaCl) aerosols with a count median diameter 75 nm or dioctyl phthalate (DOP) aerosols with a count median diameter of 185 nm [42 CFR Part 84.181(g)], which have been found to be in the most penetrating particle size range. However, as with all respirators, the greatest factor in determining their effectiveness is not penetration through the filter, but rather the face-seal leakage bypassing the device. Due to size and mobility of nanomaterials in the air, leakage may be more prevalent although no more than expected for a gas. Only limited data on face-seal leakage has been identified. Work done by researchers at the U.S. Army RDECOM on a headform showed that mask leakage (i.e., simulated respirator fit factor) measured using submicron aerosol challenges (0.72 μm polystyrene latex spheres) was representative of vapor challenges such as sulfur hexafluoride (SF6) and isoamyl acetate (IAA).

PPE may not be as effective at mitigating dermal exposure. PPE is likely to be less effective against dermal exposure to nanomaterials than macro-sized particles from both human causes (e.g., touching face with contaminated fingers) and PPE penetration (EPA, 2007). However, no studies were identified that discuss the efficiency of PPE at preventing direct penetration of nanomaterials through PPE or from failure due to human causes.

e. Quantifying exposure to nanomaterials: There is broad consensus that mass dose alone is insufficient to characterize exposure to nanomaterials (Oberdörster et al., 2005). Many studies have indicated that toxicity increases with decreased particle size and that particle surface area is a better metric for measuring exposures. This is of particular concern for nanomaterials, which typically have very high surface-area-to-mass ratios. Additionally, there currently are no convenient methods for monitoring the surface area of particles in a worker's breathing zone or ambient air. While there could be a correlation between mass and surface area, large variations in particle weight and surface area can occur within a given batch. The average particle weight and average particle surface area of the nanomaterials being assessed would also be required for any assessments based on surface area.. It has also been recommended that mass, surface area, and particle number all be measured for nanomaterials.

f. Tools for exposure assessment: Several tools exist for performing exposure assessments including monitoring data, exposure models and the use of analogous data from existing chemicals.

8.9.2 Nanohazards

The US national nanotechnology initiative has estimated that around 20,000 researchers are working in the field of nanotechnology. For the UK, the Institute of Occupational Medicine has estimated that approximately 2,000 people are employed in new nanotechnology companies and universities where they may be potentially exposed to nanoparticles. The primary production of nanoparticles takes places with the help of chemical and engineering processes that are well established. These are considered by various organisations to be relatively safe, except for accidental releases via leakages. During normal production, subsequent steps like product recovery and powder handling may result in respirable concentrations of agglomerated nanoparticles, to dermal exposure or to ingestion (mainly via hand-to-mouth contact).

So far only one study has addressed the exposure of workers to engineered nanoparticles. In the study into potential airborne and dermal exposure to carbon nanotubes samples were taken during typical activities in the production process like material removal, filling, pouring and clean up (techniques: laser ablation and high pressure carbon monoxide). The study has shown that the material only becomes airborne with a sufficient level of agitation (Christoph Lauterwasser, 2005). So-called Van der Waals forces make the nanoparticles "sticky" and imparts a strong tendency to form agglomerates. Airborne concentrations found while handling unrefined material were considered to be "very low" (between 0.7 and 53 µg/m3). The released material seems to form larger agglomerates in the size range of 1 µm rather than leading to a high number concentration of fine particles. Up to 6 mg of nanotubes were found on individual gloves. While the gloves can minimise dermal exposure, airborne clumps of material can lead to exposure of less well protected parts of the skin.

Contrary to intuition, air filtration systems such as respiratory protective equipment should be effective when used correctly. Below a particle size of 100 nm filtration efficiency for High Efficiency Particulate Arrestor (HEPA) filters even increases with decreasing particle size. The reason for this is strong Brownian motion that leads to increased probability to contact the filter elements. Once a filter element is hit, strong Van der Waals forces keep the nanoparticle stuck to the surface. It has been inferred by the UK Institute of Occupational Medicine that the primary route of exposure for many nanoparticles could be dermal exposure and subsequent ingestion exposure.

Hazard of a material translates into risk by exposure of humans and/or their environment to the agent in question, and risk is reduced by control of exposure, usually guided by regulation based on understanding of the mechanisms of harm (Seaton et al., 2010). They illustrated risks relating to the causation of diseases associated with exposure to aerosols of combustion particles and asbestos, leading to paradigms of particle toxicity, and discuss analogies with potential exposure to manufactured nanoparticles (NPs).

The health and safety hazards of nanomaterials include the potential toxicity of various types of nanomaterials, as well as fire and dust explosion hazards. Because nanotechnology is a recent development, the health and safety effects of exposures to nanomaterials, and what levels of exposure may be acceptable, are subjects of ongoing research. Of the possible hazards, inhalation exposure appears to present the most concern, with animal studies showing pulmonary effects such as inflammation, fibrosis, and carcinogenicity for some nanomaterials. Skin contact and ingestion exposure, and dust explosion hazards, are also a concern.

Guidance has been developed for hazard controls that are effective in reducing exposures to safe levels, including substitution with safer forms of a nanomaterial, engineering controls such as proper ventilation, and personal protective equipment as a last resort. For some materials, occupational exposure limits have been developed to determine a maximum safe airborne concentration of nanomaterials, and exposure assessment is possible using standard industrial hygiene sampling methods. An ongoing occupational health surveillance program can also help to protect workers.

A risk management system is composed of three parts.

i. Hazard identification involves determining what health and safety concerns are present for both the nanomaterial and its corresponding bulk material, based on a review of safety data sheets, peer-reviewed literature, and guidance documents on the material. For nanomaterials, toxicity hazards are the most important, but dust explosion hazards may also be relevant.

ii. Exposure assessment involves determining actual routes of exposure in a specific workplace, including a review of which areas and tasks are most likely to cause exposure.

iii. Exposure control involves putting procedures in places to minimize or eliminate exposures according to the hierarchy of hazard controls. Ongoing verification of hazard controls can occur through monitoring of airborne nanomaterial concentrations using standard industrial hygiene sampling methods, and an occupational health surveillance program may be instituted (Anonymous, 2012).

A recently adopted risk management method is the Safe by design (SbD) approach. It aims to eliminate or reduce risks of new technologies including nanotechnology, at the design stage of a product or production process. Anticipation of risks is challenging because some risks could emerge only after a technology is implemented (at later stages in the innovation process). In the later cases, the application of other risk management strategies based on non-design principles need to be applied. It considers the purposes and constrains for implementation of SbD approaches in the industrial innovation process and on the basis of those, establish optimal workflows to identify risks and propose solutions to reduce or mitigate them as early as possible in the innovation process called Safe by Design strategies.

The identification of hazards is the first step in determining risk and exposure. This step involves identifying chemicals or nanomaterials, and their associated processes that pose toxic, physical (e.g. high levels of noise, high pressures and vacuum, strong electromagnetic flux, etc.) and physicochemical hazards (Anonymous, 2018). When assessing the risks associated with nanomaterials, special care should be taken to identify the specific effect of surface chemistry, shape, size and morphology on toxicity caused to various organs.

The following primary hazard categories may be considered when assessing risk associated with nanomaterials.

i. **Surface charge:** The most interesting physicochemical feature of nanoparticle relating to cytotoxicity is the surface charge, with toxicity increasing in the following way: neutral < anionic < cationic.

ii. **Surface chemistry:** Surface chemistry of nanoparticles may have a role in the generation of free radicals, which influences the overall surface reactivity and toxicity of ingested particles.

iii. **Particle shape:** Studies have clearly established that exposure to fibrous particles like asbestos increases the risk of fibrosis and cancer. Similarly, the tubular structure of carbon nanotubes is believed to cause inflammation and lesions in lungs.

iv. **Particle size** : Nanoparticles have a greater chance of depositing in the lungs than micro-sized particles, and therefore have the potential to cause damage by acting directly at the site of deposition by translocating to other organs or by being absorbed through the blood. Nanoparticles can penetrate the membrane barriers resulting in significant damages. For example, silver nanoparticles with size < 9 nm can penetrate the nuclear membrane of certain human cells nucleus and cause significant DNA damage or mutation.

v. **Solubility:** Poorly soluble inhaled nanoparticles can cause oxidative stress, leading to inflammation, fibrosis, or cancer. Studies have shown significantly higher toxicity of nano metals when compared to nano ceramics, which has been attributed to higher dissolution rate in water.

vi. **Toxic hazards**

- Nanomaterials provided with the guidelines for exposure, such as threshold limit values (TLV) or permissible exposure limits (PEL) helps in determining proper safety precautions, including control measures and safety apparel.
- Materials with TLV or PEL less than 50 ppm should be handled with utmost care preferably under fume hood.
- High vapor pressure indicate potential hazard. However, low vapor pressure of even 5-10 mm Hg with low TLV (less than 10) can be highly hazardous.
- When TLV or PEL values are not available for a substance, the chemical must be used inside the fume hood.
- Whenever the chemicals have high vapor pressure, it must be used only under a fume hood. The Material Safety Data Sheet (MSDS) or container label identifies or describes the substance as toxic.

vii. **Fire Hazard** : Flammability of a chemical is determined by its flash point.

- Any chemical with a flash point below 200o F (93 °C) should be considered as a fire hazard.
- When used in quantities in excess of 100 mL, fire hazard procedures should be followed.
- When stored in quantities in excess of 500 mL, the restrictions on containers should be observed.

viii. **Explosion hazard:** The flammability and explosivity of nanomaterials can be determined by calculating the minimum ignition energy (MIE) and the minimum exposable dust concentration (MEC) also known as lower explosion limit (LEL). To assess the impact of effects and the explosion severity, parameters like the maximum pressure of explosion and maximum rate of pressure rise (Kst) need to be considered. Table 44 shows the severity of explosion based on rate of pressure rise, Kst.

Table 44: Classification according to the maximum rate of pressure

Dust explosion class	Kst (bar m/s)	Characteristics
St 0	0	No explosion
St 1	>0 - <200	Weak explosion
St 2	>200 - <300	Strong explosion
St 3	>300	very strong explosion

Correlation of Kst -Value with Maximum Pressure Rise per Unit Time (dp/dt) max.: (dp/dt) max * V (1/3) = Kst, (V: Volume of the vessel)

8.9.3 Exposure pathways and common tasks

The primary routes to exposure for nanoparticles are inhalation, dermal absorption and ingestion. The state of the nanoparticles or nanomaterials decides the exposure potential. For example, nanoparticles in powdered form present a large inhalation hazard potential when compared to nanoparticles in suspension. Table 45 summarizes the exposure pathways and suggested prevention methods.

Table 45: Exposure pathways and safety measures

Type of Exposure	Pathway	Safety measures
Dermal	Nanoparticles can migrate through skin and circulate in the body while handling nanoparticle suspensions or dry powders. Skin absorption is much less likely for solid bound or matrixed nanomaterials.	Wear gloves and lab coat while handling the nanoparticles.
Ingestion	Ingestion can occur if good hygiene practices are not followed. Nanoparticles might be absorbed and transported within the body by the circulatory system	Eating and drinking are not allowed in laboratories. Spills of nanoparticles should be quickly and properly cleaned-up.
Inhalation.	Respiratory absorption through the mucosal lining of the trachea, bronchioles or the alveoli of the lung	Nanoparticles are to be handled in a form that is not easily airborne, such as in solution or on a substrate. Use of respiratory air filters N100 or N95 is recommended.
Injection	Exposure by accidental injection (skin puncture), when working with animals or needles	Wear gloves and lab coats, and apply the standard practices for working with sharp objects
Ocular	Exposure to airborne nanoparticles placed near the eye, accidentally splashed onto the eye or transferred from hands during rubbing of eyes.	Wear safety glasses, goggles, full facepiece respirator (Recommended when there is exposure to solvent or hot material). Note: Do not wear contact lenses at work place

8.9.4 Types of toxicity in humans

i. Respiratory: Inhalation exposure is the most common route of exposure to airborne particles in the workplace. The deposition of nanoparticles in the respiratory tract is determined by the shape and size of particles or their agglomerates, and they are deposited in the alveolar compartment to a greater extent than larger respirable particles (Bair, 1995). Based on animal studies, nanoparticles may enter the bloodstream from the lungs and translocate to other organs, including the brain. The inhalation risk is affected by the dustiness of the material, the tendency of particles to become airborne in response to a stimulus. Dust generation is affected by the particle shape, size, bulk density, and inherent electrostatic forces, and whether the nanomaterial is a dry powder or incorporated into a slurry or liquid suspension .

Animal studies indicate that carbon nanotubes and carbon nanofibers can cause pulmonary effects including inflammation, granulomas, and pulmonary fibrosis, which were of similar or greater potency when compared with other known fibrogenic materials such as silica, asbestos, and ultrafine carbon black. Some studies in cells or animals have shown genotoxic or carcinogenic effects, or systemic cardiovascular effects from pulmonary exposure. Although the extent to which animal data may predict clinically significant lung effects in workers is not known, the toxicity seen in the short-term animal studies indicate a need for protective action for workers exposed to these nanomaterials. As of 2013, further research was needed in long-term animal studies and epidemiologic studies in workers. No reports of actual adverse health effects in workers using or producing these nanomaterials were known as of 2013.

Titanium dioxide (TiO_2) dust is considered a lung tumor risk, with ultrafine (nanoscale) particles having an increased mass-based potency relative to fine TiO_2, through a secondary genotoxicity mechanism that is not specific to TiO_2 but primarily related to particle size and surface area.

ii. Dermal: Some studies suggest that nanomaterials could potentially enter the body through intact skin during occupational exposure. Studies have shown that particles smaller than 1 μm in diameter may penetrate into mechanically flexed skin samples, and that nanoparticles with varying physicochemical properties were able to penetrate the intact skin of pigs. Factors such as size, shape, water solubility, and surface coating directly affect a nanoparticle's potential to penetrate the skin. At this time, it is not fully known whether skin penetration of nanoparticles would result in adverse effects in animal models, although topical application of raw SWCNT to nude mice has been shown to cause dermal irritation, and in vitro studies using primary or cultured human skin cells have shown that carbon nanotubes can enter cells and cause release of pro-inflammatory cytokines, oxidative stress, and decreased viability(Anonymous, 2017). It remains unclear, however, how these findings may be extrapolated to a potential occupational risk. In addition, nanoparticles may enter the body through wounds, with particles migrating into the blood and lymph nodes.

iii. Gastrointestinal: Ingestion can occur from unintentional hand-to-mouth transfer of materials; this has been found to happen with traditional materials, and it is scientifically reasonable to assume that it also could happen during handling of nanomaterials. Ingestion may also accompany inhalation exposure because particles that are cleared from the respiratory tract via the mucociliary escalator may be swallowed.

iv. Fire and explosion: There is concern that engineered carbon nanoparticles, when manufactured on an industrial scale, could pose a dust explosion hazard, especially for processes such as mixing, grinding, drilling, sanding, and cleaning. Knowledge remains limited about the potential explosivity of materials when subdivided down to the nanoscale. The explosion characteristics of nanoparticles are highly dependent on the manufacturer and the humidity.

For microscale particles, as particle size decreases and the specific surface area increases, the explosion severity increases. However, for dusts of organic materials such as coal, flour, methylcellulose, and polyethylene, severity ceases to increase as the particle size is reduced below ~50 μm. This is because decreasing particle size primarily increases the volatilization rate, which becomes rapid enough that that gas phase combustion becomes the rate limiting step, and further decrease in particle size will not increase the overall combustion rate.[10] While the minimum explosion concentration does not vary significantly with nanoparticle size, the minimum ignition energy and temperature have been found to decrease with particle size (Worsfoldet al., 2012).

Metal-based nanoparticles exhibit more severe explosions than do carbon nanomaterials, and their chemical reaction pathway is qualitatively different. Studies on aluminum nanoparticles and titanium nanoparticles indicate that they are explosion hazards. One study found that the likelihood of an explosion but not its severity increases significantly for nanoscale metal particles, and they can spontaneously ignite under certain conditions during laboratory testing and handling.

Radioactivity: Engineered radioactive nanoparticles have applications in medical diagnostics, medical imaging, toxicokinetics, and environmental health, and are being investigated for applications in nuclear medicine. Radioactive nanoparticles present special challenges in operational health physics and internal dosimetry that are not present for vapors or larger particles, as the nanoparticles' toxicokinetics depend on their physical and chemical properties including size, shape, and surface chemistry. In some cases, the inherent physicochemical toxicity of the nanoparticle itself may lead to lower exposure limits than those associated with the radioactivity alone, which is not the case with most radioactive materials. In general, however, most elements of a standard radiation protection program are applicable to radioactive nanomaterials, and many hazard controls for nanomaterials will be effective with the radioactive versions.

History of Nanohazards: It is appropriate to start with a warning. Two toxic substances stand out above all others in the history of peacetime technology in terms of deaths caused: quartz and asbestos. In the former case, the harmfulness of inhaling fine particles was recognized in the eighteenth century and in the latter at the end of the nineteenth century (Seaton et al., 2010). In both cases, physicians and scientists were intrigued by the possibility of understanding the mechanisms of toxicity, an area of investigation that attracted support from the industries concerned, driven by the medical mindset that if a mechanism can be found, then a cure may follow.

The concept that nanoparticles (NPs) may have unique toxic properties is a recent one, though the thinking behind it is derived from classical particle toxicology and epidemiology, especially from studies of air pollution. The harmfulness of air pollution was suspected in the reign of Charles II but its real cost in lives lost became clear in the 1930s and 1940s with two episodes, one in the Meuse valley in Belgium and the other in Donora in Pennsylvania. Coal smoke and a winter temperature inversion were a lethal combination (McDonald et al., 1951). These episodes were followed by the infamous London smog of December 1952 in which some 4000 excess deaths occurred over a week. These deaths mostly appeared to be from lung disease, chronic bronchitis being then a highly prevalent condition in urban Britain, but it was apparent that a proportion was from heart attacks. Unusually, this is an example of regulation occurring promptly (cynics would argue that this was because the episode centred on the Houses of Parliament), and the Clean Air Acts were introduced and proved very effective in reducing pollution while argument continued as to which components of pollution, gases, acid aerosols or solid particles were responsible.

The majority of deaths associated with air pollution in the epidemiological studies were from cardiac rather than respiratory disease and attempted to explain the apparent fact that toxicologically tiny doses of particulate matter (PM), mainly carbon, to the lungs could cause death from failure of another organ by proposing the hypothesis that the effect was a consequence not of the mass of particles but of the number. It was suggested that such particles evaded the lungs' primary defence, the macrophage, and caused inflammation in the interstitial space leading to release of cytokines and alteration of the coagulability of the blood (Searon et al., 1995). This rather complex idea, generally called the ultrafine hypothesis, triggered a change in the way toxicologists were to investigate air pollution, with an emphasis on the role of particles in the nanosize range.

Over the same period that scientists have investigated air pollution, there has been an analogous interest in the toxicity of asbestos. From the early twentieth century, it has been recognized that workers exposed to this mineral ran excessive risks of the development of fibrosis of the lungs and, from the 1950s and 1960s, of lung cancer and the pleural tumour mesothelioma.

The epidemiological research relating to the risks of mesothelioma led to the conclusion that it would be virtually impossible to prevent this disease unless the importation and use of asbestos were prohibited. However, asbestos was a very useful material and many materials with similar properties were necessary for industry. Thus, the toxicological research, initially carried out to understand why this chemically inert mineral had such devastating effects on the health of those inhaling it (Mossman et al., 1990), became relevant to the investigation of possible hazards from substitute fibrous minerals (Miller et al., 1999). This led to the development of what became known as the fibre paradigm.

Nanomaterials are becoming ubiquitous in medicine (e.g., molecular targeted therapies, diagnostic imaging), consumer goods (e.g., cosmetics, electronics, baby products), and building materials (e.g., concrete, steel, windows) (Joseph A. Clark, 2011). The physiochemical nature of these materials may vary depending on the intended purpose, the therapeutic value, structural support, thermal property, cosmetic appeal, durability, etc.,of the engineered product. For example, nanomaterials such as carbon nanotubes and SiO_2 nanoparticles are often used to add mechanical strength and durability, TiO_2 nanoparticles are used to increase the degree of hydration, and FeO_3 nanoparticles are used for abrasion resistance.

Currently, there are no universally accepted measurement procedures—analytical methods—established for measuring human exposure to nanomaterials precisely. This may be ascribed to the unique physicochemical properties of nanomaterials as compared to bulk materials, such as soil or dust, or volatile chemicals, such as benzene, PAHs, and TCE. The currently employed analytical methods are dynamic, mostly involving a modification of conventional methods for quantifying micro-sized materials (Joseph A. Clark, 2011).

Compared to the chemical risk assessment process, where hazards are generally identified as a function of an organism's adverse response to a particular mass of a chemical agent (a dose), characterizing the hazards of nanomaterials has been problematic. This is largely due to the many features that may contribute to the toxicity of nanomaterials. Currently, there is no consensus within the scientific community on what characteristic may be the most important in elucidating this dose-response relationship for each type of nanomaterial. This lack of agreement is understandable, given that studies evaluating the health effects of nanomaterials show a range of findings (i.e., dose-response relationships) and underscore the inappropriate generalization of responses across all types of nanomaterials (Joseph A. Clark, 2011).

Neurotoxicity: Welding fumes are a complex mixture of various compounds including nanoparticles (NPs) of metallic oxide (e.g. manganese oxide or dioxide). Especially manganese (Mn) containing particles (MnxOx) are thought to be associated with the neurotoxicity caused by this essential element. When bypassing the homeostatic regulation in the gastrointestinal tract (GT), as seen in welders exposed to Mn via inhalation, neurobehavioral debicits might occur (Nadine Hollmann et al., 2017). After the deposition of manganese NPs in the respiratory tract Mn can be directly translocated to the brain via the olfactory system or enter the blood steam from the alveolar region where approximately 14% of the total Mn mass is soluble and bioavailable. Via these routes Mn can enter the brain where it accumulated predominantly in the basal ganglia (BG) as well as frontal cortical areas. Recently, a PBPK model for Mn has been developed and combined with categorical regression modeling concluding that a concentration of 0.55 μg/g Mn in the BG is associated with an increase of 10% of risk for neurobehavioral dcicits. The underlying mechanism causing these behavioral debicits is not completely understood but various neural cell types and neurotransmitters seem to be involved.

Neither MnO_2 nor $MnCl_2$ were cytotoxic in the selected concentrations. The analyses of the first biological replicate indicate some activation of microglia after 48h exposure to $MnCl_2$ but not MnO_2 in the 135 μM condition.

Airborne pollution, especially from diesel exhaust (DE), is known to negatively affect the nervous system (NS) of exposed human populations, either through systemic or airway pathways (Bernal-Meléndez et al., 2017). However, the consequences of a gestational exposure to DE on the NS remain poorly explored. Therefore, an experimental model of gestational exposure of rabbit dams (a human hemodichorial placentation model) to DE nanoparticles that mimics conditions of exposure in human urban areas was developed to study the short- and long-term effects of such exposure on the developing NS.

Pregnant does were exposed nose-only to clean air or to diluted ($1mg/m^3$) filtered DE from gestational day 3 (GD3) to GD27, 2h/d, 5d/w [1]. At GD28, the presence of nanosize particles (20-48nm) was observed in the olfactory sensory neurons and the glomerular layer of the olfactory bulb (OB), along with cellular and axonal hypertrophy. Concomitant bulbar monoaminergic homeostasis disturbances, especially affecting the dopaminergic system, were also observed . However, the neurotransmission pathways potentially affected by this gestational exposure and their long-term consequences have yet to be further explored.

The imbalance in the dopaminergic system observed in the olfactory bulb of the exposed fetuses at the end of the gestation seems to persist in adulthood, and is associated to alteration in more central structures. Because of the known anatomical and functional continuum between the olfactory system and the rest of the brain, and the importance of dopamine homeostasis in the plasticity of neural circuits, such alterations could participate to disturbances in higher integrative structures, with possible long-term behavioral consequences.

v. Asthmatic probles: Asthma is a multifactorial disease defined by airway hyper responsiveness (AHR) and inflammation. It is associated with increased mucus secretion, epithelial injury, hyperplasia of smooth muscle cells, and deposition of excess extracellular matrix. The additional irritation of the "inblamed" epithelium with inhaled nanoparticles is supposed to aggravate the asthmatic response and, at the worst, to induce an asthmatic attack. Whether Carbon Black nanoparticles (CBNPs) aggravate an ovalbumin-induced allergic airway inflammation in mice was tested (Sina Webering et al., 2017). Printex®90 (P90) was used as a reference particle and P90 coated on the surface with Benzo[a]pyren (P90-BaP) or 9-Nitroanthracen (P90-9NA).

Nanoparticles are increasingly used also for medical therapies. The effect of nanoparticles in healthy and asthmatic subjects is still under investigation. The effect, uptake and distribution of nanoparticles in BALB/c ovalbumin (OVA) asthma mouse model were investgated (Albert J. Omlor et al., 2017). Dispersions of polyethylene-glycol-coated (PEGylated) and citrate/tannic-acid-coated (citrated) 5 nm gold nanoparticles are given intranasally to asthma and control groups. Allergic airway inflammation and airway resistance are measured. Nanoparticle uptake into extrapulmonary organs is quantified by inductively coupled plasma mass spectrometry. The asthmatic precondition increases nanoparticle uptake. Moreover, systemic uptake is higher for PEGylated gold nanoparticles compared to citrated nanoparticles. Nanoparticles inhibit both inflammatory infiltrates and airway hyperreactivity, especially citrated gold nanoparticles.

On Gene expression: The increasing use of nanomaterials in different industries consolidates the importance of nanotoxicological research. Up to now, most in vitro toxicity studies of airborne nanoparticles (NP) are performed under submerged condition (MaΣhias Hufnagel et al., 2017). However, this exposure method is a rather artificial test system to investigate inhalation toxicity. To achieve a more realistic exposure scenario, it is possible to culture cells at an air-liquid interface (ALI). Here, cells are cultured in a transwell system, being supplied with cell culture medium from the basal side and exposed to air on the apical side. Additionally, the use of appropriate ALI exposure systems, e.g. VITROCELL® Cloud, allows the quantification of the actual particle deposition, when equipped with a quartz crystal microbalance.

Within this study, A549 cells were cultured at an ALI and exposed to CuO or TiO_2 NP in the VITROCELL® Cloud. Subsequently, cytotoxic effects were determined using the colony formation assay. Additionally, gene expression profiling via high-throughput RT-qPCR was performed. CuO NP showed dose-related cytotoxic effects, while TiO_2 NP revealed no cytotoxicity. The same pattern was observed for gene expression profiling, showing no interference with gene expression by TiO_2 NP. In contrast, CuO NP exposure induced the expression of oxidative stress markers, redox-sensitive transcription factors, genes involved in DNA repair, apoptotic factors and cell cycle regulating genes. In conclusion, a new exposure system for airborne NP was established using CuO and TiO_2 NP. CuO NP affected the expression of several genes, presumably evoked by elevated levels of ROS. However, further research is needed to confirm the transcriptional results on a functional level.

Protein Corona: When a pristine nanoparticle (NP) encounters a biological fluid, biomolecules spontaneously form adsorption layers around the NP, called "protein corona". The corona composition depends on the time-dependent environmental conditions and determines the NP's fate within living organisms (Giancarlo Franzese and Oriol Vilanova, 2017). Understanding how the corona evolves is fundamental in nanotoxicology as well as medical applications. However, the process of corona formation is challenging due to the large number of molecules involved and to the large span of relevant time scales ranging from 100 μs, hard to probe in experiments, to hours, out of reach of all-atoms simulations. Here we combine experiments, simulations, and theory to study the corona kinetics (over 10−3−103 s) and its final composition for silica NPs in a model plasma made of three blood proteins (human serum albumin, transferrin, and fibrinogen). When computer simulations are calibrated by experimental protein−NP binding affinities measured in singleprotein solutions, the theoretical model correctly reproduces competitive protein replacement as proven by independent experiments. When the order of administration of the three proteins was changed, a memory effect in the final corona composition was noticed that we can explain within our model.

8.9.5 Government/NGO Recommendations

Most governmental agencies and nongovernmental organizations involved in standardization have been relatively slow in promulgating regulations and standards relating to nanotechnology. However, the National Institute of Occupational Safety and Health (NIOSH), the leading federal agency conducting research and providing guidance on the occupational safety and health implications and applications of nanotechnology, released a publication in March 2009 entitled, "Approaches to Safe Nanotechnology: Managing the

Health and Safety Concerns Associated with Engineered Nanomaterials," which sets forth a series of recommended best practices for occupational safety and health. Additionally, in November 2010, NIOSH issued, for public comment by February 18, 2011, a bulletin addressing exposure to carbon nanotubes in the workplace (Joseph A. Clark, 2011). Of paramount importance is NIOSH's recommended exposure limit (REL); this represents the first time NIOSH has provided a hard number relating to nanoparticles exposure. In arriving at the REL, NIOSH stated, "[u]ntil improved sampling and analytical methods can be developed, and until data become available to determine if an alternative exposure metric to mass may be more biologically relevant, NIOSH is recommending a REL of 7 μg/m3 elemental carbon (EC) as an 8-hr TWA respirable mass airborne concentration."

On the environmental front, the U.S. Environmental Protection Agency (EPA) has not only classified certain nanomaterials as "emerging contaminants" but has also issued Significant New Use Rules (SNURs) under the Toxic Substances Control Act (TSCA)[7] for single and multi-walled carbon nanotubes. [8]The EPA has been more active than other government agencies, and it is likely that it will be the vanguard of nanotechnology regulation for the foreseeable future.In addition to government agencies, consensus standards organizations, such as the International Organization for Standardization (ISO), the American Society for Testing and Materials (ASTM), and the American National Standards Institute (ANSI) (in conjunction with ISO) are formulating guidelines for the safe handling, use, and disposal of nanomaterials by workers and consumers. To date, progress has been slow, but the standards that do exist currently are a helpful foundation upon which to build more specific standards.

8.10 Paradigms of toxicity

8.10.1 The ultrafine hypothesis

There is not yet convincing epidemiological evidence that the toxic component of particulate air pollution resides in the nanometre-sized component, although a recent study in London reported to the Government's Committee on Medical Effects of Air Pollution has suggested that particle count, which reflects the sub-100 nm component, is the metric that best relates to the risk of heart attack. Another approach has been to investigate cardiac consequences of particulate pollution in mechanistic terms and these studies have indicated associations with rises in indices of inflammation and in some cases falls in red blood cell counts, suggesting activation of the endothelial cells that line blood vessels (Seaton et al., 2010). Human inhalation studies have used diesel particles to investigate possible cardiac effects and have shown low concentrations to be able to cause release of vasomotor factors that may be relevant to control of the coronary arteries.

Moving closer to the at-risk population, patients with stable coronary artery disease were exposed to dilute diesel exhaust (300 μg m^{-3}) during intermittent exercise and the acute release of tissue plasminogen activator, a key regulator of endogenous fibrinolytic capacity, was shown to be reduced after diesel exhaust inhalation (Seaton et al., 2010). The extent of this reduction was comparable to that seen in cigarette smokers. The clinical effect of diesel exposure was further examined in patients with coronary heart disease. When patients were exposed to diesel exhaust, myocardial ischaemia was quantified by ST-segment analysis using continuous 12-lead electrocardiography and a threefold greater increase in ischaemia was evident during exposure to diesel exhaust than during exposure to filtered air.

8.10.2 Transgression of biological membranes

An NP may for practical purposes be characterized as having at least one dimension less than 100 nm. Thus, nanotubes and nanofibres may be many micrometres in length and only a few nanometres in diameter. Many asbestos fibres and fibrils conform to this definition. The fibrogenicity and carcinogenicity of asbestos is now very well known and much research has led to the conclusion that this human toxicity is related primarily to four factors (Seaton et al., 2010).

i. A length greater than about 15 μm, below which the fibre can be removed by lung macrophages;

ii. A diameter less than about 3 μm, allowing the fibre to be inhaled to the gas-exchanging part of the lung;

iii. Insolubility or resistance to dissolution in the lung milieu;

iv. A sufficient dose to the target organ.

This has become known as the fibre paradigm and is very likely to apply to any inhaled fibre. It seems that the chemical composition of the fibre is relatively unimportant other than as a determinant of its solubility and those macrophages may ingest many short fibres without coming to harm. There is evidence, for example, that the paradigm applies to fibrous erionite, to which individuals in Turkey suffering from asbestos-type diseases have been exposed environmentally. The mechanism appears to be frustrated phagocytosis, whereby the macrophage is injured in attempting to engulf long fibres and releases cytokines, mitogens and oxidants that initiate the process of fibrosis and carcinogenesis. Experimentally, similar consequences have been shown to occur with other appropriately sized fibres of different compositions).

The concept of translocation takes the ultrafine hypothesis a step further, to propose that the toxic effect of the particles is a consequence of a direct action on the distal target organ rather than the one mediated by cytokine release and secondary effects on the target. Thus, it proposes that inhaled NPs pass through the lungs' alveolar membrane and go via the blood to the heart where they exert a directly toxic effect. This seems unlikely, ignoring as it does the known mechanisms of myocardial infarction, the amplification effect of provoking an inflammatory reaction, the dilutional effect of distribution through the blood and the likely uptake of any blood-borne particles by the reticulo-endothelial system. Nevertheless, some studies have searched for and found NPs in various organs including the heart following inhalation exposure (Seaton et al., 2010). Such translocation is small as a fraction of the deposited dose and has been focused on model and surrogate particles such as elemental carbon and iridium. So far, there are no data showing the translocation of environmental NPs such as those generated by diesel combustion.

8.10.3 Inflammation and other mechanisms

Epidemiological studies have provided ample evidence, mainly in the form of rises in C-reactive protein and white blood cell count, that exposure to air pollution is associated with a generalized inflammatory reaction in the population being studied (Mills et al., 2009). Chamber studies using concentrated ambient particles have shown these particles to be mildly inflammagenic as measured by a leucocyte response in alveolar fluid obtained by bronchoscopy. Human experimental studies have also shown increases in the release of vasoconstrictor substances with exposure to diesel exhaust, in this case without overt evidence of an inflammatory reaction. It seems to us likely that two or more mechanisms of toxicity are involved in mediating the association of heart disease and particulate air pollution: (i) indirect effects of pulmonary inflammation and oxidative stress on atherothrombosis and (ii) translocation of particles to blood vessels and a direct effect on their lining endothelial cells.

As noted above, longitudinal epidemiological studies have shown an association between historic exposures to pollution and later risk of developing a heart attack, strongly suggesting that pollution exposure is a risk factor for the development of arterial atheroma. Experimental support for this has been provided by studies of animals genetically predisposed to atheroma. The well-documented cardiovascular impact of ambient particles is considered to be a result of the combustion-derived NP component. The high fat diet rabbit and ApoE mouse model have been used extensively to demonstrate that pulmonary exposure to particles causes plaque expansion and increases in indices of plaque destabilization. Currently, there are only two studies of CNTs. One showed evidence of oxidative stress in the aorta and progression of atherosclerosis in ApoE mice after a single pulmonary aspiration of single wall CNT Seaton et al., 2010). The other showed upregulation of gene expression in aortic tissue after deposition of CNT in mouse lungs (Erdley et al., 2009)). For some genes, this was orders of magnitude greater with CNT than with carbon black, suggesting that CNT may have a special action that affects the vessel wall. It is not clear whether the CNTs themselves reach the aortic wall/atherosclerotic plaques or whether there is an indirect inflammatory/oxidative stress mechanism, although the latter is more plausible.

Both the potential benefits and the risks, associated with the application of engineered nanoparticles (ENPs) have been widely debated in recent years. In contrast to the dominating optimistic projections that nanotechnology will bring significant technological development and well-being to society, it is considered that exposure to certain ENPs may cause environmental problems and/or do harm to human health. Since the early discussions about the risks of ENPs, the chemical risk assessment (CRA) has been put forward as the most relevant approach to understand, evaluate and quantify these risks (Danail Hristozov and Ineke Malsch, 2009). Currently, a variety of methodologies are being internationally discussed and evaluated with great vengeance with the idea that, in the near future, it will be possible to perform complete and scientifically sound risk assessment of ENPs.

Hazard identification (HI): It is defined as the identification of the adverse effects, which a substance has an inherent capacity to cause. Until recently, much of the discussion about the environmental and health risks of ENPs was considered to be rather speculative than realistic. In the last few years, however, a number of experimental studies found that exposure to certain ENPs can lead to adverse health effects in living organisms.

8.11 Toxicological effects

In spite of the scientific knowledge gained in recent years in nanotoxicology, scientists have not been able to precisely anticipate the behavior and biokinetics of nanoparticles. Medical, academic, and regulatory communities, both at the national and international level, need to determine the potential threats in the workplace for manufacturers and the environment. It should be kept in mind that results are often obtained for only one type and size of NPs. Furthermore, test animals are generally exposed to high concentrations under artificial conditions (Tamara Forbe et al., 2011). Presently, there are very few data to determine which type of effects are to be expected for which type of NP.

Due to their size, nanoparticles have a large specific surface area, which may translate into increased biological activity due to different contact interactions with cells and their components and variable biokinetics. The stability of nanoparticles requires further detailed investigation; however, the possibility of Oswalt ripening and agglomeration exists. Nanoparticles favor the formation of pro-oxidants, especially under exposure to light, ultraviolet light, or transition metals, thereby, destabilizing the balance between the production of reactive oxygen species (ROS) and the biological system's ability to detoxify or repair the system. Oxidative stress induced by nanoparticles is reported to enhance inflammation through upregulation of redox-sensitive transcription factors including nuclear factor kappa B (NFKB), activating protein 1 (AP-1), extracellular signal regulated kinases (ERK) c-Jun, N-terminal kinases, JNK, and p38 mitogen-activated protein kinases pathways. Nanoparticles can modify mitochondrial function, as well as cellular redox signaling (Tamara Forbe et al., 2011).

i. **Acute toxicity :** Acute, subacute, and subchronic toxicity following oral exposure have been investigated in rodents for several different NPs (e.g., copper, selenium, zinc, zinc-oxide, and titanium dioxide NPs). The results of the available oral toxicity studies indicate that, depending on the particle size, coating, and chemical composition of the NPs, acute toxicity at high doses may occur (Bouwmeester et al., 2009).

ii. **Long term toxicity:** Information from toxicity studies indicate that several systemic effects on different organ systems may occur after long term exposure to NPs including the immune, inflammatory and cardiovascular system. Effects on the immune and inflammatory systems may include oxidative stress and/or activation of pro-inflammatory cytokines in the lungs, liver, heart, and brain. Effects on the cardiovascular system may include pro-thrombotic effects and adverse effects on the cardiac function (acute myocardial infarction and adverse effects on the heart rate). No information on the toxicity after chronic or acute low dose oral exposure is currently available. Furthermore, genotoxicity, possible carcinogenesis, and teratogenicity may occur, but no data on these endpoints are available yet (Bouwmeester et al., 2009).

Neurotoxicity: There is evidence from ADME studies that NP may pass the blood-brain barrier following systemic availability of NPs. It is not clear if this is a generic effect of all NPs or only a subgroup. This emphasizes the need for kinetic studies and increased attention of toxicologists to neurotoxicity in their search for potential effects in target tissues. Toxic effects due to the presence (or even accumulation) of NPs in the brain have not been studied so far, but risk assessors should be aware of possible neurological effects when assessing toxicology experiments. Possibly, current guideline tests will need to be adapted to render these tests more sensitive for neurotoxic effects of NPs.

Reprotoxicity: Transfer of NPs across the placenta cannot be excluded (including excretion via breast milk, the blood milk barrier), which could lead to embryotoxicity as a result of exposure to NPs. Data addressing the distribution of NPs to the reproductive cells are currently unavailable. In addition, no clear data showing the distribution of NPs in the fetus are available. This leads to the recommendation that reprotoxicity needs to be carefully considered when there is evidence for NP passage of the placenta.

Mutagenicity: Intracellular NPs do not appear to be membrane bound and might have direct access to the intracellular proteins, organelles and DNA of the cell, which might imply enhanced toxic potential. However, possible interactions of NPs with cell components are poorly understood, and validated assays with meaningful endpoints for genotoxicity are needed (Bouwmeester et al., 2009).

Allergenicity (or sensitization): Even for conventional chemicals, little is known on the induction of food allergy and the type of exposure required inducing such responses. In the case of NPs, this becomes extra prominent for two reasons. First of all, it is the possible adjuvant activity of NPs that introduces additional uncertainty. Secondly, because of the actively charged surfaces of NPs, they can absorb biomolecules as they pass through the GI tract, which might result in changed exposure of the cellular lining of the intestine. In addition, although they might also be a reason for concern, surface properties (e.g., coatings) are important determinants for the active uptake of encapsulates. For example, lectins used for coatings are highly immunogenic, can be cytotoxic, or induce inflammatory responses and gastrointestinal irritation (Bouwmeester et al., 2009).

Skin: The outer layer of the epidermis, the stratum corneum (SC), covers the entire outside of the body and only contains dead cells, which are strongly keratinized. For most chemicals the SC is the rate-limiting barrier to percutaneous absorption (penetration). The skin of most mammalian species is, on most parts of the body, covered with hair. At the sites, where hair follicles grow, the barrier capacity of the skin differs slightly from the "normal" stratified squamous epidermis. Most studies concerning penetration of materials into the skin have focussed on whether or not drugs penetrate through the skin using different formulations containing chemicals and/or particulate materials as a vehicle. The main types of particulate materials commonly used are: liposomes; solid poorly soluble materials such as TiO2 and polymer particulates and submicron emulsion particles such as solid lipid nanoparticles (Hoet et al., 2012). The penetration of these particulate carriers has not been studied in detail.

TiO_2 particles are often used in sunscreens to absorb UV light and therefore to protect skin against sunburn or genetic damage. micrometer-sized particles of TiO_2 get through the human stratum corneum and even into some hair follicles – including their deeper parts.

Penetration of non-metallic solid materials such as biodegradable poly (D,L-lactic-co-glycolic acid (PLGA) microparticles, 1 to 10 micron with a mean diameter of 4.61 ± 0.8 micron was studied after application on to porcine skin. The number of microparticles in the skin decreased with the depth (measured from the airside towards the subcutaneous layer). At 120 micron depth (where viable dermis present) a relatively high number of particles was found, at 400 micron (dermis) some micro-particles were still seen. At a depth of 500 micron no microparticles were found. In the skin of individuals, who had an impaired lymphatic drainage of the lower legs, soil microparticles, frequently 0.4–0.5 micron but as larger particles of 25 micron diameter, were found in the in the dermis of the foot in a patient with endemic elephantiasis. The particles are seen to be in the phagosomes of macrophages or in the cytoplasm of other cells. The failure to conduct lymph to the node produces a permanent deposit of silica in the dermal tissues (a parallel is drawn with similar deposits in the lung in pneumoconiosis). This indicates that soil particles penetrate through (damaged) skin, most probably in every individual, and normally are removed via the lymphatic system. Liposomes penetrate the skin in a size dependent manner. Micro-sized, and even submicron sized, liposomes do not easily penetrate into the viable epidermis, while liposomes with an average diameter of 272 nm can reach into the viable epidermis and some are found in the dermis. Smaller sized liposomes of 116 and 71 nm were found in higher concentration in the dermis.

Mechanical irritation of skin: Glass fibres and Rockwool fibres are widely distributed man-made mineral fibres because of their multiple applications, mainly as insulation materials, which have become important for replacing asbestos fibres. In contact with the skin, these fibres can induce dermatitis through the mechanical irritation(Hoet et al., 2012). Why these fibres are such strong irritant has not been examined in detail. In occlusion irritant patch tests in humans it was found that Rockwool fibres with a diameter of 4.20 ± 1.96 micron were more irritating than those with a mean diameter of

3.20 ± 1.50 micron. The fact that "small" fibres can cause strong skin irritation has been known for a long time, e.g. itching powder. It is also commonly accepted that some types of man made fibres can easily induce non-allergic dermatitis. Although this is common knowledge, it is not clear what makes these fibres irritants. In search for reports on skin irritation caused by fibres with a diameter of < 100 nm no information could be found, indicating that more research is needed.

Carbon nanotubes (CNTs): Single-walled carbon nanotubes (SWCNTs) are able to cause dose-dependent effects of interstitial inflammation and lesions in mice and rats (0–0.5 mg·kg–1 for 7 to 90 days) (Lam, 2004).A dose-dependent rise in ventilation rate, gill pathologies (oedema, altered mucocytes, hyperplasia), and mucus secretion with SWCNT precipitation on the gill mucus were noticed. They also observed a significant dose-dependent decrease in thiobarbituric acid reactive substances (TBARS), especially in the gill, brain and liver, which is an indication of oxidative stress.

C_{60} fullerenes: Most studies on the toxicological effects of C_{60} fullerenes suggest that these materials tend to induce oxidative stress in living organisms (Lai,, 2000) a significant increase in lipid peroxidation (LP) products (a sign of oxidative stress) after intravenous administration of 1 mg kg–1 C_{60} (OH)18 in male mongrel dogs ; the effects of C_{60} fullerenes in the brain of juvenile largemouth bass and observed high LP levels (0.5 and 1 ppm for 48 h).; an increase in the numbers of bronchoalveolar lavage (BAL)-recovered neutrophils (i.e., white blood cells) after intratracheal instillation of C_{60} and C_{60} (OH)24 in rats, 1 day after the exposure.

Metal and metal oxide ENPs: Metal ENPs induce more severe lung toxicity in mice than bulk particles from the same materials. Theeffects were tested on humans of exposure to zinc (Zn) ENPs. After 2 hours of exposure to 5 $mg{\cdot}m^{-3}$ of Zn ENPs, the exposed individuals started feeling sore throat, chest tightness, headache, fever and chills.environmental exposure to Zn ENPs causes pulmonary (lung) inflammatory response in mice; Zn ENPs can cause severe symptoms of lethargy, anorexia, vomiting, diarrhea, loss of body weight and even death in mice when gastrointestinally administered, whereas they observed limited effect for micro-scale Zn at equal concentrations.; hypoactivity, ataxia, emesis, exophthalmos, salivation, lacrimation, discolored and mucoid feces, injected sclera, and yellow eyes in dogs after single- dose intravenous bolus administration of 20 and 200 $mg{\cdot}kg^{-1}$ FeO ENPs and a significant increase in fetal skeletal malformations in rats and rabbits (Bourrinet, 2006).

Quantum dots (QDs): The toxicity of QDs was found to be influenced by several factors, viz., composition, size, surface charge and coating of the QDs (Hansen, 2009). CdSe/ZnS QDs (i.e., CdSe QDs in a zinc sulfide (ZnS) matrix), coated with dihydrolipoic acid (DHLA) had no effect on mammalian cells. Adverse effects were reported on mouse lymphocytes after exposure to CdSe/ZnS QDs, coated with albumin. Smaller (2.2 ± 0.1 nm), positively charged QDs exhibited stronger cytotoxicity than larger (5.2 ±0.1 nm), equally charged QDs under the same conditions. It was also found that the cytotoxicity of QDs is influenced by the exposure to light and by temperature. 56% damaged DNA after exposure to CdSe/ZnS together with UV light versus only 29% after exposure to CdSe/Zn in the absence of UV light. CdSe/CdS (i.e., CdSe QDs in a cadmium sulfide (CdS) matrix) were toxic to cancer cells at 37 °C, but at 4 °C they were not toxic at all.

Both *in vivo* and *in vitro* studies demonstrated the toxicity of AgNPs on mammalian cells. For example, AgNP exposure reduced lung function and produced inflammatory lesions in the lungs of rat (Sung et al., 2008) and resulted in the accumulation of AgNPs in the olfactory bulbs and in the brain of rats. Since AgNPs can be accumulated and transferred in the food chain, they may become dangerous to humans. Indeed, AgNPs exposure to human cells can stimulate inflammatory and immunological responses, cause oxidative stress, and lead to cellular damage. Therefore, there is an urgent need to increase our understanding of the bioaccumulation and trophic transfer of AgNPs in the food chain, which is critical for assessing and mitigating their potential harm to human health.

Antibacterial effect of silver nanoparticles (SNPs) has resulted in their widespread applications in health, electronics,medicine and home products. SNPs also possess unique optic and catalytic properties, making them highly attractive forthe creation of new advanced functional materials. Once released into the environment, SNPs may change chemically andpose toxicity to different faunal species including humans (Singh, 2018). SNPs may penetrate the gastrointestinal barrier and reach thecirculatory system resulting in severe health impacts. This review emphasized the toxicity of SNPs ondifferent human cell lines including human adipose-derived stem cells, human hepatoma cells, human mesenchymal stemcells, human intestinal cells, human epidermoid larynx carcinoma cell line and human Jurkat T cells through compilingthe available reports. SNPs were found to display toxicity to different cell lines at different concentrations. SNP exposureto humans should be minimized so as to minimize the health hazards.

Nanoparticles are reported to have various adverse effects in different animal tissue and also in human health (Elalfyet al., 2018). Various nanoparticles have toxic effect on liver, lung, skin, eye, reproductive disturbance and blood cells. The key mechanism of toxicity of nanoparticle is due to increase its concentration in non-target tissue of therapy through genotoxicity, oxidative stress or hemotoxic effects. On other word, nanoparticles could have adverse effects on healthy rather than diseased tissue.

The trace mineral elements in brain were altered after the exposure to TiO_2 NP which may responsible of the spatial recognition memory impairments reported in treated mice due to the disrupted homeostasis of trace elements, neurotransmitters and enzymes in the brain. Also Titanium dioxide (TiO_2) has been recorded to induce diverse pulmonary responses in exposed animals (Fig. 54) (Hu et al., 2010).

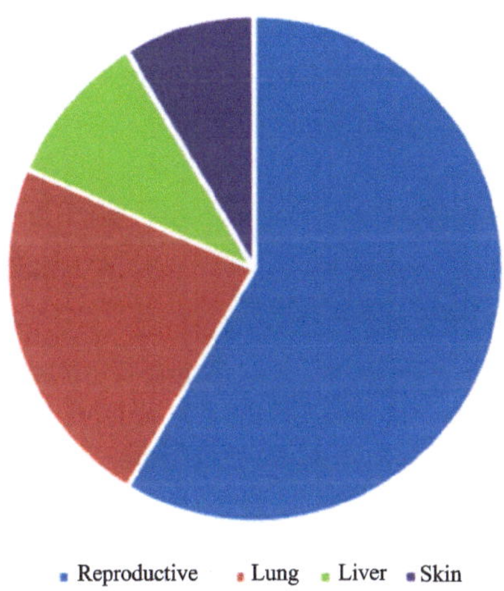

Fig. 54: Major non-target toxicity of nanoparticles

Immunity toxicity: Silicon-based materials and their oxides are widely used in drug delivery, dietary supplements, implants and dental fillers. Silica nanoparticles (SiNPs) interact with immunocompetent cells and induce immunotoxicity (Liangjiao Chen et al., 2018). However, the toxic effects of SiNPs on the immune system have been inadequately reviewed. The toxicity of SiNPs to the immune system depends on their physicochemical properties and the cell type. Assessments of immunotoxicity include determining cell dysfunctions, cytotoxicity and genotoxicity. This review focused on the immunotoxicity of SiNPs and investigates the underlying mechanisms. The main mechanisms were proinflammatory responses, oxidative stress and autophagy. Considering the toxicity of SiNPs, surface and shape modifications may mitigate the toxic effects of SiNPs, providing a new way to produce these nanomaterials with less toxic impaction.Mnocytes/macrophages/Kupffer cells and microglia. Silica nanoparticles (SiNPs) change the activity and abilities of macrophages/monocytes. The cytotoxicity of SiNPs is dependent on the physicochemical properties of particles.

i. Toxicity to immune cells

Dendritic cells (DCs) and Langerhans cells: SiNPs enhance antigen-specific cellular immune responses and change cell functions, and have size- and concentration-dependent cytotoxic effects on DCs.

Lymphocytes: The viability and proliferative activity of lymphocytes are dependent on particle size, concentration, charge and other physicochemical characteristics.

Mast cells: These cell functions were changed after exposure. No studies have focused on the genotoxicity against these cells.

Endothelial cells: Cell function was changed.

ii. Immunotoxicity to tissues & organs

Liver: Lymphocytic infiltration, granuloma formation and hydropic degeneration in liver hepatocytes were observed.

Spleen: The proliferation of immune cells such as B and T cells in spleen is decreased. Lymphocyte populations are altered in the spleen, and IgG and IgM levels in serum increase and cause histological changes.

Lung: Neutrophilic infiltration, increased mast cell abundance and pulmonary injury are greater in the lung.

Heart: SiNPs induce neutrophil-mediated cardiac inflammation in vivo.

Hemocytes and blood circulation: SiNPs initiate autophagy and apoptosis when they enter hemocytes.

iii. Mechanism of the toxicity

- Proinflammatory responses are an important element in SiNPs-induced immunotoxicity. NLRP3 inflammasome and Toll-like receptors are the main signal in inducing proinflammatory responses.
- Oxidative stress is an underlying mechanism that causes immunotoxicity. NADPH oxidase pathway and MAPK pathway are the main signal in cause oxidative stress.

- Autophagy. Autophagy dysfunction is considered a potential toxic mechanism of SiNPs.
- Other mechanisms: Activation of the intrinsic apoptotic pathway plays important role in cytotoxicity.

iv. The modification of immunotoxicity

- The toxic effects of SiNPs are dramatically attenuated by nanoparticles surface-functionalized with amino and phosphate groups.
- Shape modifications mitigate the toxic effects of SiNPs, providing a new way to produce safer nanomaterials.

8.12 Biodegradation of nanomaterials

Biodegradation of nanoparticles may result in their breakdown as typically seen in biodegradation of organic molecules, or may result in changes in the physical structure or surface characteristics of the material. The potential for and possible mechanisms of biodegradation of nanosized particles have just begun to be investigated. As is the case for other fate processes, the potential for biodegradation will depend strongly on the chemical and physical nature of the particle. Many of the nanomaterials in current use are composed of inherently nonbiodegradable inorganic chemicals, such as ceramics, metals and metal oxides, and are not expected to biodegrade. However, a recent preliminary study found that C_{60} and C_{70} fullerenes were taken up by wood decay fungi after 12 weeks, suggesting that the fullerene carbon had been metabolized (EPA, 2007). For other nanomaterials biodegradability may be integral to the material's design and function. This is the case for some biodegradable polymers being investigated for use in drug transport, for which biodegradability is mostly a function of chemical structure and not particle size.

Biodegradability in waste treatment and the environment may be influenced by a variety of factors. Recent laboratory studies on C_{60} fullerenes have indicated the development of stable colloid structures in water that demonstrate toxicity to bacteria under aerobic and anaerobic conditions . Further studies are needed to determine whether fullerenes may be toxic to microorganisms under environmental conditions. One must also consider the potential of photoreactions and other abiotic processes to alter the bioavailability and thus biodegradation rates of nanomaterials. In summary, not enough is known to enable meaningful predictions on the biodegradation of nanomaterials in the environment and much further testing and research are needed.

Bioavailability and bioaccumulation of nanomaterials: Bacteria and living cells can take up nanosized particles, providing the basis for potential bioaccumulation in the food chain. Aquatic and marine filter feeders near the base of the food chain feed on small particles, even particles down to the nanometer size fraction. The bioavailability of specific nanomaterials in the environment will depend in part on the particle. Environmental fate processes may be too slow for effective removal of persistent nanomaterials before they can be taken up by an organism. In the previous section, it was noted that some physical removal processes, such as gravitational settling, are slower for nanosized particles than for microparticles. This would lead to an increased potential for inhalation exposure to terrestrial organisms and for increased exposure of aquatic organisms to aqueous colloids. Not enough information has been generated on rates of deposition of nanomaterials from the atmosphere and surface water, or of sorption to suspended soils and sediments in the water column, to determine whether these processes could effectively sequester specific nanoparticles before they are taken up by organisms.

Complexation of metallic nanomaterials may have important interactive effects on biological availability and photochemical reactivity. For example, the biological availability of iron depends on its free ion concentrations in water and the free ion concentrations are affected by complexation (EPA, 2007). Complexation reduces biological availability by reducing free metal ion concentrations and dissolved iron is quantitatively complexed by organic ligands. Solar UV radiation can interact with these processes through photoreactions of the complexes. Further, iron and iron oxides can participate in enzymatic redox reactions that change the oxidation state, physical chemical properties and bioavailability of the metal.

Potential for toxic transformation products from nanomaterials: Certain nanomaterials are being designed for release as reactants in the environment, and therefore are expected to undergo chemical transformation. One example of this is iron nanoparticles employed as reactants for the dechlorination of organic pollutants (Zhang, 2003). As the reaction progresses, the iron is oxidized to iron oxide. Other metal particles are also converted to oxides in the presence of air and water. Whether the oxides are more or less toxic than the free metals depends on the metal. Under the right conditions, certain metal compounds could be converted to more mobile compounds. In these cases, small particle size would most likely enhance this inherent reactivity. Some types of quantum dots have been shown to degrade under photolytic and oxidative conditions, and furthermore, compromise of quantum dot coatings can reveal the metalloid core, which may be toxic. Degradation products from carbon nanomaterials (fullerenes and nanotubes) have not yet been reported.

Applicability of current environmental fate and transport models to nanomaterials: When performing exposure assessments on materials for which there are no experimental data, models are often used to generate estimated data, which can provide a basis for making regulatory decisions. It would be advantageous if such models could be applied to provide estimated properties for nanomaterials, since there is very little experimental data available for these materials. The models used by EPA's Office of Pollution Prevention and Toxics (OPPT) to assess environmental fate and exposure, are, for the most part, designed to provide estimates for organic molecules with defined and discrete structures. These models are not designed for use on inorganic materials; therefore, they cannot be applied to inorganic nanomaterials. Many models derive their estimates from structural information and require that a precise structure of the material of interest be provided. Since many of the nanomaterials in current use, such as quantum dots, ceramics and metals, are solids without discrete molecular structures, it is not possible to provide the precise chemical structures that these models need. While it is usually possible to determine distinct structures for fullerenes, the models cannot accept the complex fused-ring structures of the fullerenes. Also, the training sets of chemicals with which the quantitative structure-activity relationships (QSAR) in the models were developed do not include fullerene-type materials. Fullerenes are unique materials with unusual properties, and they cannot be reliably modeled by QSARs developed for other substantially different types of materials.

Models, used in general, to assess the environmental fate and exposure to chemicals are not applicable to intentionally produced nanomaterials. Depending on the relevance of the chemical property or transformation process, new models may have to be developed to provide estimations for these materials; however, models cannot be developed without the experimental data needed to design and validate them. Before the environmental fate, transport and multimedia partitioning of nanomaterials can be effectively modeled, reliable experimental data must be acquired for a variety of intentionally produced nanomaterials. However, models are also used which focus on the fate and distribution of particulate matter (air models) and/or colloidal materials (soil, water, landfill leachates, ground water), rather than discrete organics (EPA, 2007). For example, fate of atmospheric particulate matter (e.g., PM10) has been the subject of substantial research interest and is a principal regulatory focus of EPA's Office of Air and Radiation. Since intentionally produced nanomaterials are expected to be released to and exist in the environment as particles in most cases, it is wise to investigate applicability of these other models. In fact it can be reasoned that the most useful modeling tools for exposure assessment of nanomaterials are likely to be found not in the area of environmental fate of specific organic compounds (more precisely, prediction of their transport and transformation), rather in fields in which the focus is on media-oriented pollution issues: air pollution, water quality, ground water contamination, etc. A survey of such tools should be made and their potential utility for nanomaterials assessed.

8.13 Limitations of current research and prospects

- The biological matrices and dispersion of nanomaterials may determine the results of toxicity assessment. However, there is lack of research focusing on these points.
- The purity of SiNPs should be evaluated in details. The presence of impurities, such as the presence of surfactants (e.g., CTAB or CTAC used to make mesoporous silica) may influence the results of toxicity evaluation.
- Data on the direct toxicity of SiNPs to immune organs such as bone marrow, thymus and lymph node are very limited. Furthermore, no study has focused on the effects of SiNPs regulating the immune response; they may activate or suppress immunity.
- Different routes of administration of SiNPs in vivo have different results. The comparison of different administration route is lacked.
- The endotoxin contamination and the sterility of SiNPs affect the proinflammatory effect and reactive oxygen species. The detailed information of original articles should be clear and deeply research should consider this point.
- The interrelationships and interactions of these three mechanisms were unclear.

Bio-formation and biodistribution of mercury selenide (HgSe) nanoparticles (NPs), as the ultimate metabolic product in the Hg detoxification pathway, is an important environmental and toxicological issue not well understood nowadays (Laura Cid-Barrio et al., 2020). It wasrevealed for the first time the cellular uptake and accumulation of HgSe NPs in one embryonic and one tumour human cell line (HEK 293 and HeLa).

Due to the widespread applications of zinc oxide nanoparticles (ZnO NPs), the potential exposure of workers, consumers, and scientists to these particles has increased. This potential for exposure has attracted extensive attention in the science community (Liu et al., 2017). Many studies have examined the toxicological profile of ZnO NPs in the immune system, digestive system, however, information regarding the toxicity of ZnO NPs in the nervous system is scarce. They detected the cytotoxicity of two types of ZnO NPs of various sizes , ZnO^a NPs and ZnO^b NPs and we characterized the shedding ability of zinc ions within culture medium and the cytoplasm.

The health effects of nanoparticles (NPs) are raising considerable and growing concerns from the public and government around the world. Nano-metals are new forms of metal with special properties, but are characterized by having a particle size of less than 100 nm. Copper is an essential trace element and its deficiency leads to different diseases in humans (Marzie Hejazy et al., 2018). In general, copper NPs have been reported among the most toxic nanomaterials in mammals. The toxicity of nano-copper depends on sex. The increase in the production of reactive oxygen species (ROS) and reactive nitrogen species (RNS) plays an important role in copper-induced organic dysfunction, DNA damage as a result of oxidative stress, changes in blood biochemical indexes, and the mechanism of the nano-copper induced apoptosis.

Human oral exposure to copper oxide nanoparticles (NPs) may occur following ingestion, hand-to-mouth activity, or mucociliary transport following inhalation. Astudy byHughes et al., (2018) assessed the cytotoxicity of CuO and Cu_2O-polyvinylpyrrolidone (PVP) coated NPs and Cu_2+ ions in rat (IEC- 6) and human intestinal cells, two- and three-dimensional models, respectively.

The effect of pre-treatment of CuO NPs with simulated gastrointestinal (GI) fluids on IEC-6 cell cytotoxicity was also investigated. Both dose- and time-dependent decreases in viability of rat and human cells with CuO and Cu_2O-PVP NPs and Cu_2+ ions was observed. In the rat cells, CuO NPs had greater cytotoxicity. The rat cells were also more sensitive to CuO NPs than the human cells. Concentrations of H_2O_2 and glutathione increased and decreased, respectively, in IEC-6 cells after a 4-h exposure to CuO NPs, suggesting formation of reactive oxygen species (ROS).

Applications of nanoparticles (NP) in medicine, industry and other branches of human activities undoubtedly contribute to technology development and well-being. However, as NP are very small units in a wide range of materials, there is a lack of information related to possible side effects potentially affecting the health of organisms (Celia et al., 2014). In vitro long-term (20 wk) changes were reported in cells exposed to well- characterized gold nanoparticles (Au NPs) with varying shapes and surface coatings under both chronic (exposure to Au NPs continuously over 20 wk) and nonchronic (initial acute cell exposure to Au NPs, followed by 20 wk in NP-free cell media) conditions (Priscila Falagan-Lotsch et al., 2016).

The rapid growth of nanotechnology suggests that it will soon find widespread application in everyday consumer products and in new pharmaceutical, electronic and other industries.

Researchers from Lancaster University have discovered tiny magnetic nanoparticles from air pollution lodged in the human brain (Anonymous, 2016).

8.14 Characteristics of nanoparticles relevant for health effects

Studies specifically dealing with the toxicity of nanoparticles have only appeared recently and are still scarce. Most of the information available comes from studies on inhaled nanoparticles and from pharmaceutical studies in which nanomaterials are used, among other things to improve drug delivery. The characteristics of nanoparticles that are relevant for health effects include the following.

- **Size**: In addition to being able to cross cell membranes, reach the blood and various organs because of their very small size, nanoparticles of any material have a much greater surface to volume ratio (i.e. the surface area compared to the volume) than larger particles of that same material. Therefore, relatively more molecules of the chemical are present on the surface. This may be one of the reasons why nanoparticles are generally more toxic than larger particles of the same composition.
- **Chemical composition and surface characteristics**: The toxicity of nanoparticles depends on their chemical composition, but also on the composition of any chemicals adsorbed onto their surfaces. However, the surfaces of nanoparticles can be modified to make them less harmful to health.
- **Shape**: Although there is little definitive evidence, the health effects of nanoparticles are likely to depend also on their shape. A significant example is nanotubes, which may be of a few nanometres in diameter but with a length that could be several micrometres. A recent study showed a high toxicity of carbon nanotubes which seemed to produce harmful effects by an entirely new mechanism, different from the normal model of toxic dusts.

Inhaled nanoparticles affect health: Particulate matter present in air pollution, especially from traffic emissions, is known to affect human health, although it is not clear exactly how. Epidemiological studies on ambient air pollution have not proved conclusively that nanoparticles are more harmful than larger particles, but these studies may not be well suited to demonstrate such differences. Inhaled particulate matter can be deposited throughout the human respiratory tract, and an important fraction of inhaled nanoparticles deposit in the lungs. Nanoparticles can potentially move from the lungs to other organs such as the brain, the liver, the spleen and possibly the foetus in pregnant women. Data on these pathways is extremely limited but the actual number of particles that move from one organ to another can be considerable, depending on exposure time. Even within the nanoscale, size is important and small nanoparticles have been shown to be more able to reach secondary organs than larger ones.

Another potential route of inhaled nanoparticles within the body is the olfactory nerve; nanoparticles may cross the mucous membrane inside the nose and then reach the brain through the olfactory nerve. Out of three human studies, only one showed a passage of inhaled nanoparticles into the bloodstream..Materials which by themselves are not very harmful could be toxic if they are inhaled in the form of nanoparticles.The effects of inhaled nanoparticles in the body may include lung inflammation and heart problems. Studies in humans show that breathing in diesel soot causes a general inflammatory response and alters the system that regulates the involuntary functions in the cardiovascular system, such as control of heart rate.The pulmonary injury and inflammation resulting from the inhalation of nanosize urban particulate matter appears to be due to the oxidative stress that these particles cause in the cells

The health implications of nanoparticles used as drug carriers: Nanoparticles can be used for drug delivery purposes, either as the drug itself or as the drug carrier. The product can be administered orally, applied onto the skin, or injected.The objective of drug delivery with nanoparticles is either to get more of the drug to the target cells or to reduce the harmful effects of the free drug on other organs, or both. Nanoparticles used in this way have to circulate long distances evading the protection mechanisms of the body. To achieve this, nanoparticles are conceived to stick to cell membranes, get inside specific cells in the body or in tumours, and pass through cells. The surfaces of nanoparticles are sometimes also modified to avoid being recognized and eliminated by the immune system.

With dermal administration, it was found that particle size was less important than the total charge in terms of permeation through the skin. For instance, only negatively charged particles were found to overcome the skin barrier and only when concentration of charge was high enough.Nanoparticles may be used effectively to deliver genes to cells, to treat cancer, as well as in vaccination. The use of nanoparticles as drug carriers may reduce the toxicity of the incorporated drug but it is sometimes difficult to distinguish the toxicity of the drug from that of the nanoparticle. Toxicity of gold nanoparticles, for instance, has been shown at high concentrations. In addition, nanoparticles trapped in the liver can affect the function of this organ. Nanoparticles have the potential to cross the blood brain barrier, which makes them extremely useful as a way to deliver drugs directly to the brain. On the other hand, this is also a major drawback because nanoparticles used to carry drugs may be toxic to the brain.

How should harmful effects of nanoparticles be assessed?: Traditionally, doses are measured in terms of mass because the harmful effects of any substance depend on the mass of the substance to which the individual is exposed. However, for nanoparticles it is more reasonable to measure doses also in terms of number of particles and their surface area because these parameters further determine the interactions of nanoparticles with biological systems.Several hypotheses were proposed for the adverse health effects of nanoparticles as part of ambient air pollution. These hypotheses address nanoparticle characteristics, their distribution, and their effects on organ systems, including effects on immune and inflammatory systems.However, some of these hypotheses may be of limited or no relevance for engineered nanoparticles. For instance, the adhesion of toxic substances onto the surface of nanoparticles may be of less relevance for production and handling facilities of large volumes of engineered nanoparticles compared to the particles in ambient air.

In addition, drawing conclusions from tests on healthy animal models may be unsuitable as some of the effects of nanoparticles may only be a risk for susceptible organisms and predisposed individuals, but not to healthy people. For instance, age, respiratory tract problems and other pollutants can modify the pulmonary inflammation and oxidative stress induced by nanoparticles. Because of the specific characteristics of nanoparticles, conventional toxicity tests may not be enough to detect all their possible harmful effects. Therefore, a series of specific tests was proposed to assess the toxicity of nanoparticles used in drug delivery systems. One mechanism of toxicity of nanoparticles is likely to be the induction of oxidative stress in cells and organs. Testing for interaction of nanoparticles with proteins and various cell types should be considered as part of the toxicological evaluation.With the exception of airborne particles delivered to the lung, information on the behaviour of nanoparticles in the body including distribution, accumulation, metabolism, and organ specific toxicity is still minimal.

Knowledge gaps: Four main areas require progress

- Nanomaterail and nanoparticle specific data but also information on the use of nanomaterials/particles in products and their release from products.
- Scientific understanding of nano-toxicological behaviour to enable the step towards generalisation and abstraction.
- To address not only simple existing nanomaterials but also monitor and assess developments of new and emerging generations of nanomaterials.
- To consider aspects of risk governance and how to deal with the difference in pace between nanomaterial innovations and our scientific and regulatory capacity to assess the uncertainties and risks, and ways of dealing with these potential risks and uncertainties.
- Government, society in general, the scientific community, and the business community need to cooperateto find ways of dealing with fundamentally new and innovative developments in both materials and risks. This would add a firm foundation for increased data availability and mutual understanding.

In general, the European Commission concluded that the current EU-legislative framework, to a large extent, covers potential risks in relation to nanomaterials. Yet current legislation may have to be modified in the light of remaining gaps and new information becoming available, for example regarding thresholds.

8.15 Nanoparticles in food

Nanotechnology offers the food industry a number of new approaches for improving the quality, shelf life, safety, and healthiness of foods. Nevertheless, there is concern from consumers, regulatory agencies, and the food industry about potential adverse effects (toxicity) associated with the application of nanotechnology in foods. In particular, there is concern about the direct incorporation of engineered nanoparticles into foods, such as those used as delivery systems for colors, flavors, preservatives, nutrients, and nutraceuticals, or those used to modify the optical, rheological, or flow properties of foods or food packaging. This review article summarizes the application of both inorganic (silver, iron oxide, titanium dioxide, silicon dioxide, and zinc oxide) and organic (lipid, protein, and carbohydrate) nanoparticles in foods, highlights the most important nanoparticle characteristics that influence their behavior, discusses the importance of food matrix and gastrointestinal tract effects on nanoparticle properties, emphasizes potential toxicity mechanisms of different food-grade nanoparticles, and stresses important areas where research is still needed. The authors note that nanoparticles are already present in many natural and processed foods, and that new kinds of nanoparticles may be utilized as functional ingredients by the food industry in the future. Many of these nanoparticles are unlikely to have adverse affects on human health, but there is evidence that some of them could have harmful effects and that future studies are required. Different types of nanoscale materials that may be found in foods, and their potential origins are highlighted in Table 46.

Table 46: Examples of different kinds of nanomaterial present within foods and their origin

Nanoscale material	Origin	Characteristics	Products
Organic nanoparticles			
Casein micelles	Natural	Protein–mineral clusters	Milk, cream
Cell organelles	Natural	Ribosomes, vacuoles, lysosome etc.	Meat, fish, fruits, vegetables, spices
Oil bodies	Natural	Phospholipid/protein-coated triglyceride droplets	Plants, seeds
Lipid nanoparticles	ENP	Solid particles or liquid droplets coated by emulsifiers	Some beverages, sauces, dressings, creams
Protein nanoparticles	ENP	Clusters of protein molecules held together by physical or covalent interactions	Mainly in development
Carbohydrate nanoparticles	ENP	Small solid fragments extracted from starch, cellulose, or chitosan. Clusters of polysaccharide molecules held together by physical or covalent interactions.	Mainly in development
Inorganic nanoparticles			
Iron oxide	ENP	FeO nanoparticles used to fortify foods with iron.	Nutritional supplements, sausage casings Candies, chewing
Titanium dioxide	ENP	TiO_2 nanoparticles used as whitening agents	Candies, chewing gums, bakery goods, milk powders. Salts, icing sugar, spices, dried milk, and dry mixes
Silicon dioxide	ENP	SiO_2 nanoparticles used to control powder flowability	Meat, food packages, containers, coatings
Silver	ENP	Ag nanoparticles used as antimicrobials in foods, coatings and packaging	

Key: ENP engineered nanoparticle, which may be intentionally or unintentionally added

8.16 Physicochemical and physiological mechanisms and nanotoxicity

Ingested nanoparticles may cause toxicity due to numerous physicochemical and physiological mechanisms depending on their compositions, structures, and properties. Following are the some of the most important mechanisms of nanoparticle toxicity (McClements and Xiao, 2017).

i. Interference with normal GIT function : The presence of nanoparticles in the gastrointestinal fluids could interfere with normal GIT functions. The small size of nanoparticles means they have a high specific surface area, which offers a large area for adsorption of any surface-active components in the GIT. For example, digestive or metabolic enzymes could adsorb to nanoparticle surfaces thereby altering their normal GIT function. Many globular proteins are denatured after adsorption to particle surfaces due to the change in their thermodynamic environment, which could lead to a reduction in the catalytic activity of some enzymes. Consequently, high levels of nanoparticles could reduce the rate or extent of starch, lipid, or protein digestion within the GIT. This effect is mainly important for inorganic nanoparticles, but it may also be important for some organic nanoparticles (particularly indigestible ones). An estimate of the potential magnitude of above effect can be obtained by considering the impact of inorganic nanoparticles on the digestion of lipid droplets. In this mixed system, the lipase molecules could adsorb to the surfaces of either the lipid droplets or the nanoparticles.

The concentration of inorganic nanoparticles in the small intestine is likely to be a fraction of a percent, and so this effect is only likely to be important for relatively large lipid droplets at relatively low concentrations. In addition, the effect is likely to be less than predicted by the above equations for a number of reasons: (i) the inorganic nanoparticles may aggregate in the GIT, which reduces their exposed surface area; (ii) the lipase molecules may adsorb more strongly to the lipid droplet surfaces than to the inorganic nanoparticles surfaces; (iii) there may be other surface-active substances in the GIT that compete with the lipase for the surfaces of the inorganic nanoparticles. Indeed, recent experiments in laboratories have shown that mixing titanium dioxide nanoparticles with lipid droplets only caused a slight decrease in lipid digestion. Other substances in the GIT fluids involved in the digestion of macronutrients may also adsorb to the surfaces of inorganic nanoparticles, such as bile salts and phospholipids.

Some types of inorganic nanoparticles may also be able to physically disrupt important structures within the GIT, such as the tight junctions or microvilli, thereby altering normal nutrient absorption and the protective function of the epithelium cells. The presence of nanoparticles in the GIT may also stimulate an immune response, which could have adverse effects on human health, and so this possibility should be tested for food-grade nanoparticles.

ii. Accumulation within specific tissues : The results of animal studies suggest that certain types of ingested nanoparticles are absorbed within the GIT and accumulate in numerous tissues (Gaillet and Rouanet,, 2015). Presumably, these nanoparticles travel across the mucus layer and are then absorbed by active or passive transport mechanisms. After they have been absorbed into the cells, the nanoparticles may be metabolized, transferred out of the cells, or accumulate within the cells. These processes depend on nanoparticle characteristics such as composition, dimensions, morphology, aggregation state and interfacial properties. The accumulation of nanoparticles within specific tissues may lead to long-term problems if they exhibit toxic effects above a certain accumulation threshold. This mechanism of action is likely to be most important for inorganic nanoparticles that are bio-persistent (not normally digested or metabolized in GIT).

iii. Cytotoxicity and cellular malfunction: Nanoparticles may produce toxicity in cells through a variety of different mechanisms, depending on their composition and structure. One of the most important factors contributing to the toxicity of inorganic nanoparticles is their ability to generate ROS, such as singlet oxygen, superoxide, hydrogen peroxide and hydroxyl radicals. These ROS may then cause damage to cell membranes, organelles, and the nucleus by interacting with lipids, proteins, or nucleic acids. As a result, many biochemical functions required to maintain cell viability, such as ATP production, DNA replication, and gene expression, may be adversely affected. A number of studies have reported the ability of inorganic nanoparticles to increase the generation of ROS in cells and to produce cytotoxicity, including silicon dioxide nanoparticles, ZnO nanoparticles, and silver nanoparticles. Some inorganic nanoparticles produce toxicity by generating ions (such as Ag^+ from silver nanoparticles or Zn^{2+} from zinc oxide nanoparticles) that interact with the normal functioning cellular components (such as proteins, nucleic acids, or lipids) required to maintain biochemical processes. These mechanisms of action are most likely to be important for inorganic nanoparticles that are absorbed by the intestinal cells, since most organic nanoparticles are digested before being absorbed. However, it is still unclear about the extent to which inorganic nanoparticles would produce cytotoxicity when they are consumed as part of a complex diet under normal conditions.

iv. Altered location of bioactive release : The encapsulation of bioactive agents (nutrients or nutraceuticals) within nanoparticles may alter the location of their release and absorption within the GIT. For example, a bioactive agent that is normally released in the mouth, stomach, or small intestine could be released within the colon. As a result, the physiological response and biological impact of the bioactive agent may be altered by nanoencapsulation, which could have potentially adverse health effects. For example, the encapsulation of digestible lipids within nanolaminated dietary fiber coatings may inhibit the rate and extent of lipid digestion in the upper GIT, so that high levels of undigested lipids reach the colon. These lipids may then be fermented by the colonic bacteria, which could cause gastrointestinal problems. Alternatively, an antimicrobial agent may be encapsulated within a nanoparticle that is not digested within the upper GIT, so that it reaches the colon, where it could alter the nature of the colonic microflora, which could again have adverse health effects.

v. Enhancement of oral bioavailability: One of the most widely studied applications of nanotechnology in the food industry is for the encapsulation and delivery of hydrophobic bioactive agents, such as certain nutrients and nutraceuticals. Numerous in vitro and in vivo studies have shown that delivering these bioactive agents within nanoparticles (rather than within larger particles or in bulk phases) can greatly increase their bioavailability. For example, nanoemulsions have been shown to increase the bioaccessibility or bioavailability of carotenoids,curcumin, coenzyme Q10, ω-3 fatty acids, and fat-soluble vitamins (Walker et al., 2015).There are a number of different physicochemical mechanisms that may be responsible for this improvement. In particular, the nanoparticles may increase the bioaccessibility, chemical stability, and/or absorption of the encapsulated bioactive agents. In general, nanoparticles tend to be digested or dissolved more rapidly in the GIT and/or release any encapsulated components more rapidly because of their small size and high surface area. This will lead to differences in the pharmacokinetics of the bioactive agents within the systemic circulation. A change in the exposure level of bioactive agents within the blood could have potentially adverse health effects. The biological effects of many bioactive agents depend on their exposure levels in the blood and specific tissues. If the exposure level is too low, then the bioactive agent will have little biological impact. If the exposure level is too high, then it may be toxic.

Nanoparticles may increase the bioavailability of bioactive agents through two different approaches: delivery systems or excipient systems.. In delivery systems, the bioactive agent is encapsulated within the nanoparticles. In excipient systems, a food containing nanoparticles (excipient food) is consumed with a bioactive-rich food (such as fruits or vegetables). In both cases, the delivery or excipient system is specifically designed to increase the bioavailability of the bioactive agents by increasing the bioaccessibility or absorption, or by modulating any transformations (such as chemical or biochemical reactions) of the bioactive agents in the GIT.

vi. Enhancement of pesticide bioavailability: The ability of nanoparticles to greatly increase the oral bioavailability of hydrophobic substances could also have adverse health effects by promoting the uptake of undesirable non-polar substances in foods, such as certain pesticides and hormones. For example, a food product that contains lipid nanoparticles (such as a beverage, sauce, dressing, or cream) may increase the bioavailability of hydrophobic pesticides on fruits or vegetables consumed with them. At present there is little information on this mechanism, and there is a need for furthe studies. This mechanism is likely to be most important for foods containing lipid nanoparticles (nanoemulsions) that are consumed with foods potentially containing high levels of hydrophobic pesticides or hormones (such as some fruits and vegetables).

vii. Interference with gut microbiota: Nanoparticles that reach the colon may interact with colonic bacteria and alter their viability, thereby changing the relative proportions of different bacterial species present.The type of bacteria populating the human colon is known to play a major role in human health and wellbeing (Consortium, 2012). Consequently, any change in the gut microbiota due to the presence of food-grade nanoparticles could have adverse health effects. This is an important area that requires further research to determine the impact of specific nanoparticle characteristics on the gut microbiota and the resulting health implications.

Bibliography

Albert J. Omlor , Duc D. Le , Janine Schlicker , Raphael Ewen , Sebasθan Heck , Robert Bals, Juliane Nguyen and Q. Thai Dinh, 2017. P. 38. In : NANOSAFETY 2017. Organised by INM – Leibniz Institute for New Materials in association with Leibniz Research Alliance Nanosafety, Saarbrücken, Germany, 11–13 October 2017

Aḿerica Berenice Morales-Díaz, Hortensia Ortega-Ortíz, Antonio Juárez-Maldonado , Gregorio Cadenas-Pliego, Susana Gonzâlez-Morales and Adalberto Benavides-Mendoza, 2017. Application of nanoelements in plant nutrition and its impact in ecosystems. Adv. Nat. Sci: Nanosci. Nanotechnol., 8: 013001

An Yan and Zhong Chen, 2019. Impacts of Silver Nanoparticles on Plants: A Focus on the Phytotoxicity and Underlying Mechanism. Int. J. Mol. Sci., 20: 1003

Anne Anderson, Joan McLean, Paul McManus and David Britt, 2017. Soil chemistry influences the phytotoxicity of metal oxide nanoparticles. Intl. J. Nanotechnol.,14: 15-21

Anne Anderson, Joan McLean, Paul McManus and David Britt., 2017. Soil chemistry influences the phytotoxicity of metal oxide nanoparticles. Intl. J. Nanotechnol., 14: 15-21 Anonymous, 2013. Explainer: nanoparticles in air pollution. In: The Conversation. Editorial Anonymous, 2012. General Safe Practices for Working with Engineered Nanomaterials in Research Laboratories". U.S. National Institute for Occupational Safety and Health. doi:10.26616/NIOSHPUB2012147

Anonymous, 2017. Radiation Safety Aspects of Nanotechnology. National Council on Radiation Protection and Measurements. 2–6, 88–90, 119–130. Archived from the original on 2017- 10-31. Retrieved 2017-07-07

Anonymous, 2012.General Safe Practices for Working with Engineered Nanomaterials in Research Laboratories". U.S. National Institute for Occupational Safety and Health. doi:10.26616/NIOSHPUB2012147

Anonymous, 2016.Toxic Air Pollution Nanoparticles Found in the Human Brain.Lancaster University. Neuroscience News, 6 September 2016

Anonymous, 2017. Radiation Safety Aspects of Nanotechnology. National Council on Radiation Protection and Measurements. 2017-03-02. pp. 2–6, 88–90, 119–130. Archived from the original on 2017-10-31. Retrieved 2017-07-07

Anonymous, 2018. Guidelines and Best Practices for Safe Handling of Nanomaterials in Research Laboratories and Industries. Centre for Knowledge Management ofNanoscience & Technology (CKMNT), Nano Mission, DST, Govt of India.

Anonymous, 2018. Guidelines and Best Practices for Safe Handling of Nanomaterials in Research Laboratories and Industries. Centre for Knowledge Management ofNanoscience & Technology (CKMNT), Nano Mission, DST, Govt of India.

Bair, W. J. (1995-07-01). The ICRP Human respiratory tract model for radiological protection. Radiation Protection Dosimetry, 60 : 307–310.

Bair, W. J., 1995.The ICRP Human Respiratory Tract Model for Radiological Protection. Radiation Protection Dosimetry, 60 : 307–310.

Bernal-Meléndez E., Bouillaud P., Callebert J., Olivier B., Lacroix MC., Durieux D., Persuy MA., Tarrade A., Rousseau-Ralliard D., ChavaΣe-Palmer P. and Baly C., Schroeder H., 2017. A Daily Exposure of Rabbit Dams to Diesel Exhaust Nanoparticles During Gestation Impairs the Brain and Olfactory Dopaminergic Pathways of Offspring. P. 35. In : Nanosafety 2017. Organised by INM – Leibniz Institute for New Materials in association with Leibniz Research Alliance Nanosafety, Saarbrücken, Germany, 11–13 October 2017.

Bourrinet, P., Bengele, H., Bonnemain, B., Dencausse, A., Idee, J. M., Jacobs, P. M., Lewis, J. M.,2006. Preclinical safety and pharcokinitic profile of ferumoxtran-10, an ultrasmall superparamagnetic iron oxide magnetic resonance contrast agent. Invest. Radiol., 41: 313- 324

Bouwmeester Hans, Susan Dekkers, Maryvon Y Noordam, Werner I Hagens, Astrid S Bulder, Cees de Heer, Sandra E C G ten Voorde, Susan W P Wijnhoven, Hans J P Marvin and Adriënne J A M Sips, 2009.

Boyd, A.M., Lyon, D., Velasquez, V., Sayes, D.Y., Fortner, J. and Colvin, V.L., 2005. Photocatalytic Degradation of Organic Contaminants by Water-Soluble Nanocrystalline C60. ACS Meeting Abstracts, 229th ACS National Meeting, San Diego, CA, March 13-17, 2005.

Burklew, C.E., Ashlock, J., Winfrey, W.B. and Zhang, B.,2 012. Effects of aluminum oxide nanoparticles on the growth, development, and microRNA expression of tobacco (Nicotiana tabacum). PLoS One, 7.e34783

Branislav RuttkayNedecky, Olga Krystofova and Lukas Nejdland Vojtech Adam, 2017. Nanoparticles based on essential metals and their phytotoxicity. J. Nanobiotechnol ., 15:33

Branislav RuttkayNedecky, Olga Krystofova and Lukas Nejdland Vojtech Adam,2017. Nanoparticles based on essential metals and their phytotoxicity. J. Nanobiotechnol ., 15:33

Canivet, L., Dubot, P., Garçon, G. and Denayer, F.O., 2015. Effects of engineered iron nanoparticles on the bryophyte, Physcomitrella patens (Hedw.) Bruch & Schimp, after foliar exposure. Ecotoxicol. Environ. Saf., 113:499-505

Catrin Albrecht, Tina Wahle , Adriana Sofranko , Harm Heusinkveld, Flemming Cassee and Roel Schins, Effects of nanoparticles in the 5xfad mouse model of Alzheimer 's disease.

P. 36. In : NANOSAFETY 2017. Organised by INM – Leibniz Institute for New Materials in association with Leibniz Research Alliance Nanosafety, Saarbrücken, Germany, 11–13 October 2017.

Celá, P., Veselá, B., Matalová, E., Večeřa, Z.and Buchtová, M., 2014. Embryonic Toxicity of Nanoparticles. Cells Tissues Organs, 199:1-23

Chang Woo Lee, Shaily Mahendra, Katherine Zodrow, Dong Li, Yu-Chang Tsai, Janet Braam and Pedro J.J. Alvarez, 2010. Developmental phytotoxicity of metal oxide nanoparticles to Arabidopsis thaliana. Environ. Toxicol & Chem., 29: 1399-1399

Chen, L., Wang, C., Li, H., Qu, X., Yang, S., and Chang, X., 2017. Bioaccumulation and toxicity of ^{13}C-skeleton labeled graphene oxide in wheat. Environ. Sci. Technol., 51: 10146–10153.
Cheng, X., Kan, A.T. and Tomson, M.B., 2004. Naphthalene Adsorption and Desorption from Aqueous C60 Fullerene. J. Chem. Eng. Data, 49: 675-683.

Christoph Lauterwasser, 2005. Opportunities and risks of Nanotechnologies Report in co- operation with the OECD International Futures Programme. Allianz Center for Technology, Paris Cedex 16 France

Chu, M., Wu, Q., Yang, H., Yuan, R., Hou, S., Yang, Y., Zou, Y., Xu, S., Xu, K., Ji, A. and Sheng, L., 2010. Transfer of quantum dots from pregnant mice to pups across the placental barrier. Small, 6:670-678.

Chuanxin Ma , Jason C. White , Baoshan Xing and Om Parkash Dhankher, 2015. Phytotoxicity and Ecological Safety of Engineered Nanomaterials. Intel. J. Pl. and Environ.,10:18811

Consortium, H. M. P., 2012. Structure, function and diversity of the healthy human microbiome. Nature, 486: 207–221

Cvjetko, P., Zovko, M., Peharec Stefanic, P., Tkalec, M., Domijan, A-M., Vinkovic Vrcek, I.and Balen, B. 2017. Phytotoxic effects of silver nanoparticles in tobacco plants. Environ. Sci. & Pollu.Res. Intl., 25 : 5590-5602

Cvjetko, P., Zovko, M., Peharec, P., Sˇtefanic, R., Biba, M., Tkalec, A.M., Domijan, I., Vinkovic Vrcek, I., Letofsky-Papst, S., Sˇikic and Balen, B., 2018. Phytotoxic effects of silver nanoparticles in tobacco plants. Environ. Sci. Pollut. Res., 25 : 5590–5602.

Danail Hristozov and Ineke Malsch, 2009. Hazards and Risks of Engineered Nanoparticles for the Environment and Human Health .Sustainability,1: 1161-1194

Daohui Lin, 2017. Phytotoxicity of nanoparticles: Inhibition of seed germination and root growth. Environ. Pollu.,150 : 243-250

Di CUI, Peng ZHANG, Yu-hui MA , Xiao HE , Yuan-yuan LI , Yue-chun ZHAO and Zhi-yong ZHANG, 2014. Phytotoxicity of silver nanoparticles to cucumber (Cucumis sativus) and wheat (Triticum aestivum). J. Zhejiang Univ-Sci., A (Appl Phys & Eng), 15:662-670

Dimitrios Stampoulis,,Saion K. Sinha and Jason C. White, 2009. Assay-Dependent Phytotoxicity of Nanoparticles to Plants. Environ. Sci. Technol., 43: 9473–9479

Dimkpa, C,O., McLeanm J,E., Britt, D.W. and Anderson, A.J.,2015. Nano-CuO and interaction with nano-ZnO or soil bacterium provide evidence for the interference of nanoparticles in metal nutrition of plants. Ecotoxicol, 24:119–129

Ehsan Borzouyan Dastjerdi, Ismail Bin Sahid and Khairiah Binti Jusoh, 2016 . Phytotoxicity assessment of nano-ZnO on groundnut (Arachis hypogaea) seed germination in MS medium. Sains Malaysiana, 45: 1183–1190

Elalfy, M.M., El-hadidy, M.G. and Abouelmagd, M.M.,2018. Update of Usefulness and Adverse

Effects of Nanoparticles on Animals and Human Health. J. Vet. Med. Health, 2: 106.

EPA, 2007. U.S. Environmental Protection Agency Nanotechnology White Paper. Prepared for the U.S. Environmental Protection Agency by members of the Nanotechnology Workgroup, a group of EPA's Science Policy Council, Science Policy Council U.S. Environmental Protection Agency Washington, DC 20460

Erdely, A., Hulderman, T., Salmen, R., Liston, A., Zeidler-Erdely, P.C., Schwegler-Berry, D., Castranova, V., Koyama, S., Kim, Y,A., Endo, M. and Simeonova, P.P., 2009. Cross-talk between lung and systemic circulation during carbon nanotube respiratory exposure. Potential biomarkers. Nano. Lett., 9:36-43.

European Commission. 2004. Towards a European Strategy for Nanotechnology. Communication from the Commission COM(2004) 338 final.

Falco, W.F., Scherer, M.D., Oliveira, S.L., Wender, H., Colbeck, I. T.and Lawson, A.R.L., 2019. Caires, Phytotoxicity of silver nanoparticles on Vicia faba: evaluation of particle size effects on photosynthetic performance and leaf gas exchange. Sci.Total Environ., doi: https://doi.org/10.1016/j.scitotenv. 2019.134816

Gaillet, S. and Rouanet, J. M., 2015. Silver nanoparticles: their potential toxic effects after oral exposure and underlying mechanisms–a review. Food Chem. Toxicol., **77:** 58–63
Giancarlo Franzese and Oriol Vilanova, 2017. How the Protein Corona Evolves on a

Nanoparticle. P. 40. In : NANOSAFETY 2017. Organised by INM – Leibniz Institute for New Materials in association with Leibniz Research Alliance Nanosafety, Saarbrücken, Germany, 11–13 October 2017.

Gottschalk, F., Lassen, C., Kjoelholt, J., Christensen, F.and Nowack, B., 2015. Modeling flows and concentrations of nine engineered nanomaterials in the Danish environment. Int. J. Environ. Res. Public Health, 12: 5581–5602.

Hansen, S. Regulation and Risk Assessment of Nanomaterials-Too Little, Too Late? Technical University of Denmark (DTU): Lyngby, Denmark, 2009; Available online: http://www2.er.dtu.dk/publications/fulltext/2009/ENV2009-069.pdf (accessed 10 September 2009).

Hao,Y., Yu, F., Lv, R., Ma, C., Zhang, Z., Rui, Y., Liu, L., Cao, W., Xing, B.and Choi, J., 2016. Carbon nanotubes filled with different ferromagnetic alloys affect the growth and development of rice seedlings by changing the C:N ratio and plant hormones concentrations. PloS One, 11:e0157264.

Hardman, R., 2006. A toxicologic review of quantum dots: toxicity depends on physicochemical and environmental factors. Environ. Health Persp., 114:165–172

Hardman, R., 2006. A toxicologic review of quantum dots: toxicity depends on physicochemical and environmental factors. Environ. Health Persp., 114:165–172

Hoet, P.H., Brüske-Hohlfeld, I. and Salata, O.V., 2004. Nanoparticles – known and unknown health risks. J. Nanobiotechnol., 2: 12

Hu, R., Gong, X., Duan, Y., Li, N.and Che, Y., 2010. Neurotoxicological effects and the impairment of spatial recognition memory in mice caused by exposure to TiO2 nanoparticles. Biomaterials, 31: 8043-50.

Hughes, M., Henson, T., Navaratilova, J., Tennant, A., Bradham, K. and Rogers. K., 2018. In vitro intestinal toxicity of copper nanoparticles in rat and human cell models. QEEM II: 2nd Quantifying Exposure to Engineered Nanomaterials from Manufactured Products Workshop, Washington, DC, Washington, DC, October 09 - 10, 2018.

Jamie R. Lead,Graeme E. Batley, Pedro J. J. Alvarez, Marie-Noële Croteau, Richard D. Handy, Michael J. McLaughlin, Jonathan D. Judy and Kristin Schirmer, 2018. Nanomaterials in the environment: Behavior, fate, bioavailability, and effects-An updated review. Environ. Toxicol. Chem., 37:2029–2063

Jie Yang, Weidong Cao and Yukui Rui, 2017. Interactions between nanoparticles and plants: phytotoxicity and defense mechanisms, J. Pl. Inter., 12:158-169

John, A.C., Küpper, M., Manders-Groot, A.M.M., Debray, B., Lacome, J.M., andKuhlbusch, T.A.J., 2017. Emissions and Possible Environmental Implication of Engineered Nanomaterials (ENMs) in the Atmosphere. Atmosphere , 8: 84.

Joseph A. Clark, 2011. Potential human risks of nanomaterials. Nanotechnology, IRMI.
Kreyling, W.G., Semmler, M., Erbe, F., Mayer, P., Takenaka, S., Schulz, H., Oberdörster,

G. and Ziesenis, A., 2002. Translocation of ultrafine insoluble iridium particles from lung epithelium to extrapulmonary organs is size dependent but very low. J. Toxicol. Environ. Health , 65:1513-30

Kumar, P., Fennell, P., and Robins, A., 2010. Comparison of the behaviour of manufactured and other airborne nanoparticles and the consequences for prioritising research and regulation activities. J. Nanopart. Res, 12: 1523–1530.

Lai, H., Chen, W. and Chiang, L., 2000. Free radical scavenging activity of fullerenol on the ischemia-reperfusion intestine in dogs. World J. Surg., 24:450-454

Lam, C., James, J., McCluskey, R. and Hunter, R.,2004. Pulmonary toxicity of single-wall carbon nanotubes in mice 7 and 90 days after intratracheal instillation. Toxicol. Sci., 77: 126-134

Landa, P., Cyrusová, T., Jeřábková ,J., Drábek, O., Vaněk, T. and Podlipná, R., 2016. Effect of Metal Oxides on Plant Germination: Phytotoxicity of Nanoparticles, Bulk Materials, and Metal Ions. Water, Air, & Soil Pollution, **227**: 448

Larue, C., Castillo-Michel, H., Sobanska, S., Cecillon, L., Bureau, S., Barthe`s, V., Ouerdane, L., Carrie`re, M. and Sarret, G., 2014. Foliar exposure of the crop Lactuca sativa to silver nanoparticles: Evidence for internalization and changes in Ag speciation. J. Hazard. Mater., 264 : 98–106

Laura Cid-Barrio, Diego Bouzas- Ramos, Alfonso Salinas-Castillo, Yasunitzu Ogra, Jorge Ruiz- Encinar and Jose M Costa-Fernandez, 2020. Quantitative assessment of cellular uptake and differential toxic effects of HgSe nanoparticles in human cells.J.Analytical Atomic Spectrometry, Issue 9.

Le Van, N., Rui, Y., Cao, W., Shang, J., Liu, S., Nguyen Quang, T. and Liu, L., 2016. Toxicity and bio-effects of CuO nanoparticles on transgenic Ipt-cotton. J. Pl. Interact., 11:108–116 Lee, W.M., An, Y.J., Yoon, H. and Kweon, H.S., 2008. Toxicity and bioavailability of copper nanoparticles to the terrestrial plants mung bean (Phaseolus radiatus) and wheat (Triticum aestivum): Plant agar test for water-insoluble nanoparticles. Environ. Toxicol. Chem., 27:1915–1921

Li, W., Zheng, Y., Zhang, H., Liu, Z., Su, W., Chen, S., Liu, Y., Zhuang, J. and Lei, B., 2016. Phytotoxicity, uptake, and translocation of fluorescent carbon dots in mung bean plants. ACS Appl. Mater. Interfaces, 8:19939–19945.

Liang, L., Tang, H., Deng, Z., Liu, Y., Chen, X. and Wang, H., 2018. Ag nanoparticles inhibit the growth of the bryophyte, Physcomitrella patens. Ecotoxicol. Environ. Saf., 164: 739–748 Liangjiao Chen, Jia Liu, Yanli Zhang, Guilan Zhang, Yiyuan Kang, Aijie Chen, Xiaoli Feng and Longquan Shao, 2018 . The toxicity of silica nanoparticles to theimmune system. Nanomedicine, 13: 1939–1962

Liangjiao Chen, Jia Liu, Yanli Zhang, Guilan Zhang, Yiyuan Kang, Aijie Chen, Xiaoli Feng and Longquan Shao, 2018 . The toxicity of silica nanoparticles to theimmune system. Nanomedicine, 13: 1939–1962

Libralato, G., Costa Devoti, A., Zanella, M., Sabbioni, E., Mičetić, I., Manodori, L. A., Pigozzo, S., Manenti, F., Groppi, A. and Volpi Ghirardini , 2016. Phytotoxicity of ionic, micro- and nano-sized iron in three plant species. Ecotoxicol. & Environ. Safety, 123: 81–88

Lert, J., Hu, Q., Hudson, J.S., Reid, M.L., Ratnikova, T.A., Rao, A.M., Luo, H. and Ke, P.C., 2009. Uptake, translocation, and transmission of carbon nanomaterials in rice plants. Small, 5:1128–1132

Liu, J., Kang, Y., Yin, S., Song, B., Wei, L., Chen, L. and Shao, L., 2017. Zinc oxide nanoparticles induce toxic responses in human neuroblastoma SHSY5Y cells in a size-dependent manner. Int. J. Nanomed., 12: 8085-8099

Liu, R. and Lal, R., 2015. Potentials of engineered nanoparticles as fertilizers for increasing agronomic productions. Sci.Total Environ.,514: 131-139.

Ma, X. and Quah, B., 2016.Effects of surface charge on the fate and phytotoxicity of gold nanoparticles to Phaseolus vulgaris. J. Food Chem. Nanotechnol., 2: 57-65.

Ma, C., White, J.C., Dhankher, O.P. and Xing, B., 2015. Metal-based nanotoxicity and detoxification pathways in higher plants. Environ. Sci. Technol., 49:7109–7122

Ma, Y., Zhang, P., Zhang, Z., He, X., Zhang, J., Ding, Y., Zhang, J., Zheng, L., Guo, Z.and Zhang, L., 2015. Where does the transformation of precipitated ceria nanoparticles in hydroponic plants take place? Environ. Sci. Technol., 49:10667–10674.

Marzie Hejazy, Mohammad Kazem Koohi, Adele Bassiri Mohamad and Pourand Davod Najafi, 2018. Toxicity of manufactured copper nanoparticles - A review. Nanomed.Res.J., 3:1-9

Mafhias Hufnagel, Sarah Schoch, Dr. Betna Strauch and Prof. Dr. Andrea Hartwig, 2017. The impact of cuo and TIO2 nanoparticles on gene expression following air-liquid interface exposure. P . 39. In : NANOSAFETY 2017. Organised by INM – Leibniz Institute for

New Materials in association with Leibniz Research Alliance Nanosafety, Saarbrücken, Germany, 11–13 October 2017.

Maynard, A.D., Baron, P.A., Foley, M., Shvedova, A.A., Kisin, E.R. and Castranova, V., 2004. Exposure to Carbon Nanotube Material: Aerosol release during the handling of Unrefined Single-Walled Carbon Nanotube Material. J. Toxicol. Environ. Health, A 67: 87-107.

Mc Clements, D.J. and Xiao, H., 2017. Is nano safe in foods? Establishing the factors impacting the gastrointestinal fate and toxicity of organic and inorganic food-grade nanoparticles. npj Sci. Food, 1: 6

Mc Donald, J.C., Drinker, P. and Gordon, J.E., 1951. The epidemiology and social significance of atmospheric smoke pollution. Am. J. Med. Sci., 221:325-42

Miller, B.G., Searl, A., Davis, J.M., Donaldson, K., Cullen, R.T., Bolton, R.E., Buchanan, D.and Soutar, C.A., 1999. Influence of fibre length, dissolution and biopersistence on the production of mesothelioma in the rat peritoneal cavity. Ann. Occup. Hyg., 43:155-66.

Mills, N.L., Donaldson, K., Hadoke, P.W., Boon, N.A., MacNee, W., Cassee, F.R., Sandström, T., Blomberg, A. and Newby, D.E., 2009. Adverse cardiovascular effects of air pollution. Nat. Clin. Pract. Cardiovasc. Med., 6:36-44.

Mirta Tkaleca , Petra Peharec Štefanicb and Biljana Balen, 2019 . Phytotoxicity of silver nanoparticles and defence mechanisms. Compreh. Anal. Chem., 84 : 145-198

Mirzajani, F., Askari, H., Hamzelou, S., Farzaneh, M. and Ghassempour, A., 2013. Effect of silver nanoparticles on Oryza sativa L. and its rhizosphere bacteria. Ecotoxicol. Environ. Saf., 88: 48–54

Montillet, J.L., Chamnongpol, S., Rustérucci, C., Dat, J., Van De Cotte, B., Agnel, J.P., Battesti, C., Inzé D. Van Breusegem, F. and Triantaphylides, C., 2005. Fatty acid hydroperoxides and H_2O_2 in the execution of hypersensitive cell death in tobacco leaves. Pl. Physiol., 138:1516–1526

Mossman, B.T., Bignon, J., Corn, M., Seaton, A. and Gee, J.B., 1990. Asbestos: scientific developments and implications for public policy. Science, 247:294-301.

Nadine Hollmann, Julia Liebing, David Fluri , Jan Hengstler and Christoph van Thriel, 2017. Investigation of neurotoxic effects of manganese nanoparticles in 3d Neural Microtissue.P 33. In : NANOSAFETY 2017. Organised by INM – Leibniz Institute for New Materials in association with Leibniz Research Alliance Nanosafety, Saarbrücken, Germany, 11–13 October 2017.

National Institute for Occupational Health and Safety, 2005. Approaches to Safe nanotechnology: An Information Exchage with NIOSH.

Nhan, L., Ma, C. and Rui, Y., 2015. Phytotoxic Mechanism of Nanoparticles: Destruction of Chloroplasts and Vascular Bundles and Alteration of Nutrient Absorption. Sci.Rep., 5, 11618

Oberdörster, G., Oberdörster, E. and Oberdörster, J. 2005. Nanotoxicology: An Emerging Discipline Evolving from Studies of Ultrafine Particles. Environ. Health Perspect, 113: 823 839

Pacurari, M., Castranova, V. and Vallyathan, V., 2010. Single- and multi-wall carbon nanotubes versus asbestos: are the carbon nanotubes a new health risk to humans? J. Toxicol. Environ. Health A, 73:378-95.

Panda, K.K., Achary, V.M.M., Krishnaveni, R., Padhi, B.K., Sarangi, S.N., Sahu, S.N. and Panda, B.B., 2011. In vitro biosynthesis and genotoxicity bioassay of silver nanoparticles using plants. Toxicol. In Vitro, 25: 1097–1105

Priscila Falagan-Lotsch, Elissa M. Grzincic and Catherine J. Murphy, 2016. Gold nanoparticles alter cells at low doses. Proc. Nati.l Acad. Sci.,113 :13318-13323

Qian, H., Peng, X., Han, X., Ren, J., Sun, L. and Fu, Z., 2013. Comparison of the toxicity of silver nanoparticles and silver ions on the growth of terrestrial plant model Arabidopsis thaliana. J. Environ. Sci., 25 :1947–1956.

Qinghai Wang, Cui Li, Yu Wang and Xiaoe Que, 2019. Phytotoxicity of Graphene Family Nanomaterials and Its Mechanisms: A Review. Front. Chem., 01 May 2019 | https://doi.org/10.3389/fchem.2019.00292

Raghad, D.H, Abdul Jalill and Alyaa, M., 2015. Comparison the phytotoxicity of TiO_2 nanoparticles with bulk particles on Amber 33 variety of rice (Oryza sativa) in vitro. Sch. Acad. J. Biosci., 3: 254-262

Rao, K., Raghavendra, A. and Reddy, K., 2006. Physiology and molecular biology of stress tolerance in plants. Netherlands: Springer.

Raul Cuero, 2016. Molecular and toxicological characterization of anti-UV Yeast plasmid compound for skin protection. Front Nanosci. Nanotech., 2: 159-164

Rayhaneh Amooaghaie, Maryam Norouzi, and Mohammad Saeri, 2017. Impact of zinc and zinc oxide nanoparticles on the physiological and biochemical processes in tomato and wheat. Botany, 95: 5.

Reginald Davey, 2020. Nanomaterials - Fate of Nanomaterials in the Environment. AZoNano, 16th January 2020

Remédios, C., Rosário, F. and Bastos, V., 2012. Environmental Nanoparticles Interactions with Plants: Morphological, Physiological, and Genotoxic Aspects. J.Bot.,vol. 2012, Article ID 751686, 8 pages

Rico, C.M., Hong, J., Morales, M.I., Zhao, L., Barrios, A.C., Zhang, J.Y., Peralta-Videa, J.R. and Gardea-Torresdey, J.L.,2013. Effect of cerium oxide nanoparticles on rice: a study involving the antioxidant defense system and in vivo fluorescence imaging. Environ. Sci. Technol., 47:5635–5642

Ronaldo dos Santos Filho, Taynah Vicari, Samuel A. Santos, Karoline Felisbino, Ney Mattoso, Bruno Francisco Sant'Anna-Santo, Marta Margarete Cesta and Daniela Morais Leme,2019. Genotoxicity of titanium dioxide nanoparticles and triggering of defense mechanisms in Allium cepa. Genet. Mol. Biol., 42: No.2 On-line version ISSN 1678-4685

Rui, Y., Zhang, P., Zhang, Y., Ma, Y., He, X., Gui, X., Li, Y., Zhang, J., Zheng, L.and Chu, S., 2015. Transformation of ceria nanoparticles in cucumber plants is influenced by phosphate. Environ. Pollut., 198:8–14.

Ruttkay-Nedecky, B., Krystofova, O. and Nejdl, L., 2017. Nanoparticles based on essential metals and their phytotoxicity. J. Nanobiotechnol., 15: 33

Santner, A., Calderon-Villalobos, L.I.A. and Estelle, M., 2009. Plant hormones are versatile chemical regulators of plant growth. Nat. Chem. Biol., 5:301–307

Sarvendra Kumar, A.K., Patra, S.C., Datta, K.G., Rosin, T.J. and Purakayastha, 2015. Phytotoxicity of nanoparticles to seed germination of plants. Int. J. of Adv. Res., **3:** ISSN 2320-5407.

Scenihr, 2006. Nanotechnologies. In : Green Facts-Facts on health and environment. DG Health and Consumers of the European Commission

Seaton, A., MacNee, W., Donaldson, K. and, Godden, D., 1995. Particulate air pollution and acute health effects. Lancet, 345:176-8.

Seaton, A., Tran, L., Aitken, R. and Donaldson, K., 2010. Nanoparticles, human health hazard and regulation. J. R. Soc. Interface, Suppl 1:S119-S129. doi:10.1098/rsif.2009.0252.focus Seaton, A., Tran, L., Aitken, R.and Donaldson, K., 2010. Nanoparticles, human health hazard and regulation. J. R. Soc. Interface, Suppl 1:S119-S129. doi:10.1098/rsif.2009.0252. focus Siddiqi, K.S. and Husen, A., 2017. Plant Response to Engineered Metal Oxide Nanoparticles. Nanoscale Res. Lett., 12: 92.

Sina Webering, Karina Lindner, Peter König and Heinz Fehrenbach, 2017. Low-Dose Carbon Black Nanoparticles Exposure of the Lung Does Not Aggravate Allergic Airway Inflammation in Mice. P 41. In: NANOSAFETY 2017. Organised by INM – Leibniz Institute for New Materials in association with Leibniz Research Alliance Nanosafety, Saarbrücken, Germany, 11–13 October 2017.

Singh, Z., 2018. Toxicity of Silver Nanoparticles to Human Cell Lines. Nanomed.Nanotechnol., 3: (4):000154

Singh,Z.S.,2016. Applications and toxicity of graphene family nanomaterials and their composites. Nanotechno. Sci. & Applic., 9: 15-28

Song, U., Jun, H., Waldman, B., Roh, J., Kim, Y., Yi, J.and Lee, E.J., 2013. Functional analyses of nanoparticle toxicity: A comparative study of the effects of TiO2 and Ag on tomatoes (Lycopersicon esculentum). Ecotoxicol. Environ. Saf., 93:60–67.

Sung, J.H., Ji, J.H., Yoon, J.U., Kim, D.S., Song, M.Y., Jeong, J.,Han, B.S., Han, J.H., Chung, Y.H.and Kim, J.,2008. Lung function changes in Sprague-Dawley rats after prolonged inhalation exposure to silver nanoparticles. Inhal.Toxicol., 20: 567–574

Syeda Hafsa Ali and Syeda Ayesha Ali,2019. Nanotechnology is the potential cause of phytotoxicity. J.Biomat., 3: 1-6.

Syu, Y.Y., Hung, J.H., Chen, J.C. and Chuang, H.W., 2014. Impacts of size and shape of silver nanoparticles on Arabidopsis plant growth and gene expression. Plant Physiol. Biochem., 83 : 57-64.

Tamara Forbe, Mario García and Eric Gonzalez, 2011.Potencial risks of nanoparticles. Food Sci. & Technol., 31: No.4

Tang, Y., He, R., Zhao, J., Nie, G., Xu, L. and Xing, B., 2016. Oxidative stress-induced toxicity of CuO nanoparticles and related toxicogenomic responses in Arabidopsis thaliana. Environ. Pollut., 212: 605–614.

Thuesombat, P., Hannongbua, S., Akasit, S. and Chadchawan, S., 2014. Effect of silver nanoparticles on rice (*Oryza sativa* L. cv. KDML 105) seed germination and seedling growth. Ecotoxicol. Environ. Saf., 104: 302–309

Tsai, Y.I. Atmospheric visibility trends in an urban area in Taiwan 1961–2003. Atmos. Environ. 2005, 39, 5555–5567.

Tungittiplakorn, W., Lion, L.W., Cohen, C. and Kim, J.Y. 2004. Engineered Polymeric Nanoparticles for Soil Remediation. Environ. Sci. Technol., 38: 1605-1610.

Walker, R., Decker, E. A. and Mc Clements, D. J., 2015. Development of food-grade nanoemulsions and emulsions for delivery of omega-3 fatty acids: opportunities and obstacles in the food industry. Food Funct., 6: 42–55

Wang, D., Lin, Z., Wang, T., Yao, Z., Qin. M., Zheng, S. and Lu,W., 2016. Where does the toxicity of metal oxide nanoparticles come from: The nanoparticles, the ions, or a combination of both?. J. Hazard. Mater. 308 : 328-334

Worsfold S. Morgan, Amyotte Paul R., Khan Faisal I., Dastidar Ashok G. and Eckhoff Rolf K., 2012. Review of the explosibility of nontraditional dusts. Indus. & Engg.Chem. Res., 51: 7651–7655.

Wu, S.G., Huang, L., Head, J., Chen, D.R. and Kong, I.C., 2012. Phytotoxicity of metal oxide nanoparticles is related to both dissolved metals ions and adsorption of particles on seed surfaces. J. Pet. Environ. Biotechnol., 3:126

Yang, J., Cao, W. and Rui, Y., 2017. Interactions between nanoparticles and plants: phytotoxicity and defense mechanisms, J. Plant Interact., 12 : 158–169.

Yevheniia Konotop, Kseniia Stepanchenko, Leila-Anastasiia Karpets, Andrii Zinchenko, Mariia Kovalenko, Oleksandr Smirnov, Ludmila Batsmanova, and Nataliya Taran, 2019. Phytotoxicity of colloidal solutions of stabilized and non-stabilized nanoparticles of essential metals and their oxides.Nova Biotechnologica et Chimica,18: (I)

Yin, H.C., Liu, Q.J., Lin, Q., Chen, S.N. and Zhou, L.J.,2005. Inhibitory effects of nano-TiO_2 loaded Pd on cyanobacteria growth. Acta Bot Boreali-Occidentalia Sin., 25:1884–1887

Yinglin Liu, Lina Xu and Yanhui Dai, 2018. Phytotoxic Effects of Lanthanum Oxide Nanoparticles on Maize (*Zea mays* L.) .IOP Conf. Series: Earth and Environ. Sci., 113 :012020

Zhang, P., Ma, Y., Liu, S., Wang, G., Zhang, J., He, X., Zhang, J., Rui, Y.and Zhang, Z., 2017. Phytotoxicity, uptake and transformation of nano-CeO2 in sand cultured romaine lettuce. Environ Pollut., 220:1400–1408

Zhang, W., 2003. Nanoscale Iron Particles for Environmental Remediation: An Overview. J. Nanoparticle Res., 5: 323-332.

Zhao, J., Cao, X., Wang, Z., Dai, Y. and Xing, B., 2017.Mechanistic understanding toward the toxicity of graphene-family materials to freshwater algae. Water Res., 111: 18–27.

9

Nanosafety

Nanotechnologies are being spoken of as the driving force behind the new industrial revolution. The mass production of these nano materials and their applications mean a dramatic increase of workers dealing with Engineered Nano Materials (ENM) and becoming possibly exposed to resulting hazardous effects and increased environmental burden of the environment due to the leaks of these materials from industrial processes. It is, hence, essential to consider the importance of ENM safety and their new applications and understand that it is important for these materials and products to be safe to enable successful promotion of nanotechnology for useful applications (Eman M Osman, 2019).

The term 'ultrafine particle' has traditionally been used to describe airborne particles smaller than 100 nm in diameter that are byproducts of industry or nature. Ultra-fine particles tend to be generated through processes such as combustion and vaporization. The particles are produced in large quantities from industrial activities such as thermal spraying and welding and from domestic combustion activities like gas cooking. Ultra-fine particles are also found in the atmosphere, where they originate from combustion sources like forest fires and volcanic activity and from atmospheric gas-to-particle conversion processes, such as photo-chemically driven nucleation.

9.1 Significance of nanosafety

Research with nanomaterials has shown that the physiochemical characteristics of particles can influence their effects in biological systems. Some of these characteristics include Charge; Chemical reactivity; Degree of agglomeration; Shape; Size; Solubility; Surface area; and Surface composition. There are many unknowns as to whether the unique properties of engineered nanomaterials pose health concerns. The potential health risk following exposure to a substance is generally associated with the Magnitude and duration of the exposure; Persistence of the material in the body; Inherent toxicity of the material; and Susceptibility or health status of the person. Unfortunately, there is limited data regarding the health risks related to nanomaterials. As such, this document is to provide EHS-accepted recommendations for practicing prudent health and safety measures when working with nanomaterials.

'Nanosafety' refers to all the safety issues associated with nanotechnology. Despite the fact that nanotechnology has been around for over 20 years, it is still considered new technology and with many new technologies, the health effects of nanomaterials have not been thoroughly investigated. Consequently, the uncertainty surrounding the toxicity of nanomaterials merits a cautious approach when working with them. For this reason, discussion on nanosafety will often deal with toxicology, ecotoxicology, exposure assessment, mechanism of interaction, risk assessment and standardisation. The European Union have in the past few years focus their program activities to look at safety of manufactured nanomaterials. Thailand on the other hand is addressing nanosafety concerns in manners that embraces harmonization and enhances the overall goal of technology development. Technology development and safety awareness need to complement each other in order to achieve sustainability of technology development.

Nanosafety is a field that aims to understand the physical and chemical characteristics that can be engineered into a particle in order to maximize its safety. It deals with safety and security issues regarding nanomaterials and nanoproducts. Central topics include understanding the effects induced by nanoparticles, developing safe nanomaterials and explaining key issues in the context of nanotechnology. Nanosafety is the assessment of risks to the human health and the environment, as well as to the ecological risks arising from the use of designed nanomaterials and the evaluation of their toxicity level (Kahru and Dubourguier, 2010).

The vast majority of nanomaterials are captured by macrophages in the lungs or excreted through the feces (Aleksandra Zielińska, et.al., 2020). Safety is a term with many facets: safe production of nanomaterials, safe handling of nanomaterials, safe detection of nanomaterials, safety from exposure to nanomaterials and safety in case of incidents or accidents involving nanomaterials. BAM is part of a network of Federal Authorities (BAuA, BfR, PTB, UBA).

Information from research and animal studies on nanomaterials has identified some potential safety hazards and health effects (Anonymous, 2018). Because nanotechnology is a rapidly emerging field, more information will likely become available about potential health and safety hazards associated with some nanomaterials. The health hazard potential depends on the particular nanomaterial and a person's exposure level. For example:

- Certain inhaled nanoparticles may be deposited in the respiratory tract and may cause inflammation and damage to lung cells and tissues; e.g., carbon nanotubes and nanofibers may be capable of causing pulmonary inflammation and fibrosis.

- Titanium dioxide (TiO_2), which has many commercial applications (e.g., paint, paper, cosmetics, food), can be produced and used in varying particle sizes, including the nanoscale particle sizes (< 100 nm).
- NIOSH has determined that nanoscale TiO_2 particles have higher mass-based potency than larger particles, and that occupational exposure (by inhalation) to nanoscale TiO_2 particles should be considered a potential occupational carcinogen.
- Certain nanoparticles may penetrate cell membranes and may cause damage to intracellular structures and cellular functions.
- Some nanomaterials may act as chemical catalysts and produce unanticipated reactions, creating a risk of explosions and fires.
- Some types of nanoparticle dusts may be combustible and require less energy to ignite than larger dust particles (for instance, sugar or wood), creating a risk of explosions and fires.

Workers' Right

- Working conditions that do not pose a risk of serious harm.
- File a confidential complaint to have their workplace inspected.
- Receive information and training about hazards, methods to prevent harm, and the standards that apply to their workplace. The training must be done in a language and vocabulary workers can understand.
- Receive copies of records of work-related injuries and illnesses that occur in their workplace.
- Receive copies of the results from tests and monitoring done to find and measure hazards in their workplace.
- Exercise their rights to a safe and healthful workplace without retaliation or discrimination

9.2 Understanding nanosafety

Nanotechnology is regarded as the enabling technology of the 21st century. However, compared to the numerous inventions made during the last decades only a small number of nanoenabled products finally made their way to the market. There are several reasons why these innovative approaches got stuck in translation; one key factor being uncertainties in the safety assessment of these new materials (Cordula Hirsch et.al., 2017). Numerous examples exist where nanomaterials have deceived scientists by generating false positive or false negative results in *in vitro* studies. The parameters that are important to prevent such false results and assure reliable *in vitro* safety assessment include,

i. **Material characterization**: How well do you know the nanomaterial you are working with? How reproducible is the synthesis process? Are batch- to-batch variations to be expected? How to assess those? ii. Interference reactions with the different assay components.

iii. **Dosimetry:** sedimentation and diffusion phenomena of particulate materials make it difficult to determine the delivered dose that is relevant for a given biological response.

iv. **Benchmarking and general assay performance:** In comparison to other- known - (nano)materials: how strong is the observed effect? Did you think about the necessary control samples? Are all process control measurements within the specified and expected range? These issues add additional complexity to the well-known challenges of in vitro - in vivo comparability and finally animal-to-human predictability.

On the other hand safe by design approaches stand and fall with fast feedback loops from the material scientist (=designer) to the biologist and vice versa; precluding time consuming animal studies, longing for fast in vitro solutions. Especially in early developmental stages of a new, smart (nano)material the safety of a multitude of technically suitable candidates has to be assessed. Potential adverse biological effects need to be "designed out" to select for the most promising material; not only in terms of technical efficacy but also in terms of safety. Therefore fast in vitro tools/methods are needed. However, there is more to it than speed. Reliable, robust and reproducible results are key to success.

Nano particles are not a new phenomenon since they could be occurred naturally, incidentally or intentionally.

Normally happening nanoparticles: Numerous sorts of nanoparticles happen normally in matter or in the environment. Numerous organic materials, found in human food or food ingredients, are normally nanostructured or contain nanoparticles. Casein micelles can be considered as nanoparticles. They are the real protein segment of milk and in charge of conveying mineral supplements, for example, calcium and phosphate to neonates. It is well investigated the nearness of nanoparticles in the climate where their focus and arrangement are exceptionally factor from place to another. Common emissions from trees and different plants or soil volatiles just as from soil disintegration can dominate in certain areas, while particles from ocean splash may dominate somewhere else. Likewise, volcanic ash may convey enormous amounts of natural nanoparticles into the environment.

Unexpectedly occurring nanoparticles: Another type of atmospheric nanoparticles are the accidental results of procedures including industrials, burning, welding, and transportation activities.

Intentionally occurring nanoparticles: The well-known nanoparticles are the intentionally man-made or the engineered nanoparticles which can be made in the labs, for example nano particles of metals, metal oxides, metal hydroxides, or even polymers. These designed nanoparticles can be utilized in several fields; either industrial or therapeutic. The nearby centralizations of nanoparticles in the atmosphere are significantly influenced by environmental conditions and depend firmly on emission intensities, vicinity to sources, and meteorological conditions. All in all, the highest number of concentrations happens in urban zones while regular sources dominate in rural regions (Buseck and Adachi, 2007). Environmental and Health safety (EHS) concerns have concentrated principally on nanoscale materials that are deliberately structured and delivered frequently alluded to as Engineered Nanomaterials (ENM).

Hazards of nanoengineered materials: When assessing the risks coipled with nanomaterials, particular care must be taken to spot the specific achievement of surface chemistry, shape, size and morphology on toxicity caused to diverse organs. The successive key hazard categories could also be measured wen assessing risk linked to nanomaterials (Muthuraman Yuvraj et.al., 2020). The presence of nanomaterials isn't in itself a risk. It is just certain perspective that can make them dangerous, specifically, their versatility and their expanded reactivity. Just if certain properties of certain nanoparticles were harmful to living creatures or the environment, for this situation, it tends to be called nano pollution.

Regarding to the health and environmental impact of nanomaterials two sorts of nanostructures are to be considered.

i. **Fixed nano-particles:** where nanoscale particles are consolidated into a substance, material or device (electronic, optical, sensors and so on).

ii. **Free nanoparticles:** where individual nanoparticles of a substance are available at some phase in production or use.

There is by all accounts accord that, albeit one ought to know about materials containing fixed nanoparticles, the quick concern is with free nanoparticles. In discussing nanoparticles, it is significant that, the phase containing nanoparticles (powder or fluid) never be mono disperse, however contains a range of molecule sizes. This entangles the test investigation, as bigger nanoparticles may have various properties from littler ones. Moreover, nanoparticles demonstrate an inclination to aggregate, and such aggregates regularly act uniquely in contrast to individual nanoparticles (Eman M Osman, 2019).

Pathways and common tasks that could result in exposure: The primary routes to exposure for nanoparticles are inhalation, dermal absorption and ingestion. The state of the nanoparticles or nanomaterials decides the exposure potential. For example, nanoparticles in powdered form present a large inhalation hazard potential when compared to nanoparticles in suspension (Anonymous, 2018).

9.3 Administrative (procedural) control

Administrative means of control comprise of a supplementary approach when other methods have not achieved the expected control levels, and these control strategies include:

i. **Training:** When handling nanomaterials, effective training and instruction for the workers is critical to ensure health and safety. A number of issues should be considered while training the personnel working with nanomaterials viz., Safe handling of nanomaterials and standard operating procedures (SOP); Hazards and toxicity; Personal protective equipments(PPE); Engineering controls and equipment maintenance; Emergency procedures; Waste handling; Definition of nanoparticles; Environmental release/shipping/customer protection; Exposure monitoring; Applicable regulation; Labeling and handling of nanomaterials waste.

ii. **Housekeeping:** Good housekeeping practices in laboratories where nanomaterials are handled can minimize the risk of exposure.

Clean all working surfaces potentially contaminated with nanoparticles (i.e., benches, glassware, apparatus, exhaust hoods, support equipment) at the end of each day using industrial vacuum cleaner equipped with HEPA filters.

- The HEPA vacuums should be labeled "For use with nanoparticles only" and must be used only for this purpose.
- Do not dry sweep or use compressed air.
- Dispose of used cleaning materials in accordance with the hazárdous-waste procedure.

iii. **Work practices**

- Whenever possible, handle nanomaterials in solutions or attached to substrates to minimize airborne release.
- Consult the material safety data sheet (MSDS), if available or other appropriate references prior to using a chemical or nanomaterial.
- While working with nanomaterials in liquids.
- Avoid spreading of the liquid by working in a spill container.
- Wear gloves that are suited for the liquid being handled.
- Avoid the dispersion of liquid droplets in the lab air.
- Directly clean up spills, before evaporation or further spreading occurs.
- While working with nanomaterials in gas phase reactors.
- Work in a closed reaction vessel, preferably around atmospheric or lower than atmospheric pressure.
- Make sensitive leak checks between runs.
- When working with systems under positive pressure obey the standard safety rules for pressurized vessels and put the vessel into an enclosed safety vessel.
- For small positive pressure set-ups a closed fume hood is sufficient.
- Mount a HEPA filter (e.g. Emflon PFR filter from pall, > 3 nm) on the exhaust side of the process before leading into the fume hood and the outside air.
- While working with nanomaterial powders.

Handle free nanomaterial powder exclusively in a closed fume-hood or an enclosed vessel (glove box).

- If handling outside a closed environment cannot be avoided, wear class P3 (Filters at least 99.95% of airborne particles) certified respiratory filters.
- For characterization purposes such as XRD analysis, use a drop of oil to contain the powder and preventing it from becoming airborne.
- Clean all parts that have been in contact with nanoparticles and spills after using appropriate protection.

iv. **Medical surveillance:** The use of health surveillance program is an indicator of whether exposure is occurring, rather than in determining that levels of exposure are safe.

Several potential disease conditions such as chronic immune responses of inflammation, allergy, respiratory disorders, gastrointestinal related disorders, neurological disorders, several types of cancers resulting from oxidative damage to DNA and tissue damages are emerging with the increased use of nanomaterials.

It is recommended to have a regular health monitoring program and periodical medical surveillance of pulmonary, renal, liver and hematopoietic functions.

- The regular health monitoring helps in the early detection of any health effects which would considerably reduce the likelihood of long term harm.
- It is recommended that a basic worker health monitoring program is established. Such a program should include at a minimum
- Identifying staff exposed to engineered nanoparticles of unknown health effects.
- Conducting workplace characterization and worker exposure assessments
- Providing workers with baseline medical evaluations and including them in a non-specific routine health monitoring program.

v. **Record keeping:** In order to establish and maintain safe and healthy workplaces, effective record keeping is required. Records of the following areas need to be maintained:

Induction and training programs

- Risk assessments
- Servicing and testing of equipment
- Workplace monitoring
- Health surveillance (to be kept confidential)
- Work related injuries and illnesses
- Work place engineering control maintenance, daily checks and examinations
- Disposal records

Workplace monitoring: Reported studies have proved that the work place monitoring is important because it can significantly help the research group or organizations to describe the nanomaterial risk.

vi. Monitoring the particle concentrations at the working place using different types of particle counters, condensation particle counters, scanning mobility particle sizers or other means of particle collections, such as witness plates and particle size analysis by electron microscopy.

vii. Storage

- Storing nanoparticles might involve special protection to safeguard the products and to ensure workplace health and safety.
- Storage containers of nanoparticles should accommodate different granulometric characteristics and reactivity of particles.
- In order to avoid oxidation or explosion in case of certain metal nanoparticles, they are often needed to be protected from exposure in air and stored under inert gas/liquid bath.

viii. **Personal Protective Equipments (PPE):** Nanoparticle exposure is often due to the wearing of inadequate PPE. Typical protective clothing's should include:

Closed-toed shoes made of a low permeability material

- Long pants without cuff.
- Long sleeved shirt.
- Nitrile gloves with extended sleeves. (Note: Advisable to use two pairs of gloves, as in case of carcinogenic material, to exercise abundant caution).
- Chemical splash goggles, safety glasses and full-face shields are recommended for eyeprotection when there is potential for exposure to nanomaterials.
- Respiratory air filters (N100 recommended or N95).
- Full facepiece respirators offer eye protection in addition to respiratory protection.

ix. Waste disposal: Milligram range nanomaterials residues can be poured in sealed containers appropriately labeled and disposed off following the standard procedure in the hazard area proclaimed by the respective authority (Anonymous, 2018).

- Quantities of nanomaterials (powders, colloids) exceeding the milligram range should be treated as hazardous chemical waste, if the particle solubility in water is very small (inorganic materials like metals, metal oxides etc.).
- If the solubility is higher, the rules according to the toxicity class of the macroscopic material apply.
- Do not put engineered nanomaterials waste in the regular trash or dump it down the drain.
- All nanomaterials waste, as defined above, should be collected in labeled, enclosed hazardous waste containers with secure caps or covers. The label should include a description of the waste and prominently the words "contains nanomaterials".
- Collect paper, wipes, PPE and other items with loose contamination in a plastic bag or other sealable container and store it in a fume hood until it is full, then double-bag it, label it, securely tie or seal it, and dispose of it following appropriate procedures.
- Nanomaterial hazardous waste containers shall be collected and disposed of as hazardous waste following the standard procedures

9.4 Practices to be followed while handling nanoparticles

a. Locating emergency equipment

Know the location and proper use of emergency equipment, such as safety showers, fire extinguishers, and fire alarms.

b. Hygiene

- Do not consume or store food and beverages, or apply cosmetics where chemicals or nanomaterials are used or stored since this practice increases the likelihood of exposure by ingestion.
- Do not use mouth suction for pipetting or siphoning.
- Wash hands frequently to minimize potential chemical or nanoparticle exposure through ingestion and dermal contact.
- Remove gloves when leaving the laboratory, so as not to contaminate doorknobs, or when handling common use objects such as phones, multiuser computers, etc.

c. Labeling and signage

- Store in a well-sealed container, preferably one that can be opened with minimal agitation of the contents.

Label all chemical containers with the identity of the contents (avoid abbreviations/ acronyms) include term "nano" in descriptor (e.g., "nanozinc oxide particles" rather than just "zinc oxide." Hazard warning and chemical concentration information should also be included, if known.

- Use cautious judgment when leaving operations unattended: i) Post signs to communicate appropriate warnings and precautions, ii) Anticipate potential equipment and facility failures, and iii) Provide appropriate containment for accidental release of hazardous chemicals.

d. Cleaning procedures and spills

- Specifically watch out for exposure during cleaning operations.
- Wear gloves and work in a fume hood if the equipment allows this. Clean the fume hood afterwards.
- If needed, monitor the lab air nanomaterial concentrations during clean-up.
- Wear respiration protection when working outside a fume hood or in an open fume hood and consider overall protection.
- All contaminated material must be disposed of as chemical waste.
- Materials and surfaces can be cleaned by following techniques:

Wiped with a wet cloth (use water or solvent) where possible, rinsing off the cloth with water or disposing it off.

- Vacuum cleaned, where the exhaust of the vacuum cleaner is equipped with a HEPA filter and is not equipped with a pressure relief valve that bypasses this HEPA filter if blocked. Monitor the exhaust of the vacuum cleaner during operation. A malfunctioning filter can increase the exposure by dispersing the nanomaterial in the air.
- HEPA-filtered vacuum cleaners with combination of wet wiping are more suitable for most nanomaterial clean-up.
- Energetic cleaning methods such as dry cleaning or use of compressed air should be prohibited.
- Collect spill cleanup materials in a tightly closed container.

- Manage spill cleanup debris as hazardous waste.
- Make Nanoparticle spill kit containing the following items readily available to respond to spills
- Barricade tape
- Latex or nitrile gloves
- Disposable N100 (best rated filter by the CDC and WHO and is recommended for ultimate protection, have a minimum filter efficiency of 99.97% of particulate matter of average particle size less than 0.3 micron.) respirators
- Adsorbent material
- Wipes
- Sealable plastic bags
- Walk-off mat
- Small spills can be cleaned as described above.

e. During manufacturing and handling of nanomaterials

1. Nanomaterials must be stored and transported in sealed shatter-resistant containers. Containers must be labeled with nanomaterial name (or composition) and approximate particle size, along with any known hazard warnings.
2. Weighing and measuring of dry powders where aerosolization and release of nanomaterials are possible should be conducted in defined and closed areas .
3. Various processing steps such as dispersing, mixing, spraying, machining, gas phase processing have the potential to generate nanoparticles with a high concentration. There is a significant chance of inhalation, dermal and ingestion exposures during these steps. Hence all nanomaterial generators must have sufficient operating controls (see Table 4) to prevent employee exposure to nanomaterials during operating and maintenance procedures. Personal protective equipment should be used to minimize dermal and eye exposure
4. Employing a closed facility to process nanomaterials will considerably reduce the occupational exposure during the production and processing stages. Particular care should be taken to avoid disturbance of the closed facilities and cleaning up of residues should be carried out in wet

conditions or with a liquid medium to avoid dust dispersion and thus exposure through inhalation.

5. During working steps which cannot be performed in closed facility (e.g. filling, milling, packaging, sampling), various organizational measures such as efficient extractor systems should be used.

Working areas with potential nanoparticle emissions should be monitored by workplace or exposure measurement. A suitable method for technical monitoring of working areas is by total dust measurement.

6. Corporate responsibility is very important in the development of nanotechnology products. Various players in industry and business should take initiatives to specifically evaluate the effects of nanoproducts production and consumption with regard to health, environment and society.
7. Removal of waste and by-products generated at the production facility should be carried out with minimum exposure to humans and the environment.

9.5 Safety Practices

a. Transporting

Use sealed, double-contained container when transporting nanomaterials inside or outside the building (Anonymous, 2018).

b. Buddy system

- Communicate with others in the building when working alone in the laboratory; let them know when you arrive and leave.
- Avoid working alone in the laboratory when performing high-risk operations.

c. Explosion safety

- Due to their large surface-to-volume ratio and enhanced reactivity of a strongly curved solid surface, all oxidizable nanomaterials in the powder state must be considered as potentially pyrophoric when put in contact with air. Thus, explosive behavior is possible.
- Assess the risk of an incident in a process where a large un-oxidized sample is momentarily exposed to air. In a process that produces nanomaterial from the gas phase, the particles tend to become cleaner and cleaner while the process runs, e.g. less oxidized.

- Little is known about nano pyrophoricity, so it is recommended to test the pyrophoricity by bringing a small quantity of material in contact with air before producing or handling quantities on the level of one gram or more.
- Be certain that the test is not performed on partially oxidized particles.
- Anti-static shoes and mats should be used in areas where potentially explosive nanomaterials are handled. The shoes reduce the build-up of static charge, which could potentially ignite the materials.
- A distillation system for evaporating solvent from a colloidal dispersion being housed within an explosion-proof enclosure. This enclosure was designed with concern for the potential for these particular nanomaterials to be explosive

d. Access control

- Determine the extent of area reasonably expected to have been affected, and demarcate it with barricade tape or use another reliable means of restricting entry into the area.
- Assess the extent of the spill. Significant spills are defined as those of more than a few grams of nanoparticles.
- Smaller spills can be cleaned up by trained personnel from the lab using the procedure described under section 6d (Cleaning procedures and spills).

e. Dry materials

- Position a walk-off mat (e.g., Tacki-Mat) where clean-up personnel will exit the access controlled area.
- Do not dry sweep spilled accumulations of dry nanomaterials. Small quantities of dry materials (i.e., gram quantities) can be cleaned up with wet wipes. Dispose of the wipes as hazardous waste, as described above. Significant spills should be vacuum cleaned. Use only HEPA-filtered vacuum cleaners to clean up nanoparticles.
- Ensure that the functioning of the HEPA filters is properly tested as frequently as the manufacturer's recommends. HEPA vacuums for nanomaterials should be dedicated and labeled "For Use with Nanomaterials Only". Used HEPA filters must be appropriately characterized, collected, and disposed of as hazardous or potentially hazardous waste based on the material involved.

f. Product stewardship

Various types of guidance information are provided by the organization/ manufacturers supplying nanomaterials to customers. Different types of guidance information provided by the suppliers are listed below. MSDS is the most common form of guidance among these and should be kept at a easily accessible central location.

- Material safety data sheet (MSDS)
- Product information sheet
- Technical instructions
- Personal interaction
- Accompanying letter
- Technical data sheets
- Specification sheets
- Certificates of analysis
- Operation manuals

g. Safety considerations

Nano-safety is definitely not a basic arrangement or a convenient solution, however there should be an orderly methodology that assesses the risks and propose sensible activities to guarantee the security of the two individuals and the environment. New technologies can possibly get benefits just as inconveniences of the exposure of people to new materials. There are three fundamental settings for thinking about safety, viz., Environmental pollution, Unintentional human introduction (exposure in the working environment) and Deliberate human exposure (the intended applications)

The exposure of populaces in the working environment and customers and of the environment to nano materials is probably going to rise significantly (Zolnik et.al., 2010). In spite of the fact that, there is as of now significant learning on the effect of those environmental nano particles created unintentionally (for example through combustion), on human health basically intervened through respiratory pathways, there are still need in the fitting risk appraisal technique to assess the safety of Engineered Nano Materials. Just the same as each other new technology, Research and Development (R&D) of nano material items should be joined by safety assessment, including risk assessment and risk-management. Risk assessment and risk management require intelligent and case-explicit thought guided by potential exposure situations.

h. Green nanotechnology

Green nanotechnology refers to the utilization of either nanotechnology or the results of nanotechnology to improve the environmental maintainability of procedures delivering negative externalities. Green nanotechnology has been additionally described as the development of clean technologies, to minimize potential environmental and human health risks related with the assembling and utilization of nanotechnology items, and to encourage replacement of existing items with new nano-items that are all the more environmentally friendly all through their lifecycle (Anonymous, 2010). Green nanotechnology has two objectives: delivering nanomaterials and items without harming environment or human health and delivering nano-products that provide solution to environmental problems. It uses existing standards of green science also, green designing to make nanomaterials and nano-items without toxic ingredients, at low temperatures utilizing less energy and renewable information sources any place as possible, and utilizing lifecycle thinking in all plan and designing stages.

i. Pollution

Nano pollution is a nonexclusive name for all waste created by nano devices or during the nanomaterials assembling process. Nano waste is predominantly the gathering of particles that are discharged into the environment (air and water), or the particles that are discarded when still on their items causing disturbance of the ecosystem, as there is evidence for that MPNs are passing through the food chain from smaller to larger organisms. Since MPNs have been demonstrated to be competent to hurt organisms, it is talked about that, they can affect the functional ecosystem which is subjected to micro flora. All examined groups of manufactured nanoparticles have demonstrated to be toxic to aquatic living beings, for example, zebra fish, daphnia, algae, invertebrates and rainbow trout.

j. Social impact

Beyond the toxicity dangers to human health and environment which are associated with original nanomaterials, nanotechnology has more extensive societal impact and postures more extensive social difficulties. Numerous social researchers and associations in common society recommend that, innovation evaluation and administration ought to likewise include public cooperation. Nanotechnologies may give new answers for a huge number of individuals in developing nations who need access to fundamental services, such as , safe water, reliable energy, medicinal services, and education (Eman M Osman, 2019). The impacts of nanotechnology on: society all in all, human health and environment, the trade, security, food systems and even on the meaning of "human", have not been portrayed or politicized at this point.

k. Hazard assessment

Despite the fact that, there are many research ventures overall evaluating the potential hazards and dangers, there have been worries about their quality and significance. The absence of institutionalized materials and methodologies makes it hard to analyze results from various analysts and diverse nano products. Past the lung, there are numerous gaps in learning about the health impacts that may happen after inhalation, in connection to take-up, distribution, accumulation and biological impacts in secondary organs, in spite of the fact that it is known, for instance, that some nano particles can cross the blood–brain barrier (Eman M Osman, 2019). Generally, it may be said that, similar properties that are attractive in certain applications are likewise the properties that may offer ascent to toxicity (nano materials paradox).

There are a few issues deciding the hazard assessment of nanoparticles, as indicated below.

i. Dose must be measured dependent on the physicochemical properties of the nano material, over the long term or short term with high dosages.

ii. Standardization is basically imperative to utilize approved, institutionalized measures so that:

- Results from various analysts can be thought about, and
- Assays in vitro or creature studies can sensibly be relied upon to anticipate an impact in people.

iii. **Studies at the cellular level:** there is still little knowledge about how nano particles interact with the cell layer, how they are transported into cells, into lysosomes, mitochondria and the nucleus, and the results of these interactions

iv. **Studies at organ and system levels:** among the needs there is a must to consider toxico-kinetics studies for additional pneumonic translocation and transport after different routes of entry, especially for effect on cardiovascular, nervous, hepatic, immune and endocrine systems. It is similarly critical to do much more in assessing exposure whether intended or unintended. Few studies depicted work environment, consumer or environmental exposure or related exposure to genuine conditions. More investigations should be done to evaluate inhalation exposure in occupational settings and to increase awareness in other different settings .

Another vital issue in performing hazard identification proof is the information of the biokinetics of specific manufactured particulate nanomaterials MPN. In this way, one compiles the so-called ADME-profile which represents Absorption, Distribution, Metabolism and Excretion, depicting the nanomaterials inside the living being.

9.6 Measures to control exposure to nanomaterials

Factors affecting exposure: Every attempt should be made to prevent or minimize exposure to nanomaterials. Factors affecting exposure to nanomaterials include the amount of material being used and whether it can be easily dispersed or form airborne sprays or droplets. The degree of containment and duration of use will also influence exposure. In the case of airborne material, particle or droplet size will determine whether the material can enter the respiratory tract and where it is most likely to deposit. Inhaled particles smaller than 10 micrometers in diameter have some probability of penetrating and being deposited in the gas-exchange (i.e., alveolar) region of the lungs, but there is at least a 50% probability that particles smaller than 4 micrometers in diameter will reach the gas-exchange region.

At present there is insufficient information to predict all of the situations and workplace scenarios that are likely to lead to exposure to nanomaterials. However, there are some workplace factors that will increase the potential for exposure, including (CDC/NIOSH, 2009):

- Working with nanomaterials in liquid media without adequate protection (e.g. gloves) will increase the risk of skin exposure.
- Working with nanomaterials in liquid media during pouring or mixing operations, or where a high degree of agitation is involved, will lead to an increased likelihood of inhalable and respirable droplets being formed.
- Generating nanomaterials in the gas phase in non-enclosed systems will increase the chances of aerosol release to the workplace.
- Handling nanopowders will lead to the possibility of aerosolization.
- Maintenance on equipment and processes used to produce or fabricate nanomaterials will pose a potential exposure risk to workers performing these tasks.
- Cleaning of dust collection systems used to capture nanomaterials will pose a potential for both skin and inhalation exposure.

Since the exploration and utilization of nanomaterials keeps on extending and data about potential health impacts and exposure limits for these nanomaterials is as yet being created, managers ought to apply a mix of various measures to control potential exposures.

In general, there are three main approaches to risk and exposure control, *viz.,* engineering techniques, administrative means and personal protective equipments. These complementary approaches especially engineering techniques should be considered starting with the design stage of an industrial process. Administrative means of control constitute an additional approach when the other methods have not achieved the expected control levels. Administrative means of control must never substitute for engineering techniques, which always be performed according to standard practices. In some situations, due to insufficiently advanced technology and prohibitive costs, engineering measures can not be implemented. In these situations, performing administrative means of control through reduction of work periods, modification of work practices, personal hygiene measures, housekeeping and preventive maintenance constitute other ways of limiting the occupational exposure risks. Accordingly, to minimize the risks from known and unknown health, safety and environment hazards in research and occupational settings of the country, guideline for safe handling, use and disposal of nanoparticles has provided.

9.6.1 Engineering controls

In order to provide a safe work environment for faculty, staff, students and visitors, engineering controls must be maintained wherever nanomaterials are used or stored. At a minimum, engineering controls should include local exhaust ventilation, localized filtration, and personal protective equipment. Respiratory protection is required when working with nanomaterials when local exhaust ventilation and filtration is not available. As per NOISH, there are a few controls should be possible to catch the airborne nano particles:

Use of a chemical fume hood is recommended for all tasks with potential of aerosolizing nanomaterials in either liquid or powder form.

- A well-designed local exhaust ventilation system with a local high-efficiency particulate air (HEPA) filter should be used to effectively remove nanomaterials.
- Animals should be appropriately restrained and/or sedated prior to administering injections and other dosing methods.

- If heavy usage of aerosolized nanoparticles is in use, a proper decontamination, or buffer, area should be utilized to ensure the nanomaterials are not transported outside of the working area.
- Frequent hand washing, especially before eating, smoking, applying cosmetics, or leaving the work area should be employed.
- Laboratories and other spaces where nanomaterials are used or stored must be equipped with an eyewash station that meets American National Standards Institute (ANSI) and Occupational Safety and Health Administration (OSHA) requirements.

9.6.2 Administrative controls

Although traditional permissible exposure limits (PEL) exist for many of the substances that nanomaterials are made from, the PEL for a nanomaterial of these substances is not yet clear. Thus, it is important to incorporate the following administrative controls into all laboratory operations:

- The laboratory's safety plan should be modified to include health and safety considerations of nanomaterials used in the laboratory.
- Principal investigators should develop and implement standard operating procedures (SOPs) in the preparation and administration of nanomaterials (with minimal exposure).
- Protocols involving the in vivo use of nanomaterials must be reviewed and approved by the Animal Welfare Committee.
- Laboratory personnel must receive the appropriate training, including specific nanomaterial-related health and safety risks, standard operating procedures, and steps to be taken in event of an exposure incident, prior to working with nanomaterials.
- Laboratory personnel must be instructed to use extreme caution when performing injections involving nanomaterials since accidental needle stick presents an exposure threat.
- Exposures involving nanomaterials or any other acutely hazardous material must be reported to the office of Safety, Health, Environmental and Risk Management as soon as possible.

Work Practices: The incorporation of good work practices can help to minimize exposure to nanomaterials. Examples of good work practices include the following:

- Projects or applications with the potential for producing nanomaterial aerosols must be conducted within an approved chemical fume hood or ducted biological safety cabinet.
- Needles used for nanomaterial injection must be disposed in an approved sharps containers immediately following use. Needles used for nanomaterial injection should never be bent, sheared, or recapped.
- Bench paper utilized during preparation of nanomaterial stock should be lined with an impervious backing to limit potential for contamination of work surfaces in the event of a minor spill.
- Work areas should be cleaned at the end of each work shift (at a minimum) using either a HEPA-filtered vacuum cleaner or wet wiping methods. Dry sweeping or pressurized air should not be used to clean work areas. Bench tops, chemical fume hood interiors, biological safety cabinet interiors, equipment, and laboratory surfaces with potential for nanomaterial contamination should be routinely cleaned. Cleanup should be conducted in a manner that prevents worker contact with wastes. The disposal of all waste material should comply with all applicable federal, state, and local regulations.
- The storage and consumption of food or beverages in workplaces must be prevented where nanomaterials are handled, processed, or stored, since exposure may occur via ingestion. Wash hands carefully before eating, drinking, applying cosmetics, smoking, or using the restroom.
- Facilities for showering and changing clothes should be provided to prevent the inadvertent contamination of other areas (including take-home) caused by the transfer of nanomaterials on clothing and skin.

9.6.3 Personal Protective Equipment (PPE)

Typical chemistry laboratory apparel should be when working with nanomaterials in accordance with the University's Chemical Hygiene Plan (UTHSC-H, 2008). Always wear appropriate clothing (e.g., pants, shirts, shoes) and personal protective equipment, including safety glasses, laboratory coats, and gloves, when working with nanomaterials. Open sandals, shorts, and skirts are prohibited. Laboratory personnel involved in any task with a potential to nanomaterials must wear the following personal protective equipment

Protective gloves: Glove selection is best determined by a risk assessment and the chemicals used for the procedure. Nitrile or rubber gloves, which cover hands and wrists completely through overlapping sleeve of lab coat when working with nanomaterials, may provide adequate protection. Wearing of two sets of gloves ("double gloving") is advised whenever performing tasks involving nanomaterials and other hazardous substances. Laboratory personnel should thoroughly wash hands with soap and water before and immediately upon removal of gloves.

- Eye protection: Safety glasses or goggles are considered to be the appropriate level of eye protection for working with nanomaterials. SHERM recommends wearing a full-face shield when conducting tasks posing potential for any generation of aerosol and/or droplets.
- Protective clothing: Laboratory coats or disposable gowns that provide complete coverage of skin must be worn when working with nanomaterials. Clothing contaminated with nanomaterials should be removed immediately. Do not take contaminated work clothes home – contaminated clothing may require disposal as hazardous waste.
- Respiratory protection: If engineering controls are not adequate or are not available, and a potential aerosol exposure exists, respiratory protection is required. When working with nanomaterials, one of the following types of respirators must be worn:
- Filtering face piece (N-95 or greater)
- Elastomeric half- or full-face piece with N-100, R-100, or P-100 filters
- Powered air-purifying respirator with N-100, R-100, or P-100 filters. Anyone required to utilize respiratory protection for use with nanoparticle research must contact Chemical Safety at 713-500-5832 to be included in UTHSC-H Respiratory Protection Plan.

9.7 Labeling, Storage and Disposal of nanomaterials

Nanoparticles, due to their unique characteristics including; small size, shape, high surface area, charge, chemical properties, solubility, and degree of agglomeration can cross cell boundaries or pass directly from the lungs into the blood stream and ultimately reach to all of the organs in the body. This is the reason why they may pose higher risk than the same mass and material of larger particles. Moreover, biodegradation of nanoparticles by some kinds of fungi (like wood decay fungi) may result in metabolites which may be toxic to microorganisms under aerobic and anaerobic conditions. Bacteria and living cells can take up nanoparticles, providing the basis for potential bioaccumulation in the food chain (Amoabediny Ghassem et al., 2009).

Before begining working with nanomaterials from original or transferred containers, read the hazardous properties listed on the original container or in the Safety Data Sheet (SDS). Get a nanomaterial label like the one shown here and fill in all of the information.

- Label must include the word "nano" and indicate the chemical content and form.
- Ensure that every line is completed in case the container spills, breaks, or is opened so that others will know what to do.
- Place the label on the container so that it is visible. Add a Hazardous Waste label like the one shown here to the bottle along with the yellow label. The chemical hazards take precedence over the nanomaterial hazards. Both completed labels should be on the containers.
- Keep liquids and dry particles in closed, tightly sealed, labeled, unbreakable containers whether they are suspended in liquids or in dry particle form.
- Use secondary containment (e.g., Ziploc® bags) for dry powders.

Storage and procedures: Nanomaterials are significant for both the general population included and the effect on nature. However, with no information of the essential properties, the degree of the precautionary measures required is obscure. To comprehend the potential effect that any nanomaterial will have on the environment, it is important to indorse the lifecycle point of view. This includes surveying impacts that may happen all through an item's life, including the production, circulation, dealing with, use, storing and disposal. As indicated by the guidance created and utilized by the US Department of Energy (Anonymous, 2007) and the UK Environment Agency (EA) Guidance for the transfer of hazardous materials, an arrangement for storing and disposal of nanomaterials or nanomaterials contaminated waste ought to be created, taking in thought the hazardous idea of the materials and the amounts included and connected to nanomaterial-bearing waste streams (solid and fluid waste), including pure nanomaterials; Items contaminated with nanomaterials, for example, holders, wipes and dispensable PPE and Liquid suspensions containing nanomaterials.

Storage of nanomaterial waste prior to disposal: Coming up next are suitable methodologies for collection and storing of nanomaterial waste before disposal (Eman M Osman, 2019).

- Storage in waste container that is in great condition to anticipate the spillage of the nanomaterials. Mark the waste compartment with a depiction of the waste and incorporate accessible data describing known and suspected properties.

- Storage in plastic packs- Collect paper, wipes, PPE and different things with free contamination in a plastic pack or other sealable compartment put away in the laboratory hood. At the point when the pack is full, close it and cautiously place it into another plastic sack or other fixing holder, to maintain a strategic distance from outside contamination. Remove it from the hood and name the external sack with a suitable waste mark.

Spill Cleanup: Anyone attempting to manage any spill involving hazardous agents must be wearing the appropriate personal protective equipment. OSHA advises typically standard approaches to cleaning nanomaterial powder and liquid spills include the use of HEPA-filtered vacuum cleaners, wetting powders down, using dampened cloths to wipe up powders, and applying absorbent materials or liquid traps. (CDC/NIOSH, 2009) Energetic cleaning methods such as dry sweeping or the use of compressed air should be avoided or only be used with precautions that assure that particles suspended by the cleaning action are trapped by HEPA filters. If vacuum cleaning is employed, care should be taken that HEPA filters are installed properly, and bags and filters changed according to manufacturer's recommendations. At a minimum, the following procedures must be followed when managing an accidental spill of nanomaterials.

- Small spills (typically involving less than 5 mg of material) of nanomaterials containing powder should be wet-wiped with cloth/ gauze that is dampened with soapy water. Affected surfaces should be thoroughly wet-wiped three times over with appropriate cleaning agent and with a clean, damp cloth used for each wipe down. Following completion, all cloth and other spill clean-up materials with a potential for nanomaterial contamination must be disposed of as hazardous waste.
- Small spills (typically involving less 5 ml of material) of nanomaterial-containing solutions should be covered and absorbed with absorbent material. Areas affected by liquid spills should be triple cleaned with soap and water following removal of absorbent paper.
- For larger spills of nanomaterials, contact the Office of Environmental Health and Safety at 713-500- 5832.

As with any spill or clean-up of contaminated surfaces, handling and disposal of the waste material should follow existing Federal, State, or local regulations.

Medical screening and surveillance

- Make accessible medicinal screening and surveillance for specialists exposed to nano materials, if suitable

- Review medicinal surveillance prerequisites under OSHA measures (e.g., Cadmium, Respiratory Protection).

How to deal with the nanomaterials?: A plan to be made before using any nanomaterial for each activity and for each nanomaterial to: avoid, reduce or even eliminate the hazardous impacts of these materials on both human and environment (Eman M Osman, 2019).

Ranking the nanomaterial: Initially, it is desirable to rank the nanomaterial as per its health hazard. Three hazard classifications, 1, 2, and 3, are distinguished. The higher the classification number, the higher is the foreseen health risk. Nanoparticles of classification 1 are relatively harmless. These are required to display "just" the health hazards like the chemical mother material. Class 2 and class 3 are relied upon to display likewise explicit nano related health impacts.

Make an inventory of the activities with handling the NMP: Investigate the activities, during which nanoparticles could be discharged, this ought to be accomplished for each nanomaterial. Evaluate the entire "life cycle" of the NMP beginning from the minute the materials or items enter the organization or lab and closure when those materials or items leave the organization or lab again as "prepared for-use item" or waste material.

9.7.1 Disposal of nanomaterial waste

The existing guidelines for the disposal of nanoparticles are vague. In many cases, the laboratories are left to determine if a special disposal method is required, unless they fall under the jurisdiction of regulations for macro-sized chemicals (Katherine Moore et.al., 2016). Therefore, laboratories use the same methods for disposing of nanoparticles as they do for any other chemicals. Usually, this means they send them off to a specific company or group to deal with the disposal. 50 Very little is specialized specifically for nanoparticles and much is left to the chemical disposal team to decide. Many institutions, including Massachusetts Institute of Technology, Virginia Tech, the University of California at Berkeley, the University of Pennsylvania, and the California Institute of Technology classify nanowaste as particularly hazardous and state that it must be specially collected for disposal.

The following waste management guidance applies to nanomaterial-bearing waste streams consisting of:

- Pure nanomaterials (e.g., carbon nanotubes)
- Items contaminated with nanomaterials (e.g., wipes, PPE, bench paper).

Any material that comes into contact with nanomaterials becomes nanomaterial-bearing waste (e.g., gloves, other PPE, wipes, blotters). It must be managed as hazardous waste.

- Liquid suspensions containing nanomaterials
- Solid matrixes with nanomaterials that are easily crumbled, or have a nanostructure loosely attached to the surface, such that they can reasonably be expected to break free or leach out when in contact with air or water, or when subjected to mechanical forces. This does not apply to nanomaterials firmly bond in a solid base that will not release nanoparticles to water or when broken.

Any material coming into contact with nanomaterials becomes nanomaterial-bearing waste (e.g., gloves, other PPE, wipes, blotters).

- Never put nanomaterial-bearing waste into the regular trash or down the drain! There is a difference between nanomaterial bearing waste and embedded nanomaterial waste.
- Collect nanomaterial-bearing waste in closed, tightly sealed, labeled, unbreakable containers.

Although the major source of nanoparticles in normal waste treatment centers is from commercial products, research laboratories sometimes contribute as well. Even if they do not throw away the physical sample of nanoparticles or pour it down the drain, nanoparticles can contaminate normal rubbish such as gloves and paper towels. Whatever the outlet of disposal, the nanoparticles would ultimately arrive at one of four places: a recycling facility, incineration plant, landfill, or a wastewater treatment facility. Each place has the potential for the release of nanoparticles, with some more preventable than others. In recycling facilities there are plastics with nanoparticles in them that can be released as dust, but the dust is usually controlled as to not expose any humans or the environment. Nanoparticles can be released during the combustion in an incineration plant as ash or dust. Most plants have filters to collect pollutants, but many not rated for nanoparticles or can be damaged by them. Landfills have the most potential for nanoparticle release, be it by emission into air, water, or soil. Wastewater is turned into sludge that is then burned, or sometimes into fertilizer for agriculture. Little is known about the environmental impacts from nanoparticles in fertilizers.

When recycling materials containing nanoparticles, there are three main concerns, viz., the health of facility workers; the environmental impacts from residue that ends up in incineration, landfills, or sewage treatments;

the contamination of products made from recycled materials. There is a possibility of the nanoparticles leaching from the products and affecting their surroundings. These effects are barely studied and are poorly understood.

It is a sensible most reasonable scenario presumption to consider all nanomaterial wastes as potentially hazardous. It can consequently be discarded as synthetic waste. The List of Waste Regulations 2005 (LOWR) gives point by point data about how to deal with and discard different sorts of wastes. Two classifications are especially important: wastes from organic chemical process and wastes from inorganic chemical process (Eman M Osman, 2019). Most nanomaterial wastes can be sorted under these codes.

Prevention of fire and explosion: In the UK, issues of fire and explosion are secured by the Dangerous Substances and Explosive Atmospheres Regulations (DSEAR 2002) They set least requirements for the insurance of laborers from fire and explosion risks identified with hazardous substances and potentially explosive atmospheres and from gases under pressure and substances corrosive to metals and expect businesses to control the risks to the health of workers and consumers from these hazards. The effectiveness of methods for nanoparticle fire, explosion and catalysis prevention and control are yet to be evaluated. In some types of nanoparticle products dense clouds of nanoparticle powder are difficult to see, even though a suspension of the same product at the same concentration with coarser grade is easily visible. Explosion protection measures for dust dispersions and for hazardous quantities of larger sized materials can be applied to the handling of potentially explosive nanoparticles, as follows:

i. **For reactive or catalytically active nanoparticles:** Contact with incompatible substances should be prevented

ii. **For electrical requirements:** The design of electrical equipment protection should consider the fine granulometry and very long settling time of nanoparticles, which require dust protection. Besides, further precautions should be taken to avoid the risk of auto-ignition of nanoparticles. (Eman M Osman, 2019).

To reduce the risks of fire and deflagration, every employer shall ensure that risk is either eliminated or reduced by following measures.

- Reduction of the quantity of dangerous substances to a minimum
- Control of the release of a dangerous substance at source
- Prevention of the formation of an explosive atmosphere, including the application of appropriate ventilation

- Ensuring that any release of a dangerous substance which may give rise to risk is suitably collected, safely contained and removed to a safe place.
- Avoid ignition sources; ignition sources including electrostatic discharges. Anti-static shoes should be worn in areas where the materials are handled to reduce the build-up of static charge .

In case of fire or explosion, the selection of an extinguishing agent should take account of the compatibility or incompatibility of the nanomaterial with water. Some metallic dusts react with water to form hydrogen, which ignites very easily. Chemical powders are available to extinguish burning metallic dust powders; yet, this has the effect of putting the metallic dust in suspension, thereby increasing the risk of deflagration.

a) Using carbon dioxide, nitrogen or another inert gas as extinguisher. Sometimes, this could introduce further hazards into the system, notably the risk of asphyxiation.

The challenges to comprehend the potential ecological effect of nanomaterials: The challenges to comprehend the potential ecological effect of nano materials include choosing what nanomaterials represent the most noteworthy potential risks, what organisms are most valuable to examine, what strategies are most relevant to use, and whether conventional toxicological or emerging ecological models for surveying ecological risks are the most proper (Eman M Osman, 2019).

The present difficulties are that the accessible tools which can be utilized for surveying NMP security are frequently inappropriate, or so laborious that sufficient assessment of safety of these materials and advances remains profoundly hazardous and the assets or testing instruments are not liable to empower security evaluation of the novel nanomaterials that are flooding onto the market. This implies absolutely new safety evaluation standards should be created during the coming very long time to determine this issue.

9.8 Safety-by-Design

"Safety-by-Design" isdefinedasastartingpointconsistingontheimplementation of the principles of drug discovery and development. The concept "Safety-by-Design" appears to be a way to "ensure safety", but the superficiality and the lack of articulation with which it is treated still raises many doubts (Aleksandra Zielińska, et.al., 2020). Although the approach of "Safety-by-Design" to the principles of drug development has helped in the assessment of the toxicity of nanomaterials, a combination of scientific efforts is constantly urgent to ensure the consistency of methods and processes. This will ensure that the quality of nanomaterials is controlled and their safe development is promoted. Safety issues are considered strategies for discovering novel toxicological-related mechanisms still needed to be promoted.

Safety-by-Design (SbD) is taken as a starting point when engineering a novel NM, thereby strictly following the principles of Drug Discovery and Development (DDD) throughout the whole process of product development. The belief behind "Safety-by-Design" (SbD) is to engineer unwanted effects out of the NMs by implementing the knowledge of the NMs' adverse effects on the environment and human health into the process of designing the desired NMs.

Security is the central concept of SbD, but it is important to note that absolute security is never reached. This may lead to the question of what is sufficiently safe, who should define the levels of the security, and how this decision is made. It follows that it may be more appropriate to use the expression "safer" instead of the absolute terms "safe" and "security", for instance (Kraegeloh et.al., 2018).

Despite of all the knowledge on nanotoxicity accumulated in the last decade, it is still difficult to predict the bioavailability, biodistribution, degradation, elimination, and biological activity of nanostructures beforehand (Aleksandra Zielińska, et.al., 2020). However, safety procedures in the manufacture, commercial, and medical industry have been established. Hence, new NMs would benefit from a comprehensive understanding of their toxicological mechanisms and their interaction with the biological system. These studies can illuminate the relation between nanoparticles and its impact on cells (e.g., nanoparticle–cell interactions, endocytosis, and intracellular traffic). There are some biological models that are available to check the toxicity of substances or agents in vitro and in vivo. In vitro models are mainly based on isolated cells and can be used in test tubes exposed to the tested substance. In turn, in vivo models are applied to the life organisms in order to observe changes in their development (growth, reproduction, mortality, etc.), which are toxicity indicators. The integration of metabolomics and transcriptomics in nanotoxicity studies is emerging.

9.9 The Nanosafety Cluster Group (NSC)

The NanoSafety Cluster Group on the nanohub has been established to provide a centralised learning and education platform to which all NSC and related projects can upload all learning media to share with the community.

The NanoSafety Cluster maximises the synergies between European-level projects addressing the safety of materials and technologies enabled by the use of nanoparticles. The studied aspects include toxicology, ecotoxicology, exposure assessment, mechanisms of interaction, risk assessment and standardisation. The Cluster is an initiative of the European Commission Directorate-General for Research and Innovation (DG RTD), which sponsors these large projects. Overall, Europe targets safe and sustainable nanomaterials and nanotechnology innovations. NSC projects contribute to assuring environmental health and safety (EHS) of this Key Enabling Technology. nThe Cluster also is an open platform for dialogue and exchange. Researchers, regulators, administrators, industry, civil society representatives

These pillars work together and independently to promote nanosafety research, public engagement, create strategic collaborations and enhancing synergies between the NSC, and beyond, projects and explore potential international opportunities. Besides the continuous work taking place within the NSC's bodies, two main NSC actions also exist:

The EU Nanosafety Cluster (NSC): It is a cluster of European Commission-funded projects in the funding programs FP6 (2002–2006), FP7 (2007–2013), and Horizon 2020 aka H2020 (2014–2020) framework programmes, aimed at harmonizing the research done in these projects (Fadeel et.al., 2013). The cluster coordinates work done by the NanoSafety Cluster projects to study and establish the safety of nanomaterials. The coordination by the cluster is organized in half-yearly meetings and various working groups. An example of a result of the NanoSafety Cluster's harmonization was the prioritization of which nanomaterials to study.The NSC has become a reference actor for consumers' associations in the field. The NanoSafety Cluster followed in part from the NanoImpactNet project that ran from 2004 to 2012.

The NSC started in 2006 with FP6-funded projects such as CellNanoTox, DIPNA, NanoInteract, NANOTRANSPORT, and NANOSH. This was followed by more than 50 projects in FP7. There are currently four H2020 projects. The budget for the project is €137 million

Structure: The EU NanoSafety Cluster has nine working groups, each with different objectives and goals. WG2 (Hazard Identification) has four focus groups relating to immunosafety, marine ecotox, genotoxicity, and reprotoxicology. WG7 (Dissemination) has a sub-group dedicated to standardization and publishes the NanoSafety Cluster Newsletter. WG9 (Safe by Design and Industrial Innovation) has a sub-group called Industrial Innovation Liaison (i2L), that functions as a cross-linking working group to maximize various pilot-and innovation-led segments of the project (Fadeel et.al., 2013).

- Materials (WG1)
- Hazard (WG2)
- Exposure (WG3)
- Database (WG4)
- Risk (WG5)
- Modelling (WG6)
- Dissemination (WG7)
- Systems Biology (WG8)
- Safe by Design and Industrial Innovation (WG9)

The European Chemicals Agency updated their European Union Observatory for Nanomaterials (EUON) in 2018 with results from the Cluster, including use of software developed by the eNanoMapper project and data collected in projects like MARINA, NANoREG, and eNanoMapper.

Nanoparticle aerosol generator for nanosafety studies: A versatile system was designed to generate nanoparticulate aerosol streams with stable concentration and particle size distribution. The device was divided into a compressed air reservoir and a nanopowder container. The operational methodology consisted in storing a dried and filtered air in a 40-cm3 stainless steel reservoir under controlled pressure (Alberto Clemente et.al., 2017). Dried nanopowders were placed into a stainless steel column, with a 1.2-mm diameter reduction in the far end and a 50-µm grid on the bottom end. All nanoparticle aerosols were safely released into a state-of-the-art nanoparticle dispersion and exposure chamber in a nearly particle-free environment. The compressed air was instantaneously released impelling the nanopowder solid through column tip. Aeroxide® TiO_2 (Evonik, Germany), ZnO (Sigma-Aldrich, USA), CuO (Nanologica, Sweden) and Multi-walled carbon nanotubes (MWCNT; Colorobbia, Italy) were successfully aerosolized.

Aerosol and ofline electron microscopy tests showed the production of stable nanoparticulate streams. The air expansion in the secondary chamber and further emission using a valve produced a long-lasting aerosol plume with tunable nanoparticle concentration and narrow particle size distribution over several hours. Aerosol particle concentrations and median particle sizes depended on the pressure difference, being larger as the compressed air stream was increased. Particle concentration and size distribution in the nanoscale range was nearly independent on the mass of aerosolized powders. Finally, the proposed device is a powerful tool for generating both instantaneous clouds and continuous streams of nanoparticles with wide applications in nanotoxicology and nanosafety studies.

Morpheus – A Multiparametric platform for safety testing: Synthetic nanoparticles (NPs) are nowadays used in numerous products in technical fields and biomedicine. An evaluation of the hazardous potential of NPs at an early stage will limit reservations in public opinion as well as provide standards for preclinical development of nanoparticles (Jana Fleddermann et.al., 2017). For this purpose, a multiparametric and integrated platform for testing of NPs based on a 3D liver tissue model will be developed that include NP structure and properties, liver viability, liver functionality and morphology.

The combination of classical and new liver markers with innovative 3D cell culture technology, microscopy and PCR-based analysis of the gene expression pattern of cells in an integrated test platform will provide an important contribution to NP risk assessment.

9.10 Public perception of nanosafety

Owing to the global publicity that nanotechnology receives and the marketing success of the locally-manufactured nanotechnology-enabled functional fabrics/apparels and cosmetics, Thai consumers have slowly entered the age of nano-hypes (Tanasugarn and Tanpipat, 2020). The general public was led to believe that a product manufactured using nanotechnology must possess superior quality, regardless of any possible health or environmental risks. Possible adverse effects of nanotechnology and nanoparticles in the environment were virtually unheard of until quite recently. Since potential risks of nanotechnology for the developing world have been documented in the health, environment and social aspects, NANOTEC and CIN are both interested in the handling of nano-risks (characterization, identification, assessment, management, mitigation, prevention, etc.). For example, it is expected that a precautionary approach will be used in the eventual regulation of nanotechnology and manufactured nanomaterials. Meanwhile, owing to potential benefits of nanotechnology, the general public should be informed but not caused to panic over unsubstantiated incidents.

The advent of agri-nanotechnologies brings both increased need and opportunity to change this status quo as it introduces significant issues of intersectionality that cannot adequately be addressed by current discipline-specific approaches alone (Lombi et.al., 2019.

A complex transformation of nanomaterials occurs when released in natural aquatic environments, including stabilization, aggregation, adsorption and dissolution due to the effects of molecular interactions (from natural organic matter and proteins secreted by aquatic organisms) with nanoparticles that form a "biocorona" on the particles (Erik Nilebäck and Julie Gold, 2018). The transformation gives new characteristics to the nanomaterial that affect the toxicity towards aquatic organisms.

Nanosafety: What workers need to know: This may pose a problem if scientists do not know if different sized nanoparticles have different reactions to the human body. There is no or little research to determine if different size nanoparticles have the same properties (Dominick Fazarro, 2014). There are some unknowns about nanoparticles in terms of their characterization. Studies have found that carbon nanotubes is just as dangerous as Asbestos. It is unclear whether existing test methods for physical and chemical properties are sufficient for nanomaterials characterization in order to assess their risk and to determine their exposure and hazard. It is clear, however, that properties such as boiling point are insufficient. Single walled carbon nanotubes, for example, can be manufactured via several different processes which can generate products with different physical and chemical properties. Researchers found that carbon nano tubes (multiwalled) In addition to the early stage noncancer lung effects in animals, some studies in cells or animals have shown genotoxic or carcinogenic effects.

Gaps: There are many gaps in the knowledge still to be filled before appropriate risk assessments and workplace exposure standards can be established (Harford et.al., 2007). It is important to support the recommendations of the report commissioned for the ASCC (2006), which states that "The generally accepted approach is the application of a hierarchy of risk controls which incorporate the broad elements of elimination, substitution, engineering controls, administrative controls and – finally – the use of personal protective equipment". Specifically, until appropriate workplace exposure standards can be established, the nanotechnology industry should control exposures through the application of risk management programs using the "As Low As Reasonably Practicable" (ALARP) approach. Substitution is unlikely to be an applicable hazard reduction method because the unique properties of nanomaterials will drive their production and use. Nevertheless, traditional engineering and administrative controls and the use of PPE will be effective in reducing the exposure of workers, therefore these should be used in all nanotechnology facilities. Although there many uncertainties concerning the health effects of nanomaterials, this should not prevent the industry from moving forward because there are numerous measures that can be taken, which will limit the exposure of nanotechnology workers and minimise the harm that nanotechnologies cause to workers. Continued monitoring of current and emerging exposure and health relationships will permit early identification of unforeseen risks and the adoption of measures to mitigate them.

9.11 The DaNa 2.0 Knowledge base nanomaterials

Nanotechnology is closely related to the tailored manufacturing of nanomaterials for a huge variety of applications. However, such applications with newly developed materials are also a reason for concern. The DaNa2.0 project provides information and support for these issues on the web in condensed and easy-to-understand wording (Harald F. Krug et.al., 2018). Thus, a key challenge in the field of advanced materials safety research is access to correct and reliable studies and validated results. For nanomaterials, there is currently a continuously increasing amount of publications on toxicological issues, but criteria to evaluate the quality of these studies are necessary to use them e.g., for regulatory purposes. DaNa2.0 discusses scientific results regarding 26 nanomaterials based on actual literature that has been selected after careful evaluation following a literature criteria checklist. This checklist is publicly available, along with a selection of standardized operating protocols (SOPs) established by different projects. The spectrum of information is rounded off by further articles concerning basics or crosscutting topics in nanosafety research. This article is intended to give an overview on DaNa2.0 activities to support reliable toxicity testing and science communication alike.

The nanosafety aspect is now a topic around the world. Whereas only five nanosafety projects were funded in the 6th Framework programme of the EU (2002–2006), around 50 projects and initiatives on nanosafety were funded in the 7th EU Framework programme between 2007 and 2013 (Anonymous, 2018). During February 2018, 19 projects are running under H2020 (source EU NanoSafety Cluster . All nanosafety-related projects are consolidated in the so-called EU NanoSafety Cluster, an initiative of the European Commission Directorate-General for Research and Innovation (DG RTD). It aims to maximize synergies between research projects on a European or national level addressing all aspects of nanosafety including toxicology, ecotoxicology, exposure assessment, mechanisms of interaction, risk assessment, and standardization.

Within the first phase (2014–2015) of the European Work Programme "Horizon 2020—Topic Nanotechnologies, Advanced Materials und Production (NMP)", the nanosafety aspect gained more and more importance. One expected impact of the calls for proposals was "Promoting safe-by-design approaches in collaboration with the EU NanoSafety Cluster and contributing towards the framework of EU nanosafety and regulatory strategies..." (Harald F. Krug et.al., 2018).

In Germany, the Federal Ministry of Education and Research (BMBF) funds projects on nanosafety, together with industry, to fill knowledge gaps and to initiate measures to identify and minimize risk; it has done this since the beginning of the 1990s. Since 1998, funds made available for nanotechnology within the framework of BMBF project funding have increased by a factor of four. Competence centers were established simultaneously as supporting infrastructures (source BMBF).

In Canada and USA, several national projects and initiatives are running, for example, the National Institute for Nanotechnology (NINT) from The National Research Council of Canada or the NanoPortal as a gateway to the Government of Canada's information on nanotechnology. In the USA, NIOSH is the leading federal US agency conducting research and providing guidance on the occupational safety and health implications and applications of nanotechnology. The Organisation for Economic Co-operation and Development (OECD) also cares for the safety of manufactured nanomaterials.

With the example of nanosafety research in mind, it is obvious that not only chemicals have to be safe or labeled accordingly but also any kind of "material" used for products on the market. For more than 15 years a new philosophy called "green chemistry" has been propagated in labs and production sheds, finally generating the idea of "green toxicology" (Zimmerman et.al., 2014). One of the major doctrines of green chemistry/green toxicology is "Benign-by-Design", which means to substitute toxic compounds or to reduce them to a minimum just in the planning phase of a new product.

9.12 Safety in Research Environment

Information related to the health and environmental risk of engineered nanoparticles or nanomaterials is severely lacking and thus insufficient for completing rigorous risk assessment on their use. However, since preliminary scientific evaluations indicate that there are reasonable suspicions that activities involving nanomaterials might have damaging effects on human health; the precautionary principle must be applied (Groso et.al., 2010). Public and private institutions as well as industries have the duty to adopt preventive and protective measures proportionate to the risk intensity and the desired level of protection.

Due to the fact that these data (as well as exposure limits for each individual material) are not available, risk classes could not be determined. For each hazard level we then provide a list of required risk mitigation measures (technical, organizational and personal). The target 'users' of this safety and health methodology are researchers and safety officers. They can rapidly access the precautionary hazard class of their activities and the corresponding adequate safety and health measures. Adequate safety measures and management are promoting innovation and discoveries by ensuring a safe environment even in the case of very novel products. The proposed measures are not considered as constraints but as a support to their research. This methodology is being implemented at the Ecole Polytechnique de Lausanne in over 100 research labs dealing with nanomaterials. It would be useful to other research and academia institutions as well.

Lab safety: Nanoparticles or nanomaterials used in laboratory experiments will be likely in one of three forms: a powder, in suspension or in a solid matrix

Safety issues and concerns: The ultrasmall particle size permits the particles to be carried deeply into tissues. The molecular structure of nanoparticles and the relatively greater surface area confer on these particles different chemical reactivities than for larger structures made from the same elements or molecules. Nanoparticles present a unique challenge from an occupational health perspective as there is a limited amount of toxicological data currently available for review. The National Institute of Occupational Health and Safety (NIOSH) and other agencies are currently studying this issue, and improved guidelines for handling and disposal will be forthcoming. Currently, agencies charged with providing safety guidelines, including NIOSH, the NIH and the EPA, promote the incorporation of cautionary measures in research, with a view toward minimizing or eliminating exposures to nanoparticles.

Laboratory safety guidelines for handling engineered nanoparticles: The current practices for working with engineered nanoparticles safely are essentially the same as one would use when working with any research chemical of unknown toxicity. If handling or transporting materials outside of a glovebox, workers should wear protective equipment, including,

- Latex or nitrile gloves when handling nanoparticle powders and nanoparticles in suspension (glove changes should be performed frequently); Outer gloves should always be removed inside the hood or under the influence of local exhaust ventilation and placed into a sealed bag. This will prevent the particles from becoming airborne.
- Chemical splash goggles when working with nanomaterials in suspension or dry powdered form;
- Lab coats should be laundered on a periodic basis. Do not take lab coats home for laundering;

The use of respirators is not generally required for worker protection if you work in a ventilated area.

- When purchasing commercially available nanoscale materials, be sure to obtain the Material Safety Data Sheet (MSDS) and to review the information in the MSDS with all persons who will be working with the material. Note, however, that given the lack of extensive data on nanoparticles, the information on an MSDS may be more descriptive of the properties of the bulk material.
- Nanoparticle solutions may be handled on the lab bench once placed in solution.
- Transport of nano-materials should employ a sealed secondary containment device.
- Work surfaces should be wet-wiped regularly – daily is recommended. Because many engineered nanoparticles are not visible to the naked eye, surface contamination may not be obvious. The cleaning solution should be compatible with the vehicle in which the nanoparticles are suspended.
- To prevent ingestion, eating and drinking and chewing gum are not allowed in laboratories
- Report any safety concern

Managing spills

- All spills involving nanoparticles should be treated like a hazardous material spill and cleaned up immediately.
- The person cleaning up should wear double nitrile gloves and either vacuum up the area with a HEPA filtered vacuum or wet wipe the area with towels, or combination of the two.
- For spills that might result in airborne nanoparticles, proper respiratory protection should be worn, such as a respirator with NIOSH-approved filters that are rated as N-, R- or P-100 (HEPA).
- Do not brush or sweep spilled/dried nanoparticles.

Waste disposal procedures: Since the toxicology and environmental fate of nanoparticles is still largely unknown, all nanoparticle waste (solid material and liquids) should be conservatively managed as hazardous waste.

- Nanoparticle waste must be placed in an appropriate container and labeled for chemical content in compliance the hazardous waste management requirements.

Work practices : Health, safety and environmental (HSE) risks of a technology is an inseparable part of it which threatens all exposed employees. It has been proved for many years that exposure to particles, in an occupational setting, could be linked with the onset of lung diseases, such as pneumoconiosis, chronic obstructive pulmonary disease (COPD), and mesotelioma and lung cancer (Gh. Amoabediny et.al., 2019). Nanoparticles, due to their unique characteristics including; small size, shape, high surface area, charge, chemical properties, solubility, and degree of agglomeration can cross cell boundaries or pass directly from the lungs into the blood stream and ultimately reach to all of the organs in the body. This is the reason why they may pose higher risk than the same mass and material of larger particles. Moreover, biodegradation of nanoparticles by some kinds of fungi (like wood decay fungi) may result in metabolites which may be toxic to microorganisms under aerobic and anaerobic conditions. Bacteria and living cells can take up nanoparticles, providing the basis for potential bioaccumulation in the food chain.

Considering fast-growing in research and industrial activities, controlling nanoparticles related HSE risks should be highly considered. In general, there are three main approaches to risk and exposure control: engineering techniques, administrative means and personal protective equipments. These complementary approaches especially engineering techniques should be considered starting with the design stage of an industrial process.

Assessing worker exposures to nanomaterials : Employers should assess worker exposure to nanomaterials to identify the control measures needed and determine if the controls used are effective in reducing exposures by identifying and describing processes and job tasks where workers may be exposed to nanomaterials; determining the physical state of the nanomaterials such as dust, powder, spray, or droplets; determining routes of exposure (e.g., inhalation, skin contact or ingestion) of particulates, slurries, suspensions or solutions of nanomaterials; identifying the most appropriate sampling method to determine the quantities, airborne concentrations, durations, and frequencies of worker exposures to nanomaterials ; and determining what additional controls may be needed based on the exposure assessment results and evaluating the effectiveness of controls already in place. Employers should adopt the most effective controls available to limit worker exposure.

The incorporation of good work practices can help to minimize exposure to nanomaterials (Anonymous, 2009). Examples of good work practices include the following.

- Projects or applications with the potential for producing nanomaterial aerosols must be conducted within an approved chemical fume hood or ducted biological safety cabinet.
- Needles used for nanomaterial injection must be disposed in an approved sharps containers immediately following use. Needles used for nanomaterial injection should never be bent, sheared, or recapped.
- Bench paper utilized during preparation of nanomaterial stock should be lined with an impervious backing to limit potential for contamination of work surfaces in the event of a minor spill.
- Work areas should be cleaned at the end of each work shift (at a minimum) using either a HEPA-filtered vacuum cleaner or wet wiping methods. Dry sweeping or pressurized air should not be used to clean work areas. Bench tops, chemical fume hood interiors, biological safety cabinet interiors, equipment, and laboratory surfaces with potential for nanomaterial contamination should be routinely cleaned. Cleanup should be conducted in a manner that prevents worker contact with wastes. The disposal of all waste material should comply with all applicable federal, state, and local regulations.
- The storage and consumption of food or beverages in workplaces must be prevented where nanomaterials are handled, processed, or stored, since exposure may occur via ingestion. Wash hands carefully before eating, drinking, applying cosmetics, smoking, or using the restroom.
- Facilities for showering and changing clothes should be provided to prevent the inadvertent contamination of other areas (including take-home) caused by the transfer of nanomaterials on clothing and skin.

Personal Protective Equipment : Typical chemistry laboratory apparel should be when working with nanomaterials in accordance with the University's Chemical Hygiene Plan (UTHSC-H, 2008). Always wear appropriate clothing (e.g., pants, shirts, shoes) and personal protective equipment, including safety glasses, laboratory coats, and gloves, when working with nanomaterials. Open sandals, shorts, and skirts are prohibited. Laboratory personnel involved in any task with a potential to nanomaterials must wear the following personal protective equipment:.

Nanoparticle use in animals : Unless data exists for the use of a specific nanoparticle in an animal, the following Exposure Control methods shall be followed (Anonymous, 2009). Standard PPE for handling of dry nanoparticles when work is performed outside of chemical fume hood or ducted biological safety cabinets include: double gloves, gown, safety goggles or safety glasses, and N-95 or equivalent respirators (CDC/NIOSH, 2009). Work areas should be cleaned at the end of the procedure using either a HEPA-filtered vacuum cleaner or wet wiping methods, using a fresh cloth that is dampened with soapy water. Cleaning cloths must be disposed of. Drying and reuse of contaminated cloths may result in re-dispersion of particles. Use of nanoparticles in animals can be broken down into three segments: administration, husbandry and disposal.

Administration

Injection: Nanoparticles in suspension may be injected in to animal on a bench top covered with absorbent paper. Absorbent paper should be changed after each experiment and disposed as nanoparticles waste. Dispose of the syringe in an approved Sharps container.

Oral: if preparation is being administered via a syringe or other feeding device, a fume hood or ducted BSC (as above) must be used. If administration is by food, use of a microisolator cage is recommended.

Aerosol: fume hood or ducted BSC. Proper PPE for investigators must include an N-95 or equivalent respirator.

Husbandry

- Exposed animals must be housed under BSL-2 conditions for the first 72 hours post exposure. After 72 hrs, animals can be housed at BSL-1.
- All bedding and waste must be bagged and incinerated

Disposal

- All potentially contaminated carcasses, bedding and other materials must be disposed of through incineration.
- Any surplus nanoparticle stocks must be disposed of as hazardous waste.

As additional information is discovered concerning nanoparticles, the complexity of evaluating hazards associated with their use will probably follow a bell-curve, increasing before decreasing. Thus the information presented above must be considered subject to change.

Summary of Recommended Nanomaterial Risk Levels (NRL) is listed in table 47.

Waste disposal procedures: There are no specific EPA regulations that apply to nanomaterial waste. It is important to take a cautious approach when handling nanomaterial waste that is defined as hazardous however. The following waste management guidance applies to nanomaterial-bearing waste streams consisting of:

- Pure nanomaterials (e.g. carbon nanotubes)
- Items contaminated with nanomaterials (e.g. wipes/PPE)
- Liquid suspensions containing nanomaterials
- Solid matricies with nanomaterials that are friable or have a nanostructure loosely attached to the surface such that they can reasonably be expected to break free or leach out when in contact with air or water, or when subjected to reasonably foreseeable mechanical forces.

The guidance does not apply to nanomaterials embedded in a solid matrix that cannot reasonably be expected to break free or leach out when they contact air or water, but would apply to dusts and fines generated when cutting or milling such materials. Nanomaterial bearing waste streams should not be placed into the regular rubbish or down the drain.

Table 47: Summary of Recommended Nanomaterial Risk Levels (NRL)

S.L	Type of Nanomaterial	Practices	Engineering Controls	Personal Protective Equipment (PPE)
1.	Polymer matrix	Standard Laboratory Practices including: • Lab Safety Plan should be updated with NRL defined • Labeling of storage containers of nanomaterials with both the chemical contents and the nanostructure form	Fume hood or biological safety cabinet (Class II Type A1, A2 vented via a thimble connection, B1 or B2)	Standard PPE (lab coat, gloves, safety glasses with side shields)
2.	Liquid dispersion	NRL-1 practice plus: • Use secondary containment for containers that store nanomaterials • Wipe contaminated areas with wet disposable wipes • Dispose of contaminated cleaning materials as segregated nanomaterial waste	Fume hood or biological safety cabinet (Class II Type A1, A2 vented via a thimble connection, B1 or B2) or approved vented enclosure (e.g., Flow Sciences vented balance safety enclosure [VBSE])	NRL-1 practice plus: • Nitrile gloves • Safety goggles
3.	Dry powders or aerosols	NRL-2 practice plus: • Vacuum with HEPA-equipped hand vacuum cleaner • Label work areas with "Caution Hazardous Nanoscale Materials in Use"	Fume hoodor biological safety cabinet (Class II Type A1, A2 vented via a thimble connection, B1 or B2) or approved vented enclosure (e.g., Flow Sciences vented balance safety enclosure [VBSE]). HEPA filtered exhaust preferred for fume hoods containing particularly "dusty" operations.	NRL-2 practice plus: • N95 respirators are required if work operation must be done outside of containment

S.L	Type of Nanomaterial	Practices	Engineering Controls	Personal Protective Equipment (PPE)
4.	Dry Powders or aerosols of parent materials with known toxicity or hazards	NRL-3 practice plus: • Baseline medical evaluation or employees including physical exam, pulmonary function test (PFT) and routine blood work. • Access to the facility should be permitted only to persons who are knowledgeable about the hazards of the material and the specific control measures implemented to avoid exposures and/or environmental releases. These control measures should include work practices (SOPs), engineering controls, spill and emergency procedures, personal protective equipment, disposal procedures, and decontamination/ clean up procedures. Department procedures should address the designation and posting of the laboratory, how access will be controlled, and any required entry and exit protocols.	Fume hood or biological safety cabinet (Class II Type B1 or B2) or glove box or approved vented enclosure (e.g., Flow Sciences vented balance safety enclosure [VBSE]). HEPA filtered exhaust with Bag-In/ Bag-Out capability preferred for hoods, BSCs, and gloveboxes.	NRL-3 practice plus: • Need determined and respirator selected with reference to the engineering controls in use and potential for aerosol generation

Since the toxicology and environmental fate of nanoparticles is still largely unknown, all nanoparticle waste (solid material and liquids) should be conservatively managed as hazardous waste. This also includes any debris (i.e. PPE, plastic) that has become heavily contaminated with nanoparticles.

- Do not put material from nanomaterial-bearing waste streams into the regular waste or down the drain.
- Do not permit nanomaterial-bearing wastes to be transferred off-site to researchers' home institutions for disposal.
- Characterise and manage nanomaterial-bearing waste streams as hazardous or non-hazardous waste.
- Package nanomaterial – bearing wastes in containers that are compatible with the contents, in good condition, and that afford adequate containment to prevent the escape of nanomaterials.
- Where possible, segregate nanomaterial waste from other waste during management and disposition.
- Label the waste container with a description of the waste and the words 'contains nanomaterials'. Include available information characterising known and suspected properties.
- Collect paper, wipes, PPE and other items with loose contamination in a plastic bag or other sealable container and store in a fume hood. When the bag is full, remove from the hood and place into a second plastic bag or other sealed container. Label the outer bag with a waste label.
- Keep an inventory of all nanomaterial waste that is transferred off-site. This should contain a description of the waste, quantity and means and location of final disposition.

Potential Health Concerns

- The potential for nanomaterials to enter the body is among several factors that scientists examine in determining whether such materials may pose an occupational health hazard. Nanomaterials have the greatest potential to enter the body through the respiratory system if they are airborne and in the form of respirable-sized particles (nanoparticles). They may also come into contact with the skin or be ingested.
- Based on results from human and animal studies, airborne nanoparticles can be inhaled and deposit in the respiratory tract; and based on animal studies, nanoparticles can enter the blood stream, and translocate to other organs.

- Experimental studies in rats have shown that equivalent mass doses of insoluble incidental nanoparticles are more potent than large particles of similar composition in causing pulmonary inflammation and lung tumors. Results from in vitro cell culture studies with similar materials are generally supportive of the biological responses observed in animals
- Experimental studies in animals, cell cultures, and cell-free systems have shown that changes in the chemical composition, crystal structure, and size of particles can influence their oxidant generation properties and cytotoxicity.
- Studies in workers exposed to aerosols of some manufactured or incidental microscopic (fine) and nanoscale (ultrafine) particles have reported adverse lung effects including lung function decrements and obstructive and fibrotic lung diseases. The implications of these studies to engineered nanoparticles, which may have different particle properties, are uncertain.
- Research is needed to determine the key physical and chemical characteristics of nanoparticles that determine their hazard potential.

Potential safety concerns

- Although insufficient information exists to predict the fire and explosion risk associated with powders of nanomaterials, nanoscale combustible material could present a higher risk than coarser material with a similar mass concentration given its increased particle surface area and potentially unique properties due to the nanoscale (Anonymous, 2009)..
- Some nanomaterials may initiate catalytic reactions depending on their composition and structure that would not otherwise be anticipated based on their chemical composition

Working with engineered nanomaterials

- Nanomaterial-enabled products such as nanocomposites, surface-coated materials, and materials comprised of nanostructures, such as integrated circuits, are unlikely to pose a risk of exposure during their handling and use as materials of non-inhalable size. However, some of the processes used in their production (e.g., formulating and applying nanoscale coatings) may lead to exposure to nanomaterials, and the cutting or grinding of such products could release respirable-sized nanoparticles. (Anonymous, 2009).

- Maintenance on production systems (including cleaning and disposal of materials from dust collection systems) is likely to result in exposure to nanoparticles if deposited nanomaterials are disturbed.
- The following workplace tasks can increase the risk of exposure to nanoparticles:

Working with nanomaterials in liquid media without adequate protection (e.g., gloves)

- Working with nanomaterials in liquid during pouring or mixing operations, or where a high degree of agitation is involved
- Generating nanoparticles in nonenclosed systems
- Handling (e.g., weighing, blending, spraying) powders of nanomaterials
- Maintenance on equipment and processes used to produce or fabricate nanomaterials and the cleaning-up of spills and waste material containing nanomaterials
- Cleaning of dust collection systems used to capture nanoparticles
- Machining, sanding, drilling, or other mechanical disruptions of materials containing nanoparticles

Exposure assessment and characterization

- Until more information becomes available on the mechanisms underlying nanomaterial toxicity, it is uncertain what measurement technique should be used to monitor exposures in the workplace. Current research indicates that mass and bulk chemistry may be less important than particle size and shape, surface area, and surface chemistry (or activity) for some nanostructured materials.
- Many of the sampling techniques that are available for measuring airborne nanoaerosols vary in complexity but can provide useful information for evaluating occupational exposures with respect to particle size, mass, surface area, number concentration, and composition. Unfortunately, relatively few of these techniques are readily applicable to routine exposure monitoring. NIOSH has initiated exposure assessment studies in workplaces that manufacture or use engineered nanoparticles (Anonymous, 2009).
- Regardless of the metric or measurement method used for evaluating nanoaerosol exposures, it is critical that background nanoscale particle measurements be conducted before the production, processing, or handling of nanomaterials.

- When feasible, personal sampling is preferred to ensure an accurate representation of the worker's exposure, whereas area sampling (e.g., size-fractionated aerosol samples) and real-time (direct reading) exposure measurements may be more useful for evaluating the need for improvement of engineering controls and work practices.

Precautionary measures

- Given the limited amount of information about health risks that may be associated with nanomaterials, taking measures to minimize worker exposures is prudent (Anonymous, 2009).
- For most processes and job tasks, the control of airborne exposure to nanoaerosols can be accomplished using a variety of engineering control techniques similar to those used in reducing exposure to general aerosols.
- The implementation of a risk management program in workplaces where exposure to nanomaterials exists can help to minimize the potential for exposure to nanoparticles. Elements of such a program should include the following:
    - Evaluating the hazard posed by the nanomaterial based on available physical and chemical property data, toxicology, or health-effects data
    - Assessing the worker's job task to determine the potential for exposure
    - Educating and training workers in the proper handling of nanomaterials (e.g., good work practices)
    - Establishing criteria and procedures for installing and evaluating engineering controls (e.g., exhaust ventilation) at locations where exposure to nanomaterials might occur
    - Developing procedures for determining the need for and selecting proper personal protective equipment (e.g., clothing, gloves, respirators) — Systematically evaluating exposures to ensure that control measures are working properly and that workers are being provided the appropriate personal protective equipment.
- Engineering control techniques such as source enclosure (i.e., isolating the generation source from the worker) and local exhaust ventilation systems should be effective for capturing airborne nanoparticles. Current knowledge indicates that a well-designed exhaust ventilation system

with a high-efficiency particulate air (HEPA) filter should effectively remove nanomaterials.

- The use of good work practices can help to minimize worker exposures to nanomaterials. Examples of good practices include cleaning of work areas using HEPA vacuum pickup and wet wiping methods, preventing the consumption of food or beverages in workplaces where nanomaterials are handled, providing hand-washing facilities, and providing facilities for showering and changing clothes
- No guidelines are currently available on the selection of clothing or other apparel (e.g., gloves) for the prevention of dermal exposure to nanoaerosols. However, some clothing standards incorporate testing with nanometer-sized particles and therefore provide some indication of the effectiveness of protective clothing.
- Respirators may be necessary when engineering and administrative controls do not adequately prevent exposures. Currently, there are no specific limits for airborne exposures to engineered nanoparticles although occupational exposure limits exist for some larger particles of similar chemical composition. It should be recognized that exposure limits recommended for nonnanoscale particles may not be health protective for nanoparticle exposures (e.g., the OSHA Permissible Exposure Limit [PEL] for graphite may not be a safe exposure limit for carbon nanotubes).

9.13 Occupational Health Surveillance

Occupational health surveillance is an essential component of an effective occupational safety and health program (Anonymous, 2009). The unique physical and chemical properties of nanomaterials, the increasing growth of nanotechnology in the workplace, and information suggesting that exposure to some engineered nanomaterials can cause adverse health effects in laboratory animals all support consideration of an occupational health surveillance program for workers potentially exposed to engineered nanomaterials.

When research involves work with engineered particles for which no toxicity data is yet available, it is prudent to assume the nanoparticles may be toxic. Follow the safety guidelines below to protect yourself from possible hazards (Gilman Dr.La Jolla, 2020).

Safety training

Required: The PI must provide appropriate, rigorous, and documented safety training.

- Instruct employees about the hazards and risks of handling nanoparticles and how to protect themselves. Ensure that:
- A safety protocol is written into the research procedure
- Safety Data Sheets (formerly called *Material* Safety Data Sheets), the manufacturer's handling instructions and standard operating procedures are carefully read

Note: Given the lack of extensive data on nanoparticles, Safety Data Sheet information may be more applicable to the properties of the bulk material.

- Inform employees about the four possible routes of workplace exposure to nanoparticles:
    - Inhalation: Because of their tiny size, certain nanoparticles appear to penetrate deep into the lungs and may translocate to other organs following pathways not demonstrated in studies with larger particles.
    - Ingestion: Once ingested, some types of nanoparticles might be absorbed and transported within the body by the circulatory system.
    - Injection: Accidental injection or skin puncture is a potential route of exposure, especially when working with animals or needles.
    - Skin absorption: In some cases, nanoparticles have been shown to migrate through the skin and circulate in the body. If the particle is carcinogenic or allergenic, even tiny quantities may be biologically significant.
- Provide and train employees in the correct use of personal protective equipment and engineering controls. Find these topics in other sections of this page below:
    - Personal protective equipment
    - Control the hazards
- Make sure everyone understands what to do in case of emergency. See:
    - Emergency Guide
    - Exposure response

- Ensure that employees who work with reactive nanoparticles complete fire extinguisher training. See:
    - Fire Extinguisher Training
    - How to Use a Fire Extinguisher
- Keep training records that include:
    - Topics covered
    - Materials distributed
    - Date
    - Employee names
    - Employee signatures

Emergency Preparations

- Know the locations and how to use the nearest:
    - Eye wash and safety shower
    - Emergency Guide
    - First aid kit
    - SDS: Print and keep a copy applicable Safety Data Sheets with your emergency supplies.
    - Appropriate fire extinguisher:

Avoid incompatible extinguishing agents.

Contact EH&S Fire & Life Safety, (858) 534-7525, for questions regarding appropriate fire extinguishers.

- Prepare for spills. Clean up only very small quantities and only if you have been properly trained. (Gilman Dr.La Jolla, 2020).
    - Read How to Handle Chemical Spills in Laboratories. Keep a chemical spill ki easily accessible.
    - Wear PPE, including double nitrile gloves and a lab coat.
    - For spills that might result in airborne nanoparticles, proper respiratory protection is required (See Personal Protective Equipment above).
    - Do not brush or sweep spilled or dried nanoparticles.

    - Place Tacki-Mat® at the spill area exit to reduce the likelihood of spreading nanoparticles.
    - Dispose of spill containment materials as hazardous waste

Exposure response: Treat any exposure seriously, no matter how slight it may seem at the moment.

- **All exposures:**
  - Give first aid treatment, and then seek medical attention immediately as needed.
  - Call Police and request an ambulance, if transportation is necessary.
    - **Ingestion:** Seek medical attention immediately.
    - **Skin exposure:** Flush exposed skin with water for at least 15 minutes while removing any contaminated clothing.
    - **Eye exposure:** Flush eyes with water for at least 15 minutes. Affected individuals may need help holding their eyes open under water. Seek medical attention immediately at an emergency room.

Hazard assessment: Evaluate the hazards before beginning work:

- Consult safety resources:
  - Safety Data Sheets (formerly called *Material* Safety Data Sheets)
- **Note:** In some cases, the manufacture of nanomaterials involves hazardous or toxic chemicals. Consider the hazards of precursor materials when evaluating the process hazards or final product.
  - Approaches to Safe Nanotechnology (PDF), National Institute for Occupational Safety & Health (NIOSH)
  - General Safe Practices for Working with Engineered Nanomaterials in Research Laboratories (PDF), NIOSH
  - Nanotechnology: Guidelines for Safe Research Practices, Lab Safety Work Group, a subcommittee of the Industrial Hygienists and Safety Steering Committee
  - Nanotechnology Information Library
  - Progress Toward Safe Nanotechnology in the Workplace
- Be aware of fire risk when working with reactive nanomaterials. Prior to starting work, assess whether large quantities or high concentrations of nanoparticles will be generated. (Gilman Dr.La Jolla, 2020).

- Under certain circumstances, combustible nanomaterials may present a higher risk when exposed to air due to their large surface area and overall small size.
- Carbonaceous and metal dust can burn and explode if an oxidant such as air and an ignition source are present; self-heating may occur when reactive moieties, such as double bonds, are constituents of the carbonaceous material.

Control of hazards

- Handle nanoparticles whenever possible in a form that is not easily made airborne, such as in solution or on a substrate.
    - Alternatively, handle nanoparticles with engineering controls such as an HEPA-filtered local capture hood or glove box.

Note: Locate HEPA-filtered, local capture systems as close to the possible source of nanoparticles as possible. Installation must be properly engineered to maintain adequate ventilation capture.

- Perform work inside a fume hood if an HEPA-filtered control system or glove box is unavailable.
- Use a fume hood to expel any nanoparticles from tube furnaces or chemical reaction vessels. Do not exhaust aerosols containing engineered nanoparticles inside buildings.
- Do not eat, drink, chew gum, apply cosmetics, or handle contact lenses in the laboratory, except in clearly marked Clean Areas. Wash hands frequently and before eating.
- Follow standard sharps practices to prevent accidental injection or skin puncture.
- Clean up spills of nanoparticles promptly.
- Wet wipe work surfaces daily. Alternatively, use disposable bench paper.
- Wet wipe lab equipment and exhaust systems used with nanoparticles and HEPA-vacuum them prior to reuse, service, or disposal..
- Place sticky walk-off mat at the lab's exit to reduce the likelihood of spreading nanoparticles.

9.14 National Nanotechnology Initiative, United States (NNI)

The NNI is a U.S. Government research and development (R&D) initiative involving 20 departments and independent agencies working together toward the shared vision of "a future in which the ability to understand and control matter at the nanoscale leads to a revolution in technology and industry that benefits society." The NNI brings together the expertise needed to advance this broad and complex field, creating a framework for shared goals, priorities, and strategies that helps each participating Federal agency leverage the resources of all participating agencies (www.nano.gov, 2020). With the support of the NNI, nanotechnology R&D is taking place in academic, government, and industry laboratories across the United States.

Mihail C. Roco proposed the initiative in a 1999 presentation to the White House under the Clinton administration.The NNI was officially launched in 2000 and received funding for the first time in FY2021. The four primary goals of NNI include,

- Advance a world-class nanotechnology research and development program;
- Foster the transfer of new technologies into products for commercial and public benefit;
- Develop and sustain educational resources, a skilled workforce, and a dynamic infrastructure and toolset to advance nanotechnology; and
- Support responsible development of nanotechnology.

Coordination of the NNI: Learn about the structure, organization, and various departments involved in the Federal Government's nanotechnology initiative.

NNI Budget: The 2021 Federal Budget provides more than $1.7 billion for the National Nanotechnology Initiative (NNI), affirming the important role that nanotechnology continues to play in the Administration's innovation agenda.

Federal Legislation and Congressional Information: Learn about the annual authorization and appropriation process for the National Nanotechnology Initiative and nanotechnology-related Congressional hearings and events.

Nanotechnology Signature Initiatives: Nanotechnology Signature Initiatives (NSIs) spotlight areas of nanotechnology where significant advances in nanoscale science and technology can be made with the focus and cooperation of participating agencies. NSIs accelerate research, development, and application of nanotechnology in these critical areas (www.nano.gov, 2020).

As of December 2020, the current NSIs are,

- **NSI:** Water Sustainability through Nanotechnology – Nanoscale Solutions for a Global-Scale Challenge,
- **NSI:** Nanotechnology for Sensors and Sensors for Nanotechnology – Improving and Protecting Health, Safety, and the Environment,
- **NSI:** Sustainable Nanomanufacturing - Creating the Industries of the Future,
- **NSI:** Nanoelectronics for 2020 and Beyond.

NSIs are dynamic and are retired as they achieve their specified goals or develop an established community they no longer require the spotlight provided as a NSI. Retired NSIs are,

- **NSI:** Nanoelectronics for 2020 and Beyond,
- **NSI:** Nanotechnology for Solar Energy Collection and Conversion - Contributing to Energy Solutions for the Future,
- **NSI:** Nanotechnology Knowledge Infrastructure - Enabling National Leadership in Sustainable Design

Nanotechnology-Inspired Grand Challenges: A nanotechnology-inspired grand challenge (GC) is an ambitious goal that utilizes nanotechnology and nanoscience to solve national and global issues. The first and current GC was announced in October 2015 after receiving input and suggestions from the public. As of December 2020, the grand challenge is,

A Nanotechnology-Inspired Grand Challenge for Future Computing: Create a new type of computer that can proactively interpret and learn from data, solve unfamiliar problems using what it has learned, and operate with the energy efficiency of the human brain.

International Engagement: NNI agencies and representatives participate in many international activities, including bilateral and multilateral cooperative programs, tracking foreign progress in nanotechnology R&D, and the promotion of the trade and commercial interests of the United States. Cooperation and collaboration with other nations on nanoscale science and technology R&D can further the progress of the NNI while helping our international partners achieve their own goals.The development of a healthy global marketplace for nanotechnology products and ideas will require the establishment of consumer confidence; common approaches to nanotechnology environmental, health, and safety issues; efficient and effective regulatory schemes; and equitable trade practices for nanotechnology – not just in the United States, but worldwide.

The Global Issues Coordinator works with the Department of State and relevant NNI member agencies to coordinat and advance international programs and activities.

9.15 National Nanotechnology Coordination Office (NNCO)

The National Nanotechnology Coordination Office (NNCO) is the coordination office for a White House-led, $27 billion initiative that involves 20 Federal agencies working in the emerging field of nanotechnology (www.nano.gov, 2020). The NNCO acts as the primary point of contact for information on the NNI; provides technical and administrative support to the NSET Subcommittee, including the preparation of multiagency planning, budget, and assessment documents; develops, updates, and maintains the NNI website nano.gov; provides public outreach on behalf of the NNI; and promotes access to and early application of the technologies, innovations, and expertise derived from NNI activities.

The NNCO Director and NNCO Deputy Director are detailed from a Federal agency to the National Science and Technology Council (NSTC) and are appointed by the White House Co-Chair of the NSTC Committee on Technology (CT), in consultation with the Chair of the NSET Subcommittee. Both the NNCO Director and the NNCO Deputy Director report to the White House Co-Chair of the CT. The NNCO Director is charged in the 21st Century Nanotechnology Research and Development Act (Public Law 108-153) to arrange, on behalf of the NSET Subcommittee agencies, with the National Research Council of the National Academies to conduct a quadrennial review of the NNI and transmit the results of these evaluations to Congress.

Environmental, health, and safety implications of nanomaterials: As a goal of the NNI since inception, responsible development includes understanding potential environmental, health, and safety (EHS) implications of nanomaterials as well as the ethical, legal, and societal implications (ELSI) of nanotechnology (Anonymous, 2020). The National Nanotechnology Initiative (NNI) is committed to the responsible development of nanotechnology as one of its four main goals and as an important part of its environmental, health, and safety (EHS) research strategy. This includes sound, scientific assessment of nanotechnology's benefits and risks, and an understanding of the potential EHS impacts of nanotechnology.

Nanotechnology Environmental and Health Implications (NEHI) Working Group: The NNI addresses EHS issues through the Nanotechnology

Environmental and Health Implications (NEHI) Working Group of the Nanoscale Science, Engineering, and Technology (NSET) Subcommittee of the National Science and Technology Council. Member agencies of the NEHI Working Group have a wide range of Federal activities and responsibilities including research and regulatory missions. Representatives from the NEHI member agencies meet regularly to discuss agency research activities, to identify areas of possible collaboration and cooperation, and to efficiently leverage Federal resources.

Federal Investments in Nanotechnology-Related EHS Research: U.S. Government funding for research dedicated to nanotechnology-related EHS (nanoEHS) research is between \$58 and \$77 million per year. Cumulative nanoEHS investments since 2005 (including the 2020 request) have reached more than \$1.34 billion. These investments are coordinated through the NEHI Working Group with assistance from the National Nanotechnology Coordination Office.

Developing the NNI EHS Research Strategy and Monitoring Progress: The Federal Government's commitment to health and the environment has been part of a nanotechnology research framework since 2001. Federal agencies participating in the NNI regularly assess research progress and identify priority gaps and emerging trends through dialogue with stakeholders: academics, industry, public health advocates, and the general public. NNI agencies have incorporated these findings into its evolving EHS research strategy. In 2008, the NNI published its first formal EHS strategy. That 2008 document was updated to the current NNI EHS Research Strategy in 2011. This updated strategy included input from a wide range of stakeholder communities including academic researchers, consumer and environmental groups, manufacturers, and the general public through a series of workshops. The overall goal of the NNI EHS Research Strategy is to ensure the responsible development of nanotechnology by providing science-based guidance to the Federal agencies that produce the scientific information for risk management, regulatory decision making, product use, research planning, and public outreach. The NNI participating agencies continue to engage stakeholders through workshops and other means, to inform the ongoing implementation of the 2011 EHS Research Strategy.

The Science of EHS Research: To make federally funded nanoEHS research practical and timely, the framework underlying the 2011 nanoEHS research strategy integrates several important concepts, risk assessment, product life cycle stages, and exposure, into the basic and applied research to understand the potential EHS impacts of nanomaterials. Nanotechnology risk assessment is a process through which these advanced materials are evaluated to better understand their potential to harm humans and the environment, combined with the likelihood that there could be any exposure to the material, including the extent and duration of that exposure. Regardless of the hazard, in order for any potential effects to occur, the individual or the environment must be exposed to an agent of interest. Exposure to nanomaterials may occur unintentionally in the environment or through use of nanotechnology-enabled products. The NNI EHS Research Strategy outlines the research needed to produce the information to address these potential areas of exposure across the lifecycle and assists agencies in developing their research priorities.

International collaborations: The United States collaborates on nanoEHS research and related policy issues with other countries and international organizations. A notable example is collaboration with the European Union (EU) through the U.S.-EU Communities of Research (CoRs). The CORs serve as a platform for scientists to develop a shared repertoire of protocols and methods to overcome research gaps and barriers and to enhance their professional relationships. The CoR activities collectively complement, inform, and leverage U.S. nanoEHS efforts.

Safety issues-Indian scenario

India started a national nanotechnology mission more than a decade ago and the technology has found applications in the agriculture sector, the backbone of Indian economy (Mayank Aggarwal, 2019). The central government's department of biotechnology (DBT) has now come out with draft guidelines to regulate the use of nanotechnology in the agriculture sector. This would encourage commercialisation of the technology in products like nanofertilisers and nanopesticides, for example, while ensuring quality and safety. Nanotechnology has been supporting the Indian agricultural market to develop products and processes with higher efficiency and lower costs. With the increasing scope to commercialise this technology, the government has now proposed a set of guidelines to regulate and maintain quality and safety of the products and processes. These regulations will oversee the use and spread of hundreds of nano-agri input products (NAIP) and nano-agriproducts (NAP) which have been circulating in the Indian market for some years now, to prevent nanoparticle toxicity in humans and the environment. NAIPs and NAPs are products in a submicroscopic form used in agriculture and allied sectors for achieving varied goals like pest and disease prevention, nutrition, growth regulation, efficient packaging and storage among others.

The department of biotechnology, which comes under the Union Ministry of Science and Technology led by Harsh Vardhan, unveiled the draft 'guidelines for evaluation of nano-agri input and nano-agriproducts' on August 1, 2019 and has asked all stakeholders like different government agencies, institutions, industry, researchers and others to give their comments and suggestions by the end of August 2019. While the DBT's guidelines aim at increased quality and efficacy for commercialisation of nanotechnology-based innovations, they also focus on safety of the products. "The unique properties can also lead to nanoparticle-related toxicity in humans and the environment.

The guidelines for evaluation of nanoproducts in agriculture and food are more challenging than the existing procedures for assessment of fertilisers or safety evaluation of pesticides or toxicity evaluation in food," the guidelines said. At present, there are no unanimously acceptable international guidelines for nano-agriproducts. There exist a few provisions for nanomaterials, which are considered globally, from organisations like the World Health Organisation (WHO) with specific guidelines for quality and safety. But new innovations make it difficult to apply a universal set of evaluation parameters for different nanoproducts with different applications and thus many a time the case-by-case basis evaluation approach is advocated for NAPs, the proposed guidelines said.

General considerations of the guidelines

The European Union along with Switzerland is the only part of the world where particular provisions to deal with nanoproducts are available in the legislation. In some countries, in the absence of specific regulations for nanoproducts (Anonymous, 2019), the existing legislative and regulatory frameworks also deal with nanoproducts, many a time with necessary adaptations to account for the specific properties of nanomaterials. In India, there are different government bodies and provisions that regulate different agriproducts. Since different NAPs are considered in the guidelines, their evaluation should be conducted as per the NAP type. However, in case any specific study is not included in the suggested regulatory framework, the principles of ICH guidelines for agriproducts or OECD guidelines for chemicals may be followed. This document may also serve as useful guidelines for manufacturers, importers of NAPs and other stakeholders involved in research and development of NAPs.

The following nano-agri products are considered in these guidelines

i. **Nanofertilizers (with or without nanocarriers):** Safety, evaluation, functionality and other quality studies of nanofertilizers should be conducted under Fertilizer (Control) Order (FCO), 1985 with additional criteria for inclusion of nanofertilizers. FCO is administered by Department of Agriculture Cooperation, Government of India and issued under the Essential Commodities Act, 1955, which lays down registration requirement for fertilizers.

ii. **Nanopesticides (with or without nanocarriers):** Safety studies on chemistry, bio-efficacy and residues, toxicity, packaging and processing of molecules for registration for manufacture or import of nanopesticides should be conducted as per the regulatory aspect provisions under section 9(3) specified in the Insecticides Act, 1968 with additional criteria for inclusion of nanopesticides and plant growth regulators as per the guidelines of Central Insecticides Board & Registration Committee under the Ministry of Agriculture (Insecticide Act, 1968).

Nano-agriproducts including nanofood and nanofeed are considered in the guidelines:

Organizations for safety standard and for practice guidelines

Standards: One of the building blocks of a safe, integrated and responsible approach to the development of nanotechnologies is standardisation. Standardisation activities in the nanotechnology field are taking place at the international level and in many countries, involving a broad range of interests and organisations. At the forefront of these activities are the following five bodies.

- International Organisation for Standardization (ISO)
- European Committee for Standardization (CEN)
- British Standards Institution (BSI)
- ASTM International
- OECD Working Party on Manufactured Nanomaterials (WPMN).

A compilation of nanotechnology standards available from these organisations has been prepared by SAFENANO.

Best practice guidance : Numerous organisations have published best practice guidance for the safe handling and use of nanomaterials. Current guidance recognises the risks involved in the development, manufacture, use, clean-up and disposal of nanomaterials in order to develop and implement effective precautionary strategies and adequate control measures to minimise potential exposure. Key guidance documents include:

- Guidance on the Protection of the Health and Safety of Workers from the Potential Risks related to Nanomaterials at Work, European Commission

- Working Safely with Manufactured Nanomaterials, European Commission
- Guide to Safe Handling and Disposal of Manufactured Nanomaterials, BSI
- Health and safety practices in occupational settings relevant to nanotechnologies, ISO
- Building a Safety Program to Protect the Nanotechnology Workforce: A Guide for Small to Medium-Sized Enterprises, NIOSH[
- Using Nanomaterials at Work - Including Carbon Nanotubes (CNTs) and other High Aspect Ratio Nanomaterials (HARNs), UK Health and Safety Executive
- Working Safely with Nanomaterials in Research and Development, UK Nanosafety Group
- Safe Handling of Nanomaterials and Other Advanced Materials at Workplaces, BauA
- Best Practices Guidance for Nanomaterial Risk Management in the Workplace, IRSST

New considerations of application of nanotechnology in agriculture thus necessitates the initiation of scientific decision-making and informing platform for a cooperative national module that dedicatedly works on the policy aspects of these products before commercialisation. Thus, DBT has formulated draft guidelines for evaluation of nano-agri input and nano-agriproducts in India after a series of consultations with the stakeholders, emphasised the department of biotechnology in its draft guidelines. It explained that these guidelines are aimed to ensure the quality, safety and efficacy as well as encourage the commercialisation of nanotechnology-based innovations in agriculture and food/feed with high benefit and low-risk ratio when compared with their bulk counterparts.

Bibliorgraphy

Alberto Clemente , M. Pilar Lobera,, Francisco Balas and Jesús Santamaría, 2017. A Versatile Primary Nanoparticle Aerosol Generator for Nanosafety Studies: Instantaneous Clouds and Continuous Streams. P. 47. In : Nanosafety 2017. Organised by INM – Leibniz Institute for New Materials in association with Leibniz Research Alliance Nanosafety, Saarbrücken, Germany, 11–13 October 2017.

Aleksandra Zielińska, Beatriz Costa, Maria V. Ferreira, Diogo Miguéis, Jéssica M. S. Louros, Alessandra Durazzo, Massimo Lucarini, Piotr Eder, Marco V. Chaud, Margreet Morsink, Niels Willemen, Patrícia Severino, Antonello Santini and Eliana B. Souto,

2020. Nanotoxicology and Nanosafety: Safety-by-Design and Testing at a Glance. Int. J. Environ. Res. Public Health, 17: 4657

Amoabediny Ghassem, Naderi, A., Koohi Mohammad, Mortazavi, S., Naderi Malek and Rashedi Hamid, 2009. Guidelines for safe handling, use and disposal of nanoparticles. J. Physics: Conference Series. 170. 012037. 10.1088/1742-6596/170/1/012037.

Anonymous, 2007. Department of Energy Nanoscale Science Research Centers Approach to Nanomaterial ES&H, Revision 2, June.

Anonymous, 2008. Department of Energy Nanoscale Science Research Centers Approach to Nanomaterial ES&H, Revision 3a, May 12: 1-28.

Anonymous, 2009. Approaches to Safe Nanotechnology Managing the Health and Safety Concerns Associated with Engineered Nanomaterials. Department of Health and Human Services- Centers for Disease Control and Prevention National Institute for Occupational Safety and Health (NIOSH)

Anonymous, 2009. Approaches to Safe Nanotechnology. Department of Health and Human Services, Center for Disease Control, National Institute for Occupational Safety and Health. Anonymous, 2010. NanoSafety- Risk Governance of Manufactured Nanoparticles Interim Report - Phase II. European Technology Assessment Group ETAG, Karlsruhe and Vienna, October.

Anonymous, 2018. Guidelines and Best Practices for Safe Handling of Nanomaterials in Research Laboratories and Industries. Centre for Knowledge Management of Nanoscience & Technology. Compiled for Nano Mission, DST, Govt of India By (with inputs from Nanoregulatory Task Force), 24 pp.

Anonymous, 2018. Nanosafety Cluster-Cluster Projects-Horizon 2020 Projects. Available online: https://www. nanosafetycluster.eu/eu-nanosafety-cluster-projects/horizon-2020-projects (accessed on 28 February 2018)

Anonymous, 2018. Working Safely with Nanomaterials. OSHA (Occupational Safety and Health Administration) Fact Sheet , US Department of Labour.

Anonymous, 2019. Guidelines for Evaluation of Nanoagriinput Products and Nanoagriproducts in India. Department of Biotechnology Ministry of Science & Technology Government of India, 26 pp.

Anonymous, 2020. United States National Nanotechnology Initiative, nano.gov.

Buseck, P.R. and Adachi, K.,2007. Nanoparticles in the Atmosphere. Elements, 4: 389- 394.

Cordula Hirsch, MaΣhias Rösslein, Nils Bohmer and Peter Wick, 2017. Understanding Nanosafety – The Importance of Assay Performance in vitro. P 42. In : NANOSAFETY 2017. Organised by INM – Leibniz Institute for New Materials in association with Leibniz Research Alliance Nanosafety, Saarbrücken, Germany, 11-13october 2017

Dominick Fazarro, 2014. Nano safety: What workers need to know. CSTM University of Texas at Tayler Webinar

Eman M Osman, 2019 . Environmental and Health Safety Considerations of Nanotechnology: Nano Safety- Review Article. Biomed J. Sci.& Tech.Res., 19: 14501-14515

Erik Nilebäck and Julie Gold, 2018. MISTRA Environmental Nanosafety. Report Phase-I (2014-2019). Lunds University

Fadeel Bengt and Savolainen Kai, 2013. Broaden the discussion. Nature Nanotechnology, 8 : 71–71.

Gh. Amoabediny, Naderi , A., Malakootikhah, J., Koohi , M.K.,Mortazavi , S.A., Naderi, M.and Rashedi, H., 2019. Guidelines for Safe Handling, Use and Disposal of Nanoparticles. University of Amirkabir, Iran

Gilman Dr.La Jolla, 2020. Nanotechnology: Guidelines for Safe Research Practices. Regents of the University of California

Groso, A., Petri-Fink, A., Magrez, A., Riediker, M. and Meyer, T., 2010. Management of nanomaterials safety in research environment. Part Fibre Toxicol.,7:40.

Harald F. Krug , Nils Bohmer, Dana Kühnel, Clarissa Marquardt, Katja Nau and Christoph Steinbach, 2018. The DaNa2.0 Knowledge Base Nanomaterials-An Important Measure Accompanying Nanomaterials Development. Nanomaterials, 8: 204-218

Harford, A.J., Edwards, J.W., Priestly, B.G. and Wright, P.F.A., 2007. Nanosafe Australia. Current OHS Best Practices for the Australian Nanotechnology Industry- A Position Paper by the NanoSafe Australia Network

Jana Fleddermann, Julia Susewind , Henrike Peuschel, Julia Przibilla, Silke Kiefer, Anika Schwarz and AnneΣe Kraegeloh, 2017. P.49. In : NANOSAFETY 2017. Organised by INM – Leibniz Institute for New Materials in association with Leibniz Research Alliance Nanosafety, Saarbrücken, Germany, 11–13 October 2017

Kahru, A. and Dubourguier, H.C., 2010. From ecotoxicology to nanoecotoxicology. Toxicology, 269:105-19

Katherine Moore, Ivanna Stuart, Finn O'Brien and Andrew Lewis, 2016. Innovating Nanoparticle Safety: Storage, Handling, and Disposal Processes -An Interactive Qualifying Project Final Report. Worcester Polytechnic Institute

Kraegeloh, A., Suarez-Merino, B., Sluijters, T. and Micheletti, C, 2018. Implementation of safe-by-design for nanomaterial development and safe innovation: Why we need a comprehensive approach. Nanomaterials, 8:239

Lombi, E., Donner, E., Dusinska, M. and Wickson, F., 2019. A One Health approach to managing the applications and implications of nanotechnologies in agriculture. *Nature Nanotechnol.,* 14: 523-531

Mayank Aggarwal, 2019. New guidelines proposed to ensure safe use of nanotechnology in agriculture. MONGABAY-News & inspiration from nature's frontline in India., August 23, 2019.

Muthuraman Yuvraj, Venkatesan Yuvraj, Venugopla Arunkumar, Muthaiyan Pandiyan and Kizhaeral Sevathapandian Subramanian, 2000. Nanosafety, Intech Open, DOI: 10. 5772/ intechopen.92290

NIOSH (National Institute for Occupational Safety and Health), 2009. Patriots Plaza 1395 E Street, SW, Suite 9200, Washington, DC 20201

Saji George, Shirley S. Ho, Esther S. P. Wong, Timothy Thatt Yang Tan, Navin Kumar Verma, Robert J. Aitken, Michael Riediker, Christopher Cummings, Liya Yu, Zheng Ming Wang, Daniele Zink, Zhihan Ng, Say Chye Joachim Loo and Kee Woei Ng, 2015. The multi-facets of sustainable nanotechnology – Lessons from a nanosafety symposium, Nanotoxicology, 9: 404-406,

Tanasugarn L. and Tanpipat, N., 2020. NanoSafety in Thailand.. Intergovernmental Forum on Chemical Safety Global Partnerships for Chemical Safety Contributing to the 2020 Goal. 19 pp.

Zimmerman, J.B., Anastas, P.T. and Miller, G.W., 2014. Green chemistry as a leadership opportunity for toxicology: We must take the wheel. Toxicol. Sci., 141: 4–5

Zolnik, B.S., González Fernández, A., Sadrieh, N. and Dobrovolskaia, M.A., 2010. Nanoparticles and The Immune System. Endocrinology, 151: 458-465.

Glossary of Nanotechnological Terms

ACE Paste: Atomspheric Carbon Extractor. Harvests the greenhouse gases for Carbon, to be used for diamondoid fabrication.

Adensoine Triphosphate (ATP): A chemical compound that functions as fuel for biomolecular nanotechnology having the formula, $C_{10}H_{16}N_5O_{13}P_3$.

Acrosol: A cloud of solid or liquid particles in a gas.

Animat: This term was developed by Alan H. Goldstein. In his article I, Nanobot, he suggested that a new state of life be named after the contraction of the term "anima-materials" Ñ "animats".

Assembler: A nano-robotic device controlled by an onboard computer that can use available chemicals to manufacture nanoscale products. It has been proposed that advanced designs could communicate, cooperate, and maneuver to build macroscale products. Assemblers are much more complex, and probably less efficient, than fabricators.

Atom: The smallest unit of a chemical element, about a third of a nanometer in diameter atomic force microscope (AFM)

Atomic Force Microscope (AFM): An instrument able to image surfaces to molecular accuracy by mechanically probing their surface contours. Akind of proximal probe. Also termed a scanning force microscope.

Atomic Manipulation: Manipulating atoms, typically with the tip of an STM.

Atomic Layer Deposition(ALD) (AFM): A self-limiting, sequential surface chemistry that deposits conformal thin-films of materials onto substrates of varying compositions.

Atomic Manipulation: Manipulating atoms, typically with the tip of an STM.

Atomistic Simulations:. This new capability will allow computer simulation of mechanical devices or molecular machines using nanometer size components.

Automated Engineering: Engineering done by computer systems withoutthe input of humans.

Autoproductivity: The ability of a system, under external control, to automatically produce an identical copy of itself.

Ballistic Magnetoresistance: BMR is yet another way in which spin orientation, encoding information on a storage medium such as a hard drive, can modify electrical resistance in a nearby circuit, thereby accomplishing the sensing of that orientation.

Bio-assemblies: Containing several protein units, DNA loops, lipids, various ligands, etc.

Bio-assemblies or Biomolecular Assemblies: containing several protein units, DNA loops, lipids, various ligands, etc.

Biochauvinism: The prejudice that biolo gical systems have an intrinsic superiority that will always give them a monopoly on self-reproduction and intelligence.

BioMEMS: MEMS used in medicine, that use microchips.

Biomimetic Chemistry: Knowledge of biochemistry, analytical chemistry, polymer science, and biomimetic chemistry is linked and applied to research in designing new molecules, molecular assemblies, and macromolecules having biomimetic functions.

Biomimetic Materials: Materials that imitate, copy, or learn from nature.

Biomimetic: Imitating, copying, or learning from nature. Nanotechnology already exists in nature; thus, nanoscientists have a wide variety of components and tricks already available.

BioNEMS: Biofunctionalized nanoelectromechanical systems.

Biopolymeroptoelectromechanical Systems (BioPOEMS): combining optics and microelectromechanical systems, and used in biological applications.

Biostasis: A condition in which an organism's cell and tissue structure are preserved, allowing later restoration by cell repair machines. Applicable to cryonics.

Biovorous: From "biovore;" an organism capable of converting biological material into energy for sustenance.

Biochauvinism: The prejudice that biological systems have an intrinsic superiority that will always give them a monopoly on self-reproduction and intelligence.

Blue goo: Slang term for nanobots designed to protect against harmful nanotechnology; "blue" being a metonym for police as in the color of many police uniforms.

Bogosity Filter: A mechanism for detecting bogus ideas and propositions.

Born-Oppenheimer Approximation: Permits the use of classical mechanics in modeling and thinking about molecular and atomic motions. Needless to say, this greatly simplifies the conceptual framework required for thinking about molecular machines.

Bose-Einstein Condensates (BEC's): "...aren't like the solids, liquids and gases that we learned about in school. They are not vaporous, not hard, not fluid. Indeed, there are no ordinary words to describe them because they come from another world -- the world of quantum mechanics."

Bottom Up: Building larger objects from smaller building blocks. Nanotechnology seeks to use atoms and molecules as those building blocks. The advantage of bottom-up design is that the covalent bonds holding together a single molecule are far stronger than the weak. [NTN] Mostly done by chemists, attempting to create structure by connecting molecules.

Brownian Assembly: Brownian motion in a fluid brings molecules together in various position and orientations. If molecules have suitable complementary surfaces, they can bind, assembling to form a specific structure.

Brownian Motion: Motion of a particle in a fluid owing to thermal agitation, observed in 1827 by Robert Brown.

Buckminster fullerene: A broad term covering the variety of buckyballs and carbon nanotubes that exist. Named after the architect Buckminster Fuller, who is famous for the geodesic dome, which buckyballs resemble.

Buckyball/C60: The term "buckyball" refers only to the spherical fullerenes and is derived from the word "Buckminsterfullerene," which is the geodesic dome / soccer ball-shaped C60 molecule. C60 was the first buckyball to be discovered and remains the most common and easy to produce.

Bulk technology: Technology in which atoms and molecular are manipulated in bulk, rather than individually.

Bush robot: A concept for robots of ultimate dexterity, they utilize fractal branching to create ever-shrinking "branches," eventually ending in nanoscale "fingers."

C = Devices capable of copying themselves by molecular self-assembly. Note that any information necessary for the animat's operations cannot be stored in

DNA or RNA or any other methods that are discovered to be used naturally by life on Earth.

Carbon Nanotubes: Tubes made of carbon with nanometer dimensions

Catalyst: A substance, usually used in small amounts relative to the reactants, that modifies and increases the rate of a reaction without being consumed or changed in the process..

Cell pharmacology: Delivery of drugs by medical nanomachines to exact locations in the body.

Cell Repair Machine: Molecular and nanoscale machines with sensors, nanocomputers and tools, programmed to detect and repair damage to cells and tissues, which could even report back to and receive instructions from a human doctor if needed.

Cellular Automata: an array of identically programmed automata, or "cells," which interact with one another.

Chemical Vapour Deposition (CVD): A technique used to deposit coatings, where chemicals are first vaporized, and then applied using an inert carrier gas such as nitrogen.

Cobots: Collaborative robots designed to work alongside human operators. Prototype cobots are being used on automobile assembly lines to help guide heavy components like seats and dashboards into cars so they don't damage auto body parts as workers install them.

Cognotechnology: Convergence of nanotech, biotech and IT, for remote brain sensing and mind control.

Computational Nanotechnology: Permits the modeling and simulation of complex nanometer-scale structures. The predictive and analytical power of computation is critical to success in nanotechnology: nature required several hundred million years to evolve a functional "wet" nanotechnology; the insight provided by computation should allow us to reduce the development time of a working "dry" nanotechnology to a few decades, and it will have a major impact on the "wet" side as well.

Colloid: Nanoscale or microscale particles suspended in another medium; colloids include gels, aerosols, and emulsions.

Computronium: A highly (or optimally) efficient matrix for computation, such as dense lattices of nanocomputers or quantum dot cellular automata.

Contelligence: (Consciousness + intelligence) The combination of awareness and computational power required in an Artificially Intelligent network before we could, without loss of anything essential, upload ourselves into them.

Convergent Assembly: A process of fastening small parts to obtain larger parts, then fastening those to make still larger parts, and so on; convergent assembly can be used to build a product from many, much smaller, components.

Converging technologies : Science at the intersection of nanotechnology, biotechnology, information technology, and cognitive technologies. Cool light emitted by sources as a result of the movement of electrons from more energetic states to less energetic states.

Copolymerization: Crystalline substances that have a permanent spontaneous electric polarization (electric dipole moment per cubic centimeter) that can be reversed by an electric field.

DBERGOO: A related term to grey goo, used (jokingly) to refer to the mistaken idea that during the singularity powerful technologies would decimate non-transhumanists, and that some transhumanists would see this as desirable (which is clearly against theTranshuman Principles

Dendrimers: From the Greek word dendra - tree, a dendrimer is polymer that branches a tiny molecular structure that interacts with cells, enabling scientists to probe, diagnose, cure or manipulate them on a nanoscale." Invented by Professor Donald Tomalia from Central Michigan Design Ahead: The use of known principles of science and engineering to design systems that can only be built with tools not yet available; this permits faster exploitation of the abilities of new tools.

Design Diversity: A form of redundancy in which components of different design serve the same purpose; this can enable systems to function properly despite design flaws.

Diamondoid: Stuctures that resemble diamond in a broad sense, strong stiff structures containing dense, three dimensional networks of covalent bonds, formed chiefly from first and second row atoms with a valence of three or more. Many of the most useful diamondoid structures will in fact be rich in tetrahedrally coordinated carbon.

Dip Pen Nanolithography: An AFM-based soft-lithography technique.

Directed-Assembler: A specific type of assembler that makes use of directed-assembly, such that the assembly process requires external energy or information input.

Disassembler: An instrument able to take apart structures a few atoms at a time, recording structural information at each step. This could be used for uploading, copying objects (with an assembler), a dissolving agent or a weapon.

Disassemble: An instrument able to take apart structures a few atoms at a time, recording structural information at each step.

Disasterbation: Idly fantasizing about possible catastrophes (ecological collapse, full-blown totalitarianism) without considering their likelihood or considering their possible solutions and preventions

Disruptive Technology: Technology that is significantly cheaper than current, is much higher performing, has greater functionality, and is frequently more convenient to use. Will revolutionize markets by superseding existing technology.

Distributed Intelligence: An intelligent entity which is distributed over a large volume (or inside another system, like a computer network) with no distinct center. This is the opposite to the strategy of Concentrated intelligences.

DNA Chip : A purpose built microchip used to identify mutations or alterations in a gene's DNA.

DNA: Deoxyribonucleic acid – the double-helix molecule that provides the basis of genetic heredity, about 2 nanometers in diameter but often several millimeters in length.

Dopeyballs: Superconducting Buckyballs (they) have the highest critical temperature of any known organic compound.

Dry Nanotechnology: Derives from surface science and physical chemistry, focuses on fabrication of structures in carbon (e.g. fullerenes and nanotubes), silicon, and other inorganic materials. Unlike the "wet" technology, "dry" techniques admit use of metals and semiconductors.

Dual use: Technologies that may be used for both military and civilian purposes.

Dumb Sizing: Apealing to the least common denominator by explaining difficult concepts in such a manner so they lose meaning. Also, talking down to someone less informed or learned.

Dyson Scenario: Dyson has shown that a finite amount of energy is enough to guarantee infinite survival if it is spent sufficiently slowly.

Dyson Sphere: A shell built around a star to collect as much energy as possible, originally proposed by Freeman Dyson. In the original proposal the shell consists of many independent solar collectors and habitats in separate orbits (also known as a Type I Dyson Sphere), but later people have discussed rigid shells consisting of only one piece (called a Type II Dyson Sphere).

Ecosystem protector: A nanomachine for mechanically removing selected imported species from an ecosystem to protect native species.

Electrical Bistability: a phenomenon in which an object exhibits two states of different conductivity at the same applied voltage.

Electron beam lithography: Lithographic patterning using an electron beam, usually to induce a change in solubility in polymer films. The resulting patterns can be subsequently transferred to other metallic, semiconductor, or insulating films.

Emergence: A complex whole created by simple parts, as in the brain where billions of neurons work individually, but collectively make up our consciousness and give us the ability to think, rationalize, and create. EI - Emergent Intelligence: An intelligent system that gradually emerges from simpler systems, instead of being designed top down.

Emulation: An absolutely precise simulation of something, so exact that it is equivalent to the original (for example, many computers emulate obsolete computers to run their programs). [AS] The Star Trek replicator is an example.

Enabling science and technologies: Areas of research relevant to a particular goal, such as nanotechnology.

Engineered/manufactured nanomaterials: Nanosized materials are purposefully made. These are in contrast to incidental and naturally occurring nanosized materials. Engineering/manufacturing may be done through certain chemical and / or physical processes to create materials with specific properties.

Entanglement: From quantum mechanics, entanglement is a relationship between two objects in which they both exhibit superposition but once the state of one object is measured, the state of the other is also known.

Entropy: A measure of the disorder of a closed system. The second law of thermodynamics states that the entropy (and disorder) increases as time moves forward.

Epitaxial film : A thin crystal layer, perhaps of nanoscale thickness, deposited on the surface of another substance by processes such as vapor deposition.

Evolution: A process in which a population of self-replicating entities undergoes variation, with successful variants spreading and becoming the basis for further variation.

Exploratory engineering: Design and analysis of systems that are theoretically possible but cannot be built yet, owing to limitations in available tools.

Exponential assembly: A manufacturing architecture starting with a single tiny robotic arm on a surface. This first robotic arm makes a second robotic arm on a facing surface by picking up miniature parts ó carefully laid out in advance in exactly the right locations so the tiny robotic arm can find them ó and assembling them.

Exponential general-purpose molecular manufacturing: Exponential general-purpose molecular manufacturing means a manufacturing system capable of making a wide range of technologically advanced products, far superior to what we have today, much cheaper, much faster, and able to multiply its own source of production exponentially.

Exponential Growth: Exponential growth refers to the process of growth or replication involving doubling within a given period.

Exposure Assessment: The determination or estimation (qualitative or quantitative) of the magnitude, frequency, duration, route, and extent(number of people) of exposure to a chemical, material, or microorganism.

Fabricator: A small nano-robotic device that can use supplied chemicals to manufacture nanoscale products under external control. Fabricators could work together to build macroscale products by convergent assembly. Similar to assemblers, but less complex, easier to build, and probably more efficient.

Femtometer: [abbr: fm] A unit suitable to express the size of atomic nuclei. One quadrillionth (10 to minus 15) of a meter.

Femtosecond: It is one quadrillionth of a second, and is to a second what a second is to 32,700,000 years. At 186,000 miles per second, in one femtosecond light travels only far enough to traverse about 1,000 silicon atoms.

Femtotechnology: The art of manipulating materials on the scale of elementary particles (leptons, hadrons, and quarks).

Fluidic Self Assembly: A novel technique for accurately assembling large numbers of very small devices. The small size, planarity, and accuracy of the assembly also result in very low parasitic interconnects, comparable to on die traces. This massively parallel assembly process combines the capability and flexibility of assembly with the cost effectiveness of integration.

Foglet: A mesoscale machine. A discreet component of utility fog.

Fractal: A mathematical construct that has a fractional dimension.

Fractal Mechatronic Universal Assembler: (or Fractal Assembler) is a machine that is capable of assembling any chemical from a generic descriptions of the properties required of the chemical.The machine comprises of test tube arrays and software linked to robotic cubes and sensor arrays to implement automated mixing and testing to conduct materials research activity.

Fractal Robots: AKA: Fractal Shape Shifting Robots and Programmable "Digital Matter", are programmable machines that can do unlimited tasks in the physical world, the world of matter.

Fullerenes: Fullerenes are a molecular form of pure carbon discovered in 1985. They are cage-like structures of carbon atoms, the most abundant form produced is buckminsterfullerene (C60), with 60 carbon atoms arranged in a spherical structure.

General-Purpose molecular manufacturing: A manufacturing technology that will find many applications across many segments of society. Its extreme flexibility, precision, high capacity, and low cost will cause rapid adoption almost everywhere, and therefore will have disruptive effects in many industries

Genetic Algorithm: Any algorithm which seeks to solve a problem by considering numerous possibilities at once, ranking them according to some standard of fitness, and then combining ("breeding") the fittest in some way. In other words, any algorithm which imitates natural selection.

Genetic sequence: The ordered set of nucleotides in a particular sample of DNA or RNA.

GENIE: An AI combined with an assembler or other universal constructor, programmed to build anything the owner wishes. Sometimes called a Santa Machine. This assumes a very high level of AI and nanotechnology.

Genome : The complete set of genetic material contained in an organism, or a separately inherited portion of an organism. For instance, the mitochondrial genome is inherited maternally, whereas the nuclear genome is inherited from both parents.

Giant Magnetoresistance: (GMR). It results from subtle electron- spin effects in ultra-thin 'multilayers' of magnetic materials, which cause huge changes in their electrical resistance when a magnetic field is applied. GMR is 200 times stronger than ordinary magnetoresistance. **Golden Goo:** Another member of the grey goo family of nanotechnology disaster scenarios. The idea is to use nanomachines to filter gold from seawater.

GNR technologies: Genetic Engineering, Nanotechnology, and Robotics.

Gray Goo or Grey Goo - destructive nanobots [AKA: "gray dust"]. opposite of Blue Goo. See Star Trek scenario. Vast legions of destructive nanites. Typically, created by accident. Left unchecked, they will basically convert everthing they contact into more of themselves, or consume and digest it for energy. Either way, its pretty much bad news.

Green goo: Hypothetical use of nanomachines or bio-engineered organisms used for population control of humans, either by governments or eco-terrorist groups.

Guy Fawkes Scenario: If nanotechnology becomes widely available, it might become trivial for anyone to committ acts of terrorism (such as making nanomachines build a large amount of explosives under government buildings a la Guy Fawkes). This would either force strict control over nanotechnology (hard) or a decentralized mode of organization.

Heisenberg Uncertainty Principle: A quantum-mechanical principle with the consequence that the position and momentum of an object cannot be precisely determined. The Heisenberg principle helps determine the size of electron clouds, and hence the size of atoms. [NTN] "The more precisely the POSITION is determined, the less precisely the MOMENTUM is known" [Werner Heisenberg] Heteronuclear: Consisting of different elements.

IA: Intelligence Amplification: Technologies seeking to increase the cognitive abilities of people. Immune Machines: Medical nanomachines designed for internal use, especially in the bloodstream and digestive tract, able to identify and disable intruders such as bacteria and viruses. IMP: Electronic implant, especially in the brain.

Immune Machines: Medical nanomachines designed for internal use, especially in the bloodstream and digestive tract, able to identify and disable intruders such as bacteria and viruses.

IMP : Electronic implant, especially in the brain.

Incidental nanosized materials: Nanomaterials that are the byproducts of human activity, such as combustion, welding, or grinding.

Inline Universities: (as opposed to online universities), nanocomputer implants serving to increase intelligence and education of their owners, essentially turning them into walking universities

Intellectual property: Creations of the mind, such as inventions; literary and artistic works; designs; and symbols, names and images used in commerce. IP is protected in law by, for example, patents, copyright, trademarks and trade secrets, which enable people to earn recognition or financial benefit from what they invent or create.

Intelligent Agent: aka "software agent". Software that can do things without supervision, because it knows your patterns, history, preferences, likes, dislikes, and so forth.

Intentionally produced nanomaterials: See Engineered/manufactured nanoscale materials.

Jupiter-Brain: A posthuman being of extremely high computational power and size. This is the archetypal concentrated intelligence. The term originated due to an idea by Keith Henson that nanomachines could be used to turn the mass of Jupiter into computers running an upgraded version of himself.

Khaki goo : Slang term for hypothetical military use of nanotechnology, especially as of weaponized grey goo.

Knowbots: Knowledge robots, first developed Vinton G. Cref and Robert E. Kahn for National Research Initiatives. Knowbots are programmed by users to scan networks for various kinds of related information, regardless of the language or form in which it expressed. "Knowbots support parallel computations at different sites.

LCD (Liquid Crystal Display): It is the predominant technology used in flat panel displays. The principle that makes the display work is this: A crystalís alignment can be altered with an electric current. If the crystal is lined up one way ñ it will allow the light waves to pass through a polarized filter, but ifthe electric current alters the crystalís alignment, it will guide light so that the polarized filter blocks the light.

LED Light emitting diode: LEDs (Light Emitting Diodes) work on a completely different concept. Traditionally LEDs are created from two semiconductors.

Limited Assembler: Assembler capable of making only certain products; faster, more efficient, and less liable to abuse than a general-purpose assembler.

Linde Scenario: A scenario for indefinite survival of intelligent life. It assumes it is possible to either create basement universes connected to the original universe with a wormhole or the existence of other cosmological domains..

LMNT: An abbreviation for limited molecular nanotechnology; a narrowly specified type of MNT, using only diamondoid reactions; much easier to achieve than general MNT, but with nearly equivalent appeal and impact.

Lofstrom Loop: An beanstalk-like megaconstruction based on a stream of magnetically accelerated bars linked together. The stream is sent into space, where a station rides it using magnetic hooks, redirects it horizontally to another station, which sends it downwards to a receiving station on the ground.

Low-dimension Structures: Quantum wells, quantum wire and quantum dots. Matter as Software: "Autonomous, motile microdevices clearly are on the horizon. They may be regarded as the first step in the evolution of a technology for "programming" the structure and properties of material objects at the microscopic and the submicroscopic levels.

Macroscale: Larger than nanoscale; often implies a design that humans can directly interact with; too large to be built by a single assembler (one cubic micron of diamond contains 176 billion atoms).

Manufacturing processes: General term used to identify the variety of processes used in the production of the part. Processes may include plastic injection molding, vacuum forming, milling, stamping, casting, extruding, die-cutting, sewing, printing, packaging, polishing, grinding, metal spinning, welding, and so forth.

Manufacturing: The ability to make products, in this case ranging from clothing, to electronics, to medical devices, to books, to building materials, and much more

Mechanochemistry: Chemistry accomplished by mechanical systems directly controlling the reactant molecules; the formation or breaking of chemical bonds under direct mechanical control.

Mechanosynthesis: (where) Molecular tools with chemically specific tip structures can be used, sequentially, to modify a work piece and build a wide range of molecular structures.

Mechatronics: The study of the melding of AI and electromechanical machines to make machines that are greater than the sum of their parts.

Meme: An idea that replicates through a society as it is propagated through person-to-person interaction, both direct and indirect. Memetics is a field of study that focuses on memes' role in the evolution of a culture.

Mesoscale: A device or structure larger than the nanoscale (10^-9 m) and smaller than the megascale.

Microelectromechanical systems (MEMS): Generic term to describe micron scale electrical/mechanical devices.

Microencapsulation: Individually encapsulated small particles.

Micron: One millionth of a meter, or about 1/25,000 of an inch.

Millimeter: One thousandth of a meter, or about 1/26 of an inch.

MIMIC: [micromoulding in capillaries] one-step rapid prototyping technique.

Molecular Assembler: Also known as an assembler, a molecular assembler is a molecular machine that can build a molecular structure from its component building blocks.

MNT: An abbreviation for molecular nanotechnology; refers to the concept of building complicated machines out of precisely designed molecules. To avoid confusion between this and today's nanoscale technologies, CRN generally favors the term 'molecular manufacturing'.

MOCVD: Metal-Organic Chemical Vapor Deposition

Molecular assembler: Also known as an assembler, a molecular assembleris a molecular machine that can build a molecular structure from its component building blocks.

Molecular Beam Epitaxy (MBE): Process used to make compound (multi-layer) semiconductors. Consists of depositing alternating layers of materials, layer by layer, one type after another (such as the semiconductors gallium arsenide and aluminum gallium arsenide).

Molecular Electronics (ME) (moletronics) Any system with atomically precise electronic devices of nanometer dimensions, especially if made of discrete molecular parts rather than the continuous materials found in today's semiconductor devices.

Molecular Integrated Microsystems (MIMS): Microsystems in which functions found in biological and nanoscale systems are combined with manufacturable materials.

Molecular Manipulator: A device combining a proximal probe mechanism for atomically precise positioning with a molecule binding site on the tip; can serve as the basis for building complex structures by positional synthesis.

Molecular manufacturing: The automated building of products from the bottom up, molecule by molecule, with atomic precision. This will make products that are extremely lightweight, flexible, durable, and potentially very 'smart'

Molecular Medicine: Studying molecules as they relate to health and disease, and manipulating those molecules to improve the diagnosis, prevention, and treatment of disease

Molecular nanotechnology (MNT): The ability to construct shapes, devices, and machines with atomic precision, and to combine them into a wide range of products inexpensively.

Molecular Recognition: A chemical term referring to processes in which molecules adhere in a highly specific way, forming a larger structure; an enabling technology for nanotechnology.

Molecular Systems Engineering: Design, analysis, and construction of systems of molecular parts working together to carry out a useful purpose.

Molecular Wire: A molecular wire - the simplest electronic component - is a quasi-one-dimensional molecule that can transport charge carriers (electrons or holes) between its ends.

Monolayer: A layer, film, or coating that is only one atom or molecule thick.

Monomer: The units from which a polymer is constructed.

Monomolecular Computing: The implantation inside a single molecule of all the functional groups or circuits to realize a calculation, without any help from external artifices such as re-configuration, calculation sharing between the user and the machine, or selection of the operational devices. Monomolecular films that form or self-assemble after immersing a substrate into a solution of an active surfactant.

Moore's Law: Proposed by Intel founder Gordon Moore in the 1960s, this is a variously stated observation that the density of transistors on an integrated circuit chip has been doubling every 18 to 24 months, or the cost of a transistor has been dropping by half, or that the general capabilities of microelectronics have been improving at an exponential rate.

Nanarchist: Someone who circumvents government control to use nanotechnology, or someone who advocates this.

Nanarchy: The use of automatic law-enforcement by nanomachines or robots, without any human control.

Nano: A prefix meaning one billionth.

Nanoarray: An ultra-sensitve, ultra-miniaturized array for biomolecular analysis.

Nanobalance: A nanoscale balance for determining mass, small enough to weigh viruses and other sub-micron scale particles.

Nanobeads: Polymer beads with diameters of between 0.1 to 10 micrometers.

Nanobialys: Miniature bialy-shaped particles developed by Washington University as delivery agents for drugs and imaging agents directly to the sites of tumors and plaques.

Nanobiology: A field of study combining biology and physics which looks at how nature works on the nanometer scale, particularly how transport takes place in biological systems.

Nanobiotechnology: Applying the tools and processes of MNT to build devices for studying biosystems, in order to learn from biology how to create better nanoscale devices.

Nanobubbles: Tiny air bubbles on colloid surfaces.

Nanochemistry: A discipline focusing on the unique properties associated with the assembly of atoms or molecules on a nanometer scale. At this scale, new methods of carrying out chemical reactions are possible. Alternatively, it is the development of new tools, technologies and methodologies for doing chemistry in the nanolitre to femtolitre domains.

Nanochondria: Nanomachines existing inside living cells, participating in their biochemistry (like mitochondria) and/or assembling various structures.

Nanocombinatorics: The new analytical method utilizes a technique invented at Northwestern called polymer pen lithography, where basically a rubber stamp having as many as 11 million sharp pyramids is mounted on a transparent glass backing and precisely controlled by an atomic force microscope to generate desired patterns on a surface.

Nanocomposite: A material composed of two or more substances, of which at least one has a nanoscale dimension, such as nanoparticles dispersed throughout another solid material.

Nanocomputer: A computer made from components (mechanical, electronic, or otherwise) built at the nanometer scale.

Nanocontainers: Nanoscale polymeric containers that could be used to selectively deliver hydrophobic drugs to specific sites within individual cells.

Nanocrystals: Aggregates of anywhere from a few hundred to tens of thousands of atoms that combine into a crystalline form of matter known as a "cluster."

Nanocrystals: Also known as nanoscale semiconductor crystals. «Nanocrystals are aggregates of anywhere from a few hundred to tens of thousands of atoms that combine into a crystalline form of matter known as a «cluster.» Typically around ten nanometers in diameter, nanocrystals are larger than molecules but smaller than bulk solids and therefore frequently exhibit physical and chemical properties somewhere in between.

Nanodefenses: Any of the "good" goos, such a blue goo; protectors against grey goo, destructive nanoswarms, and the like.

Nanodisaster: Various so-called "goo" scenarios that have potentially negative outcomes.

Nanoelectronics: Electronics on a nanometer scale, whether by current techniques or nanotechnology; includes both molecular electronics and nanoscale devices resembling today's semiconductor devices.

Nanofabrication: Construction of items using assemblers and stock molecules. General terms for methods to create, assemble, or otherwise form nanoscale structures.

Nanofactory: A self-contained macroscale manufacturing system, consisting of many molecular manufacturing systems feeding a convergent assembly system.

Nanofacture: The fabrication of goods using nanotechnology.

Nanofilters: One opportunity for nanoscale filters is for the separation of molecules, such as proteins or DNA, for research in genomics.

Nanofluidics : Controlling nano-scale amounts of fluids. Science or engineering involving the flow of liquid or gas through nanoscale spaces.

Nanogate: A device that precisely meters the flow of tiny amounts of fluid.

Nanohacking: "Hacking" at the molecular level.

Nanohorns: One of the SWNT (single walled carbon nanotube) types, with an irregular horn-like shape, which may be a critical component of a new generation of fuel cells.

Nanoimprint Machine: A form of soft lithography.

Nanoimprinting: A technique that is very simple in concept, and totally analogous to traditional mould- or form-based printing technology, but that uses moulds (masters) with nanoscale features.

Nanoindentation: Similar to conventional hardness testing performed on a much smaller scale.

Nanolithography: Writing on the nanoscale.

Nanomachine: An artificial molecular machine of the sort made by molecular manufacturing.

Nanomachining: Like traditional machining, where portions of the structure are removed or modified, nanomachining involves changing the structure of nano-scale materials or molecules.

Nanomanipulation: The process of manipulating items at an atomic or molecular scale in order to produce precise structures.

Nanomanipulator: A tool for moving individual molecules or nanoscale objects, such as an atomic force microscope.

Nanomanufacturing Nanomaterials: Can be subdivided into nanoparticles, nanofilms and nanocomposites.

Nanomaterial: See Engineered/manufactured nanoscale materials

Nanomechanical: Being mechanical and very small; for example, a robot that can manipulate single molecules.

Nanomedicine: An artificial molecular machine of the sort made by molecular manufacturing.

Nanometer: One billionth of a meter; approximately the length of three to six atoms placed side-by-side, or the width of a single strand of DNA; the thickness of a human hair is between 50,000 and 100,000 nanometers.

Nano-Optics: Interaction of light and matter on the nanoscale.

Nanoparticle: Free standing nanosized material, consisting of between tens to thousands of atoms.

Nanopens & Nanopencils: Allows for drawing electronic circuits a thousand times smaller than current ones.

Nanopharmaceuticals: Nanoscale particles used to modulate drug transport for drug uptake and delivery applications.

Nanophase Carbon Materials: A form of matter in which small clusters of atoms form the building blocks of a larger structure.

Nanoplotter: A multi-tip nanopen.

Nanopores: Involves squeezing a DNA sequence between two oppositely charged fluid reservoirs, separated by an extremely small channel.

Nanoporous: Substances that have holes or pores on the nanoscale, used, for example, to separate particles or molecules by size.

Nanoprobe: Nanoscale machines used to diagnose, image, report on, and treat disease within the body.

Nanoscale: The size range roughly 1 to 100 nanometers, where many of the fundamental structures of biology are formed, composite materials may take on their distinctive characteristics, and many important physical phenomena are found.

Nanoscience: The study of unique properties of matter at the nanoscale; an interdisciplinary field of science combining physics, materials science, the chemistry of complex molecules, and related disciplines.

Nanosensor: A device for sensing radiation, forces, chemicals, or biological agents, in which some portion of a device operates at the nanoscale, for example, by having receptors into which the particular molecules to be sensed fit.

Nanostructures: Structures at the nanoscale; that is, structures of an intermediate size between molecular and microscopic (micrometer-sized) structures.

Nanotechnology: Research and technology development at the atomic, molecular or macromolecular levels, in the length scale of approximately 1 - 100 nanometer range; creating and using structures, devices and systems that have novel properties and functions because of their small and/or intermediate size; and the ability to control or manipulate on the atomic scale

Nanotube: Hollow, cylindrical structures, with a diameter usually less than 5 nanometers. They are often but not necessarily, composed of carbon, and having remarkable strength and electrical properties.

Nanowire: High aspect ratio structures with nanometer diameters that can be filled (nanorods) or hollow (nanotubes).

NEMSA: A generic term to describe nano scale electrical/mechanical devices.

OLED Organice Light Emitting Diode: OLED or Organic LED is not made of semiconductors. Itís made from carbon-based molecules. That is the key science factor that leads to potentially eliminating LEDsí biggest drawback ñ size. The carbon-based molecules are much smaller.

Omega Point: Also called the Quantum Omega Point Theory. A possible future state when intelligence controls the Universe totally, and the amountof information processed and stored goes asymptotically towards infinity. One quadrillionth of a second, and is to a second what a second is to 32,700,000 years.

Orbital Tower: Also known as a "space tether", "beanstalk" or "heavenly funicular". A cable in synchronous orbit, with one end anchored to the surface of the Earth, often with a small asteroid at the outer end to provide some extra tension and stability.

Paradigm Shift: When one conceptual world-view is replaced by another, or, a change of patterns on a massive scale..

Pervasive Computing: When computers (and sensors and actuators) become virtually invisible, and are used in almost every aspect of human commerce, interaction, and life, it will allow you full control over data and information, enabling you to send, receive, manage, and update your data from anywhere at any time.

Photolithography: A fabrication technique that utilizes light to etch patterns onto silicon wafers using a mask and light-sensitive resin. It is often used for making integrated circuits.

Picotechnology: The art of manipulating materials on a quantum scale.

Pink Goo: (humorous) Humans (in analogy with grey goo). "Pink Goo to refer to Old Testament apes who see their purpose as being fruitful and multiplying, filling up of the cosmos with lots more such apes, unmodified"
POSS Nanotechnology: Short for Polyhedral Oligomeric Silsesquioxanes Nanotechnology. POSS nanomaterials are attractive for missile and satellite launch rocket applications because they offer effective protection from collisions with space debris and the extreme thermal environments of deep space and atmospheric re-entry.

PM0.1: Particulate matter less than 0.1 micrometers in diameter **PM10:** Particulate Matter less than 10 micrometers in diameter **PM2.5:** Particulate matter less than 2.5 micrometers in diameter

Polymerz: A chemical compound, typically formed by connecting smaller molecules together, that consists of repeating structures, often arranged in a chain.

Polysilicon: Short for Polycrystalline Silicon, used in the manufacture of computer chips.

Posthuman: Persons of unprecedented physical, intellectual, and psychological capacity, self-programming, self-constituting, potentially immortal, unlimited individuals.

Positional Controlled Chemical Synthesis or Positional Synthesis: Control of chemical reactions by precisely positioning the reactive molecules, the basic principle of assemblers.

Positional Assembly: Constructing materials an atom or molecule at a time
Post Monetary Economy: After the advent of mature Nanotechnology, it is likely that our economic reality will change, possibly to the extent of eliminating currency as we know it today.

Protein Design, Protein Engineering: The design and construction of new proteins; an enabling technology for nanotechnology.

Positional Controlled Chemical Synthesis: Control of chemical reactions by precisely positioning the reactive molecules, the basic principle of assemblers.

POSS Nanotechnology: POSS nanomaterials are attractive for missile and satellite launch rocket applications because they offer effective protection from collisions with space debris and the extreme thermal environments of deep space and atmospheric re-entry. Process in which a molecule loses one or more electrons to another component of the reaction.

Protein Design: The design and construction of new proteins; an enabling technology for nanotechnology.

Protein Folding: The process by which proteins acquire their functional, preordained, three-dimensional structure after they emerge, as linear polymers of amino acids, from the ribosome.

Proteomics: The term proteome refers to all the proteins expressed by a genome, and thus proteomics involves the identification of proteins in the body and the determination of their role in physiological and pathophysiological functions.

Quantum: Describes a system of particles in terms of a wave function defined over the configuration of particles having distinct locations is implicit in the potential energy function that determines the wave function, the observable dynamics of the motion of such particles from point to point.

Quantum Computer: A computer that takes advantage of quantum mechanical properties such as superposition and entanglement resulting from nanoscale, molecular, atomic and subatomic components.

Quantum Confined Atoms (QCA): Atoms caged inside nanocrystals.

Quantum Cryptography: A system based on quantum- mechanical principles. Eavesdroppers alter the quantum state of the system and so are detected. Developed by Brassard and Bennett, only small laboratory demonstrations have been made.

Quantum Dot Nanocrystals (QDNs): Used to tag biological molecules, and "measuring between five and ten nanometres across, are made up of three components. Their cores contain paired clusters of atoms such as cadmium and selenium that combine to create a semiconductor.

Quantum Dots: Nanometer-sized semiconductor crystals, or electrostatically confined electrons. Something (usually a semiconductor island) capable of confining a single electron, or a few, and in which the electrons occupy discrete energy states just as they would in an atom (quantum dots have been called "artificial atoms").

Quantum Mechanics: A largely computational physical theory explaining the behavior of quantum phenomena, which incorporates the theory ofspecial relativity. Despite dilignet attempts, general relativity has not been sucessfully incorporated into quantum mechanics.

Quantum Tunncling: When electrons pass through a barrier, without overcoming it or breaking it down.

Quantum Well: A P-N-P junction in which the "N" layer is ~10 nm (where traditional physics leaves off and quantum effects take over) and an "electron trap" is created.

Quantum Wire: Another form of quantum dot, but unlike the single-dimension "dot," a quantum wire is confined only in two dimensions - that is it has "length," and allows the electrons to propagate in a "particle-like" fashion.

Qubit: The quantum computing analog to a bit. Qubits exhibit superposition. Thus, unlike normal bits, qubits can be both 1 and 0 at the same time

Quantum Mirage: A nanoscale property that may allow information to be transferred through use of the wave property of electrons.

Quantum Tunneling: When electrons pass through a barrier, without overcoming it or breaking it down.

Quantum Brain: [theoretical] Think of your brain. Now, think of your brain performing at vastly superior levels. Nanobots will become an as-needed addition to your existing neurons, extending your mental capacities further then you can probably now imagine.

Qubit: The quantum computing analog to a bit.

Red Goo: Deliberately designed and released destructive nanotechnology, as opposed to accidentally created grey goo

Replicator: A system able to build copies of itself when provided with raw materials and energy

Rheology: The study of the deformation and flow of matter under the influence of an applied stress, which might be, for example, a shear stress or extensional stress.

Risk assessment: A set of analytic approaches in the decision, risk, and management sciences, typically considering both the probability and the potential cost of a particular type of event.

SAMFET: (self assembled monolayer field effect transistor). Where a few molecules act as FETs, exhibiting both very strong gain, and extraordinarily rapid response.

Scanning Capacitance Microscopy: A method for mapping the local capacitance of a surface.

Scanning Electron Microscopy: see Virtual Scanning Electron Microscopy

Scanning Force Microscope (SFM): An instrument able to image surfaces to molecular accuracy by mechanically probing their surface contours. Akind of proximal probe. Adevice in which the deflection of a sharp stylus mounted on a soft spring is monitored as the stylus is moved across a surface. If the deflection is kept constant by moving the surface up and down by measured increments, the result (under favorable conditions) is an atomic-resolution topographic map of the surface. Also termed an atomic force microscope.

Scanning Near Field Optical Microscopy: A method for observing local optical properties of a surface that can be smaller than the wavelength of the light used.

Scanning Thermal Microscopy: A method for observing local temperatures and temperature gradients on a surface.

Scanning Near Field Optical Microscopy: A method for observing local optical properties of a surface that can be smaller than the wavelength of the light used.

Scanning probe lithography: A technique in which an atomic force microscope or scanning tunneling microscope scratches, indents, or heats to produce nanoscale features on a surface.

Scanning Thermal Microscopy: A method for observing local temperatures and temperature gradients on a surface.

Scanning Tunneling Microscope (STM): A scientific instrument that can make images of nanoscale details on an electrically conductive surface by moving a sharp metal probe very close to that surface, passing a low-voltage electric current across it, and measuring tiny fluctuations in the current as the probe is scanned across the surface.

Sealed Assembler Laboratory: A work space, containing assemblers, encapsulated in a way that allows information to flow in and out but does not allow the escape of assemblers or their products.

Self-assembly: In chemical solutions, self-assembly (also called Brownian assembly) results from the random motion of molecules and the affinity of their binding sites for one another. Also refers to the joining of complementary surfaces in nanomolecular interaction.

Self-Assembled Monolayers (SAMs): Self-Assembled Monolayers on Mesoporous Supports (SAMMS): nanoporous ceramic materials that have been developed to remove contaminants from environmental media.

Self-repair: Indicating ability to heal itself without outside intervention.

Self-replication: More accurately labeled "exponential replication," self-replication refers to the process of growth or replication involving doubling within a given period.

Sentience Quotient: The sentience of an intelligence is roughly directly related to the amount of data it can process per unit time and inversely to the overall mass needed to do that processing. This would be something like baud/kilograms.

Single-walled carbon nanotubes (SWNT): See Nanotubes and buckyballs

Shape Memory Alloy: A unique class of alloys which are able to "remember" their shape and are able to return to that shape even after being bent.

Shape Memory Alloys: (SMA's) are a unique class of alloys which are able to "remember" their shape and are able to return to that shape even after being bent. The ability is known as the shape memory effect.

Singularity: Defined by Vernor Vinge as the "postulated point or short period in our future when our self-guided evolutionary development accelerates enormously (powered by nanotechnology, neuroscience, AI, and perhaps uploading) so that nothing beyond that time can reliably be conceived.

Sky Hook: A long, very strong, cable in orbit around a planet which rotates around its center of mass in such a way that when one end is closest to the ground, its relative velocity is almost zero. It would function as a kind of space elevator; shuttle craft would anchor to the end and then be lifted into orbit where they would be released.

Smart Material: Materials and products capable of relatively complex behavior due to the incorporation of nanocomputers and nanomachines.

Space Fountain: A vertical stream of magnetically accelerated pellets reaching out into space, where a station held aloft by its momentum reverses the direction and directs it towards a receiver on the ground. Essentially a simpler version of a Lofstrom loop.

Smartdust: Tiny, bottle-cap-shaped micro-machines fitted with wireless communication devices - that measure light and temperature.

SNT: An abbreviation for structural nanotechnology; refers to integration of nanotech features into non-MNT products, also called nanomaterials.

Spike: Another term for the singularity, suggested by Damien Broderick since the growth curves look almost like a spike as it is approached.

Spintronics: AKA: Quantum Spintronics, Magnetoelectronics, Spin Electronics. Electronic devices that exploit the spin of electrons as well as their charge. Unlike conventional electronics which is based on number of charges and their energy, and whose performance limited in speed and dissipation, spintronics is based on the direction of electron spin, and spin coupling, and is capable of much higher speed at much lower power.

Star Trek scenario: Someone builds potentially dangerous self-replicating devices that spread disastrously.

Stewart Platforms: A positional device.

Superconductor: An object or substance that conducts electricity with zero resistance.

Superintelligence: An intellect that is much smarter than the best human brains in practically every field, including scientific creativity, general wisdom and social skills. This definition leaves open how the superintelligence is implemented: it could be a digital computer, an ensemble of networked computers, cultured cortical tissue or what have you..

Superlattice Nanowire: Interwoven bundles of nanowires using substances with different compositions and properties.

Superlattice Nanowire Pattern (transfer): [SNAP] a technique for producing "Ultra High Density Nanowire Lattices and Circuits".

Superlattice Nanowire: Interwoven bundles of nanowires using substances with different compositions and properties.

Synthespian: An artificial actor, for example a 3D model animated by motion capture from a real actor or a computer program.

Technocyte: A nanoscale artificial device (especially a nanite) in the human bloodstream used for repairs, cancer protection, as an artificial immune system or for other uses.

Technofobics: Those who have a phobia to technology, and/or to advances in technology.

Terraform: To change the properties of a planet to make it more earthlike, making it possible for humans or other terrestrial organisms to live unaided on it, for example by changing atmospheric composition, pressure, temperature or the climate and introducing a self-sustaining ecosystem.

Technocyte : A nanoscale artificial device (especially a nanite) in the human bloodstream used for repairs, cancer protection, as an artificial immune system or for other uses.

Template synthesis: Thc ability of one DNA molecule to "recognize" and attach to another molecule that has a complementary shape.

Thermal Noise: the vibration and motion of atoms and molecules caused by the fact that they have a temperature above absolute zero.

Thin film: Three-dimensional biodegradable polymers engineered for cell growth.

Toxicity: The extent to which a chemical substance is poisonous or, through chemical action, destroys living tissue.

Transhuman: Someone actively preparing for becoming posthuman. Someone who is informed enough to see radical future possibilities and plans ahead for them, and who takes every current option for self-enhancement.
Transhumanism: Philosophies of life (such as Extropianism) that seek the continuation and acceleration of the evolution of intelligent life beyond its currently human form and human limitations by means of science and technology, guided by life-promoting values.

Transistor: The basic element in an integrated circuit. An on/off switch (consisting of three layers of a semiconductor material) that consists of a source (where electrons come from), a drain (where they go) and a gate that controls the flow of electrons through a channel that connects the source and the drain.

Tribology: Study of friction, wear and lubrication of interacting surfaces.

Tubeologist: Someone who knows their nanotubes inside and out

Turing Test: Turing's proposed test for whether a machine is conscious (or intelligent, or aware

Universal Assembler: Uses raw atoms and molecules to construct consumer goods, and is pollution free. Can be programmed to build anything that is composed of atoms and consistent with the rules of chemical stability.

Universal Constructor: A machine capable of constructing anything that can be constructed. The physical analog of a "universal computer", which can perform any computation.

Uplift: To increase the intelligence and help develop a culture of a previously non- or near-intelligent species.

Upload: (a) To transfer the consciousness and mental structure of a person from a biological matrix to an electronic or informational matrix (this assumes that the strong AI postulate holds). The term "Downloading" is also sometimes used, mainly to denote transferring the mind to a slower or less spacious matrix. (b) The resulting infomorph person. [The origin of the term is uncertain, but obviously based on the computer technology term 'uploading' (loading data into a mainframe computer).]

Utility Fog: [AKA: Polymorphic Smart Materials] Objects formed of "intelligent" polymorphic (able to change shape) substances, typically having an octet truss structure. Concept concieved by Dr. J. Storrs Hall.

Vapor deposition: A chemical process commonly used in the semiconductor industry to apply thin films of one substance onto a surface composed of another substance.

Vasculoid: The vasculoid [concept] is a single, complex, multisegmented nanotechnological medical robotic system capable of duplicating all essential thermal and biochemical transport functions of the blood, including circulation of respiratory gases, glucose, hormones, cytokines, waste products, and cellular components.

Von Neumann Machine: (pronounced von noi-man) A machine which is able to build a working copy of itself using materials in its environment. This is often proposed as a cheap way to mine or colonize the entire solar system or galaxy.

Wet Nanotechnology: The study of biological systems that exist primarily in a water environment. The functional nanometer-scale structures of interest here are genetic material, membranes, enzymes and other cellular components.

Zeptosecond: One-billion-trillionth of a second.

Zettatechnology: Referring to the typical number of distinct designed parts in a product made by the systems (molecular, mature, or molecular-manufacturing-based nanotechnology).